Toxicology for the Health and Pharmaceutical Sciences

Toxicology for the Health and Pharmaceutical Sciences

Edited by
Dr. Antonio Peña-Fernández, PhD, SFHEA,
Dr. Mark D. Evans, PhD, MRSC, FIBMS and
Dr. Marcus S. Cooke, PhD, FRCPath

CRC Press
Taylor & Francis Group
Boca Raton London New York

CRC Press is an imprint of the
Taylor & Francis Group, an **informa** business

First edition published 2022
by CRC Press
6000 Broken Sound Parkway NW, Suite 300, Boca Raton, FL 33487–2742

and by CRC Press
2 Park Square, Milton Park, Abingdon, Oxon, OX14 4RN

CRC Press is an imprint of Taylor & Francis Group, LLC

Library of Congress Cataloging-in-Publication Data
Names: Peña-Fernández, Antonio, editor. | Evans, Mark D., editor. | Cooke, Marcus S., editor.
Title: Toxicology for the health and pharmaceutical sciences/edited by Dr. Antonio Peña-Fernández, PhD, SFHEA, Dr. Mark D. Evans, PhD, MRSC, FIBMS, Dr. Marcus S. Cooke, PhD, FRCPath.
Description: First edition. | Boca Raton, FL : Taylor & Francis, 2022. | Includes bibliographical references and index.
Identifiers: LCCN 2021019752 (print) | LCCN 2021019753 (ebook) | ISBN 9781138303362 (hardback) | ISBN 9781032100227 (paperback) | ISBN 9780203730584 (ebook)
Subjects: LCSH: Toxicology. | Chemicals—Safety measures.
Classification: LCC RA1190 .T697 2022 (print) | LCC RA1190 (ebook) | DDC 363.17/91—dc23
LC record available at https://lccn.loc.gov/2021019752
LC ebook record available at https://lccn.loc.gov/2021019753

ISBN: 978-1-138-30336-2 (hbk)
ISBN: 978-1-032-10022-7 (pbk)
ISBN: 978-0-203-73058-4 (ebk)

DOI: 10.1201/9780203730584

Typeset in Palatino
by Apex CoVantage, LLC

Antonio Peña-Fernández—This book is dedicated to my parents Ángela Fernández and Cástor Peña, for their unconditional support, and to my dear brothers and nephews. I love you so much.

Mark D. Evans—Dedicated to Jill for her understanding, love and support.

Marcus S. Cooke—Dedicated to Emily, Evie and Harrison, for their understanding, patience, willingness to travel, and their courage and strength wherever we arrive.

Contents

Preface

There is an increasing need for a knowledge of toxicology to safeguard the use of chemicals in industry, public and private sectors. In particular, the health sector requires workers to have a basic knowledge of toxicology due to the large burden of disease and mortality caused by chemicals and drugs of abuse. Moreover, current anthropogenic activities and the development of new techniques for chemical creation, production, manufacturing and use have made necessary the development of different tools to evaluate the safety of these substances to protect human health. Therefore, the study of toxicology should be part of the education of all fully formed professionals in the health sciences.

Knowledge of toxicology is critical for the scientific evaluation of the risks that chemicals can pose to human health and to the environment. Such knowledge is also important to manage risk, respond to any accidental or deliberate release of chemical substances to the environment and to implement remediation strategies. Moreover, a modern healthcare sector requires professionals with an education in toxicology to tackle the negative public perception of the chemical and pharmaceutical industries for the environment and public health. Thus, the subject of toxicology is of critical importance to protect human health against chemicals and/or drugs that may be present in the environment or may pose a threat to the population in the aftermath of a chemical incident (i.e. the release of one or more chemicals to the environment). For these reasons, academics should develop and offer toxicological education that will play an essential part of the education of future health workers to face these chemical threats.

Comprising a series of chapters from leading toxicology, pharmacy and public health academics and experts across Europe, the United States and beyond, *Toxicology for the Health and Pharmaceutical Sciences* provides a concise yet comprehensive volume that can be used as a relevant textbook on toxicology for the clinical, healthcare, educational and professional sectors. This book covers the fundamentals and recent developments in toxicology, to respond to local and global chemical and pharmaceutical threats due to globalization and human activities. Thus, this volume has chapters specifically designed to support the understanding of the most current, toxicology-related subjects for any undergraduate/postgraduate health programmes, as well as aiding with the delivery of continuing professional development training on up-to-date topics in toxicology for current practicing health professionals wishing to improve their background knowledge in toxicology.

The textbook begins with 10 introductory chapters that provide basic and cutting-edge information on toxicology. Chapters 11–26 were written by researchers who are experts in their fields and further cover fundamental and applied topics, together with descriptions of novel tools in relevant toxicology specialties. The final section of this volume provides practical guidance in the form of two detailed case studies based on real-world/developed scenarios for characterising human risks to environmental pollutants and on how to use innovative guidance and tools to respond to chemical incidents, which will facilitate the user to acquire and practice these relevant skills highlighted in previous chapters.

This textbook is therefore a vital, comprehensive resource and reference for students, academics, researchers and employees in the pharmaceutical-, health- and environmental health-related subject areas. Arguably, this textbook is also vital reading and reference for policymakers and others that influence and decide regulations that have an impact on the environment and human health. Although there is a particular focus on Europe and the US, reflecting the current and emerging toxicological issues in those areas, this text has global relevance.

Acknowledgments

The editors would like to thank each of the contributors for their efforts, dedication and patience in the production of this first edition of *Toxicology for Health and Pharmaceutical Sciences*. Without their input, there would be no book.

About the Editors

Antonio Peña-Fernández is a senior lecturer in toxicology and medical sciences in the Faculty of Health and Life Sciences at De Montfort University, Leicester, UK. He has a BSc in chemistry from the CEU San Pablo University and an MSc in toxicology from the University of Seville. His PhD in toxicology is from the University of Alcalá, Spain (2011), where he has held an honorary professorship since 2019. His main research work focuses on human biomonitoring as a tool for the assessment of internal doses of contaminants, and the study of emerging chemical and biological threats in urban, industrial and rural environments for the characterisation of risks and the development of protocols to protect human health and decontaminate impacted environments. Many of his current projects are international collaborations with different universities and research institutions from England, Spain and Sierra Leone. Projects relate to environmental monitoring, exposure assessment, building capabilities and the development of protocols, training materials and tools for the protection of human health to different emerging chemical and biological contaminants. He became a European Registered Toxicologist (EUROTOX) in 2013 and is a Senior Fellow of the Higher Education Academy (SFHEA, 2016).

Mark D. Evans is a senior lecturer in biomedical and medical science in the Faculty of Health and Life Sciences at De Montfort University, Leicester, UK. He is a graduate of Brunel University (BSc) and Louisiana State University (PhD), with an extensive academic and research background in the area of free radical biology and oxidative stress. During his career, he has worked in the areas of cigarette smoking–induced lung disease, autoimmune disease (rheumatoid arthritis, lupus) and cancer. Prior to coming to De Montfort University, he was a postdoctoral scientist and then lecturer at the University of Leicester. His primary interests lie in the area of DNA damage and repair and its role in disease pathogenesis. Additionally, he has worked in the area of human biomonitoring, examining excreted markers of nucleic acid oxidation and their biological sources, particularly the role of nudix hydrolases. He also has active research interests in the therapeutic potential of natural products, primarily anticancer activity and mechanisms. He has significant experience in teaching and teaching administration for undergraduate, postgraduate and research students, as well as in the design of and academic lead for MSc programmes.

Marcus S. Cooke is a professor in the Department of Cell Biology, Microbiology and Molecular Biology at the University of South Florida, Tampa, where he heads the Oxidative Stress Group. His background is in the environmental health sciences, in particular environmental toxicology. He studies oxidative stress and genomic instability: from basic mechanisms to translational application of validated biomarkers to study human health and disease; specifically, the formation and repair of damage to nuclear/mitochondrial DNA and nucleotide precursor pools, and the development and application of DNA adductomics; that is, the totality of DNA adducts in cellular DNA, and urine; and the genome-wide mapping of DNA damage. Translating such laboratory findings to the marketplace is a priority and is actively pursued. This has resulted in three US and one UK patents, license agreements, industry collaborations and an STTR award from NIEHS. As of March 2021, he has 119 peer-reviewed publications, which have attracted over 7500 citations, and he has an *h*-index of 40.

Contributors

Mª Victorina Aguilar
Biomedical Sciences Department
Alcalá University
Alcalá de Henares, Madrid, Spain

Consuelo Alvarez Herrera
Department of Molecular Biology and Biochemical Engineering, Toxicology Area
Pablo de Olavide University
Ctra. de Utrera, Sevilla, Spain

Arturo Anadón
Department of Pharmacology and Toxicology, Veterinary Faculty
Universidad Complutense de Madrid
Madrid, Spain

Irma Ares
Department of Pharmacology and Toxicology, Veterinary Faculty
Universidad Complutense de Madrid
Madrid, Spain

Ege Arzuk
Department of Pharmaceutical Toxicology, Faculty of Pharmacy
Ege University
Bornova-Izmir, Turkey

Paul C. Barrow
F. Hoffmann-La Roche Ltd
Roche Pharma Research and Early Development
Basel, Switzerland

Mª del Carmen Berrocal
Biomedical Sciences Department
Alcalá University
Alcalá de Henares, Madrid, Spain

John P. Berry
Department of Chemistry and Biochemistry, Marine Science Program
Florida International University
Miami, FL

Werner Bomann
Toxconsult
Olathe, KS

Robert P. Chilcott
Research Centre for Topical Drug Delivery and Toxicology
University of Hertfordshire
Hatfield, Hertfordshire, UK

Marcus S. Cooke
Oxidative Stress Group, Department of Cell Biology, Microbiology and Molecular Biology
University of South Florida
Tampa, FL

Sonia Dagnino
MRC Centre for Environment and Health, Department of Epidemiology and Biostatistics
Imperial College London
London, UK

Ana del Peso
National Institute of Toxicology and Forensic Sciences
Avda. Dr Fedriani
Sevilla, Spain

María del Rosario Aberturas
Department of Biomedical Sciences, Faculty of Pharmacy
University of Alcalá
Alcalá de Henares, Madrid, Spain

Raquel Duarte-Davidson
Centre for Radiation, Chemical and Environmental Hazards
Public Health England
Chilton, Didcot, UK

Ali Ergüç
Department of Pharmaceutical Toxicology, Faculty of Pharmacy
Ege University
Bornova-Izmir, Turkey

Mark D. Evans
School of Allied Health Sciences, Faculty of Health and Life Sciences
De Montfort University
Leicester, UK

María Mar Gil-Díaz
Madrid Institute for Rural Research, Agrarian and Food Research and Development (IMIDRA)
Finca El Encín
Alcalá de Henares, Madrid, Spain

Carmen González-Martín
Toxicology Section
CEU San Pablo University
Madrid, Spain

Elliott Gordon
Elliott Gordon Consulting, LLC
Princeton Junction, NJ

Esther Gramage
Toxicology Section
CEU San Pablo University
Madrid, Spain

María Haro
Department of Chemistry and Biochemistry, Faculty of Pharmacy
CEU San Pablo University
Madrid, Spain

Mª Teresa Hernández
Biomedical Sciences Department
Alcalá University
Alcalá de Henares, Madrid, Spain

Gonzalo Herradón
Department of Health and Pharmaceutical Sciences, Faculty of Pharmacy
CEU San Pablo University
Madrid, Spain

Fuat Karakuş
Department of Pharmaceutical Toxicology, Faculty of Pharmacy
Ege University
Bornova-Izmir, Turkey

Christian Khalil
Lebanese American University
Byblos, Lebanon

Jessica Laine
MRC Centre for Environment and Health, Department of Epidemiology and Biostatistics
Imperial College London
London, UK

María del Carmen Lobo-Bedmar
Madrid Institute for Rural Research, Agrarian and Food Research and Development (IMIDRA)
Finca El Encín
Alcalá de Henares, Madrid, Spain

Domingo Ly-Pen
Abbey House Medical Centre
Abbey Road
Navan, Co. Meath, Ireland

Sara Maisanaba
Department of Molecular Biology and Biochemical Engineering, Toxicology Area
Pablo de Olavide University
Ctra. de Utrera, Sevilla, Spain

María-Aránzazu Martínez
Department of Pharmacology and Toxicology, Veterinary Faculty
Universidad Complutense de Madrid
Madrid, Spain

María-Rosa Martínez-Larrañaga
Department of Pharmacology and Toxicology, Veterinary Faculty
Universidad Complutense de Madrid
Madrid, Spain

Hazem Matar
Research Centre for Topical Drug Delivery and Toxicology
University of Hertfordshire
Hatfield, Hertfordshire, UK

Jesús Molpeceres
Department of Biomedical Sciences, Faculty of Pharmacy
University of Alcalá
Alcalá de Henares, Madrid, Spain

Gary Moss
School of Pharmacy
Keele University
Keele, Staffordshire, UK

María José González-Muñoz
Department of Biomedical Sciences, Section of Toxicology, Faculty of Pharmacy
Alcalá University
Alcalá de Henares, Madrid, Spain

Hilmi Orhan
Department of Pharmaceutical Toxicology, Faculty of Pharmacy
Ege University
Bornova-Izmir, Turkey

María de los Ángeles Peña
Department of Biomedical Sciences, Faculty of Pharmacy
University of Alcalá
Alcalá de Henares, Madrid, Spain

Antonio Peña-Fernández
School of Allied Health Sciences, Faculty of Health and Life Sciences
De Montfort University
Leicester, UK

Carmen Pérez-García
Department of Health and Pharmaceutical Sciences, Faculty of Pharmacy
CEU San Pablo University
Madrid, Spain

María José Polanco
Toxicology Section
CEU San Pablo University
Madrid, Spain

Guillermo Repetto
Department of Molecular Biology and Biochemical Engineering, Toxicology Area
Pablo de Olavide University
Ctra. de Utrera, Sevilla, Spain

Sophie A. Rocks
Mechanical Engineering Sciences (SPARKlab)
University of Surrey
Guildford, Surrey, UK

Carmen Rodríguez-Rivera
Toxicology Section
CEU San Pablo University
Madrid, Spain

Raquel Rojas
Department of Molecular Biology and Biochemical Engineering
Pablo de Olavide University
Ctra. de Utrera, Sevilla, Spain

Pavel Rossner Jr.
Department of Genetic Toxicology and Nanotoxicology
Institute of Experimental Medicine of the Czech Academy of Sciences
Prague, Czech Republic

Andrea Rossnerova
Department of Genetic Toxicology and Nanotoxicology
Institute of Experimental Medicine of the Czech Academy of Sciences
Prague, Czech Republic

Francisco Javier Rupérez
Centre for Metabolomics and Bioanalysis (CEMBIO), Faculty of Pharmacy
CEU San Pablo University
Madrid, Spain

Pramila Singh
Citoxlab
Evreux, France

François Spézia
Citoxlab
Evreux, France

Radim J. Sram
Department of Genetic Toxicology and Nanotoxicology
Institute of Experimental Medicine of the Czech Academy of Sciences
Prague, Czech Republic

Christian Strupp
Gowan Crop Protection
Reading, UK

Gevorg Tepanosyan
Department of Environmental Geochemistry
Center for Ecological–Noosphere Studies of the National Academy of Sciences
Yerevan, Armenia

Carmen José Mateos Vega
Department of Biomedical Sciences, Section of Toxicology, Faculty of Pharmacy
Alcalá University
Alcalá de Henares, Madrid, Spain

Marta Vicente-Rodríguez
Department of Neuroimaging, Maurice Wohl Clinical Neuroscience Institute
King's College London
London, UK

Simon Wilkinson
School of Biomedical Sciences
University of Newcastle-upon-Tyne
Newcastle, UK

Stacey Wyke
Public Health England, Centre for Radiation, Chemical and Environmental Hazards
Chilton, Didcot, UK

1 Introduction to Toxicology

Mª Teresa Hernández, Mª del Carmen Berrocal, Domingo Ly-Pen and Mª Victorina Aguilar

CONTENTS

1.1 THE CONCEPT OF TOXICOLOGY AND AREAS OF ACTION

From the etymological point of view, the word "toxicology" is derived from two Greek words: *toxikon* (poison) and *logos* (treatise), which means "the science of poisons". According to Paracelsus (1493–1541), often called the Grandfather of Modern Toxicology, "all substances are poisons; there is none which is not poison. The right dose differentiates a poison from a remedy".

Orfila (1787–1853), the founder of the science of toxicology, in his famous *Treatise of Toxicology* (1813), defined a poison as "any substance, taken or applied in any way in small doses in a living organism, which destroys health or ends life entirely". According to this definition, substances would be classified as poisons and nonpoisons, the dose being crucial to establish their difference, and thus making the distinction between poisoning by overdose and poisoning itself.

With the advances of toxicology, this definition is now incomplete and the word "poison" is misleading. Since Paracelsus, it is well known that it is the dose that makes the poison; a chemical that is perfectly safe at one dose may be lethal at another. For example, at very high doses even table salt or drinking water can be harmful, especially if you consider the influence of other factors, such as stage of life, age, diet, diseases and concomitant exposures to various agents. Therefore, one of the most accepted definitions of toxicology today is "the study of the adverse effects of chemicals or physical agents on living organisms" (Gilbert 2012).

The change of paradigm in toxicology implies a transition from an *in vivo* science (in which the use of animals in experimental laboratory conditions were required to study parameters, such as the lethal dose for half of the population of animals) to a science in which the following are studied (even virtually): routes of exposure, mechanisms of action, events and key processes of the target molecule, cellular responses, and even the macroscopic and organographic effects on human health and the environment (Meek et al. 2014).

From a historical point of view, toxicology was consolidated as a scientific discipline, independent of medicine but related through forensic toxicology. For the development of this branch and of other more recent ones, a prerequisite was the development of increasingly sensitive and specific analysis techniques capable of analyzing toxic agents in biological samples, mainly blood and urine, of exposed subjects to confirm the presence of the toxin responsible (analytical toxicology). Analytical toxicology, with the advances in pharmacological toxicology and with the support of appropriate regulatory legislation, contributed to the current successes of forensic toxicology and clinical toxicology, which aims to diagnose and treat intoxications like any other disease that has a pathological character; that is, it can manifest in an acute and chronic way, before the death of the subject (Bello 2001).

New lifestyles, exposure to new substances and so forth have given rise to the development of different subdisciplines of toxicology. These subdisciplines are often closely related to each other, and their knowledge and application serves to protect public health.

DOI: 10.1201/9780203730584-1

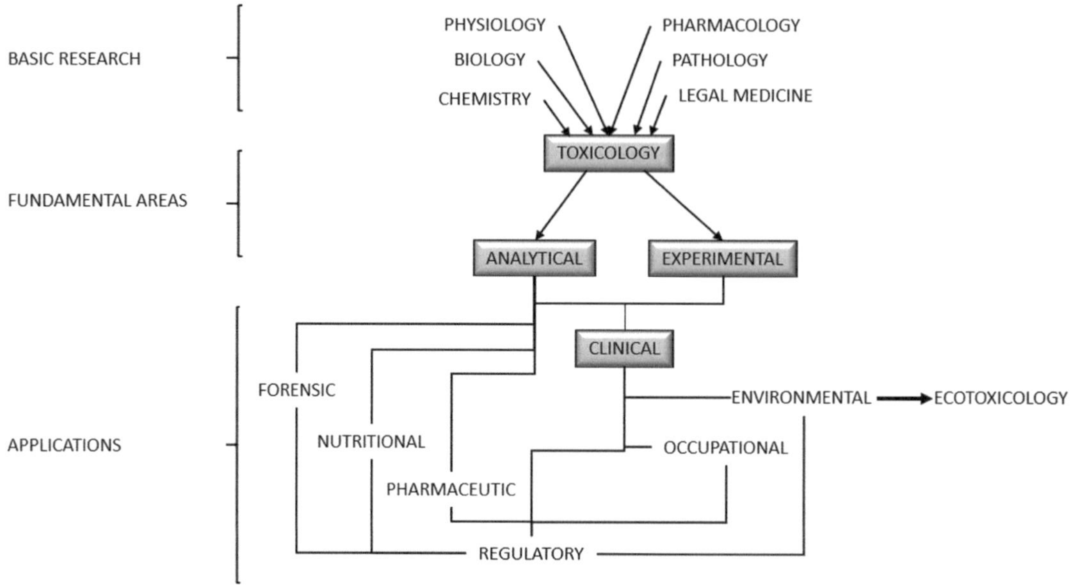

Figure 1.1 Subdisciplines of toxicology. The field of toxicology can divided into various subdisciplines, derived from the vertical and horizontal integration of toxicology with other sciences. Modern toxicology is a conglomerate of subdisciplines that cooperate to promote knowledge of physicochemical toxicity.

The development of the chemical industry has contributed to these advances in toxicology in the twentieth century, and interest has grown in increasing toxicological knowledge in relation to health in the workplace. The grouping of national toxicological societies by continental blocks has also been of crucial importance, represented by SOT in North America, EUROTOX in Europe, ASIATOX in Asia, ALATOX in Latin America and toxicological societies in Africa, such as those in South Africa and Cameroon. From the international cooperation of the different societies, the International Union of Toxicology (IUTOX) was created in 1980. It has played an important role in the applied knowledge of toxicology (Repetto and Repetto 2009).

The different subdisciplines of modern toxicology are outlined in Figure 1.1 and are established according to the way of addressing a toxic event in three areas that correspond to the type of work performed (Jaramillo et al. 2006):

1. General: the general basis of toxic actions. This includes mechanisms of action and ways of counteracting their effects.
2. Descriptive: the science of toxicity testing to provide information about safety evaluation and regulatory requirements. The different aspects and toxicological studies are grouped on toxic agents such as metal toxicology, toxicology of organic solvents, toxicology of pesticides and so forth.
3. Mechanistic: studies, identifies and attempts to understand the mechanisms by which toxic agents exert toxic effects on living beings, in order to produce safer substances and develop a rational treatment of intoxication.
4. Regulatory: integrates the information obtained from the mechanistic and descriptive areas to determine the level of risk to health and to the handling of exposure to chemical substances.

To further investigate the fundamental mechanisms and processes involved in toxic phenomena, it is necessary to integrate toxicological investigations with the knowledge of other basic sciences (Figure 1.1). This integration can be achieved by two methods (Bello 2001):

1. Vertical integration gives rise to the appearance of diverse subdisciplines: genetic, molecular toxicology, inmunotoxicology, neurotoxicology and so forth.

2. Horizontal integration is oriented towards practical applications with repercussions that affect the environment and human health. In this way, two areas appear:
   a. Retrospective toxicology: typical of the forensic branch oriented to the investigation of a toxin in corporal organs and its medicolegal aspects.
   b. Predictive toxicology: oriented towards the prediction of possible toxic effects in specific situations, including drug use, drug interactions and presence of contaminants or additives not allowed in food.

This horizontal integration of toxicology with other sciences gives rise to the appearance of different applied branches, among which we must highlight:

- Clinical toxicology, based on physiology, clinical chemistry and pharmacology, which studies the pathological changes caused by toxic agents, establishes treatments for intoxicated patients and analyzes new techniques to treat intoxications.
- Occupational or professional toxicology, based on occupational medicine and occupational hygiene. Occurrence of occupational diseases is related to toxic substances present in work environments. Therefore, toxicology investigates the harmful effects produced by substances for occupational use and determines safe exposure limits.
- Environmental toxicology, which relies on ecology and environmental chemistry to analyze the impact of pollutants present in the environment of living organisms. It is the subdiscipline responsible for evaluating the vast environmental impact produced by exposures to chemical products present in the environment, with special attention to living species other than humans in air, soil or water. A complex environment requires paying attention to the persistence of pollutants in soil, water and air, and knowing the capacity they have to join the food chain.

The joint work between ecologists and environmental toxicologists is increasing our knowledge about the impact of agrochemicals on native species. Monitoring changes of speciation, due to the effect of nearby mines or chemical plants, has led to development of the field of ecotoxicology. This is a branch of environmental toxicology, which studies the particular way(s) in which toxins impact the population dynamics of an ecosystem (Newman 2010).

- Food toxicology is a multidisciplinary approach, studying adverse effects of exposure by living organisms to chemical substances present in food. This area is supported by chemical analysis, food science and nutrition.

  It is important to know what products are safe to eat, and in what amounts; this discipline investigates the safety of the components that are added to food, deliberately or accidentally, as natural and synthetic additives or contaminants. Accidental contaminants are generally synthetic or natural environmental contaminants, such as polychlorinated biphenyls (PCBs) and methylmercury, which are found in fish; microbial toxins such as those produced by *Escherichia coli* in contaminated food; and fungal toxins, such as aflatoxins, which can contaminate grains. Recently, scientists have investigated and debated about the safety of genetically modified organisms (GMOs) as food products and the influence of new conservation technologies on food safety.
- Pharmacological toxicology studies the safety of pharmaceutical products. Toxicity testing helps ensure that pharmaceutical products are safe for humans. Advances in pharmacology and toxicological research help to ensure that the beneficial effects of therapeutic agents are not outweighed by undesired side effects.
- Forensic toxicology establishes the causes of death caused by toxins in humans and animals, their circumstances, and their medicolegal implications.

1.2 CONCEPT OF WHAT IS TOXIC

As with other fields of knowledge, toxicology has its own distinctive vocabulary: toxin, toxicant, poison and xenobiotic are often used interchangeably in the literature; however, there are subtle differences between them. The term "toxin" is best reserved for harmful substances made by living organisms (e.g. poisonous marine organisms, infectious pathogens or venomous spiders). The word "poison" is widely used for this purpose during everyday life, but it may convey a misleading interpretation, because their action depends on the dose, the individual and environmental circumstances, as mentioned previously (Burcham 2014a).

According to toxicologists themselves, the word "toxin" should be used if the foreign material came from a biological source. This means a substance is only poisonous if produced by living cells or organisms.

A "toxicant", on the other hand, should only be used if the foreign material came from man-made sources; thus it is not produced biologically. The names of toxicants are especially informative when coupled with a prefix that designates the site of toxic action for a given substance; alcohol, for example, is a hepatotoxicant because it causes liver damage at high doses (Burcham 2014b).

The term "xenobiotic" describes chemicals found but not produced in organisms or the environment. This includes numerous substances such as food additives, contaminants, drugs, recreational drugs pesticides, herbicides and industrial reagents.

The term "endobiotic", in contrast, includes chemical compounds present in the body during normal physiological processes (androgens, neurotransmitters, glucocorticoids, bilirubin, etc.). According to this definition, any substance can damage an organism because all exogenous products (xenobiotics) as the own constituents of the organism (endobiotics) when they are in a certain amount, can produce toxic effects.

This does not mean that external or synthetic chemical substances can be more toxic than natural or endobiotic ones, but simply that xenobiotics attract more attention from modern toxicology because they are widely used in industry, they are produced on a large scale and they persist in the environment for a long time (Burcham 2017).

Classifying toxic substances into endogenous and exogenous substances is complicated, because some substances come from different sources. With the development of more sensitive analytical techniques in biological fluids or tissues, it is known that many chemical compounds of industrial origin can be formed at low levels in the body. For example, humans are exposed exogenously to acrolein (combustion of tobacco and fossil hydrocarbons) but also endogenously, as a result of different biochemical reactions in the body itself (e.g. lipid oxidation; Figure 1.2). For definitions of

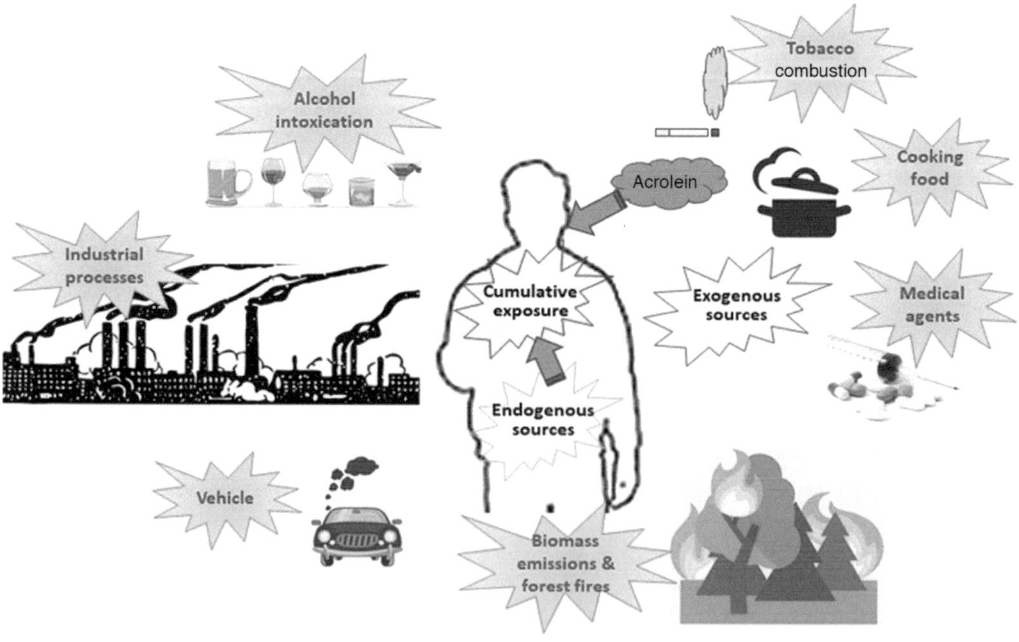

Figure 1.2 Complications of toxic substance origin based on source. Classification of toxic substances based upon their origin (endogenous or exogenous) is complicated because some substances come from different sources (e.g. acrolein). Humans are exposed to exogenous acrolein (e.g. tobacco, fossil fuels) but also to endogenous acrolein that forms via biochemical reactions within the body (e.g. lipid peroxidation).

Source: Figure drawn with inspiration from Burcham (2014a).

terms related to the study of toxicology, the reader is referred to the online resource provided by the *Encyclopaedia of Occupational Health and Safety* (Holmberg et al. 2015).

1.3 DETERMINING FACTORS OF TOXICITY

Toxicity is the activity specifically linked to the physicochemical properties of a substance due to its interaction with a receptor or receptors. This toxic activity is what determines that a substance is harmful to an organism under certain conditions. However, for each chemical agent there is a certain degree of toxicity. The range of doses necessary to produce damage in a living organism is very broad, as can be seen in Table 1.1. Toxic effect is assessed in terms of the median lethal dose (LD_{50}), which is the amount of a material, given all at once, which causes the death of 50% (one-half) of a group of test animals.

We should note that this concept is used in a relative manner, because several variables can influence the toxic activity of a substance. For a chemical compound to cause toxic effects, it is necessary for the substance or its metabolites (produced by biotransformation) to reach the appropriate place in the body and persist for a suitable amount of time at a concentration sufficient to produce a toxic manifestation. This toxic response, in turn, will depend on the chemical and physical characteristics of the substance, the size and route of exposure, the metabolism and the sensitivity of the system or individual.

1.3.1 Biological System

The biological system on which the toxic agent acts is of utmost importance, because the effect will vary markedly depending on the organism. This factor must be taken into account, because it is well known that there is a great variation in intra- and interspecies sensitivity towards toxic agents.

Therefore, we should consider that the toxic response of a substance, revealed in an animal toxicity test, does not imply that it will follow the same pattern when it is extrapolated to humans. Often the toxic effect is achieved at doses well above those of normal exposure for humans. In the history of toxicology, there are many examples of this interspecies variation. One of the most famous is the artificial sweetener saccharin, an additive widely used in foods and beverages. During the 1970s, reports of bladder tumours in rats were published after ingestion of high doses of saccharin. When investigating the mechanism related to the appearance of tumours, which only occurred in males and not in female rats, it was found that this bladder cancer was due to a specific protein in the rat called α-2u-globulin (Arnold 1983). This protein had little relevance in humans, and it was shown that the metabolism of saccharin was specific to rats. Numerous epidemiological studies provide no clear or consistent evidence to support the assertion that sodium saccharin increases the risk of bladder cancer in humans (Ellwein and Cohen 1990; NCI 2015). The US National Toxicology Program and the International Agency for Research on Cancer support the US Environmental Protection Agency's conclusion that saccharin is safe "at human levels of consumption" because, after assessing many saccharin-sweetened human foods, no association between saccharin and cancer could be established (Telišman 1998).

Table 1.1 Illustration of the Wide Order of Magnitude Observed for LD_{50} Values

Chemical	LD_{50} (mg/kg) rat
Ethanol	10,000
Sodium chloride	4000
Ferrous sulfate	1500
Morphine sulfate	900
Phenobarbital sodium	150
Nicotine	1
Botulinum toxin	0.00001

Note: The LD_{50} is one way to measure the short-term poisoning potential (acute toxicity) of a material. Toxicologists use many kinds of animals, but most often testing is done using rats and mice.

Source: Based on data in Klaassen (2005).

The different sensitivity within the same species is generally influenced, among others, by two parameters: age and sex.

Compared to adults, young children tend to be more susceptible to chemical toxicity because their inhalation volumes are relatively high compared to adults, and their gastrointestinal absorption rate is higher due to the increased permeability of the intestinal epithelium.

Children are highly susceptible to exposure to air pollutants. Minute ventilation is higher in children than in adults because children have higher basal metabolic rates and engage in more physical activity than do adults, and because children spend more time outdoors than do adults. On the basis of body weight, the volume of air passing through the airway of a child at rest is twice that of an adult under similar conditions (Künzli et al. 2010). In addition, because their enzymes for xenobiotic metabolism are immature, the excretion rate of toxic chemicals is relatively low, so the risk of toxicity may increase. In the early stages of their development, due to the immaturity of the central nervous system, children are especially susceptible to the neurotoxicity of various substances, such as lead and methylmercury, because these toxic compounds have direct (on intraneuronal regulatory mechanisms) and indirect (neurotoxic) effects. The neurological toxicity of methyl mercury at low doses in children especially affects memory (Freire et al. 2010; Oken et al. 2005; Weil et al. 2005), language and verbal skills (Lederman et al. 2008; Freire et al. 2010), and visual-motor function (Oken et al. 2008; Surkan et al. 2009). All these effects correlate with the loss of neurons from several areas of the brain (Korogi et al. 1998; Eto et al. 2010). In addition to cerebellar neurodegeneration, abnormal migration of neurons in the cerebellum and microtubule formation deficits were observed during fetal neural development (Choi et al. 1978; Castoldi et al. 2000).

Conversely, elderly people may be susceptible to adverse effects and toxicity due to age-related changes in body composition, organ-system function and consumption of a wide range of potent drugs. With advancing age, properties of functional systems that are involved in toxicokinetics process are altered (i.e. absorption metabolism, and excretion; Shibamoto and Bjeldanes 1996).

Relating to sex, there are susceptibility differences with respect to many toxic substances. These differences also occur in many mammalian species and are related to enzymatic activities, mechanisms of DNA repair and hormonal factors, and the presence of relatively greater fat deposits in women. In consequence, this produces a greater accumulation of some lipophilic toxins, such as organic solvents.

Sex differences in exposure, behavior, anatomy, physiology, biochemistry and genetics influence toxicokinetics and toxicodynamics from the molecular to whole animal level, accounting for male-female differences in responses to xenobiotics in humans and other animals. The Institute of Medicine (IOM; Wizemann and Pardue 2001) concluded that "sex matters" and exhorted: "Being male or female is an important fundamental variable that should be considered when designing and analyzing basic and clinical research". Aside from obvious differences related to sex-specific organs and reproductive events, xenobiotics can interact differently with the male and female sex hormones and their receptors. Some studies that observe male-female differences have determined that they are all due to the average body size differences (Schwartz 2003), without leading to mechanistic investigation. In other cases, statistical adjustment obscured differences. Laboratory research continued to focus on male animals or male cells, perhaps because of a mistaken belief that female development and physiology are intrinsically more variable than males (Itoh and Arnold 2015).

1.3.2 Routes of Exposure

The main ways in which toxic substances can enter the body are digestive (ingestion), pulmonary (inhalation), skin (topical, percutaneous or dermal) and other parenteral routes. In general, the most intense effect and the fastest response occurs when toxic substances are introduced directly into the bloodstream (intravenously). A decreasing order of efficacy would be inhalation, intraperitoneal, subcutaneous, intramuscular, intradermal, oral and dermal.

1.3.3 Toxicokinetic Processes

Four basic processes govern the concentrations toxicants achieve within vulnerable tissues: absorption, distribution, metabolism and excretion (Figure 1.3). These processes describe how a toxicant penetrates cell barriers to enter tissues (absorption), whether it is dispersed to particular organs and tissue compartments (distribution), how it undergoes chemical transformation within the liver (metabolism) and whether the parent compound or its metabolites are permanently eliminated in urine, faeces or both (excretion). The acronym ADME summarizes the four main processes involved in the toxicokinetic phase of xenobiotic action (Burcham 2014b).

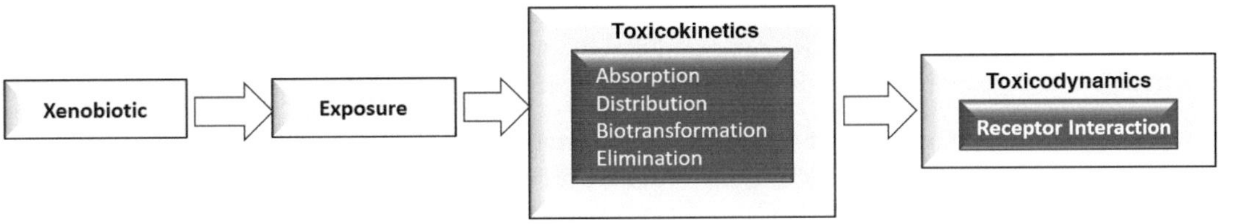

Figure 1.3 Processes affecting xenobiotic toxicity: exposure, toxicokinetics and toxicodynamics. Penetration of a xenobiotic from the environment to the sites of its toxic effect inside the organism can be divided into three phases. The exposure phase encompasses all processes occurring between various toxicants and/or the influence on them of environmental factors. The toxicokinetic phase encompasses absorption of xenobiotic into the organism and all processes which follow transport by body fluids; distribution and accumulation in tissues and organs; biotransformation (metabolism) to metabolites; and elimination (excretion) of toxicants and/or metabolites from the organism. The toxicodynamic phase refers to the interaction of toxicants with specific sites of action on or inside the cells (receptors), ultimately producing a toxic effect.

Some chemicals are readily absorbed and others are poorly absorbed. The rates and extent of absorption may vary greatly depending on the form of the chemical and the route of exposure. The behaviour of a xenobiotic during these processes is influenced by its basic physicochemical properties, including hydrophobicity, or more specifically the lipid/water partition coefficient, and the pH of the medium in which the xenobiotic is found (Ballantyne et al. 1993; Repetto 1995).

Absorption may occur through the alimentary tract, skin, lungs, eyes, mammary glands or uterus, and also from sites of injection. Toxic effects may be local, but the toxicant must be dissolved and absorbed to some extent to affect the cell. Solubility is the primary factor affecting absorption.

The skin, lungs, and alimentary canal are the main barriers that separate higher organisms from an environment containing a large number of chemicals. Toxicants must cross one or several of these incomplete barriers to exert deleterious effects. A chemical absorbed into the bloodstream through any of these three barriers is distributed throughout the body, including the site where it produces damage, which is the target organ or target tissue. A chemical may have one or several target organs; conversely, several chemicals may have the same target (Klaassen 2005). The distribution of toxicants and toxic metabolites throughout the body ultimately determines the sites where toxicity occurs.

Metabolism, also known as biotransformation, is a major factor in determining toxicity. The products of metabolism are known as metabolites, and there are two types of metabolism: detoxification and bioactivation. Detoxification is the process by which a xenobiotic is converted to a less toxic form. This is a natural defence mechanism of the organism. Generally, the detoxification process converts lipid soluble compounds to polar compounds. Except for the lungs, polar (hydrophilic) substances are more prevalent than lipid-soluble toxicants, which are excretable in bile and urine. Bioactivation is the process by which a xenobiotic may be converted to more reactive or toxic forms.

There are two phases of metabolism. Phase I includes oxidation, reduction and hydrolysis mechanisms. These reactions, catalyzed by hepatic enzymes, generally convert foreign compounds to derivatives for Phase II reactions. Products of Phase I, however, may be excreted as such, if polar solubility permits translocation. Phase II principally involves conjugation or synthesis reactions. Common conjugates include glucuronides, acetylation products, and combinations with glycine. Metabolism of xenobiotic agents seldom follows a single pathway. Usually, a fraction is excreted unchanged, and the rest is excreted or stored as metabolites (Klaassen 2005).

There are many pathways for the elimination of toxicants and/or metabolites: exhaled air via the lungs, urine via the kidneys, bile via the gastrointestinal tract, sweat via the skin, saliva via the mouth mucosa, milk via the mammary glands, and hair and nails via normal growth and cell turnover.

e) Toxicodynamic process, refers to the interaction between the molecules of the toxic substance and the cellular receptors through which the toxic effect is induced.

The toxic action of a chemical is a consequence of the physical/chemical interaction of the active form of that chemical with a molecular target within the living organism.

1.4 INTOXICATION AND TYPES

With the term "intoxication", reference is made to the damage exerted by a toxic agent on a living organism. Intoxications are classified based on different parameters (Burcham 2014b).

1.4.1 Duration of Chemical Exposure Needed to Produce Toxicity

Any intoxication usually offers a clinical picture whose toxic symptoms are usually related to the time needed to produce toxic effect. According to this criterion, they are classified as:

- Acute intoxications, characterized by the immediate appearance (usually less than 24 hours) of the pathological clinical picture. It usually occurs after exposure to a toxic agent at a high dose or after an accidental ingestion. Depending on the substance and the dose received, clinical symptoms may produce irritability, delirium, vomiting, diarrhea, seizures or even death.
- Subacute intoxication. Toxicities that manifest after repeated exposure to chemicals over several days or up to 1 month in duration are termed subacute intoxications. This broad category covers both repeated single dosing with a substance (e.g. an antibacterial drug taken every day to treat a persistent urinary tract infection) and prolonged exposures to chemicals present in our diet as food additives or contaminants of drinking water.
- Chronic intoxication. These are produced by repeated exposure to the toxic agent for a period of time, which can range from days to months or even years.

Some chronic effects caused by chemicals, such as cancer, have very long latency periods. The "latency period" is the time between the beginning of exposure and the appearance of disease caused by that exposure. The length of the latency period for chronic effects makes it difficult to establish the cause-and-effect relationship between the exposure and the illness.

The chronic effects can be varied and include neurological (damage to the nervous system), mutagenic (damage to genetic material that can be transmitted to future generations), carcinogenic (which can cause cancer), reproductive (damage to the female/male reproductive system) and teratogenic (damage to the embryo/fetus) effects.

1.4.2 Etiology of Toxic Agent

According to these criteria, intoxications can be classified into two types (Bello 2001):

- Voluntary: These intoxications are produced by an intentional action, among which are homicides, suicides, abortions, drugs, doping and aphrodisiacs.
- Accidental: These usually occur without any type of intentionality; examples are:

a. Environmental: Pollution in air, water or soil that often affects a large number of people. The causes are usually evacuation of industrial discharges, residues of pesticides in food, emissions of sulphur oxides and nitrogen to the environment by industry or use of fuels, and so forth.

b. Professional: These usually occur because of the slow but repeated action (for months or years) of a substance found in the work environment without adequate protection of ventilation and personal hygiene. Normally the legislation establishes the threshold limit value (TLV), which is the concentration of a substance whose daily exposure does not entail the development of adverse effects and thus regulates the permissible levels in the work environment (Miller et al. 1952).

c. Medicines: One of the most frequent causes of accidental poisonings due to confusion of one drug with another, overdose, drug-drug interactions or interactions between drugs and other substances, such as ethanol.

d. Food: Relatively frequent and produced by biotic or abiotic contamination of food (presence of toxic metals, pesticides, toxigenic microorganisms). In other cases it is due to certain components of a food that respond to an individual susceptibility (allergies, intolerances).

e. Domestic: These occur usually in children and the elderly, and those who inadvertently ingest medications, cleaning products, cosmetics and so forth.

Clinical manifestations can be classified as mild, moderate or severe in magnitude.

1.4.3 Characteristics of the Appearance or Location of the Toxic Effect

The intoxications can be classified as follows:

- Immediate versus delayed effects: Some substances show toxic effects in a few minutes, whereas others manifest themselves after a certain time or even after several years, as is the case with carcinogenic substances.
- Reversible versus irreversible effects: In the first case, the normal situation is restored when the toxin disappears, whereas the damage is irreparable in the second case.
- Local versus systemic effects:
  - a Local toxicity: Some toxic agents with extreme properties, such as irritants or vesicant gases, produce damage at the place of contact. Thus, very alkaline or acidic chemical compounds can cause damage directly if they contact body parts such as skin, the nasal cavity, or the eyes. Damage includes contact dermatitis, burns, blisters or irritation.
  - b Systemic toxicity (target organ toxicity): Most chemicals are absorbed and dispersed through the bloodstream and often cause damage to one or more organs. In some target organs, damage occurs due to a high accumulation of the toxin; an example is the damage to the lungs after the accidental ingestion of paraquat, a widely used herbicide (Burcham 2014b).

In addition to passive diffusion, membrane transporters have an important role in the transport of foreign substances, accumulating high amounts in target organs such as liver or kidneys. This is the case in heavy metals, such as cadmium or mercury, which can produce hepatotoxicity or nephrotoxicity.

In some cases, the specific toxicity of chemicals is localized in an organ because the tissue expresses high levels of enzymes that convert the compound into toxic metabolites that damage cells. This phenomenon is called toxicological bioactivation, and contributes to numerous pathologies induced by toxic substances; for example, German textile industry workers in the 1940s developed bladder cancer after producing the family of azo dyes, because toxic metabolites attacked DNA and proteins in exposed cells (Emiliani et al. 2014).

Other examples of toxicological bioactivation are the appearance of neurological diseases such as schizophrenia or neurological disorders such as autism related to oxidative stress. Oxidative stress is caused by an imbalance between the production of free radicals ROS (Reactive oxygen species are natural products inevitably generated along cellular metabolism. Due to their extreme reactivity, they can damage DNA, proteins and lipids) and the effectiveness of the antioxidant defences of the human body. A predominant physiological source of ROS is the mitochondria, where they are created naturally as by-products of energy generation. Mitochondrial dysfunction could lead to an overproduction of ROS and increase oxidative stress with changes in gene expression and even direct damage to DNA. If DNA repair is unsuccessful, it could lead to mutations in nuclear DNA and mitochondria, including promoting cell death (Telišman 1998). Several pathophysiological mechanisms have been implicated in these orders, including genetic predisposition, monoamine deficiency, circadian disruptions, hypercortisolemia and inflammation (Belmaker and Agam 2008). The involvement of oxidative stress mechanisms has also been suggested in some psychiatric illnesses, including depression, anxiety disorders, schizophrenia and autism spectrum disorders (Valko et al. 2007; Ng et al. 2008; Bouayed et al. 2009). Increased levels of ROS and RNS (Reactive nitrogen species are various nitric oxide–derived compounds, have been recognized as playing a crucial role in the physiologic regulation of many living cells, such as smooth muscle cells, cardiomyocytes, platelets, and nervous and juxtaglomerular cells) (Suzuki and Colasanti 2001; Dhir and Kulkarni 2011; Maes et al. 2011) and altered levels of the antioxidant glutathione (GSH) were reported in postmortem brain samples of depressed individuals (Gawryluk et al. 2011). Actually, oxidative stress mechanisms have been suggested as targets for novel antidepressants (Lee et al. 2013). This seems reasonable considering the reported occurrence of inflammation, oxidative and nitrosative stress as well as declining levels of plasma concentrations and activity of several key antioxidants in samples from depressed subjects (Maes et al. 2011). Perhaps psychologic stress disrupts the oxidant-antioxidant balance within the brain, causing impairment of antioxidant enzyme function. This leads to glutathione depletion and increases oxidative stress. Simultaneously occurring glutamate toxicity, calcium imbalance, and mitochondrial impairment intensify oxidative stress, causing biochemical distress in the brain. This disrupts neurocircuitry and weakens connections in the hippocampus, amygdala and cortex, ultimately causing behavioural and cognitive deficits. It seems reasonable to suggest that tight regulation of oxidative stress, either by enhancing the activity of enzymes of antioxidant defence or by directly quenching pro-oxidants, offers the potential to limit psychiatric symptoms (Salim 2017).

1.4.4 Dose Effect or Dose Response

"Dose" is usually defined as the amount (in units, such as mg/kg body weight) of a xenobiotic that enters an organism. According to current terminology, the adverse effect is the change in the morphology, physiology, growth, development or lifetime of an organism that results in a deterioration in the ability to compensate for an additional stress or an increase in susceptibility to harmful effects of other environmental influences. The decision of whether or not an effect is adverse is a matter that requires the judgment of the expert (Burcham 2014b). There are some toxic effects, such as death or cancer, that do not have degrees but are "all or nothing" effects.

The relationship between the response of the biological system and the amount of toxic substance administered takes such a consistent form that it is considered the most general and important concept of toxicology.

From a practical point of view, there are two types of relationships between dose and response (Burcham 2014b):

a. Individual or gradual effect dose relationship, the response of a single organism or individual at variable doses, is often called gradual because the effect measured increases according to the dose. This type of effect frequently appears in *in vitro* studies on isolated organs, tissues or cells that are very useful to study mechanisms of toxicity. It can also be observed in experimental animals but it is more difficult to interpret because the gradual effects are usually masked by the regulatory mechanisms of the organism. It should be noted that most toxic substances have several places of action and mechanisms of toxicity; each of them has its own dose-effect curve with its respective harmful effects.

b. Population or quantal dose-response relationship represents the distribution of responses to different doses in a population of organisms. Unlike the gradual dose-response relationship in a population is characterized because the individual responds or does not respond. Dose-response representations can provide important information. Experimentally it has been proven that the quantal dose-response relationship usually shows a normal distribution of frequency represented by a Gaussian, or bell-shaped, curve (Figure 1.4). In this curve, we can see how three groups of individuals can be described in relation to the action of the toxin: normal, hypersensitive or hyposensitive. This phenomenon of sensitivity to the effect of a certain toxic can occur among groups of individuals of the same population, races of the same species, species and so forth.

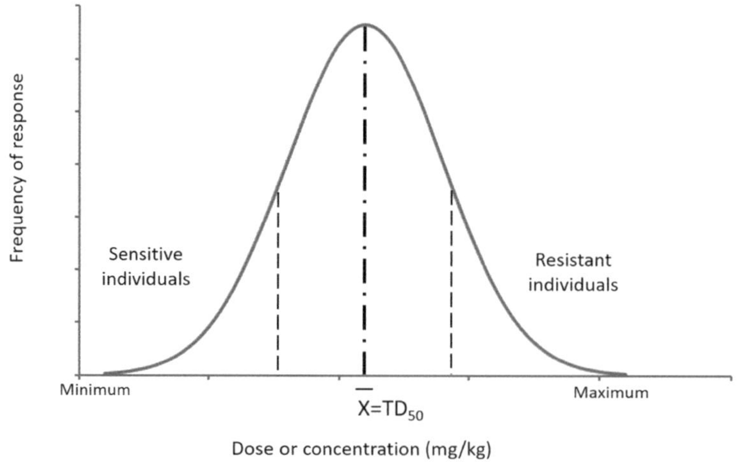

Figure 1.4 Quantal frequency dose-response relationship. This determines the dose required to produce toxic effects for each member of the population. Dose-response relationships typically proceed on the assumption that target populations are homogenous and comprise individuals who conform to bell-shaped Gaussian distributions. A median toxicant induces toxicity of comparable severity within most individuals in the population. However, the population also contains small numbers of individuals (sensitive individuals) who show toxicity at relatively low exposures (minimum dose) and other individuals (resistant individuals) who only exhibit toxicity at high exposure levels (maximum dose).

Abbreviation: TD_{50} = dose at which 50% of the exposed population experiences toxic effect.

Table 1.2 The Most Common Terms in Dose Response Curves

Term	Meaning
Dose (of a substance)	Total amount of a substance administered to, taken up, or absorbed by an organism, organ, or tissue. Often expressed as mg/kg/day.
Effective dose (ED)	Dose of a substance that causes a defined magnitude of response in a given system. Note: ED_{50} is the median dose that causes 50% of maximal response.
Dose effect	Relation between dose and the magnitude of a measured biological change.
LD_{50} (effective dose)	Dose that causes 50% lethality in an animal population.
Lethal dose (LD)	Amount of a substance or physical agent (e.g. radiation) that causes death when taken into the body.
Threshold	Dose or exposure concentration below which a defined effect will not occur.
Threshold dose (ThD 0.0)	The threshold dose (ThD 0.0) is measured as mg/kg/day.
Toxic dose (TD)	Amount of a substance that produces intoxication without lethal outcome.
Latent period	Delay between exposure to a harmful substance and the manifestations of a disease or other adverse effects. Period from disease initiation to disease detection.
NOAEL	No observed adverse effect level, or the highest dose that does not cause a toxic effect.
LOAEL	Lowest observed adverse effect level. The lowest observed effective dose on a dose-response curve, or the lowest dose that causes an effect. A safety factor is a formal, arbitrary number with which one divides the NOAEL or LOAEL derived from animal experiments to obtain a tentative permissible dose for humans. Safety factors range from 100 to 103.
Limit value (LV)	Limit concentration at or below which Member States of the European Community must set their environmental quality standard and emission standard for a particular substance according to Community Directives.
Potency (in toxicology)	Expression of relative toxicity of an agent as compared to a given or implied standard or reference.
Tolerance	Adaptive state characterized by diminished effects of a particular dose of a substance: the process leading to tolerance is called adaptation.

Source: IUPAC (2007).

A quantal dose-response curve supplies useful quantitative estimates (Table 1.2) that provide helpful insight into the toxicity of a given compound. For example, the dose eliciting the reported toxic response in 50% of the population can be easily determined (i.e. TD_{50}), as can the threshold dose at which toxicity is first observed, the lowest observed adverse effect level (LOAEL). A related concept is the no observed adverse effect level (NOAEL), which can be estimated from dose-response data of this kind (Figure 1.5).

Other parameters can also be determined, such as the threshold dose (ThD 0.0), which corresponds to the level of toxic substance below which no toxic effect is expected. The threshold is the dose below which no effect is detected or above which an effect is first observed. The threshold information is useful in extrapolating animal data to humans and calculating what may be considered a safe human dose for a given toxic substance. The threshold dose is measured as mg/kg/day. It is assumed that humans are as sensitive as the test animal used.

These parameters are insufficient, so it is necessary to establish safety margins, which are determined by the ratio between the safe dose, or threshold dose, and the lethal dose (LD). This margin of safety is very useful in environmental and food toxicology, which deals with parameters of high importance, such as NOAEL (the highest dose at which there was not an observed toxic or adverse effect), ADI (acceptable daily intake: maximum amount that an additive can be ingested in the diet

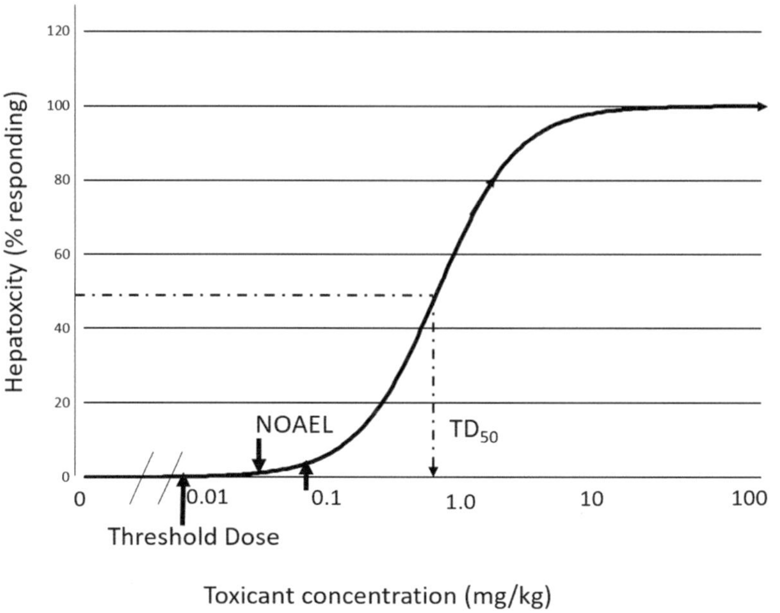

Figure 1.5 Semi-log dose response curves. The graph represents of the percentage of observed subjects with evidence of hepatotoxicity over the different dosages studied. Low doses are insufficient to generate a response, whereas high doses generate a maximal response. Toxic response in 50% of the population can be easily determined (TD_{50}), as can the threshold dose at which toxicity is first observed. Related concepts are the no observed adverse effect level (NOAEL), which can be estimated from dose-response data of this kind.

Source: Burcham (2014a).

throughout life, without adverse health effects) or VUL (limit threshold value that is the concentration of an agent to which a living organism can be exposed for 8 hours a day and 5 days a week without any damage to health).

When using a high number of doses with a high number of animals per dose, sigmoid dose-response curves can be obtained. The sigmoid curve has a relatively linear portion, and the slope of this region of the dose-response curve allows comparison between different toxins and determination of a toxin's potency and how effectively it induces a toxic effect. In Figure 1.6, the effects of three toxic compounds are compared, and we can observe that compound A shows a higher carcinogenic power than compounds B or C. This representation shows the LD_{50} or TD_{50}, which are defined as the single dose of a substance that is expected to cause death, or a certain toxic effect, in 50% of the animals subjected to the test, respectively (Burcham 2014a).

The shape of the dose-response curve has numerous and important consequences for the evaluation of toxicity. A conventional assumption of toxicology is that the dose-response relationship between a chemical and an adverse health effect will have a monotonic shape: that is, the slope of the curve does not change sign. In the case of chemicals that exhibit a U-shaped curve, effects are more prominent at low and high doses than they are at intermediate doses.

Nonmonotonic dose-response curves (NMDRCs) are mathematically defined as a change in the sign (positive/negative) of the slope of a dose-response relationship over the range of doses tested. Numerous studies have recognized the occurrence of NMDRCs in organisms' responses to nutrients, vitamins, pharmacological compounds, and other small molecules that interact with receptors including hormones (Vandenberg et al. 2012; Figure 1.7). That is, the magnitude of the harmful effects is great at low doses (or lack thereof) but decreases with increasing dose. When the dose reaches a point of nondeficiency, the harmful effects disappear, and the organism reaches a state of homeostasis. But if the dose increases to abnormally high values, an adverse response will

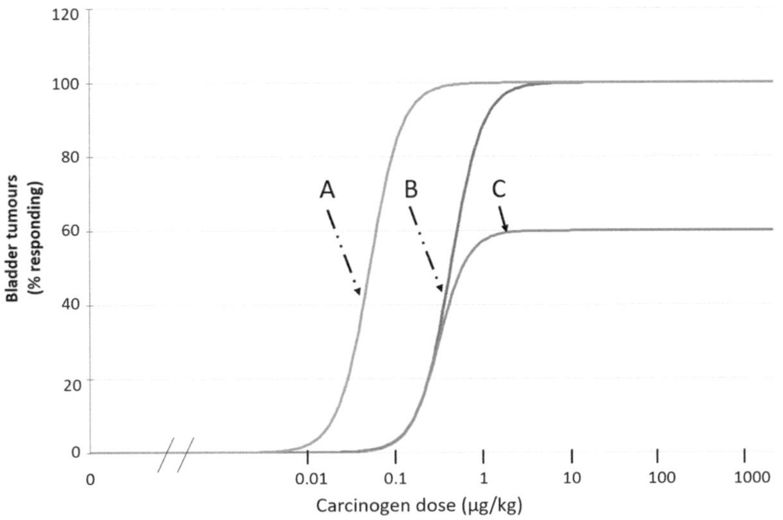

Figure 1.6 Comparison of carcinogenic potency and efficacy of three hypothetical toxicants (A, B and C) at inducing bladder cancer in a population of laboratory rats. Toxicant A has the highest potency because it increases tumour incidence most strongly at the lowest doses tested. The toxicants A and B both induce a maximal tumour response at high doses, indicating that they possess comparable efficacy. Toxicant C has comparable potency to toxicant B yet has lower efficacy.

Source: Burcham (2014a).

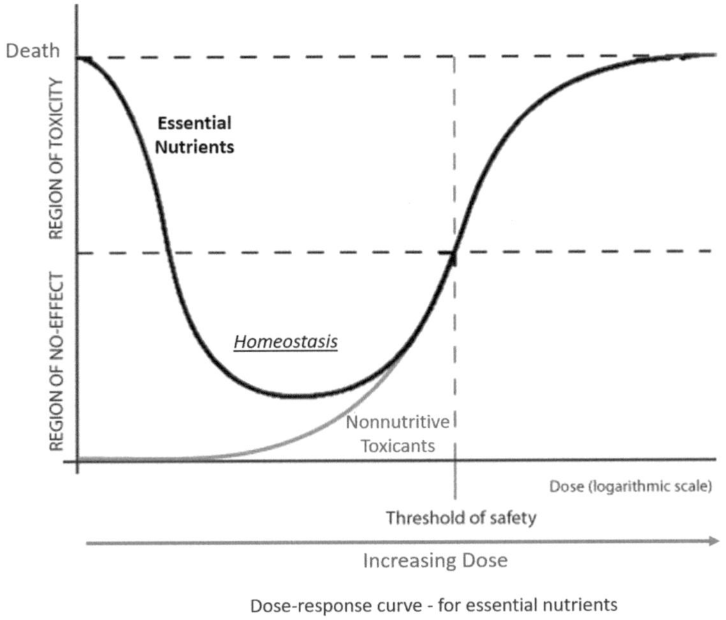

Figure 1.7 Dose-response relationships for essential vitamin or mineral nutrients and other chemical substances (black). The U-shaped hormetic response is shown with a region of homeostasis (the dose range with neither deficiency nor toxicity), which lies below the threshold for adverse response and is discontiguous with both the low-dose deficiency region (whose base is death) and the high-dose toxicity region (Hayes 2008). The grey line shows the dose-response relation.

appear, which will be different from the deficiency, and its magnitude will grow according with an increase of the dose. In the same way, some toxic substances may exert beneficial or stimulating effects at low doses, whereas at higher doses they produce toxic effects; this effect is called hormesis and the substances hormetins.

The term "hormesis" (Calabrese et al. 2007) has been most widely used in the toxicology field, where investigators use it to describe a biphasic dose response with a low-dose stimulation or beneficial effect and a high-dose inhibitory or toxic effect and those substances called hormetins. The response of the cell or organism to the low dose of some toxics substances is considered an adaptive compensatory process following an initial disruption in homeostasis. Thus, a short working definition of hormesis is: "a process in which exposure to a low dose of a chemical agent or environmental factor that is damaging at higher doses induces an adaptive beneficial effect on the cell or organism". The prevalence in the literature of hormetic dose responses to environmental toxins has been reviewed comprehensively (Calabrese and Blain 2005), as have the implications of toxin-mediated hormesis for understanding carcinogenesis and its prevention (Calabrese 2005).

These graphs also allow extrapolations, qualitative or quantitative estimations of the toxicity (extrapolations of the risk) obtained by transferring data from one species to another; or a series of dose-response data (generally in the high-dose range) to dose-response areas, for which there are no data and allows the health authorities to carry out the risk assessment for a specific toxin.

Extrapolations are theoretical qualitative or quantitative estimates of toxicity (risk extrapolations) that usually must be made to predict toxic responses outside the observation range. Mathematical modelling is used for extrapolations based upon an understanding of the behaviour of the chemical in the organism (toxicokinetic modelling) or the understanding of statistical probabilities that specific biological events will occur (biologically or mechanistically based models). Some national agencies have developed sophisticated extrapolation models as a formalized method to predict risks for regulatory purposes.

The procedures used to extrapolate from high to low doses are different for assessing carcinogenic effects and noncarcinogenic effects:

- Carcinogenic effects in general are not considered to have a threshold, and mathematical models are generally used to provide estimates of carcinogenic risk at very low dose levels.
- Noncarcinogenic effects (e.g. neurotoxicity) are considered to have dose thresholds below which the effect does not occur. The lowest dose with an effect in animal or human studies is divided by safety factors to provide a margin of safety. Based on this risk assessment, the health authorities will establish the necessary political decisions (risk management) to control the identified hazards and their communication to the population.

1.5 DOSE-RESPONSE RELATIONSHIPS TO MIXTURES OF CHEMICAL SUBSTANCES

Most of the toxicological databases of substances that produce adverse effects on health have been established using data from cell cultures, laboratory animals and even humans, but they are usually data related to exposure to a single toxin. These data can vary because, in daily life, multiple exposures usually take place simultaneously to many toxic substances that can potentially interact with each other. The latency period of the pathological response between exposure and disease is an additional complicating factor. This is the case of the appearance of tumours by accumulation of multiple genetic alterations for many years; the period of cancer manifestation can occur decades later.

The physical, chemical and biological agents can interact with each other in each phase of the toxicokinetic and/or toxicodynamic processes, with the result of three possible effects (Regl. (CE) N° 1272/2008):

- Independent: Each agent produces a different effect due to a different mechanism of action.
- Synergistic: The combined effect is greater than that of each agent separately. Here we can distinguish two types: additive, when the combined effect is equal to the sum of the effects produced separately by each agent; and enhancer, when the combined effect is greater than the sum of the individual effects.
- Antagonist: The combined effect is less than the sum of the individual effects. There are four types of antagonism: functional, chemical, pharmacokinetic and receptor. Functional antagonism assumes that two substances are counteracted by opposite effects: chemical or inactivation. It involves the chemical reaction of two toxins that leads to a less toxic product.

Pharmacokinetic antagonism occurs when absorption, biotransformation, distribution or excretion is altered, and the result is a decrease in concentration or duration of a substance or both in the target organ. The receptor antagonism, also known as a blocker, supposes that two chemical substances share the same receptor and therefore achieve less effect together.

The lack of data from toxicological studies on chemical mixtures highlights one of the main challenges for regulatory authorities. In the European Union, the regulation of the classification, labelling and packaging of chemical substances and their mixtures (Baillie-Hamilton 2008) transferred the responsibility of performing *in vivo* tests of commercial mixtures to industry, to demonstrate the existence of a toxicological hazard. However, no regulatory provision has been adopted for noncommercial artificial mixtures that represent the real-life scenario of exposure.

In addition to genetic factors and cumulative exposure to various chemical substances present in the environment, water and food also influence our individual life habits decisively. Few life factors exert such a strong influence on toxic substances as the consumption of tobacco and alcohol or the performance of physical activity. The first two significantly lower the efficiency of the liver and kidneys in the elimination of toxins from the blood and produce numerous adverse health effects. Conversely, exercise can modify the toxicokinetics of toxic agents so that the amount of toxin that arrives at the target organs or blood can be modified, regardless of the protective effect it has on oxidative stress.

Other habits such as consuming barbecued meat and processed meat have received considerable attention due to likely exposure to cooking by-products that may alter the expression of xenobiotic-metabolizing genes within the gut wall and liver. Observational studies in recent years have associated heterocyclic aromatic amines and polycyclic aromatic hydrocarbons with colorectal cancer. Strong scientific evidence has demonstrated the relationship between cancer and nitrosamines (NA), heterocyclic amines (HCAs), and polycyclic aromatic hydrocarbons (PAHs), which are the major genotoxins derived from cooking and food processing. The mechanisms of the relationship between dietary toxic xenobiotics and cancer risk are not yet well understood, but it has been suggested that differences in dietary habits affect the colonic environment by increasing or decreasing the exposure to mutagens directly and indirectly through changes in the composition and activity of the gut microbiota. Several changes in the proportions of specific microbial groups have been proposed as risk factors for the development of neoplastic lesions and the enrichment of enterotoxigenic microbial strains in stool (Nogacka et al. 2019). Likewise, the consumption of fruits and vegetables in the diet seems to be able to influence the response to exposure to different toxic substances. An increasing number of studies suggest that these diets provide protection against cancer caused by DNA-damaging chemicals (Burcham 2014b). These effects can be mediated via epigenetics. Rising interest in the role of early epigenetic programming in the health of subsequent generations stems from the recognition that the cellular phenotype is influenced by factors other than changes in the underlying genetic sequence. The three major epigenetic determinants of prenatal gene expression are histone modification, DNA methylation and noncoding RNAs.

We must consider other effects. Since the beginning of the last decade, it was pointed out that environmental cumulative toxic chemical agents could contribute to the increase in the frequency of obesity in the population. Heindel (2003) analyzed the correlation between the increase of the frequency of overweight in the adult population and the increase in the production of industrial chemical substances; in 2002, they formulated a hypothesis of the causal relationship between both events. Recently, several studies have focused on the possible repercussion of the endocrine-disrupting effect of the energy regulation system (i.e. obesogens and their relationship with the increase in the prevalence of obesity) in practically all countries (Grün and Blumberg 2006; Gluckmen and Hanson 2004; Vallverdú 2005). Soon after, the term "obesogens" was coined, which defines xenobiotics that may be present in the environment and/or in foods and inappropriately regulate and promote lipid accumulation and adipogenesis (Heindel and Levin 2005). Among the xenobiotics considered as possible obesogenic compounds in humans, the most widely studied, have been diethylstilbesterol, bisphenol A, organic compounds derived from tin, genistein and phthalates. Interest focuses on the possibility that prenatal xenobiotic exposure disrupts normal fetal programming of energy homeostasis, conferring a lifelong predisposition towards weight gain. At present, there is evidence that the period of intrauterine development is the most vulnerable for the effect of obesogens, for the specific effect on the promotion of adipogenesis in adults (Gluckmen and Hanson 2004; Bern 1992; Giusti et al. 1995).

However, more studies are needed to demonstrate the causal relationship between the concentration of different obesogens in human biological fluids and the development of obesity.

1.6 MODERN TOXICOLOGY RESEARCH

Toxicology continues to advance rapidly, supported by the development of other disciplines and methodological tools. This is the case for the toxic response in tissues, based on the broader knowledge about toxicokinetics and toxicodynamics. From the large spectrum of toxic compounds, a greater number of studies have been carried out on those that have aroused greatest interest. The conducted studies include those with chronic, neurological, mutagenic, carcinogenic, teratogenic and reproductive toxic effects with high latency periods and presence in the environment, food and cosmetics. Of great interest to health professionals is the increasingly detailed study of the toxic effects of new drugs and the interactions among them, effects that sometimes have promoted their withdrawal from the available pharmacological stock. In 2016, a British team studied the causes for the market withdrawals of weight-loss drugs. Eighty percent of the market withdrawals were based on data from spontaneous reports involving cardiac disorders for eight drugs, psychiatric disorders for seven drugs, and abuse or dependence for 13 drugs. Fenfluramine was marketed for 24 years before its worldwide withdrawal in 1997 for heart valve disease. Benfluorex was marketed in France for 33 years before its withdrawal in 2009 for cardiotoxicity; yet the heart valve disorders resembled those caused by fenfluramine and the chemical structures of these two drugs are very similar (Onakpoya et al. 2016).

The advances in instrumentation, the development of increasingly sensitive, specific assays, and lower limits of detection, have allowed the progress of analytical toxicology. With these new tools, it is easier to obtain information to characterize toxicological mechanisms or describe the interactions between the toxic agents and the different biomolecules. Chiral organic pollutants are a major trend in environmental science research and include compounds with different physical chemical properties and applications such as pesticides, herbicides, pharmaceuticals, flame retardants, and synthetic polycyclic musk. In general, agrochemicals are commercialized as racemic mixtures. However, when a racemate reaches the environment, enantiomers of the compound can differ significantly in their environmental fate and their toxicological impacts; the evaluation of the enantiomeric fraction is of critical importance to assess the environmental risk of each enantiomer (Maia et al. 2017). Therefore, we need its analytical determination, and study of the toxic effects of the configurational isomers, in order to carry out a correct analysis of the environmental risk derived from its use (Ye et al. 2015). These advances in analytical toxicology give confidence that new products entering the market, will be safer at all levels of use, because *in silico, in vitro* and *in vivo* tests are increasingly reliable.

Until now, *in vivo* tests with experimental animals have been the basis of toxicity studies of chemical compounds for the evaluation of human risk, but the use of labour animals for evaluating toxic effects of chemicals raises ethical concerns. *In vitro* methods (which include technologies such as transcriptomics, proteomics and/or metabolomics), with organ-on-a-chip or toxigenomics and computational (or *in silico*) toxicology, are the applications that allow us to understand the complexity of the biological systems. These methods, which have been used as a complement to traditional *in vivo* studies, are now the paradigm of modern testing for the identification and characterization of chemical mechanisms of toxicity in humans (Raunio 2011; Saeidnia et al. 2015). In fact, the European Food Safety Authority (EFSA), which has as one of its priority objectives the evaluation of new methodologies and technologies that serve to improve the evaluation of risks in food and feed, launched a project in 2011 to know the current situation of technologies and the application of these technologies in the assessment of risk and its future prospects.

Another important problem is the great gap in the knowledge of environmental toxins. The industrial development of the last decades has resulted in the accumulation of substances that cause damage to the environment and health There are more than 1000 industrial chemicals which have been found to be neurotoxic in experimental studies. Although more than 200 are documented to produce neurotoxic effects in humans, only 11 of them have been shown to have a toxic effect on human neurodevelopment and lead to neurodevelopmental disorders. This does not mean, however, that exposure to any of the other chemicals is harmless to fetal development but simply that they have not been studied (Ornoy et al. 2015; Emiliani et al. 2014; Grün and Blumberg 2009). Some environmental pollution can modify the epigenome by altering the pattern of gene expression at susceptible life stages such as pregnancy, neonatal life, childhood or juvenile periods. These alterations can produce harmful effects on health in later adult life, or even worse, they could be transmitted to future generations (Burcham 2014b).

Other toxic environmental compounds are the chemical compounds that act as endocrine-disrupting chemicals (EDCs); as already mentioned, these are coupled to the hormonal receptors and act as agonists, antagonists or hormonal modulators. Some EDCs frequently show non-monotonic dose-response relationships (Shanle and Xu 2011; Nadal et al. 2018). Although several molecular mechanisms have been proposed to explain NMDR relationships, they are largely undemonstrated (Nadal et al. 2018).

These environmental pollutants could be related to climate change (Europa Commission 2018). On the one hand, global warming can affect the movement and levels of chemicals, such as organochlorine pesticides in the environment, weakening the ability of animals and humans to tolerate these chemicals. On the other hand, as chemical exposure increases, sensitive populations of animals and humans may experience a reduced ability to handle extreme temperatures, severe storms, lack of food and other hazards of climate change.

The effect of toxic agents has been extended to include the study of nanomaterials, given the widespread presence in various facets of our lives: at work, pharmacological, food-related, domestic and so forth. Therefore, nanotoxicology has arisen, whose objective is to study the toxic effects of nanomaterials, of nanoparticles, compounds that can even change from a chemical to a biological nature due to their capacity to interact with biomolecules. The risk involved in their wide use has led the control agencies to balance the promotion of nanotechnology for its many beneficial effects on our way of life, with the assessment and regulation of their risks both at human and environmental level (Stone et al. 2017).

In summary, the awareness of the vulnerability of humans, and the environment, to the multiple substances to which we are exposed opens up new branches of toxicology as well as the need to study more techniques, especially at the molecular level, in order to establish prevention and diagnostic measures and treatment.

BIBLIOGRAPHY

Alexandra, S., A. S. Maia, A. R. Ribeiro, P. M. L. Castro, and M. E. Tiritan. 2017. Chiral analysis of pesticides and drugs of environmental concern: Biodegradation and enantiomeric fraction. *Symmetry* 9(9): 196.

Arnold, D. L. 1983. Two-generation saccharin bioassays. *Environ Health Perspect* 50: 27–36.

Baillie-Hamilton, P. F. 2008. Chemical toxins: A hypothesis to explain the global obesity epidemic. *J Alternat Complemen Med* 8: 185–92.

Ballantyne, B., T. Mares, and P. Turner. 1993. *General & Applied Toxicology*, Vol. 1. New York: Stockton Press.

Bello Gutiérrez, J. 2001. *Fundamentos de Ciencia Toxicológica*. Madrid: Editorial Díaz de Santos.

Belmaker, R. H., and G. Agam. 2008. Major depressive disorder. *N Engl J Med* 358: 55–68.

Bern, H. 1992. The fragile fetus. In *Chemically-Induced Alterations in Sexual and Functional Development: The Wildlife/Human Connection*, eds. T. Colborn and C. Clement, 9–15. Princeton, NJ: Princeton Scientific Publications.

Bouayed, J., H. Rammal, and R. Soulimani. 2009. Oxidative stress and anxiety: Relationship and cellular pathways. *Oxid Med Cell Longev* 2: 63–7.

Burcham, P. C. 2014a. The emergence of modern toxicology. In *An Introduction to Toxicology*, 1–27. London: Springer-Verlag.

Burcham, P. C. 2014b. Core concepts in toxicology. In *An Introduction to Toxicology*, 29–51. London: Springer-Verlag.

Burcham, P. C. 2017. Acrolein and human disease: Untangling the knotty exposure scenarios accompanying several diverse disorders. *Chem Res Toxicol* 30(1): 145–61.

Calabrese, E. J. 2005. Historical blunders: how toxicology got the dose-response relationship half right. *Cell Mol Biol* 51: 643–654.

Calabrese, E. J., K. A. Bachmann, A. J. Bailer, et al. 2007. Biological stress response terminology: Integrating the concepts of adaptive response and preconditioning stress within a hormetic dose-response framework. *Toxicol Appl Pharmacol* 222(1): 122–8.

Calabrese E. J., and Blain R. 2005. The occurrence of hormetic dose responses in the toxicological literature, the hormesis database: an overview. *Toxicol Appl Pharmacol* 202: 289–301.

Castoldi, A. F., T. Coccini, S. Ceccatelli, and L. Manzo. 2000. Neurotoxicity and molecular effects of methylmercury. *Brain Res Bull* 55(2): 197–203.

Choi, B. H., L. W. Lapham, L. Amin-Zaki, and T. Saleem. 1978. Anormal neuronal migration, deranged cerebral cortical organization, and diffuse White matter astrocytosis of human fetal brain: A major effect of methylmercury poisoning in utero. *J Neuropathol* 37: 719–33.

Dhir, A., and S. K. Kulkarni. 2011. Nitric oxide and major depression. *Nitric Oxide* 24: 125–31.

European Commission. Communication from the Commission to the European Parliament, the Council, the European Economic and Social Committee and the Committee of the Regions. Towards a comprehensive European Union framework on endocrine disruptors. COM (2018) 734 final. Available at: <http://ec.europa.eu/transparency/regdoc/rep/1/2018/EN/COM-2018-734-F1-EN-MAIN-PART-1.PDF. Accessed 30 May 2019>

Ellwein, L. B., and S. M. Cohen. 1990. The health risks of saccharin revisited. *Crit Rev Toxicol* 20(5): 311–26.

Emiliani, F., T. Sedlak, and A. Sawa. 2014. Oxidative stress and schizophrenia: Recent breakthroughs from an old story. *Curr Opin Psychiatry* 27: 185–90.

Eto, K., M. Marumoto, and M. Takeya. 2010. The pathology of methylmercury poisoning (Minamata disease). *Neuropathology* 30: 471–9.

Freire, C., R. Ramos, M. J. Lopez-Espinosa, et al. 2010. Hair mercury levels, fish consumption, and cognitive development in preschool children from Granada, Spain. *Environ Res* 110: 96–104.

Gawryluk, J. W., J. F. Wang, A. C. Andreazza, L. Shao, and L. T. Young. 2011. Decreased levels of glutathione, the major brain antioxidant, in post-mortem prefrontal cortex from patients with psychiatric disorders. *Int J Neuropsychopharmacol* 14: 123–30.

Gilbert, S. G. 2012. *A Small Dose of Toxicology*. Healthy World Press.

Giusti, R. M., K. Iwamoto, and E. E. Hatch. 1995. Diethylstilbestrol revisited: A review of the long-term health effects. *Ann Intern Med* 122: 778–88.

Gluckmen, P. D., and M. A. Hanson. 2004. Developmental origins of disease paradigm: A mechanistic and evolutionary perspective. *Pediatr Res* 56: 311–17.

Grün, F., and B. Blumberg. 2006. Environmental obesogens: Organotins and endocrine disruption via nuclear receptor signaling. *Endocrinology* 147: S50–S55.

Grün, F., and B. Blumberg. 2009. Endocrine disrupters as obesogens. *Mol Cell Endocrinol* 304: 19–29.

Hayes, D. P. 2008. Adverse effects of nutritional inadequacy and excess: A hormetic model. *Am J Clin Nutr* 88(Suppl): 578S–581S.

Heindel, J. J. 2003. Endocrine disruptors and the obesity epidemic. *Toxicol Sci* 76: 247–9.

Heindel, J. J., and E. Levin. 2005. Developmental origins and environmental influences. *Birth Defects Res A: Clin Mol Teratol* 73: 469.

Holmberg, B., J. Högberg, and G. Johanson. 2015. Chapter 33—Toxicology: Definitions and concepts. In *Encyclopaedia of Occupational Health and Safety*, ed. E. K. Silbergeld. www.iloencyclopaedia.org/part-iv-66769/toxicology-57951/general-principles-of-toxicology/item/4 (accessed October 23, 2020).

Itoh, Y., and J. P. Arnold. 2015. Are females more variable than males in gene expression? Meta-analysis of microarray datasets. *Biol Sex Diff* 6: 18–27.

IUPAC. 2007. Glossary of terms used in toxicology (2nd edition). *Pure Appl Chem* 79(7): 1153–344.

Jaramillo González, F., A. R. Rincón Sánchez, and F. A. Posadas del Río (Coords.) 2006. *Toxicología básica*. Mexico: Universidad Autónoma de Aguascalientes.

Klaassen, C. D. 2005. *Casarett & Doull's Toxicology: The Basic Science of Poisons*. Madrid: McGraw-Hill.

Korogi, Y., M. Takahashi, T. Okajima, and K. Eto. 1998. MR findings of Minamata disease-organic mercury poisoning. *J Magn Reson Imaging* 8: 308–16.

Künzli, N., L. Pérez, and R. Rapp. 2010. *Air Quality and Health*. Lausanne, Switzerland: European Respiratory Society.

Lederman, S. A., R. L. Jones, K. L. Caldwell, et al. 2008. Relation between cord blood mercury levels and early child development in a World Trade Center cohort. *Environ Health Perspect* 116: 1085–91.

Lee, S. Y., S. J. Lee, C. Han, A. A. Patkar, P. S. Masand, and C. U. Pae. 2013. Oxidative/nitrosative stress and antidepressants: Targets for novel antidepressants. *Prog Neuropsychopharmacol Biol Psychiatry* 46: 224–35.

Maes, M., P. Galecki, Y. S. Chang, and M. Berk. 2011. A review on the oxidative and nitrosative stress (O&NS) pathways in major depression and their possible contribution to the (neuro) degenerative processes in that illness. *Prog Neuropsychopharmacol Biol Psychiatry* 35: 676–92.

Maia, A. S., Ribeiro, A. R., Castro, P. M. L., and Tiritan, M. E. 2017. Chiral Analysis of Pesticides and Drugs of Environmental Concern: Biodegradation and Enantiomeric Fraction. *Symmetry*, 9 (9): 196.

Meek, M. E., A. Boobis, I. Cote, et al. 2014. New developments in the evolution and application of the WHO/IPCS framework on mode of action/species concordance analysis. *J Appl Toxicol* 34: 1–18.

Miller, E., A. M. Plescia, J. A. Miller, and C. Heidelberger. 1952. The metabolism of methylated aminoazo dyes: I. The demethylation of 3′-methyl-4-dimethyl-C14-aminoazobenzene in vivo. *J Biol Chem* 196: 863–74.

Miodovnik, A. 2011. Environmental neurotoxicants and developing brain. *Mt Sinai J Med* 78(1): 58–77.
Nadal, A., E. Fuentes, C. Ripoll, et al. 2018. Extranuclear-initiated estrogenic actions of endocrine disrupting chemicals: Is there toxicology beyond Paracelsus? *Steroid Biochem Mol Biol* 176: 16–22.
National Cancer Institute. 2015. *Artificial Sweeteners and Cancer.* www.cancer.gov/about-cancer/causes-prevention/risk/diet/artificial-sweeteners-fact-sheet (accessed October 23, 2020).
Newman, M. C. 2010. *Fundamentals of Ecotoxicology.* Boca Raton: CRC Press, Taylor & Francis Group.
Ng, F., M. Berk, O. Dean, and A. I. Bush. 2008. Oxidative stress in psychiatric disorders: Evidence base and therapeutic implications. *Int J Neuropsychopharmacol* 11: 851–76.
Nogacka, A. M., M. Gómez-Martín, A. Suárez, O. González-Bernardo, C. G. de Los Reyes-Gavilán, and S. González. 2019. Xenobiotics formed during food processing: Their relation with the intestinal microbiota and colorectal cancer. *Int J Mol Sci* 20(8): 2051.
Oken, E., J. S. Radeskky, R. O. Wright, et al. 2008. Maternal fish intake during pregnancy, blood mercury levels, and child cognition at age 3 years in a US cohort. *Am J Epidemiol* 167: 1171–81.
Oken, E., R. D. Wright, K. P. Kleinman, et al. 2005. Maternal fish consumption, hair mercury, and infant cognition in a U.S. cohort. *Environ Health Perspect* 13: 1376–80.
Onakpoya, I. J., C. J. Heneghan, and J. K. Aronson. 2016. Post-marketing withdrawal of anti-obesity medicinal products because of adverse drug reactions: A systematic review. *BMC Med* 14: 191.
Ornoy, A., L. Weinstein-Fudim, and Z. Ergaz. 2015. Prenatal factors associated with autism spectrum disorder (ASD). *Reprod Toxicol* 56: 155–69.
Raies, A. B., and V. B. Bajic. 2016. In silico toxicology: Computational methods for the prediction of chemical toxicity. *WIREs Comput Mol Sci* 6: 147–72.
Raunio, H. 2011. In silico toxicology—Non-testing. *Methods Pharmacol* 1(2): 33.
Reglamento (CE) N ° 1272/2008 del Parlamento Europeo y del Consejo de 16 de diciembre de 2008 sobre clasificación, etiquetado y envasado de sustancias y mezclas, y por el que se modifican y derogan las Directivas 67/548/CEE y 1999/45/CE y se modifica el Reglamento (CE) nº 1907/2006.
Repetto, M., and P. Sanz. 1995. *Glosario de Términos toxicológicos.* Asociación Española de Toxicología.
Repetto Jiménez, M., and G. Repetto Kuhn. 2009. *Toxicología Fundamental.* Madrid: Editorial Díaz de Santos.
Saeidnia, S., A. Manayi, and M. Abdollahi. 2015. Curr Drug Discov. From in vitro experiments to in vivo and clinical studies. *Pros Cons Technol* 12(4): 218–24.
Salim, S. 2017. Oxidative stress and the central nervous system. *J Pharmacol Exp Thera* 360(1): 201–5.
Schwartz, J. B. 2003. The influence of sex on pharmacokinetics. *Clin Pharmacokinet* 42: 107–21.
Shanle, E. K., and W. Xu. 2011. Endocrine disrupting chemicals targeting estrogen receptor signaling: Identification and mechanisms of action. *Chem Res Toxicol* 24(1): 6–19.
Shibamoto, T., and L. F. Bjeldanes. 1996. *Introducción a la toxicología de los alimentos.* Zaragoza: Editorial Acribia.
Stone, V., M. R. Miller, M. J. D. Clift, et al. 2017. Nanomaterials versus ambient ultrafine particles: An opportunity to exchange toxicology knowledge. *Environ Health Perspect* 125(10): 106002.
Surkan, P. J., D. Wypij, F. Trachtenberg, et al. 2009. Neuropsychological function in schoolage children with low mercury exposures. *Environ Res* 109: 728–33.
Suzuki, H., and M. Colasanti. 2001. NO: A molecule with two masks of 'NO' theatre. *Biofactors* 15: 123–5.
Telišman, S. 1998. Chapter 33—Toxicology: Effects of age, sex and other factors. In *Encyclopaedia of Occupational Health and Safety*, ed. E. K. Silbergeld. www.iloencyclopaedia.org/part-iv-66769/toxicology-57951/general-principles-of-toxicology/item/7 (accessed October 23, 2020).
Valko, M., D. Leibfritz, J. Moncol, M. T. Cronin, M. Mazur, and J. Telser. 2007. Free radicals and antioxidants in normal physiological functions and human disease. *Int J Biochem Cell Biol* 39: 44–84.
Vallverdú, J. 2005. La evolución de la Toxicología: De los venenos a la evaluación de riesgos. *Rev Toxicol* 22: 153–16.
Vandenberg, L. N., T. Colborn, T. B. Hayes, et al. 2012. Hormones and endocrine disrupting chemicals: Low dose effects and non-monotonic dose responses. *Endocr Rev* 33(3): 378–455.
Weil, M., J. Bressler, P. Parsons, K. Bolla, T. Glass, and B. Schwartz. 2005. Blood mercury levels and neurobehavioral function. *JAMA* 293: 1875–82.
Wizemann, T. M., and M. Pardue. 2001. *Exploring the Biological Contributions to Human Health: Does Sex Matter*? Washington, DC. National Academy Press.
Ye, J., M. L. Zhao, L. Niu, and W. Liu. 2015. Enantioselective environmental toxicology of chiral pesticides. *Chem Res Toxicol* 28: 325–38.

2 Toxicokinetics

Jesús Molpeceres, María de los Ángeles Peña and María del Rosario Aberturas

CONTENTS

2.1 INTRODUCTION

When referring to toxicokinetics (TK), the reader will notice that the concepts and basis described herein are the same as those described in pharmacokinetics (PK). Indeed, both TK and PK characterize the absorption, distribution, metabolism and excretion (ADME) processes of xenobiotics, either drugs or toxins, from the time course of their concentrations (kinetics) as parent chemical(s) and/or metabolite(s) in biological media, tissues and excreta. Moreover, ADME studies identify factors influencing the different steps involved and try to represent it mathematically, to quantitatively estimate and predict temporal concentrations of chemicals in target tissues, where pharmacological or toxicological responses may be observed (El-Masri et al. 2015). However, it should be noted that PK and TK differ in terms of goals and technology as well as in the philosophical emphasis in the two approaches (Welling 1995). TK studies mainly refer to animals, either to determine toxicity mechanisms, appropriate species, study design and treatment regimen in subsequent nonclinical toxicity studies, or to assess the relevance of these findings to human safety. In veterinary medicine, TK will contribute to the avoidance of undesirable xenobiotic residues in animal tissues. It is also remarkable that TK focuses more on systemic *exposure*, and typically doses are higher relative to the therapeutic or effective dose (Van der Merwe and Buur 2018; Singh 2018; Rang and Hill 2013). Then, TK is defined as "the generation of pharmacokinetic data, either as an integral component in the conduct of non-clinical toxicity studies or specially designed supportive studies, to assess systemic exposure" (ICH 1995). However, its application towards risk assessment of industrial chemicals (fertilizers, pesticides, biocides, carcinogens, nanomaterials, etc.) and safety of nonpharmaceutical ingredients added to food, cosmetics and personal care products have brought a more clear definition: "the study of the time course of absorption, distribution, elimination (i.e., excretion and metabolism), and uptake of potentially harmful xenobiotics leading to a

 DOI: 10.1201/9780203730584-2

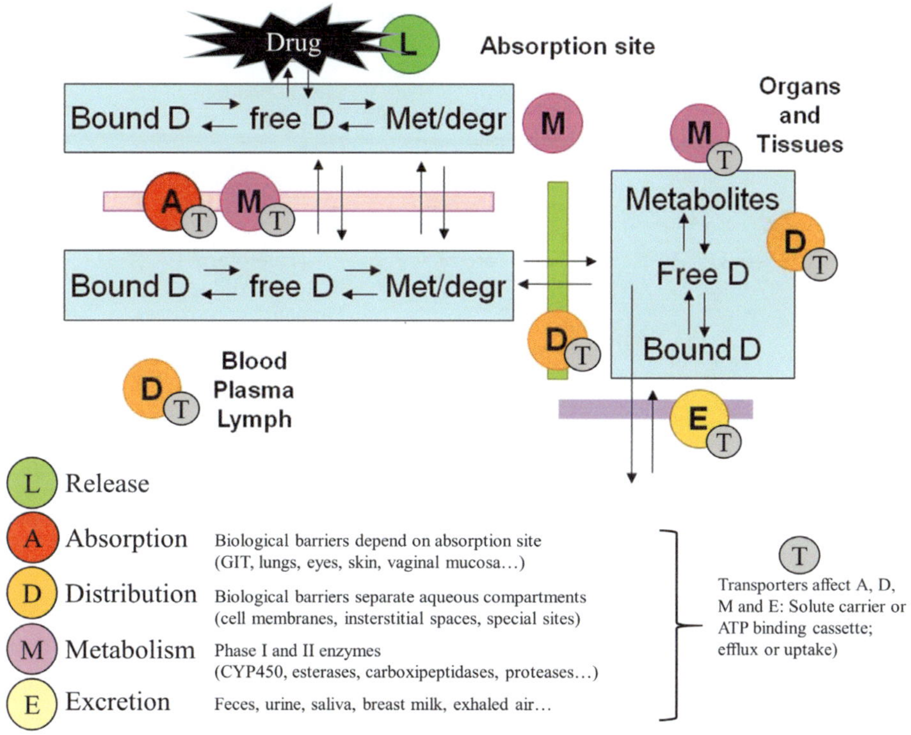

Figure 2.1 General overview of the LADME process.

toxic response" (Boroujerdi 2015). Then, the biological response, either toxicological or pharmacological, is a function of the ADME process associated with the time course evolution of the xenobiotic and/or its metabolite concentrations. It is usually assumed that the higher the dose, the higher the concentration and the longer the time the xenobiotic compound stays in blood, so that a higher body exposure and a more intense and long-lasting response will be attained. This principle was recognized by Paracelsus (1493–1541): "All substances are poisons; there is none which is not a poison. The right dose differentiates a poison from a remedy". Figure 2.1 shows the general scheme of the LADME process. L represents liberation or release of the toxicant from the matrix where it may be contained or the source where it is created.

2.2 DATA IN TOXICOKINETIC STUDIES

To obtain the data needed for TK studies, blood, plasma and urine are the most commonly used sample matrices. However, depending on the goal of the study, tissues, nails, hair or other samples might be relevant. The design of all TK experiments implies the selection of a test species and the response to be measured, an exposure period, the length of the observation period and the different dosing levels to be tested.

Data for TK studies can be obtained from occupational human epidemiology studies, clinical exposure studies, environmentally exposed epidemiologic studies, acute accidental poisonings, animal toxicity tests or other alternatives, such as the use of nonmammalian or nonavian species or *in vitro* systems. Because all have advantages and disadvantages, they should be selected according to the study objective (James et al. 2000). A proper selection and design of the approach best fitted to the study objectives should consider the factors affecting the ADME process (Table 2.1).

As shown in Table 2.1, these factors can be grouped into organism related factors or physicochemical specific factors. Some authors also consider intrinsic and extrinsic factors affecting susceptibility, referring to the differences in toxicity risks resulting from variation in the toxicological response (sensitivity) and exposure. Lifestyle, socioeconomic status, geographical factors and diet can affect exposure. Some ethnic groups eat fish or shellfish frequently or use ethnic medicines, both of which might increase ingestion of mercury. Certain population groups can be highly

Table 2.1 Factors Affecting the ADME Process in TK

Source	Factors	Facts	Example	
Organism	Route of exposure	Differences in epithelial barriers and physiological environment: Parenteral, GIT, skin, lungs, eyes, rectal, vaginal.	LD_{50} methadone in rats (mg/kg)	
			Oral	90
			SC	43
			IV	10
	Species (and strains)	The rate of production of reactive metabolites is different in humans, rabbits, rats, mice.	LD_{50} chloroform (mg/kg/day)	
			Human	602
			Rabbit	100
			Wistar rat	2180
			Sprague Dawley rat	908
			CD 1 mouse	250
			Swiss mouse	1100
	Sex	Metabolic differences, percentage fat of total BW.	Sensitivity to strychnine in rats: more toxic to females than to males when administered SC or IP due to higher rates of metabolism by males.	
	Genetics (ethnic group)	The largest contribution is metabolic differences. Extensive and poor metabolizers.	Acetylation of isoniazid; polymorphic CYP2C19, CYP2D6.	
	Age or life stage	Immature organ development, percentage fat, muscle or water of total BW. "Children are not little adults": chloramphenicol, cisapride.	Differential expression of enzymes during gestation: CYP3A, FMO and SULT1 and 2.	
	Diseases	GIT oedema in congestive heart failure, altered protein synthesis in renal or hepatic impairment.	Asthmatic individuals are more susceptible to air pollutants.	
Physicochemical	Chemical composition	Particle size, chirality, solubility, lipophilicity, pKa, product type, formulation.	LD_{50} (mg/kg) in rats	
			DDT	40
			morphine	265
			fentanyl	3.1
	Exposure conditions	Dose, single or repeated, time, substrate where toxic is included.	Morphine toxicity in rats	
			Single IV dose	LD_{50} 265 mg/kg
			Repeated dosing (35 mg/kg/day)	No toxic effects

Abbreviations: GIT = Gastrointestinal tract; SC = subcutaneous; IV = intravenous; BW = body weight; IP = intraperitoneal; SULT = sulfotransferase; FMO = flavin monooxygenase.

Source: Adapted from James et al. (2000) and Hines et al. (2010).

exposed to contaminants because of their geographic proximity to industrial plants, those near and/or using a polluted water body, or those living or working near roads with dense vehicular traffic. Smoking or drinking alcohol can modify metabolism as well. Indeed, this is another important source of variation because two chemicals playing simultaneously might exhibit additive, synergistic, potentiated or antagonistic effects. These can be based on functional, chemical, dispositional or response mediated interactions. For example, ethanol increases gastritis associated with aspirin (functional), decreases the toxicity of ethylene glycol, thereby inhibiting the production of toxic metabolites (dispositional), the additive effects of ammonia and cyanide (chemical), or the induction of CYP1A2 in smokers, affecting antipyrine or theophylline metabolism (dispositional). A good example of several factors simultaneously affecting the toxicokinetics and metabolism of polychlorinated dibenzo-p-dioxins (PCDDs) and dibenzofurans (PCDFs) has been described by Van den Berg et al. (1994). The absorption, body distribution and metabolism can vary greatly between species and also may depend on the congener (any individual of a specified genus) and dose. Besides, exposure to complex mixtures has the potential to alter the TK of each compound.

2.3 ROUTES OF EXPOSURE

As mentioned above, considering absorption is the first step after release, necessary for a xenobiotic to undergo disposition in the body; the different routes available to reach the systemic circulation will be described next. The administration routes have been classified into four categories (Boroujerdi 2015):

- Category 1 groups various routes that represent a biological barrier that may have different characteristics. The gastrointestinal tract (oral), the respiratory tract and parenteral or mucosal (rectal, vaginal nasal) administration are included here. Depending on the physicochemical and reactive properties of the compound, local and systemic effects can be obtained.
- In category 2, administration takes place directly in the bloodstream, circumventing all barriers (intravenous or intraarterial).
- Category 3 involves percutaneous absorption, given the particular multilayered structure of the skin.
- Category 4 includes all administration routes seeking targeted or local exposure (e.g. intracardiac, epidural, intrathecal, intraarticular, intracerebral, intravesical).

However, it should also be mentioned that in TK, very often the entry of xenobiotics into the body follows environmental exposure; skin (topical), lung (inhalation), eye (ophthalmic) and gastrointestinal tract (oral) are the major routes. Also, it is often challenging to determine the exact dose and the exposure pattern (frequency, duration and extent). Most accidental or autolytic acute poisonings occur after oral ingestion. Toxic reactions associated with gases happen through inhalation, and exposure to toxic industrial chemicals is often due to skin contact. The eyes represent a less frequent route, whereas toxicity after rectal and vaginal absorption is usually associated with illicit drug trafficking. Oral, parenteral and inhalation routes are predominantly linked to fatal poisoning (García et al. 2002).

2.3.1 Gastrointestinal Tract (GIT)

The oral route is preferred for drug product administration because of physiological reasons. It is also very important in TK because many toxicants can be ingested with food. Absorption in the gastrointestinal tract (GIT) is a complex process affected by several factors mainly classified as route related (physiology and dynamics of the GIT) and drug (product) related (physicochemical properties). Important route-related factors include pH, surface, secretory processes (e.g. bile, pancreatic juice), microflora, enzymes, transport proteins, gastric emptying and intestinal motility. Drug-related factors that can significantly affect oral absorption include polymorphism, solubility, chirality, particle size, porosity, partition coefficient, molecular weight, wettability, dosage form (excipients: pharmacologically inactive ingredients in the dosage form). Genetic differences or disease conditions might also affect the GIT.

The whole GIT is 20–25 feet long, and five absorption sites can be distinguished: buccal and sublingual mucosa, the stomach, small intestine, colon and rectum.

Buccal and sublingual absorption: The permeability of xenobiotics in the buccal and sublingual mucosa is ranked between that of the intestine and the skin. There are several epithelial layers with various degrees of differentiation, but the limiting step is permeability through the upper oral epithelium that shows different anatomical characteristics depending on absorption sites (Figure 2.2 and Table 2.2).

Therefore, sublingual absorption is faster than buccal and palatal (the roof of the mouth posterior to the ridge of bone behind the upper teeth). The absorption mechanism is either transcellular passive diffusion or intercellular pathways, which means solubility and pKa are important determinants for absorption. Saliva volume and pH might affect ionization and accordingly modify absorption. Saliva is secreted at a highly variable rate depending on several factors such as stimulation of the taste buds (0.5–7 ml/min); it is hypotonic and has a pH close to neutral conditions (6.2 to 7.6) with buffering potential. Toxicants in the mouth can exert acute local reactions or systemic effects because of sublingual absorption or ingestion if they are lipophilic.

Gastric absorption: The gastric mucosa is a folded surface of 0.1 m^2, and according to the pH partition theory, weak acids at gastric pH (pH 1–3) will be in their unionized form and might be absorbed by passive diffusion. However, physiological gastric transit times are relatively short (half-life of 10–30 min), although various factors affect gastric emptying such as volume, viscosity, tonicity, drugs, food composition, pH, body position and temperature. The structure of the absorbing membrane in the stomach is heterogeneous, as shown in Figure 2.2. Then, regardless of the

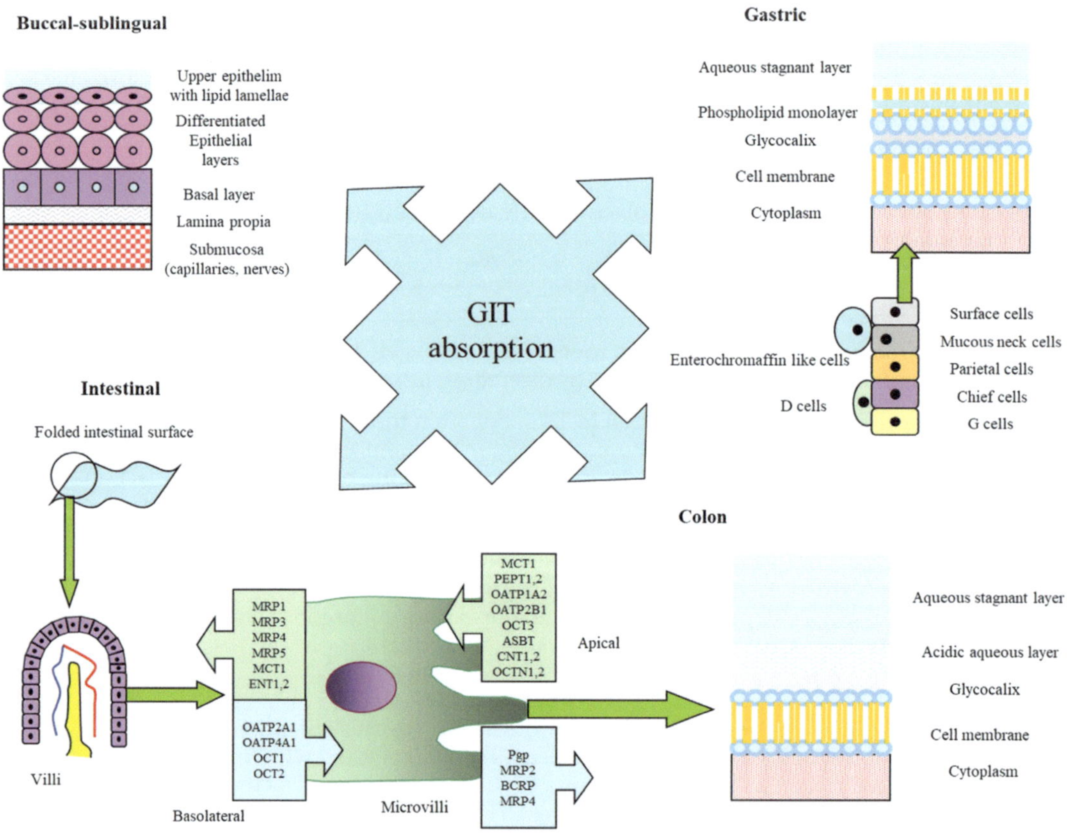

Figure 2.2 Absorption sites in the gastrointestinal tract.

Table 2.2 Differences in Absorption Site within the Oral Cavity

Absorption site	Thickness (μm)	Keratinized surface?	Lipids*
Cheeks and lips (buccal)	500–600	N	Polar
Under the tongue (sublingual)	100–200	N	Polar
Gums (gingival)	200	Y	Apolar
Buccal ceiling (palatal)	250	Y	Apolar

*Typical lipid composition.

Source: Adapted from Bermejo and Garrigues (2013).

potential absorption of acid chemicals in the stomach, the main absorption site in the GIT is the small intestine due to the huge surface area available.

Small bowel absorption: the small intestine is the biological site physiologically and anatomically adapted for absorption after ingestion, with a large surface area close to 200 m^2. The absorption process is conditioned by the absorbing membrane structure. Usually three steps are involved, passing from the lumen (1) through the aqueous luminal layer in the apical side, (2) through the lipid cell bilayer and (3) from the cytoplasm to the basolateral side and lymph or blood capillaries. The lipid membrane has the characteristic sandwich architecture (water–lipid bilayer–water). However, on top of the head groups and glycocalyx facing the luminal side there is a thin and acidic aqueous layer (20 µm, pH 5.3) covered by a thicker (300 µm) stagnant layer acting as a barrier to diffusion of hydrophobic molecules. The main mechanism responsible for absorption is diffusion, either transmembrane or through pores, but influx and efflux transporters are also expressed both in the luminal and basolateral side of enterocytes (Figure 2.2). The permeability of xenobiotics once in the cytoplasm is considered as a nonlimiting step because of the thin and porous structure of capillary and lacteal walls. Nevertheless, the presence of intracellular enzymes such as CYP450 or esterases may affect the bioavailability through a first-pass effect. The first-pass effect (also known as first-pass metabolism or presystemic metabolism) is a phenomenon of drug metabolism whereby the concentration of a drug is reduced before it reaches the systemic circulation. It is the fraction of lost drug during the process of absorption which is generally related to the liver and gut wall. All along the intestine, P450 enzymes and efflux transporters play a combined role to prevent the absorption of certain xenobiotics. However, a first-pass effect may occur in all administration routes depending on the xenobiotic compound. For example, Chang et al. (1994) demonstrated a significant metabolic first-pass effect for the pesticides carbaryl and parathion upon contact with porcine skin; both metabolized to naphthol and paraoxon and para-nitrophenol, respectively.

Absorption in the colon: As compared to the small intestinal site, the colonic mucosa is devoid of microvilli and the absorbing surface is reduced to 1/30. The number of transport proteins is also decreased, but in contrast, the longer residence time and a slightly higher pH of the acidic layer make the diffusion of basic compounds with pKa between 9 and 11 favoured. Besides, the presence of microflora with enzymes showing the ability to cleave specific bonds such as azoreductases, β-lyases, β-glucuronidases, nitroreductases and sulfatases may affect absorption and toxicity. There is clear evidence that air and food pollutants interact with GI microbiota. Liver metabolites of PAHs are deconjugated or transformed into CH_3S^-, nitro-PAHs and nitrotoluenes are reduced to amine metabolites, pesticides are dechlorinated or deconjugated, polychlorinated biphenyls (PCBs) are metabolized to $MeSO_2$ derivatives, metals are demethylated (Hg) or methylated (As, Bi), azo dyes are transformed into aromatic amines or melamine metabolized to cyanuric acid (Claus et al. 2016).

Rectal absorption: The absorption of toxicants in the rectum is mainly associated with illegal trafficking of drugs of abuse. The rectum is a richly perfused area divided into inferior, middle and upper segments according to venous drainage. The epithelium resembles that of the GIT but there is no serosa layer. The volume contained in the rectal ampoule is low (1–3 mL), viscous and holds a neutral pH (6–8) but without buffer capacity. The main absorption mechanism is transcellular passive diffusion and pKa becomes a determinant factor in rectal absorption, which is considered to be slow and erratic depending on the absorptive segment and the facts mentioned above.

2.3.2 Pulmonary Tract

The airways begin with a nasal mucosa of about 100–150 cm^2 and 20 mL connected to the trachea through the nasopharynx, oropharynx and larynx. The inner epithelium is folded into three turbinates (these are shell-shaped networks of bones coated with a thin mucosal layer, containing vessels and tissue located in the external wall of the nasal passageways) provoking a turbulent flow of inhaled air that enables it to intimately contact the mucosa. The nasal epithelium is made of ciliated and nonciliated cells, goblet mucus-secreting cells and basal cells. Mucociliary clearance acts as a physical barrier against the absorption of xenobiotics and prevents the inhalation of particles. The nasal fluid is a mixture of tears and local fluid, with the latter made of 96% water, 2% electrolytes and 2%–3% secretions containing proteins (e.g. albumin, α-macroglobulin, IgA, lactoferrin and aminopeptidases) or mucin (neutral and acidic glycoproteins). Nasal mucus is 2–5 µm thick, and it is cleared to the pharynx with a linear velocity of 5–6 mm/min, which means particle contact with the mucosa will last 20–30 minutes.

Nasal absorption can take place by passive diffusion, either through the membrane or pores, but a plethora of ABC transporters (P-gp, MRP and CFTR), solute carrier (SLC) transporters (POT, DAT, OAT, OATP, OCT, EAAT2/GLT1, and GLUT), amino acids and metal transport proteins have been identified as well (Anand et al. 2014). Also, phase I and II metabolic enzymes such as P450 (family 2 and 3), de-ethylases, hydroxylases, amino- and carboxypeptidases, esterases, glutathione or glucuronosyl transferases, dehydrogenases and oxidases are present with high interspecies variability and eventually higher activity than the liver (Chapter 3). Anilines and nitrosamines are metabolized in the nasal mucosa (Dhamankar 2013). Remarkably, certain disease conditions or toxic compounds can provoke local effects affecting the physiology of nasal mucosa, modifying mucociliary clearance and pH of mucus.

The pulmonary route is further divided into a conductive tract and a respiratory tract. Considering the anatomy of the airways, the upper conductive tract includes the trachea, bronchi and bronchioli up to generation number 16 (the upper airways branch into two or more smaller airways, and each division point or level is called a generation number). Then the respiratory tract goes deeper into the lungs to the alveolar region, corresponding to generation numbers 20–24 (Figure 2.3). The epithelial surface of the respiratory tract shares some features with the nasal epithelium, such as mucus secretion and mucociliary clearance to the pharynx. In contrast, the alveolar region shows a specialized epithelium for gas exchange and absorption with a huge surface (100 m^2). Type I alveolar cells are cells for O_2 and CO_2 exchange, whereas type II cells are type I precursors and produce the luminal surfactant coating containing phosphatidylcholine. The alveolar epithelium is not considered a restrictive barrier to absorption. Then, the lungs constitute an absorption site where local and systemic effects can be attained. Some toxicants can cause irritation and constriction of bronchi and bronchioli, altering mucociliary clearance and mucus production. In contrast, soluble and small particles can be absorbed in alveoli. From a toxicology

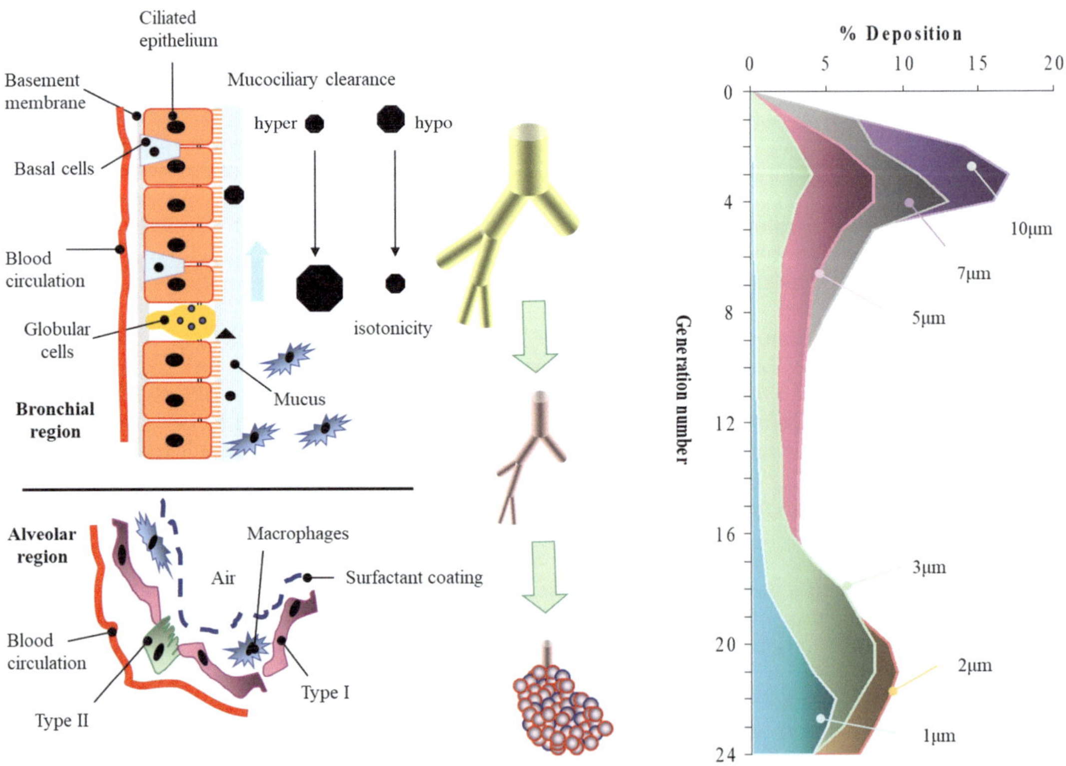

Figure 2.3 The inhalatory route and pattern for particle deposition according to aerodynamic diameter.

standpoint, the lungs can be exposed to gases or vapours and particles. Under normal breathing conditions, inflow of 10,000 L per day in the upper airways is fast, whereas in the deeper lungs, the narrower lumen and higher generation number favour gas exchange by diffusion and particle deposition. Particle disposition in the pulmonary tract depends on the particle aerodynamic diameter, relating particle size to the diameter of a sphere of unit density that has the same settling velocity as the particle of interest regardless of its shape or density. According to their aerodynamic size, particles larger than 10 µm undergo inertial impact in the mouth and throat, and they are removed by nose-wiping, blowing, or by sneezing. Smaller sizes show diffusion and settling, causing 5–10 µm particles to be trapped in the upper conductive tract, which are removed by the mucociliary escalator. Those smaller than 1 µm are light and exhaled. Only 2–3 µm particles will reach and settle down in the alveolar region for further absorption (Figure 2.3). Some particulates (asbestos, fiberglass) that are phagocytized remain in the lungs, where they may have adverse effects and result in the development of respiratory disease.

Absorption in the lungs is related to solubility because only those chemical compounds that can get dissolved in the alveoli quickly will be able to penetrate the epithelium with the help of the surfactant coating. Gases and vapours of volatile compounds diffuse readily across the alveolar cell membrane. Although lipid solubility is important in determining the rate of absorption, the solubility of the toxic substance in the blood and its interaction with components of the blood is also crucial. The mechanisms for absorption are transcellular passive diffusion for lipophilic molecules and diffusion through pores (40–400 nm) for hydrophilic ones. Moreover, many protein transporters including organic cation transporters (e.g. OCT1, OCT2, OCT3, OCTN1), organic anion transporters (e.g. OATP2B1, OATP2B1, OATP3A1, OATP4C1), ABC transporters (e.g. MRP1, MRP3) and peptide transporters (e.g. PEPT2) are expressed. Also, a broad spectrum of P450 (CYP1A1, CYP3A4, CYP3A5, CYP1B1, CYP2B6, CYP2D6, CYP2E1 and CYP2J2), non-P450 (peptidases, flavin mono-oxygenases, esterases and cyclo-oxygenases) or phase II enzymes (UDP glucuronosyl transferases, glutathione S-transferases and sulfotransferases) have been identified.

2.3.3 Ocular Route

The eye is a highly protected organ. In humans, general toxic effects on the eyes are usually considered as mainly external due to the direct ocular exposure. Usually, upon contact with chemicals, the eyes show an acute local irritation reaction, with increased secretion of tear fluid outweighing the tear film volume stably associated with the corneal epithelium (7–30 µL), thus eliminating the chemical from the surface. However, some compounds can permeate the corneal epithelium to further diffuse through the stroma and endothelium into the aqueous humour. Then the trabecular meshwork, iris and ciliary body can be reached. The clearance of chemicals from these structures depends on molecular size. Also, chemicals can be absorbed from the conjunctiva and sclera to the iris and/or ciliary body without entering the aqueous humour due to the larger pore sizes than those found in the cornea. Very low local exposure of the anterior eye chamber is achieved after chemical contact with the eye surface (1%–5%). Factors such as short retention time on the eye surface (less than 2 min), low corneal permeability and systemic absorption from lacrimal fluid (tear fluid differs, in that it consists not only of fluid from the lacrimal gland but also contains components from ocular surface epithelial cells, stromal immune cells and meibomian gland acinar cells) in the palpebral and bulbar conjunctiva explain this fact. Small molecules permeate across the cornea and conjunctiva by passive diffusion (trans-cellularly and/or paracellularly) depending on physicochemical properties, such as logD, hydrogen bonding and polar surface area. Although transporters such as P-glycoprotein (P-gp) and multidrug-associated proteins (MRP-1 and MRP-4) have been identified in the cornea, their influence on ocular exposure remains to be determined (Subrizi et al. 2019). Ocular toxicity is usually associated with nonadverted exposure to environmental or occupational chemicals or accidental splashing. Jaga and Dharmani (2006) reviewed the ocular toxicity from pesticide exposure. They concluded that ocular toxicity resulted from inhalation, ingestion, dermal contact and ocular exposure. Effects of pesticides have been observed in the anterior chamber (conjunctiva, cornea, lens) but also retina and the optic nerve, suggesting permeation to the posterior chamber and/or systemic toxicity after absorption in the lungs, GIT or skin. Pesticide exposure has also been associated with retinopathy, Saku disease, optic-autonomic peripheral neuropathy and abnormal ocular movements. Ocular toxicity has been reported after systemic administration of certain drugs such as antineoplastic agents, ethambutol or hydroxychloroquine.

2.3.4 Dermal Route

The skin is a relatively large organ (1.73 m^2 surface in adults) with a main protective role. It is a multilayered and nonpermeable barrier, where the outermost layer of the epidermis, stratum corneum, is considered as the rate-limiting step in skin absorption. It consists of a cell layer packed with keratin and devoid of blood vessels. The cell walls of the keratinized cells are double in thickness due to the presence of the keratin. Blood vessels are usually about 100 μm below the skin surface.

Drug permeability through the stratum corneum can occur via paracellular or transcellular pathways, but the first (involving the lamellar intercellular space) is the main one. The intercellular route is through the lipids of the stratum corneum and the transcellular route is through the corneocytes. In both cases, the drug must diffuse into the intercellular lipid matrix, which is recognized as the major determinant of drug absorption by the skin. Mainly, small lipophilic molecules can pass through the stratum corneum in small amounts. No active transport mechanisms functioning within the epidermis have been reported so far, but in 2014, Fujiwara et al. described the expression of human solute carrier family transporters in the skin.

Also, a minor absorption pathway is associated with sweat glands, sebaceous glands and hair follicles. Because these structures represent only a very small percentage of the skin's total surface area (0.1%), they are not ordinarily viewed as important contributors to dermal absorption. However, this route enables the permeation of charged molecules and large polar compounds (e.g. peptide-based drugs; Morais et al. 2016).

Once a substance penetrates the stratum corneum, it enters the viable epidermis, the dermis and subcutaneous tissue. These contain a porous, nonselective aqueous diffusion medium where toxicants can readily penetrate by simple diffusion into the circulatory system via the large numbers of venous and lymphatic capillaries in the dermis.

Skin absorption is influenced by several factors. The thickness of the stratum corneum shows regional variability. In the palms and soles, it is very thick (400–600 μm) but is much thinner in the arms, back, legs and abdomen (8–15 μm). The stratum corneum of the neck behind the ear, scrotum and the axillary and inguinal regions is the thinnest. Assuming diffusion is the absorption mechanism, skin permeability inversely relates to the thickness of the epidermis. Then any skin injuries, abrasion, scratching or cuts—either mechanically or chemically induced—will make it easier for toxicants to penetrate this layer. Certain chemicals such as methyl and ethyl alcohol, hexane, and acetone are lipid soluble and can degrade the lipid barrier of the cell membrane. Skin burns and dermatitis are the most prevalent conditions.

The skin's basal level of hydration (7% by weight) favours the absorption of polar substances 10-fold, and additional hydration increases penetration by 3–5 times.

Skin penetration can vary by species: monkey, pig, and guinea pig skin permeability is similar to that of humans; that of the rat and rabbit are generally more permeable; and the skin of the cat is generally less permeable. For safety reasons, the rat and rabbit have been used for preclinical dermal toxicity tests after shaving.

Some toxicants can gain entry into the body after skin contamination. Examples include poisoning of agricultural workers by organophosphate pesticides; death after skin contact with the neurological warfare agent sarin; and systemic toxicity from several industrial solvents such as carbon tetrachloride (hepatotoxic) and hexane (nerve damage). Dermal exposure to semivolatile organic compounds (SVOCs) has often been underestimated but it can occur even to a larger extent than the amount taken in via inhalation. (SVOCs include polybrominated flame retardants, endocrine-disrupting chemicals [EDCs], polycyclic aromatic hydrocarbons, phthalates, plasticizers, pesticides, and antimicrobials and are encountered indoors in cleaning products, detergents, aerosol cans, brushes and sponges.) Also, dermal exposure to particles and dust, the role of clothing and bedding as transport vectors, and the relevance of a potential absorption through hair follicles are all areas of research interest.

As happens in other routes of exposure (lungs, rectal, buccal, sublingual), chemicals entering the blood through the skin escape the hepatic first-pass effect and do not encounter the same detoxification pathways found when ingested and processed by the stomach, intestines and liver, making them potentially more toxic.

2.3.5 Intravaginal Route

The vaginal mucosa is a highly perfused region, usually acidic due to the presence of *Lactobacillus acidophilus* which buffers the pH between 3.8 and 4.2. Regional but also systemic toxicological effects can be expected after vaginal absorption. It takes place mainly by transcellular and paracellular passive diffusion. Although the protein expression of P-gp, BCRP and MRP-2 in endocervical and vaginal tissue of premenopausal women has been confirmed, no uptake transporters have been reported for this site (Grammen et al. 2014; Nicol et al. 2014). It should also be pointed out that the vaginal epithelium possesses protease enzymatic activity. Proteases are likely to be the prominent barrier for the absorption of intact peptide and protein molecules into the systemic circulation (Ashok et al. 2012).

Mucosal irritation induced by some chemicals, vaginal changes during the menstrual cycle, and menopausal changes modify the regional vascularity and thickness of the mucosa, which leads to significant differences in the permeability of the barrier, with inter- and intraindividual absorption variability. The vaginal absorption of therapeutic agents is influenced by their lipophilicity, degree of ionization, chemical structure and molecular weight, and interaction with vaginal barrier and secretion. The TK of a vaginally absorbed compound depends to some extent on the type of product used. For example, the absorption of a compound surrounded by a plastic film depends on the release rate. In general, the absorption through mucosal barriers of drugs and their related pharmacokinetic analysis are the same as other extravascular routes of administration.

2.4 MECHANISMS OF PASSAGE THROUGH BIOLOGICAL BARRIERS

The passage of a compound through physiological barriers can take place by different mechanisms. The GIT is the only site where all these absorption mechanisms are present and in some instances some play simultaneously: (1) passive diffusion, either transcellular or paracellular; (2) carrier-mediated transcellular diffusion or facilitated diffusion, or passive-mediated transport; (3) transcellular diffusion subject to P-gp efflux; (4) active transport; (5) pinocytosis and receptor-mediated endocytosis; (6) solvent drag, osmosis and two-pore theory; and (7) ion-pair absorption. The most important are passive diffusion, facilitated diffusion and active transport.

2.4.1 Passive Diffusion

Passive diffusion is the main mechanism for the transfer of xenobiotics through biological barriers. It can take place through the cell membrane (transcellular) or using pores between the cells (paracellular). Drug molecules on one side of the membrane start crossing because a concentration gradient acts as the driving force. According to Fick's first law, flux (J_{sm}) by passive diffusion can be written as

$$J_{sm}\ (\text{mass length}^{-2}\ \text{time}^{-1}) = \text{P}\ (\text{length time}^{-1})\ \Delta\text{C}\ (\text{mass length}^{-3}) \qquad (2.1)$$

where P is the permeability constant and ΔC is the concentration gradient across the membrane corresponding to dissolved solutes. Therefore, to have a nonlimited mass transfer rate, the concentration gradient should always be in favour of the absorption. This usually happens in most biological barriers because blood flow withdraws absorbed drug molecules from the basolateral side. *In vitro* experiments designed to estimate mass transfer rates refer to this situation as a sink condition, and it means that solute concentrations passing through should never exceed 10%–30% of the solute solubility in the receiving chamber after drug passage. Equation 2.1 shows a clear dependency of drug diffusion on concentration following first-order kinetics. P depends on the diffusion coefficient (D), the partition coefficient (P_c) and the length of the diffusion pathway (x) as shown in Equation 2.2:

$$P\ (\text{length time}^{-1}) = D\ (\text{length}^2\ \text{time}^{-1})\ P_c\ (\text{unitless})\ /x\ (\text{length}) \qquad (2.2)$$

Partition coefficient: P_c is an index of lipid solubility. It is very useful to classify chemical compounds in a rank order, because it has been shown that their values for nonionized forms of several series of representative chemicals and drugs can be correlated with their transfer rates through biological membrane systems—from intestinal lumen into blood, from plasma into brain and cerebrospinal fluid, and from lung into blood. In general, the higher the lipid solubility, the higher the P_c and the easier the passage through the membrane. In many instances, P_c is referred to as log P_c to make values more understandable. P_c can be determined experimentally by the shake flask method using hexane, heptane or n-octanol, or by chromatography. The classical shake flask method is based on the ratio of the equilibrium concentrations at 20–25°C (tolerance ±1°C) of a dissolved substance in a two-phase system consisting of two largely immiscible solvents. Different volumes ratio of n-octanol to water are used, and after shaking both phases are separated by centrifugation and the concentrations in each one determined by appropriate analytical techniques. It can also be predicted using software packages such as ADMET Predictor's PCB, ACD/LogP from the individual contribution of functional groups or molecular structure.

When molecules contain ionizable functional groups like monoprotic weakly acidic or basic xenobiotics or polyprotic ampholytes, their true partition coefficient depends on the pH, and thus, the distribution coefficient D_c is estimated based on the presence of both ionized and unionized forms in the lipid and aqueous phases.

Diffusion coefficient: D is also a determinant feature for passive diffusion. As shown in the Stokes-Einstein equation (Equation 2.3) it is inversely proportional to molecular size:

$$D = k_B T/(N_A 6\eta r) \text{ (length}^2 \text{ time}^{-1}) \tag{2.3}$$

where K and T are the Boltzmann constant (1.38×10^{-23} units J/K) and absolute temperature, respectively; N_A is Avogadro's number; η is viscosity; and r is hydrodynamic radius or molecular size, assuming a spherical shape. In this expression, the energy per molecule is kBT/N_A; the friction per molecule is $6\pi\eta r$ and is equivalent to a force divided by velocity: (g cm s^{-2})/(cm s^{-1}). Larger molecules show slower diffusion because of frictional resistance and in some cases steric hindrance. In contrast, very small molecules pass through membranes faster than expected based on their P_c. Smaller molecules are usually associated with lower P_c values than their larger homologs, and they may be able to move through membrane pores. Pores are present in all membranes, but their size varies with the nature and function of the membrane. For example, GIT cell membranes will not allow passage of water-soluble molecules larger than about 0.4 nm in diameter, whereas renal or hepatic blood capillary walls are permeable to molecules up to about 100 nm in diameter.

Diffusion through pores follows Equation 2.4:

$$J_{sp} = D_p \frac{A_p}{A_m} \frac{\Delta C}{x} \tag{2.4}$$

where Dp is the diffusion coefficient through aqueous pores and A_p/A_m represents the fractional area of pores in the whole membrane surface. In small pores, the effective pore area A_p is reduced because of entrance effects, friction between pore wall and solute, and solute-solute and solute-solvent interactions.

When transcellular and paracellular diffusion occur simultaneously, total flux is the sum of the two processes (Equation 2.5):

$$J = J_{sm} + J_{sp} = \frac{P_c D_m}{x}\left(1 - \frac{A_p}{A_m}\right)\Delta C + D_p \frac{A_p}{A_m}\frac{\Delta C}{x} \tag{2.5}$$

Ionization: When chemicals are ionized at physiological pH, only the nonionized forms will cross the membrane by transcellular diffusion, because small electrolytes will be able to diffuse through pores regardless of their ionization state and ionized larger molecules will show very low diffusion and partition coefficients to pass through the membrane (Figure 2.4).

The Henderson-Hasselbalch equation (Equation 2.6) predicts the fraction of the drug that remains unionized at a given pH depending on pKa.

$$\text{for acids } pK_a - pH = \log \frac{non - ion}{ion} \tag{2.6}$$

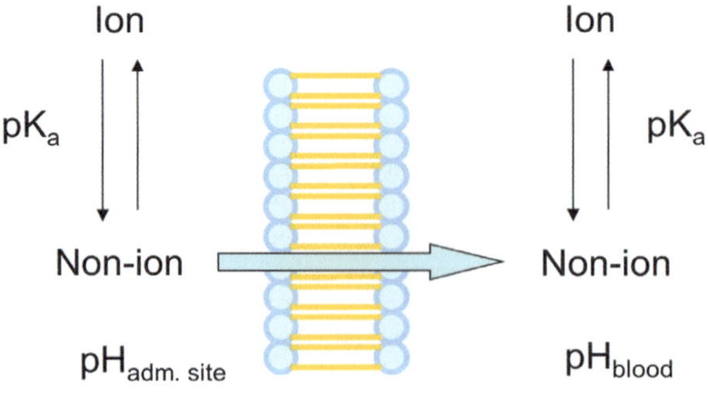

Figure 2.4 The pH-partition theory.

$$\text{for bases } pK_a - pH = \log \frac{ion}{non - ion}$$

Thus, acidic compounds (benzoic acid pKa = 4) will be more ionized at pH values above pKa (1% nonionized at pH 5 and 90% nonionized at pH 3), whereas bases will show an opposite behavior (aniline pKa = 5; 1% nonionized at pH 3 and 90% nonionized at pH 6). Most toxicants move across the cell membrane by diffusion. Organic substances such as nitrous oxide, ethylene and divinyl ether diffuse across the cell membrane of the alveoli easily because they do not have an electrical charge and are lipid soluble.

2.4.2 Facilitated Diffusion

Facilitated diffusion, also called passive mediated transport or carrier-mediated transcellular diffusion, is a mechanism to enable permeation through biological membranes using carrier proteins or ion channels in the membrane that facilitate transport. In contrast to active transport, it is a passive diffusion through a passageway without requiring energy or conformational change of the protein. Therefore, molecules do not cross the barrier against the concentration gradient. Lipid insoluble compounds or those larger than pores allowing for paracellular diffusion may use this mechanism. In facilitated diffusion, substances are transported at a much faster rate than expected based on the molecular size and polarity of the molecule. As opposed to passive diffusion, mass flux (J) is not just driven by the concentration gradient but rather by the amount of the protein expressed on the membrane, the diffusion coefficient and the affinity between the carrier and the substrate. Thus, saturation and/or competitive phenomena may occur, and mass transfer kinetics may deviate from linearity at high concentrations, influencing TK behavior. Compounds such as sugars, amino acids, steroids or vitamin B12 have been shown to use facilitated transport.

2.4.3 Transcellular Diffusion Subject to P-Glycoprotein Efflux

Nowadays, many proteins expressed on membranes have been identified, and their roles in the disposition of chemicals have been characterized. They are usually considered as phase III enzymes (phase III transporters affect drug disposition, including drug metabolism) because they affect the passage of many chemical compounds through biological barriers. These are both expressed in the apical and basolateral side of cells so that they can help in permeability across the membrane or work against it. These proteins are grouped into adenosine triphosphate (ATP)-binding cassette and soluble proteins. One of the most important proteins affecting the absorption of xenobiotics is P-gp, a 170 KDa protein embedded in the cell membrane of enterocytes, hepatocytes, renal tubular cells and endothelial brain cells. The net effect of the P-gp efflux on the kinetics of xenobiotics passage through biological barriers has been well established *in vitro* by using *in vitro* Caco-2 cells culture systems.

2.4.4 Active Transport

Active transport shares some features with facilitated transport, because a carrier protein is needed to help the xenobiotic to pass through a biological barrier. Then, saturation and competitive processes might occur as well. However, active transport also demands energy input and conformational changes of the carrier. It maintains transport against a concentration gradient, and the main energy source comes from ATP hydrolysis. These two steps of energy production or protein interaction can be affected by toxicants like fluorides or dinitrophenols.

Active transport can be referred to as primary transport when ions permeate the membrane because of protein interaction and energy expenditure. Examples are Ca^{2+}, Na^+ or H^+ pumps. Toxic substances such as cadmium, lead or strontium will also bind to the calcium-binding protein (CaBP) because their physicochemical characteristics are similar to calcium. However, in secondary transport, primary transport generates ion gradients that make other ions pass through either in the same or the opposite direction. This is also called cotransport. An example is the intestinal absorption of Na^+ and glucose. Usually, when xenobiotics interact with proteins, the model that best explains how they pass through membranes is the well-known Michaelis-Menten kinetics equation (Equation 2.7):

$$T = T_{max}\, C/K_m + C \tag{2.7}$$

where T_{max} is the maximum transport rate, K_m the xenobiotic concentration when half of the maximum transport rate is achieved and C the xenobiotic concentration.

Active transport is an essential mechanism for nutrient supply and to detoxify the organism. Accordingly, transport proteins are expressed in body sites such as the liver, kidneys, intestine, placenta or brain. More than 49 ABC transporter subtypes have been reported in humans, and they are divided into seven subfamilies: ABCA, ABCB, ABCC, ABCD, ABCE, ABCF and ABCG1. In contrast, more than 384 unique protein sequences have been identified, classified into 52 distinct solute carrier families (SLC1 to SLC52; Fujiwara et al. 2014).

Chedik et al. (2018, 2019) have shown that some pesticides, such as organochlorine, pyrethroid and organophosphorus pesticides, interact with various uptake and efflux drug transporters, including the efflux pump P-gp and the uptake organic cation transporters (OCTs). Table 2.3 is a classification of the different transport proteins.

2.4.5 Pinocytosis and Receptor-Mediated Endocytosis

These mechanisms refer to the surrounding of particles or large molecules by the cell membrane to form vesicles within the cell cytoplasm which are further processed. The size and properties of engulfed material make a difference: if large entities are included in vesicles, the process is called phagocytosis; smaller molecules associated with the extracellular fluid are taken up by pinocytosis. Both are nonselective but require energy for vesicle formation and processing to phagosomes

Table 2.3 List of Transporter Proteins

Organ	Efflux	Uptake
Brain	Apical: P-gp, BCRP, MRP4, MRP5, OAT3, OATP-A, MRP1, MRP3, MCT1	Apical: OATP1A2, OATP2B1
Liver	MRP1, MRP2, MRP3–6, P-gp (MDR1), BCRP, MDR3, OAT7, BSEP, MATE1, OSTα-β	OCT1, OATP-C, OATP-B, OATP8, NTCP, OAT2, OAT7, OSTα-β
Intestine	Apical: BCRP, P-gp, MRP2 Basolateral: OCT1, OSTα-β	Apical: OATP, PEPT1, ASBT, MCT1 Basolateral: MRP3, OSTα-β
Kidney	Apical: OAT4, P-gp, MATE 1 and 2, MRP2 and 4, OCTN1 and 2 Basolateral: OCT1, OCT2, OAT1, OAT2, OAT3, MRP1	Apical: OAT4, URAT1, PEPT1 and 2, OCTN1 and 2
Eyes	MRP1–7, MDR1, LRP in the corneal epithelium MRP1–7, MDR1, LRP and BCRP in the conjunctival epithelium MRP1–2, MRP6–7, MDR1 and LRP in the iris-ciliary body MRP1–3, MRP6–7, MDR1 and LRP in the retina MRP1–3, MRP6–7, MDR1 and LRP in human corneal epithelial cell line (HCEC) MRP7, MDR1, LRP, and BCRP in the human retinal pigment epithelial cell line (ARPE-19)	PEPT1 and 2, OCT1 to 3
Placental membrane	Apical: P-gp, BCRP, MRP2 Basolateral: OCT3, OAT4, OATP2B1	Apical: NET, SERT, OATP4A1, OCTN2 and 1 Basolateral: MRP5 and 1

Transporters (some of the following acronyms are followed by numbers/letters indicating polymorphism, families and subfamilies of transporter) listed in order of occurrence: P-gp = P-glycoprotein (also known as MDR1); MRP = multidrug-resistant proteins; BCRP = breast cancer resistant protein; OAT = organic anion transporter; OATP = organic anion transporter polypeptide; MCT = monocarboxylate transporter; OCT = organic cation transporter; NTCP = Na^+-taurocholate cotransporting polypeptide; OST = organic solute and steroid transporter; BSEP = bile salt–exporting pump; MATE = multidrug and toxin exclusion protein; PEPT = peptide transporter; ASBT = apical sodium dependent bile acid transporter; URAT = uric acid transporter; OCTN = organic cation transporter, novel; LRP = lung resistance protein; NET = norepinephrine transporter; SERT = sodium-dependent serotonin transporter.

Transporter families: ABC transporters (ATP-binding cassette); family members include P-gp, MRP; SLC (solute carrier) family members include OCT, OAT, URAT, MCT.

Source: Adapted from Li et al. (2012), Chen et al. (2013) and the International Transporter Consortium et al. (2010).

and lysosomes. Proteins, polypeptides or particles (e.g. nanoparticles, liposomes, microemulsions) may permeate membranes by pinocytosis. Also, there is a highly selective transport, when endocytosis takes places after the molecular entity binds to a specific receptor or ligand on the cell surface. Phagocytosis happens in the alveoli of the lung and the reticuloendothelial system of the liver and spleen. Clathrin-mediated endocytosis, clathrin-independent, caveolar endocytosis, macropinocytosis or mixed pathways have been identified. The receptors commonly involved are transferrin receptor (TfR), low-density lipoprotein receptor (LDLR), epidermal growth factor receptor (EGFR), G-protein coupled receptor, integrins (αvβ3, αvβ5, α5β1, α6β4, α4β1 and αvβ6), folate receptor (FR), CD44, ICAM, CD55, CD59, CD90, the heavy chain of amino acid transporters CD98; CD147; or the glucose transporter Glut1 (Xu et al. 2013).

2.4.6 Solvent Drag and Osmosis

The passage of large molecules like proteins through membranes can be contributed by another mechanism related to the presence of pores that can be crossed by solvents as a function of hydrostatic and osmotic pressures. Then, solvent permeability may show a significant dragging effect for large molecules. However, this effect is not so important for small molecules because they permeate membranes faster by other mechanisms.

2.4.7 Ion-Pair Formation

Some chemicals can cross biological barriers to a significant extent and at a faster rate even when ionized at the pH of the absorption site. This fact cannot be explained by the pH-partition theory mentioned above for passive diffusion. However, it has been hypothesized that this ionized chemical species might form a neutral complex with a counter-ion in the absorption site with new physicochemical properties better suited for membrane penetration. It is accepted that the ion-pair formation increases the lipophilicity of hydrophilic ionized compounds and enhances their partition coefficient. The formation of the complex will be the rate-limiting step in the absorption process, and once absorbed its breakdown will release the charged molecule in the basolateral side of the membrane. Some counter-ion compounds for positively charged molecules are alkyl carboxylate, cholate, n-alkyl sulphates, n-alkyl carbonates and trichloroacetate. This hypothesis has been used to promote the absorption of peptides, prodrugs and antiviral compounds.

2.5 DISPOSITION OF XENOBIOTIC COMPOUNDS

Disposition is a term describing the combined processes of distribution, biotransformation and elimination (excretion).

2.5.1 Intravascular Distribution

After absorption or intravenous administration through the capillary walls into the blood, xenobiotics undergo first intravascular and then extravascular distribution to other organs and tissues (Figure 2.1). Within the blood compartment, binding to plasma proteins such as albumin, glycoproteins transferrin, globulin, and lipoproteins oppose distribution to other body sites. A dynamic equilibrium exists between the bound and unbound forms of a toxicant in plasma. The equilibrium will be determined by factors such as concentrations or affinity constant and only unbound toxicant passes through the endothelial cells of the capillaries into the extravascular space. Some examples of protein binding include metals such as mercury (95%), ochratoxin (80%), chlorinated phenoxy acid herbicides (Hagelberg et al. 1989; Roberts et al. 2011), or perfluorinated carboxylic acids (PFCAs; 98%) (Ohmori et al. 2003). El-Moneim and Afify (2010) described the protein binding of xenobiotics (especially pesticides) from a general standpoint, including potential displacement and competition between chemically related structures for the same binding sites. Sometimes binding involves different plasma elements, such as aluminium binding to small molecule plasma binders (citrate, fluoride, phosphate, bicarbonate, and low molecular weight proteins), albumin, transferrin and other larger molecular weight proteins (Wilhelm et al. 1990). Also, protein binding has been suggested as a potential biomarker of exposure (i.e. albumin binding of organophosphorus pesticides is considered a complement to the widely used measure of acetylcholinesterase (AChE) inhibition; Tarhoni et al. 2008).

2.5.2 Extravascular Distribution

After extravasation the toxicants may be stored in the target tissue, possibly resulting in an adverse response on other tissue types, preventing exposure of target organs which may not be readily affected. Uptake of xenobiotics may occur either by passive diffusion or by special transport processes. Organs or tissues differ in the amount of a chemical that they receive or to which they are exposed. This is primarily due to blood flow or permeability through a specific tissue barrier. Organs with larger perfusion rates, such as the liver and the kidneys, can potentially accumulate more of a given toxicant. These organs together receive almost 50% of cardiac output and are the main elimination pathways for xenobiotic compounds through metabolism and/or excretion. However, some poorly perfused tissues can be the primary storage sites for many toxicants because of affinity. For example, adipose tissue, which has a meagre blood supply, concentrates fat-soluble toxicants whereas others form complexes with minerals commonly found in bone. Once deposited in these storage tissues, toxicants may remain for long periods due to their solubility in the tissue and the relatively low blood flow. Examples of storage depots for specific chemicals in mammalian organisms are bones (lead, strontium, fluoride), kidneys (cadmium), transferrin (iron), fat (polychlorinated pesticides such as DDT), nerve tissue (lead) and skin or hair (arsenic; Savković-Stevanović 2011). A potential irreversible interaction between the xenobiotic and the tissue might occur (Pumford et al. 1997).

Distribution of toxicants to extravascular sites means mass transfer from and to different aqueous compartments: vascular, interstitial and intracellular. Human plasma accounts for about 5%–7% of the total body weight in comparison to interstitial tissue fluids (15%–20% of body weight) and intracellular fluids (35%). The key parameter characterizing the magnitude of the distribution process is the apparent volume of distribution. It is the total volume (in litres) of body fluids in which a toxicant is distributed. It is a virtual value because very often it exceeds the total volume of the aqueous compartments mentioned above (60% of body weight). However, it is a relatively constant value for given xenobiotics and species and is very useful for upscaling in interspecies comparisons.

During distribution, there are key sites with specific structural barriers that restrict the entrance of toxicants. The primary barriers are those of the brain, placenta and testes.

In the blood-brain barrier, specialized cells called astrocytes have many small branches, which form a barrier between the capillary endothelium and the neurons of the brain. Membrane astrocyte lipids, very tight junctions between adjacent endothelial cells and efflux transport proteins expressed on their surface, limit the passage of molecules. Nevertheless, the blood-brain barrier shows variable permeability with health status and disease state, mainly to very small lipophilic compounds. Thus, the rate at which toxicants cross into brain tissue is slowed down while allowing essential nutrients, including oxygen, to pass through. The placental barrier has also a protective role and effectively slows down the diffusion of most toxicants from the mother into the foetus. This barrier consists of several cell layers combining lipids, efflux transport proteins, and enzymes to limit the diffusion of water-soluble toxicants. In contrast, nutrients, gases and waste from the developing foetus can pass through the placental barrier.

Metabolism is one of the options to eliminate xenobiotics from the body, but it will not be discussed in this section because it is the focus of Chapter 3 in this book.

Finally, the excretion of xenobiotics can take place in different organs and biological fluids as parent or biotransformed chemical compounds. The most common are urine, faeces, saliva, breast milk, bile, sweat or exhaled air. Usually, biotransformed polar (hydrophilic) substances are more likely excreted from the body. Excretion shares with other ADME steps the passage of chemicals through biological barriers, and the same chemical and physical principles are also applicable.

2.5.3 Renal Excretion

The kidneys are the primary route of excretion, and there are three primary regions involved: the glomerulus, proximal convoluted tubule and distal convoluted tubule. There are also three processes: filtration, secretion and reabsorption.

Filtration takes place in the glomerulus, and an average value of 125 mL/min (10% of the blood flow) filters through the glomerulus into the nephron tubule. This results from the large pores (40 angstroms) in the glomerular capillaries and the hydrostatic pressure of the blood. Small molecules, both lipid-soluble and polar substances, will pass through the glomerulus into the tubule filtrate. The physiological volume of urine excreted daily is around 1.5 L, thus, about 99% of the aqueous filtrate is reabsorbed downstream in the nephron tubule.

Molecules with molecular weights greater than 60,000 (which include large protein molecules and blood cells) cannot pass through the capillary pores of healthy glomeruli. Therefore, binding to plasma proteins will influence urinary excretion. Polar substances usually do not bind with the plasma proteins and thus can be filtered out of the blood into the tubule filtrate. In contrast, substances extensively bound to plasma proteins remain in the blood.

Secretion occurs in the proximal convoluted tubule, affecting potassium ions, hydrogen ions and some xenobiotics. Secretion occurs by active transport mechanisms that are capable of differentiating among compounds based on polarity. Two systems exist: one that transports weak acids (such as many conjugated drugs and penicillins) and the other that transports basic substances (such as histamine and choline).

Reabsorption also happens in the proximal convoluted tubule of the nephron. The renal tubules reuptake nearly all the water, glucose, potassium and amino acids lost during glomerular filtration. Reabsorption occurs primarily by passive transfer, thereby urine pH has a great influence. If the urine is alkaline, weak acids (such as glucuronide and sulfate conjugates) are more ionized and excretion is increased. Because the urinary pH varies in humans (due to diet or drugs), the urinary excretion rates of weak electrolytes also vary. Examples of drugs are phenobarbital (acidic) and amphetamine (basic). Treatment of barbiturate poisoning (such as an overdose of phenobarbital) may include changing the pH of the urine to facilitate excretion. Then, molecular size and polarity are the main determinants of renal excretion. In some cases, large molecules (including some that are protein bound) may be secreted (by passive transfer) to enter the urine. Lipid-soluble toxicants can be reabsorbed, which lengthens their half-life in the body and potential for toxicity. Nephrotoxicity caused by some metals (Cd, As, Pb, Hg or U) and chemicals (e.g. melamine, herbal Chinese drugs, vancomycin) diminish the ability to excrete toxicants, thus making those individuals more susceptible upon exposure (Vervaet et al. 2017).

2.5.4 Excretion into Faeces

Excretion of toxicants in the faeces is contributed by biliary and/or intestinal excretion. The biliary route generally involves active secretion rather than passive diffusion. Specific transport systems exist for organic bases, organic acids, and neutral substances. Some heavy metals (As, Pb and Hg) and large, ionized molecules such as conjugates greater than 300 daltons are excreted in the bile. Bile excretion makes intestinal reabsorption possible, but not for water-soluble xenobiotics. However, enzymes in the intestinal flora can cleave glucuronide and sulfate conjugates, allowing the parent compound to undergo enterohepatic recycling.

Enterohepatic recycling prolongs the half-life of the xenobiotic or its phase I metabolites, and in some cases the metabolite is more toxic than the excreted conjugate. Continuous enterohepatic recycling can occur and lead to very long half-lives of some substances. Dimethylmercury is secreted in bile, and chelators or resins taken orally can bind this compound to prevent recycling. Changes in the production and flow of bile into the liver affect the efficiency of biliary excretion. Bile flow is a determinant factor of bile excretion, and some drugs can affect it. Phenobarbital increases bile flow rate, thereby enhancing the excretion of methylmercury.

Intestinal excretion is not a major route of elimination, but several substances, especially those that are poorly ionized in plasma (such as weak bases), may passively diffuse from blood capillaries through the intestinal submucosa and into the intestinal lumen. This process is only relevant for xenobiotics with large circulation times due to slow biotransformation, or slow urinary or biliary excretion. The presence of lipids in the intestinal tract favours intestinal excretion of lipophilic substances. Mineral oil (liquid paraffin, derived from petroleum) is sometimes added to the diet to help eliminate toxic substances, which are known to be excreted directly into the intestinal tract.

2.5.5 Pulmonary Excretion

The lungs are an important route of excretion for volatile xenobiotics (and metabolites) in the blood. Gases are excreted by passive diffusion according to a concentration gradient. The lower the gas solubility, the faster the excretion by alveoli. Vapor pressure of the volatile toxicant also has a significant influence on the amount excreted. Well-known examples of pulmonary excretion are the measurement of alcohol, metals (Pb, Cd, Al), solvents or toxic elements from cigarette smoking in exhaled air or exhaled breath condensates. In some cases, these have been suggested as biomarkers of disease or exposure to noxious pneumotoxic substances (Corradi and Muti 2005).

2.5.6 Excretion into Breast Milk

Alternative but minor routes of excretion are mother's milk, sweat, saliva, tears, and semen.

Excretion into milk is relevant from a toxicological point of view because toxicants can be passed through orally by breastfeeding to babies or by adults consuming dairy products manufactured with contaminated milk from cows, goats or sheep. Toxicants are excreted into milk by simple diffusion. Thus, basic substances and lipid-soluble compounds can be excreted into milk because of the more acidic milk pH (~6.5) and lipid content 3%–4%. Substances such as lead, mercury, bisphenol A (BPA) and phthalates that are chemically similar to calcium can also be excreted into milk along with it. Volatile organic compounds found in indoor air can also be found in breast milk. Examples include methyl tert-butyl ether (MTBE), chloroform, benzene and toluene. However, average exposure of the infant population by inhalation of indoor air exceeded exposure by ingestion rates 25- to 135-fold.

Under conditions of great sweat production, excretion in sweat may reach a significant degree. Some metals, including cadmium, copper, iron, lead, nickel and zinc, may be eliminated in sweat to some extent. Xenobiotics that passively diffuse into saliva may be swallowed and absorbed by the gastrointestinal system. The excretion of some substances into saliva is responsible for the unpleasant taste that sometimes occurs with time after exposure to a substance. Chronically high intakes of the organic and inorganic forms of selenium provoke a garlic odour in the breath and a metallic taste in the mouth, similar to arsenic toxicity (ToxTutor 2018).

2.6 MODELS AND METHODS

A major reason to build models from TK data relates to the fact that toxicity only arises when a chemical and/or its metabolites interact with specific targets at predetermined concentrations, and quantitative correlations between toxic effects and target dose can be attained (the target dose concept). In TK models, the potential factors affecting the relationship between external exposure or administered dose and target dose are characterized. As mentioned before, these are the route of exposure, exposure pattern, exposure duration, physical exercise, body build, enzyme induction, enzyme inhibition, species differences in size, physiology and metabolism. TK models can use data obtained from different experimental conditions, such as dosing levels or administration routes, to predict the results for other administration routes or groups of related compounds. Also, TK models can help to investigate the relationship between external exposure and amount, concentration or excretion rate of the parent chemical or its metabolites in sampling media. Among the pharmacokinetic approaches to quantitatively analyze the ADME of toxic compounds, compartmental, noncompartmental, physiologically based and population models are the most used.

Classical TK models, also called empirical or compartmental models, represent the body using three connected compartments or fewer, where the chemical is distributed homogenously. Compartments do not necessarily reflect the anatomy of the species of interest. The number of compartments and parameter values (e.g. exchange rates between compartments or volumes) are estimated from fitting the model to TK data. In Figures 2.5 and 2.6, the schemes of the one-compartment and two-compartment open model are shown, respectively.

Physiologically based TK (PBTK) modelling involves mathematically describing the complex interplay of the critical physicochemical (e.g. partition coefficients) and biological determinants (e.g. metabolic processes, breathing rate, blood flows, tissue volumes) involved in the disposition of chemicals. In PBTK models, the body is subdivided into a series of anatomical or physiological compartments. These depend on the objective of the study and of the possible mode(s) of action and site(s) of toxicity of the chemical studied. PBTK modelling improves the scientific basis

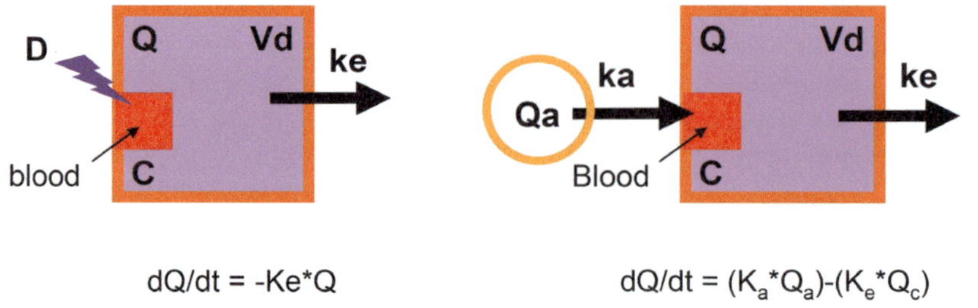

Figure 2.5 One-compartment open model for bolus IV exposure and EV exposure.

extrapolating animal data to humans. The fitting uses the available physiological prior information on parameters (Table 2.4), where population means or variances and population modelling offer a natural setting for inclusion of such information (Andersen and Krishnan 1994; Bois 2001). Figure 2.7 shows the scheme of a PBTK model.

TK is a critical component in understanding toxicological responses by linking chemical exposure to internal tissue concentrations. High-throughput TK (HTTK) *in vitro* or *in silico* approaches

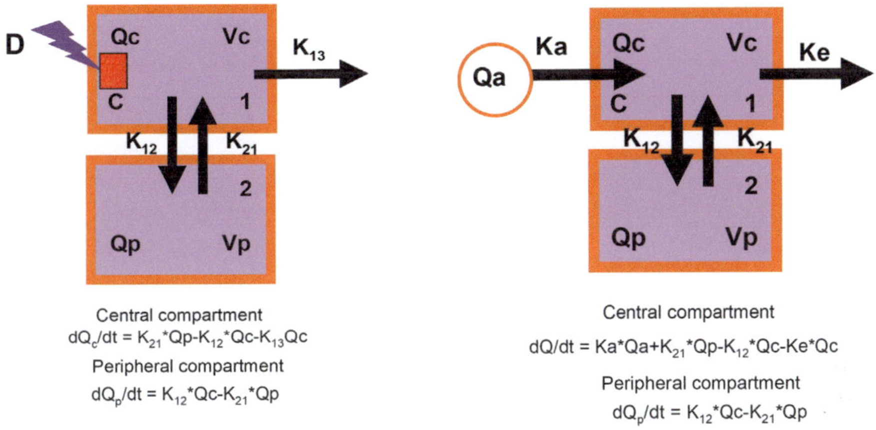

Figure 2.6 Two-compartment open model for bolus IV exposure and EV exposure.

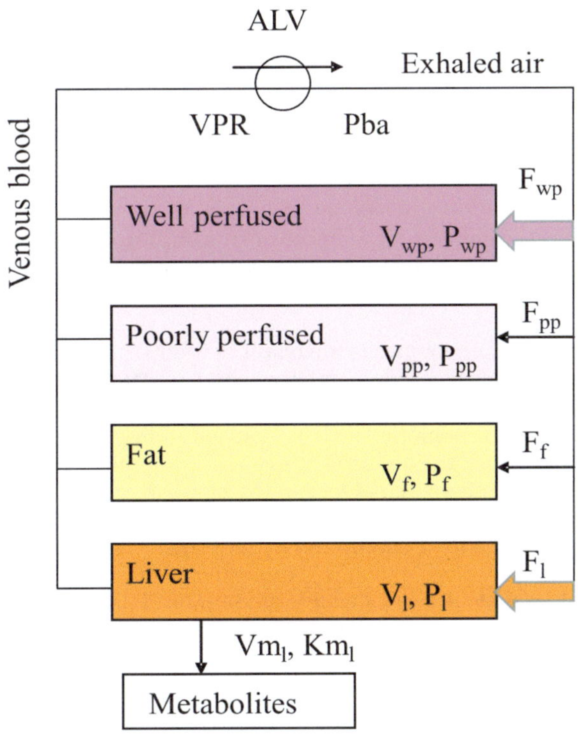

Figure 2.7 Scheme of the physiological model used to simulate the distribution and metabolism of tetrachloroethylene. The symbols are: V, volumes; F: blood flows; P, partition coefficients; Vml and Kml, Michaelis-Menten coefficients; ALV, alveolar ventilation rate; VPR, ventilation over perfusion ratio. (Adapted from Bois et al., 1996)

Table 2.4 Physiological Parameters Used in Physiologically Based Pharmacokinetic (PBPK) Models

Physiological parameters	Mouse		Rat		Human	
Body weight (kg) (BW)	0.025		0.25		70	
Tissue	Volume (% BW)	Perfusion (% CO)	Volume (% BW)	Perfusion (% CO)	Volume (% BW)	Perfusion (% CO)
Adipose	5–14		4.6–12		19	3.5–11.8
Brain	1.35–2.03	3.1–3.5	0.38–0.83	1.5–2.6		8.6–20.4
Heart	0.4–0.6	5.9–7.2	0.27–0.4	4.5–5.1		3.8–8
Kidneys	1.35–1.88	7–11.1	0.49–0.91	9.5–19		12.2–22.9
Liver	4.19–7.98		2.14–5.16	13.1–22.1	2.6	11–34.2
Lungs	0.66–0.86		0.37–0.61	11.1–17.8		
Muscle	35.8–39.9	12.2–19.6	35.4–45.5		62	5.7–42.2
Skin	15.9–20.8	3.3–8.3	15.8–23.6			3.3–8.6
Cardiac output (CO) (L/min)	0.017		0.083		6.2	
Alveolar ventilation (L/min)	0.025		0.117		5	

offer a promising approach to estimating the *in vivo* TK properties for hundreds of chemicals. The HTTK methods include *in vitro* assays that assess metabolic clearance using *in vitro* liver systems (i.e. hepatocytes) and binding to plasma proteins. These chemical-specific HTTK data (i.e. hepatocyte metabolic clearance, protein binding) rely on *in vitro* to *in vivo* extrapolation (IVIVE) to scale the *in vitro* data and parameterize TK models of varying sophistication (Wambaugh et al. 2018).

As adopted for other studies within the drug discovery and development process, regulatory bodies have published guidelines including information and strategies to carry out TK studies as far as the development of pharmaceutical products intended for use in human subjects is concerned. Released in 1994, the ICH topic S3A describes the general principles to be considered but also more specific aspects to be addressed in the different areas of toxicity testing. More recently, the US Food and Drug Administration (FDA) has published additional guidance to the S3A (2017) that contains nonbinding recommendations focusing on microsampling. The microsampling approach refers to the use of blood or blood-derived samples less than 50 μL and contributes to reducing animal stress (refinement) and satellite groups or the number of animals (reduction) in experimental TK. The main scientific advantage of microsampling is that the relationship between the safety data and drug exposure can be directly evaluated in the same animal.

2.7 CONCLUSION

The toxic response exerted by xenobiotics is the result of their LADME process. All these steps involve the passage of molecules through membranes by passive transfer or diffusion, or by other transfer processes such as facilitated diffusion, active transport, phagocytosis and pinocytosis. Physicochemical properties of xenobiotics such as lipid solubility, molecular size and degree of ionization affect the rate at which the molecule moves through membranes. Toxicants are mainly absorbed in the GI tract, lung and/or skin. Once in the bloodstream, the compound will undergo distribution to different organs. Some may act as depot tissues (e.g. fat, bones), but others will show the ability to eliminate the toxic compound by biotransformation and/or excretion. Excretion takes place mainly from kidney, liver (including enterohepatic circulation) and lung, and by less general routes such as skin, hair, sweat, nails or milk. Simple classical and physiologically based kinetic models can be used to study xenobiotics disposition.

When rate of absorption exceeds the rate of elimination, the xenobiotic may accumulate and reach critical concentration at a target site, when toxicity may ensue. Hence, whether a xenobiotic elicits toxicity depends not only on its inherent potency and site specificity but also on how an organism can dispose of that toxicant.

BIBLIOGRAPHY

Anand, U., A. Parikh, M. C. Ugwu, and R. Agu. 2014. Drug transporters in the nasal epithelium: An overview of strategies in targeted drug delivery. *Future Med Chem* 6(12): 1381–97.

Andersen, M. E., and K. Krishnan. 1994. Physiologically based pharmacokinetics and cancer risk assessment. *Env Health Perspect (suppl)* 102(1): 103–8.

Ashok, V., R. M. Kumar, D. Murali, and V. S. Brahmaji. 2012. A review on vaginal route as a systemic drug delivery. *Critical Rev Pharmaceut Sci* 1: 1–19.

Bermejo, M. V., and M. T. Garrigues. 2013. Administración de fármacos por vía bucal y perlingual. In *Tratado general de Biofarmacia y Farmacocinética vol II*, eds. J. Doménech, J. Martínez, and C. Peraire, 77–91. Barcelona: Síntesis.

Bois, F. Y. 2001. Applications of population approaches in toxicology. *Toxicol Lett* 120(1–3): 385–94.

Bois, F. Y., Gelman, A., Jiang, J., D. R. Maszle, L. Zeise, and G. Alexeef. 1996. Population toxicokinetics of tetrachloroethylene. *Arch Toxicol* 70(6): 347–55.

Boroujerdi, M. 2015. The influence of routes of administration on ADME. In *Pharmacokinetics and Toxicokinetics*, ed. M. Boroujerdi, 1–33. Boca Raton: CRC Press.

Chang, S. K., P. L. Williams, W. C. Dauterman, and J. E. Riviere. 1994. Percutaneous absorption, dermato-pharmacokinetics and related biotransformation studies of carbaryl, lindane, malathion, and parathion in isolated perfused porcine skin. *Toxicol* 91(3): 269–80.

Chedik, L., A. Bruyere, A. Bacle, S. Pontin, M. Le Vée, and O. Fardel. 2018. Interactions of pesticides with membrane drug transporters: Implications for toxicokinetics and toxicity. *Expert Opin Drug Metab Toxicol* 14(7): 739–52.

Chedik, L., A. Bruyere, and O. Fardel. 2019. Interactions of organophosphorus pesticides with solute carrier (SLC) drug transporters. *Xenobiotica* 49(3): 363–74.

Chen, P. L., H. Chen, X. Zang, et al. 2013. Expression of efflux transporters in human ocular tissues. *Drug Metab Dispos* 41(11): 1934–48.

Claus, S. P., H. Guillou, and S. Ellero-Simatos. 2016. The gut microbiota: A major player in the toxicity of environmental pollutants? *Npj Biofilms Microbiomes* 16003 2: 1–11.

Corradi, M., and A. Mutti. 2005. Exhaled breath analysis: From occupational to respiratory medicine. *Acta Biomedica* 76(Suppl 2): 20–9.

Dhamankar, V. S. 2013. *Cytochrome P450-Mediated Drug-Metabolizing Activity in the Nasal Mucosa*. PhD (Doctor of Philosophy) thesis, University of Iowa, http://ir.uiowa.edu/etd/1585 (accessed October 24, 2020).

El-Masri, H., E. McLanahan, and S. Martin. 2015. Chapter 5: Principles of toxicokinetics and predictive toxicokinetics modeling. In *Mammalian Toxicology*, ed. A. D. Mohamed, 85–98. New York: John Wiley & Sons.

El-Moneim, A., and M. R. Afify. 2010. Biological function of xenobiotics through protein binding and transportation in living cells. *Int J Agric Res* 5(8): 562–75.

Fujiwara, R., S. Takenaka, M. Hashimoto, T. Narawa, and T. Itoh. 2014. Expression of human solute carrier family transporters in skin: Possible contributor to drug-induced skin disorders. *Sci Rep* 4(2014/06/11 article 5251) https://doi.org/10.1038/srep05251 (accessed October 24, 2020).

García, E., E. Valverde, M. A. Agudo, J. Novales, and M. I. Luque. 2002. Toxicología clínica cap 2.13. In *Farmacia Hospitalaria—Tomo I*, 667–711. SEFH Coordinadora: Mª C. Gamundi Planas.

Grammen, C., M. Baes, S. Haenen, et al. 2014. Vaginal expression of efflux transporters and the potential impact on the disposition of microbicides in vitro and in rabbits. *Mol Pharmaceut* 11(12): 4405–14.

Hagelberg, S., K. Hult, and R. Fuchs. 1989. Toxicokinetics of ochratoxin A in several species and its plasma-binding properties. *J Appl Toxicol* 9(2): 91–6.

Hines, R. N., D. Sargent, H. Autrup, et al. 2010. Approaches for assessing risks to sensitive populations: Lessons learned from evaluating risks in the pediatric population. *Toxicol Sci* 113(1): 4–26.

ICH. 1995. *Topic S3A Toxicokinetics: A Guidance for Assessing Systemic Exposure in Toxicology Studies.* CPMP/ICH/384/95. https://www.ema.europa.eu/en/documents/scientific-guideline/ich-s-3-toxicokinetics-guidance-assessing-systemic-exposure-toxicology-studies-step-5_en.pdf (accessed October 24, 2020).

International Council for Harmonisation of Technical Requirements for Pharmaceuticals for Human Use. 2017. *S3A Guidance: Note for Guidance on Toxicokinetics: The Assessment of Systemic Exposure in Toxicity Studies: Focus on Microsampling. Questions and Answers.* https://database.ich.org/sites/default/files/S3A_Q%26As_Q%26As.pdf (accessed October 24, 2020).

International Transporter Consortium, K. M. Giacomini, S. M. Huang, D. J. Tweedie, et al. 2010. Membrane transporters in drug development. *Nat Rev Drug Discov* 9(3): 215–36.

Jaga, K., and C. Dharmani. 2006. Ocular toxicity from pesticide exposure: A recent review. *Env Health Prev Med* 11: 102–7.

James, R. C., S. M. Roberts, and P. L. Williams. 2000. General principles of toxicology. In *Principles of Toxicology. Environmental and Industrial applications*, eds. P. L. Williams, R. C. James, and S. M. Roberts, 3–32. Boca Raton: CRC Press.

Li, Y., J. Lu, and J. W. Paxton. 2012. The role of ABC and SLC transporters in the pharmacokinetics of dietary and herbal phytochemicals and their interactions with xenobiotics. *Curr Drug Metab* 13(5): 624–39.

Morais Ruela, A. L., A. Gravinez Perissinato, M. Esselin de Sousa Lino, P. S. Mudrik, and G. Ribeiro Pereira. 2016. Evaluation of skin absorption of drugs from topical and transdermal formulations. *Brazilian J Pharmaceut Sci* 52(3).

Nicol, M. R., Y. Fedoriw, M. Mathews, et al. 2014. Expression of six drug transporters in vaginal, cervical, and colorectal tissues: Implications for drug disposition in HIV prevention. *J Clin Pharmacol* 54(5): 574–83.

Ohmori, K., N. Kudo, K. Katayama, and Y. Kawashima. 2003. Comparison of the toxicokinetics between perfluoro carboxylic acids with different carbon chain length. *Toxicology* 184: 135–40.

Pumford, N. R., N. C. Halmes, and J. A. Hinson. 1997. Covalent binding of xenobiotics to specific proteins in the liver. *Drug Metab Rev* 29(1–2): 39–57.

Rang, H. P., and R. G. Hill. 2013. Chapter 15: Assessing drug safety. In *Drug Discovery and Development: Technology in Transition*, eds. R. G. Hill and H. P. Rang, 211–26. London: Churchill Livingstone.

Roberts, D. M., A. H. Dawson, L. Senarathna, et al. 2011. Toxicokinetics, including saturable protein binding, of 4-chloro-2-methyl phenoxyacetic acid (MCPA) in patients with acute poisoning. *Toxicol Lett* 201(3): 270–6.

Savković-Stevanović, J. 2011. Xenobiotics in tissue and organs. In *Recent Researches in Sociology, Financing, Environment and Health Sciences*, eds. N. Mastorakis, V. Mladenov, J. Savkovic-Stevanovic, 275–80. WSEAS Press. www.wseas.org

Singh, G. 2018. Chapter 4: Preclinical drug development. In *Pharmaceutical Medicine and Translational Clinical Research*, eds. D. Vohora, and G. Singh, 47–63. Cambridge, MA: Academic Press.

Subrizi, A., E. M. del Amo, V. Korzhikov-Vlakh, T. Tennikova, M. Ruponen, and A. Urtti. 2019. Design principles of ocular drug delivery systems: Importance of drug payload, release rate, and material properties. *Drug Discov Today* 24(8): 1446–57.

Tarhoni, M. H., T. Lister, D. E. Ray, and W. G. Carter. 2008. Albumin binding as a potential biomarker of exposure to moderately low levels of organophosphorus pesticides. *Biomarkers* 13(4): 343–63.

ToxTutor: Learn Essential Principles of Toxicology. August 2018 Release. U.S. National Library of Medicine. Environmental Health and Toxicology Information. https://toxtutor.nlm.nih.gov/ (accessed October 24, 2020).

U.S. Environmental Protection Agency (EPA). 2006. *Approaches for the Application of Physiologically Based Pharmacokinetic (PBPK) Models and Supporting Data in Risk Assessment*. National Center for Environmental Assessment, Washington, DC; EPA/600/R-05/043F. National Technical Information Service, Springfield, VA. http://epa.gov/ncea (accessed October 24, 2020).

Van den Berg, M., J. De Jongh, H., Poiger, and J. R. Olson. 1994. The toxicokinetics and metabolism of Polychlorinated Dibenzo-p-Dioxins (PCDDs) and Dibenzofurans (PCDFs) and their relevance for toxicity. *Crit Rev Toxicol* 24(1): 1–74.

Van der Merwe, D., R. Gehring, and J. L. Buur. 2018. Chapter 8—Toxicokinetics in veterinary toxicology. In *Veterinary Toxicology. Basic and Clinical Principles*, ed. R. C. Gupta, 133–43. Cambridge, MA: Academic Press.

Vervaet, B. A., P. C. D'Haese, and A. Verhulst. 2017. Environmental toxin-induced acute kidney injury. *Clin Kidney J* 10(6): 747–58.

Wambaugh, J. F., M. F. Hughes, C. L., et al. 2018. Evaluating in vitro-in vivo extrapolation of toxicokinetics. *Toxicol Sci* 163(1): 152–69.

Welling, P. G. 1995. Differences between pharmacokinetics and toxicokinetics. *Toxicol Pathol* 23(2): 143–7.

Wilhelm, M., D. E. Jager, and F. K. Ohnesorge. 1990. Aluminium toxicokinetics. *Pharmacol Toxicol* 66: 4–9.

Xu, S., B. Z. Olenyuk, C. T. Okamoto, and S. F. Hamm-Alvarez. 2013. Targeting receptor-mediated endocytotic pathways with nanoparticles: Rationale and advances. *Adv Drug Deliv Rev* 65(1): 121–38.

3 Metabolism and Biotransformation of Xenobiotics

Hilmi Orhan

CONTENTS

3.1 INTRODUCTION

3.1.1 Metabolism versus Biotransformation

The fundamental instinct of life is to preserve the existence of the cell, for maintenance of its future generations by self-reproducing. This is accomplished by the inevitable interaction between the cell and other living and/or nonliving things. The selfish nature of life entails taking advantage of the interacted entities first by degrading them to their building blocks, second by utilising them as much as (or if) possible and finally by getting rid of the useless leftovers without causing any harm. The term "degradation" varies from being broken into small particles to molecular levels, and it is contemporarily expressed as "metabolism" or "biotransformation". Therefore, this process is maintained both for nutrition and for protection. Current data show that this is gradually become more and more complex, like other biochemical events, as ancestral monocellular eukaryotic organisms have evolved into multicellular eukaryotes throughout life's history. Figure 3.1 summarises the interaction between monocellular organisms and any living or nonliving things, and the similarity of this interaction to the transformed versions in a multicellular organism like humans.

Traditionally, the term "metabolism" refers to a broad concept which comprises the wide range of biochemical reactions of endogenous components in an organism, together with digestion of food and the total fate of the exposed chemical in the body from entering to leaving the organism, whereas the term "biotransformation" is rather used for the chemical modification of exogenous compounds that entered the organism. Basically, both are cascades of chemical reactions involving either the components of the organism (endogen molecules/building blocks) or chemicals entering the organism intentionally (food, drugs, etc.) and/or unintentionally (contaminants, pollutants, etc.) from outside, so-called xenobiotics which means "foreign to the body" (*xenos* means "unknown" in Greek). Although metabolism and biotransformation have been used to describe the events above, differences between the two are still not clear, though they are used interchangeably in scientific jargon. More importantly, the cascade of reactions and their catalytic mechanisms for both endogenous chemicals and xenobiotics are similar. The major distinction is in substrate specificities of the catalytic enzymes, where a substantial number of enzymes seem to have little or no affinity towards xenobiotics. Biotransformation will be used in this chapter to describe the molecular structural changes of xenobiotics (including drugs) by mammalian enzymes that catalyse chemical reactions, whereas metabolism will be used mainly to describe the other processes of interaction between chemical and organism, as well as the absorption, distribution and excretion of xenobiotics. Traditionally, these processes are abbreviated as ADME (*M* stands for metabolism, but here it refers to biotransformation). The

DOI: 10.1201/9780203730584-3

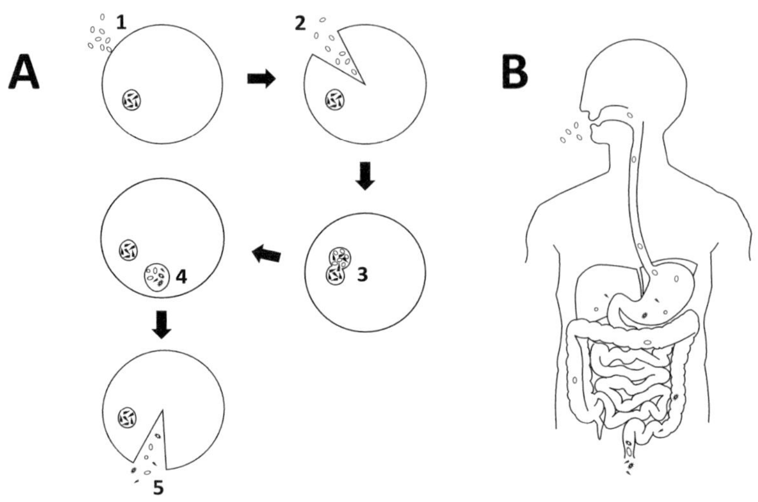

Figure 3.1 Interaction of (A) a monocellular organism with living and/or nonliving entities, the primitive processes of "metabolism", and the similarity of these events to (B) the multicellular organism (human). (A1) Entities approach cell; (A2) cellular membrane forms invagination; (A3) internalised entity bag/sack is merged with cellular organelle containing digestive enzymes; (A4) entities are digested and utilised by the cell; and (A5) the debris/leftovers are excreted.

entire ADME process is also named "disposition". The term "pharmacokinetics" (or "toxicokinetics") defines the numerical characterisation of ADME, such as the time course of absorption, distribution, biotransformation and excretion of chemicals (Shen 2013). The systemic effect of any xenobiotic on an organism depends on two major factors: the amount (dose) delivered to the site(s) of action (target) and the time of interaction. These two factors are exclusively determined by the processes of ADME; thus, toxicokinetics is essential for the assessment of the ultimate toxicity of an exposed compound.

3.2 ADME

3.2.1 Absorption

Absorption of a chemical in the life sciences is defined as crossing body membranes and entering the blood circulation. The lipid bilayer nature of almost all mammalian cell membranes (Gorter and Grendel 1925) is the reason that any chemical should have relatively high lipid solubility to be readily absorbed. Thus, lipid solubility is a prerequisite for chemicals to reach their targets of biological activity, either in deeper membranes (membrane-bound organelle receptors) or other subcellular structures (soluble receptors, messenger cofactors, DNA, etc.). Unlike distribution, absorption is irreversible.

Membrane crossing can be categorised into two main types: *passive transport* and *special transport*. Passive transport, or simple diffusion, occurs in two distinct ways: *paracellular diffusion* or *filtration*, where water-soluble midsize chemicals (100–200 daltons [Da], up to a maximum of 600 Da) cross the membrane via aqueous pores (with an average width of 40 angstroms) (Schanker 1962; Benz et al. 1980; Lu 1996); and *transcellular diffusion*, where hydrophobic molecules diffuse across the lipid bilayer. In both forms, chemicals move from one side of the membrane, where the concentration is high, to the other side, where the concentration is low; therefore, cellular energy is not consumed. In the latter, the compound must be neutral (uncharged) and lipid soluble. Many synthetic chemicals including pesticides, flame retardant compounds such as polychlorinated biphenyls (PCBs) or polybrominated diphenyl ethers (PBDEs), and drugs such as thiopental cross cell membranes by transcellular diffusion (Hjelmborg et al. 2008). In contrast, small hydrophilic compounds such as urea cross membranes by filtration (Harvey et al. 2000). Insoluble particulate matter is taken into the cells by endocytosis, which is called phagocytosis for solid matter and pinocytosis for liquid droplets (Underhill et al. 2016).

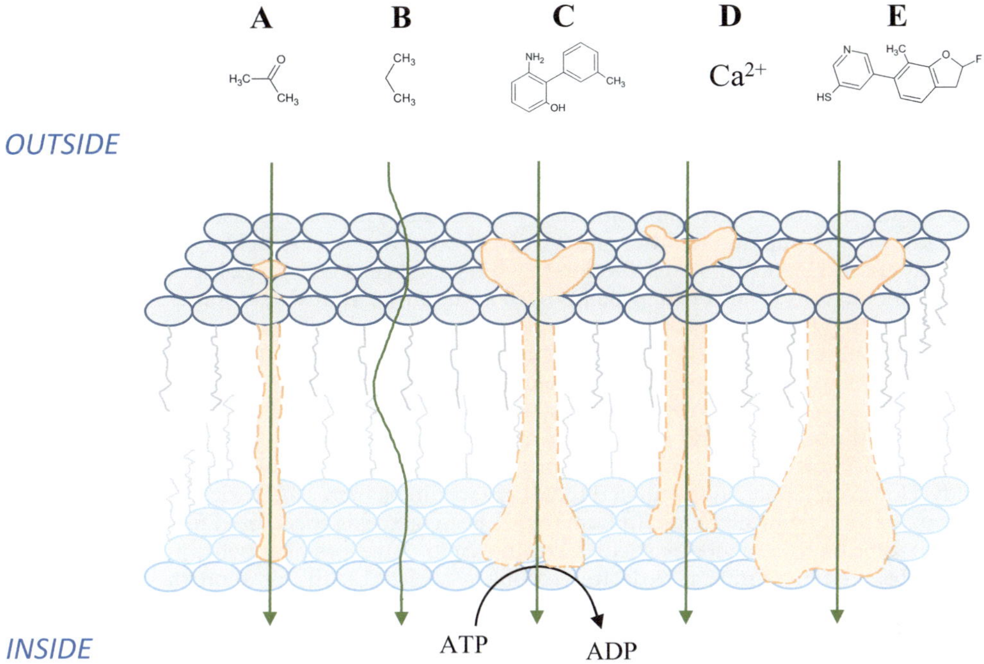

Figure 3.2 The major membrane crossing mechanisms: (A) paracellular diffusion via aqueous pores (small, water-soluble molecules); (B) transcellular diffusion via lipid membrane (lipid-soluble, nonionised molecules); (C) active transport via carrier proteins, energy required (mostly water-soluble molecules); (D) ion channels; and (E) facilitated diffusion via carrier proteins, energy not required (mostly water-soluble molecules). Indicated structures are hypothetical.

For chemicals that do not have appropriate lipid solubility or are not suitable for passing through the aqueous pores, there are carrier-mediated processes classified as *special transport*. This is accomplished by a number of transport or carrier proteins on cytoplasmic and some organelle membranes. Two major types of special transport have been described: *active transport* and *facilitated diffusion*. Figure 3.2 illustrates the major types of membrane crossing mechanisms.

In contrast to passive transport, special transport progresses against an electrochemical (ions) or concentration (molecules) gradient. Another difference is that *active transport* requires cellular energy (ATP) for this task, and therefore metabolic poisons can inhibit this process. In contrast, *facilitated transport* does not require energy, although it is still mediated by carrier proteins. Therefore, facilitated transport is not affected by metabolic poisons that interfere with the production of ATP. Although it is similar to active transport, it does not transport molecules against a concentration gradient.

One should keep in mind that the structure of the membranes and transport proteins in the organism varies on a large scale qualitatively and has become highly specialised in different organs during evolution in order to accomplish numerous physiological tasks. For instance, gaps between endothelial cells are relatively larger in the intestine, kidneys and liver. Special transport is also relatively more active therein, as these sites are the main entrance and exit pathways for xenobiotics (food, drug, contaminants and their metabolites) in the body. In contrast, transport is highly restricted in the central nervous system, placenta, testes, eyes and ovaries (Fröhlich 2002; Wen et al. 2018), as there are no gaps between the endothelial cells, termed a "tight junction". The same transport proteins function in excluding the metabolites of drugs and other xenobiotics from cells, therefore they are also considered to be essential components of phase III metabolism, where the water-soluble conjugates are transported by specialised proteins from inside of the cell to the outside. Phase III metabolism is discussed in detail in Section 3.3.5.3.

3.2.2 Distribution

Absorption is followed by distribution. Once the chemical enters systemic circulation, it initially distributes to various tissues depending on the local blood flow and endothelium permeability. However, accumulation in a particular organ is dictated by the existence of both specific transport proteins and binding structures in the cells. This is the basis for the process called redistribution. For example, the chemical thiopental initially distributes to the liver, kidneys and central nervous system, where blood flow is relatively high, and then it is redistributed and concentrated in adipose tissues due to its high lipid solubility (Enna and David 2008). Considering that blood is a more fluidic organ compared to other organs, distribution first occurs in blood between plasma and mainly the red blood cells (distribution to white blood cells is negligible due to their relatively much smaller quantity) and serum proteins such as albumin, α1-acid glycoprotein and so forth. In this sense, binding to plasma proteins at a high rate serves as a reservoir for the biological effect (either therapeutic or toxic) of the absorbed chemical in the body, which may affect its toxicokinetics significantly. Nonionised and highly lipid-soluble compounds, such as dichlorodiphenyltrichloroethane (1,1′-(2,2,2-trichloroethane-1,1-diyl)bis(4-chlorobenzene), or DDT), tend to concentrate in adipose tissues. Adipose tissue and fat fractions in the organs serve as depot for such chemicals, where there is no indication of a toxic outcome as long as lipid-soluble chemicals stay in this fat fraction of the body. However, it may be speculated that there would be increased toxicity of fat-stored chemicals in case of a sudden weight loss in humans. This, in fact, was speculated for migratory birds indirectly; some environmental pesticides are known to cause loss of appetite in birds. Several large bird kills observed in migrating birds are probably associated with this anorexia occurring at a time when most of the birds' fat reserves have been used (Grue 1982; Harvey 1967). Rapid release of stored fat and persistent organic pollutants from adipose tissue into the bloodstream may contribute to deaths. Bone is another site of accumulation for xenobiotics without indicating apparent toxic signs. Especially, metals such as lead, strontium and fluoride are accumulated in bones by substituting for calcium in hydroxyapatite (Boivin 2007). Such storage in bones or adipose tissues decreases the availability of corresponding metals or chemicals for their target sites in the body.

As mentioned above, distribution of xenobiotics to some sites in the organism is restricted. These corresponding restrictions are made up of various combined structural and functional features and are named according to the protected tissue, such as blood-brain, blood-testis, blood-follicle and blood-placenta barriers. At the blood-brain barrier, for instance, there are no gaps (fenestrae) between capillary endothelial cells, and therefore possess tight junctions; they are also surrounded with astrocytes, which lack special transport except for several targeted substrates (glucose transport to the brain for instance), and present various enzymes to detoxify chemicals (Daneman and Prat 2015; Keaney and Campbell 2018). Although very efficient against hydrophilic chemicals, these barriers are not effective against lipid-soluble chemicals.

Transport proteins are obviously one of the major players in distribution, and their types and quantities vary significantly between organs. These proteins also play an essential role in the absorption and excretion of xenobiotics. As discussed in detail below, transport proteins are considered to be fully responsible for transferring the metabolites of phase I and/or phase II biotransformation from the cell. Figure 3.3 represents how ADME changes the biological effect of a xenobiotic.

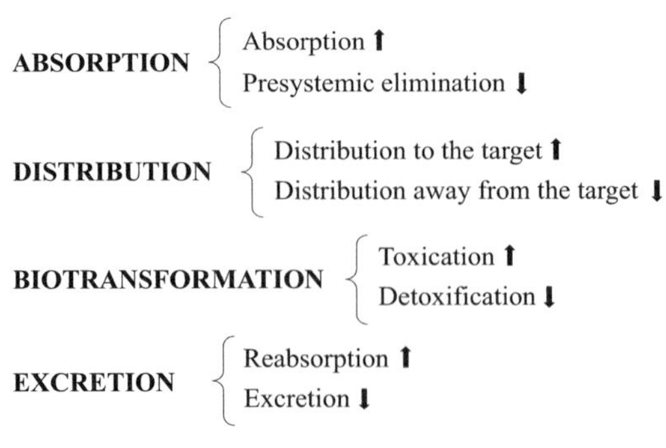

Figure 3.3 Kinetic factors that change the biological effect of a xenobiotic *in vivo*. Up arrow represents increase and down arrow represents decrease in the biological effect.

Source: Adapted from Gregus (2013).

3.3 BIOTRANSFORMATION (METABOLISM)

3.3.1 History of Drug Metabolism and Relevance to Toxicology

Despite there being plenty of data on xenobiotic metabolising enzymes (XMEs) and biotransformation reactions accumulated to date, until the beginning of the 19th century it was believed that chemicals entering an organism were excreted unchanged, and that man would be unable to synthesise constituent compounds of the organism because of its divine nature. The German chemist Friedrich Woehler was either very bull-headed, or not enough of a believer, to accept this dogma without questioning it. As a result of many investigations on his dog and himself, Woehler succeeded in synthesising urea in 1828, and he notified his former boss and scientific mentor Berzelius about this exciting news: "I must tell you that I can prepare urea without requiring a kidney of an animal, either man or dog" (Jaffe 1942). The resulting paradigm shift caused a significant impact on the science of those days, demonstrating that organic chemistry could replace the need for biological tissues. Following this and several subsequent observations, it was understood that, as xenobiotics change the organism, the organism can change (the structure of) the xenobiotic. Alexander Ure was the first to show the metabolic capacity of the human body in 1841 when he detected hippuric acid in urine after ingestion of benzoic acid (Ure 1841). This metabolic conversion of benzoic acid to hippuric acid could not be proved by Woehler despite his extensive works in dogs, because hippuric acid had not been fully characterised until 1829 by Liebig (ISSX 2020). Following Ure's report, formation and excretion of hippuric acid in horse urine and his own after ingestion of benzoic acid were confirmed by Keller in Woehler's lab (Keller 1842). (Incidentally, hippuric acid acquired its name from the Greek word *hippo*, which means "horse".) Further studies of chemical biotransformation demonstrated the formation of phenol and phenol sulphate after administration of benzene. About 30 years later, phase II conjugation studies were extended when a urinary excretion product of bromobenzene was isolated and shown to be an *N*-acetylcysteine conjugate of bromobenzene (mercapturic acid derivative; Baumann et al. 1879; Jaffee 1879). By the end of the 19th century, the discipline of drug metabolism was well underway (Williams 1959). Although it was realised after Woehler's and others' observations that xenobiotics entering the body are not excreted but only unchanged, the role of biotransformation in toxicity was only understood much later. A hundred years after Ure and Keller, Miller showed that the rat carcinogen dimethylaminobenzene is metabolised by the liver, binds to cellular components (e.g. DNA) and initiates cancer (Miller and Miller 1947; Mueller and Miller 1948; Mueller and Miller 1949; Orhan 2015).

3.3.2 Enzymes as the Key Components of Biotransformation

Changes to the chemical structure of the parent molecule can generate several products called "metabolites", in accordance with interchangeable use of the two terms of "biotransformation" and "metabolism". The chemical reactivity of metabolites towards endogenous components of an organism varies on a broad scale: they may not react with any component and are excreted from the body, or they may react instantly with cellular molecules in their immediate proximity or at the distance of diffusion. Therefore, biotransformation reactions result in changes not only to "movement" of the xenobiotics (kinetics) but also in their biological effect (dynamics) either in favour for and/or against the organism. As can easily be inferred, the organism's attempt to take advantage of xenobiotics and avoid any harm associated with this process does not always work ideally. The outcome of biotransformation may be harmful for the organism, depending on the properties of the xenobiotic and the organism, the nature of the metabolites formed and conditions of exposure. This is the basis for why the term "detoxification" is incorrect, and metabolism or biotransformation is favoured, as the metabolites may be more toxic than the parent compound (see below). This is actually one of the major reasons for the existence of the discipline of toxicology, wherein formation of reactive metabolites may result in adverse reactions when protective mechanisms are overwhelmed. Biotransformation is one the most important modifiers of the toxicity of xenobiotics.

Although it is a relative term, enzymes involve in xenobiotic biotransformation are called drug- or xenobiotic-metabolising enzymes (DME or XME). The major enzymes under this term are given in Table 3.1 (Guengerich and Macdonald 1984; Ortiz de Montellano 1986; Nebert and Gonzalez 1985; Guengerich 1987; Gonzales 1988; Guengerich 1982; Schladt et al. 1988; Thomas and Oesch 1988; Mantle et al. 1987; Sies and Ketterer 1988; Mannervik and Danielson 1988; Bock 1987; Tephly et al. 1988; Guengerich and Macdonald 1984; Ortiz de Montellano 1986; Guengerich 1987;

Table 3.1 The Major Drug- or Xenobiotic-Metabolising Enzymes (DME, XME)[1]

Phase I: Xenobiotic metabolising enzymes	Phase II: Xenobiotic metabolising enzymes
Cytochrome P450 (CYP450) superfamily	Uridine 5′-diphospho-glucuronosyltransferases (UGTs)
Epoxide hydrolyse	*N*-acetyltransferases (NATs)
Flavine-containing monooxygenase	Glutathione *S*-transferases (GSTs)
Prostaglandin synthase	*N*- and *O*-methyl transferases
Quinone reductase	Sulphotransferase (SULT)
Alcohol dehydrogenase	Etc.
Aldehyde dehydrogenase	
Ketone dehydrogenase	
Cysteine conjugate β-lyase	
Monoamine oxidase	
Etc.	

1 Refer to appropriate section of the chapter above for references to each metabolic process.

Guengerich 1982; Schladt et al. 1988; Tomisawa et al. 1988; Mantle et al. 1987; Sies and Ketterer 1988; Mannervik and Danielson 1988; Bock 1987; Burchell and Coughtrie 1989; Mckenzie et al. 1985; Weber 1986; Hein 1988; Andres et al. 1988; Ziegler 1988a, 1988b; Marnett and Eling 1983; Marnett 1985; Krauss and Eling 1984; Prochaska et al. 1985; Emster et al. 1987; Jaiswal et al. 1988; von Bahr-Lindstrom et al. 1986; Bosron and Li 1986; Hempel et al. 1984; Wermuth et al. 1988; Ansher and Jakoby 1986; Scott et al. 1988; Weinshilboum 1988; Shinohara et al. 1986; Weinshilboum 1986; Ramaswamy and Jakoby 1987; Weinshilboum 1990; Stevens and Jakoby 1983; Lash et al. 1986; Tomisawa et al. 1988; Tipton et al. 1984).

On the other hand, such categorisations depend on how one conceptualises the multiple processes in the "biotransformation of xenobiotics" (Guengerich 1990). For instance, certain drugs, or other chemicals, can lead to an increased production of reactive oxygen species (ROS), leading to oxidative stress. Enzymes that catalyse detoxification of ROS, namely superoxide dismutase, catalase and glutathione peroxidase, have also been considered as XME (Sies and Cadenas 1983). Another important feature is that enzymes in gut microflora contribute to biotransformation reactions too (Adamson et al. 1970; Goldin 1990; Naranjo et al. 2018) Although biotransformation involves reactions catalysed by various enzymes, only a few biotransformation reactions occur spontaneously without involving enzymes such as the conjugation of various xenobiotics with glutathione (an example of a phase II reaction), the hydrolysis of some ester conjugates in the stomach (catalysed by stomach acid) and so forth. Biotransformation reactions are important in both drug development and the clinical use of drugs. Figure 3.4 shows the contribution of phase I and phase II enzymes to the biotransformation of contemporarily used drugs in clinic.

3.3.3 Outcomes of Biotransformation (Detoxification versus Bioactivation)

The usual outcome of biotransformation of a xenobiotic is to convert it into a more water-soluble metabolite and at the same time to diminish the potential harmfulness of both the parent compound and the metabolite, which is historically called "detoxification". However, although the sequential phase reactions suggest, and generally occur towards converting lipophilic compounds into hydrophilic compounds, and simultaneously decreasing the toxic potential of compounds, this is not always the case. Increasing the toxic potential via biotransformation is called "bioactivation". This can occur via several different ways. The parent xenobiotic may be converted into electrophilic intermediate(s) via oxidation catalysed by CYP450s, NADPH oxidase, myeloperoxidase, thyroid peroxidase and so forth, or into a free radical via one electron reduction, free electrons via redox cycle or proton subtraction (Orhan 2015). There are many examples of formation of electrophilic or radical intermediates upon biotransformation reactions, which will be exemplified in the coming subsections. Formation of nucleophilic intermediates is also possible although very rare, such as formation of cyanide from amygdalin catalysed by β-glucosidase in gut microflora (Gregus 2013). The same nucleophile is also formed from the industrial compound acrylonitrile after phase I epoxidation reaction, followed by glutathione conjugation (El Hadri et al. 2005). Table 3.2 summarises various examples of drugs that undergo

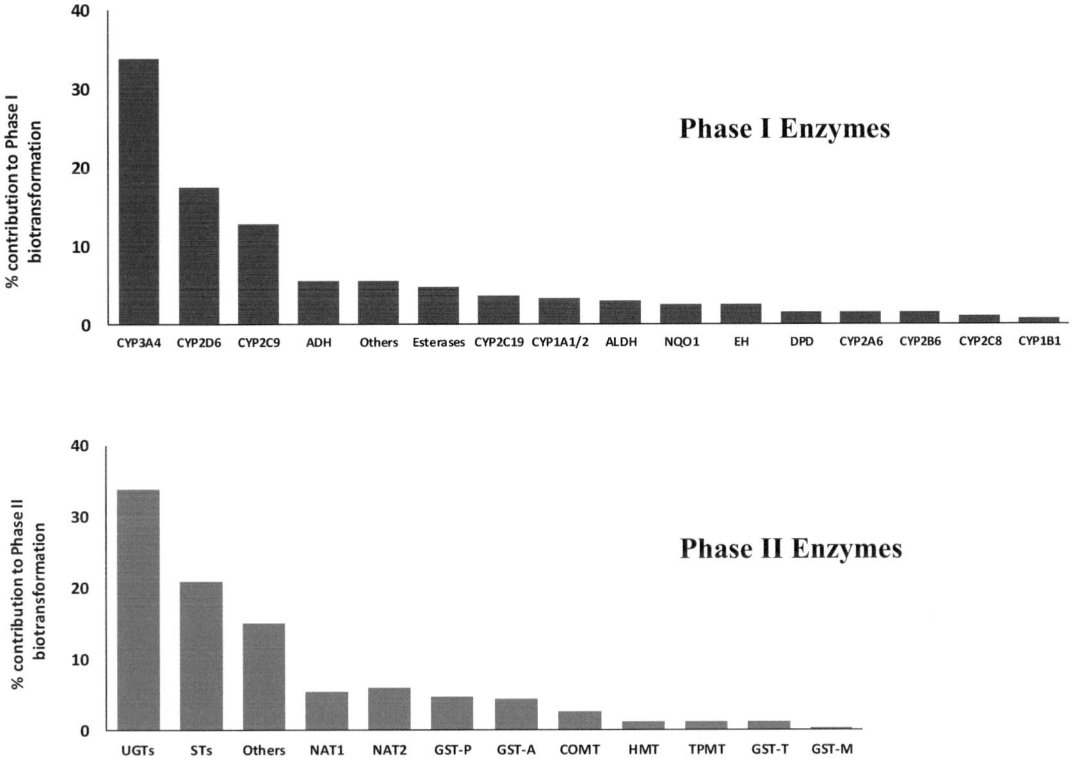

Figure 3.4 The relative contributions of phase I and phase II biotransformation enzymes involved in the metabolism of clinical drugs.

Abbreviations: ADH = alcohol dehydrogenase; ALDH = aldehyde dehydrogenase; COMT = catechol *O*-methyltransferase; CYP = cytochrome P450; DPD = dihydropyrimidine dehydrogenase; EH = epoxide hydrolyse; GSTs = glutathione *S*-transferases; HMT = histamine methyltransferase; NATs = *N*-acetyltransferases; NQO1 = quinone oxidoreductase or DT diaphorase; STs = sulfotransferases; TPMT = thiopurine methyltransferase; UGTs = uridine 59-triphosphate glucuronosyltransferases.

Source: Reproduced from Evans and Relling (1999).

bioactivation, as well as the reactive metabolites formed, mechanisms of toxicity and toxic outcome (Lu 1996; Mazel and Pessayre 1976; Wirth et al. 1975; Williams et al. 2007).

3.3.4 The Site of Biotransformation at the Organ and Subcellular Levels

Because of its central role in the overall metabolism of the organism, the liver is the primary organ for biotransformation of both endogenous and exogenous chemicals. This is the reason why the liver is the most frequent target for drug- or xenobiotic-induced toxicities. Nevertheless, non-liver tissues such as kidney, lung, heart, brain and so forth also possess various enzyme systems responsible for xenobiotic biotransformation. This feature even forms one of the principles of organ-specific toxicity (target organ concept; Orhan 2015). Three main mechanisms contribute to the organ-specific toxicity of chemicals. First, particular organs may have specific uptake mechanisms for certain drugs and thereby may concentrate them. For instance, there is a chemical-specific uptake mechanism for polyamines such as spermine and spermidine in the lungs that also concentrates structurally similar paraquat in pulmonary tissue (Orhan 2015). This relative increase in the concentration of drug alone may result in toxicity, either directly or by its metabolites. Second, the drug may undergo bioactivation by the action of specific enzymes that are either exclusively expressed in that organ, or show higher affinity for the drug compared to the same enzyme in other tissues. The toxin of *Fusarium solani*, 4-ipomeanol causes specifically lung damage

Table 3.2 Representative Drugs and Nondrug Chemicals That Undergo Bioactivation, Chemical Structure of the Corresponding Reactive Metabolites, Mechanisms of Toxic Reaction with Organismal/Cellular Components and Toxic Outcome

Drug/xenobiotic	Toxic metabolite	Mechanism	Toxicity
Paracetamol	N-OH derivative (NAPQI)	Covalent protein binding	Hepatic necrosis
Aflatoxin B1	Aflatoxin-exo-8,9-epoxide	Covalent DNA binding	Hepatic tumours
Benzene	Benzene epoxide	Covalent bone marrow DNA binding and damage	Bone marrow depression
Bromobenzene	Bromobenzene epoxide	Covalent binding in liver, kidney and lung	Hepatic, renal and bronchial necrosis
Carbon tetrachloride	Trichloromethyl radical	Covalent DNA binding in liver	Hepatic necrosis, tumours
Halothane	Trifluoroacetyl chloride	Covalent binding in liver	Hepatic necrosis
Parathione	Paraoxon	Covalent binding to Cholinesterase	Neuromuscular paralysis
Furosemide	Epoxide and enedione	Covalent protein binding Disruption in Ca^{2+} homeostasis	Lethal synthesis, hepatic and renal necrosis
Isoniazid	Acetylhydrazine	Covalent binding	Hepatic necrosis
Methoxyflurane	Inorganic fluoride	Covalent binding Protein phosphatase inhibition	Renal failure Bladder cancer
Amygdalin	Mandelonitrile	Cyanide formation	Cytotoxic hypoxia
Cyclophosphamide	Many proposed	Alkylation	Cytotoxicity
Nitrates	Nitrites	Hemoglobin oxidation	Methemoglobin
Cycasin	Methylazoxymethanol	Alkylation	Cancer, hepatic necrosis
Safrole	1-Hydroxysafrole	Covalent DNA binding	Various cancers

independent of its administration route in test animals. It is activated specifically by the lung specific CYP4B1 and produces selective necrosis in Clara cells where bioactivation occurs (Orhan 2015). Third, relatively more stable reactive metabolite may be formed in one organ and then transported to the target organ where it exerts its toxicity (Orhan 2015). An example of this type is monocrotaline-induced pulmonary toxicity. This alkaloid is activated to a reactive metabolite in liver, is transported by the red blood cells and exerts its toxicity in the lungs (Wilson et al. 1992).

At the subcellular levels, XMEs are located primarily in smooth endoplasmic reticulum (sER), in terms of quantity and quality for both phase I and phase II reactions. Scientific interest in this subcellular membrane fraction has a long history, as more specified *in vitro* biotransformation studies started several decades later than the isolation of this fraction from the homogenate of animal tissues by fractional centrifugation (Claude 1943), which was later identified to be fragmented membrane vesicles of the sER and named "microsomes" (Palade and Siekevitz 1956). Nevertheless, microsomes are not the only site where XMEs are located. Cytosolic, and usually water-soluble, enzymes also catalyse various biotransformation reactions. Some groups of enzymes have both microsomal and cytosolic forms. A good example is the glutathione *S*-transferase (GST). Finally, mitochondria have both phase I and phase II enzymes in the inner membrane as insoluble enzymes and in matrix fraction as soluble enzymes (Devlin 1992). It has been proposed that there may also be soluble P450s in mitochondria (Anandatheerthavarada et al. 1997; Neve and Ingelman-Sundberg 1999). Should mitochondrial P450s be involved in drug metabolism, then the presence of soluble P450s may cause the generation of metabolites relatively closer to the components of mitochondrial matrix such as mitochondrial DNA and so forth, with potential detrimental effects. Notably, GSTs have a mitochondrial soluble form, GST kappa (Morel and Aninat 2011).

3.3.5 Biotransformation Reactions

Although a molecule may be foreign to the organism, it may still be a substrate for an enzyme of the endogenous metabolic pathways, provided that its chemical structure is similar to that of an endogenous compound. There are two major hypotheses to explain why mammals have large amounts of enzymes that function in xenobiotic metabolism. The first is that endogenous compounds are the original substrates for these enzymes, and they show activity towards certain xenobiotics because of the structural similarity mentioned above (Guengerich 1990). The second is that these enzymes evolved in order to provide protection against xenobiotics they were exposed to, mainly via food, such as constituents of plants (Nebert and Gonzalez 1985; Jakoby 1980). Thus, the ability of animals to detoxify xenobiotic chemical compounds, including pharmaceutical drugs, is likely part of a heritage from ancestral animals to detoxify various noxious compounds in their food (Omura 2013). On the other hand, there is an overlap between biotransformation of some endogenous and exogenous chemicals. The same enzymes implicated in xenobiotic biotransformation also contribute to the hepatic catabolism of steroid hormones, phase II conjugation of bilirubin, thyroid hormones and so forth (Parkinson et al. 2013).

The biotransformation of xenobiotics can be divided into four categories based on the chemical reactions: oxidation, reduction, hydrolysis and conjugation. Due to their nature, biotransformation reactions were classified traditionally as phase I and phase II. The first three chemical reactions (oxidation, reduction, hydrolysis) are classified as phase I biotransformations, as these reactions expose or introduce a functional group to the parent xenobiotic. The product (metabolite) of phase I reactions is usually converted to a water-soluble and mostly final metabolite by the phase II (conjugation) reactions. Phase I reactions do not result in a significant increase in the size or molecular weight from the parent molecule, with some instead decreasing after hydrolysis reactions. In contrast, phase II conjugation reactions generally lead to a substantial increase in molecular weight. Although this reaction cascade perfectly follows a plain logic in theory, total disposition of a compound may not follow a tiered process. That is, the parent xenobiotic may be excreted unchanged or may undergo direct conjugation without phase I when there is relevant functional group(s) to be conjugated, or in some cases, one or more phase I reactions may occur even after phase II conjugation has completed (Ogilvie et al. 2006). Figure 3.5 illustrates these biotransformation possibilities of a drug in the organism.

3.3.5.1 Phase I Biotransformation

Phase I biotransformation reactions are oxidation, reduction and hydrolysis. Oxidation exposes or introduces a functional group to the parent xenobiotic. All biotransformation reactions aim to be processes where chemical reactivity is reduced, and water solubility is enhanced in order to excrete the chemical from the body in water-based media such as urine and faeces. Lipophilicity is a prerequisite for a xenobiotic to pass through the bilayer lipid of membranes without requiring energy-consuming transporters. Reducing the reactivity of the exposed chemical is the net result of the sum of phase I and phase II. However, reactivity of the parent molecule may slightly be increased during phase I to facilitate phase II, if the inherent/intrinsic reactivity of the parent molecule is not sufficient for interacting with the endogenous cofactors of phase II.

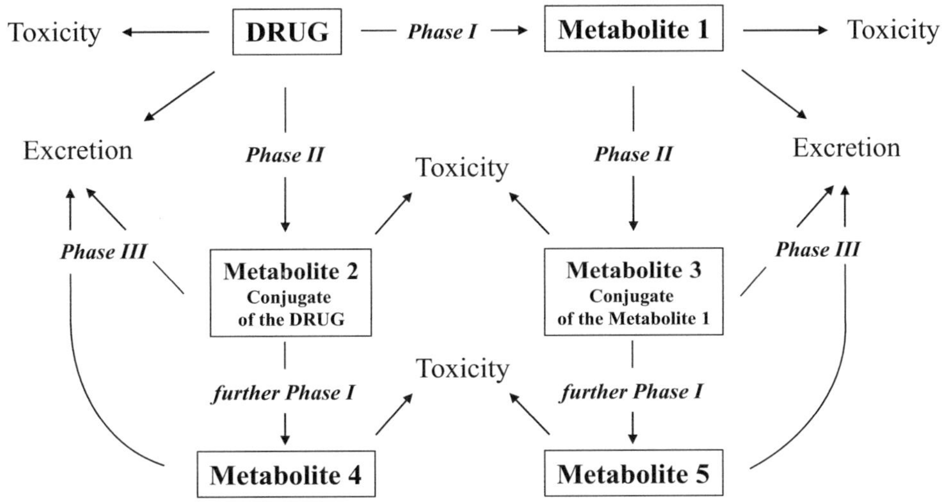

Figure 3.5 Possible biotransformation reactions and outcome of a drug in the organism.

3.3.5.1.1 Oxidation

Oxidation is the most frequently occurring phase I reaction on exogenously (i.e. food, drug or contaminant) and occupationally exposed chemicals. Perhaps because life on earth has evolved under an oxygenated atmosphere, after plants advanced from the sea onto dry land, metabolic reactions utilised oxygen in changing the structure of chemicals to readily remove them from the organism. Oxygen itself is reduced and detoxified to water along an *electron transport chain* system in the inner membrane of mitochondria. Oxidation of a molecule can simply be defined as the addition of oxygen to, or the removal of hydrogen from, the molecule, which means losing an electron. Electron removal from a molecule was first discovered following the addition of an oxygen molecule, and therefore it has been termed "oxidation". Moreover, as in removal of hydrogen from a molecule, oxygen is not involved at all in some oxidation reactions. Three types of oxidation reactions are summarised in Figure 3.6: oxygenation, dehydrogenation and electron transfer.

The most abundant enzymes in oxidation reactions, both in type and amount, are the cytochrome P450s (CYP450). As will be described in detail in the coming section, CYP450s are also widely distributed to all five kingdoms of life: animals, plants, fungi, bacteria and archaea (Omura 1962; Murphy and West 1969; Ferris et al. 1973; Appleby 1967; Wright et al. 1996). Accumulated data on the subject strongly suggest that all CYP450s have diversified from a single ancestral protein to many forms during the course of biological evolution and are distributed in various forms of life, from which it can be inferred that the origin of life is common (Omura 1999, 2010, 2013). The cytosolic locations of CYPs vary. All CYPs are membrane-associated enzymes that can be found in the endoplasmic reticulum, where they are also known as microsomal fraction, and the inner membrane of mitochondria. Moreover, several soluble CYPs have been suggested to be in the mitochondrial matrix by some research groups (Anandatheerthavarada et al. 1997; Neve and Ingelman-Sundberg 1999).

Oxidation reactions involve both an aromatic ring and aliphatic side chain of a molecule. Some examples shown in parentheses that undergo biotransformation via each type of oxidation reactions are aromatic ring hydroxylation (phenytoin, propranolol, warfarin), aliphatic side chain oxidation (barbiturates, ibuprofen, digoxin), alcohol and aldehyde oxidation (ethanol, benzyl alcohol, pyridoxine), oxidative *N*- and *O*-dealkylation (caffeine, lidocaine), oxidative deamination (histamine, noradrenaline), aromatic amine oxidation (aniline derivatives), *N*- and *S*-oxidation (imipramine, paracetamol, phenothiazine) and desulphuration (thiopental, parathion). Some of these oxidation reactions are catalysed by non-CYP-dependent oxidases. These are flavin adenine dinucleotide (FAD)–containing monooxygenase (FMO), alcohol dehydrogenase (ADH), aldehyde dehydrogenase (ALDH), aldehyde oxidase, xanthine oxidase, monoamine oxidase (MAO), diamine oxidase (DAO) and a number of peroxidases (prostaglandin synthase, lactoperoxidase, myeloperoxidase and uterine peroxidase). Two examples of clinically used drugs (nifedipine and *p*-nitrophenol) that undergo oxidation catalysed by CYPs are shown in Figure 3.7.

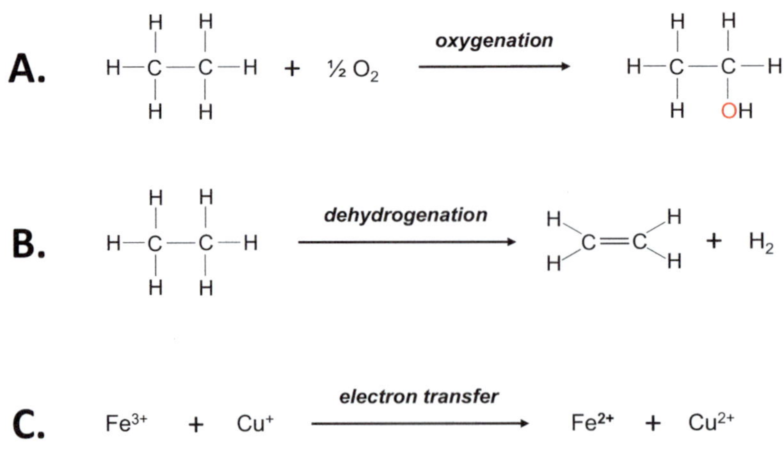

Figure 3.6 Three types of CYP450-catalysed oxidation reactions: (A) oxygenation; (B) dehydrogenation; (C) electron transfer.

Source: NIH (2018).

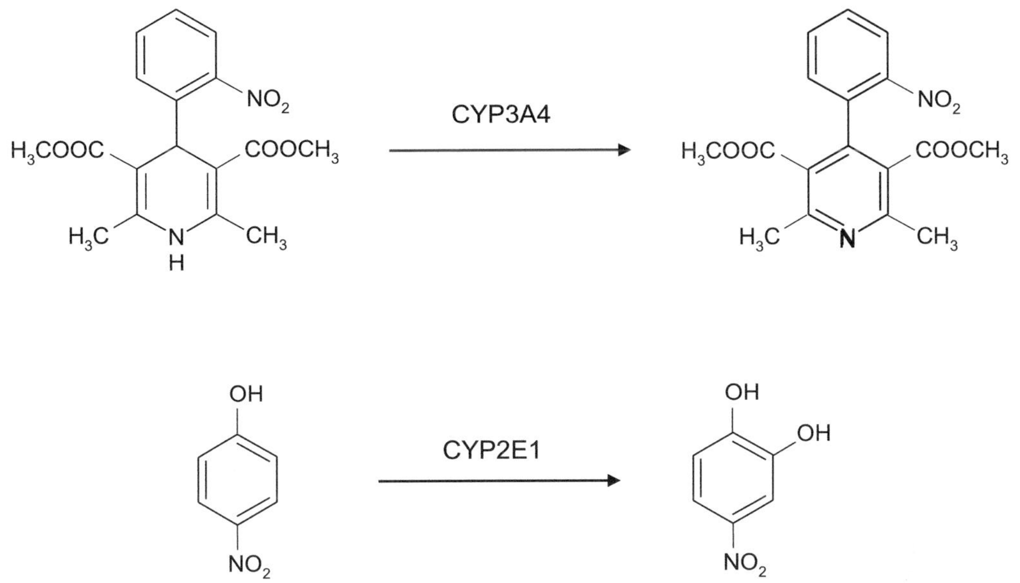

Figure 3.7 Dehydrogenation of nifedipine by CYP3A4 to its corresponding pyridine derivative (upper) and aromatic hydroxylation of *p*-nitrophenol by CYP2E1 to 4-nitrocatechol (lower).

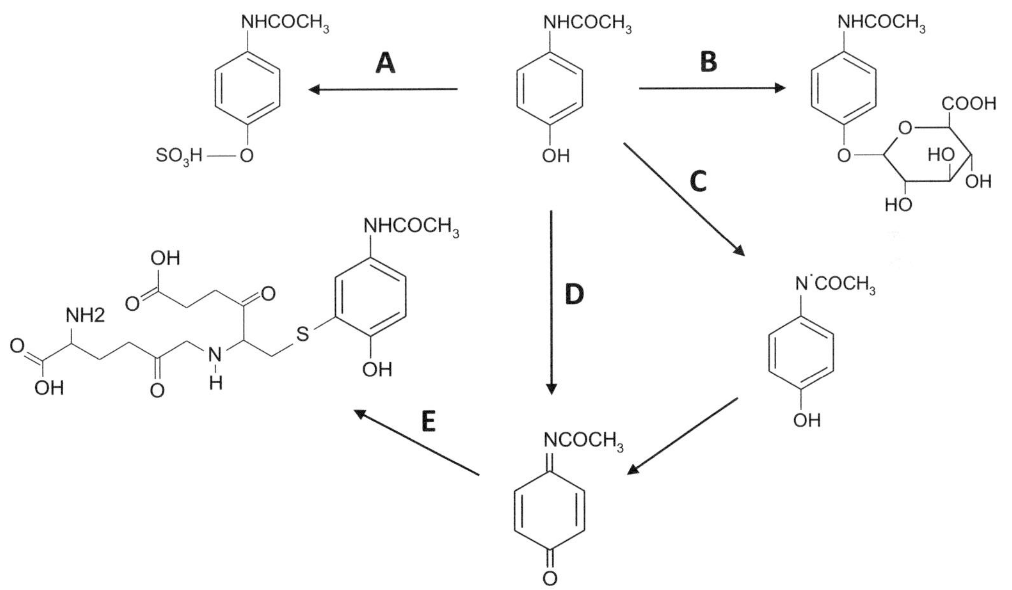

Figure 3.8 Metabolism and bioactivation of paracetamol leading to hepatic and renal toxicity: (A) sulphotransferase (SULT); (B) uridindiphosphoglucuronyltransferase 1A9 (UGT1A9); (C) prostaglandin H synthase (PHS); (D) cytochrome P450 2E1 (CYP2E1); (E) glutathione *S*-transferase (GST).

Source: Adapted from Parkinson et al. (2013).

Paracetamol is the best and most referred example in the textbooks that exerts its toxicity via biotransformation, and its hepatotoxic effect is directly proportional to the dose administered. Because paracetamol overdose is one of the frequent reasons of deaths globally, great effort has been invested to understand its toxic mechanism. Therefore, it is perhaps the most well-known mechanism among the drug-induced adverse effects. As seen in Figure 3.8, sulphation followed by glutathione conjugation is saturated by the overdose of paracetamol, and the free *N*-acetyl-*p*-benzoquinoneimine (NAPQI), formed mainly by CYP2E1-catalysed oxidation, attacks the thiol groups of cellular proteins primarily found in the liver (because of the highest rate of metabolism in this organ), specifically mitochondrial proteins (Xie et al. 2015). Subsequent mitochondrial damage and nuclear DNA fragmentation result in necrotic cell death in the liver (McGill et al. 2012).

3.3.5.1.2 Cytochrome P450s

Isolation of microsomes from animal tissues (Claude 1943) and discovery of the presence of various oxygenase reactions in this novel subcellular fraction (Hogeboom 1949; Strittmatter and Velick 1956; Horecker 1950; Yoshikawa 1951; Strittmatter and Ball 1951) in the 1940s and 1950s led to the discovery of CYP450s. Klingenberg reported an unknown carbon monoxide–binding pigment in rat liver in 1958 (Klingenberg 1958), of which the haemoprotein nature of this pigment was elucidated by Omura and Sato (1962), and the name "cytochrome P450" was proposed by the same authors (Omura and Sato 1964; Omura and Sato 1964), where "P" stands for pigment or protein and "450" stands for the unique absorption maximum at 450 nm of the cytochrome enzymes when reduced by carbon monoxide binding. This intense progress on the subject led to the identification of the catalytic function of CYP450s in the hydroxylation of progesterone (Estabrook et al. 1963) and in the oxidation of various drugs (Cooper et al. 1965). Endogenous substrates of CYPs include fatty acids, steroids, eicosanoids and fat-soluble vitamins (Coon and Koop 1983). CYPs are also the most abundant enzymes in humans and are the most active in biotransformation reactions. So far, 57 genes have been identified coding for CYP450s in humans: seven for mitochondrial forms (Type-I enzymes) and 50 for microsomal forms (Type-II enzymes; Omura 2006). The families of 11 (A, B1, B2), 24 (A) and 27 (A, B, C) are the mitochondrial forms and play essential roles in the biosynthesis of steroid hormones from cholesterol (Omura and Morohashi 1995). CYP1, CYP2 and CYP3 seem to have relatively higher activity towards xenobiotics, whereas CYP4 appears to catalyse endogenous fatty acid metabolism.

As mentioned earlier, the CYP450 gene superfamily has most likely evolved and diversified from a single ancestral gene to higher organisms in response to the necessity of changing chemical structures of plant-derived alkaloids, terpenes and so forth to exclude them from the organism (Omura 2010). Because of the diversity of the oxidation reactions they catalyse, CYP450s are not limited to transferring oxygen to a molecule and are therefore called *mixed function oxidases*. Thus, their major function is the monooxygenation of various substrates, while intramolecular oxygen transfer (DeWitt and Smith 1983) without requirement of molecular oxygen and supply of reducing equivalents from NADPH has also been reported (Rahimtula and O'Brien 1975; Hrycay et al. 1975). Oxidised substrates are also subjected to further rearrangement, hence products are not limited to only alcohols and/or phenols. For their catalytic action, CYPs require molecular oxygen, NADPH, magnesium ions and NADPH-cytochrome P450 reductase; the last provides electrons to CYPs. To a lesser extent, electrons can also be supplied by cytochrome b5. A molecule of oxygen binds to the iron atom (Fe^{3+}, ferric form) of haem, while the substrate binds to the protein near the haem. After this, one oxygen atom is incorporated into the substrate, while the other is reduced by NAD(P)H. Figure 3.9 depicts the general catalytic cycle of CYPs. Thus, it is obvious that the iron-oxygen complex plays an essential role in CYP450-catalysed reactions. Another important feature of the haem iron is that, in the absence of oxygen, it can be a reductant (Fe^{2+}) as well (Guengerich 1996).

3.3.5.1.3 Reduction

In contrast to oxidation, reduction is a reaction where the chemical gains electrons. In fact, reduction and oxidation proceed together always, as there should be a counterpart to be oxidised or reduced in each type of reaction. Depending on the substrate and conditions, such as the availability of sufficient oxygen, azo and nitro reductions can be catalysed by CYPs or by NADPH-cytochrome c reductases; reduction reactions are generally inhibited by oxygen. Gut microflora reductases contribute to azo and nitro reduction significantly, especially for orally taken drugs or chemicals. Examples of drugs that undergo reduction catalysed by CYPs or microflora enzymes are depicted in Figure 3.10.

The reduction reactions, with examples in the parentheses, are aldehyde (chloral hydrate), nitro group (chloramphenicol), azo group (prontosil), carbonyl (haloperidol), disulphide (disulfiram), sulphoxide and *N*-oxide (sulindac), quinone (menadione), dihydropyrimidine dehydrogenase (5-fluorouracyl) and reductive dehalogenation (halothane, methoxyflurane, DDT). Azo and nitro reduction, catalysed largely by intestinal microflora (Chung et al. 1992), frequently generate

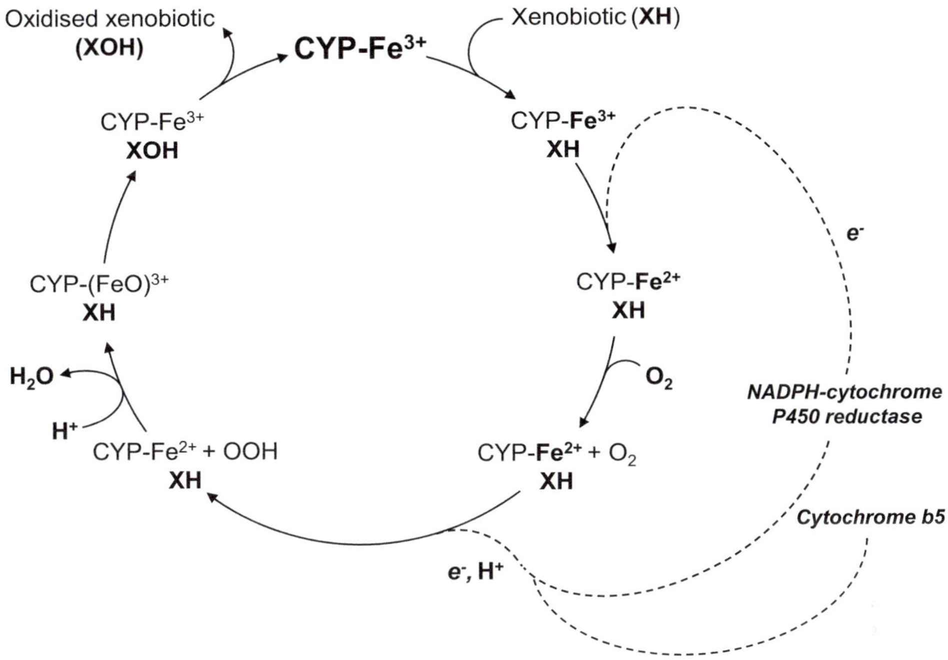

Figure 3.9 The catalytic cycle of cytochrome P450 enzymes. Hypothetical hydroxylation of a xenobiotic is shown.

Source: Adapted from Parkinson et al. (2013).

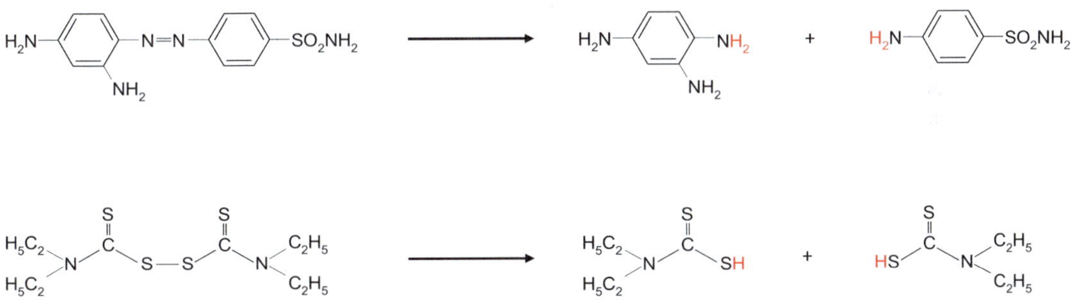

Figure 3.10 Representative azo group reduction of prodrug prontosil (upper) and disulphide reduction of disulfiram (lower).

corresponding amino compounds, which subsequently are oxidised to toxic metabolites. It can be said that reduction of a substrate generally results in bioactivation rather than detoxification.

3.3.5.1.4 Hydrolysis

A number of esterases and amidases present in mammalian blood and other tissues hydrolyse carboxylic acid esters (acetylsalicylic acid, cocaine, procaine, acetylcholine, etc.), amides (indomethacin, lidocaine, etc.), thioesters (spironolactone, etc.), acetals—glycoside—(digital glycosides, etc.), epoxides (carbamazepine, etc.) and phosphoric acid esters (paraoxon, etc.). Although most esterases are present in the cytoplasm, microsomal forms of both esterases and amidases have also been identified. These enzymes have been classified according to their interaction with organophosphates as A, B and C class. Class A enzymes hydrolyse organophosphate compounds, class B are inhibited by organophosphates and class C do not interact with organophosphates (Aldridge

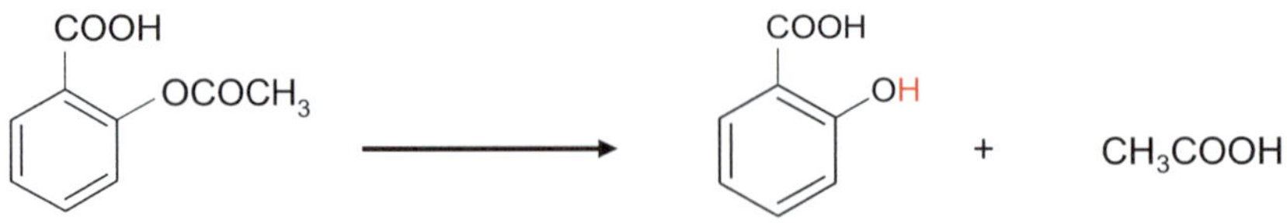

Figure 3.11 Hydrolysis reaction of acetylsalicylic acid (aspirin) to salicylic acid.

1953a, 1953b). Paraoxonase (PON) is in class A, whereas carboxylesterases and cholinesterases are in class B. Because of the high toxic potency of organophosphate compounds that initially were produced as warfare agents and subsequently used as pesticides, esterases have been studied in depth in the literature (Ramesh 2015). PON is another member of esterases that has attracted a high degree of interest. This enzyme not only catalyses the hydrolysis of organophosphates but also plays an important role in lipid metabolism (Mackness et al. 2002). On the other hand, amidases catalyse the hydrolysis of amides, hydrazides and carbamates. These enzymes play an important role in isoniazid-induced hepatotoxicity and neuropathy (Mitchell et al. 1975; Mandel 1959; Thomas et al. 1981). As an example, the hydrolysis reaction of acetylsalicylic acid (aspirin) is shown in Figure 3.11.

3.3.5.2 Phase II Biotransformation

Because an appropriate/relevant endogenous molecule attaches to the xenobiotic, the phase II reactions are called conjugation reactions. Direct conjugation of the parent xenobiotic, or its relatively more reactive metabolite, yields a highly water-soluble compound that can easily be excreted via urine and/or bile secretion (in faeces), as well as via other minor means such as sweat and saliva. For example, there is an approximately 10-fold increase in the excretion rate upon biotransformation of paracetamol, where the parent drug has a renal clearance value of 13 mL min^{-1}, whereas its sulphate conjugate has a rate of 130 mL min^{-1} (Prescott 1980). Excretion via lungs exhibits a different nature. Relatively more lipid-soluble compounds are transported from the lungs via exhaled air in contrast to the general propensity of increased water solubility. The major conjugation reactions are glucuronidation, sulphonation (sulphation), conjugation with glutathione, conjugation with amino acids, acetylation and methylation. The last two, however, result in a decrease in chemical water solubility in contrast to the general rule.

3.3.5.2.1 Glucuronidation

Conjugation of xenobiotics and/or their metabolites with the endogenous cofactor, uridine-5′-diphospho-α-D-glucuronic acid (UDPGA), is one of the major phase II reactions in humans. UDPGA reacts with acceptor groups such as hydroxyl (hydroxylamines, phenol, enol and primary, secondary and tertiary alcohols) or carboxyl groups (aromatic and aliphatic acids) on the parent compound to form corresponding glucuronic acid conjugates. Other groups such as aromatic amino and imino (carbamate and sulphonamide) and sulfhydryl (heterocyclic, thiol and carbamodithioic acid) may also react with UDPGA, although to a much lower extent (Abou-Donia 1995). The corresponding conjugates are ether, ester, thiolic, *N*- and *C*-glucuronides (named according to whether they bind to N- or C- groups in the substrate). In addition to xenobiotics having these groups, glucuronidation involves endogenous substrates such as bilirubin, steroid hormones and thyroid hormones. These reactions are catalysed by UDP-glucuronyl transferases (UGTs) located in the sER fraction of the cell (microsomal fraction). There are two main families of enzymes in humans, UGT1A and UGT2B, which have many members in each. Similar to CYPs, UGTs are also induced by phenobarbital (Burchell and Coughtrie 1989). Depending on the parent molecule, ether or ester glucuronides are formed. In rats, these are excreted by the kidneys (urine) or liver (bile), depending on the size of the aglycone part of the conjugate. Urine is preferred if the molecular weight is smaller than 250 Da, and bile if the molecular weight is larger than 350 Da (Yang et al. 2009). The bond between UDP and glucuronic acid is in α-configuration, hence it is resistant to hydrolysis catalysed by β-glucuronidase. However, glucuronic acid conjugates are formed in β-configuration and subjected to hydrolysis by this enzyme (hence its name).

The sites of glucuronidation in the body are the liver, intestinal mucosa and the kidneys. Other organs have a very limited capacity. Acyl glucuronides of nonsteroidal anti-inflammatory drugs (NSAIDs) may undergo hydrolysis and isomerisation that may bind to proteins. Some of these adducts may cause hypersensitivity (Xie et al. 2015) or idiosyncratic liver toxicity (Boelsterli et al.

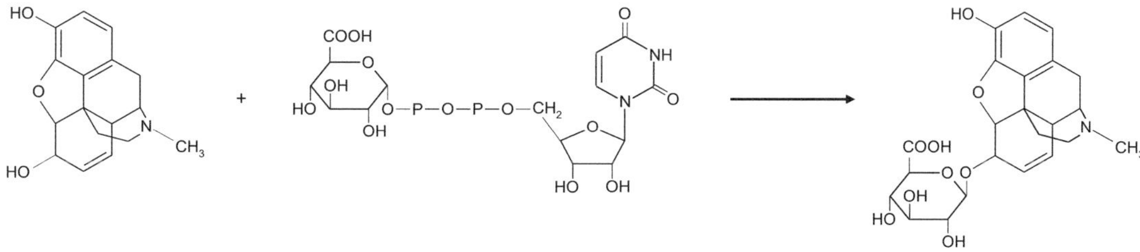

Figure 3.12 Glucuronidation of morphine.

1995). In fact, this acyl migration (see below) happens also for other drugs or xenobiotics containing carboxylic acid groups, such as isoxepac, zomepirac, clofibrate, valproate and oxaprozin (Smith et al. 1986).

Two mechanisms have been proposed for the initial drug-protein covalent binding associated with glucuronide-induced toxicity. One is nucleophilic displacement of glucuronosyl group by the $-NH_2$, –SH or –OH group of the protein (van Breemen and Fenselau 1985; McDonagh et al. 1984; Wells et al. 1987; Ruelius et al. 1986); the other, termed "acyl migration", involves migration of the acyl group to the other –OH group of the glycone moiety (Smith et al. 1986, 1990). As a result, the rearranged aldehyde moiety of the molecule binds to lysine residues on proteins which, in turn, forms an imine derivative (Hodge 1955). Hypolipidemic clofibrate is another drug example that exerts its toxicity via this pathway (Faed 1984). It has been reported that the reactive glucuronide of zomepirac covalently binds to cellular proteins, and the resulting adduct acts as a hapten to produce an immune response which may result in anaphylaxis (Smith et al. 1986). A similar mechanism has been suggested/proposed for cholestasis or conjugated hyperbilirubinemia, causing liver damage (Blanckaert et al. 1978). A UDPGT evolved to catalyse bilirubin glucuronidation may be "duped" by a modern chemical to participate in an adverse reaction (Burchell and Coughtrie 1989). Morphine-glucuronide formation is illustrated in Figure 3.12. Finally, although glucuronidation has low affinity, it has a high capacity for phenolic substrates.

3.3.5.2.2 Sulphation

Phenols, aliphatic alcohols and aromatic amines are conjugated with the sulphate group from endogenous 3′-phosphoadenosine-5′-phosphosulphate (PAPS), a reaction that is catalysed by sulphotransferases (SULTs). In contrast to microsomal CYPs and UDPGTs, SULTs are cytosolic enzymes. In general, conjugation with a sulphate group results in the detoxification of xenobiotics; however, few bioactivations also occur following sulphation. For example, conjugation of *N*-hydroxylated 2-acetylaminofluorene with sulphate yields an unstable N-O-sulphate ester. This ester decomposes to a nitrenium-carbonium ion that covalently binds to cellular macromolecules and initiates a toxic pathway. Highly polar sulphate conjugates are readily excreted in urine. Among all conjugation reactions, sulphation is not inducible by the actions of microsomal XMEs, probably due to the cytosolic nature of these proteins. In addition to structurally appropriate xenobiotics, steroids may also undergo sulphation. As with all other conjugation reactions, sulphation requires consumption of cellular energy (ATP). In addition, its cofactor PAPS is synthesised from inorganic sulphate ion and ATP. Because several other ions may replace sulphate, this type of replacement causes both ATP depletion and PAPS shortage, which eventually results in cellular damage.

The capacity of sulphate conjugation is low compared to glucuronidation. Therefore, it is saturated relatively faster by ingestion of large doses of sulphate-consuming metabolites and/or parent chemicals such as paracetamol. In case of paracetamol overdose, saturated sulphate conjugation diverts the biotransformation to the other two pathways: glucuronidation and bioactivation to NAPQI, which are catalysed by UGT1A9 (Mutlib et al. 2006) and CYP2E1 (Arzuk et al. 2018), respectively (Figure 3.8).

Animals, invertebrates, fungi and bacteria all have the capacity for sulphation. Among mammals, the rate is different between species (e.g. it is high in cats but low in rabbits and pigs). In contrast to glucuronidation, sulphation has high affinity but low capacity for phenols. This conjugation primarily occurs in the liver, lungs, gastrointestinal system and the kidneys. Examples of drugs that undergo sulphation are aniline, paracetamol and sulphonamides (aniline sulphation is shown in Figure 3.13).

3.3.5.2.3 Glutathione Conjugation

Glutathione (GSH) represents one of the most important antioxidant/antielectrophilic cofactors in all tissues, with the highest concentration of 5–10 mM in hepatocytes (liver). GSH may react with electrophilic substrates either spontaneously or by glutathione *S*-transferase (GST)–catalysed reactions. Interestingly, the configuration of the corresponding GSH conjugate may be different in the presence of GSTs (Dragovic et al. 2010), which may catalyse conjugation of GSH to a different position compared to spontaneous binding, and may therefore have a significant influence on further metabolism. GSTs are a large family of enzymes with seven cytosolic (GSTA, GSTM, GSTT, GSTP, GSTS, GSTZ, GSTO), one mitochondrial (GSTK) and three microsomal forms (MGSTs; Hayes et al. 2005; Higgins and Hayes 2011; Mannervik et al. 2005). GSTA, in addition to other functions, binds several endogenous substrates, including haem and bilirubin, and serves as transfer protein.

Although GSH is a tripeptide composed of three amino acids, its conjugates are structurally different. Thioethers are formed upon GSH conjugation, whereas amides are the final products upon amino acid conjugation. Analogous to UDPGA, GSH also provides an evolutionary protection against hydrolysis by corresponding enzymes (peptidases in this case). For example, glutamine binds to cysteine from γ-carbon instead of α-carbon (Figure 3.14).

Therefore, GSH is protected from peptidase-catalysed hydrolysis because these enzymes have affinity to peptides that bind the α-carbon.

GSH conjugation is an important detoxification pathway, although sometimes it results in toxicity. GSH conjugates of certain halogenated alkanes and alkenes, such as hexachlorobutadiene, chlorotrifluoroethylene and dibromoethylene, are further bioactivated by the cysteine conjugate β-lyase in kidney tissue. Conjugation of sevoflurane with GSH and subsequent proposed toxic pathways (Orhan et al. 2004) is shown in Figure 3.15.

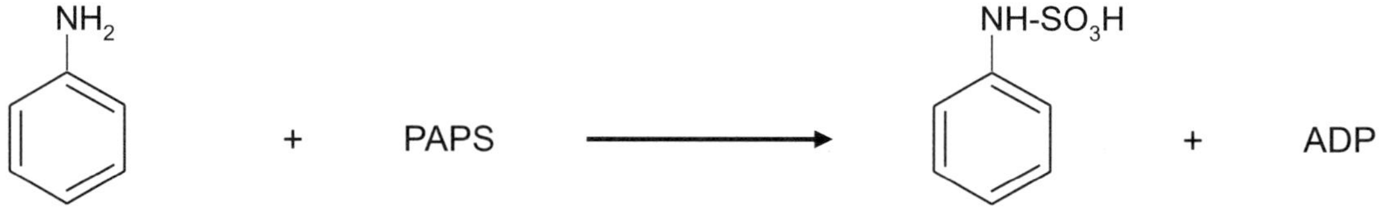

Figure 3.13 Sulfation of aniline.

Abbreviations: PAPS = 3′-phosphoadenosine-5′-phosphosulphate; ADP, adenosine diphosphate.

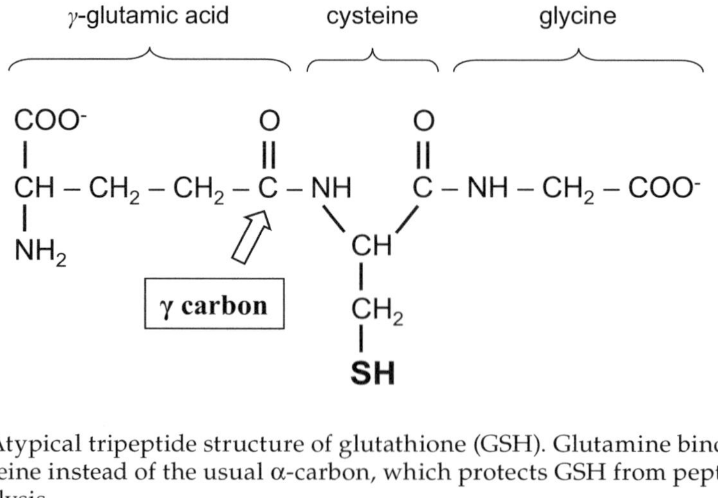

Figure 3.14 Atypical tripeptide structure of glutathione (GSH). Glutamine binds to the γ-carbon of cysteine instead of the usual α-carbon, which protects GSH from peptidase-catalysed hydrolysis.

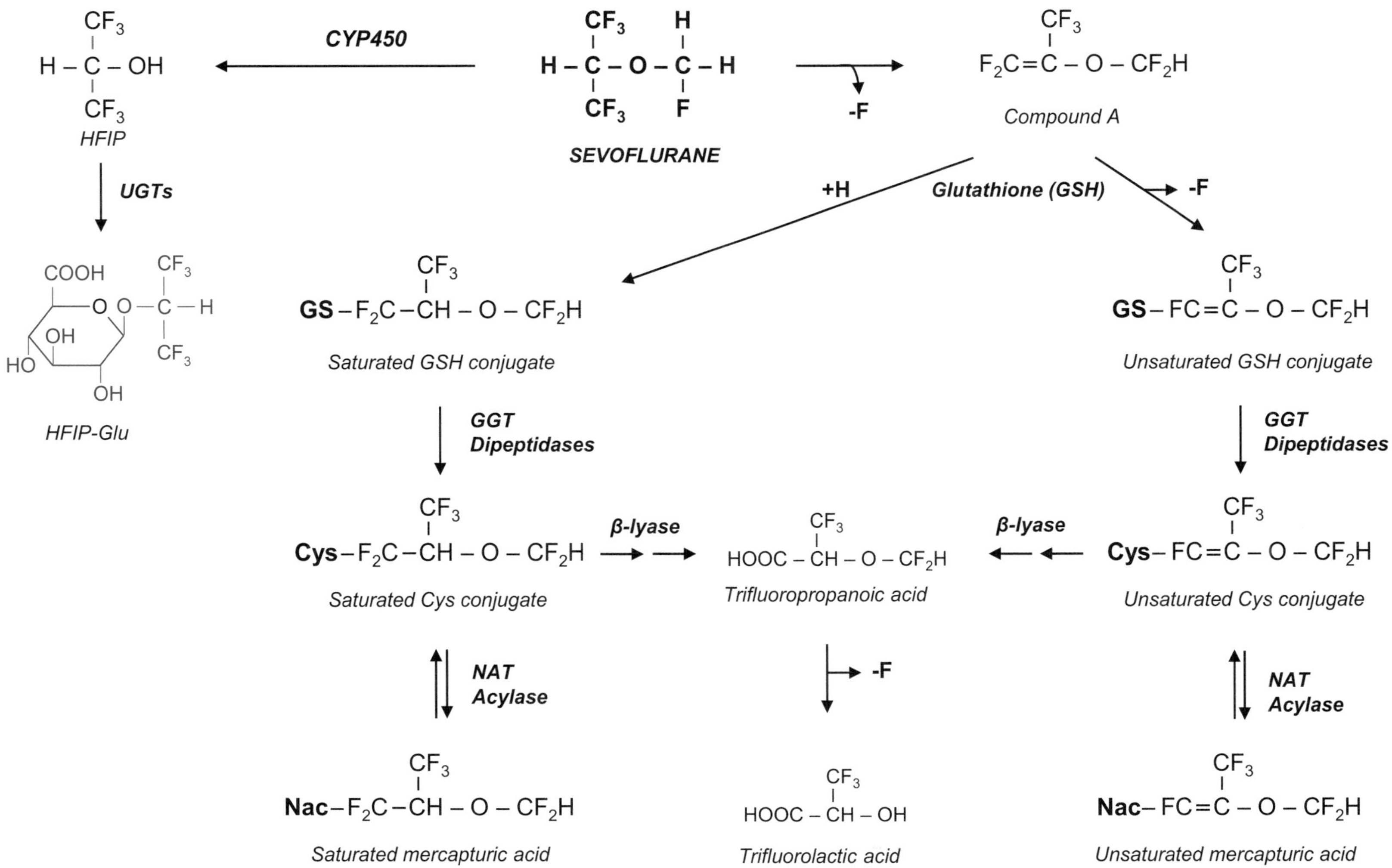

Figure 3.15 Glutathione conjugation, mercapturic acid formation and proposed cysteine conjugate β-lyase-mediated bioactivation of sevoflurane (Orhan et al. 2004). Phase I oxidation and glucuronide conjugation are shown in fade tone to emphasise the glutathione conjugation pathway.

Abbreviations: Compound A = reactive degradation product of sevoflurane; CYP450 = cytochrome P450; GGT = gammaglutamyltranspeptidase; HFIP = hexafluoroisopropanol; NAT = *N*-acetyltransferase; UGTs = uridindiphosphoglucuronic acid transferases.

3.3.5.2.4 *Amino Acid Conjugation*

Aromatic carboxylic acids, arylacetic acids and heterocyclisc carboxylic acids are conjugated with α-amino acids. The amino acids that are utilised vary among species; glycine and glutamine are the most common ones used for conjugation with the acids above in mammals, whereas taurine is used in conjugation with bile acids. In some birds and reptiles, ornithine is preferred (Idle et al. 1976; J. Smith 1958), whereas insects utilise mostly arginine. Ticks (which are arachnids) also use arginine, and fish use taurine for conjugation. As such, amino acid conjugation is the most variable interspecies phase II conjugation reaction (Dauterman 1994). Amino acids are conjugated with xenobiotics in two ways: carboxylic acid moiety of xenobiotics can react with amino groups of glycine, glutamine or taurine, or aromatic hydroxylamine group of xenobiotics can bind to carboxylic acid group of serine or proline. Bile acids are the endogenous substrates of glycine and taurine and are excreted in bile, whereas the other amino acid conjugates of xenobiotics are exclusively excreted in urine. The conjugation of amino acids with carboxylic acids are catalysed by two different forms of Acyl-CoA:*N*-acetyl transferases in mitochondrial fraction. There is an interesting cross-correlation between amino acid and glucuronic acid conjugations. Amino acid conjugates of carboxylic acid containing/bearing xenobiotics are nontoxic, whereas aromatic amines are usually toxic because they can be bioactivated by the corresponding reactive nitrenium or carbonium ions. On the other hand, as already discussed, glucuronic acid conjugates of $–NH_2$ groups are nontoxic, whereas conjugates of –COOHs, such as NSAIDs, are frequently subjected to bioactivation. Amino acid conjugation of NSAIDs is important, as it may limit its acyl migration–mediated toxicity. Examples of drugs for conjugation with glycine are benzoic acid and acetylsalicylic acid (ASA), and with glutamine are diphenhydramine, phenylacetic acid and indole acetic acid. Conjugation of ASA with glycine and formation of the product salicyluric acid are shown in Figure 3.16.

There are differences in the conjugation of organic acids among mammalians; carnivores favour glucuronic acid conjugation, herbivores favour amino acid conjugation, and omnivores utilise both (Calabrese 1983).

3.3.5.2.5 *Acetylation*

In this type of reaction, the acetyl group of acetyl coenzyme A is conjugated with primary aromatic amines, sulphonamides, hydrazines and hydrazides. Acetylation reactions are catalysed by the cytosolic *N*-acetyl transferases (NAT), mainly in liver, gastric mucosa and white blood cells. Initially, the NAT enzyme is acetylated by the cofactor acetyl coenzyme A (acetyl CoA), and subsequently the acetyl group is transferred to the amino group of the substrate. Primary aliphatic amines are not substrates for this type of reaction, except cysteine conjugates. Following sequential hydrolyses of glutamine and glycine from GSH conjugates, the subsequent acetylation

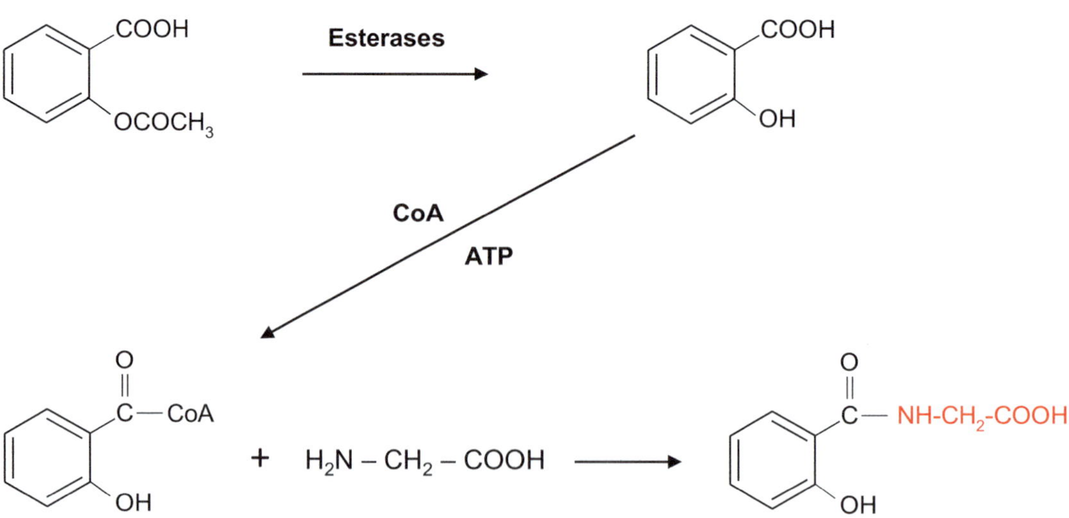

Figure 3.16 Conjugation of acetylsalicylic acid (ASA) with glycine, and formation of salicyluric acid.

generates a cysteine conjugate of "soft" electrophiles which, in turn, is converted into the corresponding mercapturic acid derivative upon *N*-acetylation. NATs are cytosolic enzymes with a limited number of copies compared to other XMEs; humans express only two enzymes, NAT1 and NAT2. An important feature is that, similar to some other XMEs, NAT2 is genetically polymorphic, which has significant outcomes regarding clinically used drugs that are biotransformed by *N*-acetylation. This is the case for the hydralazine drug which may cause lupus erythematosus in slow acetylators, where NAT2 functions slower with frequencies of 70% in Moroccans, 90% in Middle Easterners, 10% in Japanese and 40% in Caucasian Americans and Europeans (Parkinson et al. 2013). This significant difference in the rate of acetylation between individuals has been established by a series of clinical observations with the antitubercular drug isoniazid; half of the patients, namely slow acetylators, treated with this drug suffered from peripheral neuropathy and hepatotoxicity (Parkinson et al. 2013).

A large interspecies variation, similar to amino acid conjugation, exists for acetylation; the hamster and the rabbit possess relatively high *N*-acetylating capacities, whereas the dog has little or no acetylating capacity. Rodents such as rats and mice have moderate capacity (Dauterman 1994).

A toxicologically important feature is acetyl conjugates of several xenobiotics, which may precipitate in kidney tubules due to diminished water solubility upon acetylation and eventually exert local toxicity. *N*-acetylated sulphonamides are examples of this crystalluria (Dutton 1978). Dapsone, isoniazid and sulphonamides are some examples that undergo acetylation (Figure 3.17 illustrates acetylation of isoniazid).

3.3.5.2.6 Methylation

Methylation is a common conjugation reaction in mammals involving the transfer of a methyl group from *S*-adenosylmethionine (SAM) to a variety of substrates: proteins, nucleic acids, phospholipids and so forth. The reaction is catalysed by a number of methyl transferases. SAM is synthesised from methionine and ATP, similar to PAPS. As a less important alternative, a methyl group can also be supplied by N^5-methyltetrahydrofolic acid. SAM is the essential coenzyme in the methylation of oxygen, nitrogen or sulphur nucleophiles (Lin 2011). Two enzymes catalyse methylation reactions: phenol *O*-methyltransferase (POMT) and catechol-*O*-methyltransferase (COMT) (160). COMT catalyses the degradation of catecholamines such as adrenaline, noradrenaline and dopamine. Methylation masks functional groups and reduces the ability to be conjugated; on the other hand, it is a relatively unimportant conjugation for xenobiotics. Except for the tertiary amines, methylation generates fewer water-soluble conjugates compared to the parent chemical together with acetylation, in contrast to the prevalent tendency of conjugation reactions. Nevertheless, methylation of xenobiotics generally results in detoxification. Catecholamines, histidine and methadone are some examples metabolised by methylation (epinephrine is shown in Figure 3.18).

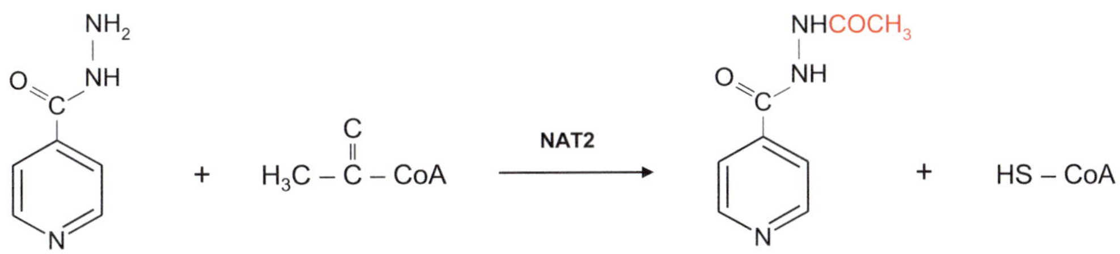

Figure 3.17 Acetylation of the antitubercular drug isoniazid.

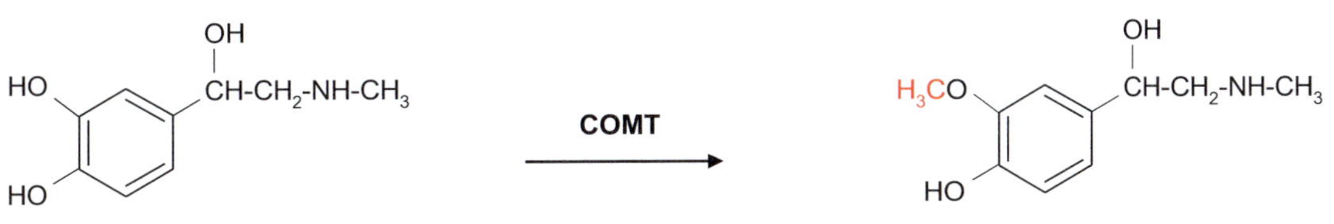

Figure 3.18 Methylation of epinephrine.

Abbreviation: COMT, catechol *O*-methyl transferase.

3.3.5.3 Phase III Biotransformation

Both phase I and II reactions may be followed by phase III metabolism, which is actually not a process that involves catalytic enzymes. The water-soluble conjugates should be transported from inside of the cells where they are formed to the interstitial space between cells, enter the capillary system and move to the sites of excretion (glomeruli, bile, alveoli, etc.). This proves difficult, as lipid membranes represent a barrier against highly water-soluble conjugates. Hence the process is still not completed for the conjugate after being transported to the interstitial space. As mentioned above, conjugated and/or nonconjugated metabolites must pass several more membranes before reaching the excretion sites. This is an essential process because any failure in transporting water-soluble conjugates from cells can lead toxicity to cellular components and/or functions by hydrolysis of the conjugate back to the reactive intermediate.

The difficult membrane crossing is accomplished by a wide range of membrane-bound transport systems. These systems have been evolved to actively remove water-soluble metabolites and other low molecular weight compounds from cells. It has been estimated that around 5% of human genes are associated with transporters, indicating the importance of transport processes for both endogenous and exogenous chemicals (Hediger et al. 2004).

The first xenobiotic transporter identified was the phosphoglycoprotein (P-gp, or multidrug resistance protein [MDR1]) that conferred multidrug resistance to tumour cells (Ambudkar et al. 1999). P-gp or MDR1 pumps anticancer drugs out from the tumour cell, which represents a clinical problem in cancer therapy. MDR1 is a member of ATP binding cassette (ABC) transporters, of which there are 49 genes in humans (Ueda 2011). Some members of these seven subfamilies, from A to G, take charge in transporting endogen substrates such as cholesterol (ABCA1) or phospholipid substrates (ABCB4). ABCC, which is also known as multidrug resistance protein (MRP), is the second important subfamily. Members of MRP transport UDP-glucuronic acid or glutathione conjugates from cells. The breast cancer resistance protein (BCRP) is a member of ABCG subfamily. Apart from breast cancer cells, this protein is expressed in normal tissues and plays a role in efflux of sulphate conjugates of both endogenous and exogenous origin (Ni et al. 2010). ABC transporters take place in absorption from the alimentary tract and excretion into the bile and urine of many xenobiotics. They contribute also to the barrier function of several specialised anatomical/histological structures, such as blood-brain, blood-testis, blood-placenta and blood-follicle barriers (Hutson et al. 2010; Iqbal et al. 2012; Bloise et al. 2013; Baello et al. 2014; Guerreiro et al. 2018).

Next to the ABC family, the second major family of xenobiotic transporters is the solute carriers (SLCs) that facilitate transport process. Despite their important roles in transporting xenobiotic and endogenous chemicals especially in the liver and kidneys, data so far suggest that SLCs do not involve in transfer/transport of metabolite/drug conjugates as much as ABC transporters do (Hediger et al. 2004; Perland and Fredriksson 2017).

An important feature of transporters is that some enzymes and transporters are coregulated, as inducers increase the capacity of both systems (Cummins et al. 2002). One example is CYP3A4 and P-gp, wherein flucloxacillin induced the expression of both proteins simultaneously (Huwyler et al. 2006).

3.4 EXCRETION

After absorption through different routes, such as the alimentary tract, skin and lungs, and being subjected to distribution and biotransformation reactions, xenobiotics are excreted from different sites in the body. The main site of excretion is via the kidneys, where urine serves as the highest volume for water-based secretion. It has previously been emphasised that lipid-soluble xenobiotics can readily enter the organism and that this physicochemical feature is, in general, converted by biotransformation reactions to more water-soluble metabolites to facilitate excretion. Kidneys are one of the most perfused organs, receiving 25% of the cardiac output. The functional unit of the kidney, the nephron, is anatomically composed of glomeruli, proximal and distal tubules that contain the loop of Henle. Distal tubules open into collecting ducts which ultimately end up in the bladder. Figure 3.19 illustrates the functional units of the major excretion organs (kidneys, liver and lungs). On average, there are about a million nephrons in an adult human. Two major mechanisms exist for renal excretion of endogenous and exogenous chemicals: glomerular filtration and tubular secretion. Although it functions against excretion, tubular reabsorption is equally important.

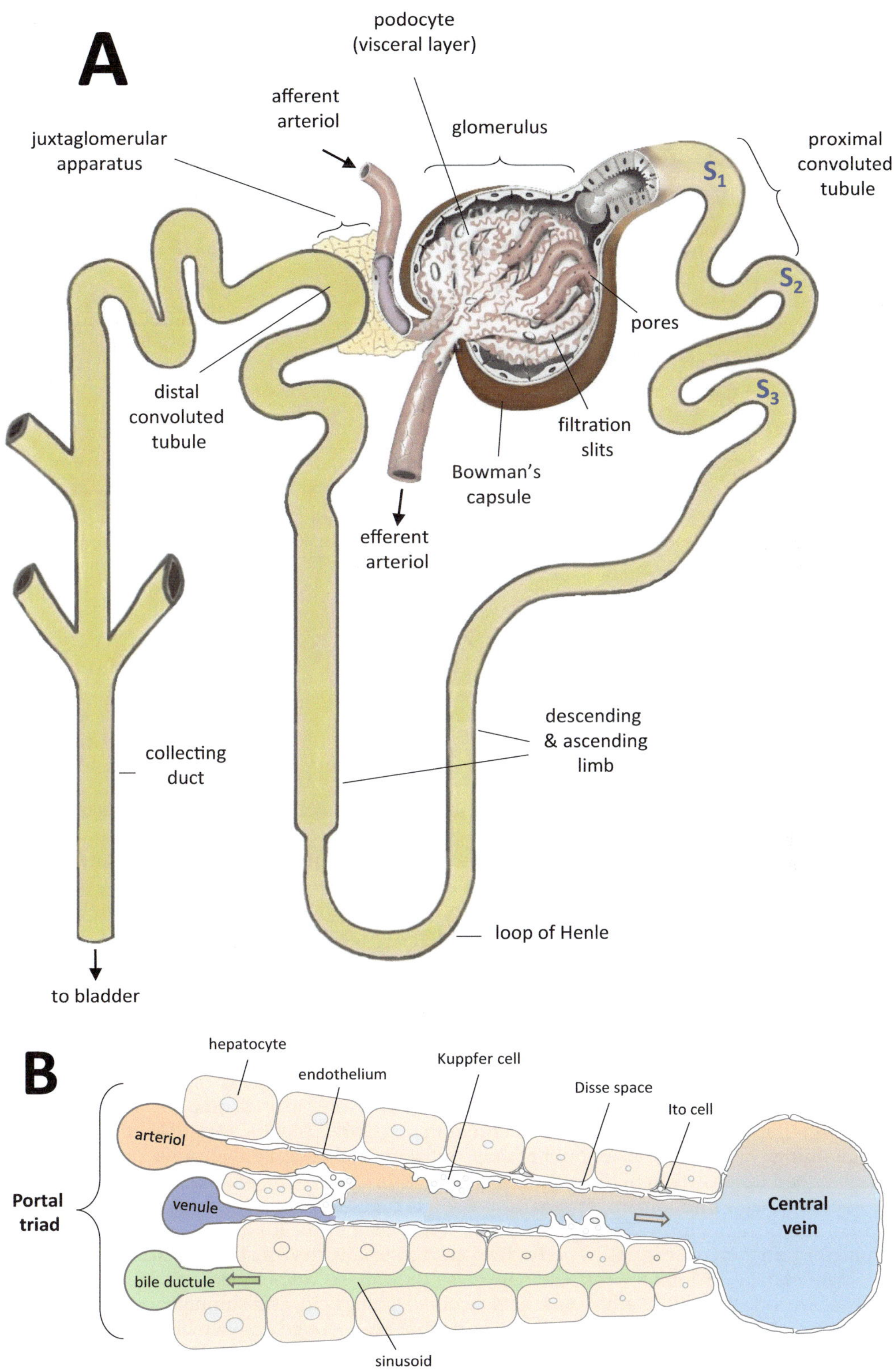

Figure 3.19 The functional units of major excretion organs: (A) nephron (kidneys); (B) hepatic acinus (liver); (C) alveoli (lungs).

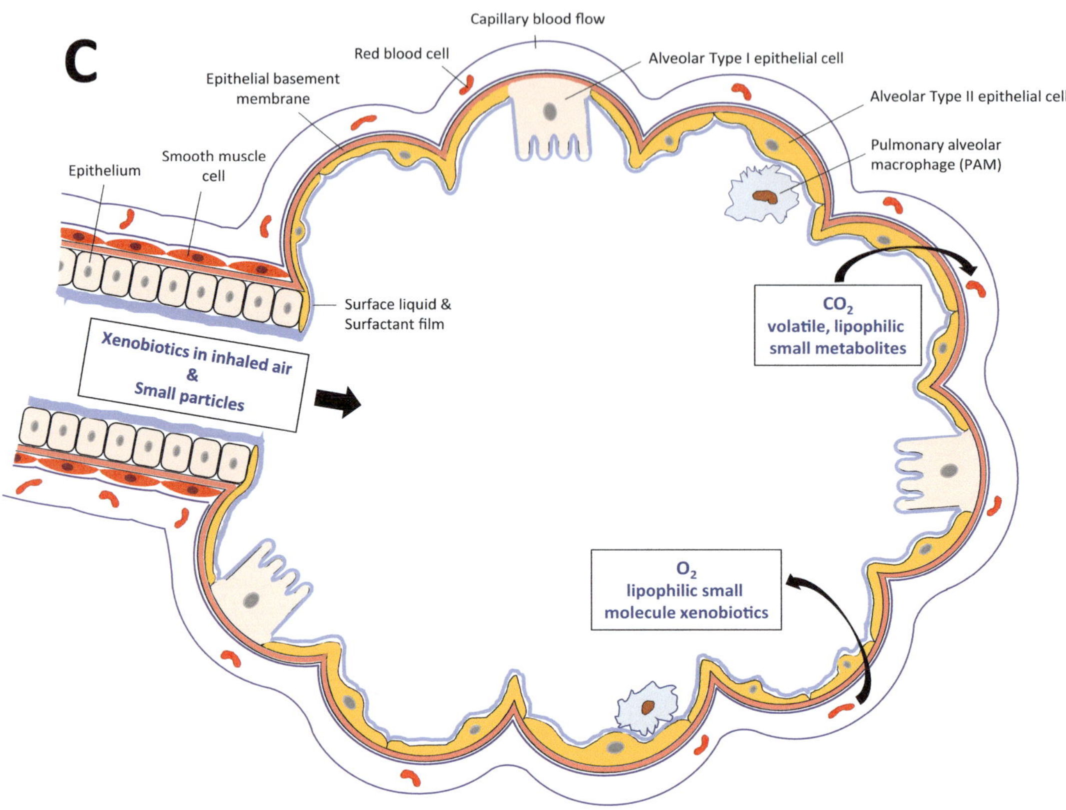

Figure 3.19 (Continued)

Glomerular filtration, which is similar to paracellular diffusion from aqueous pores on the cell membrane, involves the hydrostatic or osmotic pressure of the blood through highly porous glomeruli to filter excretable end products (Figure 3.4). Sufficiently polar and relatively small molecules, including water, are readily filtered through the pores of the glomerular capillaries (70 nm in diameter). As the diameter of the pores is the largest among the pores at other sites in the organism, molecules with the size of up to 60 kDa are filtered in general. On the other hand, molecular weight is not the only determining factor in what materials can be filtered through glomerular pores. Haemoglobin, with a molecular weight of 68 kDa, has a diameter of only 6 nm and should be able to pass the glomerulus, however the filtered proportion is only 3%, compared to inulin which is completely filtered (Rickert 1997). The holes and epithelial cells of the basal membrane of the glomerulus also play an important role in filtration. Drugs and other xenobiotics bound to plasma proteins (albumin, α1-acid glycoprotein, etc.) are not filtered; only the unbound free portion is available for filtration through the glomerulus. The volume and composition of the first filtrate are significantly different from the eventual urine. Only 1% of the filtrate is excreted as urine. The remaining water, required amount of ions, nutrients and sufficiently lipid-soluble compounds are reabsorbed downstream of the nephron tubule.

The second major mechanism for renal excretion is tubular secretion, which occurs at the proximal tubule of nephron. There are two different systems: one for transporting organic anions and one for transporting cations by active transport mechanisms into tubular lumen. Many free or conjugated drugs exhibit a weak acid nature, so they are secreted by organic anion transporters. These include penicillin, ethacrynic acid, acetylsalicylic acid, and sulphate and glucuronide conjugates of drugs. The endogenous compounds that are secreted actively include 5-hydroxyindolacetic acid, the main metabolite of serotonin (Pritchard and Miller 1992).

Reabsorption takes place in the convoluted proximal tubule of the nephron. In contrast to tubular secretion that involves active transport of chemicals into the tubular lumen, reabsorption occurs primarily by passive transport (transcellular diffusion) from high concentration in the proximal tubule to low concentration of the chemical in the blood around the tubule (concentration gradient). In terms of toxic outcome via reabsorption, one significant example is the nephrotoxicity of several halogenated alkenes via the cysteine conjugate β-lyase, which is discussed in more detail in Section 3.3.5.2.3.

The second major site of excretion is the liver. Mostly, glucuronide and glutathione conjugates, parent chemicals and endogenous compounds are actively transferred into bile. The anatomical structure of the liver represents a unique feature: one side of the hepatocytes faces the sinusoids (blood circulation), whereas the other faces the bile canaliculus (Figure 3.19). Xenobiotics in the systemic circulation enter the liver via the hepatic artery or through the lymph. The entrance of chemicals from the blood circulation into hepatocytes occurs not exclusively via active transport but by combination of passive diffusion and facilitated transport. Bile is produced at an average rate of 600 mL/day by hepatocytes in adults and secreted into the bile canaliculus. It flows in the opposite direction to the blood into hepatic bile ducts, which end at the gallbladder. The gallbladder empties its contents, containing endogenous and exogenous end products, into the duodenum following a stimulus from food in the stomach.

In contrast to entrance of chemicals into hepatocytes via passive diffusion and facilitated transport, biliary excretion generally involves the active transport of various xenobiotics and their metabolites. Several transport proteins present on the canalicular membrane mediate excretion of bile (discussed in detail in Section 3.3.5.3). Therein, the multiresistance drug protein-2 (MRP2) has the greatest role. It is responsible for the transport of organic anions, including previously mentioned glucuronide and glutathione conjugates of both drugs and other xenobiotics, as well as endogenous compounds. Highly polar and plasma protein-bound chemicals, as well as those with molecular weights greater than 300 Da, are likely to be excreted in bile. In most cases, bile is the ultimate route of excretion for many conjugates that are not reabsorbed into the blood. Bile flows down the bile duct towards the gallbladder in the opposite direction of the blood. Six times a day, the gallbladder empties its contents into the duodenum where xenobiotics and/or their metabolites, together with endogenous compounds, are excreted in faeces. However, some glucuronide conjugates are hydrolysed in the intestine by glucuronidases of the microflora, and then the free metabolite returns to the liver via hepatic portal vein and partially enters back to the systemic circulation, while a fraction may be conjugated and secreted to bile again. This phenomenon is known as enterohepatic recirculation. In addition to biliary excretion, direct excretion of chemicals into the gastrointestinal tract also occurs, either from the stomach, although at trace amounts (morphine, etc.; Hatcher and Davis 1925); or the intestine (digitoxin, dinitrobenzamide, hexachlorobenzene, ochratoxin A, etc.; Rozman 1986).

The third major organ of excretion is the lung. In contrast to the xenobiotics with increased water solubility, which are excreted in urine or faeces, lipid-soluble compounds and gases are mainly excreted via lungs in exhaled air. Compounds which are volatile at physiological temperature are also eliminated in exhaled air. Specialised transport for both endogenous and exogenous chemicals does not exist in pulmonary tissue; therefore, the chemical transfer from capillary blood to alveolar air and vice versa is governed by the concentration gradient of a particular compound (i.e. by simple diffusion). In terms of drugs, highly volatile nitrous oxide is excreted by the lungs, for instance.

There are some other minor routes of excretion such as sweat, tears, milk, cerebrospinal fluid and saliva. Given their relatively much less volume produced daily, excretion into these media is practically negligible/insignificant.

BIBLIOGRAPHY

Abou-Donia, M. B. 1995. Metabolism and toxicokinetics of xenobiotics. In *CRC Handbook of Toxicology*, eds. M. J. Derelanko and M. A. Hollinger, 568. CRC Press.

Adamson, R. H., J. W. Bridges, M. E. Evans, and R. T. Williams. 1970. Species differences in the aromatization of quinic acid in vivo and the role of gut bacteria. *Biochem J* 116(3): 437–43.

Aldridge, W. N. 1953a. Serum esterases. I. Two types of esterase (A and B) hydrolysing p-nitrophenyl acetate, propionate and butyrate, and a method for their determination. *Biochem J* 53(1): 110–7.

Aldridge, W. N. 1953b. Serum esterases. II. An enzyme hydrolysing diethyl p-nitrophenyl phosphate (E600) and its identity with the A-esterase of mammalian sera. *Biochem J* 53(1): 117–24.

Ambudkar, S. V., S. Dey, C. A. Hrycyna, M. Ramachandra, I. Pastan, and M. M. Gottesman. 1999. Biochemical, cellular, and pharmacological aspects of the multidrug transporter. *Annu Rev Pharmacol Toxicol* 39: 361–98.

Anandatheerthavarada, H. K., S. Addya, R. S. Dwivedi, G. Biswas, J. Mullick, and N. G. Avadhani. 1997. Localization of multiple forms of inducible cytochromes P450 in rat liver mitochondria: Immunological characteristics and patterns of xenobiotic substrate metabolism. *Arch Biochem Biophys* 339(1): 136–50.

Andres, H. H., A. J. Klem, L. M. Schopfer, J. K. Harrison, and W. W. Weber. 1988. On the active site of liver acetyl-CoA. Arylamine N-acetyltransferase from rapid acetylator rabbits (IWJ). *J Biol Chem* 263: 7521–7.

Ansher, S. S., and W. B. Jakoby. 1986. Amine N-methyltransferases from rabbit liver. *J Biol Chem* 261: 3996–4001.

Appleby, C. A. 1967. A soluble hemeprotein P-450 from N2-fixing Rhizobium bacteroids. *Biochim Biophys Acta* 147: 399–402.

Arzuk, E., B. Turna, M. Sözbilen, and H. Orhan. 2018. Inter-individual and inter-organ variability in the bioactivation of paracetamol by human liver and kidney tissues. *Environ Toxicol Pharmacol* 61: 8–17.

Baello, S., M. Iqbal, E. Bloise, M. Javam, W. Gibb, and S. G. Matthews. 2014. TGF-beta1 regulation of multidrug resistance P-glycoprotein in the developing male blood-brain barrier. *Endocrinology* 155: 475–84.

Baumann, E. Preusse, and C. Ueber. 1879. Bromphenylmercaptursäure. *Chemische Berichte* 12: 806–10.

Benz, R., K. Janko, and P. Läuger. 1980. Pore formation by the matrix protein (porin) of Escherichia coli in planar bilayer membranes. *Ann NY Acad Sci* 358: 13–24.

Blanckaert, N., F. Compernolle, P. Leroy, R. Van Houtte, J. Fevery, and K. P. Heirwegh. 1978. The fate of bilirubin-IX alpha glucuronide in cholestasis and during storage in vitro. Intramolecular rearrangement to positional isomers of glucuronic acid. *Biochem J* 171(1): 203–14.

Bloise, E., M. Bhuiyan, M. C. Audette, et al. 2013. Prenatal endotoxemia and placental drug transport in the mouse: Placental size-specific effects. *PLoS One* 8: e65728.

Bock, K. W. 1987. Drug glucuronidation and sulfation in rat and human liver. *ISI Atlas Sci Pharmacol* 1: 5–7.

Boelsterli, U. A., H. J. Zimmerman, and R. A. Kretz. 1995. Idiosyncratic liver toxicity of nonsteroidal antiinflammatory drugs: Molecular mechanisms and pathology. *Crit Rev Toxicol* 25: 207–35.

Boivin, G. 2007. The hydroxyapatite crystal: A closer look. *Medicographia* 29(2): 126–32.

Bosron, W. F., and T. K. Li. 1986. Genetic polymorphism of human liver alcohol and aldehyde dehydrogenases, and their relationships to alcohol metabolism and alcoholism. *Hepatology* 6: 502–10.

Burchell, B., and M. W. H. Coughtrie. 1989. UDP-glucuronosyltransferases. *Pharmacol Ther* 43(2): 261–89.

Calabrese, E. J. 1983. *Principles of Animal Extrapolation*, 220. Wiley-Interscience.

Chung, K. T., S. E. Jr. Stevens, and C. E. Cerniglia. 1992. The reduction of azo dyes by the intestinal microflora. *Crit Rev Microbiol* 18(3): 175–90.

Claude, A. 1943. The constitution of protoplasm. *Science* 97(2525): 451–6.

Coon, M. J., and D. R. Koop. 1983. P450 oxygenases in lipid biotransformation. In *The Enzymes*, Vol. 16, 645. New York: Academic Press.

Cooper, D. Y., S. Levin, S. Narasimhulu, and O. Rosenthal. 1965. Photochemical action spectrum of the terminal oxidase of mixed function oxidase systems. *Science* 147(3656): 400–2.

Cummins, C. L., W. Jacobsen, and L. Z. Benet. 2002. Unmasking the dynamic interplay between intestinal P-glycoprotein and CYP3A4. *J Pharmacol Exp Ther* 300: 1036–45.

Daneman, R., and A. Prat. 2015. The blood-brain barrier. *Cold Spring Harb Perspect Biol* 7(1): a020412.

Dauterman, W. C. 1994. Metabolism of toxicants: Phase II reactions. In *Introduction to Biochemical Toxicology*, eds. E. Hodgson and P. Levi, 113. Hartford, CT: Appleton & Lounge.

DeWitt, D. L., and W. L. Smith. 1983. Purification of prostacyclin synthase from bovine aorta by immunoaffinity chromatography. Evidence that the enzyme is a hemoprotein. *J Biol Chem* 258(5): 3285–93.

Dragovic, S., J. S. Boerma, L. van Bergen, N. P. Vermeulen, and J. N. Commandeur. 2010. Role of human glutathione S-transferases in the inactivation of reactive metabolites of clozapine. *Chem Res Toxicol* 23(9): 1467–76.

Dutton, G. J. 1978. Phase II metabolic reactions in man. In *Drug Metabolism in Man*, eds. J. W. Gorrod and A. H. Beckett, 81. London: Taylor & Francis Group.

El Hadri, L., B. Chanas, and B. I. Ghanayem. 2005. Comparative metabolism of methacrylonitrile and acrylonitrile to cyanide using cytochrome P4502E1 and microsomal epoxide hydrolase-null mice. *Toxicol Appl Pharmacol* 205(2): 116–25.

Emster, L., R. W. Estabrook, P. Hochstein, and S. Orrenius. 1987. *DT-Diaphorase*. Cambridge: Cambridge University Press.

Enna, S. J., and B. B. David. 2008. *xPharm: The Comprehensive Pharmacology Reference.* Amsterdam and Boston: Elsevier.
Estabrook, R. W., D. Y. Cooper, and O. Rosenthal. 1963. The light reversible carbon monoxide inhibition of the steroid C21-hydroxylase system of the adrenal cortex. *Biochem Z* 338: 741–55.
Evans, W. E., and M. V. Relling. 1999. Pharmacogenomics: Translating functional genomics into rational therapeutics. *Science* 286(5439): 487–91.
Faed, E. M. 1984. Properties of acyl glucuronides: Implications for studies of the pharmacokinetics and metabolism of acidic drugs. *Drug Metab Rev* 15(5–6): 1213–49.
Ferris, J. P., M. J. Fasco, F. L. Stylianopoulou, D. M. Jerina, J. W. Daly, and A. M. Jeffrey. 1973. Monooxygenase activity in Cunninghamella bainieri: Evidence for a fungal system similar to liver microsomes. *Arch Biochem Biophys* 156(1): 97–103.
Fröhlich, E. 2002. Structure and function of blood-tissue barriers. *Dtsch Med Wochenschr* 127(49): 2629–34.
Goldin, B. R. 1990. Intestinal microflora: Metabolism of drugs and carcinogens. *Ann Med* 22(1): 43–8.
Gonzalez, F. J. 1988. The molecular biology of cytochrome P-450s. *Pharmacol Rev* 40(4): 243–88.
Gorter, E., and F. Grendel. 1925. On bimolecular layers of lipoids on the chromocytes of the blood. *J Exp Med* 41(4): 439–43.
Gregus, Z. 2013. Mechanisms of toxicity. In *Casarett & Doull's Toxicology*, ed. Curtis D. Klaassen, 57. McGraw-Hill.
Grue, C. E. 1982. Response of common grackles to dietary concentration of four organophosphate pesticides. *Arch Environ Contam Toxicol* 11: 617–26.
Guengerich, F. P. 1982. Epoxide hydrolase: Properties and metabolic roles. *Rev Biochem Toxicol* 4: 5–30.
Guengerich, F. P. 1987. *Mammalian Cytochromes*, Vols. 1 and 2. Boca Raton, FL: CRC Press.
Guengerich, F. P. 1990. Enzymatic oxidation of xenobiotic chemicals. *Crit Rev Biochem Mol Biol* 25(2): 97–153.
Guengerich, F. P. 1996. The chemistry of cytochrome P450 reactions. In *Cytochromes P450: Metabolic and Toxicological Aspects*, eds. C. Ioannides and D. V. Parke, 68. New York: CRC Press.
Guengerich, F. P., and T. L. Macdonald. 1984. Chemical mechanisms of catalysis by cytochromes P-450: A unified view. *Acc Chem Res* 17: 9–16.
Guerreiro, D. D., L. F. de Lima, G. T. Mbemya, et al. 2018. ATP-binding cassette (ABC) transporters in caprine preantral follicles: Gene and protein expression. *Cell Tissue Res* 372(3): 611–20.
Harvey, J. M. 1967. Excretion of DDT by migratory birds. *Can J Zool* 45(5): 629–33.
Harvey, L., B. S. Arnold, Z. Lawrence, M. Paul, B. David, and D. James. 2000. *Molecular Cell Biology*, 4th ed. New York: W. H. Freeman.
Hatcher, R. A., and D. Davis. 1925. The excretion of morphine into the stomach. *J Pharmacol Exp Ther* 26(1): 49–60.
Hayes, J. D., J. U. Flanagan, and I. R. Jowsey. 2005. Glutathione transferases. *Annu Rev Pharmacol Toxicol* 45: 51–88.
Hediger, M. A., M. F. Romero, J. B. Peng, A. Rolfs, H. Takanaga, and E. A. Bruford. 2004. The ABCs of solute carriers: Physiological, pathological and therapeutic implications of human membrane transport proteins. *Pflugers Arch* 447(5): 465–8.
Hein, D. W. 1988. Acetylator genotype and arylamine-induced carcinogenesis. *Biochim Biophys Acta* 948: 37–66.
Hempel, J., H. von Bahr-Lindström, and H. Jörnvall. 1984. Aldehyde dehydrogenase from human liver. Primary structure of the cytoplasmic isoenzyme. *Eur J Biochem* 141: 21–35.
Higgins, L. G., and J. D. Hayes. 2011. Mechanisms of induction of cytosolic and microsomal glutathione transferase (GST) genes by xenobiotics and pro-inflammatory agents. *Drug Metab Rev* 43(2): 92–137.
Hjelmborg, P. S., T. K. Andreassen, and E. C. Bonefeld-Jørgensen. 2008. Cellular uptake of lipoproteins and persistent organic compounds: An update and new data. *Environ Res* 108(2): 192–8.
Hodge, J. E. 1955. The Amadori rearrangement. *Adv Carbohydr Chem* 10: 169–205.
Hogeboom, G. H. 1949. Cytochemical studies of mammalian tissues. II. The distribution of diphosphopyridine nucleotide cytochrome c reductase in rat liver fractions. *J Biol Chem* 177: 847–58.
Horecker, B. L. 1950. Triphosphopyridine nucleotide-cytochrome c reductase in liver. *J Biol Chem* 183: 593–605.
Hrycay, E. G., J. A. Gustafsson, M. Ingelman-Sundberg, and L. Ernster. 1975. Sodium periodate, sodium chloride, organic hydroperoxides, and H2O2 as hydroxylating agents in steroid hydroxylation reactions catalyzed by partially purified cytochrome P-450. *Biochem Biophys Res Commun* 66(1): 209–16.

Hutson, J. R., G. Koren, and S. G. Matthews. 2010. Placental P-glycoprotein and breast cancer resistance protein: Influence of polymorphisms on fetal drug exposure and physiology. *Placenta* 31: 351–7.

Huwyler, J., M. B. Wright, H. Gutmann, and J. Drewe. 2006. Induction of cytochrome P450 3A4 and P-glycoprotein by the isoxazolyl-penicillin antibiotic flucloxacillin. *Curr Drug Metab* 7(2): 119–26.

Idle, J. R., P. Millburn, R. T. Williams, and G. Zini. 1976. The conjugation of arylacetic acids in the pigeon compared with the hen. *Biochem Soc Trans* 4(1): 141–3.

Iqbal, M., H. L. Ho, S. Petropoulos, V. G. Moisiadis, W. Gibb, and S. G. Matthews. 2012. Pro-inflammatory cytokine regulation of P-glycoprotein in the developing blood-brain barrier. *PLoS ONE* 7: e43022.

ISSX webpage. *Woehler: The Birth of Metabolism*. 2020. www.issx.org/page/Woehler (accessed April 20, 2020).

Jaffe, B. 1992. *Crucibles: The Stories of Great Chemist*. Cleveland and New York: The World Publishing Company.

Jaffee, M. 1879. Ueber die nach Einführung von Brombenzol und Chlorbenzol im Organismus entstehenden schwefelhaltigen Säuren. *Chem Ber* 12: 1092–8.

Jaiswal, A. K., O. W. McBride, M. Adesnik, and D. W. Nebert. 1988. Human dioxin-inducible cytosolic NAD(P)H: Menadione oxidoreductase. cDNA sequence and localization to chromosome 16. *J Biol Chem* 263: 13572–8.

Jakoby, W. B. 1980. Detoxication enzymes. In *Enzymatic Basis of Detoxication*, ed. W. B. Jakoby, 1. New York: Academic Press.

Keaney, J., and M. Campbell. 2018. The dynamic blood-brain barrier. *FEBS J* 282(21): 4067–79.

Keller, W. 1842. Ueber Verwandlung der Benzoësäure in Hippursäure. *Ann der Chemie* 43: 108–11.

Klingenberg, M. 1958. Pigments of rat liver microsomes. *Arch Biochem Biophys* 75(2): 376–86.

Krauss, R. S., and T. E. Eling. 1984. Arachidonic acid-dependent cooxidation: A potential pathway for the activation of chemical carcinogens in vivo. *Biochem Pharmacol* 33: 3319–24.

Lash, L. H., A. A. Elfarra, and M. W. Anders. 1986. Renal cysteine conjugate β-lyase. Bioactivation of nephrotoxic cysteine S conjugates in mitochondrial outer membrane. *J Biol Chem* 261: 5930–5.

Lin, H. 2011. S-Adenosylmethionine-dependent alkylation reactions: When are radical reactions used? *Bioorg Chem* 39(5–6): 161–70.

Lu, F. C. 1996. *Basic Toxicology: Fundamental Target Organs, and Risk Assessment*, 3rd ed. Washington, DC: Taylor & Francis Group.

Mackness, M. I., P. N. Durrington, and B. Mackness. 2002. *Paraoxonase (PON1) in Health and Disease: Basic and Clinical Aspects*. Kluwer.

Mandel, W. 1959. Pyridoxine and the isoniazid-induced neuropathy. *Dis Chest* 36: 293–6.

Mannervik, B., P. G. Board, J. D. Hayes, I. Listowsky, and W. R. Pearson. 2005. Nomenclature for mammalian soluble glutathione transferases. *Methods Enzymol* 401: 1–8.

Mannervik, B., and U. H. Danielson. 1988. Glutathione transferases: Structure and catalytic activity. *CRC Crit Rev Biochem* 23: 283–337.

Mantle, T. J., C. B. Pickett, and J. D. Hayes. 1987. *Glutathione S-Transferase and Carcinogenesis*. London: Taylor & Francis Group.

Marnett, L. J. 1985. Arachidonic acid metabolism and tumor initiation. In *Arachidonic Acid Metabolism and Tumor Initiation*, ed. L. J. Marnett, 39. Boston: Martinus Nijhoff.

Marnett, L. J., and T. E. Eling. 1983. Cooxidation during prostaglandin biosynthesis. In *A Pathway for the Metabolic Activation of Xenobiotics, in Reviews in Biochemical Toxicology*, eds. E. Hodgson, J. R. Bend, and R. M. Philpot, 135. New York: Elsevier.

Mazel, P., and D. Pessayre. 1976. Significance of metabolite-mediated toxicities in the safety evaluation of drugs and chemicals. In *New Concepts in Safety Evaluation*, Vol. 1, Part 1, eds. M. A. Mehlmen, R. E. Shapiro, and H. Blumenthal. Washington, DC: Hemisphere.

McDonagh, A. F., L. A. Palma, J. J. Lauff, and T. W. Wu. 1984. Origin of mammalian biliprotein and rearrangement of bilirubin glucuronides in vivo in the rat. *J Clin Invest* 74: 763–70.

McGill, M. R., M. R. Sharpe, C. D. Williams, M. Taha, S. C. Curry, and H. Jaeschke. 2012. The mechanism underlying acetaminophen-induced hepatotoxicity in humans and mice involves mitochondrial damage and nuclear DNA fragmentation. *J Clin Invest* 122(4): 1574–83.

Mckenzie, P. I. n.d. Structure and function of UDP glucuronosyltransferases. In *Principles, Mechanisms, and Biological Consequences of Induction*, Vol. 2, eds. K. Ruckpaul and H. Rein, 211–43. Berlin: Academie Verlag.

Miller, J. A., and E. C. Miller. 1947. The metabolism and carcinogenicity of p-dimethylaminoazobenzene and related compounds in the rat. *Cancer Res* 7(1): 39–41.
Mitchell, J. R., U. P. Thorgeirsson, M. Black, et al. 1975. Increased incidence of isoniazid hepatitis in rapid acetylators: Possible relation to hydranize metabolites. *Clin Pharmacol Ther* 18(1): 70–9.
Morel, F., and C. Aninat. 2011. The glutathione transferase kappa family. *Drug Metab Rev* 43(2): 281–91.
Mueller, G. C., and J. A. Miller. 1948. The metabolism of 4-dimethylaminoazobenzene by rat liver homogenates. *J Biol Chem* 176(2): 535–44.
Mueller, G. C., and J. A. Miller. 1949. The reductive cleavage of 4-dimethylaminoazobenzene by rat liver; the intracellular distribution of the enzyme system and its requirement for triphosphopyridine nucleotide. *J Biol Chem* 180(3): 1125–236.
Murphy, P. J., and C. A. West. 1969. The role of mixed function oxidases in kaurene metabolism in Echinocystis macrocarpa Greene endosperm. *Arch Biochem Biophys* 133(2): 395–407.
Mutlib, A. E., T. C. Goosen, J. N. Bauman, J. A. Williams, S. Kulkarni, and S. Kostrubsky. 2006. Kinetics of acetaminophen glucuronidation by UDP glucuronosyltransferases 1A1, 1A6, 1A9 and 2B15. Potential implications in Acetaminophen-induced hepatotoxicity. *Chem Res Toxicol* 19(5): 701–9.
Naranjo, P. M., I. Montoliu, A. M. Aura, T. Seppänen-Laakso, D. Barron, and S. Moco. 2018. In vitro gut metabolism of [U-13 C]-quinic acid, the other hydrolysis product of chlorogenic acid. *Mol Nutr Food Res* 16: e1800396.
Nebert, D. W., and F. J. Gonzalez. 1985. P-450 genes: Structure, evolution, and regulation. *Annu Rev Biochem* 56: 945–93.
Neve, E. P., and M. Ingelman-Sundberg. 1999. A soluble NH(2)-terminally truncated catalytically active form of rat cytochrome P450 2E1 targeted to liver mitochondria. *FEBS Lett* 460(2): 309–14.
Ni, Z., Z. Bikadi, M. F. Rosenberg, and Q. Mao. 2010. Structure and function of the human breast cancer resistance protein (BCRP/ABCG2). *Curr Drug Metab* 11(7): 603–17.
NIH, National Institute of Health. 2018. *ToxTutor, Section 12. Biotransformation.* https://toxtutor.nlm.nih.gov/12-002.html#Four (accessed April 2020).
Nishimura, M., and S. Naito. 2006. Tissue-specific mRNA expression profiles of human phase I metabolizing enzymes except for cytochrome P450 and phase II metabolizing enzymes. *Drug Metab Pharmacokinet* 21(5): 357–74.
Ogilvie, B. W., D. Zhang, W. Li, et al. 2006. Glucuronidation converts gemfibrozil to a potent, metabolism-dependent inhibitor of CYP2C8: Implications for drug-drug interactions. *Drug Metab Dispos* 34(1): 191–7.
Omura, T. 1999. Forty years of cytochrome P450. *Biochem Biophys Res Commun* 266(3): 690–8.
Omura, T. 2006. Mitochondrial P450s. *Chem Biol Interact* 163(1–2): 86–93.
Omura, T. 2010. Structural diversity of cytochrome P450 enzyme system. *J Biochem* 147(3): 297–306.
Omura, T. 2013. Contribution of cytochrome P450 to the diversification of eukaryotic organisms. *Biotechnol Appl Biochem* 60(1): 4–8.
Omura, T., and K. Morohashi. 1995. Gene regulation of steroidogenesis. *J Steroid Biochem Mol Biol* 53: 19–25.
Omura, T., and R. Sato. 1962. A new cytochrome in liver microsomes. *J Biol Chem* 237: 1375–6.
Omura, T., and R. Sato. 1964. The carbon monoxide-binding pigment of liver microsomes. I. Evidence for its hemoprotein nature. *J Biol Chem* 239: 2370–8.
Omura, T., and R. Sato. 1964. The carbon monoxide-binding pigment of liver microsomes. II. Solubilization, purification, and properties. *J Biol Chem* 239: 2379–85.
Orhan, H. 2015. Extrahepatic targets and cellular reactivity of drug metabolites. *Curr Med Chem* 22(4): 408–37.
Orhan, H., J. N. Commandeur, G. Sahin, U. Aypar, A. Sahin, and N. P. Vermeulen. 2004. Use of 19F-nuclear magnetic resonance and gas chromatography-electron capture detection in the quantitative analysis of fluorine-containing metabolites in urine of sevoflurane-anaesthetized patients. *Xenobiotica* 34(3): 301–16.
Ortiz de Montellano, P. R. 1986. *Cytochrome P450.* New York: Plenum Press.
Palade, G. E., and P. Siekevitz. 1956. Liver microsomes; an integrated morphological and biochemical study. *J Biophys Biochem Cytol* 2(2): 171–200.
Parkinson, A., B. W. Ogilvie, D. B. Buckley, F. Kazmi, M. Czerwinski, and O. Parkinson. 2013. Biotransformation of xenobiotics. In *Casarett & Doull's Toxicology*, ed. K. Curtis, 190. McGraw-Hill.

Perland, E., and R. Fredriksson. 2017. Classification systems of secondary active transporters. *Trends Pharmacol Sci* 38(3): 305–15.

Prescott, L. F. 1980. Kinetics and metabolism of paracetamol and phenacetin. *Br J Clin Pharmacol* 10(Suppl 2): 291S-8S.

Pritchard, J. B., and D. S. Miller. 1992. Proximal tubular transport of organic anions and cations. In *The Kidney: Physiology and Pathophysiology*, 2nd ed., Vol. 3, eds. DW Seldin and G. Giebisch, 2921–45. New York: Raven Press.

Prochaska, H. J., M. J. DeLong, and P. Talalay. 1985. On the mechanisms of induction of cancer-protective enzymes: A unifying proposal. *Proc Natl Acad Sci USA* 82: 8232–6.

Rahimtula, A. D., and P. J. O'Brien. 1975. Hydroperoxide dependent O-dealkylation reactions catalyzed by liver microsomal cytochrome P450. *Biochem Biophys Res Commun* 62(2): 268–75.

Ramaswamy, S. G., and W. B. Jakoby. 1987. Amine N-sulfotransferase. *J Biol Chem* 262: 10039–43.

Ramesh, C. G. 2015. *Handbook of Toxicology of Chemical Warfare Agents*, 2nd ed. Academic Press.

Rozman, K. 1986. Fecal excretion of toxic substances. In *Gastrointestinal Toxicology*, eds. K. Rozman and O. Hanninen, 119–45. Amsterdam, New York and Oxford: Elsevier.

Ruelius, H. W., S. K. Kirkman, E. M. Young, and F. W. Janssen. 1986. Reactions of oxaprozin-1-O-acyl glucuronide in solutions of human plasma and albumin. *Adv Exp Med Biol* 197: 431–41.

Schanker, L. S. 1962. Passage of drugs across body membranes. *Pharmacol Rev* 14: 501–30.

Schladt, L., H. Thomas, R. Hartmann, and F. Oesch. 1988. Human liver cytosolic epoxide hydrolases. *Eur J Biochem* 176: 715–23.

Scott, M. C., J. A. Van Loon, and R. M. Weinshilboum. 1988. Pharmacogenetics of N-methylation: Heritability of human erythrocyte histamine N-methyltransferase activity. *Clin Pharmacol Ther* 43(3): 256–62.

Shen, D. D. 2013. Toxicokinetics. In *Casarett & Doull's Toxicology*, 8th ed., ed. Curtis D. Klaassen, 367. McGraw-Hill.

Shinohara, A., K. Saito, Y. Yamazoe, T. Kamataki, and R. Kato. 1986. Acetyl coenzyme A dependent activation N-hydroxy derivatives of carcinogenic arylamines. Mechanism of activation, species difference, tissue distribution, and acetyl donor specificity. *Cancer Res* 46: 4362–7.

Sies, H., and E. Cadenas. 1983. Biological basis of detoxication of oxygen free radicals. In *Biological Basis of Detoxication*, eds. J. Caldwell and W. B. Jakoby, 181. New York: Academic Press.

Sies, H., and B. Ketterer. 1988. *Glutathione Conjugation: Its Mechanisms and Biological Significance.* London: Academic Press.

Smith, J. N. 1958. Comparative detoxication. 5. Conjugation of aromatic acids in reptiles: Formation of ornithuric acid, hippuric acid and glucuronides. *Biochem J* 69(4): 509–16.

Smith, P. C., L. Z. Benet, and A. F. McDonagh. 1990. Covalent binding of zomepirac glucuronide to proteins: Evidence for a Schiff base mechanism. *Drug Metab Dispos* 18: 639–44.

Smith, P. C., A. F. McDonagh, and L. Z. Benet. 1986. Irreversible binding of zomepirac to plasma protein in vitro and in vivo. *J Clin Invest* 77: 934–9.

Stevens, J., and W. B. Jakoby. 1983. Cysteine conjugate β-lyase. *Mol Pharmacol* 23: 761–5.

Strittmatter, C. F., and E. G. Ball. 1951. A hemochromogen component of liver microsomes. *Proc Natl Acad Sci USA* 38: 19–25.

Strittmatter, P., and S. F. Velick. 1956. A microsomal cytochrome reductase specific for diphosphopyridine nucleotide. *J Biol Chem* 221: 277–86.

Tephly, T., M. Green, J. Puig, and Y. Irshaid. 1988. Endogenous substrates for UDP-glucuronosyltransferases. *Xenobiotica* 18: 1201–10.

Thomas, B. H., L. T. Wong, W. Zeitz, and G. Solomonraj. 1981. Isoniazid metabolism in the rabbit, and the effect of rifampin pretreatment. *Res Commun Chem Pathol Pharmacol* 33(2): 235–47.

Thomas, H., and F. Oesch. 1988. Functions of epoxide hydrolases. *ISI Atlas Sci Biochem* 1: 287–91.

Thomas, M. D. 1992. *Textbook of Biochemistry*. New York: Wiley-Liss.

Tipton, K. F., P. Dostert, and M. Strolin-Benedetti. 1984. *Monoamine Oxidase and Disease: Prospects for Therapy with Reversible Inhibitors*. New York: Academic Press.

Tomisawa, H., N. Ichimoto, Y. Takanohashi, S. Ichihara, H. Fukazawa, and M. Tateishi. 1988. Purification and characterization of cysteine conjugate transaminases from rat liver. *Xenobiotica* 18: 1015.

Ueda, K. 2011. ABC proteins protect the human body and maintain optimal health. *Biosci Biotechnol Biochem* 75(3): 401–9.

Underhill, D. M., S. Gordon, B. A. Imhof, G. Núñez, and P. Bousso. 2016. Élie Metchnikoff (1845–1916): Celebrating 100 years of cellular immunology and beyond. *Nat Rev Immunol* 16(10): 651–6.

Ure, A. 1841. On gouty concretions, with a new method of treatment. *Pharm J Transact* 24: 30–5.
van Breemen, R. B., and C. Fenselau. 1985. Acylation of albumin by 1-O-acyl glucuronides. *Drug Metab Dispos* 13: 318–20.
van Breemen, R. B., C. Fenselau, W. Mogilevsky, and G. B. Odell. 1986. Reaction of bilirubin glucuronides with serum albumin. *J Chromatogr* 383: 387–92.
von Bahr-Lindstrom, H., J. O. Höög, L. O. Hedén, et al. 1986. cDNA and protein structure for the α subunit of human liver alcohol dehydrogenase. *Biochemistry* 25: 2465–70.
Weber, W. W. 1986. Commentary: The molecular basis of hereditary acetylation polymorphisms. *Drug Metab Dispos* 14: 377–94.
Weber, W. W., and D. W. Hein. 1985. N-Acetylation pharmacogenetics. *Pharmacol Rev* 37: 25–79.
Weinshilboum, R. M. 1986. Phenol sulfotransferase in humans: Properties, regulation, and function. *Fed Proc* 45(8): 2223–8.
Weinshilboum, R. M. 1988. Pharmacogenetics of methylation: Relationship to drug metabolism. *Clin Biochem* 21: 201–10.
Weinshilboum, R. M. 1990. Sulfotransferase pharmacogenetics. *Pharmacol Ther* 45(1): 93–107.
Wells, D. S., F. W. Janssen, and H. W. Ruelius. 1987. Interactions between oxaprozin glucuronide and human serum albumin. *Xenobiotica* 17: 1437–49.
Wen, Q., E. I. Tang, Y. Gao, et al. 2018. Signaling pathways regulating blood-tissue barriers—Lesson from the testis. *Biochim Biophys Acta Biomembr* 1860(1): 141–53.
Wermuth, B., K. M. Bohren, G. Heineman, J. P. von Wartburg, and K. H., Gubbay. 1988. Human carbonyl reductase. Nucleotide sequence analysis of a cDNA and amino acid sequence of the encoded protein. *J Biol Chem* 263: 16185–8.
Williams, D. P., D. J. Antoine, P. J. Butler, et al. 2007. The metabolism and toxicity of furosemide in the Wistar rat and CD-1 mouse: A chemical and biochemical definition of the toxicophore. *J Pharmacol Exp Ther* 322: 1208–20.
Williams, R. T. 1959. *Detoxication Mechanisms*. New York: John Wiley & Sons.
Wilson, D. W., H. J. Segall, L. C. Pan, M. W. Lamé, J. E. Estep, and D. Morin. 1992. Mechanisms and pathology of monocrotaline pulmonary toxicity. *Crit Rev Toxicol* 22(5–6): 307–25.
Wirth, P. J., P. J. Bettis, and W. L. Nelson. 1975. Microsomal metabolism of furosemide evidence for the nature of the reactive intermediate involved in covalent binding. *Mol Pharmacol* 12: 759–68.
Wright, R. L., K. Harris, B. Solow, R. H. White, and P. J. Kennelly. 1996. Cloning of a potential cytochrome P450 from the archaeon Sulfolobus solfataricus. *FEBS Lett* 384(3): 235–9.
Xie, Y., M. R. McGill, K. Du, et al. 2015. Mitochondrial protein adducts formation and mitochondrial dysfunction during N-acetyl-m-aminophenol (AMAP)-induced hepatotoxicity in primary human hepatocytes. *Toxicol Appl Pharmacol* 289(2): 213–22.
Yang, X., Y. A. Gandhi, D. B. Duignan, and M. E. Morris. 2009. Prediction of biliary excretion in rats and humans using molecular weight and quantitative structure-pharmacokinetic relationships. *AAPS J* 11(3): 511–25.
Yoshikawa, H. 1951. The oxidation reduction potential of cytochrome b′, a hemochromogen-like component of cytochrome in liver. *J Biochem* 38: 1–5.
Zia-Amirhosseini, P., R. Z. Harris, F. M. Brodsky, and L. Z. Benet. 1995. Hypersensitivity to nonsteroidal anti-inflammatory drugs. *Nature Med* 1: 2–4.
Ziegler, D. M. 1988a. Detoxication. Oxidation and reduction. In *The Liver: Biology and Pathobiology*, 2nd ed., eds. I. M. Arias, W. B. Jakoby, H. Popper, D. Schacter, and D. A. Shafritz, 363. New York: Raven Press.
Ziegler, D. M. 1988b. Flavin-containing monooxygenases. Catalytic mechanism and substrate specificities. *Drug Metab Rev* 19: 1–32.

4 Mechanistic Toxicology

Pramila Singh, Christian Strupp and Werner Bomann

CONTENTS

4.1 INTRODUCTION TO MECHANISTIC TOXICOLOGY

4.1.1 Background and Context for Understanding Mechanistic Toxicology

Mechanistic toxicology has been a propeller of the multidisciplinary science of toxicology and is possibly the most rapidly evolving subdiscipline due to the availability of ever-improving technologies, methods and testing platforms in molecular and cellular biology. It is this close relationship between basic scientific investigations and the broad scope of applications that make mechanistic toxicology an indispensable tool in the medical and product safety sciences, including risk assessment. Mechanistic toxicology involves a detailed and systematic approach to the investigation of the molecular, biochemical, cellular and tissue-level interactions and pathways underlying the toxicological effects caused by xenobiotics and other foreign substances when they are introduced into a biological system.

For practical purposes, throughout this chapter the term "substance" is used to refer to any foreign agent coming into contact with or entering the body, including drugs, chemicals and other exogenous materials. The term "mechanism of toxicity" refers to the cascade of key events, produced as a consequence of exposure to a substance at sufficiently active concentrations and through whatever route of entry into the body, that leads to toxicologically significant outcomes. The mechanism of toxicity is what links a given exposure to a substance with the inherent potential that substance may have to produce toxicological effects.

Toxic effects have been categorized based on definitions of the types of pathology that ensue such as inflammation, degeneration of tissues and cell death, immune-mediated mechanisms,

 DOI: 10.1201/9780203730584-4

functional toxicity, mutagenesis and carcinogenesis, developmental or reproductive toxicity, metabolic dysfunction and harmful endocrine-altering effects. These pathologies can be caused by multiple mechanisms, but it is also possible that one molecular mechanism can cause different types of toxicity. Substances may change both the functions and structure of tissue and organ systems, which making it important to evaluate the intrinsic hazard potential of substances by both qualitative and quantitative methods.

Characterization of the potential effects of a substance once it comes into contact with the biological environment of the host is a critical part of risk assessment that relies on well-characterized hazard data to determine the probable impact associated with the degree of exposure. Many toxicological mechanisms are now known, thanks to the discipline of mechanistic toxicology and to the improved methods now available and in use. The study of mechanistic toxicology has enabled advances in antidotes to poisons and identification of biomarkers of toxicity that can be monitored for indications of disease progression.

4.1.2 Distinction between Mechanistic Toxicology and Mode of Action Studies in Human Health Risk Assessment

Regulatory toxicology and human health risk assessment are areas of applied science directly impacted by mechanistic toxicology. Within this domain, a related but distinct term, mode of action (MOA), has emerged as a methodological approach for safety and risk assessments. The MOA of a given substance refers to a well-described and reproducible pathway beginning with a molecular initiating event, an initiating interaction between the substance and a sensitive receptor or other molecular target of the host, subsequently followed by a cascade of key events that ultimately leads to an adverse toxicological outcome (Figure 4.1). The described cascade of events or pathway is typically established in a rodent testing model and may not lead to toxicity in an exposed human, depending on a number of different factors. A later section of this chapter will focus on the application of MOA in regulatory toxicology.

In some cases, knowledge of the mechanisms of toxicity can be helpful in demonstrating that certain toxicological effects identified in laboratory animal species are nonrelevant to humans. In general, with a better understanding of mechanisms of toxicity it is possible to improve the precision in human health risk assessments of substances. Some examples of mechanisms of toxicity induced by a variety of substances are summarized in a later section of this chapter. These can also be referred to as adverse outcome pathways (AOPs) when they lead to defined adverse endpoints.

4.1.3 Regulatory Environment for Pharmaceuticals, Agricultural Products and Chemicals Registrations

Regulatory data requirements prior to first-in-human clinical trials and market approvals for pharmaceuticals as well as for chemical and agrochemical registrations and market authorizations include extensive safety testing in both *in vitro* models and laboratory animals. In pharmaceutical development, efficacy testing is done using many of the same *in vitro* and animal models. Most of these data requirements are fixed and based on international product safety testing guidelines; however, some approaches to the data evaluations and risk assessments vary somewhat from region to region. During the regulatory review of product safety dossiers for medicines, agricultural products and chemicals, mechanistic data play an essential role in the characterization of biological activity and the potential risks following exposure to the active substance. Recent trends in the regulatory environment have increased the requirement of MOA data in line with both the

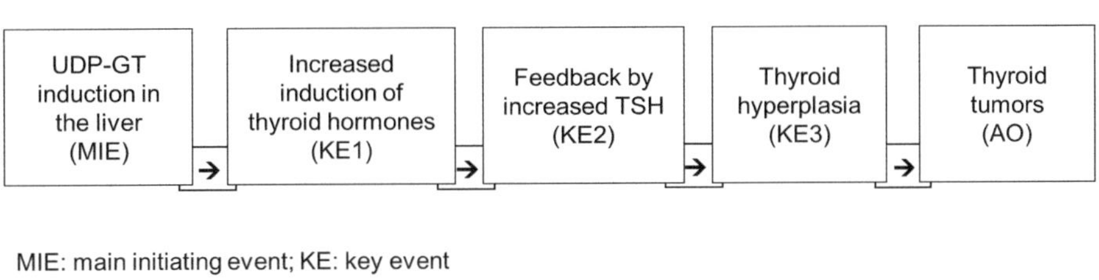

Figure 4.1

need for more detailed hazard characterization of substances and with the 3Rs approach to animal testing: calling for *reduction, refinement* and *replacement* of the use of animals in product testing. The principles of mechanistic toxicology and MOA investigations for use in product safety evaluations and risk assessments are largely aligned with 3R principles to arrive at conclusions on human risk. It has been discussed in numerous regulatory guidance documents that a paradigm shift is eventually foreseen in toxicology, moving away from a prescriptive list of hazard characterization studies in vertebrate animals to a more targeted approach using increasingly refined data to complement a thorough evaluation of all relevant existing data.

4.2 PRINCIPLES OF MECHANISTIC TOXICOLOGY

4.2.1 Portals of Entry

The entry points of foreign substances into an organism, referred to as portals of entry (Table 4.1), may be the first site of toxicological effects. Each portal of entry has distinct characteristics in its microenvironment, including the presence of metabolizing enzymes and physicochemical properties, such as pH level, that may impact the fate of the foreign substances and the types of toxicities that may occur. Substances may have toxicological effects at the portal of entry or may require absorption into systemic circulation before having the potential to exert adverse effects.

Substances and their metabolites are characterized by their physical and chemical properties, which contribute to their toxicological activity and potential. The types of effects can be differentiated by their site of action. Local effects will be exerted on the site of first contact (e.g. the skin or eyes), whereas systemic effects will occur in the body at sites that are different from the entry location.

The physical and chemical properties of substances play a role in their toxicokinetic processing once they have entered the body. Some chemicals are weak organic acids or bases and thus are ionized in solutions with usually low lipid solubility and poor ability to permeate through membranes. For nonionized forms, the lipid solubility is relevant for their diffusion potential. For certain substances, special transport systems have been found to support rapid transport through membranes, even against concentration gradients.

With regard to systemic effects, before a substance is able to exert toxic effects, a certain amount of exposure at the toxicological target site (receptors, DNA, proteins, etc.) is required. The dose, and thus the systemic concentration and persistence, of a substance is a determinant of whether the substance can cause toxicity and will also affect the degree of toxicity. Therefore, toxicokinetic processes such as absorption, distribution, metabolism and elimination (ADME) play important roles in the effects on an organism and should be considered in safety testing when *in vitro* approaches are used or the relevance of animal data for humans is discussed.

4.2.2 Interspecies Differences in Absorption, Distribution, Metabolism and Elimination (ADME)

Although animal testing still, in many cases, generates the most accurate predictions for human toxicity, there are many important differences between laboratory species and humans. In the case of systemic exposure, a simple correction for body weight is not appropriate, as food intake or respiration rate relative to body weight is very different between animal species. The situation can be improved by allometric scaling, which takes these factors into account. Nevertheless, it does not correct for intrinsic differences, such as the substantially higher permeability of rabbit and rat skin compared to human skin.

Another very important aspect is metabolism. Rodent metabolism is much quicker and more extensive than that of humans. Although the basic types of reaction and metabolic enzymes may be conserved, the localization of these, their compound specificity and their levels of induction are different. This affects general toxicity but also tumorigenicity or tumor-causing potential. Many rodent carcinogens, where exposures at high dose levels lead to tumors in the liver, lungs and thyroid, are rodent-specific and may have no or little relevance as human carcinogens.

Table 4.1 Common Portals of Entry

Portal of entry	Method of entry
Ingestion	Via dietary admixture or by oral gavage directly into the stomach
Injection	Intramuscular, intravenous, intraperitoneal, intraocular, etc.
Dermal contact	A cream or gel gently massaged on the skin, a patch positioned and held in place on the skin, etc.
Inhalation	As nasal or oropharyngeal aerosol exposure

Development is also very different (between laboratory species and humans). Simplified approaches such as "1 dog year is equivalent to 7 human years" are not suitable. There are distinct differences; for example, humans are born with an intact blood-brain barrier and a much more advanced stage of central nervous system development when compared to rats, in which these develop in the first weeks after birth.

Sensitivity to and reserves of hormones are very different in rodents and humans. Stresses due to handling and circadian rhythms have to be controlled when working with laboratory animals, particularly during hormonal investigations. Pharmaceutical research has shown that immune reactions in humans are complex and can be quite distinct from those predicted by animal models. Such differences are very important and have to be understood and accounted for in mechanistic toxicology investigations in order to avoid spurious or unpredictive results.

4.2.3 Molecular Target Site

In general, after initial contact and entry into the system, and once a substance has reached its target or site of action, actions may consist of either direct interaction with target molecules or alteration of the biological microenvironment, both of which may lead to cellular injury or functional changes. There are many well-described mechanisms of toxicity that have facilitated a better understanding of the toxic effects of compounds, including exaggerated pharmacological activities. Before a substance reaches its target within the host, there are many ways it can exert effects due to its physical and chemical properties at the site of entry, prior to absorption into the systemic circulation or distribution into various tissues and organs. The ultimate impact or toxicological outcome may depend on whether or not the chemical is bioactivated or deactivated and eliminated by interaction with the host's metabolic system. Subsequent steps may consist of adaptive mechanisms, which could reduce the impact on tissues by changing their cellular structures or by activating repair mechanisms that can improve recovery from disturbed functions and morphological alterations. If the repair mechanisms are overwhelmed or lead to uncontrolled injury and dysfunction, toxicity occurs in the affected tissues (e.g. tissue necrosis), and if there is more extensive damage over a longer duration, cancer may result.

4.2.4 Off-Target Effects

Contrary to the expected or even intended targeted effects of different substances, there may be different unintentional effects, referred to as off-target effects, which could remain unknown until toxicity studies are performed. In drug development, off-target effects may be responsible for unwanted side effects, as an alternate explanation to exaggerated pharmacological effects. Off-target effects may also increase with increasing dose levels as the specificity of activity may be overwhelmed, leading to saturation of clearance, gross toxicological effects associated with high dose levels or cross-reactivity with nonspecific molecular targets.

4.3 WELL-DESCRIBED MECHANISMS OF TOXICITY

4.3.1 Metabolism and Bioactivation

Normally the organism is able to reduce the toxic effects of chemicals via biotransformation to metabolites with lower toxicity, but in some cases chemicals are metabolized to more potent and toxic molecules by metabolic reactions, mainly in the liver. This phenomenon is called bioactivation. Cytochrome P450 (CYP) is the most important enzyme family involved in these mechanisms; more than 400 CYP genes and many isoforms have been identified. Some well-known examples include cocaine, which is metabolized in the liver to several reactive metabolites that can cause liver effects in laboratory species and in humans and parathion, an organophosphate insecticide, which is converted into the active cholinesterase inhibitor paraoxon.

Another example of a general mechanism of bioactivation is the formation of electrophiles, which are molecules containing an electron-deficient atom with a partial or full-positive charge, so they can share electrons with electron-rich molecules. This formation of electrophiles is involved in the bioactivation of many compounds, such as ethanol (converted to acetaldehyde), benzene (converted to muconic aldehyde) and acetaminophen (converted to N-acetyl-p-benzoquinone imine).

Peroxidases, another family of enzymes, can catalyze (in the presence of hydrogen peroxide) the one-electron oxidation of some chemicals in a reaction over many steps. The human peroxidase family contains prostaglandin H synthase, peroxidase and myeloperoxidase (MPO), among others. MPO is abundant in blood cells, which can cause peroxidase-dependent bioactivation of substances in humans.

These examples show that the biotransformation and elimination processes that normally reduce the toxic effects of substances can, in some cases, increase their toxic effects by producing more potent and toxic molecules (Bornheim 1998; Guengerich 2001).

4.3.2 Covalent Binding of Reactive Metabolites to Cellular Molecules

Binding of substances to proteins or intracellular molecules, for example, is normally a noncovalent, reversible interaction, but some substances have covalent binding potential and can thus irreversibly bind to molecules. Some substances are known to convert into reactive metabolites, after which they can covalently bind to DNA. This DNA adduct formation is regarded as an important step in chemical-induced carcinogenesis. In addition to covalent binding to DNA, some compounds can covalently bind to proteins or other large molecules.

Selective binding of a substance to a specific protein is the underlying mechanism in the classic example of acetaminophen, an antipyretic and analgesic compound that induces toxicity in the liver. Acetaminophen (paracetamol) is primarily metabolized by glucuronidation and sulfation, but its hepatotoxicity is mediated by its toxic metabolite, N-acetyl-p-benzoquinone imine, which is generated by liver cytochrome P450s (CYP2E1 and CYP1A2) and detoxified by conjugation with hepatic glutathione (GSH). During an acetaminophen overdose, cellular GSH is depleted, and cellular functions are impaired by the binding of N-acetyl-p-benzoquinone imine to macromolecules, which can lead to cell death. Other examples with generally similar mechanisms exist, such as the immune-mediated effects of diclofenac (a nonsteroidal anti-inflammatory drug that binds to proteins), n-hexane (present in cleaning fluids and can lead to neuropathy) and benzo(a)pyrene (a polycyclic aromatic hydrocarbon—which can be found in coal tar, tobacco smoke and many foods, especially grilled meats—that leads to DNA adduct formation and mutagenesis).

4.3.3 Signal Transduction and Gene Regulation

Knowledge about the activation or disruption of cellular signal transduction by chemicals has increased recently due to the rapid development of molecular techniques and models, such as transgenic animal models or gene-silencing methods. Signaling consists of sensing (by receptors) and responding. The connecting mechanism is the signal transduction pathway, which can be affected by xenobiotics, either by activation or inhibition, thereby causing toxic effects.

Sensing is undertaken by receptors, which can be soluble in the cytosol and/or nucleus, or membrane bound. Among the membrane-bound receptors, the main types are G-protein-coupled (GPCR), kinase-linked receptors and ligand-gated ion channels. One toxicologically important group of signaling pathways are those of the protein kinase type.

Protein kinases can covalently attach a phosphate to a serine, threonine or tyrosine on specific proteins which can be other kinases being converted from an inactive to an active state. Such pathways can be activated by xenobiotics, but also by growth factors, inflammatory cytokines, reactive oxygen species and bacterial liposaccharides. These pathways can end in the transcriptional activation of genes that are involved in cell survival and proliferation or in cell toxicity and death, which can lead to imbalanced cell proliferation, apoptosis or neoplastic transformation.

4.3.4 Nuclear Receptor-Mediated Mechanisms

Nuclear receptor-mediated mechanisms are triggered by the soluble receptors already mentioned, which are in the cytosol and/or nucleus. The possible effects triggered by xenobiotics are typically tissue specific and provoke a rapid transcriptional response for which the activation of specific genes can be demonstrated. Two families of nuclear-signaling molecules are of particular toxicological interest, the Per-Arnt-Sim (PAS) receptors and the steroid hormones. The PAS receptor family includes the aryl hydrocarbon receptor (AHR). The steroid hormone family includes, among others, the peroxisome proliferator-activated receptor (PPAR), thyroid hormone receptor (TR), estrogen receptor (ER) and androgen receptor (AR). The constitutive androstane receptor (CAR) and pregnane X receptor (PXR) are slightly different as they belong to the NR1I nuclear receptor family.

4.3.5 Mechanisms of Interactions with Cell Maintenance and Cell Death

Certain functions are essential for cell survival, for example, energy production, intracellular environment maintenance and macromolecule biosynthesis, which are important functions in membranes and cellular organelles. Interaction between chemicals and essential functions can impair cell survival and lead to cell death. Such toxicant-induced events include, among others, membrane-damaging actions and increased intracellular Ca^{2+} and adenosine triphosphate (ATP) depletion. Depending of the duration and severity of the chemical impact, the cell can either recover or die. Cell death can occur as either necrosis or apoptosis (Table 4.2).

Table 4.2 Differences between Apoptosis and Necrosis

Parameter	Necrosis	Apoptosis
Cell size	Enlargement (swelling)	Reduction (shrinkage)
Nucleus	Pyknosis → karyorrhexis → karyolysis	Fragmentation
Plasma membrane	Disrupted	Intact
Cellular contents	Enzymatic digestion; content may leak out of cell	Intact; may be released in apoptotic bodies
Accompanied by inflammation	Frequent	No
Physiologic or pathologic role	Pathologic	Physiologic and pathologic

Necrosis is characterized by swollen organelles, changes in the nuclei and membranes, membrane rupture and lysis. Although these steps are well researched and known (in principle), for single toxicants, the exact steps leading to necrosis may be complex, can differ according to the substance and are not always known in detail.

Apoptosis is a more organized process of cell death, and it follows well-regulated pathway steps and is linked to the mechanisms that regulate cell proliferation and DNA repair. It is often described as programmed cell death because it is genetically controlled and involved in many physiological processes, such as cell turnover, where it helps to eliminate cells that could develop into cancer or proliferating cells; this programmed process can also be initiated by chemical injury. The process normally starts with the activation of caspases (i.e. proteases), but other pathways not involving caspases also exist. During apoptosis, the membranes remain intact so that no intracellular molecules leak through and no inflammatory reactions occur.

Among the substances that can cause apoptosis are heavy metals, organotin compounds and dithiocarbamates. Research seems to indicate that severe toxicity leads more often to necrosis, whereas milder levels of toxicity lead to apoptosis (Orrenius et al. 2011).

4.3.6 Oxidative Stress

Oxidative stress is an unavoidable consequence of life in an oxygen-rich atmosphere (Davies 2000). It is usually described as an imbalance between oxidizing molecules and antioxidants in the cells; when there is a dominance of oxidizing molecules; this imbalance can cause cell injury.

The production of molecular oxygen initiates events such as the formation of singlet oxygen and reactive oxygen species (ROS), which cause oxidoreductive stress. For example, small amounts of ROS are produced in the mitochondria under normal conditions controlled by inactivation processes, but if substances are able to enter the mitochondria and interact with the electron-chain transport mechanisms, this balance can be disturbed, causing increased ROS formation and finally NADPH (nicotinamide adenine dinucleotide phosphate) depletion. Examples of substances involved in this mechanism are, among others, benzene, quinones and paraquat (a herbicide), all of which have different toxicological effects. Oxidative stress is often claimed to be a key event in the MOA, however such claims should be critically reviewed, as measurable oxidative stress can be an associated event as a consequence of primary toxicity.

4.3.7 Impairment of Cell Proliferation and Tissue Repair

Mitogenesis is the normal biological process of replacing cells that have been lost during physiological turnover in the cell cycle. This ensures the balance between the regeneration and elimination of cells and their components. If injury to the cell occurs, cell proliferation processes are activated to replace lost tissue. In contrast to this continuous process, when cell proliferation occurs after chemical injury, the cells are most likely to be entering mitosis and starting cell division following the release of mediators from damaged cells (Schafer 1998). Although this process usually results in repair and adaptation to the chemical injury, sometimes these mechanisms are not efficient or even lead to toxicity after inappropriate repair or adaptation, with the potential to result in carcinogenesis.

4.4 MECHANISTIC TOXICOLOGY IN THE REGULATORY ENVIRONMENT

4.4.1 Evolving Perspectives in Toxicology and the 3Rs

Regulatory toxicology has to manage the challenge that at best only approximately 70% of human toxicities can be predicted by animal models (Shanks et al. 2009). These limitations are due to species differences, mainly in metabolism, but also because the same toxicological pathways are not always preserved in the phylogenetic tree. This has variable implications on the true risk assessment for human safety when it is based on studies conducted using laboratory animals. Depending on the degree of exposure, the safety margin can be relatively low (i.e. for pharmaceuticals, which are intentionally used at biologically active doses, or in cases of accidental exposure to chemicals) or relatively high (i.e. for exposure of the general public to trace levels of agrochemicals in food or to cleaning products in the home or to pesticides in public spaces in most instances).

Regulatory agencies must understand and integrate various domains of the science in order to make decisions regarding the acceptability of the risk-to-benefit ratio when reviewing safety dossiers of registered chemicals and pharmaceutical products. In addition, these evaluations are often performed against the backdrop of public interest or perspectives that may complicate the distinction between facts and perception.

Regulatory decisions have traditionally been risk-driven in regions with a longer history of robust regulatory science and tend to be more hazard-driven in regions with less history of regulatory science. One other reason for hazard-driven approaches is communication: hazard-based classifications are easier to understand and communicate to the public, whereas risk-based regulatory evaluations require an understanding of and accounting for the quantitative dose-response relationships, differences between humans and experimental animals (interspecies differences), and the levels of exposure resulting from product use in a variety of representative scenarios, including indicated uses and unapproved uses (e.g. abuse and unintentional or accidental exposures). This can lead to different conclusions for different product uses, which require specific skills and tools in communication to the general public.

Recently, some regions that have historically used more integrated, risk-based regulatory science have been tending towards hazard-based regulations. This is a concern if it results from influences driven by and responding to the general public's fear of uncertainty, as most of the public do not have access to the technical data, evaluation tools and perspective to understand the associated risks or to weigh them against the intended benefits. In some cases, political or public interests have entered into the scientific community, calling for a paradigm change that threatens the fundamental principles of toxicology, for example, "The dose does not make the poison". The implication of this idea is that any amount of a substance could be toxic, regardless of the dose, exposure level or available detoxifying mechanisms. This trend should cause concern, as despite the absence of scientific proof, regulatory scientists and risk managers alike may become overly cautious, which can have damaging consequences on innovation, science and the time it takes for new, improved products to reach the market. Publication bias in the media towards sensational reporting and disinterest in reporting of studies that discount more alarming, sensationalized scientific hypotheses further fuel this trend.

A constructive progressive trend has been the implementation of the 3Rs (reduce, refine, replace) and the "*in vitro* and alternative methods" movement in regulatory toxicology. This trend has been particularly active and has made a positive contribution to the way scientific research has been conducted since the 1980s. Many achievements have been made in terms of a better understanding of mechanistic pathways, refinement and replacement of animal tests, more appropriate and humane housing of experimental animals, development of less invasive methods in *in vivo* testing, appropriate anesthesia and analgesia (pain management), and the trimming of regulatory requirements for excess studies that do not lead to an increase in human safety.

Recently, several governments and safety agencies have issued plans to work free of vertebrate tests within a few years and many academic institutes are leaning towards *in vitro* methods in order to defend and promote animal welfare. Although the ethical aspect of promoting *in vitro* methods is very welcome and needed, an independent shift towards *in vitro* away from *in vivo* is in some ways problematic to the integrity of the discipline.

In vitro studies have helped in many areas, particularly in understanding the relevance of animal findings for humans. Thus, *in vitro* studies are an essential element of mechanistic toxicology. Simple, well-understood modes of action (such as genotoxicity, cholinesterase inhibition and local irritation) can be addressed effectively in standalone *in vitro* studies. However, this becomes more difficult when attempts are made to replace complex, not yet well-understood toxicities in animal models by *in vitro* standalone methods without human data to validate the substitution.

The limitations of *in vitro* systems should always be assessed (i.e. solubility and bioavailability issues, absence/limitation of metabolism, absence of effective compound or metabolite removal, absence of complex interactions with other cell types), and the relationship between an *in vitro* test concentration and an *in vivo* exposure level is never simple but must be considered. In particular, the results of *in vivo* testing up to the maximum tolerated dose level (e.g. safety testing at very high dose levels, as required for registered chemical substances) are very difficult to predict using *in vitro* methods, as toxicity is often a consequence of nonspecific metabolism and overloading of elimination systems in the whole organism. Exposure-based *in vivo* testing, that is several times above the human exposure level, as for pharmaceuticals, is often more suitable for *in vitro–in vivo* extrapolations.

Mechanistic toxicology has a firm place in each of these areas. Risk assessments are typically done using large safety factors, and decades of this practice have shown that they are health protective. However, when overly conservative safety margins are applied, this can lead towards a reduction in the number of products being approved depriving society of some of the essential solutions and advances needed in the face of increasing challenges such as food supply, disease and drought.

In order to properly perform or refine a quantitative risk assessment, detailed understanding of the qualitative and quantitative mechanistic aspects of toxicity is required. Mechanistic toxicology is also an essential component of hazard-based regulation, as many chemical effects in experimental animals, particularly at high dose levels, are of variable relevance to humans due to inherent species differences, and require deeper, more detailed analyses and interpretations.

4.4.2 Total Exposure Assessment

Exposure differs according to product use. In pharmaceutical settings, exposure is, in general, intentional and regulated (i.e. available when the drug is approved and usually controlled by a doctor's prescription). Uptake, metabolism and clearance are well understood for therapeutic drugs and often continuously monitored during clinical use. Predominant routes of exposure include oral, dermal, inhalation and injection. The duration of exposure varies from acute (e.g. aseptic and hydration treatments) to subchronic (e.g. antibiotics and chemotherapy) to chronic (e.g. statins and beta blockers).

Safety assessments for unintentional exposure (e.g. to chemicals and agrochemicals or freely available substances such as food additives, over-the-counter medication and cosmetics) are often more complex. Routes of exposure may be oral for food additives, over-the-counter medication and agrochemical residues but are mainly limited to dermal and inhalation exposure for cosmetics and occupational exposure to chemicals and agrochemicals. In these cases, assumptions have to be made in the process of risk assessment, based on data such as dietary intake in combination with measured residues, or measured exposure as a consequence of general activities. These general factors are often extracted from the scientific literature, combined with the results of regulatory exposure studies, and captured in various guidance documents and mathematical models.

Important corrective factors have to be taken into account, based on the relevant route of exposure. Oral bioavailability, dermal penetration, and deposition characteristics in the lungs need to be well understood in order to correctly estimate the potential systemic exposure via these routes. Although traditionally well understood for pharmaceuticals due to the availability of human data, this can be a complicated endeavor for substances where intentional exposure to test for potential effects in humans would be unethical. Consequently, this area of study still operates with many assumptions and safety factors to address residual uncertainties.

Exposure to single substances is well understood, based on the available models, but regulatory agencies are struggling with a lack of suitable models and the mandate for safety regulation concerning exposure to multiple substances, often called complex mixtures. In these instances, additional regulatory tools are needed and should be developed to more accurately conclude on the safety and risk of these types of exposures in humans. Of further consideration are the epidemiological data indicating that social and behavioral factors including access to medical treatment and care, or the results of moving from manual work to more sedentary work impacting lifestyle conditions (e.g. burnout, musculoskeletal issues), also affect public health outcomes.

4.4.3 Mode of Action in Toxicology Studies

Toxicology still relies heavily on *in vivo* safety studies in animals, mainly in rodents, but there have been many attempts since the 1980s to develop *in vitro* and *in silico* models designed to reduce and finally replace *in vivo* testing, as discussed above in the context of the 3Rs.

Although successful for simple, nonspecific endpoints (cytotoxicity, genotoxicity and corrosiveness), no combination of fully *in vitro* models or test batteries are ready to address the types of toxicity that depend on complex interdependent responses of a whole organism in order to develop and manifest. Thus toxicology will have to continue relying on ethically sensible *in vivo* testing until the necessary progress is made in predictive and *in silico* tools to guarantee sufficient safety margins while respecting and building upon the scientific knowledge that has evolved over decades if not centuries of biomedical and toxicology research.

A good understanding and adequate presentation of results in their correct context are thus key skills for toxicologists, and even more relevant when explaining the MOA, which typically involves substantial amounts of complex data, both qualitative and quantitative, in safety and risk assessments. Simplification of information into a manageable format, including qualified statements on the strengths, weaknesses and relevance of the research, has proven to be very useful in regulatory decision-making. This approach was developed by the World Health Organization and the concept has since been further developed and discussed as the MOA-Human Relevance Framework (Boobis et al. 2016; Meek et al. 2014; EFSA Guidance).

Although it is true that large *in vivo* studies typically have improved statistical power, it must be accepted that the results obtained in animal models often diverge from the responses that would occur in humans (Ames and Gold 1990; Smith and Perfetti 2018). Quite often, the source of the difference is the much more active metabolism in rodents and other differences (e.g. pharmacokinetics and pharmacodynamics) between experimental animal species and humans. In rodents, the generation of reactive metabolites at excessive doses can lead to local cytotoxicity with repair-induced cellular proliferation or direct mitogenicity, both of which lead to more cell division, which means a greater statistical chance of the occurrence of errors during gene amplification, thus an inheritable mutation of the cell, and if uncontrolled then potentially the development of cancer (Tomasetti and Vogelstein 2015; Cohen 2010). The same distinction can be noted in the species-specific differences in tumor targets and rates, such as when background tumor patterns in unexposed healthy populations of humans and rodents are compared. Hormonal cascades can also be affected by interspecies differences, as metabolic enzymes do not convert only xenobiotics but also messenger molecules such as hormones and proteins.

To overcome these gaps, MOA research is indicated when nongenotoxic tumors or reproductive toxic effects are observed. The *in vitro* methods currently used, despite their limitations, are very helpful tools in MOA research. They have a big advantage in that the response in humans can be directly investigated often without ethical concerns because the provenance of samples may be from either deceased persons or surgical resection materials. When these methods are used in connection with *in vivo* models, and in particular with genetically modified models (knock-out, knock-in, humanized, surgically preconditioned), they become powerful tools for understanding the animal response and, most importantly the human relevance of effects observed in animals.

It is important to remember that *in vivo* testing should be done at dose levels reflecting human end-user exposure.

Well-conducted dose range-finding studies, in which unnecessary suffering is prevented by minimizing animals that are overdosed and wherein toxicokinetics are integrated for investigating how systemic exposure increases with the administered dose are essential in the design of main studies. This is particularly important for MOA investigations because generalized, nonspecific responses may skew the data and make it more difficult to answer the key research questions.

MOA investigations typically start with a review of the effects reported in required toxicology studies (e.g. international regulatory testing guidelines on repeated dose, carcinogenicity, reproductive toxicity or developmental toxicity studies). Patterns of effects (e.g. increased liver weights with hypertrophy and specific metabolic enzyme induction in short-term studies, combined with increased rates of liver tumors in carcinogenicity assessments) may suggest a known MOA. If not, histopathology is the endpoint that provides the most relevant data as a starting point for postulation of a hypothesis. The observed alterations in tissues can often be tracked to a specific cell population, which gives a first hint of the MOA, i.e. if a metabolically competent cell population (e.g. hepatocytes, club cells) is the target, metabolic activation is likely to play a role. If the lesion is located where there is either first contact (e.g. esophagus, stomach or duodenum after oral administration; airways upon inhalation exposure, skin at dermal exposure) or where, due to physicochemical processes, the test substance concentrates (e.g. kidney or bladder), cytotoxicity or local irritancy due to crystal growth may play a role. A thorough literature search should be performed before designing further investigations, as inclusion of controls (e.g. at least one negative control and, ideally, one to several positive controls) can help to either bridge to existing data on substances with the same target or distinguish from substances that cause effects at the same toxicological target via an alternate MOA.

Important milestones in successful MOA research are (1) postulating clear hypotheses, also known as problem-formulation (Sauve-Ciencewicki et al. 2018); (2) conducting a series of studies to determine key events and their temporal development in a robust *in vivo* model; and (3) evaluating the plausible human relevance using all available *in vitro* and *in vivo* data in multiple species, including humans. Alternative modes of action should be excluded, and this is best when considered early during experimental design. Ideally, the postulated hypotheses and investigations at each milestone should allow determination of the likelihood of whether the established rodent MOA would be applicable in humans to be determined.

4.4.4 Human Relevance Framework

There is a strong trend towards better integration of mechanistic investigations in today's safety assessments. Regulatory processes largely rely on high-quality data, generated under good laboratory practice (GLP) conditions and following internationally agreed standards. This helps to create global harmonization, which in turn avoids having to repeat similar animal tests in different areas of the world. Internationally agreed protocols (OECD and ICH) are proven tools for detecting safety concerns in humans, although often the study design in the guidelines can be behind state-of-the-art designs reflecting current scientific advances.

In the mid-2000s, the OECD and WHO proposed MOA and human relevance frameworks that can help to integrate and weigh scientific evidence from nonguideline work, within the regulatory processes. This has been picked up by regulatory agencies worldwide and further developed, and most regulatory systems now have the possibility to include such data. The general principle is the same: relevant data (e.g. from literature or purpose generated) is evaluated for relevance and quality, and if it is of suitable quality it is integrated into the assessment, weighed against possible alternative modes of action and then assessed by experts. One of the key challenges is communication of MOA results because such data are complex, nonstandard and require a multidisciplinary scientific background to appropriately evaluate. Simplification is key, followed by an organized, detailed discussion of the different lines of evidence linking exposure to the cascade of observed molecular, biochemical, cellular and tissue-level events leading to an adverse outcome, with repeatable causal, associative, temporal and dose relationships.

4.5 APPLIED TOXICOLOGY AND MODE OF ACTION INVESTIGATIONS

4.5.1 Improved Understanding of Targeted Pharmacology versus Off-Target Effects

In pharmaceutical discovery and development, characterization of the pharmacological mechanisms of a new drug or biological therapy is a key component of the application for Investigational New Drug (IND) status. Different testing methods are employed to describe the pharmacological activity and the specific interactions of the substance with its intended biological target. In the study of toxicology, similar testing methods and platforms can be applied to determine off-target effects, unintended cross-reactivity and nonspecific binding or interactions that may lead to unintended toxicological effects. One objective of standard toxicological testing of regulated substances is the identification of targeted organs, where toxicities may occur preferentially with increasing dose levels. Once these target organ effects are identified, further mechanistic studies may be performed to better understand the dose-dependency of the molecular initiating and causal key events that determine the outcomes observed in the off-target organs. MOA and mechanistic studies may be used to justify safe starting dose levels in first-in-human clinical trials together with toxicokinetic and metabolism data generated in the toxicology test species.

4.5.2 Relevant Species Selection in Nonclinical Drug Development

The standard rodent test species is the rat or mouse, and for nonrodent testing, the dog or nonhuman primate has usually been the standard test species for small molecules and biologicals, respectively. An additional nonrodent model used in drug or medical device development is the minipig or pig. Final selection of the rodent and nonrodent test species must be justified by the sponsor, and part of this justification includes the considered relevance to human safety and targeted screening as well as the availability of historical data in a given laboratory species. The use of MOA studies and targeted screening in species selection and justification has become increasingly important in drug development to avoid unwarranted expense and use of animals that will not generate relevant data for human safety and risk evaluations. Some testing platforms that may be used for species selection justification include hepatocyte activity profiling, drug-receptor profiling and tissue reactivity in different laboratory species compared to humans.

4.5.3 Determination of Human Relevance of Rodent Liver Tumors

One well-known example of human relevance determination in regulatory toxicology is the MOA of rodent liver tumor induction by phenobarbital via activation of CAR, as demonstrated for a number of pesticides, human drugs and industrial chemicals. To demonstrate nonrelevance in humans, the putative rodent MOA was investigated and confirmed by performing studies to demonstrate CAR activation as the molecular initiating event, followed by cell proliferation in rodent livers. Using *in vitro* models, the generation of less or no cell proliferation in human hepatocytes was demonstrated compared to clear induction of cell proliferation in rodent hepatocytes.

Additional studies have been helpful for demonstrating hepatocellular proliferation in wild-type rats versus its absence in CAR-knockout rats. Together, these data support a phenobarbital-like MOA for liver tumor induction, comprising activation of CAR/PXR followed by altered gene expression specific to CAR activation, increased cell proliferation, formation of altered liver foci and eventually, the development of liver tumors. As no evidence of liver tumor induction by phenobarbital exists in humans, this MOA is considered not relevant to humans (Elcombe et al. 2014).

Another example is the group of peroxisome-proliferating chemicals, including some herbicides, plasticizers, drugs and natural products, which are able to induce peroxisome proliferation in rodent liver and other tissues. Liver tumors were observed in long-term studies with some of these compounds in rodents, although the compounds were not genotoxic. This mechanism is characterized by increased volume, density and fatty acid oxidation of the peroxisomes, which contain oxidation systems that are important in lipid metabolism.

The activation of peroxisome proliferator-activated receptor alpha (PPARα), a nuclear hormone receptor implicated in energy homeostasis, is the first step in peroxisome proliferation in rodent hepatocytes, and it leads to subsequent steps and eventually to liver tumors. In rodent liver studies, several key events, including the activation of PPARα, perturbation of cell proliferation and apoptosis, selective clonal expansion and a series of associated events were demonstrated. Furthermore, studies in wild-type rodents and PPARα-knockout animals demonstrated that the proliferation observed in wild-type animals did not occur in PPARα-knockout animals. By using the human relevance framework, it was determined that the data supported the conclusion that this MOA for liver tumors was not relevant to humans, as the key events do not lead to liver tumor formation in humans following longer-term exposures (Klaunig et al. 2003; Strupp et al. 2018).

BIBLIOGRAPHY

Ames, B. N., and L. S. Gold. 1990. Chemical carcinogenesis: Too many rodent carcinogens. *Proc Natl Acad Sci USA* 87: 7772–6.

Boobis, A. R., S. M. Cohen, V. L. Dellarco, et al. 2016. Classification schemes for carcinogenicity based on hazard-identification have become outmoded and serve neither science nor society. *Regul Tox Pharmacol* 82: 158–66.

Bornheim, L. M. 1998. Effect of cytochrome P450 inducers on cocaine-mediated hepatotoxicity. *Toxicol Appl Pharmacol* 150(1): 158–65.

Cohen, S. M. 2010. Evaluation of possible carcinogenic risk to humans based on liver tumors in rodent assays: The two-year bioassay is no longer necessary. *Toxicol Pathol* 38: 487–501.

Davies, K. J. 2000. Oxidative stress, antioxidant defenses, and damage removal, repair, and replacement systems. *IUBMB Life* 50(4–5): 279–89.

EFSA Guidance. www.efsa.europa.eu/ (accessed October 24, 2020).

Elcombe, C. R., R. C. Peffer, D. C. Wolf, et al. 2014. Mode of action and human relevance analysis for nuclear receptor-mediated liver toxicity: A case study with phenobarbital as a model constitutive androstane receptor (CAR) activator. *Crit Rev Toxicol* 44(1): 64–82.

Guengerich, F. P. 2001. Common and uncommon cytochrome P450 reactions related to metabolism and chemical toxicity. *Chem Res Toxicol* 14(6): 611–50.

Klaunig, J. E., M. A. Babich, K. P. Baetcke, et al. 2003. PPAR alpha agonist-induced rodent tumors: Modes of action and human relevance. *Crit Rev Toxicol* 33(6): 655–780.

Meek, M. E., C. M. Palermo, A. N. Bachman, et al. 2014. Mode of action human relevance (species concordance) framework: Evolution of the Bradford Hill considerations and comparative analysis of weight of evidence. *J Appl Toxicol* 34(6): 595–606.

Orrenius, S., P. Nicotera, and B. Zhivotovsky. 2011. Cell death mechanisms and their implications in toxicology. *Toxicol Sci* 119(1): 3–19.

Sauve-Ciencewicki, A., K. P. Davis, J. McDonald, et al. 2018. A simple problem formulation framework to create the right solution to the right problem. *Regul Toxicol Pharmacol* 101: 187–93.

Schafer, K. A. 1998. The cell cycle: A review. *Vet Path* 35(6): 471–8.

Shanks, N., R. Greek, and J. Greek. 2009. Are animal models predictive for humans? *Philos Ethics Humanit Med* 4: 2. https://doi.org/10.1186/1747-5341-4-2.

Smith, C. J., and T. A. Perfetti. 2018. The "false-positive" conundrum in the NTP 2-year rodent cancer study database. *Toxicol Res Applic* 2: 1–13.

Strupp, C., W. H. Bomann, F. Spézia, et al. 2018. A human relevance investigation of PPARα-mediated key events in the hepatocarcinogenic mode of action of propaquizafop in rats. *Regul Toxicol Pharmacol* 95: 348–61.

Tomasetti, C., and B. Vogelstein. 2015. Variation in cancer risk among tissues can be explained by the number of stem cell divisions. *Science* 347(6217): 78–81.

5 Genetic Toxicology and Carcinogenesis

Pavel Rossner Jr., Radim J. Sram and Andrea Rossnerova

CONTENTS

5.1 INTRODUCTION

Genetic toxicology studies the effects of chemical, biological or physical agents on genetic information. Although changes on the level of DNA are primarily investigated, modifications of other macromolecules (lipids and proteins) are also of interest. This chapter focuses mostly on DNA modifications induced by chemical exposures and implications of these changes for the process of carcinogenesis.

Upon entering the organism, chemical agents (xenobiotics) may interact with DNA and cause damage. For some xenobiotics, the effects are direct, not requiring metabolic activation; other compounds are first converted to reactive intermediates that bind to DNA. DNA modifications may be manifested as chemical structures bound to building blocks of DNA (nucleobases, 2-deoxy-D-glucose) or as single-strand (ss) or double-strand (ds) DNA breaks (Figure 5.1). Modifications of DNA are called adducts, and their presence may cause mispairing of nucleobases potentially leading to the induction of DNA mutations. DNA breaks may result in chromosome aberrations. Both events represent a risk factor for cancer initiation. Damage to DNA is corrected by repair processes that are required to maintain genomic integrity and stability. These processes remove modified nucleobases and/or nucleotides, repair base mispairing and correct DNA breaks. The following DNA repair pathways have been described: base excision repair (BER), nucleotide excision repair (NER), mismatch repair, nonhomologous end-joining (NHEJ) and homology-directed repair (Figure 5.2; Jalal et al. 2011).

Genetic toxicology applies a battery of biomarkers that are used as an indicator of potentially harmful processes in the organism. Biomarkers are broadly classified into three groups: biomarkers of exposure, biomarkers of effect and biomarkers of susceptibility (Figure 5.3). To be suitable for human studies, biomarkers should meet certain criteria. They should be sensitive and specific, and their detection should be standardized, validated and relatively inexpensive. Sample collection should be simple, noninvasive or minimally invasive, and the detection method should be high throughput (Rossner et al. 2015). Biomarkers of exposure include the xenobiotics themselves or their metabolites detected within the organism as a result of exposure. They are directly related to biomarkers of effect, which are detected as changes in the levels of certain molecules induced in association with the exposure to xenobiotics. Finally, the response of the organism to xenobiotics exposure is modulated by biomarkers of susceptibility, mostly single nucleotide polymorphisms (SNPs), that affect activity of enzymes, often those involved in metabolism of xenobiotics (Demetriou et al. 2012; Watson and Mutti 2004). Here we will concentrate on biomarkers of effect in relation to carcinogenesis, specifically on oxidative stress markers, bulky DNA adducts and chromosomal aberrations.

DOI: 10.1201/9780203730584-5

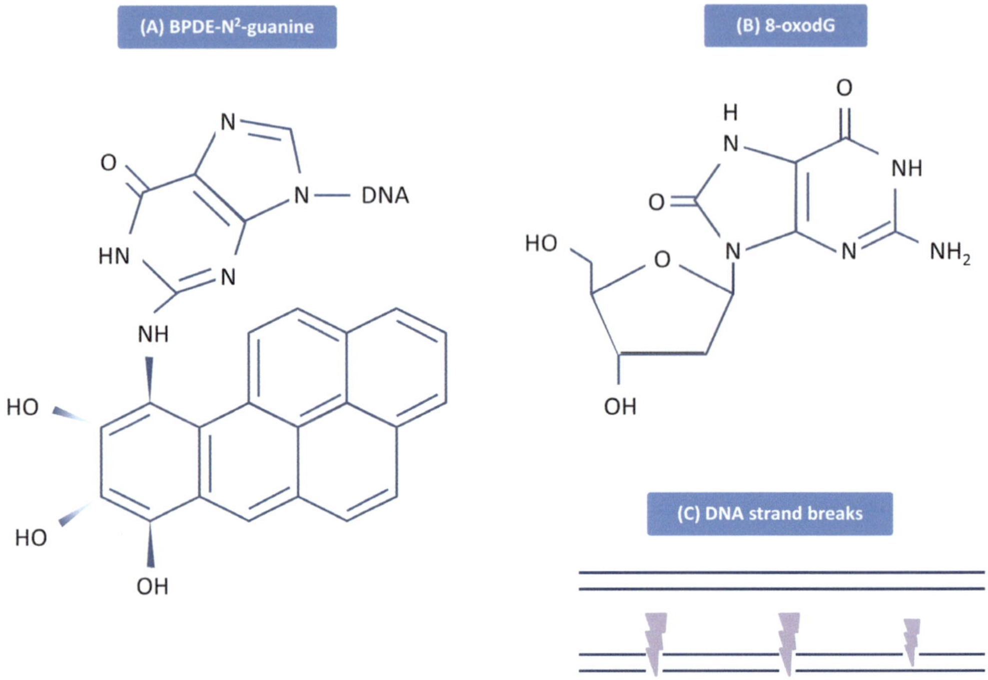

Figure 5.1 Examples of DNA modifications: (A) benzo[a]pyrene-7,8-dihydrodiol-9,10-epoxide (BPDE-N^2-guanine, a bulky DNA adduct); (B) 8-oxo-2′-deoxyguanosine (8-oxodG, a product of reactive oxygen species attack); and (C) single-strand or double-strand DNA breaks.

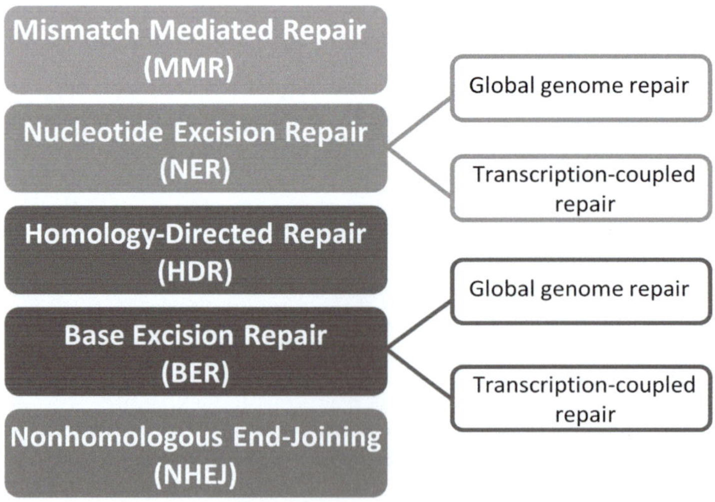

Figure 5.2 An overview of DNA repair pathways. Base excision repair (BER) and nucleotide excision repair (NER) are involved in removal of oxidatively damaged DNA bases. Bulky DNA adducts are repaired by NER. Repair of double-strand DNA breaks by nonhomologous end-joining repair (NHEJ) or homology-directed repair (HDR) may result in formation of stable chromosomal aberrations.

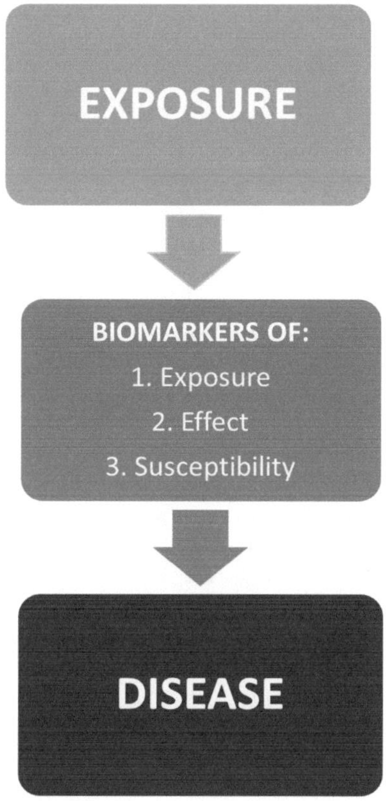

Figure 5.3 The relationship between exposure, biomarkers used in genetic toxicology, and disease. Biomarkers are usually classified into three groups: biomarkers of exposure, biomarkers of effect and biomarkers of susceptibility.

5.2 METABOLIC ACTIVATION OF XENOBIOTICS AND THE PROCESS OF CARCINOGENESIS

The primary aim of metabolic activation is to convert lipophilic xenobiotics into hydrophilic derivatives that can be readily excreted from the organism (see Chapter 3). Briefly, the organism possesses two groups of enzymes that are involved in phase I and phase II of xenobiotic metabolism (Croom 2012). Phase I enzymes are responsible for hydrolysis, reduction or oxidation of xenobiotics. In phase II, xenobiotics are conjugated usually with glucuronide, sulfate, glutathione or amino acids, resulting in formation of water-soluble compounds that are eliminated from the body (Figure 5.4). This process, which primarily serves for detoxification of xenobiotics, may have negative consequences: during the reactions, some compounds are converted to reactive intermediates that may bind to macromolecules and cause damage.

We will illustrate this latter process using benzo[a]pyrene (BaP) as an example. BaP is one of the polycyclic aromatic hydrocarbons (PAHs), by-products of incomplete combustion of organic matter. PAHs are commonly found in polluted air, but they also contaminate water, soil and food. Although BaP is classified as carcinogenic to humans (Group 1, according to the International Agency for Research on Cancer [IARC 2012]), it requires metabolic activation to become reactive and cause damage to DNA or other macromolecules. During phase I xenobiotic metabolism, BaP is converted to BaP-diol epoxide by cytochrome P450 (CYP) enzymes by adding an epoxide group. This metabolite is further converted to dihydrodiols by epoxide hydrolase and by CYPs to diol epoxides. Diol epoxides may be either detoxified by phase II reactions (conjugation with glutathione or glucuronide, for example) or bind to DNA via amino groups of purines (Baird et al. 2005). If unrepaired, these structures may cause mutations, or DNA breaks indirectly as a result of blocking of DNA synthesis and uncoupling of the other strand (Bi et al. 2005).

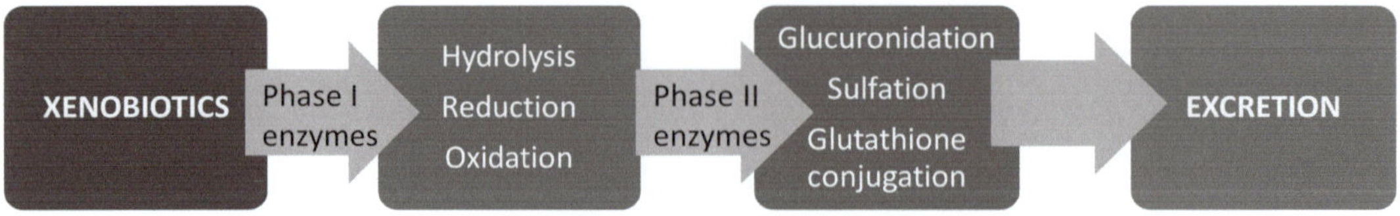

Figure 5.4 Metabolic activation of xenobiotics. The process is responsible for conversion of lipophilic compounds to hydrophilic derivatives that can be excreted from the organism. It involves phase I and phase II enzymes with specific metabolic activities.

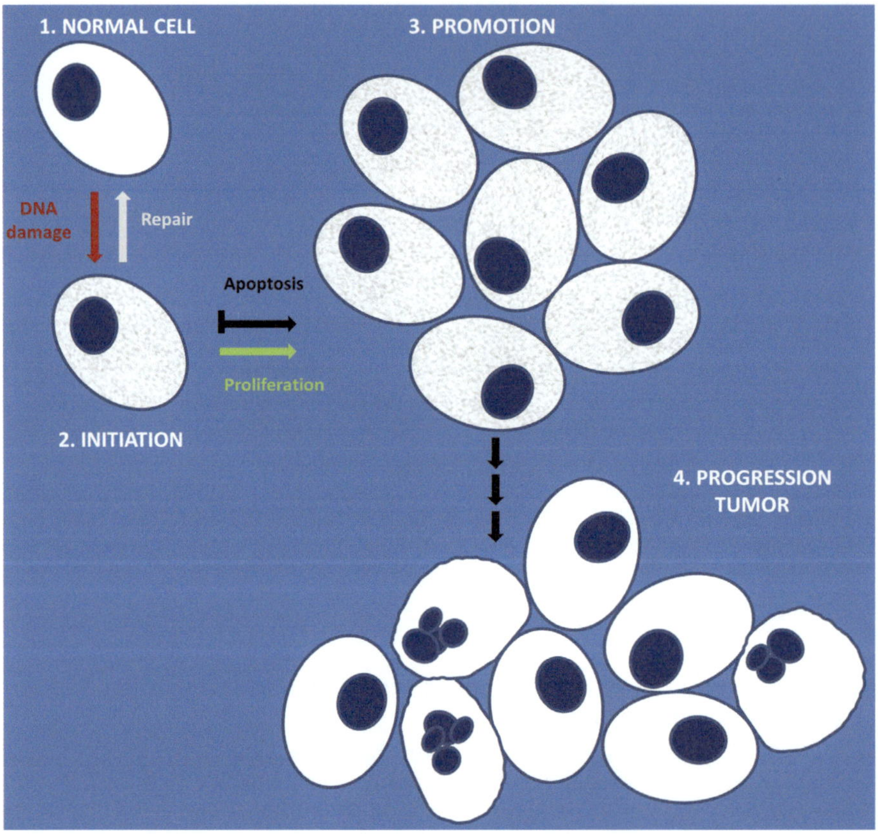

Figure 5.5 Steps involved in the process of carcinogenesis.

DNA damage may potentially induce the process of carcinogenesis, which has three distinct stages: initiation, promotion and progression (Figure 5.5; Klaunig and Kamendulis 2004). During initiation, DNA damage caused by genotoxic events results in formation of preneoplastic cells. The initiation step is irreversible and leads to the second stage, tumor promotion, which is characterized by selective clonal expansion of the initiated cells. Promotion is usually accompanied by an increased cell proliferation and/or a decrease of apoptosis; it is reversible provided that the promotion stimulus is removed. The last step, progression, is associated with irreversible cellular and molecular changes that cause a shift from the preneoplastic to the neoplastic state. In this stage, genetic instability and disruption of chromosome integrity also occurs.

5.3 OXIDATIVELY DAMAGED DNA AND CARCINOGENESIS

Reactive oxygen species (ROS) are a group of molecules and radicals derived from molecular oxygen (Galadari et al. 2017). They include superoxide anion ($O_2^{\bullet-}$), hydroxyl radical ($^{\bullet}OH$), hydroperoxyl radical ($HOO^{\bullet}$), alkoxyl radical ($RO^{\bullet}$), organic peroxy radical ($ROO^{\bullet}$) and hydrogen peroxide

(H_2O_2). Superoxide radical, regarded to be the primary ROS, can be converted to H_2O_2 or, in the presence of Fe^{2+} or Cu^{2+}, by a Fenton reaction to the highly reactive hydroxyl radical. Due to their reactivity, both superoxide and hydroxyl radicals do not diffuse in the cells but cause local damage. In contrast, hydrogen peroxide is more stable and easily diffuses through cellular membranes to other parts of the organism. ROS may be generated by endogenous and exogenous processes. In cells, ROS are predominantly formed by the mitochondrial electron transport chain, but the activities of some enzymes (NADPH oxidases, xanthine oxidase, lipoxygenases, CYP450) may also cause ROS generation. Further, there are numerous exogenous factors that affect ROS levels (e.g. air pollution, radiation, food or drug intake).

ROS occur naturally in all aerobic organisms at low levels. They play an important physiological role in bacterial inactivation via respiratory burst, cell signaling, regulation of activities of cytokines, growth factors or hormones, as well as in transcription, ion transportation, immunomodulation or apoptosis (Gupta et al. 2014). However, in excessive amounts they cause oxidative stress leading to damage of macromolecules, including DNA. The organisms possess antioxidant mechanisms that regulate negative biological effects of ROS. These mechanisms are commonly divided into enzymatic (activities of enzymes, such as superoxide dismutase, catalase, glutathione peroxidase) and nonenzymatic (e.g. vitamins, glutathione, albumin; Gupta et al. 2014). Ideally, there is a balance between pro-oxidant activities of ROS and activities of antioxidants. If the balance is disturbed, oxidative stress accompanied by oxidatively generated damage to macromolecules occurs (Figure 5.6).

Oxidatively generated damage to DNA results in several types of lesions including apurinic/apyrimidinic sites, ssDNA and dsDNA breaks, and oxidized purines and pyrimidines (Cooke et al. 2003; Kryston et al. 2011). Among oxidized nucleobases, 8-oxo-7,8-dihydroguanine (8-oxoGua) and 2,6-diamino-4-hydroxy-5-formamidopyrimidine (FapyGua) are the most commonly studied products (Figure 5.7). Both lesions are formed by the addition of the hydroxyl radical to the C8 position of the guanine ring. Oxidation of pyrimidines at positions 5 or 6 generates other oxidized nucleobases, for example, 5,6-dihydroxy-5,6-dihydrothymine (thymine glycol, Tg) which is one of the most abundant (Figure 5.7). Tg and 8-OxoGua are commonly used markers of oxidative stress in various organisms including humans. Oxidatively damaged DNA is primarily repaired by base

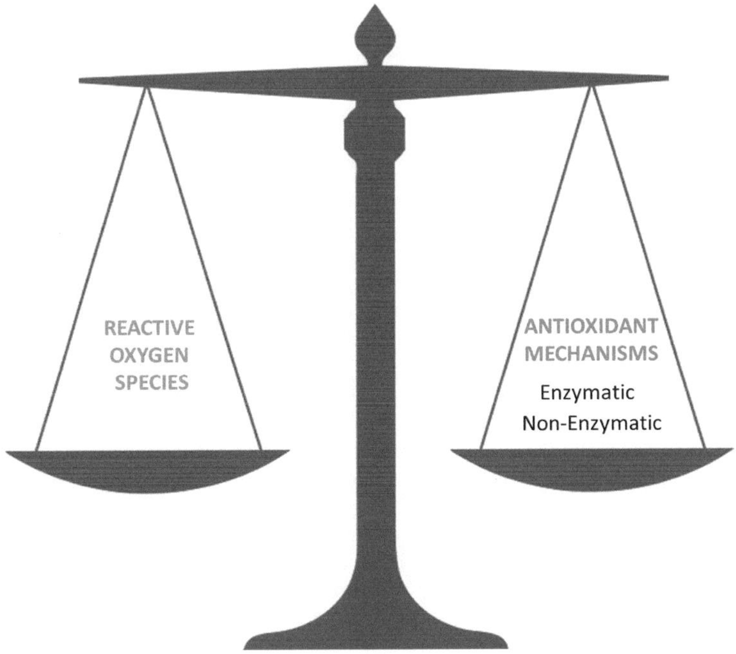

Figure 5.6 A balance between reactive oxygen species (ROS) generation and activities of antioxidant mechanisms. If production of ROS exceeds antioxidant activity of the organism, oxidative stress and subsequent oxidative damage occurs.

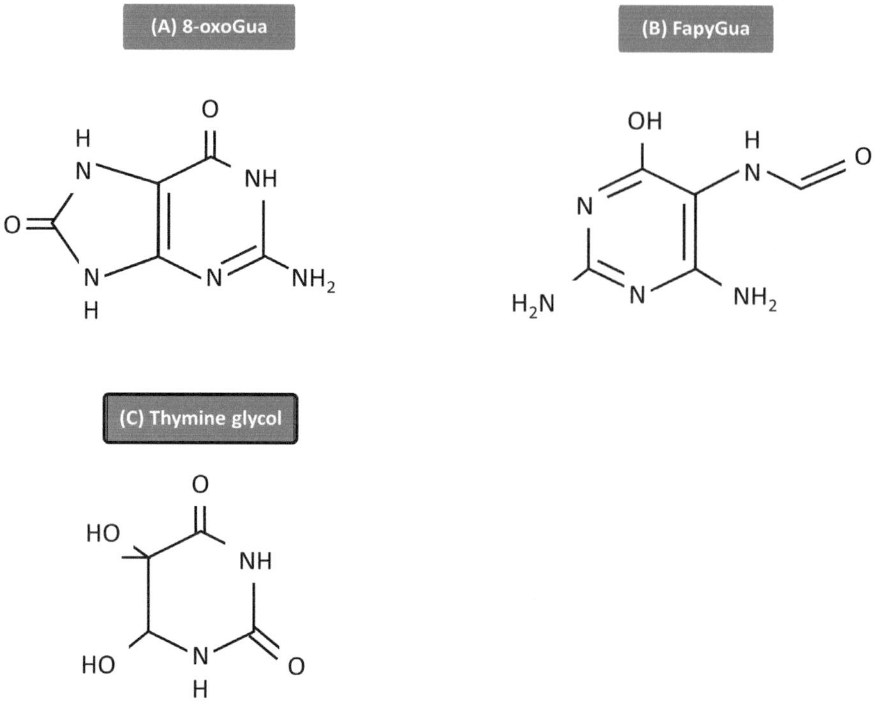

Figure 5.7 Examples of oxidized bases derived from guanine [(A) 8-oxo-7,8-dihydroguanine (8-oxoGua), (B) 2,6-diamino-4-hydroxy-5-formamidopyrimidine (FapyGua)] and thymine [(C) 5,6-dihydroxy-5,6-dihydrothymine (thymine glycol)].

excision repair, but nucleotide excision repair, both global genome and transcription coupled, may be also involved (Figure 5.2; Melis et al. 2013). If unrepaired, oxidized bases may cause mutations, which are the first step in the process of carcinogenesis (e.g. the presence of 8-oxoGua leads to GC to TA transversions; Markkanen 2017).

Apart from induction of mutations, ROS may contribute to carcinogenesis by tumor initiation, malignant transformation and induction of resistance to chemotherapy (Galadari et al. 2017). Tumor initiation may be started by DNA damage caused by ROS. This leads to activation of oncogenes or inactivation of tumor suppressor genes, which in turn increases ROS production. For example, mutant K-ras enhances ROS production, and loss of function of p53 protein increases transcription of cytokine genes, leading to an increased production of ROS. Apart from DNA damage, ROS may modulate tumorigenesis by induction of epigenetic modifications (DNA methylation and/or demethylation) or modulation of activities of proteins involved in tumor cell proliferation (e.g. phosphoinositide 3-kinase/protein kinase B, mitogen-activated protein kinase; Galadari et al. 2017).

Although ROS may start the process of tumorigenesis, the actual role of oxidative stress in human cancers is difficult to confirm experimentally. Tumor tissues have naturally higher levels of oxidatively damaged DNA caused by increased production of ROS due to defective mitochondrial metabolism (Galadari et al. 2017), and thus measurement of DNA oxidation or other parameters of ROS production in cancer patients may reflect consequences rather than causes of tumorigenesis. Indeed, increased oxidative stress levels have been documented in many cancers (e.g. breast, cervical, colorectal, gastric, hepatocellular or lung carcinoma; melanoma or leukemia; in patients and in experimental cell lines; Kryston et al. 2011). However, to prove that oxidatively damaged DNA is involved in cancer etiology, prospective studies are needed. These studies are difficult to perform because they require large sample sets and a long time for observation of the study populations. Thus, the number of such studies is rather limited. Nevertheless, the associations between various oxidative stress markers and cancer have been reported. In a study investigating associations between 8-oxoGua urinary excretion and lung cancer incidence, a significant cancer risk was found among men, never smokers and former smokers with increased 8-oxoGua levels (Loft et al. 2012). Prediagnostic serum levels of reactive oxygen metabolites (hydroperoxides) and increased colorectal cancer risk were reported (Leufkens et al. 2012). Levels of 8-epi-prostaglandin F2α, a marker of lipid peroxidation, were associated with risk of lung (Yuan et al. 2018) and liver cancer (Ma et al. 2018). Markers of oxidatively damaged DNA, lipid peroxidation and levels of plasma fluorescent oxidation products were associated with increased risk of breast cancer (Fortner et al. 2013; Lee et al. 2017). Risk of colorectal cancer was reduced by antioxidant lifestyle exposures over pro-oxidant exposures (Dash et al. 2015). In summary, although oxidative stress causes DNA damage and should thus contribute to the process of carcinogenesis, large prospective studies are still needed to confirm associations between oxidative stress markers and cancer risk.

5.4 METHODS OF OXIDATIVELY DAMAGED DNA DETECTION

Although oxidatively damaged DNA may be analyzed in various biological matrices (e.g. whole blood, plasma, serum, DNA, exhaled breath condensate or tissues), urine represents a convenient, commonly used material for the assessment of oxidative state of the organism. Unlike other sample types, urine can be easily obtained in sufficient quantity, its collection is noninvasive and does not pose ethical issues (e.g. in case the samples are to be obtained from newborns). Oxidized nucleobases (e.g. 8-oxo-7,8-dihydroguanine, 8-oxoGua) and 2′-deoxyribonucleosides (e.g. 8-oxo-7,8-dihydro-2′-deoxyguanosine, 8-oxodG; Figure 5.1B; Cooke et al. 2010) in urine are relatively stable; therefore the samples do not need to be frozen immediately after collection and the risk of artifactual generation of oxidized bases during storage or analysis is minimal. Finally, the detection of 8-oxodG is considered an approach reflecting whole-body oxidative stress. The commonly used methods of 8-oxodG analysis in urine include chromatography (high-performance liquid chromatography–tandem mass spectrometry [HPLC-MS/MS], gas chromatography–mass spectrometry [GC/MS]), ELISA, or ^{32}P-postlabeling. For interpretation of the results obtained in various laboratories, standardization of analytical methods is very important. Interlaboratory comparisons of chromatographic methods and ELISA have been performed with the aim to standardize 8-oxodG analysis (Barregard et al. 2013; Evans et al. 2010; Rossner et al. 2016). The results showed good agreement for laboratories that used chromatography, whereas for ELISA greater variability has been observed. Moreover, ELISA tended to overestimate urinary 8-oxodG levels. HPLC-MS/MS is a very sensitive and specific method that is currently considered the gold standard of 8-oxodG analysis, but because it relies on expensive laboratory equipment and requires trained laboratory personnel, it may thus be difficult to perform for some laboratories. ELISA represents a cheaper, high-throughput alternative (Rossner and Sram 2012), although it suffers from lower specificity and high interlaboratory variability. These drawbacks originate in the nature of the primary anti-8-oxodG antibody and the presence of compounds in the urine samples that may cross-react with the antibody. Recently, specificity and reproducibility of ELISA have been improved (Rossner et al. 2013a, 2016) by application of solid phase extraction that helped to remove some cross-reacting compounds from the samples. Incubation of the purified samples with the primary antibody at lower temperatures (4°C) together with correcting for creatinine levels further helped to improve the assay. Despite these modifications. some discrepancy between ELISA and HPLC-MS/MS still persists, and the ELISA data need to be interpreted with caution (Rossner et al. 2016).

5.5 BULKY DNA ADDUCTS AND THEIR DETECTION

Bulky DNA adducts are usually derived from large organic molecules, including PAHs and their derivatives, aflatoxins, nitrosamines, aromatic amines or heterocyclic aromatic amines (Hwa Yun et al. 2018). They may be detected using ^{32}P-postlabeling, antibody-based methods (ELISA, immunohistochemistry), gas chromatography–mass spectrometry (GC/MS) or liquid chromatography–mass spectrometry (LC/MS) (Hwa Yun et al. 2018). The antibody-based methods are relatively easy to perform and cheap and are thus suitable for the analysis of large sample sets. The limitation is lower sensitivity and cross-reactivity of the antibody with structurally similar compounds. A PAH-DNA ELISA was first used by Poirier et al. (1977), and the method was later used to explore DNA damage by carcinogenic PAHs in human organs (e.g. Perera et al. 1982). Immunohistochemistry can be used to detect DNA adducts in tissues in paraffin blocks. This can help to localize chemical carcinogens actions in the nuclei of specific cells, informing about intracellular DNA damage (Santella et al. 1999; Pratt et al. 2011).

A highly sensitive method for detection of chemically stable DNA damage is ^{32}P-postlabeling (Randerath and Randerath 1994). The detection limit is 1 adduct/10^8 nucleotides with DNA input of 1–5 μg. It detects a wide range of DNA adducts simultaneously but requires application

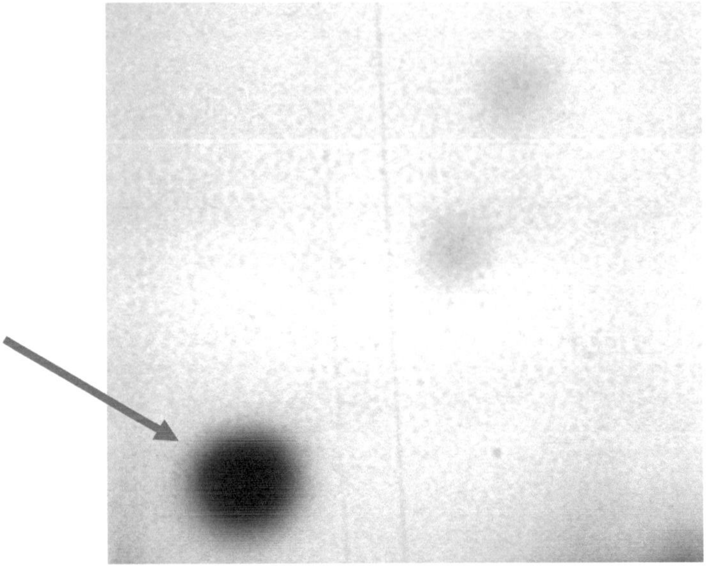

Figure 5.8 An autoradiogram of ^{32}P-postlabeled DNA. The arrow indicates the spot representing benzo[a]pyrene-7,8-dihydrodiol-9,10-epoxide (BPDE)-N^2-guanine, a bulky DNA adduct formed by interaction of benzo[a]pyrene (BaP) with DNA.

of radioactive isotopes and is very labor-intensive. The method does not allow identification of unknown adducts. Due to its sensitivity, it is suitable for measuring DNA adduct levels in samples exposed to environmental complex mixtures, especially combustion emissions, where DNA damage is generally relatively low (Figure 5.8; Phillips and Arlt 2007).

Both GC/MS and LC/MS can elucidate DNA adduct structure and are thus highly specific. They require low DNA quantities and allow absolute quantitation of bulky DNA adduct levels. The limitation is the high cost of the instruments (Boysen and Hecht 2003; Farmer and Singh 2008). Despite these facts, LC/MS is the most broadly used technique to detect bulky DNA adducts (Hwa Yun et al. 2018).

5.6 THE ROLE OF BULKY DNA ADDUCTS IN CARCINOGENESIS

Although bulky DNA adducts may be formed at different nitrogen and oxygen atoms of both the purines and pyrimidines, the *N3* and *N7* positions of the guanine (Figure 5.1) and adenine are preferentially modified due to their nucleophilic character. Some chemical carcinogens (e.g. BaP) may also induce adducts on the phosphate group of DNA (Hwa Yun et al. 2018). The stability of bulky DNA adducts differs, and some compounds even form labile glycosidic bonds which cause depurination of the adducted nucleobase and formation of an apurinic site. Bulky DNA adducts are removed by nucleotide excision repair pathways, either global genomic NER or transcription-coupled NER (Figure 5.2; Zhong et al. 2010). If not repaired, bulky DNA adducts can interfere with processes of DNA transcription and replication, and their presence may lead to formation of mutations. Thus, the major BaP-derived bulky DNA adduct, BPDE-N^2-guanine, induces GC to TA, GC to AT, GC to CG and frameshift mutations. Mutations in oncogenes (*H-ras, K-ras*) and tumor suppressor genes (*p53*) are critical for tumor initiation. Mutations in *p53*, particularly GC to TA transversions, are commonly found in almost 40% of human lung cancers, especially in smokers. This observation suggests an association between exposure to PAHs in tobacco smoke and lung cancer development in smokers.

The proof of a direct link between bulky DNA adduct levels and cancer incidence depends on the compounds associated with adduct formation. A clear association has been established for aflatoxin B1 (AFB1) and aristolochic acid exposure. AFB1 is a mycotoxin that contaminates food, particularly peanuts, corn and other grains. AFB1 is metabolized in the liver to aflatoxin B1 exo-8,9-epoxide that binds to DNA, preferentially to *N7* guanine. Exposure to AFB1 is a risk factor for hepatocellular carcinoma (HCC) in humans. The incidence of HCC associated with food AFB1 intake is increased in populations in Asia and Africa (Groopman et al. 1992a, 1992b). Aristolochic acid (AA) is a compound naturally found in *Aristolochia* plants. The compound, structurally related to nitrophenanthrenes, is nephrotoxic and carcinogenic. Humans are exposed to AA through the use of herbs found in traditional Chinese medicine or the ingestion of homemade bread contaminated with *Aristolochia* seeds (Hwa Yun et al. 2018). The exposure is associated with increased incidence of upper tract urothelial carcinoma. Human urothelial tissue is rich in peroxidases required for metabolic activation of AA. Aristolactam DNA adducts formed upon AA exposure were shown to be directly linked to AT to TA transversions in *p53* (Barnes et al. 2018).

PAHs are a class of ubiquitous compounds that contaminate air, water, soil and food. Some of these compounds are potentially carcinogenic to humans; BaP is classified as a human carcinogen (IARC 2012). Due to health concerns associated with PAH exposure, many authors have investigated the effect of environmental PAH exposure on bulky DNA adduct levels. It has been suggested that bulky DNA adducts are not only markers of exposure to PAHs, but their levels also reflect the ability of an organism to metabolically activate carcinogens and to repair DNA damage (Phillips 2005; Poirier 2016). According to Phillips (2005), levels of bulky DNA adducts as a measure of exposure several years prior to the onset (or clinical manifestation) of disease can identify individuals at higher probability of subsequent cancer development. Ricceri et al. (2010) performed a pooled analysis of bulky DNA adducts in leukocytes of 3600 subjects and observed weak positive associations between bulky DNA adduct levels and various exposure variables. The effect was more pronounced if the DNA adducts were determined in peripheral lymphocytes. DeMarini (2013) reviewed 18 studies for traffic-associated bulky DNA adducts, which also included exposure assessment. In all studies, he observed higher bulky DNA adduct levels in exposed versus control subjects; in nine studies, he found an association between levels of bulky DNA adducts and exposure. The relationship between the exposure to PAHs and bulky DNA adduct levels is not linear at high concentrations of PAHs (Lewtas et al. 2007; Rossner et al. 2013b). This response may be explained by saturation of enzyme activity, as would be expected for carcinogens that require metabolic activation, or by adaptation to long-term exposure to the compounds (Rossnerova et al. 2017).

Despite the reports mentioned above, the direct link between PAH-DNA adducts and cancer is uncertain (Veglia et al. 2008). Although the elevated levels of bulky DNA adducts have been detected in subjects suffering from liver (Chen et al. 2002), stomach (Agudo et al. 2012), colon (Gunter et al. 2007; Agudo et al. 2012), lung (Tang et al. 1995, 2001), prostate (Tang et al. 2013), and breast cancer (Gammon et al. 2004), prospective studies linking bulky DNA adducts with cancer development are rare, and a clear relationship has been shown only for the association between DNA adduct levels in white blood cells and the risk of lung cancer (Veglia et al. 2008).

5.7 CHROMOSOMAL ABERRATIONS AND CARCINOGENESIS

Single-strand or double-strand DNA breaks induces e.g. by ROS, ionizing radiation or bulky DNA adduct formation, may result in chromosomal aberrations (CA). If repaired by nonhomologous end-joining or homology-directed repair (Figure 5.2), dsDNA breaks may form stable chromosomal aberrations as translocations and reciprocal translocations.

Current research repeatedly shows a clear relationship between the increased levels of CA in peripheral blood lymphocytes (PBL) of healthy subjects and risk of cancer. This observation, first reported already in 1998 (Hagmar et al. 1998), suggested that the frequency of CAs in PBL is a relevant biomarker for cancer risk in humans. Moreover, results from Nordic, Italian and Czech cohorts indicate an approximately twofold cancer risk among subjects with high frequencies of CAs (Hagmar et al. 2004). Later studies confirmed this observation in databases of 11,834, 6430 and 22,358 healthy individuals from 1, 6 and 11 countries, respectively, and added novel information on the role of aberration subclass and cancer type (Rossner et al. 2005; Boffetta et al. 2007; Bonassi et al. 2008). The association between alternative cytogenetic biomarkers with increased risk of cancer has been investigated in a number of other studies. In contrast to CAs, the relationship with the risk of cancer was not confirmed for sister chromatid exchanges (SCE; Norppa et al. 2006). A total of 6718 subjects selected from the database of the Human MicroNucleus (HUMN) international collaborative project provided preliminary evidence that micronuclei (MN) frequency in PBL is a predictive biomarker of cancer risk within a population of healthy subjects (Bonassi et al. 2007). Although many laboratories focus on the investigation of stable chromosomal aberrations in healthy subjects (Sigurdson et al. 2008), no relevant representative study concerning increased frequency of this CA type in PBL and risk of cancer has been published. Despite this fact, numerous chromosomal rearrangements, including stable translocations, are typically and frequently found

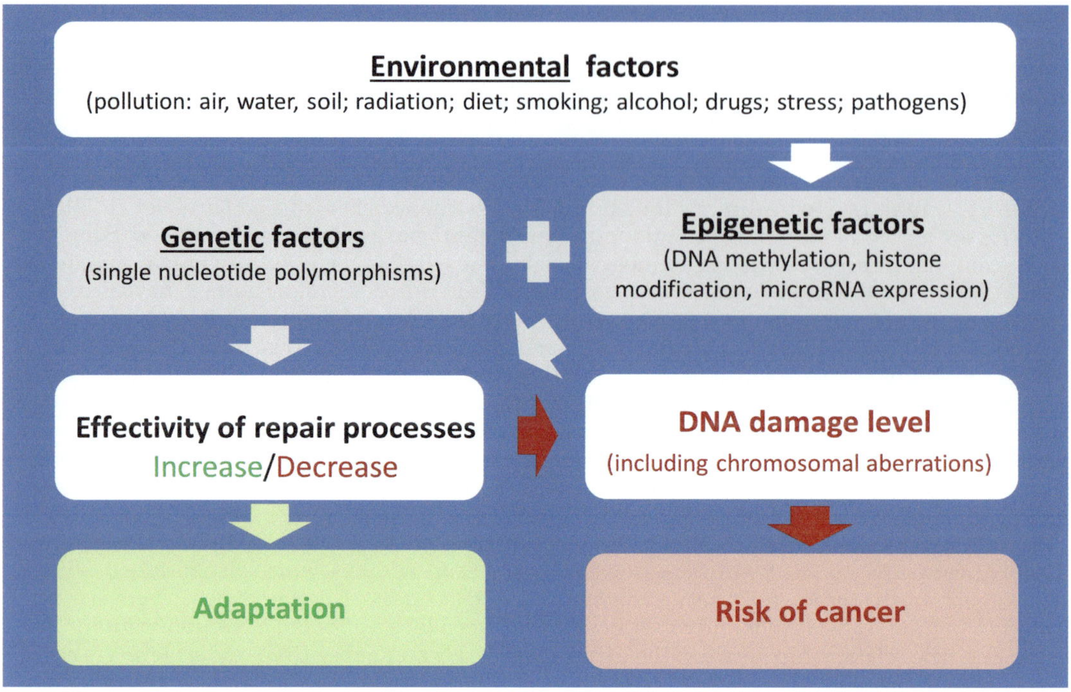

Figure 5.9 Factors affecting the process of carcinogenesis.

in tumor cells of cancer patients (reported in more than 50,000 records in a PubMed database search for keywords "translocation" + "cancer" + "human").

On the other hand, the relationship between the exposure to xenobiotics and frequency of CA is not nearly as clear as the relationship between CA increase and the risk of cancer (Bonassi et al. 2010). Even though many studies reported effects of exposure to particular chemicals on chromosomal damage, including micronuclei formation, some authors found no effect or even an opposite effect of the studied compounds (Natarajan et al. 1996; Rossner Jr. et al. 2015; Nersesyan et al. 2016; Sram et al. 2016). Broad toxicological analyses revealed the critical importance of other factors, including genetic (SNPs) and epigenetic (DNA methylation, histone modification or microRNA expression) factors. Both play an important role in effectivity of repair processes affecting DNA damage levels, including chromosomal aberrations as a potential signal of carcinogenesis initiation (Figure 5.9). Moreover, type of exposure and exposure history are other important factors significantly affecting CA levels, mainly due to epigenetic adaptation of an organism to the environment (Rossnerova et al. 2017).

5.8 METHODS FOR CHROMOSOMAL ABERRATION INVESTIGATION

Cytogenetic analyses are among the basic tools of genetic toxicologists for investigation of chromosomal aberrations. The following cytogenetic methods are most commonly used: (1) conventional cytogenetic analysis (CCA), (2) sister chromatid exchanges, (3) analysis of micronuclei and (4) fluorescence *in situ* hybridization (FISH).

5.8.1 Conventional Cytogenetic Analysis

CCA is the oldest cytogenetic biomonitoring method (Buckton and Evans 1973), standardized and accepted for analysis of clastogenic effects of exposure in PBL (Carrano and Natarajan 1988; Sorsa et al. 1992). This method originated in 1965, when the first protocol for whole-blood cultivation and metaphase chromosome preparation was published (Hungerford 1965). An almost identical protocol is used now, which uses fetal bovine serum, phytohemagglutinin and glutamine during cultivation, colchicine to accumulate metaphases during cell division, and KCl (hypotonic solution) followed by methanol-acetic acid fixation. Moreover, other cytogenetic methods (analysis of micronuclei and preparation of samples for FISH analysis) are also based on these cultivation and cell-harvesting conditions. The modified conventional Hungerford method is used for short-term,

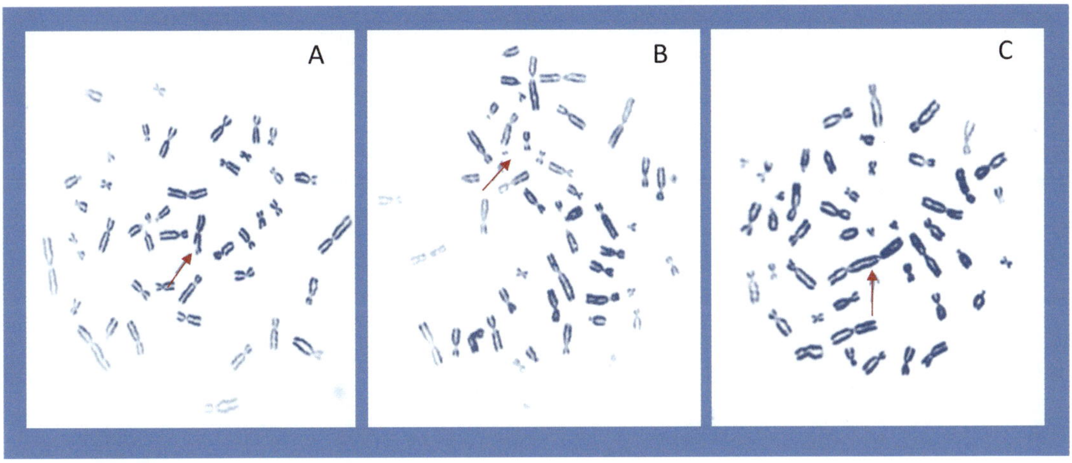

Figure 5.10 Examples of chromosomal aberrations (highlighted with arrow) analyzed by the conventional cytogenetic method: (A) chromatid break; (B) chromosome break; (C) dicentric chromosome.

48-hour cultures, with all cells being in the first division. After cell cultivation and harvesting, the microscopic slides are prepared and stained by Giemsa solution. Typically, 100 or 200 well-spread metaphases with 46 ± 1 centromeres are examined per donor.

Low price is the main advantage of the CCA method, but the evaluation of individual metaphases is relatively laborious. Chromatid and chromosome breaks are main classes of the chromosomal aberrations that are recognized (Figure 5.10A, B). Thus, this approach is presented mainly as a method for detection of unstable chromosomal aberrations. Stable chromosomal rearrangements (e.g. translocations or insertions) are rarely recognized, even though it is possible in combination with the G-banding technique. Other types of aberrations recognized by CCA are rings, gaps, minute, dicentric or tricentric chromosomes (Figure 5.10C). Moreover, the cells with an aberrant number of chromosomes (aneuploid or polyploid cells) can be recorded. Even though individual metaphases can be found and scanned by automated systems, recognition of individual broad spectrum of various types of chromosomal rearrangements still requires input from researchers. Advances were reached only in the case of separate evaluation of dicentric chromosomes (Schunck et al. 2004). Taken together, this technique is still used by many laboratories, as it was suggested for monitoring of human exposure to occupational genotoxicants (Rossner et al. 1995), and given the publication record, it remains the cytogenetic method yielding the most convincing results related to risk of cancer (Hagmar et al. 2004; Rossner et al. 2005; Boffetta et al. 2007; Bonassi et al. 2008).

5.8.2 Micronuclei Analysis

The first attempt to use a micronucleus test was made in 1959 (Evans et al. 1959) in roots of *Vicia faba*. A shift to the cultivation of PBL and application of cytochalasin-B (to inhibit cytokinesis) in 1985 were important milestones in the development and future routine use of the method in human biomonitoring studies (Countryman and Heddle 1976; Fenech and Morley 1985). This method, which analyzes DNA damage in binucleated cells (BNC), is one of the preferred cytogenetic methods frequently used in molecular epidemiology. MN represent a measure of both chromosome breakage and chromosome loss, and reflect exposure to agents with clastogenic or aneugenic modes of action, respectively (Fenech and Morley 1985). Moreover, the method offers various variants for both *in vitro* and *in vivo* applications related to (1) evaluated objects, (2) type of staining or (3) procedure of evaluation.

Related to evaluated objects, the cytokinesis-block micronucleus assay (CBMN), later described as a "cytome" assay, implies that every cell studied in the system is scored for its viability status, mitotic status and chromosomal instability or damage status (Fenech 2006). Some of the most important events that can be a result of various exposure episodes are presented in Figure 5.11.

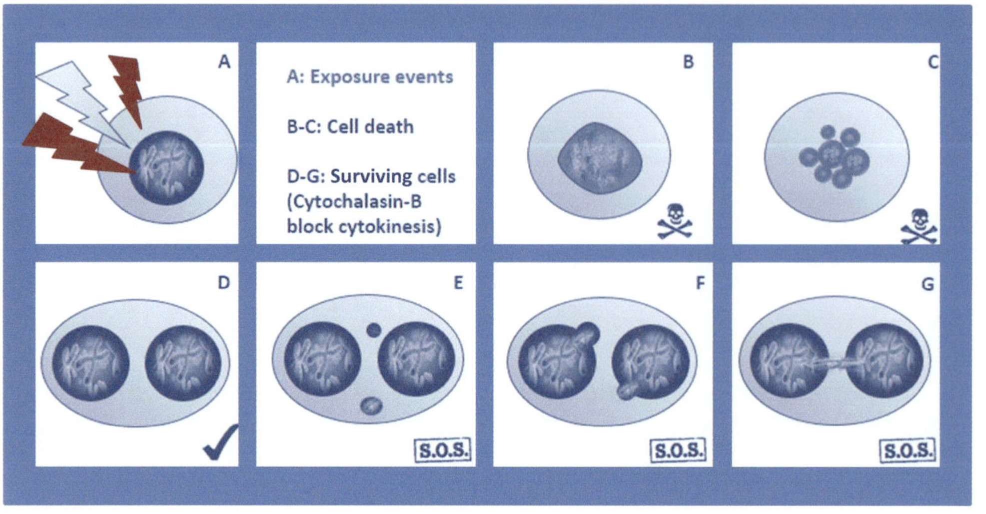

Figure 5.11 Summary of various cell exposure effects after exposure events: (A) exposure events; (B–G) possible exposure consequences, including (B) cell death (necrosis); (C) cell death (apoptosis); (D) binucleated cell (BNC) without micronucleus assay (MN); (E) BNC with two MN; (F) BNC with two nuclear buds (NBUDs); and (G) BNC with nucleoplasmic bridge (NPB).

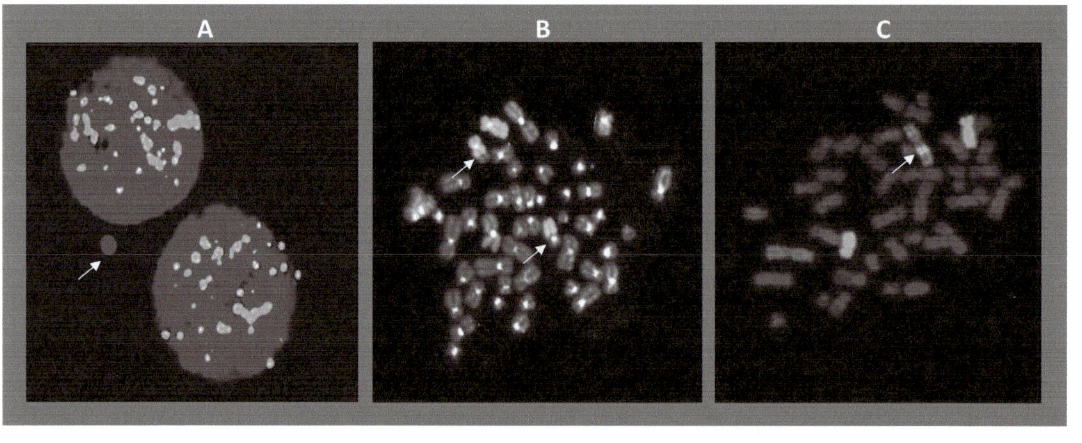

Figure 5.12 Examples of various cytogenetic findings analyzed by the FISH method: (A) cytochalasin-B-blocked binucleated cell (BNC) with centromere negative micronucleus assay (MN; arrow); (B) combination of centromere and whole chromosome painting of chromosomes #1 (darker staining) and #4 (brighter staining) in metaphase cell detecting a reciprocal translocation (arrows); and (C) whole chromosome painting of chromosomes #1 (darker staining) and #4 (bright staining) in metaphase cell detecting an insertion (arrow).

The type of staining can affect the output of the method. Besides the basic methodological approach, which mostly concerns the frequency of total MN (including chromosomal breaks and losses) and in some cases also nucleoplasmic bridges (NPBs) or nuclear buds (NBUDs), detailed identification of the content of micronuclei can be investigated in combination with immunostaining techniques (e.g. antikinetochore antibody staining [CREST staining]; Henning et al. 1988) or fluorescence hybridization methods (e.g. FISH of centromeres, telomeres or even specific chromosomes; Moore et al. 1993). These approaches have been applied as follow-up techniques that significantly improve the output of the studies despite being laborious and costly (examples in Figure 5.12A).

Other methodological variants relate to the type of scoring of evaluated objects. Up to now, most studies have been performed utilizing visual scoring. Although this approach is relatively easy, the scoring of thousands of cells is very time-consuming. Owing to this fact, efforts to automate the counting of cells have been made. Hutter and Stöhr reported the automation of micronucleus analysis by flow cytometry as early as 1982 (Hutter and Stöhr 1982), followed by the first automatic technique with computerized image analysis in 1990 (Tates et al. 1990). Recently, a few new systems have been developed, and their abilities have been compared (Fenech et al. 2013). The current leading system was introduced by MetaSystems in 2004 (Schunck et al. 2004; Rossnerova et al. 2011). Detailed testing and a comparison between samples stained with Giemsa for visual counting and with DAPI (4′-6-diamidino-2-phenylindole) for automated analysis were performed in a group of breast cancer patients and controls to evaluate the performance of the automated analysis (Varga et al. 2004). The results of this comparison indicated that the system appears to be more reliable for DAPI-stained samples using fluorescence. Another system was introduced in 2009 (Decordier et al. 2009). This system, suggested for biomonitoring, was developed for scoring Giemsa-stained slides. The current state of the art shows the relatively broad utilization of automated scanning systems in many studies not only related to human PBL. Moreover, automation allows evaluation of more objects, up to several thousand, instead of the 1000 or 2000 BNC analyzed by visual scoring.

5.8.3 Fluorescence *In Situ* Hybridization

FISH, frequently used in both clinical and tumor cytogenetics as well as in human biomonitoring, is based on the ability of DNA strands to anneal to complementary base pairs. The method was introduced by Pinkel in 1986. We have mentioned the FISH method in connection with the analysis of micronuclei as a way to get more information on chromosomal rearrangements. Besides the possibility to stain regions of interest in interphase chromosomes, this fluorescent staining is preferably applied on well-spread metaphase chromosomes. The procedure of cell cultivation and harvesting is the same as in CCA, with only prolonged time of cultivation (up to 72 h) due to the increased number of evaluated objects. The main advantage of this method is the possibility to evaluate stable CA.

Today, a broad spectrum of fluorescent probes, consisting of a set of DNA sequences labeled by various fluorochromes, are available for staining of individual whole chromosomes, their arms, selected genes, centromeres or telomeres. Moreover, custom mixtures for painting of selected chromosomes or regions can be also prepared based on the requests of individual laboratories. Depending on the purpose of staining, even the whole karyotype can be painted, but due to the cost and relatively laborious evaluation of the samples, whole-chromosome painting of selected chromosomes is the most frequently used approach in genetic toxicology. The frequently recognized combinations of stained chromosomes include the following sets: #1, #4 or #1, #2, #4 or #3, #5, #6. Also, other combinations of chromosome adding (e.g. #8, #11 or #12) have been painted by some laboratories (Sigurdson et al. 2008). As only a subset of chromosomes was painted, the fraction of exchanges detected was calculated using equations appropriate for one- to three-color paints (Tucker et al. 1997). The calculations are based on the DNA content of the painted chromosomes and gender (Morton 1991). The final genomic frequency of stable chromosome exchanges is calculated according to the published formula (Lucas and Sachs 1993), which allows the comparison of results between laboratories staining the different sets of chromosomes.

As already mentioned, the main advantage of this methodological approach is the ability to analyze stable chromosomal aberrations (e.g. translocations, including one-way and two-way reciprocal translocations) and insertions. Also, acentric fragments and dicentric chromosomes are frequently recorded during evaluation. Thus, many types of chromosomal rearrangements are observed. The aberrant cells are classified according to the Protocol for Aberration Identification and Nomenclature (PAINT), describing their various individual types (Figure 5.12B, C; Tucker et al. 1995).

5.9 CONCLUSIONS

For successful cancer treatment, early detection of the disease is crucial. For this purpose, application of biomarkers may be a convenient approach, provided that the changes in biomarker levels can predict the onset of the disease. In this chapter, we summarized current knowledge on associations between biomarkers of oxidatively damaged DNA, bulky DNA adducts and chromosomal aberrations and cancer risk. Although all the biomarkers discussed here have some predictive value, the frequency of chromosomal aberrations is the only parameter for which the link with cancer risk is well established. Although elevated levels of all biomarkers indicate DNA damage with potential consequences for cancer initiation, the causal link between the process of carcinogenesis and elevated oxidative stress and bulky DNA adduct levels is less clear.

ACKNOWLEDGMENTS

Supported by the Czech Science Foundation (18–02079S) and the Ministry of Youth, Education, and Sports of the Czech Republic (LO1508). The excellent technical help of Ms. Zuzana Novakova is appreciated.

BIBLIOGRAPHY

Agudo, A., M. Peluso, A. Munnia, et al. 2012. Aromatic DNA adducts and risk of gastrointestinal cancers: A case-cohort study within EPIC-Spain. *Cancer Epidemiol Biomarkers Prev* 21: 685–92.

Baird, W. M., L. A. Hooven, and B. Mahadevan. 2005. Carcinogenic polycyclic aromatic hydrocarbon-DNA adducts and mechanism of action. *Env Mol Mutagen* 45: 106–14.

Barnes, J. L., M. Zubair, K. John, M. C. Poirier, and F. L. Martin. 2018. Carcinogens and DNA damage. *Biochem Soc Trans* 46: 1213–24.

Barregard, L., P. Moller, T. Henriksen, et al. 2013. Human and methodological sources of variability in the measurement of urinary 8-Oxo-7,8-dihydro-2′-deoxyguanosine. *Antiox Redox Signal* 18: 2377–91.

Bi, X., D. M. Slater, H. Ohmori, and C. Vaziri. 2005. DNA polymerase kappa is specifically required for recovery from the benzo[a]pyrene-dihydrodiol epoxide (BPDE)-induced S-phase checkpoint. *J Biol Chem* 280: 22343–55.

Boffetta, P., O. van der Hel, H. Norppa, et al. 2007. Chromosomal aberrations and cancer risk: Results of a cohort study from Central Europe. *Am J Epidemiol* 165: 36–43.

Bonassi, S., L. Hagmar, U. Strömberg, et al. 2010. Chromosomal aberrations in lymphocytes predict human cancer independently of exposure to carcinogens. European Study Group on Cytogenetic biomarkers and Health. *Cancer Res* 60: 1619–25.

Bonassi, S., H. Norppa, M. Ceppi, et al. 2008. Chromosomal aberration frequency in lymphocytes predicts the risk of cancer: Results from a pooled cohort study of 22 358 subjects in 11 countries. *Carcinogenesis* 29: 1178–83.

Bonassi, S., A. Znaor, M. Ceppi, et al. 2007. An increased micronucleus frequency in peripheral blood lymphocytes predicts the risk of cancer in humans. *Carcinogenesis* 28: 625–31.

Boysen, G., and S. S. Hecht. 2003. Analysis of DNA and protein adducts of benzo[a]pyrene in human tissues using structure-specific methods. *Mutat Res* 543: 17–30.

Buckton, K. E., and H. J. Evans. 1973. *Methods for the Analysis of Human Chromosome Aberrations.* Geneva: WHO.

Carrano, A. V., and A. T. Natarajan. 1988. International Commission for Protection Against Environmental Mutagens and Carcinogens. ICPEMC publication no. 14. Considerations for publication monitoring using cytogenetic techniques. *Mutat Res* 204: 379–406.

Chen, S. Y., L. Y. Wang, R. M. Lunn, et al. 2002. Polycyclic aromatic hydrocarbon-DNA adducts in liver tissues of hepatocellular carcinoma patients and controls. *Int J Cancer* 99: 14–21.

Cooke, M. S., M. D. Evans, M. Dizdaroglu, and J. Lunec. 2003. Oxidative DNA damage: Mechanisms, mutation, and disease. *FASEB J* 17: 1195–214.

Cooke, M. S., S. Loft, R. Olinski, et al. 2010. Recommendations for standardized description of and nomenclature concerning oxidatively damaged nucleobases in DNA. *Chem Res Toxicol* 23: 705–7.

Countryman, P. I., and J. A. Heddle. 1976. The production of micronuclei from chromosome aberrations in irradiated cultures of human lymphocytes. *Mutat Res* 41: 321–32.

Croom, E. 2012. Metabolism of xenobiotics of human environments. *Prog Mol Biol Transl Sci* 112: 31–88.

Dash, C., R. M. Bostick, M. Goodman, et al. 2015. Oxidative balance scores and risk of incident colorectal cancer in a US prospective cohort study. *Am J Epidemiol* 181: 584–94.

Decordier, I., A. Papine, G. Plas, et al. 2009. Automated image analysis of cytokinesis-blocked micronuclei: An adapted protocol and a validated scoring procedure for biomonitoring. *Mutagenesis* 24: 85–93.

DeMarini, D. M. 2013. Genotoxicity biomarkers associated with exposure to traffic and near-road atmosphere: A review. *Mutagenesis* 28: 485–505.

Demetriou, C. A., O. Raaschou-Nielsen, S. Loft, et al. 2012. Biomarkers of ambient air pollution and lung cancer: A systematic review. *Occup Env Med* 69: 619–27.

Evans, H. J., G. J Neary, and F. S. Williamson. 1959. The relative biological efficiency of single doses of fast neutrons and gamma-rays on *Vicia faba* roots and the effect of oxygen. Part II. Chromosome damage: The production of micronuclei. *Int J Radiat Biol* 1: 216–29.

Evans, M. D., R. Olinski, S. Loft, and M. S. Cooke. 2010. Toward consensus in the analysis of urinary 8-oxo-7,8-dihydro-2′-deoxyguanosine as a noninvasive biomarker of oxidative stress. *FASEB J* 24: 1249–60.

Farmer, P. B., and R. Singh. 2008. Use of DNA adducts to identify human health risk from exposure to hazardous environmental pollutants: The increasing role of mass spectrometry in assessing biologically effective doses of genotoxic carcinogens. *Mutat Res* 659: 68–76.

Fenech, M. 2006. Cytokinesis-block micronucleus assay evolves into a "cytome" assay of chromosomal instability, mitotic dysfunction and cell death. *Mutat Res* 600: 58–66.

Fenech, M., M. Kirsch-Volders, A. Rossnerova, et al. 2013. HUMN project initiative and review of validation, quality control and prospects for further development of automated micronucleus assays using image cytometry systems. *Int J Hyg Environ Health* 216: 541–52.

Fenech, M., and A. A. Morley. 1985. Measurement of micronuclei in lymphocytes. *Mutat Res* 147: 29–36.

Fortner, R. T., S. S. Tworoger, T. Wu, and A. H. Eliassen. 2013. Plasma florescent oxidation products and breast cancer risk: Repeated measures in the Nurses' Health Study. *Breast Cancer Res Treat* 141: 307–16.

Galadari, S., A. Rahman, S. Pallichankandy, and F. Thayyullathil. 2017. Reactive oxygen species and cancer paradox: To promote or to suppress? *Free Radical Biol Med* 104: 144–64.

Gammon, M. D., S. K. Sagiv, S. M. Eng, et al. 2004. Polycyclic aromatic hydrocarbon-DNA adducts and breast cancer: A pooled analysis. *Arch Environ Health* 59: 640–9.

Groopman, J. D., A. J. Hall, H. Whittle, et al. 1992a. Molecular dosimetry of aflatoxin-N7-guanine in human urine obtained in The Gambia, West Africa. *Cancer Epidemiol Biomarkers Prev* 1: 221–7.

Groopman, J. D., Z. Jiaqui, Z., P. R. Donahue, et al. 1992b. Molecular dosimetry of urinary aflatoxin-DNA adducts in people living in Guangxi Autonomous Region, People's Republic of China. *Cancer Res* 52: 45–52.

Gunter, M. J., R. L. Divi, M. Kulldorff, et al. 2007. Leukocyte polycyclic aromatic hydrocarbon-DNA adduct formation and colorectal adenoma. *Carcinogenesis* 28: 1426–9.

Gupta, R. K., A. K. Patel, N. Shah, et al. 2014. Oxidative stress and antioxidants in disease and cancer: A review. *Asian Pac J Cancer Prev APJCP* 15: 4405–9.

Hagmar, L., S. Bonassi, U. Strömberg, et al. 1998. Chromosomal aberrations in lymphocytes predict human cancer: A report from the European Study Group on Cytogenetic Biomarkers and Health (ESCH). *Cancer Res* 58: 4117–21.

Hagmar, L., U. Strömberg, H. Tinnerberg, and Z. Mikoczy. 2004. Epidemiological evaluation of cytogenetic biomarkers as potential surrogate end-points for cancer. *IARC Sci Publ* 157: 207–15.

Henning, U. G., N. L. Rudd, and D. I. Hoar. 1988. Kinetochore immunofluorescence in micronuclei: A rapid method for the in situ detection of aneuploidy and chromosome breakage in human fibroblasts. *Mutat Res* 203: 405–14.

Hungerford, D. A. 1965. Leukocytes cultured from small inocula of whole blood and preparation of metaphase chromosomes by treatment with hypotonic KCl. *Stain Technol* 40: 333–38.

Hutter, K. J., and M. Stöhr. 1982. Rapid detection of mutagen induced micronucleated erythrocytes by flow cytometry. *Histochemistry* 75: 353–62.

IARC. 2012. *IARC Monographs on the Evaluation of the Carcinogenic Risk of Chemicals to Humans. Chemical Agents and Related Occupations.* Lyon, France: IARC Publications.

Jalal, S., J. N. Earley, and J. J. Turchi. 2011. DNA repair: From genome maintenance to biomarker and therapeutic target. *Clin Cancer Res* 17: 6973–84.

Klaunig, J. E., and L. M. Kamendulis. 2004. The role of oxidative stress in carcinogenesis. *Ann Rev Pharmacol Toxicol* 44: 239–67.

Kryston, T. B., A. B. Georgiev, P. Pissis, and A. G. Georgakilas. 2011. Role of oxidative stress and DNA damage in human carcinogenesis. *Mutat Res Mol Mech Mutagen* 711: 193–201.

Lee, J. D., Q. Cai, X. O. Shu, and S. J. Nechuta. 2017. The role of biomarkers of oxidative stress in breast cancer risk and prognosis: A systematic review of the epidemiologic literature. *J Womens Health* 26: 467–82.

Leufkens, A. M., F. J. B. van Duijnhoven, S. H. S. Woudt, et al. 2012. Biomarkers of oxidative stress and risk of developing colorectal cancer: A cohort-nested case-control study in the European prospective investigation into cancer and nutrition. *Am J Epidemiol* 175: 653–63.

Lewtas, J. 2007. Air pollution combustion emissions: Characterization of causative agents and mechanisms associated with cancer, reproductive, and cardiovascular effects. *Mutat Res* 636: 95–133.

Loft, S., P. Svoboda, K., Kawai, et al. 2012. Association between 8-oxo-7,8-dihydroguanine excretion and risk of lung cancer in a prospective study. *Free Radical Biol Med* 52: 167–72.

Lucas, J. N., and R. K. Sachs. 1993. Using three-color chromosome painting to test chromosome aberration models. *Proc Natl Acad Sci USA* 90: 1484–87.

Ma, X., Y.-T. Tan, Y. Yang, et al. 2018. Pre-diagnostic urinary 15-F $_{2t}$ -isoprostane level and liver cancer risk: Results from the Shanghai Women's and Men's Health Studies: Urinary 15-F $_{2t}$ -isoprostane and liver cancer risk. *Int J Cancer* 143: 1896–903.

Markkanen, E. 2017. Not breathing is not an option: How to deal with oxidative DNA damage. *DNA Repair* 59: 82–105.

Melis, J. P., H. van Steeg, and M. Luijten. 2013. Oxidative DNA damage and nucleotide excision repair. *Antiox Redox Signal* 18: 2409–19.

Moore, L. E., N. Titenko-Holland, P. J. Quintana, and M. T. Smith. 1993. Novel biomarkers of genetic damage in humans: Use of fluorescence in situ hybridization to detect aneuploidy and micronuclei in exfoliated cells. *J Toxicol Environ Health* 40: 349–57.

Morton, N. 1991. Parameters of the human genome. *Proc Natl Acad Sci USA* 88: 7474–76.

Natarajan, A., J. J. Boei, F. Darroudi, et al. 1996. Current cytogenetic methods for detecting exposure and effects of mutagens and carcinogens. *Environ Health Perspect* 104: 445–8.

Nersesyan, A., M. Fenech, C. Bolognesi, et al. 2016. Use of the lymphocyte cytokinesis-block micronucleus assay in occupational biomonitoring of genome damage caused by in vivo exposure to chemical genotoxins: Past, present and future. *Mutat Res* 770: 1–11.

Norppa, H., S. Bonassi, I. L. Hansteel, et al. 2006. Chromosomal aberrations and SCEs as biomarker of cancer risk. *Mutat Res* 600: 37–45.

Perera, F. P., M. C. Poirier, S. H. Yuspa, et al. 1982. A pilot project in molecular cancer epidemiology: Determination of benzo[a]pyrene-DNA adducts in animal and human tissues by immunoassays. *Carcinogenesis* 3: 1405–10.

Phillips, D. H. 2005. DNA adducts as markers of exposure and risk. *Mutat Res* 577: 384–92.

Phillips, D. H., and V. M. Arlt. 2007. The 32P-postlabeling assay for DNA adducts. *Nature Protocols* 2: 2772–81.

Pinkel, D., T. Straume, and J. W. Gray. 1986. Cytogenetic analysis using quantitative, high-sensitivity, fluorescence hybridization. *Proc Natl Acad Sci USA* 83: 2934–38.

Poirier, M. C. 2016. Linking DNA adduct formation and human cancer risk of chemical carcinogenesis. *Environ Mol Mutagen* 57: 499–507.

Poirier, M. C., S. H. Yuspa, I. B. Weinstein, and S. Blobstein. 1977. Detection of carcinogen-DNA adducts by radioimmunoassay. *Nature* 270: 186–8.

Pratt, M. M., K. John, A. B. Maclean, et al. 2011. Polycyclic aromatic hydrocarbon (PAH) exposure and DNA adduct semi-quantitation in archived human tissues. *Int J Environ Res Public Health* 8: 2675–91.

Randerath, K., and E. Randerath. 1994. 32P-postlabeling methods for DNA adduct detection: Overview and critical evaluation. *Drug Metab Rev* 26: 67–85.

Ricceri, F., R. W. Godschalk, M. Peluso, et al. 2010. Bulky DNA adducts in white blood cells: A pooled analysis of 3,600 subjects. *Cancer Epidemiol Biomarkers Prev* 19: 3174–81.

Rossner Jr, P., P. Boffetta, M. Ceppi, et al. 2005. Chromosomal aberrations in lymphocytes of healthy subjects and risk of cancer. *Environ Health Perspect* 113: 517–20.

Rossner Jr, P., M. Cerna, H. Bavorova, A. Pastorkova, and D. Ocadlikova. 1995. Monitoring of human exposure to occupational genotoxicants. *Cent Eur J Public Health* 3: 219–23.

Rossner Jr, P., V. Mistry, R. Singh, R. J. Sram, and M. S. Cooke. 2013a. Urinary 8-oxo-7,8-dihydro-2′-deoxyguanosine values determined by a modified ELISA improves agreement with HPLC-MS/MS. *Biochem Biophys Res Commun* 440: 725–30.

Rossner Jr, P., H. Orhan, G. Koppen, et al. 2016. Urinary 8-oxo-7,8-dihydro-2′-deoxyguanosine analysis by an improved ELISA: An inter-laboratory comparison study. *Free Radical Biol Med* 95: 169–79.

Rossner Jr, P., and R. J. Sram. 2012. Immunochemical detection of oxidatively damaged DNA. *Free Radical Res* 46: 492–522.

Rossner Jr, P., V. Svecova, J. Schmuczerova, et al. 2013b. Analysis of biomarkers in a Czech population exposed to heavy air pollution. Part I: Bulky DNA adducts. *Mutagenesis* 28: 89–95.
Rossner Jr, P., E. Tulupova, A. Rossnerova, et al. 2015. Reduced gene expression levels after chronic exposure to high concentrations of air pollutants. *Mutat Res* 780: 60–70.
Rossnerova, A., M. Pokorna, V. Svecova, et al. 2017. Adaptation of the human population to the environment: Current knowledge, clues from Czech cytogenetic and "omics" biomonitoring studies and possible mechanisms. *Mutat Res* 773: 188–203.
Rossnerova, A., M. Spatova, C. Schunck, and R. J. Sram. 2011. Automated scoring of lymphocyte micronuclei by the MetaSystems Metafer image cytometry system and its application in studies of human mutagen sensitivity and biodosimetry of genotoxin exposure. *Mutagenesis* 26: 169–75.
Santella, R. M. 1999. Immunological methods for detection of carcinogen-DNA damage in humans. *Cancer Epidemiol Biomarkers Prev* 8: 733–9.
Schunck, C., T. Johannes, D. Varga, T. Lorch, and A. Plesch. 2004. New developments in automated cytogenetic imaging: Unattended scoring of dicentric chromosomes, micronuclei, single cell gel electrophoresis, and fluorescence signals. *Cytogenet Genome Res* 104: 383–89.
Sigurdson, A. J., M. Ha, M. Hauptmann, et al. 2008. International study of factors affecting human chromosome translocations. *Mutat Res* 652: 112–21.
Sorsa, M., J. Willbourn, and H. Vainio. 1992. Human cytogenetic damage as a protector of cancer risk. *IARC Sci Publ* 543–54.
Sram, J. R., V. Svecova, A. Rossnerova, et al. 2016. Systematic review of the use of the lymphocyte cytokinesis-block micronucleus assay to measure DNA damage induced by exposure to polycyclic aromatic hydrocarbons. *Mutat Res* 770: 162–9.
Tang, D., O. N. Kryvenko, Y. Wang, et al. 2013. Elevated polycyclic aromatic hydrocarbon-DNA adducts in benign prostate and risk of prostate cancer in African Americans. *Carcinogenesis* 34: 113–20.
Tang, D., D. H. Phillips, M. Stampfer, et al. 2001. Association between carcinogen-DNA adducts in white blood cells and lung cancer risk in the physicians health study. *Cancer Res* 61: 6708–12.
Tang, D., R. M. Santella, A. M. Blackwood, et al. 1995. A molecular epidemiological case-control study of lung cancer. *Cancer Epidemiol Biomarkers Prev* 4: 341–6.
Tates, A. D., M. T. van Welie, and J. S. Ploem. 1990. The present state of the automated micronucleus test for lymphocytes. *Int J Radiat Biol* 58: 813–25.
Tucker, J. D., J. W. Breneman, J. F. Briner, et al. 1997. Persistence of radiation-induced translocations in rat peripheral blood determined by chromosome painting. *Environ Mol Mutagen* 30: 264–72.
Tucker, J. D., W. F. Morgan, A. A. Awa, et al. 1995. A proposed system for scoring structural aberrations detected by chromosome painting. *Cytogenet Cell Genet* 68: 211–21.
Varga, D., T. Johannes, S. Jainta, et al. 2004. An automated scoring procedure for the micronucleus test by image analysis. *Mutagenesis* 19: 391–97.
Veglia, F., S. Loft, G. Matullo, et al. 2008. DNA adducts and cancer risk in prospective studies: A pooled analysis and a meta-analysis. *Carcinogenesis* 29: 932–6.
Watson, W. P., and A. Mutti. 2004. Role of biomarkers in monitoring exposures to chemicals: Present position, future prospects. *Biomarkers* 9: 211–42.
Yuan, J.-M., S. G. Carmella, R. Wang, et al. 2018. Relationship of the oxidative damage biomarker 8-epi-prostaglandin F2α to risk of lung cancer development in the Shanghai Cohort Study. *Carcinogenesis* 39: 948–54.
Yun, B. H., J. Guo, M. Bellamri, and R. J. Turesky. 2018. DNA adducts: Formation, biological effects, and new biospecimens for mass spectrometric measurements in humans. *Mass Spectrom Rev* 39: 55–82.
Zhong, Q., S. Amin, P. Lazarus, and T. E. Spratt. 2010. Differential repair of polycyclic aromatic hydrocarbon DNA adducts from an actively transcribed gene. *DNA Repair* 9: 1011–6.

6 Reproductive and Developmental Toxicology

François Spézia, Paul C. Barrow and Pramila Singh

CONTENTS

6.1 INTRODUCTION

Reproductive and developmental toxicology is a discipline dealing with methods aimed at identifying potential adverse effects of chemical, drug, biological or physical agents on human reproduction, including all steps that allow perpetuation of the species.

More specifically, developmental toxicology is the study of adverse effects on the developing organism that may result from exposure to xenobiotics prior to conception, during prenatal development, or during postnatal development up to the time of sexual maturity. The major

DOI: 10.1201/9780203730584-6

manifestations of developmental toxicity include intrauterine death (miscarriage), morphological abnormality, altered growth and functional deficiency.

The term "reproductive toxicology" is used inconsistently. It is sometimes used to describe the study of adverse effects on the reproductive systems of adult males or females, including alterations to male or female reproductive organs, the endocrine system or pregnancy outcomes (Kimmel et al. 2006). The ICH (International Council for Harmonisation) for the testing of pharmaceuticals uses the term to describe any effect on reproductive capacity in males or females in addition to any manifestation of developmental toxicity (ICH 2005). The most recent ICH guidance on reproductive and developmental toxicology emphasizes use of current principles in regulatory toxicology, including the qualification and use of alternative assays and integrated testing strategies, to refine the risk evaluation while reducing the overall use of animals.

For chemicals and pharmaceuticals, the aim of reproductive and developmental toxicology is to reveal any potential adverse effects on mammalian reproduction and/or embryo-fetal or postnatal health and development.

6.2 BASIC PRINCIPLES OF DEVELOPMENTAL TOXICOLOGY

The most appropriate testing strategy will consider any known pharmacological and toxicological activities of the test substance and the specific stages of development that may be impacted. Using this approach will ensure that all existing data are taken into account, including any investigative studies that may have been performed revealing the inherent activity of the compound as well as the known activities of similar substances. There are four recognized manifestations of prenatal toxicity: malformation, embryonic/fetal death, growth retardation, and functional deficiency (Schardein 2000; Hood and Parker 2007), which may occur individually or in combination.

6.2.1 Malformation

Teratology studies, termed embryo-fetal development (EFD) studies for pharmaceuticals or prenatal development studies for chemicals, are designed to detect any potential to induce congenital abnormalities. These are morphological defects (i.e. dysmorphogenesis) that can be detected at hysterectomy or birth and which result from injury during the intrauterine period.

Congenital abnormalities are changes relative to the normal state for the species and developmental stage. The normal state can change over time, so careful monitoring of in-house historical control data is necessary. Congenital abnormalities should be described according to the internationally harmonized terminology (Makris et al. 2009) and are typically classified as either malformations or variations (Chahoud et al. 1999):

- A *malformation* is a permanent structural change that is likely to adversely affect survival or health. A compound is classed as teratogenic if it has the potential to induce malformations under the conditions of use.
- A *variation* is a deviation from normal that is unlikely to adversely affect survival or health. Variations may include manifestations of delayed growth or an alternative morphological form.

The induction of a teratogenic response depends upon the administration of a specific agent at a particular dose to a genetically susceptible species when the embryo is at a susceptible stage of development (Karnofsky 1965).

- *Susceptible species*: Not all species are equally susceptible or sensitive to teratogenic compounds. Differences may be due to genetic, anatomic and/or metabolic factors. Variability of response is commonly observed, even between fetuses in the same litter.
- *Susceptible stage of development*: The nature and incidence of teratogenic effects depend on the affected developmental stage. The most highly susceptible stage of development to teratogenic insult is during organogenesis, which is a period of continuous organ development and which would be the most appropriate period for identifying substances with teratogenic potential. During the preimplantation period (i.e. blastogenesis), the conceptus is generally resistant to congenital malformations, but embryonic death or implantation failure may occur. During the late development stage (i.e. histogenesis), the fetus becomes progressively less susceptible to teratogenicity. Nonetheless, some organs, such as the brain and gonads, show significant morphological development during the fetal period and may still be susceptible to dysmorphogenesis in the later stage of gestation.

- *Dose dependency*: Teratogenic effects are expected to show a dose response. A "no adverse effect level" (NOEL) should be demonstrated for all known teratogenic agents. The severity of malformations and the incidence of fetuses affected increase with dose. At higher exposure levels, only resorptions may be seen. The current concept is that chemicals are considered to present a teratogenic risk only when they disrupt embryonic development at doses that are not toxic to the mother (Schardein 2000). Note, however, that many drugs of abuse, such as alcohol, would probably not be considered as teratogenic if this definition was strictly applied. Also, a recent review (Danielsson 2013) cited several examples where chemicals or pharmaceuticals were labeled as teratogens or developmental toxicants even though malformations were only observed at dose levels that induced maternal toxicity.

6.2.2 Growth Retardation

Intrauterine growth restriction (IUGR) occurs when fetuses do not develop at a normal rate and are much smaller than normal fetuses of the same gestational age. IUGR may occur as the result of maternal toxicity, embryo-fetal toxicity or a combination of the two.

6.2.3 Lethality

Intrauterine death is a manifestation of developmental toxicity. It may be due to a direct effect of the chemical on the embryo or fetus (which may be malformed or not) or as a consequence of maternal toxicity. As for IUGR, it is often difficult or impossible to confirm or dismiss a direct embryo/fetal effect at dose levels that are toxic for the mother.

6.2.4 Functional Deficiency

Birth defects can be of a structural or functional nature. Functional defects typically affect the organ systems that have a prolonged period of maturation (e.g. central nervous system [CNS], immune system and reproductive system). Some functional deficits may only be detectable when the affected individual reaches maturity. Structural or functional impairment of the CNS may result in behavioral or learning difficulties.

6.3 PHARMACEUTICALS

Nearly all health authorities worldwide apply the ICH S5 guideline for the testing of pharmaceuticals for reproductive and developmental toxicity. The first version of the guideline was issued in 1993. The current version, ICH S5 (R3 was issued in February 2020 and became available in July 2020 (ICH 2020; Wise 2013).

ICH S5 (R3) applies to pharmaceuticals, including biotechnology-derived pharmaceuticals, preventative vaccines for infectious diseases and novel ingredients or excipients that are part of a final pharmaceutical product. The guideline gives recommendations on the use of information concerning the pharmaceutical and the patient population, so that only studies that are essential for evaluation of the stages where current knowledge about the risk to reproduction and development is insufficient are performed. Observations through a complete life cycle, from conception in one generation through to conception in the following generation, allow the detection of immediate and latent adverse effects. The following reproduction stages are usually assessed:

A. Premating to conception (adult male and female reproductive functions, development and maturation of gametes, mating behavior, fertilization).

B Conception to implantation (adult female reproductive functions, preimplantation development, implantation).

C Implantation to closure of the hard palate (adult female reproductive functions, embryonic development, major organ formation).

D Closure of the hard palate to the end of pregnancy (adult female reproductive functions, fetal development and growth, organ development and growth).

E Birth to weaning (adult female reproductive functions, neonate adaptation to extrauterine life, preweaning development and growth).

F Weaning to sexual maturity (postweaning development and growth, adaptation to independent life, attainment of full sexual function).

6.3.1 Three Separate Studies Design

The testing evaluation is most often made using three protocols (referred to as the *most probable option/approach* or the *three separate studies* design), which subdivide the above-mentioned stages for the rodent into three study types as follows:

- *Fertility and Early Embryo Development (FEED)*: If effects on fertility are suspected, based on mode of action or on the results of repeat-dose studies, it can be advisable to treat males and females in separate arms of a study or in separate studies where treated animals are mated with untreated animals (stages A and B).
- *Embryo-Fetal Development (EFD)* in a rodent and nonrodent species (stage C).
- *Pre- and Postnatal Development, including maternal function (PPND)* (stages D, E and F).

6.3.2 Single Study Design

A combination single study including fertility, gestation and postnatal development (stages A–F) in rodents can be appropriate if reproductive toxicity is not anticipated. If the results of such a study are clearly negative at appropriate dose levels, no further reproduction studies are warranted in the species. In this study design, all newborns, pups, stillbirths and culled pups are examined for morphological abnormalities. If reproductive or developmental toxicity is observed, the observed effects are assessed in detail. Although specified in the guideline, this study design is almost never used in practice, due to the earlier requirement for the FEED and EFD studies (in support of clinical trials) compared to the timing of the PPND study (only required in support of marketing applications; ICH 2009).

6.3.3 Two Studies Design

6.3.3.1 Combined FEED and EFD Study (Stages A–D) Plus a PPND Study (Stages C–F)

This combination of the FEED and EFD investigations in one study provides all the same information as in separate studies but requires fewer animals.

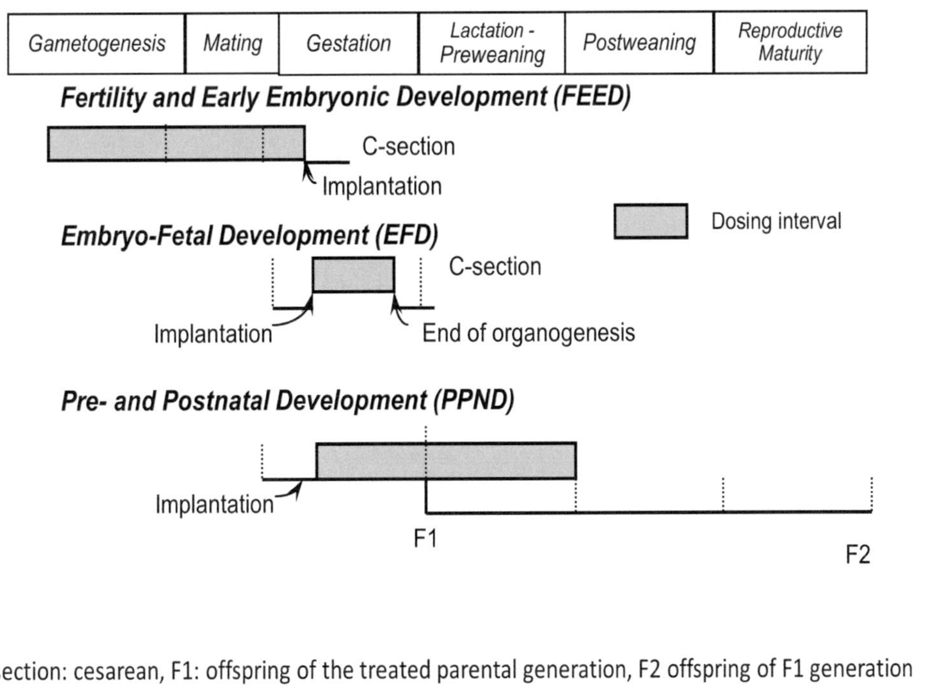

Figure 6.1

6.3.3.2 EFD Study (Stages C–D) Plus a Combined FEED and PPND Study (Stages A–C + D–F)

No assessment of external, soft tissue or skeletal morphology is included in the combined FEED and PPND study, so it could be useful when no treatment-related effects have been observed in the EFD study. Like the single study design, this study design is almost never used in practice.

6.3.4 Study of Fertility and Early Embryonic Development to Implantation (FEED)

6.3.4.1 Study Objectives

The aim of the FEED study is to determine whether there are disturbances in male and/or female fertility and early embryonic development resulting from treatment of young, sexually mature adults before mating, through mating and up to implantation. Stages A and B of the reproductive process are evaluated (see Section 6.2). FEED studies are usually performed in rodents and more rarely in rabbits. Mating studies are not usually performed in nonrodents such as dogs and nonhuman primates (NHPs). Health and safety concerns have increased interest in evaluating the potential effects of pharmaceutical and chemical agents on sperm cycle and motility as well as female estrous cyclicity, both of which are influenced by induced changes in hormone levels or disruption of hormone synthesis.

A dedicated fertility study includes a mating phase for the detection of effects that cannot be assessed by histopathology of the reproductive organs. But if the drug has clinically important adverse effects on male or female reproductive organs in a repeat-dose toxicity study, a fertility study in the affected sex may be unwarranted. A fertility study is also generally unwarranted for pharmaceuticals not destined for use in patients of reproductive age. The repeated-dose toxicity study results can generally be used to help design a fertility study without requiring further dose range-finding studies. It is also possible to include fertility evaluations such as spermatograms and healthy estrous cyclicity within a repeat-dose toxicity study, as long as the animals have reached sexual maturity.

Male and female rodents can be evaluated in the same fertility study when no adverse effects on fertility are expected. However, when effects on fertility are identified in a combined male and female study, it is necessary to determine the extent to which each sex is affected.

If NHPs are the only pharmacologically relevant species, as is the case for many monoclonal antibodies, the evaluation of fertility is based on the results of repeat-dose toxicity studies. Most of the effects on male and female fertility can be detected at histopathological examination of the reproductive organs, provided the animals are sexually mature. If the repeat-dose studies are intended to be used for this purpose, the males and females should be sexually mature from the start of the study, which is typically not common practice for repeat-dose studies in dogs, minipigs and NHPs. If effects on fertility are assessed in repeat-dose toxicity studies, a comprehensive histopathological examination should be performed on reproductive organs from both males and females. When there is concern over the mode of action or effects in previous study data, additional investigations of the reproductive organs can be included in the repeat-dose toxicity studies in addition to the routine histopathology examinations (e.g. sperm collection or monitoring of the estrous or menstrual cycle). Two- to four-week studies can provide an initial evaluation of the effects on reproductive organs. This information can be supplemented later with similar evaluations in subchronic and chronic toxicity studies.

6.3.4.2 Male-Only, Female-Only and Combined Male and Female FEED Study

In most FEED studies (see Figure 6.2), males and females are treated and then paired. If potential paternally mediated or maternally mediated effects are anticipated, FEED studies can be performed by treating one sex only and pairing with untreated animals. The males are euthanatized only after hysterectomies have been performed on most of the females, thus allowing the possibility of re-mating the males with untreated females if the initial results indicate that this is advisable.

Males are treated for 2, 4 or 9/10 weeks before mating and then until termination. Males are treated for at least 9 weeks overall, even in the case of a 2- or 4-week premating period, to ensure exposure throughout spermatogenesis (which takes about 63 days in rats).

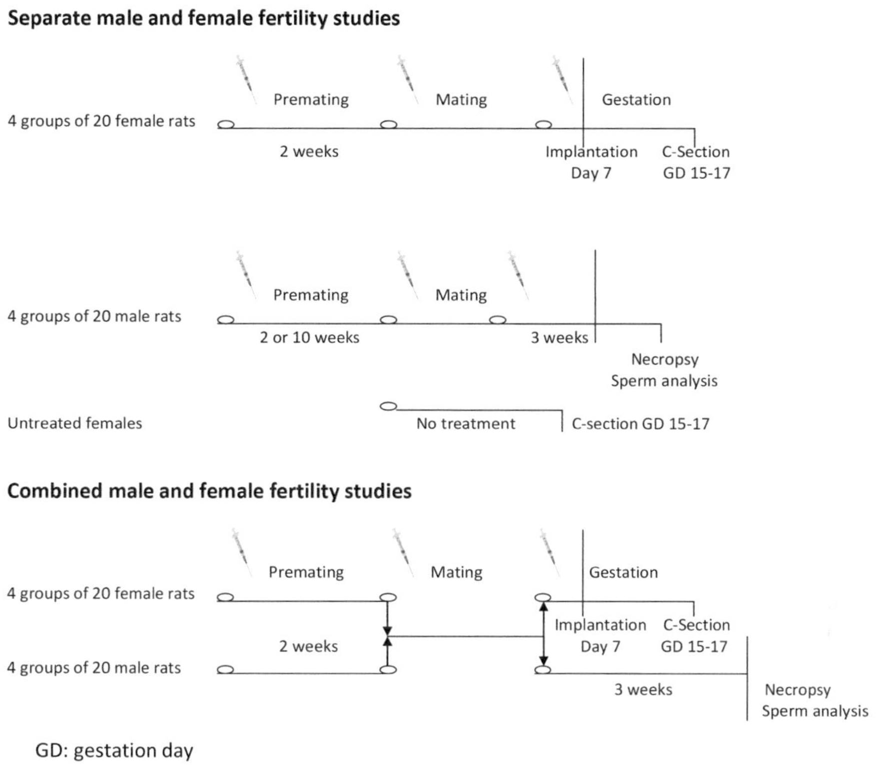

Figure 6.2

6.3.5 Pre- and Postnatal Development

6.3.5.1 Study Objectives

The aim of a pre- and postnatal developmental (PPND) toxicity study is to detect adverse effects on the pregnant or lactating female and on development of the conceptus and offspring (referred to as the F1 generation) up to maturity. Females are treated from implantation through to weaning. The manifestation of effects induced during this period can be delayed, so observations are continued up to sexual maturity (i.e. stages C to F, inclusive, of the reproductive process). PPND studies are designed to assess toxicity in pregnant females, pre- and postnatal mortality, altered growth/development and functional deficits in offspring. Examinations in the offspring include sexual maturation (puberty), reproductive capacity at maturity, sensory functions, motor activity, learning ability and memory.

Females deliver and rear their offspring until weaning. At this time, at least one male and one female pup per litter are selected for rearing to adulthood. Then the mature offspring are mated to assess reproductive competence.

6.3.5.2 Modification of Rodent PPND Study to Assess Juvenile Toxicity Endpoints

When a juvenile animal study is warranted to address a specific area of concern, a PPND study can be modified to add juvenile toxicity endpoints, which can potentially avoid the need for a dedicated juvenile study (see Chapter 7). The following should be taken into consideration in the study design:

- In order to support the pediatric therapeutic application, the appropriate exposure period of the juvenile rodents should be determined.
- In order to demonstrate adequate exposure in the pups, either samples of treated dams' milk can be collected and/or over the period of interest, toxicokinetic sampling can provide exposure data that may help to avoid direct administration of the test item to the pups during the preweaning period (by gavage).

Endpoints in a modified PPND study should be chosen according to the principles appropriate for juvenile animal study designs in support of human pediatric use (see Chapter 7).

6.3.5.3 Enhanced Pre- and Postnatal Developmental Toxicity Study (ePPND) in NHPs

The ePPND study is the only reproductive and developmental toxicity study performed for most monoclonal antibodies that only exert their intended pharmacological activity in the human and NHP, as defined in the ICH M3(R1) guideline (ICH 2009). The ePPND combines endpoints from EFD and PPND studies. The dams are treated throughout the gestation period until parturition (i.e. GD20 to parturition). Monoclonal antibodies that cross-react in rodents and/or rabbits are tested in those species, avoiding NHP studies whenever possible.

6.3.6 EFD Toxicity Study

The aim of the embryo-fetal developmental (EFD) toxicity study is to detect any adverse effects on pregnant females and the development of the embryo and fetus following exposure of the dam from implantation until closure of the hard palate (stages C to D of the reproductive process). This study is designed to assess enhanced maternal toxicity versus nonpregnant females, embryo-fetal death, altered growth and structural changes. EFD studies are most often performed in rats (Leroy and Allais 2013) and rabbits (Allais and Reynaud 2013), although mice (Marsden and Leroy 2013) are also used.

6.3.6.1 DRF Study

Dose range-finding (DRF) studies in mated females are used to select appropriate dose levels or dose schedules for the definitive EFD studies, but tolerability and TK data from existing repeat-dose toxicity can be sufficient for this purpose.

6.3.6.2 pEFD Study

The preliminary embryo-fetal developmental (pEFD) toxicity study is similar in design to the definitive embryo-fetal developmental toxicity study but uses fewer animals. Such studies can be used to support clinical trials in Europe and Japan, including a limited number of women of child-bearing potential for a short period of treatment. A typical pEFD study design includes dosing over the period of organogenesis, has adequate dose levels covering the range of expected human exposure (including a safety margin, i.e. the ratio between the toxic and pharmacologically active dose), evaluates a minimum of six pregnant females per group, and includes assessments of fetal survival and weight as well as external and soft tissue examinations.

6.3.6.3 Definitive Embryo-Fetal Developmental Toxicity Study

The pregnant females are submitted to cesarean section near term for assessments of fetal survival and weight as well as external, soft tissue (Critchell 2013; Leroy and Jocteur-Monrozier 2013) and skeletal (Reynaud and Jocteur-Monrozier 2013) examinations.

6.3.7 Combination Studies

6.3.7.1 Fertility and Embryo-Fetal Development (FEFD)

The objective of the combined FEFD study is to test for toxic effects/disturbances that result from treatment before mating (males/females) through mating and implantation, until the end of organ-ogenesis (stages A to C of the reproductive process; Reynaud and Marsden 2013). A combined

Table 6.1 Embryo-Fetal Developmental Toxicity Studies

Species	Group size	Treatment period	Hysterectomy	Fetal examinations
Mouse	4 groups: more than 24 mated females/group	Day 6 to 15 p.c.	Day 18 p.c.	100% external, 50% visceral and 50% skeletal
Rat	4 groups 20 mated females/group	Day 6 to 15 p.c.	Day 29 p.c.	100% external, 50% visceral and 50% skeletal
Rabbit	4 groups 20 mated females/group	Day 6/7 to 18/19 p.c.	Day 29 p.c.	100% external, visceral and skeletal

Abbreviation: p.c. = post-coitum.

male/female FEFD is most commonly used, but a female-only option is possible. In this case male fertility is assessed in a separate study of suitable duration (e.g. repeat-dose toxicity study). For the specific study designs, see Sections 6.3.4.2 and 6.3.6.3 (FEED and EFD).

6.4 CHEMICALS

The testing and registration of chemicals is controlled by national laws and regional interpretations, most of which are based on the international test guidelines of the Organisation for Economic Co-operation and Development (OECD). In this section, the term "chemical" is used to refer to a broad category of substances including agrochemicals, industrial chemicals and food additives that are subject to OECD-based testing requirements to obtain authorizations for market registration.

The OECD, an intergovernmental economic organization with 36 member countries, is committed to sharing good practices and coordinating the domestic and international policies of its members.

OECD governments have comprehensive regulatory frameworks for preventing and/or minimizing health and environmental risks linked to chemicals. The frameworks ensure that marketed chemical products are handled safely and that new chemicals are correctly assessed before being marketed. The different national chemical control policies can, however, lead to the duplication of testing and government assessments. Implementation of the Mutual Acceptance of Data system—under which chemical safety data developed in one member country using the OECD Test Guidelines and OECD Principles of Good Laboratory Practices must be accepted by all member countries—is helping to minimize unnecessary divergences across regulatory frameworks and to make work-sharing easier for governments.

The OECD Guidelines for the Testing of Chemicals is a collection of around 150 of the most relevant internationally agreed testing methods that are used by governments, industry and independent laboratories to identify and characterize potentially dangerous chemicals. They are used to select and rank candidate chemicals in toxicology research and during the development of new chemicals and products.

The following OECD guidelines are of crucial importance for the assessment of potential reproductive and developmental toxicities:

- Prenatal Developmental Toxicity Study (OECD 414, 2018) (OECD 2018a)
- Two-Generation Reproduction Toxicity Study (OECD 416, 2001) (OECD 2001)
- Reproduction/Developmental Toxicity Screening Test (OECD 421, 2018) (OECD 2016)
- Combined Repeated Dose Toxicity Study with Reproduction/Developmental Toxicity Screening Test (OECD 422, 2018) (OECD 2018b)
- Developmental Neurotoxicity Study (OECD 426, 2007) (OECD 2007a)
- Uterotrophic Bioassay in Rodents (OECD 440, 2007) (OECD 2007b)
- Hershberger Bioassay in Rats (OECD 441, 2009) (OECD 2009)
- Extended One-Generation Reproductive Toxicity Study (OECD 443, 2018) (OECD 2018c)
- Testing for Endocrine Disruption (OECD 150 guidance document 2018) (OECD 2018d)

6.4.1 Prenatal Developmental Toxicity Study (OECD 414, 2018)

The latest version of the OECD 414 test Guideline for Developmental Testing was adopted on June 25, 2018. The guideline is designed to provide information about the effects of prenatal exposure on the pregnant test animal and the developing conceptus. The test substance is administered to pregnant animals from implantation until the day before scheduled euthanasia (close to the normal day of delivery). Rodents (preferably rats) and nonrodents (preferably rabbits) are tested, and each test treatment and control group should contain a sufficient number of females to obtain 20 animals with uterine implantation sites at necropsy. At least three dose levels should be tested, and the test compound or vehicle is usually administered by oral gavage (esophageal administration directly into the stomach via a tube) but may also be administered in the diet or drinking water. A single dose level of 1000 mg/kg/day (referred to as a limit test) may be used if expected to result in no observable toxicity. The animals are weighed regularly and daily clinical observations are made. The females are euthanized 1 day prior to the expected day of delivery and a cesarean section is performed. The uterine contents are examined and the fetuses are evaluated for soft tissue (Critchell 2013; Leroy and Jocteur-Monrozier 2013) and skeletal (Schneider 2013) changes. Certain endocrine-related measurements in dams and fetuses were added in 2018. If no effects indicative of toxicity are noted, further investigations should be considered to investigate test substance absorption and bioavailability before ruling out toxicological potential.

6.4.2 Two-Generation Reproduction Toxicity Study (OECD 416, 2001)

The latest version of the OECD 416 test Guideline for Two-Generation Reproduction Toxicity testing was adopted on January 22, 2001. The guideline is designed to provide information on the effects of the test substance on the integrity and performance of male and female reproductive systems, and on offspring growth and development. The OCED 416 study design has been updated to include estrous cycling, measurement of anogenital distance and the presences of nipples. The test substance is administered daily to at least three groups of males and females at graduated dose levels, plus a concurrent control group, by the oral route (diet, drinking water or gavage). Parent generation males and females (5–9 weeks old) are treated until and during mating, during the resulting pregnancies and through the weaning period of their offspring. Administration is continued to the first-generation offspring during their growth to adulthood, then through mating, production and weaning of a second generation. The rat is the preferred species. Each test and control group should contain sufficient animals to yield no fewer than 20 pregnant females. A limit test may be performed if no effects are expected at 1000 mg/kg/day. Study results include the following: body weight, sperm parameters, estrus cycle parameters, offspring parameters, daily clinical observations, necropsy and histopathology. The findings of this two-generation reproduction toxicity study are evaluated in terms of the observed effects, including necropsy and microscopic findings. A correctly conducted study should provide a satisfactory estimation of the no-effect level and an understanding of any adverse effects on reproduction, parturition, lactation and postnatal development (including growth and sexual development).

6.4.3 Reproduction/Developmental Toxicity Screening Test (OECD 421, 2018)

The latest version of the OECD 421 test Guideline for Reproduction/Developmental Toxicity Screening was adopted on July 29, 2016. This guideline concerns the effects of test chemicals on male and female reproductive performance. Endocrine disruptor endpoints were subsequently added, in particular the measurement of anogenital distance in pups, male nipple retention in pups and thyroid evaluations. The test substance is administered daily by the oral route to at least three test groups at graduated dose levels and to a control group of males and females (with at least 10 animals per sex/group). Males are treated for at least 4 weeks. Females are treated throughout the study (i.e. for approximately 63 days). The guideline is designed for use with rats. The dose levels may be chosen according to information obtained in acute toxicity or repeated dose studies. Study results include clinical observations, body weight, food/water consumption, estrous cycle monitoring, offspring observations and measurements, thyroid hormone measurement, necropsy and histopathology. The findings should be evaluated according to the observed effects and the necropsy and microscopic findings. As the male treatment period is relatively short, testis and epididymis histopathology should be considered along with fertility data to assess male reproductive effects.

6.4.4 Combined Repeated Dose Toxicity Study with Reproduction/ Developmental Toxicity Screening Test (OECD 422, 2018)

The latest version of the OECD 422 test Guideline for the Combined Repeated Dose Toxicity Study with Reproduction/Developmental Toxicity Screening was adopted on July 29, 2016. This guideline combines reproduction/developmental toxicity screening with repeated dose toxicity (including histopathological endpoints) in concordance with Test Guideline 407.

6.4.5 Uterotrophic Bioassay in Rodents (OECD 440, 2007)

The latest version of the OECD 440 test Guideline was adopted on October 16, 2007. The Uterotrophic Bioassay in Rodents is a short-term screening test for estrogenic properties, based on an induced increase in uterine weight (uterotrophic response). The sensitivity of the uterotrophic bioassay relies on a nonfunctional hypothalamic-pituitary-ovarian axis in the female animal test system during two estrogen-sensitive states in female rodents: (1) immature females after weaning and prior to puberty and (2) young adult females after ovariectomy (after adequate time for uterine tissue regression). The test substance is administered daily by the oral route or by subcutaneous injection. Each treated and control group should include at least six females. The test substance is administered to at least two treatment groups at graduated dose levels for 3 consecutive days for immature females or for a minimum of 3 consecutive days for ovariectomized adults. The animals are euthanized approximately 24 hours after the last administration. An increase in uterine weight in the treated group compared with the vehicle group is indicative of estrogenic activity in this bioassay. Study reporting should include daily body weights, daily health status, wet and blotted uterine weights, and daily food consumption.

6.4.6 Hershberger Bioassay in Rats (OECD 441, 2009)

The latest version of the OECD 441 test Guideline was adopted on September 7, 2009. The Hershberger Bioassay is a short-term screening assay for the determination of (anti)androgenic properties. The ability of a chemical to elicit biological activities consistent with androgen agonists, antagonists or 5 α-reductase inhibitors is evaluated, based on weight changes in five androgen-dependent tissues in the castrated peripubertal male rat: ventral prostate, seminal vesicle (plus fluids and coagulating glands), levator ani-bulbocavernosus muscle, paired Cowper's glands and glans penis. In order to detect androgenic or anti-androgenic action, two or three test substance groups, plus positive and vehicle (negative) controls with at least six animals per group are normally sufficient. The test substance is administered daily by gavage or subcutaneous injection for 10 consecutive days. To test for anti-androgens, the test substance is administered along with a reference androgen agonist. The animals are euthanized approximately 24 hours after the last treatment. Tissues are excised and weighed (fresh). A statistically significant increase (androgenic) or decrease (anti-androgenic) in 2/5 tissue weights indicates a positive response.

6.4.7 Extended One-Generation Reproductive Toxicity Study (OECD 443, 2018)

The latest version of the OECD 443 test Guideline was adopted on June 25, 2018. The Extended One-Generation Reproductive Toxicity Study is used for the evaluation of reproductive and developmental effects that may occur as a result of pre- and postnatal chemical exposure. Systemic toxicity in pregnant and lactating females or in young and adult offspring is also evaluated. Sexually mature male and female rodents (parental [P] generation) are exposed to graduated dose levels of the test substance. Treatment of the P generation starts 2 weeks before mating and continues through mating and gestation until weaning of the F1 generation pups. Pups are assigned to one of three cohorts (but cohorts 2 and 3 may be requested or may not be included). Cohort 1 is used for reproductive/developmental investigations, cohort 2 is used for developmental neurotoxicity testing and cohort 3 serves for developmental immunotoxicity testing. The F1 generation receives the test compound from weaning until adulthood. Clinical observations and pathology examinations are performed on all animals to detect any signs indicative of toxicity; special emphasis is placed on the integrity and performance of male and female reproductive systems and the offspring's health, growth, and development. Part of cohort 1 (cohort 1B) may be extended to include an F2 generation, in which case the procedures for F1 animals will be similar to those of the P generation.

6.4.8 Testing for Endocrine Disruption (OECD 150 Guidance Document, 2018)

The OECD 150 guideline (OECD 2018d) was originally published in 2012 and updated in 2018 (OECD 2018d) to reflect other newer OECD test guidelines and scientific progress in the use of test methods to assess potential endocrine disruption activity. The document provides guidance for the evaluation of chemicals using standardized testing together with newer methods as well as proposed methods still under development. The document presents the OECD Conceptual Framework for evaluating chemicals for endocrine disruption, organizing all available assays and methods into five levels of testing, from physicochemical characterization to *in vitro* and *in vivo* assays in both mammalian and nonmammalian testing systems. This Conceptual Framework integrates OECD assays with the endocrine screening tests implemented by the US Environmental Protection Agency (EPA) under the Endocrine Disruptor Screening Program (EPA 1998a, 1998b), providing a useful reference for international harmonization.

A case-by-case approach to testing is encouraged based on the known properties and biological activities of each test chemical. The document also provides guidance on how test results can be interpreted, based on assay outcomes. Some of the key questions addressed concern the possible mechanisms of endocrine action and the possible apical effects or final toxicological outcomes that can result from such actions. The guideline is not prescriptive but offers some suggestions for the possible steps in testing, and it focuses primarily on endocrine modalities including estrogen-, androgen- and thyroid-mediated endocrine disruption and chemicals that interfere with steroidogenesis. OECD 150 is considered as a living document that is expected to undergo revisions and updates in accordance with the latest developments in this area of testing. One recent OECD document that is complementary to the Conceptual Framework on endocrine disruption testing is the New Scoping Document on In Vitro and Ex Vivo Assays for the Identification of Modulators of Thyroid Hormone Signaling (OECD 2017). The aim of this scoping document is to highlight the relevant *in vitro* and *ex vivo* thyroid assays that could be used and further developed by OECD member countries in case of gaps that are not yet covered in thyroid signaling pathways by existing tests.

6.5 ALTERNATIVE TESTING METHODS

The term "alternative testing methods" has been used to refer to the *in vitro, ex vivo* and nonmammalian *in vivo* assays that can be used for reproductive and developmental toxicity testing. With the emphasis now being on integrated testing strategies for both pharmaceuticals and chemicals, there is a need for new approaches and alternative testing methods that specifically address the relevant knowledge gaps for a given substance, taking into account all the data already available and the intended use. Both ICH S5(R3) and OCED 150 discuss the use of alternative methods and give guidance on appropriate ways to qualify new alternative assays for reproductive and developmental toxicology risk characterization and/or regulatory acceptance for use in risk assessments.

BIBLIOGRAPHY

Allais, L., and L. Reynaud. 2013. Teratology studies in the rabbit. In *Teratogenicity Testing: Methods and Protocols*, ed. P. C. Barrow, 139–56. Totowa, NJ: Humana Press.

Chahoud, I., J. Buschmann, R. Clark, et al. 1999. Classification terms in developmental toxicology: Need for harmonisation 1. *Reprod Toxicol* 13(1): 77–82.

Critchell, K. 2013. Fetal soft tissue examination by serial sectioning. In *Teratogenicity Testing: Methods and Protocols*, ed. P. C. Barrow, 233–42. Totowa, NJ: Humana Press.

Danielsson, B. R. 2013. Maternal toxicity. In *Teratogenicity Testing: Methods and Protocols*, ed. P. C. Barrow, 311–25. Totowa, NJ: Humana Press.

EPA. 1998a. Endocrine disruptor screening program. *Federal Register* 63(154): 42852–5.

EPA. 1998b. Endocrine disruptor screening program; Proposed statement of policy. *Federal Register* 63(248): 71542–68.

Hood, R. D., and R. M. Parker. 2007. Reproductive and developmental toxicology. In *Preclinical Development Handbook*, ed. S. C. Gad, 363–422. Hoboken: John Wiley & Sons, Inc.

ICH. 2005. *Harmonised Tripartite Guideline: Detection of Toxicity to Reproduction for Medicinal Products & Toxicity to Male Fertility S5(R2).* www.ich.org/fileadmin/Public_Web_Site/ICH_Products/Guidelines/Safety/S5/Step4/S5_R2__Guideline.pdf (accessed December 6, 2018).

ICH. 2009. *Guideline M3(R2) on Non-Clinical Safety Studies for the Conduct of Human Clinical Trials and Marketing Authorisation for Pharmaceuticals.* www.ich.org/fileadmin/Public_Web_Site/ICH_Products/Guidelines/Multidisciplinary/M3_R2/Step4/M3_R2__Guideline.pdf (accessed December 7, 2018).

ICH. 2015. *Final Concept Paper S5(R3): Detection of Toxicity to Reproduction for Human Pharmaceuticals dated 9 February 2015.* Endorsed by the ICH Steering Committee on 27 March 2015. www.ich.org/fileadmin/Public_Web_Site/ICH_Products/Guidelines/Safety/S5/S5_R3__Final_Concept_Paper_27Mar2015.pdf (accessed December 6, 2018).

ICH. 2020 Guideline on reproductive toxicology: Detection of toxicity to reproduction for human pharmaceuticals - step 5. https://www.ema.europa.eu/en/ich-s5-r3-guideline-reproductive-toxicology-detection-toxicity-reproduction-human-pharmaceuticals

Karnofsky, D. A. 1965. Drugs as teratogens in animals and man. *Annu Rev Pharmacol* 5: 447–72.

Kimmel, C. A., G. L. Kimmel, and S. Y. Euling. 2006. Developmental and reproductive toxicity risk assessment for environmental agents. In *Developmental and Reproductive Toxicology, a Practical Approach.* Boca Raton, FL: CRC Press.

Leroy, M., and L. Allais. 2013. Teratology studies in the rat. In *Teratogenicity Testing: Methods and Protocols*, ed. P. Barrow, 95–109. Totowa, NJ: Humana Press.

Leroy, M., and A. Jocteur-Monrozier. 2013. Fetal soft tissue examinations by microdissection. In *Teratogenicity Testing: Methods and Protocols*, ed. P. C. Barrow, 243–53. Totowa, NJ: Humana Press.

Makris, S. L., H. M. Solomon, R. Clark, et al. 2009. Terminology of developmental abnormalities in common laboratory mammals (version 2). *Reprod Toxicol* 28(3): 371–34.

Marsden, E., and M. Leroy. 2013. Teratology studies in the mouse. In *Teratogenicity Testing: Methods and Protocols*, ed. P. Barrow, 111–23. Totowa, NJ: Humana Press.

OECD. 1983. *OECD Guideline for Testing of Chemicals. OECD 415. One-Generation Reproduction Toxicity Study*. www.oecd-ilibrary.org/docserver/9789264070844-en.pdf?expires=1544181741&id=id&accname=guest&checksum=E270722AA6F3CB2E3D465CC8FC3B2B22 (accessed October 22, 2020).

OECD. 2001. *OECD Guideline for Testing of Chemicals. OECD 416. Two-Generation Reproduction Toxicity Study*. https://read.oecd-ilibrary.org/environment/test-no-416-two-generation-reproduction-toxicity_9789264070868-en#page1 (accessed October 22, 2020).

OECD. 2007a. *OECD Guideline for Testing of Chemicals. OECD 426. Developmental Neurotoxicity Study*. https://read.oecd-ilibrary.org/environment/test-no-426-developmental-neurotoxicity-study_9789264067394-en#page1 (accessed October 22, 2020).

OECD. 2007b. *OECD Guideline for Testing of Chemicals. OECD 440. Uterotrophic Bioassay in Rodents*. https://read.oecd-ilibrary.org/environment/test-no-440-uterotrophic-bioassay-in-rodents_9789264067417-en#page1 (accessed October 22, 2020).

OECD. 2009. *OECD Guideline for Testing of Chemicals. OECD 441. Hershberger Bioassay in Rats*. https://read.oecd-ilibrary.org/environment/test-no-441-hershberger-bioassay-in-rats_9789264076334-en#page1 (accessed October 22, 2020).

OECD. 2016. *OECD Guideline for Testing of Chemicals. OECD 421. Reproduction/Developmental Toxicity Screening Test*. https://read.oecd-ilibrary.org/environment/test-no-421-reproduction-developmental-toxicity-screening-test_9789264264380-en#page1 (accessed October 22, 2020).

OECD. 2017. *New Scoping Document on in vitro and ex vivo Assays for the Identification of Modulators of Thyroid Hormone Signalling*. www.oecd-ilibrary.org/content/publication/9789264274716-en (accessed October 22, 2020).

OECD. 2018a. *OECD Guideline for Testing of Chemicals. OECD 414. Prenatal Developmental Toxicity Study*. https://read.oecd-ilibrary.org/environment/test-no-414-prenatal-development-toxicity-study_9789264070820-en#page1 (accessed October 22, 2020.

OECD. 2018b. *OECD Guideline for Testing of Chemicals. OECD 422. Combined Repeated Dose Toxicity Study with Reproduction/Developmental Toxicity Screening Test*. https://read.oecd-ilibrary.org/environment/test-no-422-combined-repeated-dose-toxicity-study-with-the-reproduction-developmental-toxicity-screening-test_9789264264403-en#page1 (accessed October 22, 2020).

OECD. 2018c. *OECD Guideline for Testing of Chemicals. OECD 443. Extended One-Generation Reproductive Toxicity Study*. https://read.oecd-ilibrary.org/environment/test-no-443-extended-one-generation-reproductive-toxicity-study_9789264185371-en#page1 (accessed October 22, 2020).

OECD. 2018d. *Revised Guidance Document 150 on Standardised Test Guidelines for Evaluating Chemicals for Endocrine Disruption*. www.oecd.org/publications/guidance-document-on-standardised-test-guidelines-for-evaluating-chemicals-for-endocrine-disruption-2nd-edition-9789264304741-en.htm (accessed October 22, 2020).

OECD. *OECD Test Guidelines for the Chemicals*. www.oecd.org/chemicalsafety/testing/oecdguidelinesforthetestingofchemicals.htm (accessed October 22, 2020).

Reynaud, L., and A. Jocteur-Monrozier. 2013. Skeletal examination by alizarin staining. In *Teratogenicity Testing: Methods and Protocols*, ed. P. C. Barrow, 201–13. Totowa, NJ: Humana Press.

Reynaud, L., and E. Marsden. 2013. Combined fertility and embryotoxicity study. *Methods Mol Biol* 947: 125–37.

Schardein, J. 2000. *Chemically Induced Birth Defects*. Boca Raton, FL: CRC Press.

Schneider, S. 2013. Skeletal examination by double staining for ossified bone and cartilaginous tissue. In *Teratogenicity Testing: Methods and Protocols*, ed. P. C. Barrow, 215–21. Totowa, NJ: Humana Press.

Wise, L. D. 2013. The ICH S5(R2) guideline for the testing of medicinal agents. In *Teratogenicity Testing: Methods and Protocols*, ed. P. Barrow, 1–11. Totowa, NJ: Humana Press.

7 Juvenile Toxicology

Paul C. Barrow, Pramila Singh and François Spézia

CONTENTS

7.1 REGULATORY CONTEXT

A pediatric assessment is required for all new pharmaceutical marketing applications in most regions of the world unless a waiver has been requested and granted. The critical issues in clinical pediatric drug development are outlined in the recently revised ICH E11 guideline (ICH 2017). The guideline aims to encourage and facilitate the timely development of safe, efficient therapies for use in the pediatric population. Juvenile animal studies (JAS) are performed only when they can provide useful information to fill important gaps in the knowledge (between safety in the adult and safety in the child). Often, the data available from clinical trials in adults and from safety studies in adult animals are sufficient for this purpose. When JAS are warranted, the studies in juvenile laboratory animals (usually rodents) are designed to ensure the safety of pediatric patients who will be exposed to the drug (Barrow and Schmitt 2017).

Advice on the need for JAS and on the design of the studies is given in a recent ICH guideline titled "Nonclinical Safety Testing in Support of the Development of Pediatric Medicines—S11" (ICH 2020). The ICH S11 guideline replaced previous guidance documents applicable in the United States (FDA 2006), Europe (EMA 2013) and Japan (MHLW Japan). The ICH M3(R2) guideline (ICH 2009) gives advice on the timing of JAS with respect to clinical trials. Additional considerations on the need for JAS for biopharmaceuticals and for anticancer agents are discussed in the ICH S6(R1) (ICH 2011) and S9 (ICH 2010) guidelines, respectively. The ICH S11 guideline recommends a customized approach to JAS, with fewer standardized study designs than specified in the previous regional guidelines.

The need for a nonclinical evaluation in juvenile animals should be discussed with the regulatory authorities before starting any studies. Likewise, it is advisable to request regulatory feedback on JAS study designs. The choice of species and the design of juvenile studies are based on a series of complex considerations, including the therapeutic use of the drug, the age at which pediatric patients will be treated, the duration of treatment and the potential age- or species-specific differences in pharmacokinetics or toxicity (Barrow et al. 2011). The ICH S11 guideline (ICH 2020) includes guidance on the development of safe pediatric therapies, preparation for conducting clinical trials in pediatric populations and support of the use and implementation of the 3R principles (replacement, reduction and refinement) in animal research.

DOI: 10.1201/9780203730584-7

7.2 POSSIBLE INCREASED SUSCEPTIBILITY OF PEDIATRIC PATIENTS

The efficacy and safety of a pharmaceutical may be different in pediatric patients and in adults. Many organs undergo growth, development or functional maturation after birth. The immaturity of these organs may render them more susceptible to toxicity. Differences in body composition between the juvenile and the adult, differences in drug metabolizing enzyme activities and incomplete activity of elimination mechanisms can also affect the pharmacokinetics of a drug. Differences in pharmacodynamics or efficacy may occur as the result of maturing receptor density or affinity. In addition, the biomedical community and the government oversight committees have become increasingly aware of the vulnerability among infants and children, who generally lack the ability to give consent or understand the potential risks of participation in clinical research (Presidential Committee for the study of Bioethical Issues 2016). A nonclinical juvenile strategy can only be devised with a complete understanding of the clinical pediatric study plan.

7.3 IS A JUVENILE ANIMAL STUDY NEEDED?

The clinical context is a crucial consideration. The possible need for a JAS is influenced by the intended clinical use of the drug, the intended patient age groups, the degree of maturity of the organs and tissues most affected, and the intended duration of dosing. Data from initial clinical trials in adults are often the most important information to consider when determining the need for JAS. An integrated assessment is performed following a weight-of-evidence (WOE) approach, which takes into account all available clinical data, knowledge of pharmacology, data on pharmacokinetics and the results of previous nonclinical studies. The following key factors influence the consideration of whether a JAS is needed or not (ICH 2020):

- Youngest intended patient age: A JAS is more likely to be needed for younger ages.
- Known or suspected adverse effects on developing organ systems: A JAS is more likely to be needed if effects on developing organs are suspected.
- The pharmacological target: A JAS is more likely to be needed if the pharmacological target has a role in organ development.
- Pharmacological selectivity: A JAS is more likely to be needed for compounds with broader off-target effects.
- Clinical treatment duration: A JAS is more likely to be needed for indications requiring long-term therapy.
- Amount/type of existing data: A JAS is most likely to be needed if no nonclinical or clinical data have been generated. The likelihood of a requirement for JAS is reduced if adult nonclinical data are available and further reduced if adult clinical data are available. A JAS is least likely to be needed if pediatric clinical data are available.

The first two items in the above list carry more weight in the WOE evaluation than the remaining four. A review of the literature can provide valuable information on the ontogeny of the drug target and on a potential role of the target in organ development.

A JAS is not warranted when the existing data are sufficient to support pediatric use or when the identified concerns can be managed clinically (e.g. by monitoring predictive biomarkers of toxicity). Pharmacokinetic and simulation systems can be used to predict drug levels in pediatric patients (particularly neonates and infants; Yang et al. 2014; Villiger et al. 2016) and may be more reliable for this purpose than JAS. The existing nonclinical toxicity data are evaluated for indications of effects in organs that will be undergoing development in the pediatric patients at the time of treatment.

7.4 STUDY DESIGN

The ICH S11 guideline defines core elements to be included in all JAS, to which additional endpoints can be added to address particular concerns. If the need for a JAS is due to a specific safety concern, the design of the study should focus on elucidating that concern. A broader study will be needed if the JAS is required to compensate for a lack of relevant knowledge in the pediatric population.

An understanding of the relative rates of development and maturity of organ systems across species is needed both to design an appropriate JAS and to interpret findings in juvenile animals with respect to their significance in human pediatric patients. Many reviews on the comparative

Table 7.1 Comparison of the Developmental Windows of Selected Postnatal Events in Humans and Rats

Postnatal event	Developmental windows	
	Human (years)	Rat (days)
Glutamate receptors (maximal binding)	1–2 cortex Decline to adult 2–16	28 cortex Decline to adult >28
Monoamine system	2–4 Maximum receptor density	21–30 Adult levels
Ocular dominance	0–3	21–35
Cerebellum persistent external germinal layer	0.6–2	0–21
Rapid phase of myelination ends	2	25–30
Cognitive development: delayed response learning	1–2	10–35
Puberty	11–12	40–60

Source: Adapted from the FDA Guidance for Industry Nonclinical Safety Evaluation of Pediatric Drug Products (2006).

development of various organ systems in laboratory animals and humans have been published. Organ systems that undergo significant postnatal development and are therefore of particular interest with regard to the design of JAS include the following: central nervous system (CNS; Watson et al. 2006; Wood et al. 2003; Ridings 2014); immune system (Holsapple et al. 2003; Sewald et al. 2015); bone (Zoetis et al. 2003; DeSesso and Scialli 2018); kidney (Zoetis and Hurtt 2003a); lung (Zoetis and Hurtt 2003b; Lewin and Hurtt 2017); heart (Hew and Keller 2003); female reproductive system (Beckman and Feuston 2003; Laffan et al. 2018); male reproductive system (Marty et al. 2003; Picut et al. 2018); gastrointestinal system (Walthall et al. 2005; Downes 2018); and ear (Powles-Glover and Maconochie 2018). A comparison of the developmental windows of selected postnatal events in humans and rats is shown in Table 7.1.

7.4.1 Selection of Species

A single species, usually the rat or mouse, is required. In exceptional circumstances, a nonrodent species (i.e. dog [Robinson et al. 2012], minipig [Barrow 2012] or nonhuman primate [NHP; Weinbauer et al. 2012]) may be considered, either alone or in addition to a rodent species. JAS in two species are generally only required when the JAS are intended to replace the repeat-dose studies in pediatric-first or pediatric-only indications (see below).

The following factors are considered when selecting a species for JAS:

- Ontogeny of the pharmacological target in animals when compared to humans
- Species used in previous adult toxicology studies
- Species that have previously been shown to be susceptible to the effects of the drug
- The chronology of development of the identified target organs in animals when compared to humans
- The technical feasibility of dosing and handling of animals at the required age
- Similarity of pharmacokinetics and metabolism in animals when compared to humans.

A table of the advantages and disadvantages of various species is presented in the S11 guideline (ICH 2020). NHPs are rarely suitable for JAS because of the reliance of the infant on the mother, and should only be considered as a last resort (Weinbauer et al. 2012). However, in most repeat-dose studies using NHPs, treatment is started at a juvenile age (i.e. three years old or less), so the general toxicology package will often provide some information on the effects of juvenile exposure in cases where the nonrodent species evaluated is the NHP.

7.4.2 Age of Animals

The age at which dosing commences in the JAS should correspond to the earliest developmental stage at which paediatric patients will be treated. Because the chronology of development of different organs is not consistent across species, the age of the animals at the start of dosing should

Table 7.2 General Comparison of the Postnatal Stages of Development in Various Species as Compared to Humans

Species	Preterm newborn	Term newborn	Infants and toddlers	Children	Adolescents
Human	–	0–27 days	28 days to 23 months	2–11 years	12–16 or 18* years
Monkey	–	0–15 days	0.5–6 months	0.5–3 years	3–4 years
Dog	–	0–21 days	3–6 weeks	6–20 weeks	5–7 months
Minipig	–	0–15 days	2–4 weeks	4–14 weeks	4–6 months
Rabbit	0–4 days	4–10 days	1.5–5 weeks	5–12 weeks	3–6 months
Rat	0–4 days	4–12 days	1.5–3 weeks	3–5 weeks	5–10 weeks
Mouse	0–4 days	0–10 days	1.5–3 weeks	3–5 weeks	5–7 weeks

* Dependent on region

Source: Adapted from Barrow and Schmitt (2017).

be based on a consideration of the timing of development of those organs targeted by the pharmacological action of the drug and those organs considered likely to be susceptible to the toxicity of the drug. Table 7.2 shows a general comparison of the postnatal stages of development in various species as compared to humans. This table is intended only as a simplified guide. The appropriate age of exposure in JAS must be based on the chronology of development of the particular organs concerned by the drug in the selected species.

7.4.3 Route of Administration

The route of administration in JAS is generally that by which the drug will be given to human patients. For drugs intended to be administered to pediatric patients by several routes, a single JAS using the route that results in the highest exposure is usually sufficient. When the human route of administration is not feasible in very young animals (e.g. intrathecal), an alternative route that results in the same or higher exposure may be used. Methods have been developed over recent years allowing drugs to be given to juvenile animals by routes previously considered as nonfeasible at early ages (e.g. intravenous administration in neonatal rats).

7.4.4 Dose Level Selection

There should be some overlap between the exposures attained in the JAS with those obtained in repeat-dose toxicology studies in young-adult animals. It is advisable to investigate the dose-response relationship of any effects, but severe toxicity should be avoided. The low dose level should preferably achieve an exposure in juvenile animals that is similar to the anticipated exposure in pediatric patients. When there are large age-related changes in exposure due to metabolism maturation, it may be preferable to adjust the administered doses as the animals get older to ensure consistent and maintained exposure levels.

7.4.5 Number of Animals Per Group

The ICH S11 guideline gives no clear guidance on the number of animals to be used, but it is recommended that both male and female animals are included. For rodents, a minimum group size of 10 males and 10 females would be appropriate for the evaluation of most endpoints, based on the recommendations for repeat-dose studies. More animals are needed if the various endpoints are to be evaluated in separate subgroups, such as when additional endpoints are triggered based on safety concerns. Investigations that typically generate highly variable data, such as behavioural tests (see Section 7.4.10), require larger group sizes in order to draw valid conclusions. These behavioural tests include functional observation battery (FOB), reactivity to handling and different stimuli in behavioural, neurological and autonomic tests performed in the cage, in the hand and in a standard area, along with motor activity measurements.

7.4.6 Allocation of Animals to Treatment Groups

For studies in which the animals are treated during the preweaning period, whole-litter allocation to treatment groups is preferred so that all pups in the litter receive the same treatment, thus avoiding cross-contamination between littermates. Cross-fostering is often used to provide litters of the required size with an equal number of male and female pups (Richmond and Clode 2014).

The maternal and/or litter origin of the pups should be taken into account in the allocation to groups, even for studies in which treatment is started after weaning. Thus, pups from the same mother or litter should be equally distributed across the control and treatment groups. Likewise, if subgroups are used for various types of examinations, each subgroup should comprise pups drawn from as many different litters as possible to minimize genetic or maternal bias (Hoberman and Barnett 2012).

7.4.7 Duration of Treatment

The duration of treatment should cover all sensitive developmental periods during which pediatric patients will be exposed. This often leads to comparatively long treatment periods in animals for drugs that are intended for only short-term administration in human patients due to the need for covering the sensitive periods of both early- and late-developing organs in the same animals. For instance, if the reproductive system is of potential concern, rodents in JAS need to be exposed up to sexual maturity. In practice, it is rarely practical to treat nonrodent species up to sexual maturity because of their long maturation period, although this can be feasible in rabbits (Allais et al. 2014) and minipigs (Barrow 2012).

7.4.8 Treatment-Free Period

A treatment-free period may or may not be required depending on the nature of safety concerns. If, for instance, organ toxicity is expected based on the findings in adult animals, a treatment-free recovery period may be needed to investigate whether the effects are more persistent following juvenile exposure than in adults. A second purpose of a treatment-free period is to detect latent effects induced early in development that only become visible as the organ reaches maturity. For example, an inhibition of Sertoli cell division in neonate testes may remain undetected throughout childhood and may only manifest as a reduced sperm count after puberty. Additional subgroups of dedicated animals are usually required for investigations during the treatment-free period.

7.4.9 Core Endpoints

The following endpoints are generally included as a minimum in all JAS and are referred to as core endpoints. Clinical condition is monitored. Body weight is recorded at least twice weekly for rodents. Long bone (e.g. femur) length is measured at least at necropsy (or at intervals during the study in case of specific concern). Food consumption is monitored after weaning. The onset of puberty is recorded; for example, in rodents, the age of vaginal opening in females and balanopreputial separation (cleavage of the prepuce from the glans penis) in males. Blood samples are taken for standard clinical pathology at the end of the treatment period. This includes determination of hematology, blood biochemistry and urinary parameters (possibly along with coagulation parameters). Organ weights are recorded at necropsy and histopathology is performed on all major organs (e.g., bone/marrow, brain, gastrointestinal tract, heart, kidney, liver, lung, ovary, testis, target organs of toxicity and any tissues with macroscopic lesions). Blood samples for toxicokinetic investigations are taken at least at the beginning and end of the treatment period.

7.4.10 Additional Endpoints

The following endpoints are only included to address specific safety concerns. Imaging techniques may be used to monitor bone development. Bone mineral density and/or bone biomarkers may be determined at intervals during the study or at necropsy. Additional clinical pathology parameters and/or time points may be added, taking into consideration that only small volumes of blood can be taken from small rodents. Ophthalmology examinations are not a routine requirement in JAS, as eye development in humans is mostly prenatal, but they may be included to address specific concerns. Additional tissues may be submitted to histopathology and additional examinations may be performed (e.g. immunohistochemistry).

CNS examinations may be included to address potential changes in behaviour, learning or memory (FDA Guidance for Industry 2006). Common behavioural tests in rodent JAS include a modified Irwin test (evaluation of qualitative effects on behaviour and physiological CNS function), FOB (see 7.4.5), locomotor activity (number of horizontal and vertical movements), coordination and reflex tests, and the habituation and prepulse inhibition test (acoustic startle reflex response). Learning and memory tests (e.g. complex maze performance), are usually conducted during a treatment-free period to avoid any confounding pharmacological influences.

Neurohistopathology examinations (e.g. sections in specific brain regions), specialized stains and/or histomorphometry can also be added.

The ICH S11 guideline (ICH 2020) states that mating assessments are no longer recommended in JAS. For female rodents, estrous cyclicity is recommended as an initial assessment of reproductive and endocrine function. For males, sperm analysis and/or testicular immunochemistry are recommended if they can contribute useful data not already collected in other studies. Hormonal assessments are not generally recommended because of the high hormonal variability in juvenile animals and the difficulty of obtaining consistent samples.

Immunotoxicity assessments may be included if the pharmacological class or existing toxicology findings give rise to concern. Immune function assays, such as a T cell–dependent antibody response, should only be performed once the immune system has reached relative maturity (e.g. 45 days of age for rats).

7.5 DOSE RANGE-FINDING (DRF) STUDIES

Preliminary or dose range-finding (DRF) studies are recommended for the determination of tolerability in juvenile animals and for the collection of toxicokinetic data. The DRF study should start at the youngest age at which the animals will be treated in the main study. The duration of the DRF study is typically shorter than that of the main study (e.g. four weeks), as it is the younger animals that are less likely to tolerate the drug. Blood samples for toxicokinetics should be taken after the first administration and at later time points if this is considered to be useful. Additional investigations, such as plasma protein binding and brain tissue concentration, may be useful for explaining differences in tolerability and sensitivity between juvenile and adult animals.

7.6 TOXICOKINETICS

Because of the small size of juvenile animals and the limited volume of blood that can be taken, it is often not feasible to investigate the full kinetic profile of the drug in juvenile animals (particularly neonates). Sampling at a few time points, using pooled samples if necessary, should be performed to at least compare exposure levels with those in adult studies. Microsampling (Verhaeghe et al. 2017; Nilsson et al. 2013) should be used whenever possible, and this may allow samples to be collected in "main/principal" study animals, thus avoiding the inclusion of satellite animals, and may also permit longitudinal sampling (i.e. multiple samples from the same individual to obtain a complete profile, as opposed to composite sampling, where samples are collected from different subjects at each time point).

7.7 PEDIATRIC-ONLY AND PEDIATRIC-FIRST DEVELOPMENT

For drugs being developed exclusively for a pediatric indications (e.g. neonatal respiratory distress syndrome) with no foreseeable use in adults (i.e. pediatric-only), the initial clinical trials may be conducted in either healthy adult volunteers or in pediatric patients (Schmitt et al. 2016). Pediatric-first drugs are developed initially for pediatric indications, with pediatric clinical trials being run before any clinical trials in adults have been conducted. In these cases, it may be possible to avoid performing separate repeat-dose (as required in ICH M3(R2)) and juvenile studies. Juvenile rodents and/or nonrodents are allocated to the study and the treatment is continued into adulthood so that all the usual repeat-dose toxicity assessment endpoints and juvenile endpoints can thus be investigated.

7.8 FURTHER TRENDS IN JUVENILE TOXICOLOGY TESTING OF ENVIRONMENTAL CHEMICALS

Recently, toxicologists and pathologists have come together to discuss trends and challenges in the field of juvenile toxicology. Changes in the assessment of juvenile toxicity suggested by the US National Toxicology Program (an interagency program composed of government agencies within the US Department of Health and Human Services) have been made subsequent to more extensive evaluations of gestational exposures to hormonally active compounds showing potential to produce latent effects at the juvenile life stage (Remick et al. 2015). This is consistent with regulatory trends over the last 10 years that have called for increased testing of environmental chemicals for their potential to induce endocrine disrupting effects. Among pathologists evaluating juvenile toxicity, one of the greatest challenges is the combination of the inherently dynamic tissue morphologies associated with growing and developing organs together with a general lack of historical control data, especially for intermediate ages not typically included in the standard developmental and reproductive toxicology study designs.

BIBLIOGRAPHY

Allais, L., L. Reynaud, L. Crouille, M.-C. Dufour, S. Pilet, and T. Gros. 2014. Postnatal evaluations in juvenile rabbits. *Reprod Toxicol* 48: 23–4.

Barrow, P. 2012. Use of the swine pediatric model. In *Pediatric Nonclinical Drug Testing*, ed. A. Hoberman, and E. Lewis, 213–29. Hoboken NJ: John Wiley & Sons, Inc.

Barrow, P. C., and G. Schmitt. 2017. Juvenile nonclinical safety studies in support of pediatric drug development. In *Drug Safety Evaluation: Methods and Protocols*, ed. J.-C. Gautier, 25–67. New York: Springer New York.

Barrow, P. C., S. Barbellion, and J. Stadler. 2011. Preclinical evaluation of juvenile toxicity. *Methods Mol Biol* 691: 17–35.

Beckman, D. A., and M. Feuston. 2003. Landmarks in the development of the female reproductive system. *Birth Defects Res Part B: Dev Reprod Toxicol* 68(2): 137–43.

DeSesso, J. M., and A. R. Scialli. 2018. Bone development in laboratory mammals used in developmental toxicity studies. *Birth Defects Res* 110(15): 1157–87.

Downes, N. J. 2018. Consideration of the development of the gastrointestinal tract in the choice of species for regulatory juvenile studies. *Birth Defects Res* 110(1): n/a–n/a.

EMA. 2013. *Guideline on Pharmaceutical Development of Medicines for Paediatric Use*. www.ema.europa.eu/docs/en_GB/document_library/Scientific_guideline/2013/07/WC500147002.pdf (accessed October 27, 2020).

FDA. 2006. *Guidance for Industry Nonclinical Safety Evaluation of Pediatric Drug Products*. www.fda.gov/media/119658/download (accessed October 27, 2020).

FDA. 2006. *Non-clinical Testing in Juvenile Animals on Human Pharmaceuticals for Pediatric Indications*. www.fda.gov/downloads/drugs/guidancecomplianceregulatoryinformation/guidances/ucm079247.pdf (accessed October 27, 2020).

Hew, K. W., and K. A. Keller. 2003. Postnatal anatomical and functional development of the heart: A species comparison. *Birth Defects Res Part B: Dev Reprod Toxicol* 68(4): 309–20.

Hoberman, A. M., and J. F. Barnett. 2012. Juvenile toxicity study design for the rodent and rabbit. In *Pediatric Nonclinical Drug Testing: Principles, Requirements, and Practices*, eds. A. Hoberman and E. Lewis, 141–82. Hoboken, NJ: Wiley.

Holsapple, M. P., L. J. West, and K. S. Landreth. 2003. Species comparison of anatomical and functional immune system development. *Birth Defects Res B: Dev Reprod Toxicol* 68(4): 321–34.

ICH. 2009. *Guideline M3(R2) on Non-Clinical Safety Studies for the Conduct of Human Clinical Trials and Marketing Authorisation for Pharmaceuticals*. www.ich.org/fileadmin/Public_Web_Site/ICH_Products/Guidelines/Multidisciplinary/M3_R2/Step4/M3_R2__Guideline.pdf (accessed October 27, 2020).

ICH. 2010. *Guidance for Industry: S9 Nonclinical Evaluation for Anticancer Pharmaceuticals*. www.fda.gov/downloads/Drugs/.../Guidances/ucm085389.pdf (accessed October 27, 2020).

ICH. 2011. *Harmonised Tripartite Guideline: Preclinical Safety Evaluation of Biotechnology-Derived Pharmaceuticals*. S6(R1). www.ich.org/fileadmin/Public_Web_Site/ICH_Products/Guidelines/Safety/S6_R1/Step4/S6_R1_Guideline.pdf (accessed October 27, 2020).

ICH. 2017. *ICH Harmonised Guideline Addendum to ICH E11: Clinical Investigation of Medicinal Products in the Pediatric Population E11 (R1)*. www.ich.org/fileadmin/Public_Web_Site/ICH_Products/Guidelines/Efficacy/E11/E11_R1_GuidelineWithAddendum_Step4.pdf (accessed October 27, 2020).

ICH. 2020. *Nonclinical Safety Testing in Support of Development of Paediatric Medicines*. S11. https://database.ich.org/sites/default/files/S11_Step4_FinalGuideline_2020_0310.pdf (accessed September 30, 2021).

Laffan, S. B., L. M. Posobiec, J. E. Uhl, and J. D. Vidal. 2018. Species comparison of postnatal development of the female reproductive system. *Birth Defects Res* 110(3): 163–89.

Lewin, G., and M. E. Hurtt. 2017. Pre- and postnatal lung development: An updated species comparison. *Birth Defects Res* 109(19): 1519–39.

Marty, M. S., R. E. Chapin, L. G. Parks, and B. A. Thorsrud. 2003. Development and maturation of the male reproductive system. *Birth Defects Res Part B: Dev Reprod Toxicol* 68(2): 125–36.

MHLW Japan. 2012. *Guideline on the Nonclinical Safety Study in Juvenile Animals for Pediatric Drugs and Questions and Answers Regarding Guideline on the Non-clinical Safety Study in Juvenile Animals for Pediatric Drugs*. www.pref.kagawa.lg.jp/yakumukansen/yakujinotice/listH24/161_241015.pdf (accessed October 27, 2020).

National Toxicology Program. https://ntp.niehs.nih.gov/whoweare/organization/index.html.

Nilsson, L. B., M. Ahnoff, and O. Jonsson. 2013. Capillary microsampling in the regulatory environment: Validation and use of bioanalytical capillary microsampling methods. *Bioanalysis* 5(6): 731–8.

Picut, C. A., M. K. Ziejewski, and D. Stanislaus. 2018. Comparative aspects of pre- and postnatal development of the male reproductive system. *Birth Defects Res* 110(3): 190–227.

Powles-Glover, N., and M. Maconochie. 2018. Prenatal and postnatal development of the mammalian ear. *Birth Defects Res* 110(3): 228–45.

Presidential Committee for the Study of Bioethical Issues. 2016. *Vulnerable Populations in Safeguarding Children: Pediatric Medical Countermeasure Research.* https://bioethicsarchive.georgetown.edu/pcsbi/sites/default/files/6%20Vulnerable%20Populations%20 Safeguarding%20Children%209.30.16.pdf (accessed October 27, 2020).

Remick, A. K., N. R. Catlin, E. M. Quist, T. J. Steinbach, and D. Dixon. 2015. Juvenile toxicology: Relevance and challenges for toxicologists and pathologists. *Toxicologic Pathol* 43(8): 1166–71.

Richmond, E. S., and S. A. Clode. 2014. Reduction and refinement of rodent juvenile toxicity studies: The simple approach to cross-fostering. *Reprod Toxicol* 48: 27–8.

Ridings, J. E. 2014. Assessment of CNS safety in juvenile animals. *Reprod Toxicol* 48: 18–19.

Robinson, K., S. Y. Smith, and A. Viau. 2012. Dog juvenile toxicity. In *Pediatric Nonclinical Drug Testing: Principles, Requirements, and Practices,* eds. A. Hoberman and E. Lewis, 183–212. Hoboken, NJ: John Wiley & Sons, Inc.

Schmitt, G., J. Ridings, L. De Schaepdrijver, F. L. C. van Doesum-Wolters, G. D. Cappon, and A. Hartmann. 2016. Nonclinical safety considerations for the development of pediatric-first drugs: An industry view. *Ther Innov Regul Sci* 50(5): 632–38.

Sewald, K., M. Mueller, J. Buschmann, T. Hansen, and G. Lewin. 2015. Development of hematological and immunological characteristics in neonatal rats. *Reprod Toxicol* 56: 109–17.

Verhaeghe, T., L. Dillen, H. Stieltjes, et al. 2017. The application of capillary microsampling in GLP toxicology studies. *Bioanalysis* 9(7): 531–40.

Villiger, A., C. Stillhart, N. Parrott, and M. Kuentz. 2016. Using physiologically based Pharmacokinetic (PBPK) modelling to gain insights into the effect of physiological factors on oral absorption in paediatric populations. *The AAPS J* 18(4): 933–47.

Walthall, K., G. D. Cappon, M. E. Hurtt, and T. Zoetis. 2005. Postnatal development of the gastrointestinal system: A species comparison. *Birth Defects Res Part B: Dev Reprod Toxicol* 74(2): 132–56.

Watson, R. E., J. M. DeSesso, M. E. Hurtt, and G. D. Cappon. 2006. Postnatal growth and morphological development of the brain: A species comparison. *Birth Defects Res Part B: Dev Reprod Toxicol* 77(5): 471–84.

Weinbauer, G. F., G. J. Chellman, A. D. Rasmussen, and E. Vogelwedde. 2012. Use of primate pediatric model. In *Pediatric Nonclinical Drug Testing: Principles, Requirements, and Practices,* eds. A. Hoberman and E. Lewis, 255–79. Hoboken, NJ: John Wiley & Sons, Inc.

Wood, S. L., B. K. Beyer, and G. D. Cappon. 2003. Species comparison of postnatal CNS development: Functional measures. *Birth Defects Res Part B: Dev Reprod Toxicol* 68(5): 391–407.

Yang, X., S. M. Morris, J. M. Gearhart, et al. 2014. Development of a physiologically based model to describe the pharmacokinetics of methylphenidate in juvenile and adult humans and nonhuman primates. *PLoS ONE* 9(9): e106101.

Zoetis, T., and M. E. Hurtt. 2003a. Species comparison of anatomical and functional renal development. *Birth Defects Res Part B: Dev Reprod Toxicol* 68(2): 111–20.

Zoetis, T., and M. E. Hurtt. 2003b. Species comparison of lung development. *Birth Defects Res Part B: Dev Reprod Toxicol* 68(2): 121–4.

Zoetis, T., M. S. Tassinari, C. Bagi, K. Walthall, and M. E. Hurtt. 2003. Species comparison of postnatal bone growth and development. *Birth Defects Res Part B: Dev Reprod Toxicol* 68(2): 86–110.

8 *In Vitro* Toxicity Testing

Guillermo Repetto, Consuelo Alvarez Herrera, Raquel Rojas, Ana del Peso and Sara Maisanaba

CONTENTS

DOI: 10.1201/9780203730584-8

8.1 TYPES OF EXPERIMENTAL TOXICOLOGICAL STUDIES

Although risk assessment is the most relevant activity related to toxicological research, there are others that should also be taken into account. From the practical point of view, according to its application, the experimental toxicological studies that can be carried out are, apart from training and education objectives, to fulfil regulatory requirements or to increase toxicological knowledge.

8.1.1 Regulated Research

Regulatory research, commonly known as toxicity testing, investigate the main potentially adverse effects of chemicals or products to obtain information with the objective of protecting human beings and the environment. Toxicity testing is performed to satisfy legislative requirements in order to establish the basis for the safe use of chemicals and products. Toxicity testing is mainly carried out directly by industry or by contract laboratories, to support the dossiers submitted to the regulatory agencies for registration or approval of the application of substances or products (Repetto and Repetto 2009; Worth 2019).

The number and type of tests required depends on the proposed use of the chemicals. Medicines are particularly demanding for toxicology studies. Less information is required for food additives and minimum for cosmetics. Pesticides require, in addition to health assays, ecotoxicological tests that are also needed for industrial chemicals, proportional to the quantity produced. Other products subjected to testing include medical device materials.

The hazard identification of the substances (i.e. the toxic or intrinsic toxicity) is mainly carried out through the acute, subchronic and chronic exposure of laboratory animals to the chemicals. This is based on the assumption that exposure of animals to high doses of chemicals could predict the risk of low doses in humans (Worth et al. 2019). The procedure basically consists of establishing the dose-effect relationship and to calculate the mean lethal dose (LD_{50}) in acute cases. The "no observable adverse effects level" (NOAEL) in repeated exposure has been classically used as a dose useful for the risk assessment calculations of safety levels. The kinetics and metabolism of the chemical in the organism and its environmental degradation complete the safety information. Another regulated area is the control and monitoring of pollution, evaluating the toxic potency of environmental samples.

The valid information for regulatory purposes should be obtained using standardized approved protocols, mainly according to the testing guidelines (TG) and testing documents of the Organisation for Economic Co-operation and Development (OECD 2020a, 2020b); the European Council Regulation (EC) No 440/2008, laying down test methods pursuant to Regulation (EC) No 1907/2006 of the European Parliament and of the Council on the Registration, Evaluation, Authorisation and Restriction of Chemicals (REACH); the European Pharmacopoeia; the International Council for Harmonisation of Technical Requirements for Pharmaceuticals for Human Use (ICH); and others and carried out under good laboratory practice (OECD 2018). In some cases, integrated approaches for testing and assessment are applied. They are flexible approaches for chemical safety assessment based on the integration and translation of the data derived from multiple methods and sources.

In summary, the objectives of most of the regulated studies are to evaluate the hazard of chemicals by the identification of the adverse effects caused, the target organs on which the action of the toxic compound is mainly produced, the kinetics of the substance in the body, the metabolism towards more or less active products, and the variable susceptibility between the different species in ecotoxicity tests.

8.1.2 Nonregulated Research

There are also studies not imposed by regulations that can have two aims: basic or applied research.

8.1.2.1 Basic Research

Basic research has the essential purpose of increasing our knowledge without direct practical implications. It includes the study of the structures, physiology, diseases, mechanisms of actions, interactions and so forth. Investigative toxicology arises from a desire to study toxic effects more deeply and specifically than in regulated routine trials. Although regulatory toxicology is primarily focused on screening and main adverse effects, the basic research work emerges from these studies.

Included among basic studies in toxicology are the investigation of the mechanisms of pathophysiological action; the molecular targets and the homeostatic responses of the organism; the treatment of the adverse changes and the reversibility of the effects; the variable susceptibility between the different sexes, ages, ethnic groups and so forth (i.e. those factors that modify the toxicity and that will serve to identify groups of population with more risk); and the interactions with other substances, both nutritional and medicinal, toxic or theoretically inactive, as well as the conditions that may modify their absorption or elimination.

8.1.2.2 Applied Research

Applied research is directed to answer questions with a practical objective, such as the selection or screening of new substances, comparing biological activities. The classic example is the development of medicines, but applied research is also used for pesticide or biocide products. Other applied toxicological research studies include the monitoring and control of the presence of harmful compounds in the environment; the use of bioassays to detect the presence of a toxin, as in the case of the detection of botulinum toxin; or the quality control of pharmaceutical products, vaccines and so forth.

8.1.3 Education and Training

In other cases, the main objective of the activity consists of the instruction on biomedical disciplines of aspects related to anatomy, physiology or pathology using animals or the training of surgical procedures, administration of pharmaceuticals and extraction of samples or even the handling of patients or animals.

8.2 THE CONCEPT OF ALTERNATIVES

The three Rs correspond to the initial letters of the three basic principles that identify alternative methods to laboratory animals: *R*eplacement of the procedures that use animals by others that do not require them; *R*eduction in the number of animals used; and *R*efinement of the methods applied (Russell and Burch 1959).

The concept of alternatives acquires a special meaning in this context and includes all the methods or techniques that could be used: *replace* the experiments carried out with animals, *reduce* the number of animals used in each test, or *refine* the existing procedures in order to reduce stress and to avoid suffering to the animals (Balls et al. 2019). The most relevant fields in which the alternative methods are applied are education, biomedical research and evaluation of the toxicity of chemical compounds.

About 10%–15% of all laboratory animals are used for toxicity testing, because millions of new chemicals are synthesized every year, including industrial chemicals or active molecules for medicines, cosmetics, pesticides, food additives and so forth. The need to experimentally evaluate the toxicity of a substance for the regulatory authorities in order to authorize its commercialization and use has provoked a series of reactions. The opposition of animal defence associations has been motivated by, on the one hand, the requirement to guarantee the public health of the human population, and on the other hand, the large number of animals used (1,500 on average per compound) and the stress to which they are subjected. Moreover, manufacturers complain about the excessive cost of toxicological studies. Furthermore, there are the toxicologists who consider that current methods are ethically and scientifically improvable and, in many cases, replaceable.

All this is encouraging interest in the development and use of alternative methods to the official protocols for the evaluation of toxicity in animals, either by refining the *in vivo* procedures or by designing new assays completely *in vitro.*

8.3 CLASSIFICATION OF ALTERNATIVE METHODS

The wide range of possible alternative test procedures in practical terms can be structured as follows (see Table 8.1).

Table 8.1 Main Alternative Experimental Methods

Type of approach	Example
1. Avoidance of unnecessary *in vivo* and *in vitro* experiments	• Protocols and previous studies: availability of information, exchange • Grouping of similar substances and read across and evaluation of the weight of evidence
2. Computational systems (*in silico*)	• Predictive models: quantitative structure-activity relationship (QSAR); in chemico studies; drug-toxicokinetics (physiologically based pharmacokinetic [PB-PK]); environmental kinetics of chemical compounds; *in silico* clinical trials • Exploitation of big data • Artificial intelligence
3. Improved animal studies	• Reduction: number of animals used • Refinement: minimization of pain and distress; new models
4. Lower organisms	Bacteria, fungi, protozoa, algae, plants and invertebrate animals
5. Vertebrate embryos and larvae	Fish, amphibians, reptiles, birds and mammals
6. *In vitro* and *ex vivo* models	• Organs: baths, perfusion, culture, slides, reconstituted organs, organoids microchip (microfluidic 3D co-culture models) • Cellular reaggregates, cocultures, spheroids • Primary culture of dispersed cells • Cell lines/transgenesis/stem cells • Cell-free systems
7. Human information	• Microdosing experiments with volunteers • Epidemiological, surveillance, occupational, clinical and forensic studies
8. Models in education and training	Mechanical models, audiovisual systems, computer simulations and virtual reality systems

8.3.1 Avoidance of Experiments

Improvements in the availability, use and exchange of toxicological information are necessary to avoid unnecessary performance or repetition of *in vivo* and *in vitro* studies. This implies not only the existence of available updated international databases but also the legal mutual acceptance of the data obtained in other countries. Several regulations, particularly REACH, includes the obligation among companies to share, at least, the information obtained with laboratory animals.

The strategies of assessment should integrate all the available information, including *in chemico* data, which identifies the intrinsic reactivity of a substance, using physicochemical properties or simple *in vitro* procedures carried out in small containers. It is necessary to allow enough flexibility to avoid assays, on the basis of weight-of-evidence approaches and association with similar chemicals (read-across or categorization).

8.3.2 Computational Systems (*In Silico*)

Computational or *in silico* models were first developed as predictive systems through the correlation of the chemical structures and the biochemical and physiological activity of compounds (quantitative structure-activity relationship [QSAR]). This nontesting approach is very interesting, and the current knowledge allows the prediction with increasing certainty of the toxicity of xenobiotics based on theoretical data. Models that predict the environmental kinetics of chemical compounds and their toxicokinetics (physiologically based pharmacokinetic [PB-PK]) have also been developed. Predictive procedures are currently used, for example, as part of weight-of-evidence approaches or to support information to facilitate the acceptance of data generated in laboratories without good laboratory practices (GLPs); to develop testing strategies; to support exemptions of animal assays from higher level studies; to provide data when a test is not technically possible; to predict the toxicity of new molecules; and to prioritize chemical substances that require further evaluation. There are also new applications such as *in silico* clinical trials.

In addition, extremely large datasets from different sources, usually known as big data, can be analyzed computationally to reveal patterns, trends and associations. The modern models are based in artificial intelligence to predict hazards and to make complex decisions better than human experts. They integrate information of different types from multiple sources, including *in silico, in chemico, in vitro*, and animal and human data (Myatt et al. 2018).

8.3.3 Improved Animal Studies

It is necessary to improve the design of experiments with animals in order to increase their validity and reproducibility and to reduce the stress and the number of animals used; statistical test procedures should be applied not only in the evaluation phase of the data but also in the experimental design. New models should be carefully evaluated in terms of animal welfare and biological relevance.

8.3.4 Lower Organisms

The current legislation for the protection of experimental animals does not protect living organisms with less evolved sensory systems such as invertebrates (with the exception of some cephalopods) or bacteria, algae, fungi, protozoa and plants, which can be used *in vitro*.

8.3.5 Vertebrate Embryos and Larvae

The use of vertebrates in the initial stages of their development has ethical advantages over the use of adult animals for the same reasons stated in the previous section, having less evolved sensory systems and therefore not being included, in most countries, in the protection regulations. In this sense, the use of embryos/larvae of fish (e.g. zebrafish), amphibians, reptiles, birds and mammals is very useful.

8.3.6 *In Vitro* Models

In vitro studies are so named because this type of study was initially carried out in glass containers. Furthermore, a test performed *in vitro* means that it is done outside of a living organism and it usually involves isolated tissues, organs or cells. *Ex vivo* tests are *in vitro* assays carried out with materials from humans or animals.

According to their purpose, *in vitro* tests can replace animals, or complement them, improving the sensitivity and specificity of animal studies. On the other hand, we must bear in mind that *in vitro* procedures have their own entity among experimental methods and provide more in-depth information on the mechanisms of toxic action than *in vivo* tests. In addition, regardless of their specific purpose, *in vitro* tests are usually grouped to form a battery of tests, the results of which, taken together, will allow the interpretation of the proposed objective. The main types of *in vitro* systems will be described later.

8.3.7 Human Information; Microdosing

The human toxicological data of interest come from epidemiological studies, toxic surveillance after the introduction of products on the market; from occupational exposure, clinical poisonings and forensic cases; and, with many limitations, from the use of volunteers. Microdosing is a promising way to assess, by means of positron emission tomography and accelerator mass spectrometry, pharmacokinetics and pharmacodynamics of tested drugs administered at very low doses to humans (Herrmann et al. 2019). This facilitates exploratory studies in humans without complete preclinical toxicological studies in animals (Bergstrom 2017). Microdosing is based on two main principles: first, that human data are superior to animal and *in vitro* data when modelling the effects of human disease and therapeutics, and second, that data obtained with limited pharmacological exposure can be reliably extrapolated to the therapeutic dose range. The use of microdosing has been accepted by the International Conference for Harmonisation (ICH 2009).

8.3.8 Education and Training

Many systems are used in education and training, including mechanical models, audiovisuals, computer simulations and virtual reality systems. Computer systems present a number of advantages: they force students to be active; many factors can be studied simultaneously and the parameters can be quantitatively varied; the program can provide feedback, give suggestions and offer help; the speed of the processes is adapted to the suitable for the learning of that student; experiments can be repeated at any time and almost anywhere; effects that can make it difficult to understand the basic principles can be simplified; less time is needed to present information and students to master it; students control the pace of learning, as teaching is personalized; activities can appear in different orders; fewer resources are required; and they stimulate the teacher's creativity compared to conventional methods such as dissection.

8.4 BASIS FOR THE USE OF *IN VITRO* TESTS

The main reasons why *in vitro* research methods should replace laboratory animals as much as possible are as follows:

Moral and ethical: Laboratory animals are subjected to stress and suffering. It is estimated that up to 1,500 animals are used for the toxicological evaluation of each chemical compound.

Economic: The cost of animal toxicity tests, especially those that use nonrodent species such as dogs, is so high that it is an obstacle to the development of many drugs and chemical compounds. In addition, as these studies requires a long time to complete (2–3 years), they produce an important delay in the commercial launch of the products.

Scientific: Traditional research with animals is fundamentally empirical; many of the effects that are observed originated in complex pathways sometimes not sufficiently known or understood. On the contrary, the experiences with cells or cellular fractions allow to limit or pull apart the components of the pathophysiological processes, which favours the understanding of the phases, even of the molecular interaction and the measurement in quantitative terms of actions and effects, making possible the objective comparison among substances.

In addition, animal studies present many limitations in the prediction of human risks. For example, many pharmaceutical products without hazards in animals have presented risks to humans (e.g. mibefradil, troglitazone, cerivastatin, valdecoxib). Furthermore, a less empirical and more scientific approach than animal testing is necessary, which includes the technical and toxicological advances produced and which improves the methods of research and the extrapolation procedures.

Logistic: There is not a sufficient number of animals, laboratories, technicians and toxicologists to evaluate a minimum of the more than 1,000 new compounds that are introduced every year into the market.

Sociopolitical: There is pressure, based on ethical reasons, from different social groups that demand the abolition of animal research. Paradoxically, other sectors are demanding improvements in regulations and in protecting the health of the human population when exposed to chemical compounds.

Legal: European legislation establishes that "wherever possible, a scientifically satisfactory method or testing strategy, not entailing the use of live animals, shall be used instead of a procedure with animals" (Art. 4.1, Directive 2010/63/EU). This legislation is, in theory, clearly in favour of abolishing animal experimentation as a whole.

8.5 ADVANTAGES AND DISADVANTAGES OF *IN VITRO* TESTS

In a very simplified way, the main advantages of *in vitro* toxicological research methods with respect to those employed *in vivo* include the following:

- They are ethically more acceptable than *in vivo* tests, because it is not necessary to put live animals in contact with the test agents.
- In cases where *in vitro* tests require animals as a source of materials, they need a considerably smaller number of animals, because with the organs of a single animal, hundreds (even thousands) of *in vitro* tests can be performed.
- They can use human tissue taken from cadavers or by biopsy in living subjects, which can greatly simplify the process of extrapolation of the data.
- They can use homogeneous biological material obtained with standardized techniques.
- They allow to study the action of the compound in the main target in a specific cellular population or isolated subcellular fraction. This avoids the interference produced on the studied target by other organs, cells or organelles.
- They facilitate certain specialized studies, such as electrophysiological assays.
- They are more easily objectifiable and quantifiable than *in vivo* tests, because they can be directly measured by physicochemical techniques (spectrophotometry, fluorimetry, etc.).
- The results are more reproducible than animal assays, because each test consists of reactions of a few components, under controlled conditions, without interference from other factors such as animal stress, seasonal changes, weather influences and so forth.

- They present greater precision in the direct control of the dosage (i.e. the toxicity in interaction with the target organ).
- They allow the absolute control of the duration of the period of contact with the evaluated product, because the potentially toxic agent can be applied and removed as required.
- They facilitate the study of the mechanisms of toxic action, differentiating the primary lesion from those that are triggered next, in line with the adverse outcome pathways promoted by the OECD (2017). The reason is because it is possible to interrupt or modify the process at any time. In addition, they reduce the latency period and facilitate the chronological study of toxic actions.
- Being simpler models, they allow the study of the toxicokinetic influences in the uptake, transformation and elimination of substance more easily and in greater detail.
- They provide important time savings, facilitating more rapid generation of results.
- In general, they have a much lower economic cost and require less complex facilities than those legally required to conduct animal studies.
- They require smaller quantities of the tested products than the *in vivo* methods. This is particularly important for newly synthesized substances or during a trial period.

In contrast, *in vitro* toxicological experimental methods have a series of drawbacks:

- The need to use more than one test system. That is, the alternatives to animal methods are usually batteries of *in vitro* tests that combine several biological substrates and indicators of toxicity.
- The lack of complex interactions between the different organs of a whole animal, unless three-dimensional tissues, microfluidic systems or organ-on-a-chip systems are used.
- The generally limited duration of the physiological activity, mainly of specialized cells, and consequently, the difficulties to study delayed or chronic toxicities. In addition, they are usually relatively static models.
- The absence, in most models, of certain toxicokinetic phenomena such as the first-pass effect and the ability to activate and detoxify the chemicals.
- The interference of solvents and the difficulties in testing substances that are not very soluble, volatile, gaseous or that decompose rapidly or explode in contact with the culture medium.
- The complexity of the extrapolation to the *in vivo* situation.

8.6 TYPES OF *IN VITRO* MODEL SYSTEMS

There are many biological alternatives to the use of higher (vertebrate) animals for studying the toxicity produced by chemical compounds. These include invertebrates, microorganisms, plants, microalgae and *in vitro* preparations. The characteristics of the main *in vitro* models are briefly described below and in Table 8.2.

8.6.1 Cell Free Systems

At the minimum level of biological organization are the different individual components of cells, which can be separated by centrifugation techniques. Research with an isolated subcellular fraction has the great advantage of avoiding tissue, cellular and other organelle and enzyme interactions, making it suitable for testing very specific effects on subcellular structures. This is also a drawback because the target may be different from that of the whole-organism system and because most of the cellular processes are interrelated and the different stages take place in different cellular locations. The most used subcellular fractions are the purified components (e.g. enzymes, DNA fragments, receptors, ion channels, synaptosomes), the complete fraction or total homogenate, the soluble fraction, the postmitochondrial, the microsomal, the nuclear, and the membrane fraction.

The typical pharmacological receptor binding assays use a source of receptor (membranes, soluble/purified) and analyze receptor-ligand interactions. Other options are, for example, using a biomembrane as a model of skin and detecting the deterioration caused by corrosive chemicals.

Table 8.2 Characteristics of *In Vitro* Models

Model	Advantages	Disadvantages
Lower organism	Easy culture and handling. Very useful in different types of studies: planaria (neurobiology), nematodes (endocrinology), molluscs (environmental biology and ecotoxicology), insects (mutagenesis).	Complex extrapolation of results to higher organisms.
Embryos	Easy control of the treatments applied. Direct observation in eggs with transparent chorionic shells.	Difficult for certain organisms. According to stage of development, they can be considered animal experimentation.
Organs in baths or in culture	Very good correlation with *in vivo* behaviour (ileum, brain, pituitary gland, liver). Good interaction between cell types. 3D.	Difficult to grow. Direct extraction of animals. Very short life.
Perfused organs	Even better than the previous ones to study the uptake and secretion of substances (liver, kidney, pancreas, gills). 3D.	More difficult methodology. Require researcher experience.
Reconstructed and printed organs	The use of an extracellular matrix allows 3D growth, for example of the skin. They can be generated using 3D printers. Better correlation with behaviour *in vivo*. Easy comparison between species. Excellent interaction between different cell types. Allows histological and biochemical studies. Effects dependent on organ area.	Technically challenging. Poorly reproducible. Different activity according to the zones.
Organ-on-a-chip, microfluidic 3D cell cultures	3D models on a plate that allows fluid control to and from the reconstituted organs of nutrients and compounds tested. Can be multiorgan chips.	Labour-intensive and technically challenging/complex. Cost.
Organoids	Miniaturized versions of an organ produced *in vitro*. Three-dimensional architecture and physiology, better than in reconstituted organs, more similar to that existing *in vivo*. Good intercellular communication. Derived from tissue cells, embryonic stem cells or induced pluripotent stem cells.	Variability between experiments. Technically complex and difficult to grow.
Embryoids	These are 3D miniature versions containing embryonic tissues, blastocyst-like structures, similar to organoids.	Similar to organoids.
Spheroids	By constant agitation, 3D microaggregates are formed. The original three-dimensional architecture is largely restored. The differentiation is more complete than in the growth of a monolayer and in matrices. Depending on the case, they may be better or worse than organoids.	Hard to grow. Short life. Some cell types do not form spheroids easily. Variability between experiments.
Primary culture of dispersed cells	They are obtained from different tissues, isolating cells of interest from animals or humans from biopsies. They maintain quite a few specific tissue characteristics.	Short life. Rapid loss of its specific characteristics. 2D.

Culture of established cell lines; clones	Cells from primary cultures that transform to have infinite replicative life span. Continuous source of fairly homogeneous cells. Easy cultivation, storage and production control. Available for most human organs and for many species.	Cells lose some specific tissue characteristics, although others can be differentiated to more mature cells. Cells age and change after a number of subcultures depending on the cell type. 2D.
Genetically modified cells	Genetic modification, now widespread with editing systems, but also available through other types of systems, gives cells stability or encodes structural or functional proteins. Very useful for generating polymorphic cells in various enzymes and for introducing reporting systems.	Very specialized technique. Changes not always stable. 2D.
Cocultures	The situation is better reproduced *in vivo*, as in the blood-brain barrier.	Difficult by the requirements of each cell type. 2D.
Stem cells	Obtained from mesenchymal (MSCs), embryonic (ESCs) or induced pluripotent stem cells (iPSCs) from adult tissues of animal or human origin. They can be differentiated into many specific cell types very similar to those *in vivo*.	Ethical problems for human embryonic origin and consent for adult. Limitations of cell types and species. Complex growth. 2D.
Unicellular organisms	Easy culture and handling. Very useful in different types of studies: bacteria (genetics, genotoxicity, ecotoxicity), protozoa (embryology), algae (ecotoxicology, photosynthesis).	Complex extrapolation of results to higher organisms.
Cell-free models: postmitochondrial fraction, microsomal, synaptosomes, ion channels, enzymes, DNA, protein solutions, biomembranes, etc.	Very useful for very specific studies such as receptor binding, binding to subunits of channels, inhibition of enzymes, adducts to DNA, changes in membrane stability, protein denaturation, changes of biomembranes that simulate the skin.	They do not reproduce the interactions between cells and tissues.

8.6.2 1D- and 2D-Cell Cultures

There are many options, from unicellular organisms to use suspension (1D) or adherent cultures. The limitations of 2D-cultures include that 2D geometry constrains morphogenesis and lacks cell-substrate interactions; the high stiffness of the surface of culture provides supraphysiological mechanical signals; a continuous, flat surface available for unencumbered adhesion, spreading, and migration; and soluble gradients are absent without microfluidics.

8.6.2.1 Unicellular Organisms

Unicellular organisms, such as bacteria (e.g. *Escherichia coli*), fungi (*Saccharomyces cerevisiae*), protozoa and algae (*Chlorella*), can be cultured and employed as simple but useful models due to ease of application and genetic manipulation.

8.6.2.2 Culture of Primary Cells

Animal, human or vegetal tissues are dissociated enzymatically or mechanically and the seeded cells grown in a monolayer. The histochemical characteristics can be easily visualized but remain only partly if all type of cells from the tissue are seeded. The reproducibility of electrophysiological, morphological and biochemical studies is adequate but time limited. Apart from the complexity of separation and culture requirements, the main drawback is that these are two-dimensional structures in which a large proportion of the original tissue architecture is lost.

8.6.2.3 Culture of Cell Lines; Cellular Clones

A single cell type, usually from a primary culture, can be immortalized to live in suspension or adherent cultures. By the process called transformation—which can occur spontaneously or induced by chemicals, radiation or viruses—the cells acquire continuous life, becoming an established cell line. Continuous life means the ability of the cells to multiply *in vitro* for long periods of time. The mitotic, continuous or established cell lines present the advantage of supplying large, homogenous amounts of a single cell type being very suitable for quantitative biochemical or electrophysiological studies. In addition, some of them can be differentiated to a mature form. A cell clone is a cell line that comes from a single cell, so the morphological, biochemical and electrophysiological characteristics of its population are absolutely homogeneous. The drawbacks of cell lines are the loss of normal tissue structures. In addition, some lines may not reflect all the functions of the cells of origin.

Cell cultures are useful for many different types of studies, including physiological (e.g. receptor activation, electrophysiology, channels), pharmacological, toxicological, microbiological, genetic, biochemical, molecular and so forth.

8.6.2.4 Stem Cells

The three commonly used stem cell types are mesenchymal stem cells (MSCs), embryonic stem cells (ESCs) and induced pluripotent stem cells (iPSCs). They are obtained from embryonic or adult tissues of animal or human origin. Human induced pluripotent stem cells (hiPSCs) and their differentiated derivatives come from healthy subjects and also from patients to model their diseases (Skeldon et al. 2018).

MSCs, which are adult multipotent stem cells often isolated from bone marrow or adipose tissues, have the potential to differentiate into osteoblasts, adipocytes or chondroblasts *in vitro*. However, they can also become other cell types, including myocytes, tendocytes, ligament cells, smooth muscle cells, endothelial cells, cardiomyocytes, hepatocytes and neural-like cells.

Both ESCs and iPSCs can differentiate into somatic cells of all three germ layers: ectoderm, mesoderm and endoderm. Thus, they offer greater multipotency compared to adult stem cells. ESCs originate from within the inner cell mass of a blastocyst, whereas iPSCs represent somatic cells that have been reprogrammed into pluripotency. Concerns with ESCs include the controversy over their derivation from embryos and the limited number of cell lines, which could induce an immune response depending on the patient. But, iPSCs offer a feasible alternative to ESCs without the ethical and immunogenic drawbacks of the latter.

Mouse ESCs are used more often than human pluripotent stem cells because they have a short developmental period; the duration of mouse foetal development is about 20 days, which is much shorter than the 300 days of human development.

8.6.3 3D-Culture or Complex Systems

The 3D cell culture is a step toward mimicking normal cell-matrix and cell-cell interactions. These cultures are mainly established by using a cell matrix or a scaffold, or by the agitation of the culture medium. They recreate *in vitro* the physiological and spatial arrangements of cells; the matrix surfaces guide and hinder cell motility; there are gradients of soluble factors, nutrients and oxygen; cells are free to self-organize in 3D; embedded cells generate apical-basal polarity; it is easier to grow difficult-to-culture cells, including primary cells and tumours; they improve cell type–specific function and gene expression; and they facilitate the generation of 3D structures (blood vessels, branching morphogenesis and lumenization) and the accumulation of proteins, as occur in some diseases. Furthermore, 3D systems are particularly useful for studying functions poorly reproduced in 2D cell cultures, such as amyloid accumulation, chondrogenesis, angiogenesis, cancer metastasis and morphogenesis. The main limitations of 3D cell cultures are high cost, the difficulty of use and microscopic analysis, limitations for some biochemical assays and the difficulties in replicating the systems.

8.6.3.1 Reconstructed Organs

Natural or artificial extracellular matrices or other systems can be used to support the growth of the cells that reproduce part of an organ (e.g. the skin).

8.6.3.2 3D Printed Organs

Modified inkjet printers can produce 3D biological tissues. Three-dimensional printing allows the layer-by-layer construction of a particular organ structure to form a cell scaffold, followed by cell seeding.

8.6.3.3 3D Spheroids

When cells grow in a medium subjected to constant agitation, they tend to regroup forming microaggregates, in which the original three-dimensional architecture is restored in large part. Although the multiplication of some cell types is limited, the differentiation in the spheroids is more complete than in a monolayer (2D) culture.

8.6.3.4 Organoids

An organoid is a miniaturized and simplified version of an organ produced *in vitro* in three dimensions that shows realistic micro-anatomy. They are derived from one or a few cells from a tissue, either ESCs or iPSCs. They can self-organize in 3D culture owing to their self-renewal and differentiation capacities.

8.6.3.5 Embryoids

Blastocyst-like structures are generated from a single stem-cell type as well as from somatic reprogrammed cells, and these artificial embryo-like structures can implant, induce decidualization and give rise to cells of the three embryonic founding tissues: epiblast, trophectoderm and primitive endoderm. They are miniature versions containing embryonic tissues, similar to organoids and spheroids.

8.6.3.6 Microfluidic 3D Cell Cultures: Organs-on-a-Chip

Microfluidic 3D cell cultures combine the growth of cells in an extracellular matrix in a multichannel 3D microfluidic cell culture chip, which simulates the activities, mechanics and physiological responses of entire organs and organ systems. There is a perfect control of nutrients, growth factors, oxygen and chemicals delivered to the 3D structures and of the factors generated. For example, in the gut-on-a-chip, Caco-2 cells cultured in a fluid flow (but not in static membrane culture) displayed villus-like differentiation and morphology and increased mucus secretion, glucose uptake and drug metabolism. Several human organs can be interconnected in a chip in multiorgan chips.

8.6.4 Organs in Baths or in Culture

Liver, brain, pituitary gland, ileum, and other organs from animals or humans can be used. They are maintained, both intact or in slices, in baths or in culture. The spatial relationships between the different types of cells are conserved. They can present problems of hypoxia in the centre of the organ and difficulties in the microscopic visualization.

8.6.5 Perfused Organs

Certain organs from animals such as liver, kidney or pancreas can be maintained in optimal conditions by continuous perfusion. Culture facilitates toxicokinetic studies of uptake, biotransformation and secretion of substances.

8.6.6 Embryos

The use of vertebrates in the initial stages of their development presents ethical advantages over the use of adult animals. Embryos (pre- or postimplanted) of rodents, birds, amphibians, reptiles and fish are mainly used to examine alterations in development and differentiation. They are characterized by maintaining their metabolism intact and physiological cellular interactions, and can be exposed at any stage of embryonic development. For example, zebrafish embryo and larva facilitate the visualization of morphological and cellular changes. However, the researcher should know the precise moment in which the species used is considered an animal and becomes protected by legislation.

8.6.7 Lower Organisms

Less evolved living organisms such as invertebrates and plants are useful experimental models, particularly the nematode *Caenorhabditis elegans*, the vinegar fly *Drosophila melanogaster*, the Cladocera *Daphnia magna*, planaria or molluscs.

8.7 BIOINDICATORS USED IN *IN VITRO* MODELS

Many types of indicators can be applied to *in vitro* systems to quantify the changes induced by chemical exposure (see Table 8.3). Indicators range from the classical cytomic techniques (biochemical, microscopic and flow-cytometrical investigations) to molecular approaches comprising

Table 8.3 Main Types of Bioindicators Used in *In Vitro* Systems

	Type of biomarker	Example
1	Cellular and tissue morphology (optical and electronic microscopy, high content analysis)	Shape, size, differentiation Membranes, organelles
2	Cell viability (photometry, fluorimetry, bioluminescence, cytometry)	Dye uptake, adhesion, phagocytosis, vacuolization.
3	Cell proliferation	Protein markers, DNA content, cell cycle
4	Metabolic activity (+ calorimetry)	Regulatory substances (e.g. ATP, Ca^{2+}) Use of energy, enzyme activities Synthesis of macromolecules
5	Cytoskeleton/membranes	Composition and stability Ionic permeability Transporter systems
6	Signalling pathways	Gap junction intercellular communication, cytokines
7	Nucleic acids (transcriptomics, genomics)	Gene expression/inhibition Mutation/degradation of nucleic acid/necrosis/apoptosis
8	Proteins (proteomics)	Selective protein induction
9	Biotransformation systems	Mixed function oxidase (MFO, cytochrome P450s)
10	Defence systems (antioxidative, antimicrobial, etc.)	Glutathione (GSH), glucose-6-phosphate dehydrogenase (G6PDH), metallothionein, mediators of inflammation, oxidative stress
11	Specific indicators	Binding and activation of receptors, contractility, electrophysiology, secretion and removal of substances (e.g. acetylcholine, norepinephrine)

genomic (DNA sequence), transcriptomic (RNA), proteomic (amino acid), metabolomic (nutrients, detritus, changes in molecules), biosensor and reporter gene (genes activated by the contact with one molecule induce a visible change) assays. Most of the systems have been miniaturised and adapted to robot-assisted high-throughput screening.

8.8 MAIN REGULATED *IN VITRO* TOXICOLOGICAL TESTS

As previously considered, there are different regulations that require toxicological information, such as REACH (EC 2008a); the United Nations Globally Harmonized System (GHS); the EU Classification, Labelling and Packaging (CLP) system (EC 2008b); and the cosmetics (Adler et al. 2011; EU 2019) and pharmaceutical (ICH 2000, 2005, 2009) regulations. Table 8.4 shows the main *in vitro* assays accepted to be used in safety regulatory testing, grouped according to different endpoints. The tests should be carried out under special requirements according to GLPs (OECD 2018).

8.8.1 Acute Toxicity (Oral, Dermal, Inhalation Routes)

The investigation of acute effects aims to classify substances by their hazard properties. The classification of a substance means its inclusion in the corresponding group according to its toxic capacity. Acute toxicity is the ability of a substance to produce adverse effects after a single exposure. These effects can vary from a mere skin irritation to death. As it is an obvious parameter, the mean lethal dose (LD_{50}, the dose of a test substance that is lethal for 50% of the animals in a dose group) has traditionally been considered the best indicator of the acute toxic capacity of a substance.

The replacement of animals in acute systemic toxicity testing remains a considerable challenge (Prieto et al. 2019). Nevertheless, *in vitro* studies should be used at least to select the starting doses for the acute *in vivo* assays (Guidance Document [GD] 129). An evaluation based on the weight of the evidence can be applied to avoid the oral LD_{50} test in mammals by means of a cytotoxicity test using cultures of fibroblasts and evaluating the decrease in viability caused by the chemicals by measuring the capacity of live cells to take up/accumulate the dye neutral red. It can be used for the identification of substances that do not require classification for acute oral toxicity (LD_{50} >2000 mg/kg) according to the classification and labelling requirements. In addition, if it is available a

Table 8.4 Main Assays Alternative to Animals Accepted in Safety Regulatory Testing, Grouped according to Different Endpoints

Endpoint	*In vitro*	Comments
Acute systemic toxicity (oral, dermal, inhalation)	3T3 neutral red uptake cytotoxicity test	For *in vivo* dose selection (129)1 To avoid LD_{50} for the identification of substances that do not require classification for acute oral toxicity (Classification, Labelling and Packaging [CLP]) If it is available, a repeated-dose study with an NOAEL ≥1000 mg/kg, avoid the LD_{50} test and use the same classification (REACH).
Skin corrosivity	Reconstructed human epidermis (RHE) (431)	Allow the classification and subcategorization of the corrosivity of the chemicals (1A, 1B, 1C) (see text)
	Ex vivo transcutaneous electrical resistance (TER) (430)	Only for classification
	Membrane barrier test (435)	Only for classification for some products
Skin irritation	Reconstructed human epidermis (439)	Classification and subcategorization 1, 2, 3
Eye irritation	Short exposure *in vitro* method in monolayers of rabbit cornea (SIRC) cell line 491 Isolated eye of chicken test *ex vivo* (438)	Identify (1) chemicals inducing serious eye damage and (2) chemicals not requiring classification for eye irritation or serious eye damage
	Fluorescein leakage test in the canine kidney cell line MDCK 460 Opacity Permeability test in *ex vivo* bovine cornea (BCOP) (437)	Identify ocular corrosives and severe irritants
	Reconstructed human cornea-like epithelium (RhCE) test (492) The vitrigel test with hCE cells (494)	Identify compounds not requiring classification and labelling for eye irritation or serious eye damage
Sensitization	*In chemico* skin sensitisation: direct peptide reactivity test DPRA (442C)	Replacement
	Skin sensitisation *in vitro*: ARE-Nrf2 luciferase test: KeratinoSens (442D)	Replacement
	Dermal sensitization *in vitro* (442E): activation of dendritic cells by (1) activation surface markers; (2) activation test of the U937 or U-SENS cell line; and (3) assay of the interleukin-8 reporter gene or IL-8 luc assay	Replacement
Repeat-dose toxicity		Not available
Genotoxicity		
(a) Gene mutation	Reversion of bacterial mutation (471)	Many false positives
	Mammalian cell gene mutation using the Hprt or xprt genes (476) or the thymidine kinase gene (490)	Many false positives
(b) chromosomal abnormalities	*In vitro* mammalian cell micronucleus test (487)	Detect clastogens and aneugens
	In vitro mammalian chromosomal aberration test (473)	Detect clastogens
(c) Primary DNA damage		Not available. Somatic cell *in vivo* tests to investigate gene mutations are TGR-transgenic rodent somatic and germ cell gene mutation assays (488); the *in vivo* mammalian alkaline comet assay (489); or the liver unscheduled DNA synthesis (UDS) *in vivo* (486)

(*Continued*)

Table 8.4 (Continued)

Endpoint	*In vitro*	Comments
Carcinogenicity	Cell transformation assay (CTA): Syrian hamster embryo (214) Bhas (BALB 3T3 A31–1–1 cell line transfected with H-ras gene) (231)	Can be used as part of a weight of evidence approach in the testing of substances for carcinogenic potential
Developmental toxicity	Mouse embryonic stem cells (mEST) Whole rat embryo test (WEC) Micro mass (MM) assay	None accepted
Reproductive toxicity	Steroidogenesis in H295R cells (456)	Complementary
Endocrine disruption	Estrogenic agonists/antagonists: performance-based test guideline for stably transfected transactivation *in vitro* assays using stably transfected TA (STTA) assay and the BG1Luc ER (455) Estrogen-binding affinity HERα (493)	Complementary
	Androgenic agonists/antagonists: AR-STTA stably transfected human androgen receptor transcriptional activation assay (458)	Complementary
Aquatic acute toxicity	Threshold approach for acute fish toxicity testing	In combination with fish acute toxicity test (203)
	Fish embryo acute toxicity (FET) test (236) and RTgill-W1 fish cell line test (249)	Replacement in some cases
ADME/Kinetics	Dermal absorption (428)	
	Distribution	Not available
	Metabolism	Not available
	Bioaccumulation in fish: *In vitro* intrinsic clearance using cryopreserved rainbow trout hepatocytes (RT-HEP) (319A) *In vitro* intrinsic clearance using rainbow trout liver S9 subcellular fraction (RT-S9) (319B)	Can be used to improve *in silico* predictions of the bioaccumulation in fish
Pharmacokinetics and pharmacodynamics	Microdosing in human volunteers	ICH (2009)
Others	*In vitro* 3T3 NRU phototoxicity test (432) ROS (reactive oxygen species) assay for photoreactivity (495), ICH S10	Partial replacement
	Pyrogenicity: *in vitro* release of interleukins	Accepted in the European Pharmacopoeia
	Human vaccines: —Tetanus vaccine for human use: ELISA and the inhibition of the binding of the toxin (ToBI) —Erysipelas vaccines potency: ELISA	European Pharmacopoeia
	Medical device materials: several cytotoxicity tests	ISO 10993–5
	Marine toxins in live bivalve molluscs: liquid chromatography–mass spectrometry (LC-MS/MS) technique	Replacement of rodent bioassay EU. Com. Regulation 15/2011

1 The number corresponding to the OECD test guideline code is enclosed in parentheses; for example, the OECD test guideline 129 is the 3T3 neutral red uptake cytotoxicity test.

repeated dose study yielding a NOAEL ≥1000 mg/kg, it is possible to avoid the requirement of the LD_{50} test and use the same classification.

8.8.2 Corrosive and Irritative Dermal Capacity

The corrosive effect of a substance is the irreversible destruction of the tissues after its application. The EU and the OECD recommend following an integrated strategy to detect corrosives or severe irritants on eyes and skin. After considering the physicochemical properties, the pH and the result of validated *in vitro* tests, the compounds can be directly classified as severe irritants without being tested in animals. First, strongly alkaline (pH = 11.5) or acidic (pH = 2) substances should not be subjected to skin studies because this property is sufficiently predictable to classify it as corrosive. In addition, the inclusion of a predictive system of structure activity relationship in the first stage greatly simplifies the procedure (Doe and Botham 2019).

In the evaluation of the corrosive capacity to the skin, there are three accepted approaches. First, the *in vitro* skin corrosion: reconstructed human epidermis (RHE) (TG431) allows the subcategorization of the corrosivity of the chemical products (UN GHS/CLP UE subcategories 1A, 1B and 1C, being corrosive after very short, short or more than 1 hour exposure, respectively), in addition to classifying the noncorrosive ones. There are several commercial systems very similar, both morphologically and functionally, to skin (EpiSkin, EpiDerm SCT, SkinEthic RHE, epiCS); these systems rely on the metabolism of a dye by viable cells to produce a colour with absorbance read using a spectrophotometer or plate reader.

Second, the transcutaneous electrical resistance (TER, TG430) test is an *ex vivo* procedure that employs rat or human skin and correctly identifies corrosives and noncorrosives. It is based on the decrease in electrical resistance that occurs between two compartments separated by a dermal membrane when damaged by the compound.

Third, the *in vitro* membrane barrier test method for skin corrosion controls the time required for a compound to destroy a collagen biomembrane that separates two compartments from a bottle, allowing the passage of a dye from one compartment to another. It is only allowed in some legislations (United States) and for certain types of OECD substances (CORROSITEX, TG435). Only when it was not possible with the previous procedures to classify the substance as corrosive could the dermal and ocular corrosive action in rabbit be exceptionally evaluated.

Skin irritation refers to the production of a reversible skin lesion as a consequence of the application of a test substance to rabbits for a period of up to 4 hours. Currently, only exceptionally are animal tests required for the assessment of dermal irritative capacity. Various *in vitro* procedures, measuring the reductive capacity on the vital dye MTT of reconstructed human epidermis (B.46, TG439), can be used to identify the hazard of irritant chemicals (substances and mixtures) according to category 2 of the UN GHS and the CLP. In the EU, and in other regions that have not adopted the optional category 3 of the UN GHS (mild irritants), they can also be used to identify unclassified chemicals (i.e. the "uncategorized" of the UN GHS and the CLP of the EU). They can be used to determine the skin irritancy capacity of the chemicals, such as an independent test to replace the skin irritation test *in vivo*, within a sequential evaluation strategy.

The models of reconstructed human skin (EpiSkin, EpiDerm, SkinEthic RHE, epiCS) are therefore a substitution alternative for the dermal irritation test in rabbit (EU B.46). They use the reduction of the MTT dye as a biomarker, although it can also rely on the release of IL-1α to increase its sensitivity or confirm negative results. The EpiDerm procedure requires the confirmation of negative results by another procedure.

8.8.3 Eye Irritant or Corrosive Ability

Currently, only exceptionally are rabbit tests required for the evaluation of ocular irritative capacity, because several OECD *in vitro* procedures are applied. First, there are two *ex vivo* systems using animal material from slaughterhouses: the opacity and permeability test in bovine cornea identifies ocular corrosives and severe irritants (BCOP, TG437); and the test of the isolated eye of chickens identifies (1) severe eye irritants and (2) compounds that do not require classification by eye irritation or severe damage (TG438).

Second, the fluorescein release test in the canine kidney cell line MDCK identifies severe ocular corrosives and severe irritants (TG460); the short exposure *in vitro* method in monolayers of Statens Seruminstitut rabbit cornea (SIRC) cells to identify (1) compounds that induce severe eye damage and (2) compounds that do not require classification for ocular irritation or severe ocular damage (STE, TG491); and the method of reconstituted human corneal epithelium (RhCE, TG492) or the vitrigel (TG494) to identify compounds that do not require classification for ocular irritation or severe ocular damage (which should be used in combination to replace the Draize rabbit assay).

The US Environmental Protection Agency has established an *in vitro* irritation testing approach for antimicrobial cleaning products using the cytosensor microphysiometer, the BCOP and the EpiOcular model.

The sequential procedure for the ocular evaluation starts from the application of the dermal test. Therefore, the substances that demonstrate corrosive power or severe irritation should not be subjected to the ocular test and are classified directly as corrosive or ocular irritants.

8.8.4 Sensitizing Capacity

A skin sensitizer is a substance that induces an allergic response after contact with the skin. Since 2016, REACH considers *in vitro* or *in chemico* systems as the trials of choice. (*In chemico* means the use of an abiotic system to quantify chemical reactivity.)

The mechanistic information of the pathway of adverse effects is provided with *in vitro/in chemico* methods of each of the key events of skin sensitization: (a) molecular interaction with skin proteins, (b) inflammatory response in the keratinocyte and (c) activation of dendritic cells.

Defined Approaches (DA) on Skin Sensitisation have been published (TG497, OECD, 2021). A DA consists of a fixed data interpretation procedure applied to data (*e.g. in silico* predictions, *in chemico, in vitro* data) generated with a defined set of information sources to derive a prediction without the need for expert judgment. The DAs included in this Guideline have shown to either provide the same level of information or be more informative than the murine Local Lymph Node Assay for hazard identification. In addition, two of the DAs provide information for sensitisation potency categorisation that is equivalent to the potency categorisation information provided by the LLNA. The DA is valid if at least 2 of the 3 of the following *in vitro* or *in chemico* tests are performed: (1) *in chemico* dermal sensitization direct peptide reactivity test (DPRA, TG442C); (2) ARE-Nrf2 luciferase test KeratinoSens (TG442D); (3) activation of dendritic cells by means of the activation test in human cell line or h-CLAT method, measuring surface markers by loose cytometry. If it is necessary to apply an *in vivo* test, the local lymph node test (LLNA) in the mouse is used.

8.8.5 Repeated-Dose Toxicity

Several models have been developed to mimic *in vitro* prolonged exposure, particularly using human stem cells and organs-on-a-chip and bioreactors. However, it is not possible at the moment to predict adequately *in vitro* repeated-dose toxicity.

8.8.6 Genotoxicity/Mutagenicity

Genotoxicity is defined as the deleterious effect of any chemical that could potentially damage the cell's own genetic material, its DNA or RNA, thus affecting its integrity and normal function (Dixit and Kumar 2018). Genotoxic compounds can be carcinogens, teratogens or mutagens. A *carcinogen* is any substance or agent that promotes cancer and is often (but not necessarily) a teratogen or mutagen. A *teratogen* is any agent that can disturb the development of an embryo or foetus. A *mutagen* is a chemical compound or form of radiation (such as ultraviolet [UV] radiation) that causes irreversible and heritable changes (mutations) in the cellular genetic material. Genotoxicity is often confused with mutagenicity, but it is important to note that all mutagens are genotoxic agents, but all genotoxic substances need not be mutagens (Dixit and Kumar 2018).

Genotoxicity assessment is an essential component of the safety assessment of all types of substances, ranging from pharmaceuticals, industrial chemicals, pesticides, biocides, food additives and cosmetics ingredients to veterinary drugs, relevant in the context of international legislations aiming at the protection of human and animal health. A variety of well-established *in vitro* assays are in place and have been used successfully to predict genotoxicity; however, at present they cannot be considered to fully replace animal tests currently used to evaluate the safety of substances (Corvi et al. 2019).

A battery of *in vitro* and *in vivo* tests is chosen for the toxicological evaluation process. The most common *in vitro* genotoxicity tests are described below.

8.8.6.1 The Bacterial Mutation Test or Ames Test

The bacterial mutation test, or Ames test, is developed to determine if the chemical tested has mutagen capacity. This assay is based on the principle of reverse mutation or back mutation. The bacterial reverse mutation test uses amino acid–requiring strains of *Salmonella typhimurium* and *Escherichia coli* to detect point mutations, which involve substitution, addition or deletion of one or a few DNA base pairs. The principle of this bacterial reverse mutation test is that it detects mutations which revert mutations present in the test strains and restore the functional capability of the bacteria to synthesize an essential amino acid. The revertant bacteria are detected by their ability to grow in the absence of the amino acid required by the parent test strain (TG471).

8.8.6.2 *The* In Vitro *Mammalian Cell Gene Mutation Test*

The purpose of the *in vitro* mammalian cell gene mutation test is to detect gene mutations induced by chemicals. Several suitable cell lines are available; the most commonly used genetic endpoints measure mutation at thymidine kinase (TK) (TG490) and hypoxanthine-guanine phosphoribosyl transferase (HPRT) and a transgene of xanthine-guanine phosphoribosyl transferase (XPRT) (TG476). With respect to the *in vitro* mammalian cell gene mutation tests using the TK gene, the cells with a deficiency in the TK enzyme (due to a mutation from TK± to TK–/–) often arise due to chromosomal aberration or gene mutation. Gene mutations include point mutations, frameshift mutations and deletions, whereas chromosomal aberrations can be due to the lack of a large portion of chromosomal DNA or rearrangements, either of which leads to abnormal chromosome numbers.

It is really useful to recognize the aneugen compound, which is any substance or process that, by interacting with the components of the mitotic and meiotic cell division cycle, leads to aneuploidy (any deviation from the normal diploid number of chromosomes by a single chromosome or more than one, but not by entire sets of chromosomes) in cells or organisms. In respect to the *in vitro* mammalian cell gene mutation tests using the Hprt and xprt genes, this test detects mutation at the locus of hypoxanthine-guanine phosphoribosyl transferase (HPRT) and xanthine-guanine phosphoribosyl transferase transgene (gpt) (XPRT). There is a mutation in the X chromosome of the gene at the HPRT locus. This locus is normally essential for encoding the HPRT enzyme, a transferase that aids the conversion of hypoxanthine and guanine to inosine monophosphate and guanosine monophosphate, respectively. This enzyme is crucial in the wild pathway of purine synthesis. Purines and pyrimidines make up the two groups of nitrogenous bases, including the two groups of nucleotide bases. Two of the four deoxyribonucleotides (deoxyadenosine and deoxyguanosine) and two of the four ribonucleotides (adenosine [AMP] and guanosine [GMP]), the respective building blocks of DNA and RNA, are purines. A mutation in the gene at the HPRT locus could lead to serious syndromes (Nguyen and Nyhan 2013).

8.8.6.3 *The* In Vitro *Mammalian Cell Micronucleus Test*

The *in vitro* micronucleus (MNvit) test is a genotoxicity test for the detection of micronuclei (MN) in the cytoplasm of interphase cells. Micronuclei may originate from acentric chromosome fragments (i.e. lacking a centromere) or whole chromosomes that are unable to migrate to the poles during the anaphase stage of cell division. Therefore, the micronucleus test is an *in vitro* method that provides a comprehensive basis for investigating chromosome damaging potential *in vitro* because both aneugens and clastogens can be detected in cells that have undergone cell division during or after exposure to the test chemical. Micronuclei represent damage that has been transmitted to daughter cells, whereas chromosome aberrations scored in metaphase cells may not be transmitted. In either case, the changes may not be compatible with cell survival (TG487).

8.8.6.4 *The Mammalian Chromosome Aberration Test*

The purpose of the *in vitro* chromosome aberration test is to identify agents that cause structural chromosome aberrations in cultured mammalian cells. Structural aberrations may be of two types: chromosome or chromatid. Chromosome aberrations and related events are the cause of many human genetic diseases, and there is substantial evidence that chromosome mutations and related events causing alterations in oncogenes and tumour suppressor genes of somatic cells are involved in cancer induction in humans and experimental animals (TG473).

Mutagenicity tests were the first *in vitro* assays included in the regulation. However, the *in vitro* tests are not considered a full replacement of animal tests due to their low specificity. Consequently, at least positive results should be evaluated *in vivo*, mainly with the mammalian erythrocyte micronucleus test (TG474), the mammalian bone marrow chromosomal aberration test (TG475), the transgenic rodent somatic and germ cell gene mutation assay (TG488) and the *in vivo* mammalian alkaline comet assay (TG489).

8.8.7 Carcinogenicity

Repeated-dose testing of both sexes in mice and rats with complete pathological analysis is the basic procedure for assessing carcinogenic capacity. However, these assays need a long time, high dedication, are very expensive and require more than 400 animals for each of the two species used.

Several mutagenicity models with genetically modified animals have been proposed, which have the advantage of requiring short exposure time (only 28 days) to develop tumours. The main gene mutation assays are lacZ bacteriophage mouse (MutaMouse); lacZ plasmid mouse; gpt delta (gpt and Spi–) mouse and rat; lacI mouse and rat (Big Blue). In addition, the cII positive selection assay can be used for evaluating mutations in the Big Blue and MutaMouse models (TG488).

These rodent *in vivo* gene mutation tests are especially relevant to assessing mutagenic hazard, in that the assays' responses are dependent upon *in vivo* metabolism, pharmacokinetics, DNA repair processes, and translesion DNA synthesis. An *in vivo* assay for gene mutations is useful for further investigation of a mutagenic effect detected by an *in vitro* system and for following up on results of tests using other *in vivo* endpoints. In addition to being causally associated with the induction of cancer, gene mutation is a relevant endpoint for the prediction of mutation-based noncancer diseases in somatic tissues and diseases transmitted through the germline. Pharmaceutical regulatory agencies accepted the models to be used in lieu of a second species full carcinogenicity bioassay (Corvi et al. 2019).

Although there are not complete *in vitro* tests for carcinogenicity, the *in vitro* Bhas 42 transformation assay (BALB 3T3 A31–1–1 cell line transfected with H-ras gene) can be used for hazard identification of potential carcinogenicity of chemicals with initiating and/or promoting activity and is able to differentiate between genotoxic and nongenotoxic carcinogens (GD231). The *in vitro* Syrian hamster embryo (SHE) cell transformation assay (CTA) can be used as part of a weight of evidence approach in the testing of substances for carcinogenic potential (GD214).

8.8.8 Developmental and Reproductive Toxicity

The studies for reproductive and developmental toxicity can be grouped into three types: studies of reproduction (multigenerational), development and endocrine disruption. Developmental toxicity is evaluated with the prenatal development study: alterations in embryogenesis (TG414). The extended one-generation reproductive toxicity study (TG443) is preferred because it uses approximately half the number of animals (1400 rats) than the two-generation reproductive test (TG416) (Worth 2019).

There are three embryotoxicity procedures validated but not accepted: the test of mouse embryonic stem cells test (mEST) and the whole rat embryo test (WEC) distinguish moderate, strong and nonembryotoxic. The mammalian micro mass (MM) assay identifies potent embryo toxicants. Other useful tests are the frog embryo (FETAX) and the RETox.

The *in vivo* procedures for endocrine disruption include the uterotrophic assay in rodents to detect estrogenic properties (TG440); the Hershberger bioassay in the rat for androgenic agonists, antagonists and 5α-reductase inhibitors (TG441); and the oral repeated 28-day doses in rodents. There are *in vitro* transcriptional options to detect estrogenic agonists and antagonists with the performance-based test guideline for stably transfected transactivation *in vitro* assays using stably transfected TA (STTA) assay and the BG1Luc ER assay (TG455), estrogen binding affinity HERα (TG493) and for androgenic agonist and antagonist activity (AR-STTA) (TG458). In addition, the steroidogenesis test can be carried out in H295R cells (TG456).

8.8.9 Toxicity to the Environment

Acute and chronic aquatic toxicity is required for environmental hazard classification under the GHS; hazardous product legislation also requires classification as a marine pollutant.

Acute toxicity is required in fish, *Daphnia magna* and algae for the classification of substances for the aquatic environment. The most commonly used fish species are *Brachydanio rerio* (zebrafish), *Pimephales promelas, Cyprinus carpio* (carp), *Poecilia reticulata, Lepomis macrochirus, Onchorhynchus mykiss* (rainbow trout), *Leuciscus idus* and *Oryzias latipes* (medaka). The fish test should not be performed if the substance is poorly soluble or does not penetrate the membranes.

In a second stage, prolonged toxicity is usually evaluated by means of full-cycle tests on fish. For plant protection products, acute toxicity, long-term data and alteration of reproduction in birds are also required.

In addition to the use of a threshold of toxicological concern and the limit test (GD 126), another reduction alternative is available. The threshold approach for acute fish toxicity test is a tiered testing strategy based on the fact that the LC_{50} value of the most sensitive of the three test species (fish, algae and invertebrates) is commonly used for hazard and risk assessment and that fish is often not the most sensitive test species (Zuang et al. 2018). When fish acute toxicity data is required, the fish limit test should be carried out at only one concentration. The threshold concentration corresponds to the lowest LC_{50}/EC_{50} derived from reliable acute invertebrate (e.g. *Daphnia*, TG202) and algae (e.g. TG201) data. If no fish mortality is observed at the threshold concentration, it demonstrates that the fish is not the most sensitive test species. No further testing is required and the TC value can be used as LC_{50} fish. If mortality occurs at the threshold concentration, a full LC_{50} test should be conducted.

The alternatives for the replacement or further reduction of fish use are mainly focused on the fish embryo acute toxicity test (FET) (TG236) and the application of fish cell lines, such as the RTgill-W1 (TG249) (Scholz et al. 2013; Zurita et al. 2019).

8.8.10 Toxicokinetics in the Body and Degradation in the Environment

Although official toxicokinetic protocols of absorption, distribution, metabolism and excretion still only include *in vivo* studies (TG417, EU B.37), with the exception of dermal absorption, numerous *in vitro* procedures have been developed and are routinely applied as election procedures (e.g. in the development of medicines).

Generally, the results obtained from different systems are integrated in hierarchical schemes in predictive models of kinetics and metabolism. These latter computer models are based on two types of parameters: first, the interspecific differences in anatomy and physiology, such as the pulmonary ventilatory frequency; and second, the specific parameters of the compound, such as the blood/tissue partition coefficients, blood/air, and kinetic parameters such as the constants of Michaelis-Menten, Vmax and Km. With all this, it is possible to extrapolate route to route, dose to dose and between species.

The skin absorption is quantified *in vitro* with human or rat dermal membranes in a diffusion cell, studying the proportion of the substance that crosses them. This robust *in vitro* procedure (TG428) was accepted before the *in vivo* procedure (TG427).

Gastrointestinal absorption used to be studied with the *ex vivo* method of isolated colon, but can be investigated in monolayer cultures of Caco-2 colon cells grown for 3 or 21 days on a semipermeable membrane, which also allows studying the metabolism and effects.

The passage of the blood-brain barrier is usually modelled by monolayers of endothelial cells of brain capillaries, or MDCK or CaCo-2 cells grown on filters.

Although the systems must be improved, the Caco cell line expressing the multiple drug resistance protein 1 (MDCKmdr-1) distinguishes the compounds that cross the blood-brain barrier by passive diffusion from those that require active transport. For membrane absorption, membrane cultures are also used. The passage of membranes by passive transport is determined by the physicochemical properties such as lipophilicity (LogP, LogD), pKa, solubility and molecular weight.

In the metabolism and biotransformation phases, the intrinsic metabolic clearance (CLint), the Km and Vmax activities of metabolizing enzymes, the stability and the metabolic profile are usually investigated in hepatocytes (isolated, cultured or cryopreserved) and slices. The use of organelles (microsomes, S9) has decreased because biotransformer systems in phase II are poorly represented, although the use of human liver material has become popular. The *in vitro* procedures are particularly useful for the study of metabolites formed, which shows the predominant metabolic pathways and the active metabolites.

The existence of genetic polymorphisms in biotransformation can be investigated *in vitro* by several batteries of cells genetically engineered to express different human isoenzymes of cytochrome P450 or N-acetyltransferase. This is intended to know the stability of the compound, if the toxicity changes with metabolization and which is the isoenzyme responsible for biotransformation. This makes it possible to predict risk of human groups, which would be difficult to identify through *in vivo* tests, to know the differences between species and to optimize the design of animal tests in (selection of species and strain) and clinical trials (selection of individuals), as well as to predict interactions by studying the modification of the isoenzymes.

There are also specific protocols to investigate degradation and bioconcentration. *In vitro*–derived clearance data can be used in REACH to improve the prediction of *in silico* methods to derive a bioconcentration factor as a means for bioaccumulation. Cryopreserved rainbow trout hepatocytes (TG319A) and rainbow trout liver S9 subcellular fraction (TG319B) can be used.

8.8.11 Phototoxicity

In the evaluation of phototoxic capacity, of great interest for the compounds applied to the skin, there is no validated or accepted *in vivo* method, but an *in vitro* assay has been accepted in fibroblast cultures of the mouse cell line (3T3 NRU PT), which are exposed to the substance tested and irradiated or not with ultraviolet light, to then compare the uptake of the vital dye red neutral. In the case of phototoxic substances, there is an increase in the toxicity in the irradiated cells due to activation by the light. This was the first *in vitro* toxicity test experimentally validated and accepted for regulatory applications (TG432, ICH S10), apart from the mutagenicity tests. However, this procedure does not detect all the possible phototoxic mechanisms, so it must be complemented with other systems to detect reactive oxygen species (ROS), photostability (ICH S10, TG495) and so forth (Zuang et al. 2018).

8.8.12 Pyrogenicity

In the specific case of medicines, as substitution alternatives to the rabbit assay for gram-negative endotoxin-mediated pyrogenicity, five *in vitro* monocyte activation (MAT) assays were validated by ECVAM in 2006: the release of IL-1 in blood complete human; the release of IL-6 in whole human blood; the release of IL-6 in peripheral blood mononuclear cells (PBMC); the release of IL-6 in the human monocytic cell line MonoMac-6 (MM6); and the release of IL-1 in whole human blood cryopreserved. These procedures replace the use of rabbits by 2026 and also the *in vitro* system that has been used with great utility for many years consisting in the coagulation of the amoebocyte lysate of the horseshoe crab (*Limulus polyphemus*) (LAL test).

8.8.13 Potency and Safety of Vaccines and Biological Products

A wide variety of products require, the evaluation of the safety of each of their batches produced. They include vaccines for human and veterinary use, products obtained from blood and plasma and those obtained by biotechnology. General safety tests for batch release of human and veterinary vaccines are no longer required in Europe and were deleted from European Pharmacopoeia monographs (abnormal toxicity test, target animal batch safety test) but not in other countries (Zuang et al. 2018).

Since 2003, the European Pharmacopoeia accepts as valid the technique of ELISA and an assay of inhibition of the binding of the toxin (ToBI) for the evaluation of the potency of the batches of the tetanus vaccine for human use. The European Pharmacopoeia has also accepted since 2004 the use of an ELISA test to evaluate the potency of batches of human erysipelas vaccines (EC 2018). Various techniques are being promoted, including one by PCR, to replace the mouse bioassay for the evaluation of botulinum toxin. ESAC-ECVAM already recommended in 2002 the elimination of the evaluation test of the potency of the concentrated erythropoietin solution and also the routine safety testing of each batch of vaccines for veterinary use, considering them unnecessary as long as there were no changes in their manufacture.

For veterinary vaccines, there are several VICH guidelines for waiving the target animal batch safety test for inactivated (VICH GL50(R); VICH 2017a) and live vaccines (VICH GL55; VICH 2017b) and the general batch safety test in laboratory animals (e.g. abnormal toxicity test).

8.8.14 Other

Many other specific endpoints are applied, especially for safety pharmacology studies, particularly for cardiovascular, respiratory and central nervous systems but also for neutropenia, hepatotoxicity and so forth (ICH 2000).

One of the most relevant causes of cardiotoxicity is the prolongation of the duration of the action potential and thus the QT interval. This syndrome is due to the interaction of the drugs with the ether-a-go-go-related gene (HERG), which codes for the potassium channel, increasing the duration of the action potential in cardiac cells, inducing arrhythmias and death. Several types of *in vitro* assays are used to screen compounds, which may exert this adverse effect on potassium channels, particularly using stem cell–derived cardiomyocytes (hPSC-CM) and automated patch clamp electrophysiology (ICH 2005).

Numerous drugs cause hepatotoxicity in humans, and even only in certain susceptible individuals but not in animal toxicity studies. Different individual assays have been proposed to investigate drug-induced liver injury (DILI). For example, HepaRG cells retain the expression of many liver-specific functions as well as many CYP450s, nuclear receptors, membrane transporters and phase II enzymes. High DILI sensitivity (>70%) was achieved when toxicity studies were undertaken for several days *in vitro* by using liver spheroids or when data provided by multiple mechanistically relevant assays were combined. Useful combinations include reactive metabolite formation, cell cytotoxicity, bile salt export pump inhibition and mitochondrial injury (Kenna and Ram 2019).

The CFU-GM granulocyte-monocyte colony formation assay was validated by ECVAM in 2006 as an *in vitro* substitution alternative for the second animal species (dog) to predict acute human neutropenia, recommending its use as a complement to the HTD assay in the mouse to detect bone marrow toxicity of antitumorals.

Medical device materials can be investigated by several cytotoxicity tests according to ISO 10993–5. For other special studies, the most appropriate species and system is used according to the characteristics of the compound, using more sophisticated indicators, such as neurotoxicity. Although *in vitro* tests of the inhibition of neuropathy target esterase (NTE) can be used to detect substances capable of producing delayed polyneuropathy, the negative results do not exclude that the substance studied is neurotoxic. Other systems are developed for neurotoxicity and development neurotoxicity (Ríos et al. 2003).

In the area of food safety, since 2014 the rodent bioassay for the detection of marine toxins in live bivalve molluscs was replaced in the EU by the liquid chromatography–mass spectrometry technique (CR 15/2011; EU 2011).

In conclusion, many *in vitro* methods are efficiently applied in biomedical and environmental research. In regulatory studies, *in vitro* systems are progressively replacing the use of animals in specific assays and are being implemented in integrated testing strategies. The most complex endpoints to replace seem to be repeat-dose toxicity tests, reproductive and developmental toxicity and carcinogenicity assays.

ACKNOWLEDGMENTS

This work was partially supported by the grants CTM2012–31344, UNPO13–1E-2444 and EQC2018–005220-P from the Spanish Ministry of Economy and Competitiveness-FEDER and Proyecto Puente CTM2016–76304-C2–1-R and 2019-PPI1901 (VPPI-UPO) from the Consejería de Economía y Conocimiento de la Junta de Andalucía 2017, Spain. CA was recipient of a contract from the Programa Operativo de Empleo Juvenil, Consejería de Economía y Conocimiento (Junta de Andalucía, Spain) and Fondo Social Europeo, EJP48, 2017.

ONLINE RESOURCES

The following are some useful websites:

- EU-RL-ECVAM. EU Reference Laboratory for alternatives to animal testing (https://ec.europa.eu/jrc/en/eurl/ecvam).
- US EPA. Pesticides (www.epa.gov/test-guidelines-pesticides-and-toxic-substances/final-test-guidelines-pesticides-and-toxic).
- US EPA. OPPTS Harmonized Test Guidelines (www.epa.gov/test-guidelines-pesticides-and-toxic-substances).
- US EPA. Water (CWA) (www.epa.gov/waterscience/methods/method/index.html).
- FDA guidances and guidelines: CBER (www.fda.gov/cber/guidelines.htm).
- ICH. International Council for Harmonisation of Technical Requirements for Pharmaceuticals for Human Use (www.ich.org/).
- ICH. Safety Guidelines (www.ich.org/products/guidelines/safety/article/safety-guidelines.html)
- IPCS. The International Programme on Chemical Safety (www.who.int/pcs/).
- Johns Hopkins University. Center for Alternatives to Animal Testing (https://caat.jhsph.edu/)
- OECD Testing Guidelines for testing of chemicals (TG) (www.oecd.org/chemicalsafety/testing/oecdguidelinesforthetestingofchemicals.htm).
- OECD Guidance Documents (GD). Series on Testing and Assessment (www.oecd.org/chemicalsafety/testing/series-testing-assessment-publications-number.htm).
- Buscaalternativas.com (http://buscaalternativas.com/).

BIBLIOGRAPHY

Adler, S., Basketter, D., Creton, S., et al. 2011. Alternative (non-animal) methods for cosmetics testing: Current status and future prospects-2010. *Arch Toxicol* 85: 367–85.

Balls, M., R. Combes, and A. Worth. 2019. *The History of Alternative Test Methods in Toxicology*. London: Academic Press.

Bergstrom, M. 2017. The use of microdosing in the development of small organic and protein therapeutics. *J Nucl Med* 58: 1188–95.

Burt, T., and R. D. Combes. 2019. The use of imaging, biomonitoring and microdosing in human volunteers to improve safety assessments and clinical development. Chapter 3.8. In *The History of Alternative Test Methods in Toxicology*, eds. M. Balls, R. Combes, and A. Worth, 229–40. London: Academic Press.

Corvi, R., H. Spielmann, and T. Hartung. 2019. Alternative approaches for carcinogenicity and reproductive toxicity. Chapter 3.6. In *The History of Alternative Test Methods in Toxicology*, eds. M. Balls, R. Combes, and A. Worth, 209–17. London: Academic Press.

Dixit, M., and A. Kumar. 2018. *In vitro* gene genotoxicity test methods. In *In Vitro Toxicology*, eds. A. Dhawan and S. Kwon, 67–89. London: Academic Press.

Doe, J., and P. Botham. 2019. Chemicals and pesticides: A long way. Chapter 3.3. In *The History of Alternative Test Methods in Toxicology*, eds. M. Balls, R. Combes, and A. Worth, 177–84. London: Academic Press.

EC. Council of Europe, 2018 & European Pharmacopoeia 2018. General chapter 5.2.14—Substitution of *in vivo* methods by *in vitro* methods for the quality control of vaccines. In *European Pharmacopoeia*, 9th ed.

EC. Council Regulation (EC 2008a) No 440/2008 of 30 May 2008 Laying Down Test Methods Pursuant to Regulation (EC) No 1907/2006 of the European Parliament and of the council on the registration, evaluation, authorisation and restriction of chemicals (REACH) (Text with EEA relevance) OJ L 142, 31.5.2008, p. 1–739.

EC. Council Regulation (EC 2008b) No 1272/2008 of the European Parliament and of the Council of 16 December 2008 on classification, labelling and packaging of substances and mixtures, amending and repealing Directives 67/548/EEC and 1999/45/EC, and amending Regulation (EC) No 1907/2006 (Text with EEA relevance). OJ L 353, 31.12.2008, p. 1–1355.

EU. 2019. Report from the Commission to the European Parliament and The Council on the Development, Validation and Legal Acceptance of Methods Alternative to Animal Testing in the Field of Cosmetics (2018). Brussels, 15.10.2019. COM(2019) 479 final.

EU. Commission Regulation (EU) No 15/2011 of 10 January 2011 amending Regulation (EC) No 2074/2005 as regards recognised testing methods for detecting marine biotoxins in live bivalve molluscs text with EEA relevance. OJ L 6, 11.1.2011, p. 3–6.

EU. Directive 2010/63/EU of the European Parliament and of the Council of 22 September 2010 on the protection of animals used for scientific purposes text with EEA relevance OJ L 276, 20.10.2010, p. 33–79.

Herrmann, K., F. Pistollato, and M. L. Stephens. 2019. Beyond the 3Rs: Expanding the use of human-relevant replacement methods in biomedical research. *ALTEX* 36: 343–52.

ICH. 2000. *ICH Harmonised Tripartite Guideline Safety Pharmacology Studies for Human Pharmaceuticals S7A*. Current Step 4 version. www.ema.europa.eu/en/ich-s7a-safety-pharmacology-studies-human-pharmaceuticals (accessed October 31, 2020).

ICH. 2005. *Harmonised Tripartite Guideline the Non-Clinical Evaluation of the Potential for Delayed Ventricular Repolarization (qt Interval Prolongation) by Human Pharmaceuticals S7B*. Current Step 4 version dated 12 May 2005. www.ema.europa.eu/en/ich-s7b-non-clinical-evaluation-potential-delayed-ventricular-repolarization-qt-interval (accessed October 31, 2020).

ICH. 2009. *Guideline M3(R2) on Non-Clinical Safety Studies for the Conduct of Human Clinical Trials and Marketing Authorisation for Pharmaceuticals*. www.ema.europa.eu/en/ich-m3-r2-non-clinical-safety-studies-conduct-human-clinical-trials-pharmaceuticals (accessed October 31, 2020).

Kenna, J. G., and R. Ram. 2019. Safety assessment of pharmaceuticals. Chapter 3.2. In *The History of Alternative Test Methods in Toxicology*, eds. M. Balls, R. Combes, and A. Worth, 167–76. London: Academic Press.

Myatt, G. J., E. Ahlberg, Y. Akahori, et al. 2018. In silico toxicology protocols. *Regul Toxicol Pharmacol* 96: 1–17.

Nguyen, K. V., and W. L. Nyhan. 2013. Identification of novel mutations in the human HPRT gene. *Nucleosides Nucleotides Nucleic Acids* 32: 155–60.

OECD. 2017. *Revised Guidance Document on Developing and Assessing Adverse Outcome Pathways*. Series on Testing & Assessment, No. 184.

OECD. 2018. *Guidance Document on Good in Vitro Method Practices (GIVIMP)*.

OECD. 2021. *Guidelines for the Testing of Chemicals*. Organisation for Economic Co-operation and Development.

OECD. 2021. *Guideline No. 497: Defined Approaches on Skin Sensitisation*. OECD Guidelines for the Testing of Chemicals, Section 4. Paris: OECD Publishing. https://doi.org/10.1787/b92879a4-en

Prieto, P., R. Graepel, K. Gerloff, et al. 2019. Investigating cell type specific mechanisms contributing to acute oral toxicity. *ALTEX* 36: 39–64.

Repetto, M., and G. Repetto. 2022. *Toxicología Fundamental*, 5th ed. Madrid: Díaz de Santos.

Ríos, J. C., G. Repetto, A. Jos, et al. 2003. Tribromophenol induces the differentiation of SH-SY5Y human neuroblastoma cells in vitro. *Toxicol in Vitro* 17: 635–41.

Russell, W. M. S., and R. L. Burch. 1959. *The Principles of Human Experimental Technique*. London: Methuen.

Scholz, S., E. Sela, L. Blaha, et al. 2013. A European perspective on alternatives to animal testing for environmental hazard identification and risk assessment. *Regul Toxicol Pharmacol.* 67: 506–30.

Skeldon, G., B. Lucendo-Villarin, and W. Shu. 2018. Three-dimensional bioprinting of stem-cell derived tissues for human regenerative medicine. *Philos Trans R Soc Lond B Biol Sci* 373(1750). pii: 20170224.

VICH. 2017a. *VICH GL50(R) Harmonisation of Criteria to Waive Target Animal Batch Safety Testing for Inactivated Vaccines for Veterinary Use*. www.ema.europa.eu/en/vich-gl50-harmonisation-criteria-waive-target-animal-batch-safety-testing-inactivated-vaccines (accessed October 31, 2020).

VICH. 2017b. *VICH GL55 Harmonisation of Criteria to Waive Target Animal Batch Safety Testing for Live Vaccines for Veterinary Use*. www.ema.europa.eu/en/vich-gl55-harmonisation-criteria-waive-target-animal-batch-safety-testing-live-vaccines-veterinary (accessed October 31, 2020).

Worth, A. P. 2019. Types of toxicity and applications of toxicity testing. Chapter 1.2. In *The History of Alternative Test Methods in Toxicology*, eds. M. Balls, R. Combes, and A. Worth, 7–10. London: Academic Press.

Zuang, V., A. Dura, D. Asturiol Bofill et al. 2018. *EURL ECVAM Status Report on the Development, Validation and Regulatory Acceptance of Alternative Methods and Approaches (2018), EUR 29455*. Publications Office of the European Union, Luxembourg. http://publications.jrc.ec.europa.eu/repository/handle/JRC113594 (accessed October 28, 2020).

Zurita, J., A. del Peso, R. Rojas, et al. 2019. Integration of fish cell cultures in the toxicological assessment of effluents. *Ecotoxicol Environ Safety* 176: 309–20.

9 *In Vivo* Toxicity Testing

Carmen González-Martín, Esther Gramage, María José Polanco and Carmen Rodríguez-Rivera

CONTENTS

9.1 INTRODUCTION

The aim of *in vivo* toxicity studies is to assess the toxicological risk of xenobiotics to humans: what is the mechanism of action of the toxic chemical on health and the concentration which, if not exceeded, will not generally cause adverse effects to the exposed person. Generally, laboratory animals are necessary to perform these toxicity assays. The maximum exposure dose that does not cause any detectable adverse effects is determined in animals and, after applying a safety factor, the dose or exposure level estimated to be safe for humans is established. Toxicological studies in *in vivo* experimental models allow the reproduction of a toxic case in different contexts (e.g. occupational exposure to chemical substances, environmental issues, drugs of abuse, food additives, household products). It is important to note that the European Regulation (EC) No 1223/2009 of November 30, 2009, on cosmetic products banned from 2013 the use of animals to assess the safety of ingredients used for cosmetics in the European Union (EU). However, their use is still allowed in 80% of countries worldwide.

Regarding the *in vivo* preclinical studies of pharmaceutical drugs, these studies can reduce and predict possible unwanted effects on human health before starting the clinical trial. Throughout history, catastrophic events have occurred due to drug-induced toxicity in humans, such as that of thalidomide in the 1960s, despite prior testing in animals. These facts motivated the relevant authorities to increase the safeguards so a greater number of toxicological *in vivo* assays, especially in teratogenicity, are required to be passed before approving the administration to humans.

Regulatory agencies, such as the US Food and Drug Administration (FDA), the Ministry of Health, Labour, and Welfare (MHLW) in Japan, and the European Medicines Agency (EMA), have published guidelines with all these requirements. Moreover, studies must be in accordance with good laboratory practices (GLPs), guides considered as a reference for the surveillance (design and endpoints) and execution of preclinical trials. Especially relevant for chemical testing is the

DOI: 10.1201/9780203730584-9

Organisation for Economic Co-operation and Development (OECD), which provides specific periodically reviewed guidelines (not only for animal testing). Those documents are available at the chemical safety site of the OECD (OECD iLibrary 2020).

9.2 ANIMAL WELFARE CONSIDERATIONS REGARDING *IN VIVO* TESTING

After the 1970s, the 3Rs (Russell and Burch 2009) became one of the basic concepts for protection of experimental animals. The 3Rs are based on (a) replacement of animals with other research options; (b) reduction of the number of animals used, usually possible due to the development and application of statistical approaches to clearly establish the minimum number of animals required, or due to the use of the better species for each study; and (c) refinement of the methods to minimize animal suffering and improve welfare. Environmental enrichment for animals used in research (by identifying and providing the environmental stimuli necessary for optimal psychological and physiological wellbeing), the proper use of anaesthesia and analgesia when needed, and the establishment of an early finishing point based on the classification of the severity of the procedures are some examples of refined processes.

To this end, different specific laws were developed to regulate the use of animals for research purposes, to further protect them, ensure their welfare and decrease their suffering. These norms include, among others, inspections of animal facilities or revisions to the procedures by new ethics committees for animal welfare. One of these legislations is Directive 2010/63/EU, which establishes strict control over animal welfare and implements the 3Rs, promoting the use of alternative methods and applying the replacement of animals whenever it is scientifically possible. When replacement is not possible, reduction of the number of animals and refinement of the methods are mandatory. The directive also recommends the establishment of international programs to share organs and tissues among the states' members.

Based on this, before starting any toxicological study, especially acute ones, it is fundamental to study all the published data of the experimental substance to assure the minimum but necessary number of animals needed to obtain reliable conclusions from the results of the study.

9.3 RISK QUANTIFICATION

The dose/response ratio of a substance is used to quantify the risk of exposure to a substance—in other words, the exact dose of a substance that produces a determined response. To quantify this risk in a laboratory, different animal species are exposed to increasing doses of the given substance, through different routes of administration and during different periods of time, even including "the lifetime of an animal". To do so, different species of laboratory animals are used to which the chemical studied in increasing doses is presented through different routes of administration and for different periods of time, including the life of an animal if necessary. It is important to note that critical phases such as growth and development (e.g. *in utero*, neonate, adolescent) may represent periods of specific toxicological sensitivity for certain substances. Any kind of adverse effect detected is collected, and the percentage of animals affected with each effect is determined. Afterwards, these results are extrapolated to obtain quantitative data for human health. The quantification of the risk is usually represented in a dose/response curve (generally monotonic) as the accumulated frequency versus the dose or the log dose. It allows the following:

a. To know the toxic doses for the different adverse effects identified (lethal doses, doses that induce a 10% reduction of the body weight (bw), neurotoxic and hepatotoxic doses, etc.).

b. To render information of the critical effect (the one that appears at the lowest concentrations), which is normally more relevant for the chronical exposure. This information may include the severity of the effect, if it is a primary adverse effect or an adaptative one due to another alteration, and if it is a precursor of a more significant or severe toxic effect. It is also recommended to estimate the correlation between the alterations in parameters (clinical or biochemical), the weight of the organs and the histopathological effects.

c. To identify the smallest toxic dose below which apparently no adverse effects are observed. The NOAEL (no observed adverse effect level) is the highest dose at which there is not an observed toxic or adverse effect (in morphology, physiology, growth, development or life expectancy of the exposed animal). NOAEL is the parameter of reference most widely used in risk assessments and research.

In some cases (e.g. in effects with a threshold dose), the LOAEL (lowest observed adverse effect level) is also useful, which is the lowest dose at which there was an observed toxic or adverse effect. When the NOAEL/LOAEL approach presented some limitations, such as unknown NOAEL, the US Environmental Protection Agency (EPA) proposed the benchmark dose (BMD) approach as an alternative. The BMD is defined as the lower statistical confidence limit for the dose corresponding to a specified increase in effect over background (US EPA 2012). This approach considers sample size and the shape of the dose-response curve. It is important to notice that there are some products, such as endocrine disruptors, that are characterized by chronic exposure at low doses. The toxicity of endocrine disruptors does not follow monotonic curves. For this reason, it is especially relevant to study the effects of these substances at low doses chronically maintained (Fagin 2012). They will be studied in more detail at the end of this chapter.

Some factors need to be taken into consideration when designing an *in vivo* experimental toxicology protocol:

- The animal species to be used must be suitable for the effects assessed, and the results will be able to be extrapolated to humans. In general, species such as mice, rats, rabbits, guinea pigs, dogs, and nonhuman primates are chosen according to specific susceptibilities, but some studies require that the test species phylogenetically resembles humans as much as possible.
- Appropriate controls must be included.
- The same administration route as the one for human exposure (oral, dermal, inhalatory, etc.) must be used.
- The number and distribution of the animals per experimental group must be reasonable to allow a correct statistical analysis and interpretation of results.
- The dose or doses used in the study, the parameters and results determined during the study, and the duration of the evaluation period must be properly designed to guarantee the detection of whatever toxicity may occur. As toxicological studies evolve in time and complexity, so do the experimental animals in number and species.

9.4 CLASSIFICATION OF *IN VIVO* TOXICOLOGICAL ASSAYS

This review will be based on the guidelines recommended by the OECD (2020) and in the Globally Harmonized System of Classification and Labeling of Chemicals (GHS; United Nations 2017). Readers are encouraged to view these test guidelines by following the link to the OECD (2020); test guideline (TG) numbers are given in the text and selected tables to supplement the information provided in the chapter. To determine the exposure hazard of a chemical substance, acute and repeated dose studies are traditionally developed. A general classification of these studies is given in Table 9.1.

Table 9.1 Classification of *In Vivo* Toxicity Studies

Short-duration studies	Acute toxicity	Acute oral toxicity
		Acute dermal toxicity
		Acute inhalation toxicity
	Irritation/corrosion studies	Acute dermal irritation/corrosion*
		Acute eye irritation/corrosion*
	Skin sensitization	
Toxicokinetic studies		
Long-term toxicity studies (repeat-dose toxicity studies)	Subchronic toxicity	
	Chronic toxicity	
	Mutagenicity/genotoxicity studies	
	Carcinogenicity studies	
	Evaluation of toxicity during reproduction and development	
Other studies		
Toxic effects on the environment	Acute aquatic toxicity	
	Chronic aquatic toxicity	

* Irritation/corrosion: Reversible/irreversible injury after applying the test substance.

9.4.1 Short-Duration Studies

Acute toxicity is defined by the GHS as the adverse effects that appear after a unique oral or dermal administration of one or a mixture of substances, multiple administrations within 24 hours, or an inhalatory exposure of 4 hours. The acute toxicity is expressed in milligrams per kilograms of animal body weight (mg/kg/bw) and is the initial step for the design of studies of repeated administration. An important consideration in the toxicity assays performed by administration of a single dose of a test substance is that proper design is needed to ensure the appearance of signs of toxicity and the determination of death conditions, should they occur.

The objectives of acute toxicity studies are:

a. To establish the oral, inhalatory, dermal, or ocular toxicities, as well as the dermal irritation and cutaneous hypersensibility to a substance.
b. To determine the lethal dose 50 (LD_{50}), which is the dose that causes the death of 50% (one-half) of a group of test animals.
c. To obtain information about the dose-effect relationship and the toxicological profile of the substance by determining the slope of its curve.
d. To establish if sex differences in toxicity exist.
e. To know the probable effects of an overdose in humans.

For this kind of study, it is especially important to compile and analyze all the information from the literature about the effects of every substance before proceeding to animal experimental procedures following the 3Rs strategy. A good design for acute toxicity testing attempts to derive a maximum amount of information from a minimum number of animals.

9.4.1.1 Acute Oral Toxicity

The acute oral toxicity trial design renders information about the possible adverse effects following a short exposure of a substance through the oral route. Three animal species must be used (one species must not be a rodent). A single dose is administered in 24 hours together with food or by oral gavage; three dose levels are checked.

The results obtained can be used as a basis to classify the substance in categories ranging from 1 to 5, or unclassified, according to the GHS (United Nations 2017). The lower the LD_{50}, the higher the substance toxicity. The assessment of the acute toxicity must include the relationship, if any, between the exposure of the animals to the tested substances and the appearance and severity of all the alterations/symptoms, including abnormalities in behaviour and clinical parameters (e.g. changes in body weight and any other toxic effect, reversible or not, during the study). A necropsy must be carried out whenever animals die during observational periods, and to the surviving animals once the study has finished. Macroscopic lesions and weights of target organs will be examined, and a detailed histopathological study of the lesioned organs will be performed (Gad 2014).

For those substances initially considered as innocuous, a limit dose test is used. A single dose of 2000 mg/kg/bw is usually administered to five animals per sex. This test is recommended when toxicity is expected to be low and mortality is unlikely at the limit dose.

To determine the LD_{50} using an acute toxicity assay, the classical method (TG 401; OECD 2020) was traditionally used until it was rescinded in 2002 following a OECD Council decision. After that, sequential methods of increasing doses were recommended, and the term "evident toxicity" started to be used to describe clear signs of toxicity after the administration of a substance. These signs must be enough for the risk assessment, and they must evolve from superficial to severe with a probable mortality as the administered dose increases. The sequential studies that are currently used are shown in Table 9.2.

- Acute toxic class (ATC) method (TG 423; OECD 2020): Prespecified fixed doses of 5, 50, 300 and 2000 mg/kg are normally used. Groups of three animals of one sex are dosed in a stepwise procedure, with the initial dose being selected as the dose expected to produce mortality in some animals. Further groups of animals may be treated at higher or lower fixed doses, depending on the presence of mortality, until the study objective is achieved—that is, the classification of the test substance based on the identification of the dose(s) causing mortality—except when there are no effects at the highest fixed dose.
- Fixed dose procedure (TG 420; OECD 2020): A principle of this method is that only moderately toxic doses are used in the main study, and the administration of doses that are expected to be

Table 9.2 Acute Oral Toxicity: OECD Test Guidelines (TG)

	Acute toxic class method (ATC) (TG 423)	Fixed-dose procedure (FDP) (TG420)	Up-and-down procedure (UDP) (TG 425)
Animals	Young adult rats (one sex)		
Dosing type	Single bolus dose		
Dosing route	Oral gavage with constant volume and concentration		
Observations	Clinical observations, body weight follow-up		
Sighting study*	Yes	No	
Methodology	Fixed doses of 5, 50, 500, and 2000 mg/kg		Starting at estimation for LD_{50} (175 mg/kg dose)
Aim	Determination of lowest fixed dose that causes evident toxicity		Estimation of LD_{50}
Estimation	Range estimation of LD_{50}		Point estimation of LD_{50} with confidence intervals
Observations	Necropsy.** Macroscopic lesions, weight of target organs, and a detailed histopathological study of the lesioned organs.		

* To allow the selection of the appropriate starting dose for the main study.

** Must be carried out whenever animals die during observational periods and to the surviving animals once the study has finished.

lethal should be avoided. This guideline is intended primarily for use with rats. For the main study, groups of animals of a single sex (normally females) are treated by oral gavage in a stepwise procedure using the fixed doses of 5, 50, 300 and 2000 mg/kg (exceptionally 5000 mg/kg). Animals should be fasted prior the treatment. The initial dose level is selected based on a sighting study as the dose expected to produce some signs of toxicity without causing severe toxic effects or mortality (single animals used per step). Further groups of animals may be treated at higher or lower fixed doses, depending on the presence or absence of signs of toxicity or mortality. This procedure continues until the dose causing evident toxicity or death is identified, or when no effects are seen at the highest dose or when deaths occur at the lowest dose. The results of this study include measurements (weighing at least weekly) and daily detailed observations as well as gross necropsy. The method provides information on the hazardous properties of the substance and allows it to be classified for acute toxicity according to the GHS.

- Up-and-down procedure (TG 425; OECD 2020): In this procedure, animals are dosed one at a time, at a minimum of 48-hour intervals. If an animal survives, the dose for the next animal is increased; if it dies, the dose is decreased usually by a factor of 3.2. This sequence continues until there is a reversal of the initial outcome (i.e. the point where an increasing dose results in death, rather than survival, or decreasing dose results in survival rather than death). Then, additional animals are dosed following the up-down principle until a stopping criterion is met. If there is no reversal before reaching the selected upper-limit dose (2000 or 5000 mg/kg), then no more than a specified number of animals are dosed at the limit dose. The LD_{50} must always be considered together with the observed toxic effects, classified as evident toxicity or pathological findings. The symptoms most commonly observed are weight loss, nausea, salivation, abdominal distension, equilibrium loss, respiratory problems, seizures, and weakness and debilitation. The three tests approved by OECD are summarized in Table 9.2.

9.4.1.2 Acute Dermal Toxicity

Acute dermal toxicity is the adverse effects occurring within a short time of dermal application of a single dose of a substance or multiple doses given within a 24-hour period. These tests give information to classify and label products, and they are an initial step for other studies concerning longer exposure times and also to gather information about the absorption through this route (TG 402; OECD 2020). Rats and rabbits are used for these purposes, assaying three levels of doses in their dorsal area. The evaluation of the dermal acute toxicity must include the relationship, if it exists, between the exposure of the animals to the substance and the incidence and severity of all abnormalities, including behavioural and clinical abnormalities, changes in body weight, serious injuries and any other toxic effects that cause mortality.

9.4.1.3 Acute Inhalation Toxicity

In the risk assessment of an inhalable material, such as a gas, volatile substance or aerosol particulate matter, determination of acute inhalation toxicity is an initial step (TG 403; OECD 2020). The test provides information on health hazards likely to arise from short-term exposure by the inhalation route. Data from an acute study may serve as a basis for classification and labelling. It is an initial step in establishing a dosage regimen in subchronic and other studies, and may provide additional information on the substance's mechanism of action. Rats are recommended to be used for these purposes; the period of exposure is 4 hours, and three levels of dose are applied. The percentage of mortality, the clinical symptoms, the weight variations, and the food intake are registered. At the end of the assay, necropsies are performed to observe any possible macroscopic alterations. The lethal concentration 50 (LC_{50}) is obtained, expressed as the weight of the substance per unit of volume of normal air in milligrams per litter (mg/L).

9.4.2 Irritation/Corrosion Studies

9.4.2.1 Acute Dermal Irritation/Corrosion

Dermal irritation is defined as "reversible damage" to the skin that appears after a period of 4 hours since the administration of a substance. Dermal corrosion is an "irreversible lesion" produced in the skin with necrosis affecting epidermis and dermis after exposure to the substance under study for 4 hours, including the appearance of ulcers, haemorrhages, or haemorrhagic blisters. In this case, after a total observational period of 14 days, complete decolouration, complete alopecia and crusts may appear.

Young adult albino rabbits are used for this test (TG 404; OECD 2020). The test substance is administered at one only dose, 0.5 mL (liquid) or 0.5 g (solid), in a small area (approximately 6 cm^2) of the skin of the dorsal region of the animal. The time of exposure is 4 hours and the observational period varies from 72 hours to 14 days. To avoid the unnecessary use of animals and to protect their welfare, it is essential to review all the available information regarding the irritant and corrosive potential of the chemical substances before starting this test. In this sense, it is important to note that this assay must not be performed with strong acids and bases (pH $\leq$ 2 or pH $\geq$ 11.5); with products that have an LD_{50} smaller than 200 mg/kg through the dermal route or that have not induced irritation in a limit test method assay with a dose of 2000 mg/kg body weight; with products that might be corrosive by predictions using their structure-activity relationships (SAR), the relationship between the chemical structure of a molecule and its biological activity; or with products that have shown corrosive properties using *in vitro* validated tests.

9.4.2.2 Acute Eye Irritation/Corrosion

The evaluation of the irritant or corrosive potential of a substance in the eye of an animal is an initial test to assess the toxic effects of a substance, because the obtained information may be used to predict possible risks after eye exposure and associated mucous membranes. This test must not be carried out when there is previously available information on humans or animals, or results coming from *in vitro* tests, that suggests a high probability of eye hazard. Examples include if the substance has previously caused ocular irritation, has previously caused corrosion or intense irritation, has a pH $\leq$ 2 or pH $\geq$ 11.5 or when there is existing evidence of ocular lesions provided by SAR models. Applying the 3Rs principle, this test should not be performed if the same information can be obtained using an *in vitro* irritation test or an *ex vivo* ocular test.

This test is conducted in adult healthy albino rabbits that present no signs of ocular irritation and have no ocular defects or previous corneal lesions (TG 405; OECD 2020). The test substance is initially applied as a unique dose in the eye of only one animal, using the nonexposed eye as a control. The possible effects are evaluated and graded at specified intervals. The substance is applied in the conjunctival sachet, maintaining the eyelids in a closed position after administration to limit the substance loss. Eyes must not be washed within the following 24 hours, but after that they may be washed if it is considered necessary. Liquids, solids, pastes, particulate substances and aerosols can be administered; if they are solid or granular, they must be ground into a fine powder before administration. After an observation period of 21 days, this test allows the detection of both ocular reversible changes (irritation) and irreversible effects in the eye (corrosion).

9.4.3 Skin Sensitization

Dermal contact sensitivity is an immunological process that may occur after a repeated exposure of the substance to the skin. The host animal usually acquires an allergic sensitivity to that specific substance. The recommended species is the guinea pig, used in the guinea pig maximization test (GPMT) and the delayed contact dermal sensitization test, or Buehler test (TG 406; OECD 2020). In GPMT, the substance is intradermally administered and the time of exposure is 4 hours. However, the induction-observation period, during which the immune response may develop, differs depending on the method (from 23 to 25 days in the GPMT and from 30 to 32 days in the Buehler test). The dermal contact sensitivity test is positive when an erythema appears, considering one or more points of reddening as a positive reaction. The sensitivity potential of the substance is determined depending on the percentage of animals showing a positive reaction. The Buehler test renders information about the health hazards that can arise through topical application of the test substance.

For pure compounds, mice are used in tests such as the local lymph node assay (LLNA). The rationale for the test is that the sensitizers induce lymphocyte proliferation in the lymph node to which the place of application of the substance drains. Proliferation is proportional to the dose applied, and the potency of the allergen and can be measured quantitatively. In the LLNA:DA (nonradioactive) test, the content of adenosine triphosphate (ATP) is measured by bioluminescence as an indicator of lymphocyte proliferation (TG 442a; OECD 2020). In the LLNA-BrdU, measurement of proliferation is based on the quantification of 5-bromo-2′-deoxyuridine as a thymidine analogue that is determined by an ELISA technique using a commercially available kit and flow cytometry (TG 442b; OECD 2020).

9.5 TOXICOKINETIC STUDIES

Together with the acute toxicity studies, toxicokinetic (TK) studies (see Chapter 2) are highly relevant to the toxicological characterization of a chemical substance. Data obtained by toxicokinetic studies aid to select further toxicity studies. The quantity of the substance that enters the body, its interaction with different tissues and organs, and the manner in which the substance is excreted are some of the parameters obtained by toxicokinetic studies. The principal objective of these studies is to understand the kinetics of the chemical substance, named ADME (absorption, distribution, metabolism and elimination) in the body:

- Absorption is the process by which a given substance enters the body.
- Distribution refers to how this substance and its metabolites are transported or disseminated throughout the body.
- Metabolism is the process of transformation of the original substance to another (metabolite) by a chemical reaction within the body.
- Elimination or excretion describes the process by which a substance leaves the body. The chemical structure and physicochemical properties of the substance help to predict the absorption, distribution, metabolism and elimination of the substance.

In vivo toxicokinetics tests provide information about the absorption, bioavailability, tissular distribution, metabolism, excretion, and basic toxicokinetic parameters of a substance, as well as complementary approaches (metabolite/TK data) that may give useful information about the toxicokinetics of the substance. Normally, the substance under study (unlabelled or labelled using ^{14}C) is administered orally, although other routes can be used. The administration of a single dose of the substance (preferably with at least two dosage levels) can be sufficient for the test, but sometimes a repeated dose is required. The toxicokinetic studies should be performed preferably in the same species that is used for the other toxicological studies of the substance (usually rats, and minimum four animals of each sex per dose). The initial determination of the absorption rate can be obtained by measuring the mass balance assessment (amount of substance absorbed), but more exhaustive studies might be necessary, such as those performed by intravenous administration and biliar excretion rate. Bioavailability can be determined by examining the plasma or blood kinetics assessed after oral or intravenous administration. The percentage of the total dose that reaches the tissues must be measured at the end of the experiment at a minimum; additional time points might be required. Those metabolites exceeding 5% of the administered dose must be identified. The excretion rate of the administered dose is determined by measuring the percentage of that dose detected in urine, feces and exhaled air (TG 417; OECD 2020).

In vivo percutaneous absorption may be used to determine the penetration of a given substance through the skin and its correlation to absorption by the oral route. The substance under study, again preferably radio-labelled, is applied to the skin of the animals for a fixed time in one single dose or different dosages in the form of a representative preparation of its intended use. As mentioned above, rats are commonly used for these studies, at a minimum of four animals of each sex per preparation and time point. A known dose of a given substance is applied uniformly to the skin. This dose must mimic the exposure dose for a human (e.g. 1–5 mg/cm^2 for solids and up to 10 μL/cm^2 for liquids). A relevant exposure time must be used (generally 6 or 24 hours), based on the predictable human exposure time. Animals should be observed periodically during the whole exposure period for signs of toxicity or abnormal reactions. This kind of studies include daily measurements, detailed regular observations (urine and feces) and animal euthanasia and collection of blood samples for further analysis at established time points (TG 427; OECD 2020).

9.6 LONG-TERM TOXICITY STUDIES/REPEAT-DOSE TOXICITY STUDIES

Long-term and repeated-dose toxicity studies allow the determination of the effects of accumulation of a given substance and to identify the target tissues. In order to accurately predict these effects, administration in humans is reproduced, and the dose and the exposure time is increased to mimic the effects of an intoxication. The principal long-term toxicity studies are those on subchronic toxicity, chronic toxicity and carcinogenesis. Their objective is to learn the effects of a given substance after repeated administration in animal models and to extrapolate these effects to humans.

9.6.1 Subchronic Toxicity

The GHS system defines subchronic toxicity as the toxicity caused by repeated exposure to a given substance at the systemic or tissue-specific level. The aim of subchronic toxicity studies is to establish the adverse effects occurring after repeated or continuous administration of a given substance not exceeding the 10% of the animal's life span. Three different daily repeated doses are administered by two different routes, one of them being the oral route. These studies allow to further estimate the dose to administer for chronic and carcinogenic studies.

The animal species used for these studies are:

a. Rats, for the evaluation of toxicity after administration of repeated doses from 14 to 28 days by oral, dermal, and inhalation routes (TG 407, TG 410, TG 412; OECD 2020). Animals are kept in metabolic cages, and the assessment of the toxic effects is based on observation of animal behaviour and changes in body weight. Food and water intake measures and ophthalmologic analysis is performed every 7 days. At the end of the experimental period, blood and urine biochemical analysis, and necropsy and macroscopic and histopathologic analysis of animal tissues and organs are performed. The final goal of these studies is to learn the adverse effects after repeated doses and to identify the target tissues and organs.

b. Dogs, for the evaluation of toxicity after administration of repeated doses for up to 90 days by oral, dermal, and inhalation routes. Mortality, clinical symptoms, changes in body weight and food intake, blood and urine analysis, necropsy, and macroscopic and histopathologic analysis of tissues and organs will be assessed (TG 409). The main goal of these studies is to establish the systemic toxicity of a given substance, to identify the target tissues or organs and the principal toxic effect (toxicodynamics). This allows to know the LOAEL and NOAEL, the reversible nature of the adverse effects, the latency and the bioaccumulation.

9.6.2 Chronic Toxicity

Chronic toxicity and carcinogenic studies aim to determine the adverse effects of a given substance in mammals after a chronic repeated exposure. The exposure time must be at least 10% of the animal's life span and can cover the entire life of the animal.

The objective of chronic toxicity studies is to characterize the toxicological profile of a given substance in mammalian species (primarily rodents) following prolonged and repeated exposure.

Chronic toxicity studies use both sexes of two species of rodents (mouse and rat) and a nonrodent (normally dog). For rodents, at least 20 animals per group and sex must be used, whereas for nonrodents a minimum of 4 animals per sex and group is recommended. Three different daily doses are administered normally by the oral route, although the dosage may change depending on the route of administration (oral, dermal or inhalation). The exposure time will be at least 6 to 18 months (TG 452; OECD 2020).

The objectives of these studies include:

a. Identifying the chronic toxicity of a given substance
b. Identifying target organs and tissues
c. Characterizing the dose-response relationship
d. Identifying the NOAEL or starting point for the establishment of the BMD
e. Predicting chronic toxicity effects for human exposure
f. Obtaining the necessary data to hypothesize a mechanism of action for the substance.

Mortality, clinical symptoms, and variations in body weight and food intake are established, and blood and urine analysis are performed. At the end of the study, following the necropsy of the animals, macroscopy analysis of the organs and a histopathological study are performed, and the NOAEL is obtained. The lowest NOAEL obtained from all the studies is normally the most adequate one to calculate the admissible daily intake (ADI) for humans (ADI = long-term NOAEL (lowest value)/100).

9.6.3 Mutagenicity/Genotoxicity Studies

Genotoxicology is the area of toxicology that studies the mutagenic, carcinogenic and teratogenic effects of a substance. These studies provide information about the mutagenic and genotoxic potential of a given substance. Mutation is defined as a permanent change in the quantity or structure of the genetic material of a cell. The terms "mutagenic" or "mutagen" are used to refer to the substances that cause an increase in the incidence of mutations in populations of cells and organisms. A mutagen is a physicochemical agent that causes mutations in human germinal cells that can be transmitted to the offspring. The term "genotoxin" is more generic and is generally used for substances or processes that alter the structure, the content, or the segregation of the DNA by interfering with the normal processes of duplication, or those that alter temporally its replication in a nonphysiological way. Teratogenesis is a prenatal toxicity characterized by structural or functional defects in the developing embryo or fetus. The OECD, the European Commission (Joint Research Centre), and the International Council for Harmonisation of Technical Requirements for Pharmaceuticals for Human Use (ICH) have defined the studies that must be performed for the assessment of genotoxicity (Benigni and Bossa 2008). The most common mutagenicity and genotoxicity assays are compiled in Table 9.3.

In vivo genotoxicity assays are performed to confirm a positive result from an *in vitro* genotoxicity assay. Commonly used *in vivo* genotoxicity tests are the micronucleus test and the comet assay. The micronucleus test, also known as the chromosomal aberration test, is the most widely used *in vivo* genotoxicity test, in which mice are exposed to a substance and their bone marrow cells or lymphocytes are examined microscopically. When the DNA fragments or the chromosomes do not separate properly during mitosis, nuclei smaller than the original nucleus (micronuclei) may appear. Treatment of the cells with a genotoxic agent can facilitate the increase in the number of these micronuclei; this increase is an indicative parameter of genotoxic cell damage (Fenech et al. 2011). The comet assay, given its name because of the shape that mammalian cell DNA acquires under a fluorescence microscope when treated with a genotoxic agent, measures DNA fragmentation induced by a genotoxic agent. The cells are immobilized on an agarose matrix, lysed *in situ* and subjected to an electric field. The DNA fragmentation releases supercoiling and allows migration of DNA in the electric field. DNA remains tethered to what is left of the nuclear matrix after lysis and smaller DNA fragments migrate into the "tail", giving the overall appearance of a comet. The length and/or intensity of the tail after staining with a fluorescent dye serves to quantify the genotoxic potential of the agent under study. The assay is highly sensitive, requires a very small number of cells and can be completed in a relatively short period of time (Hartmann et al. 2003).

9.6.4 Carcinogenicity Studies

During carcinogenicity studies, the animals are observed during most of their lives in order to detect the development of neoplastic lesions during or after the exposure to several doses of a

Table 9.3 Genotoxicity Assays: OECD Test Guidelines (TG)

Type of test	Assay
DNA strand breakage assays	The comet assay for DNA damage
In vivo tests of mutagenicity of hereditary germ cells	Rodent dominant lethal test (TG 478) Mouse heritable translocation assay (TG 485) Specific local mutation assay in mice
In vivo assays of mutagenicity of somatic cells: structural and numerical chromosome aberrations	Mammalian bone marrow chromosome aberration test (TG 475) Mouse spot test (TG 484) Mammalian erythrocyte micronucleus (TG 474)
Germ cell mutagenicity assays	Mammalian spermatogonial chromosome aberration test (TG 483) Test of micronuclei in spermatids
Germ cell genotoxicity	Analysis of sister chromatid exchange (SCE) in spermatogonia Test of unscheduled DNA synthesis (UDS) in testicular cells
Genotoxicity assays in somatic cells	Unscheduled DNA synthesis (UDS) test with mammalian liver cells *in vivo* (TG 486) Sister chromatid exchange (SCE) of mammalian bone marrow
Mutagenicity assays	*In vitro* mammalian chromosome aberration test (TG 473) *In vitro* mammalian cell gene mutation test (TG 476) Bacterial reverse mutation test (TG 471)

given substance through a specific route. The study will provide information about the toxic effects of the substance, including potential carcinogenicity, and may indicate target organs and the possibility of bioaccumulation. Carcinogenic toxicity studies use both sexes of two species of rodents (mouse and rat) and a nonrodent (normally dog). At least 50 animals per sex per group must be used, and the duration of the treatment can last up to 24 months. Daily doses are administered orally, dermally or inhalatory, depending on the toxicokinetic profile of the substance. Mortality higher than 75% of the animals in the lower dose group is considered the endpoint of the study. The results of carcinogenic studies include measurements and quantifications of body weight and food intake, daily observations of animal behaviour and phenotype, haematological studies, detection of tumour formation, and necropsies and histopathologic studies. These studies may provide an estimation of the NOAEL for toxic effects, and the NOAEL for the tumour processes in the case of nongenotoxic carcinogens, which can be used for establishing the safety criteria for human exposure. Oncogenic NOAEL is obtained if tumours are developed. A substance is considered as carcinogen when it is able to produce an increase in the number of tumours (benign and malignant), an increase in the incidence or malignancy of the tumour, or an increase in the speed of cancer development.

The objectives of carcinogenicity studies covered by this test guideline include:

a. Identifying the carcinogenic properties of a substance that can increase the incidence of neoplasms, the proportion of malignant neoplasms or the reduction in time of appearance of neoplasms, compared with concurrent control groups
b. Identifying target organs of carcinogenicity
c. Identifying the time to appearance of neoplasms
d. Characterizing the tumour dose-response relationship
e. Identifying the NOAEL or starting point for the establishment of a BMD
f. Extrapolating carcinogenic effects for human exposure
g. Obtaining the data necessary data to hypothesize a mechanism of action of the substance.

9.6.5 Reproductive and Developmental Toxicity Studies

The study of the toxicity during reproduction and development determines the adverse effects of a given substance on the progenitors' reproduction system and on their offspring. The evaluation of the toxicity is determined during three reproduction phases: (I) fertility and early embryo development; (II) embryofoetal development (organogenesis period); and (III) peri- and postnatal period.

For phase I (fertility and early embryo development) studies, rats are normally used. Exposure to the substance starts 60 days before mating for males and 15 days before mating for females and continues until the weaning of the offspring. The main goal of phase I studies is to determine the adverse effects occurring during gametogenesis, mating, pregnancy and postnatal development. The toxic effects during phase II (embryofoetal development [organogenesis period]) determine the teratogenicity of a given substance when administered during gestation. The route of administration is generally oral, although other routes can be used depending on the physical properties of the substance and the most likely route of exposure in humans. The disaster caused by thalidomide during the 1950s and 1960s demonstrated the need to detect the toxicity of a substance at any stage of the reproductive cycle, due to its potential to cause malformations in the embryo, delayed growth or even death. The toxicological studies during the phase III (peri- and postnatal toxicity studies) aim to determine the potential adverse effects of a given substance during embryo development, birth, and lactancy, as well as peri- and postnatal development.

The goal of the studies of toxicity in reproduction and development is to evaluate the adverse effects of a given substance in the reproductive organs and their functions, along with the alterations that might happen to the offspring. They are normally carried out in rats and rabbits. The study of the alterations on the parents' fertility and offspring development comprises the administration of the substance from fertilization to delivery. Parameters must be established, such as the number of embryos and resorptions; fetal weight; relationship between sexes; and external, visceral and skeletal morphology. These studies determine the NOAEL for maternal, reproductive and litter toxicity, and they establish whether a substance is teratogenic.

Although these tests have only limited predictive value, extrapolation to human beings is highly relevant. A substance defined as "reprotoxic" exerts adverse effects in the offspring at doses that are not toxic for the parents and that are close to those used in humans. Reprotoxic assays are described in the Table 9.4.

9.6.6 Endocrine Disruptors

In this chapter, the authors wished include particular mention of endocrine disruptors due to the current concern they generate and because, as previously mentioned, they do not follow monotonic curves. The World Health Organization (WHO) has defined an endocrine disruptor as an exogenous substance or mixture that alters functions(s) of the endocrine system and consequently causes adverse health effects in an intact organism, its progeny, or its (sub)populations (Damstra

Table 9.4 Reproductive/Deelopmental Toxicity Tests: OECD Test Guidelines (TG)

OECD TG	Basis	Aim of study
414	Prenatal development toxicity study	Effects of prenatal exposure on the pregnant test animal and on the developing organism
415	One-generation reproduction toxicity study	Male and female reproductive performance
416	Two-generation reproduction toxicity	Integrity and performance of the male and female reproductive systems and on the growth and development of the offspring
421	Reproduction/developmental toxicity screening test	Effects of a test substance on male and female reproductive performance. Studies include oestrous cycle monitoring, anogenital distance, and male nipple retention in pups, among others.
422	Combined repeated dose toxicity study with the reproduction/developmental toxicity screening test	To obtain specific information concerning the effects of a chemical substance on male and female reproductive capacity, as well as gonadal function, mating behaviour, conception, fetal development and parturition
426	Developmental neurotoxicity study	Effects of repeated exposure to a substance during *in utero* and early postnatal development. Study of gross neurological effects and behavioural abnormalities.
443	Extended one-generation reproductive toxicity study	Reproductive and developmental effects that may occur as a result of pre- and postnatal chemical exposure, as well as evaluation of systemic toxicity in pregnant and lactating females and young and adult offspring

et al. 2002). The increasing incidence of disorders such as obesity, diabetes, metabolic syndrome, cardiovascular diseases, reproductive dysfunction, hormone-sensitive cancers, and neurodevelopmental abnormalities in some human populations have created an alarm towards environmental endocrine disruptors and the possibility that exposure to them may be involved (Gore et al. 2015). Both natural and artificial substances are thought to cause endocrine disruption. Among natural endocrine disruptors are substances of edible plant origin found in soy, such as phyto-oestrogens (genistein and daidzein), and in other edible plants such as apples, legumes and almonds (cyanogenic glycosides; Bolarinwa et al. 2015; Jargin 2014; Rietjens et al. 2017; Thodberg et al. 2018). Artificial substances include pharmaceuticals, phthalates, bisphenol A (BPA), dioxin and dioxin-like compounds, polychlorinated biphenyls (PCB), and herbicides and pesticides such as dichlorodiphenyltrichloroethane (DDT; Beszterda and Franski 2018; Onuzulu et al. 2019). Endocrine disruptors may be found in many everyday products, including plastic bottles, the lining of metal food cans, detergents, flame retardants, food, toys, cosmetics, and pesticides. The greatest risk of endocrine disruptors occurs during prenatal and early postnatal development, when organ and neural systems are forming (Walley and Roepke 2018).

The mechanisms through which endocrine disruptors influence the endocrine system and alter hormonal functions are:

- To mimic or partly mimic naturally occurring hormones in the body, like estrogens (genistein), androgens (tetrabromethylcyclohexane, TBECH), and thyroid hormones (perchlorates), potentially producing overstimulation.
- To bind to a receptor within a cell and block the endogenous hormone from binding. The normal signal then fails to occur, and the body fails to respond properly. Examples of chemicals that block or antagonize hormones (anti-estrogens and anti-androgens) are phthalates, tamoxifen and BPA.
- To interfere or block the way natural hormones or their receptors are made or controlled, for example, by altering their metabolism in the liver.

The effects of endocrine disruptors are evaluated via studying the hypothalamus-pituitary-adrenocortical (HPA) axis, the hypothalamus-pituitary-gonadal (HPG) axis, the somatotropic axis, the retinoid signalling pathway, the hypothalamus-pituitary-thyroid (HPT) axis, the vitamin D signalling pathway, and the peroxisome proliferator-activated receptor (PPAR) signalling pathway. The tests that are carried out are described in Table 9.5. The OECD guidelines, from 2018, compile scientific advances in the use of test methods and the evaluation of the endocrine activity of chemical products (OECD 2018).

Table 9.5 *In Vivo* Tests for Evaluating the Endocrine Disrupting Activity of Substances: OECD Test Guidelines (TG)

	HPA axis		HPG axis		Somatotropic axis		Retinoid signalling pathway		Vitamin D signalling pathway		PPAR signalling pathway	
Levels	*Mam*	*nMam*	*Mam*	*nMam*	*Mam*	*nMam*	*Mam*	*nMam*	*Mam*	*nMam*	*Mam*	*nMam*
3	440, 441	229, 230, 231	441	229	N/A	N/A	N/A	N/A	N/A	N/A	440, 441	229, 230
4	408, 415	206	N/A	N/A	N/A	N/A	N/A	N/A	N/A	N/A	415	206
5	N/A	N/A	N/A	231	416, 443	N/A	416, 443	N/A	416, 443	N/A	416, 443	N/A

Abbreviations: HPA = hypothalamus-pituitary-adrenocortical; HPG = hypothalamus-pituitary-gonadal; N/A = not applicable; PPAR = peroxisome proliferator-activated receptor signaling.

Note: OECD tests for evaluation in mammalian (*Mam*) and nonmammalian (*nMam*) animal models of endocrine disruption toxicity. Numbers indicate the OECD test guidelines, which are classified based on the main hormonal axis affected. Levels 1 and 2 are excluded from the table because they are not measuring *in vivo* effects. Level 1: existing data and nontest information (*in silico* predictions, ADME model predictions, physical and chemical properties, etc.); Level 2: *in vitro* assays; Level 3: *in vivo* assays providing data about selected endocrine mechanism(s)/pathway(s); Level 4: *in vivo* assays providing data on adverse effects on endocrine relevant endpoints; Level 5: *in vivo* assays providing more comprehensive data on adverse effects on endocrine relevant endpoints over more extensive parts of the life cycle of the organism.

9.7 OTHER STUDIES

Specific studies can be performed for certain substances, such as the local tolerance test (normally cutaneous or muscular tolerance) which uses rats, rabbits and dogs. Special studies aim to determine the toxic mechanisms of a certain substance on specific organs (e.g. hepatotoxicity studies) or systems (e.g. neurotoxicity studies). The potential neurotoxicity of a substance can be detected by many different means, such as functional changes, structural changes, or biochemical alterations in the central or peripheral nervous system. Neurotoxicity assays use rodents exposed to single or repeated doses of the substance, and they evaluate the motor phenotype and the animal behaviour as consequences of an alteration of neurophysiological state, neuron degeneration or blood-brain barrier damage. Based on the data obtained from acute toxicity studies, a preliminary idea of the possible neurotoxicity of a given substance can be elucidated. Repeated-dose toxicity studies include the evaluation of the neurotoxicological effects of a substance, and they highlight the importance of daily observations of animals to obtain the maximum amount of information (TG 407; OECD 2020). Neurotoxicity studies are especially important to detect neurotoxic effects of certain substances that cannot be detected by any other toxicity assay. This is the case for some organophosphates that produce delayed neurotoxicity and must be evaluated by specific assays after one unique exposure (TG 418; OECD 2020) or after a repeated administration (TG 419; OECD 2020).

A substance causing immunotoxicity can either decrease the viability or the function of the immune system cells (immunosuppression) or trigger an aberrantly increased response against an innocuous stimulus (e.g. hypersensitivity or autoimmunity). Repeated-dose toxicity studies include evaluation of immunotoxic effects (TG 407; OECD 2020). These studies must detect substances with immunotoxic potential, and they must determine whether additional studies are required. The evaluation of immunotoxic effects is especially relevant for biological medicines, such as recombinant proteins and monoclonal antibodies, due to their potential to induce immunogenicity and immunological cross-reactions with endogenous molecules.

The vascular tolerance studies must be performed for those substances to be administered intravenously. Parameters to be evaluated are arterial dilation, endothelial lesions, alterations in red blood cell morphology (assessing the formation of echinocytes), oxygen supply, and coagulation disorders.

Furthermore, specific renal tolerance assays must be performed for those substances that, could produce renal alterations during their excretion that may lead to renal insufficiency.

9.8 TOXIC EFFECTS ON THE ENVIRONMENT

The nature of environmental effects of toxic substances is explained in more detail elsewhere in this book. In this chapter, we are focused exclusively on the *in vivo* methods for the assessment of environmental toxicity. In terms of the aquatic environment, GHS classifies hazardous substances as those producing acute or chronic aquatic toxicity (United Nations 2017).

9.8.1 Acute Aquatic Toxicity

Acute aquatic toxicity means the intrinsic property of a substance to cause injury to an aquatic organism after a short-term exposure. Substances and mixtures of this hazard class are assigned to one of three toxicity categories based on acute toxicity data: LC_{50} (fish), EC_{50} (crustacea) or ErC_{50} (median effect concentration of growth rate reduction for algae or other aquatic plants; TG 201, TG 202, TG 203; OECD 2020).

9.8.2 Chronic Aquatic Toxicity

Chronic aquatic toxicity means the potential or actual properties of a material to cause adverse effects to aquatic organisms during exposures that are determined in relation to the life cycle of the organism. The harmonized classification system for substances consists of three acute classification categories and four chronic classification categories. The acute and the chronic classification categories are applied independently. The criteria for classification of a substance in acute categories I to III are defined on the basis of the acute toxicity data only (LC_{50}, EC_{50} or ErC_{50}). The criteria for classification of a substance into chronic categories combine two types of information: acute toxicity data and environmental fate data (degradability and bioaccumulation data). By measurement of aquatic toxicity, chronic category IV is considered safe, even if some concerns are emerging regarding their safety for aquatic life.

To sum up, the authors would like to specify that the studies carried out in animals, although appropriate, present a limited biological variability compared to that of the worldwide human population. On the other hand, it would be inaccurate to directly compare data from animal testing to those results in humans due to interspecies differences. Moreover, several adverse reactions, such as headache or depression, are very hard to determine in animals; therefore extrapolation to human beings should be done very cautiously.

BIBLIOGRAPHY

Benigni, R., and C. Bossa. 2008. Structure alerts for carcinogenicity, and the Salmonella assay system: A novel insight through the chemical relational databases technology. *Mutat Res* 659(3): 248–61.

Beszterda, M., and R. Franski. 2018. Endocrine disruptor compounds in environment: As a danger for children health. *Pediatr Endocrinol Diabetes Metab* 24(2): 88–95.

Bolarinwa, I. F., C. Orfila, and M. R. Morgan. 2015. Determination of amygdalin in apple seeds, fresh apples and processed apple juices. *Food Chem* 170: 437–42.

Damstra, T., S. Barlow, A. Bergman, R. Kavlock, and G. Van Der Kraak. 2002. *Global Assessment of the State-of-the-Science of Endocrine Disruptors.* www.who.int/ipcs/publications/new_issues/endocrine_disruptors/en/ (accessed October 28, 2020).

Fagin, D. 2012. Toxicology: The learning curve. *Nature* 490(7421): 462–5.

Fenech, M., M. Kirsch-Volders, A. T. Natarajan, et al. 2011. Molecular mechanisms of micronucleus, nucleoplasmic bridge and nuclear bud formation in mammalian and human cells. *Mutagenesis* 26(1): 125–2.

Gad, S. 2014. Rodents model for toxicity testing and biomarkers. In *Biomarkers in Toxicology*, ed. R. Gupta, 7–69. London: Academic Press.

Gore, A., V. Chappell, S. Fenton, et al. 2015. Executive summary to EDC-2: The Endocrine Society's second scientific statement on endocrine-disrupting chemicals. *Endocrine Rev* 36(6): 593–602.

Hartmann, A., E. Agurell, C. Beevers, et al. 2003. Recommendations for conducting the in vivo alkaline Comet assay. 4th International Comet Assay Workshop. *Mutagenesis* 18(1): 45–51.

Jargin, S. V. 2014. Soy and phytoestrogens: Possible side effects. *Ger Med Sci* 12. www.egms.de/static/en/journals/gms/2014-12/000203.shtml (accessed October 28, 2020).

OECD. 2018. *Revised Guidance Document 150 on Standardised Test Guidelines for Evaluating Chemicals for Endocrine Disruption.*

OECD. 2020. *OECD Guidelines for the Testing of Chemicals.* www.oecd.org/env/ehs/testing/oecd-guidelinesforthetestingofchemicals.htm (accessed October 28, 2020).

OECD iLibrary. 2020. www.oecd-ilibrary.org/ (accessed October 28, 2020).

Onuzulu, C. D., O. A. Rotimi, and S. O. Rotimi. 2019. Epigenetic modifications associated with in utero exposure to endocrine disrupting chemicals BPA, DDT and Pb. *Rev Environ Health.* www.degruyter.com/view/journals/reveh/34/4/article-p309.xml (accessed October 28, 2020).

Regulation (EC) No 1223/2009 of the European Parliament and of the Council of 30 November 2009 (L 342/59–209) (2009). *Official Journal of the European Union.*

Rietjens, I., J. Louisse, and K. Beekmann. 2017. The potential health effects of dietary phytoestrogens. *Br J Pharmacol* 174(11): 1263–80.

Russell, W. M. S., and R. L. Burch. 2009. *The Three Rs and the Humanity Criterion, An Abridged Version of: The Principles of the Humane Experimental Technique*, ed. M. Balls.

Thodberg, S., J. Del Cueto, R. Mazzeo, et al. 2018. Elucidation of the amygdalin pathway reveals the metabolic basis of bitter and sweet almonds (Prunus dulcis). *Plant Physiol* 178(3): 1096–111.

United Nations. 2017. *A Guide to the Globally Harmonized System of Classification and Labeling of Chemicals.* www.osha.gov/dsg/hazcom/ghsguideoct05.pdf (accessed October 28, 2020).

US EPA. 2012. *Benchmark Dose Technical Guidance.* www.epa.gov/risk/benchmark-dose-technical-guidance (accessed November 9, 2020).

Walley, S. N., and T. A. Roepke. 2018. Perinatal exposure to endocrine disrupting compounds and the control of feeding behavior: An overview. *Horm Behav* 101: 22–8.

10 Toxicological Applications of Genomic, Proteomic and Metabolomic Technologies

Carmen Pérez-García, Francisco Javier Rupérez, María Haro and Gonzalo Herradón

CONTENTS

10.1 INTRODUCTION

'Omics technologies are high-throughput technologies used to analyze all the molecules that make up a cell, tissue or organism. Their aim is to detect all the genes (genomics), mRNA (transcriptomics), proteins (proteomics) and metabolites (metabolomics) in a specific biological sample in a non-targeted and non-biased manner. Traditional analytical techniques are focused in one or a few genes, proteins or metabolites (e.g. RT-qPCR, Northern blotting, Western blotting, ELISA), and they are mainly used in hypothesis-driven studies. In contrast, in some 'omics studies no hypothesis is known or assumed, but all data are analyzed to define a hypothesis that will be further validated. This nonbiased approach (i.e. not studying a specific molecule of interest) provides a great potential to decipher the molecular basis of physiological processes but also of pathological ones, such the ones derived from toxicant exposure. Every day we are exposed to a continuously growing number of potentially toxic substances; if we consider only those of a chemical nature, more than 80,000 molecules are listed on the US Environmental Protection Agency (EPA) Toxic Substances Control Act Inventory (US EPA 2018). Therefore, the high-throughput 'omics technologies, which allow the assessment of responses of tens of thousands of genes and their products in a single sample, are highly valuable tools for the risk assessment, diagnosis and treatment related to toxicant exposure. In this chapter, we provide a review of the current technologies used in genomics, proteomics and metabolomics with a particular emphasis on their applications in toxicology.

10.2 GENOMICS

10.2.1 Definition

A genome is the complete DNA sequence of an organism that is considered to be haploid. That sequence comprises both the coding regions, that is, the entire set of genes with all their regulatory sequences and the noncoding regions. The genomic sequence contains all the instructions for the cell's function. It also determines the characteristics of the cell, the groups of genes that the cell will express, the cell type in which it will be differentiated and, ultimately, the role that the cell will play in the organism. It contains all the necessary information for developing an organism and for its particular phenotype. A phenotype is the result of the interaction between the genotype and the environment and comprises the observable characteristics or traits of the cell,

DOI: 10.1201/9780203730584-10

Table 10.1 Genomics-Related Definitions

Term	Definition
Gene	A DNA sequence that codifies one or more set of RNA or proteins; it is not a single entity but it is part of a gene regulatory network
Genome	The complete DNA sequence of an organism considered in a haploid manner
Genomics	The study of genomes
Phenotype	An observable trait of an individual that is determined by its DNA sequence
Genotype	A specific DNA sequence in a determined locus
Epigenetics	The mechanism of gene expression regulation that yields phenotype changes but does not involve alterations in the DNA sequence

such as its function, structure or morphology. Thus, the phenotype of an individual depends on its genotype, on the environment and on epigenetic factors. Epigenetics is the existence of different phenotypes with the same genotype. This is due to the existence of epigenetic factors that modify the expression of genes giving rise to different phenotypes without modifying the DNA sequence (Berger et al. 2009). Therefore, the biological response to a certain situation depends on the genotype plus the environment and the epigenetic factors.

Genomics is the study of the genomes, that is, the study of the contents, structure, organization and function of genetic material of an organism. Genomics describes the exact nucleotides that appear in a sequence; the regions involved in gene codification, expression and regulation; and the regions lacking known functions. Genomics science is focused on the study of the cartography, sequencing and analysis of complete genomes. It is divided into structural genomics, which aims to identify and define all the elements of a whole genome, and functional genomics, which determines the functions of those elements. The structural genomic field was developed from the improvement of sequencing techniques, especially next generation sequencing.

Genome variations are natural differences in the sequence of DNA of two random people (Venter et al. 2011). Point substitutions of just one nucleotide in the DNA sequence are called single nucleotide polymorphisms (SNPs). In the event that more than one nucleotide can potentially be present at a certain nucleotide position, it is said to be a polymorphism or a polymorphic site (polymorphic locus). One of the applications of the study of genomic variability is to obtain useful information by performing genome-wide association studies (GWAS). For instance, this allows establishing a phenotype/genotype relationship by comparing the nucleotide sequences of individuals carrying a specific disease to identify the existence of common polymorphisms which may be related to that disease. In the same manner, this procedure is used to identify genome variations that determine different individuals' responses to toxicant exposure, so-called gene × environment interactions (i.e. when two different genotypes respond to environmental exposures in different ways).

10.2.2 Genomic Techniques

High-throughput sequencing or next-generation sequencing (NGS) allows scientists to sequence DNA and RNA faster and more competitively than the previously used Sanger sequencing. It allowed to obtain the complete sequence of a whole genome in a shorter period of time and started the genomic medicine era. The first genome sequenced was the bacteriophage X-174 by Frederick Sanger and colleagues in 1977, which has 5386 base pairs. Since then, longer and more complex genomes have been sequenced. In 2001, the complete sequence of the human genome was published by the Human Genome Project Consortium; this was an inflection point in the rapid advance of understanding of the elements of the genome. The human genome has 3.2 billion base pairs, but only 1.1% of them are coding regions. However, the large noncoding DNA sequence is far from having no purpose, as reflected in the results of the Encyclopaedia of DNA Elements (ENCODE) project. The project emerged as an international collaboration aiming to determine the functions of the DNA sequence and, particularly, of the regions which do not code for proteins. ENCODE shows what DNA elements correspond with coding sequence, regulatory regions or protein binding sites; all of these remain as conserved DNA elements. ENCODE also describes the existence of SNPs. Most noncoding DNA is located between genes, serving as separators, and has no known function. Other noncoding sequences called introns are found within genes. Unexpectedly, two-thirds of noncoding DNA turns out to be transcribed and seems to play a regulatory role in gene expression (e.g. small interfering RNAs).

Massive techniques are applied to the gene expression analysis. This analysis is made through transcriptomics studies. Transcriptomics is the science, between genomics and proteomics, that studies the transcriptome, which is defined as the full set of messenger RNA (mRNA) molecules (or transcripts) that is in a specific single cell and that, at a certain time, determines the genes that the cell is expressing. The development of technologies such as DNA arrays has allowed the simultaneous assessment of the levels of mRNA expression of thousands of genes. Using this technique, all the mRNA in a cell or a tissue is transformed into complementary DNA (cDNA) and then hybridized to the microarray surface in which a collection of microscopic DNA spots are attached. Each DNA spot contains a known DNA sequence called a probe. Probe-target hybridization is usually quantified by detection of fluorescence or chemiluminescence-labelled targets, determining the relative levels of nucleic acid sequences. Applied to the field of toxicology, this microarray-based gene expression profiling is important to identify genes whose levels of expression are regulated in response to toxicants (Figure 10.1). The same approach is used in genotyping DNA arrays for massive SNP analysis (Ates et al. 2018), in which each target represents a different SNP; therefore, binding of the probe to the target means the sample carries that SNP.

Transcriptional profiling analyses can be performed as well with sequence-based techniques, like RNA-Seq, which render the identification of mRNA molecules and quantification of their levels of expression (Wang et al. 2009). Massive techniques, such as DNA array and RNA-Seq, generate results which are very informative, but they need to be validated by other traditional and tested techniques, such as quantitative PCR (qPCR). Additionally, a set of criteria have been established that define the minimum information that should be included when describing a microarray or sequencing study: Minimum Information About a Microarray Experiment (MIAME) (and MINSEQE [Minimum Information About a Next-Generation Sequencing Experiment]), which includes such validation steps (www.ncbi.nlm.nih.gov/geo/info/MIAME.html). Indeed, many journals and funding agencies require microarray data to comply with MIAME and MINSEQE standards.

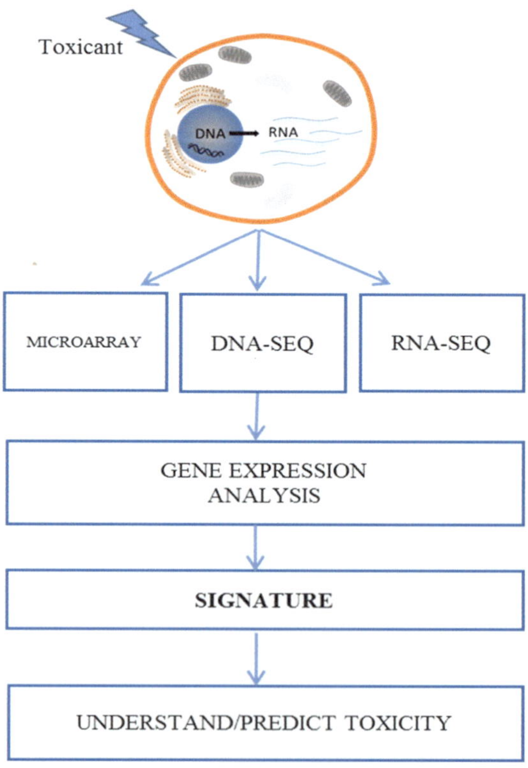

Figure 10.1 Procedure usually followed in genomics: The arrival of a toxic agent to the cell produces a toxic effect that leads to the modification in gene expression. This expression pattern will be a trace or a footprint of the toxic effect that can be used as a signature to predict the effect of other similar substances, as well as to understand the mechanism of action.

In order to process the data derived from genomics technologies, like NGS or DNA arrays, bioinformatics is essential to analyze the raw sequence data. Significant efforts are being performed to create databases that could allow, in the near future, robust correlational analyses between studies conducted all around the world.

10.2.3 Applications

Toxicogenomics aims to establish a genotype-phenotype correlation, in which the gene expression pattern is associated with the individual's response to a potentially toxic compound. Toxicogenomics also studies changes in the pattern of gene expression induced by the exposure to a toxicant, which often renders relevant information about the pathways implicated in the effects of the toxicants (i.e. the mechanism or mode of action). This information could allow making future predictions about an organism's response to a toxicant easier, because similar chemicals may induce similar changes in the pattern of gene expression. Nonetheless, it is important to note that significant alterations in the pattern of gene expression do not always translate into significantly different phenotypes (Corvi 2002).

An example of the application of toxicogenomics to the study of environmental pollutants toxicity was published by Laine and Fry in 2016. This study described how prenatal exposure to inorganic arsenic is associated with dysregulated foetal gene expression, as shown by the results obtained in different cohorts in which the same type of biological response is repeated. This study related the accidental arsenic environmental exposure, for example through contaminated drinking water, with certain alterations in the gene expression pattern and with adverse birth outcomes and susceptibility to certain diseases in adult life. For instance, prenatal arsenic exposure has been associated with preterm birth, reduced height and head circumference, lower birth weight, higher susceptibility to infection and later life cancers (Naujokas et al. 2013). It is known that arsenic is a carcinogen associated with skin, lung, bladder, kidney, and liver cancers. Moreover, arsenic exposure changes the foetal cord leukocyte CpG methylation leading to dysregulation at the epigenome level, adding complexity to the genetic alterations induced by exposure to this toxicant.

Another common application of toxicogenomics deals with the study of genetic polymorphisms that influence the toxic responses to therapeutic drugs. The SNPs most extensively studied are those affecting the enzymes involved in drugs metabolism, such as CYP2D6 and CYP2C19 genes (members of cytochrome P450 family 2). CYP2C19 and CYP2D6 show over 30 and 100 allelic variants respectively, ranging from nonfunctional alleles (which correspond to poor metabolizer phenotype) to alleles with enhanced activity (ultra-rapid metabolizers). CYP2D6 and CYP2C19 are two of the most common enzymes involved in the metabolism of a wide variety of therapeutic drugs, such as the selective serotonin reuptake inhibitors (SSRIs). These drugs are the first option for the treatment of depressive and anxiety disorders. CYP2C19 or CYP2D6 poor metabolizers suffer exacerbated adverse reactions to SSRIs, including cardiovascular and central nervous system toxicity (Hicks et al. 2015). Analysis of CYP2C19 or CYP2D6 polymorphisms of patients prior to SSRIs use could help to prevent those adverse events. A different example, concerning the drug's ability to induce alterations in the genetic pattern of expression, is the anticancer drugs gefitinib and erlotinib, which are inhibitors of the epidermal growth factor receptor (EGFR; Verbist et al. 2015). They were used as reference compounds to analyze the effects of 35 compounds potentially displaying similar effects. One of these compounds was selected because the expression profile analysis demonstrated its capacity to downregulate the levels of fibroblast growth factor carrier protein (FGFBP1), an effect related to the inhibition of the proliferation of cancer cells. However, it was also found that mitochondrial membrane genes such as MT1X were also downregulated by this compound, which predicted an unfortunately relevant toxic profile.

These examples confirm the importance of toxicogenomics in the prediction of drug-induced toxic responses in early phases of drug development. Moreover, they illustrate how toxicogenomics can be useful to dissect the molecular mechanisms involved in adverse drug reactions (ADR), which is of great interest because ADR are one of the most important causes of drug withdrawal. Moreover, toxicogenomic studies can also help prevent ADR in patients with specific gene polymorphisms, which is one of the pillars of personalized medicine or precision medicine. Precision medicine combines genetic and environmental information to classify patients according to their susceptibility to particular diseases and/or their responses to particular treatments. Its aim is to optimize the therapies for each person (Chambliss and Chan 2016).

In summary, the field of toxicogenomics has emerged as a powerful tool for clarifying a toxicant's mode of action and predicting an individual's response to exposure to a given toxicant. This is the beginning of a revolution in classical toxicology, but it needs further significant developments to improve the understanding of the relationship between altered genes and the appearance of the toxic effect, approaching the study from a multidisciplinary, integrated or holistic standpoint.

10.3 PROTEOMICS

10.3.1 Definition

Proteomics can be briefly defined as the science that studies the proteome, a word first used in 1994 to describe the entire set of proteins expressed by a genome (Wilkins et al. 1996). However, one single genome may result in different proteome profiles, because gene expression can be differently regulated during transcription, processing and translation. Moreover, proteins can be also highly modified by many posttranslational modifications, such as phosphorylation, glycosylation, acetylation and sulphation, which may alter protein structure, stability, location, interaction and function (Csizmok and Forman-Kay 2018). Therefore, it is impossible to predict all the characteristics of the proteome just by studying the genome. Consequently, given the high variability of the proteome, there is a more precise definition of proteomics as the science that studies the identity, quantity, structure, and biochemical and cellular functions of all proteins in an organism, organ or organelle, and their modifications depending on the time, space and physiological state (Kenyon et al. 2002). Toxic compounds can also produce changes in the proteome, which are studied by toxicoproteomics, a term used to define toxicology research using proteomic approaches (Wetmore and Merrick 2004). Toxicoproteomics evaluates the alterations of all the proteins present in cells, tissues and biofluids following exposure to toxic substances. This approach is highly useful to identify the mechanism of action of the toxicants, biomarkers of toxicant exposure and new drug targets for the treatment of intoxications (Rabilloud and Lescuyer 2015).

10.3.2 Proteomic Techniques

The complexity of the study of the proteome requires the use of highly reliable techniques with high capacity of processing; these techniques, especially regarding their application in toxicoproteomics, have been reviewed and described in detail (Miller et al. 2014; Titz et al. 2014; Rietjens et al. 2018) and will be summarized in this section. In general, the proteomic workflows include sample preparation by different methods; protein separation, quantification and identification by high-throughput techniques and data analysis by efficient bioinformatics tools (Figure 10.2).

Sample preparation is a key step in proteomic studies that requires the extraction and solubilization of the proteins of the sample, and the protocols vary depending on the type of sample (tissues, isolated cells, biofluids) and the aim of the study. In some studies, the aim is to analyze all

Table 10.2 Proteomics-Related Definitions

Term	Definition
Protein	Large molecule composed of one or more chains of amino acids in a specific order determined by the base sequence of nucleotides in the DNA coding for the protein
Proteome	The entire set of proteins found (or expressed) in a cell, tissue or organism at a particular time and under defined environmental conditions
Complete proteome	The entire set of proteins found in an organism over its entire life cycle
Proteomics	Science that studies the identity, quantity, structure, and biochemical and cellular functions of all proteins in an organism, organ or organelle, and their modifications depending on the time, space and physiological state
Total expression proteomics	Identification of all the components of a proteome and their classification to generate protein databases
Differential expression proteomics	Identification of the components of the proteome that suffer changes in their expression as a consequence of physiological or pathological processes. The aim of these studies is the identification of proteins involved in those processes, which may act as biological markers and/or therapeutic targets.
Interactions proteomics	Characterization of subcellular interactions between proteins. This type of study is intended to investigate protein functions characterizing their interactions inside the cell.

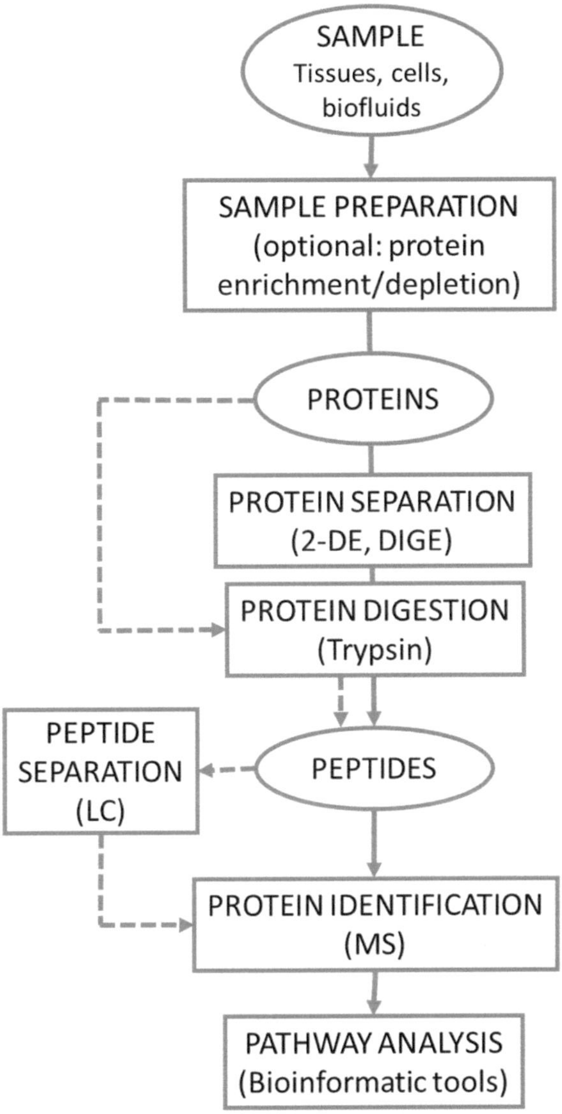

Figure 10.2 Main steps of the most common proteomic approaches. 2-DE: two-dimensional electrophoresis, DIGE: differential gel electrophoresis, LC: liquid chromatography, MS: mass spectrometry

the proteins of the sample, but in some cases enrichment or depletion techniques (mainly antibody purification or chromatographic separation) are included in sample preparation protocols in order to increase or decrease the concentration of the least and most abundant proteins of the sample, respectively. Enrichment techniques are used, for instance, to increase the concentration of proteins with a specific posttranslational modification (i.e. phosphorylated or glycosylated proteins). Depletion techniques are typically used when analyzing plasma samples to eliminate the most abundant proteins (mainly albumin and globulins) that would otherwise interfere with the study of the less abundant ones.

Regarding protein separation and identification, classic proteomic approaches are gel based; these approaches couple two-dimensional electrophoresis (2-DE) for protein separation with mass spectrometry (MS) for protein identification. In 2-DE, proteins are separated in two steps. In the first step, or first dimension, proteins are separated according to their charge (i.e. isoelectric point) by isoelectric focusing (IEF) in immobilized pH gradient (IPG) strips. In the second step, or second

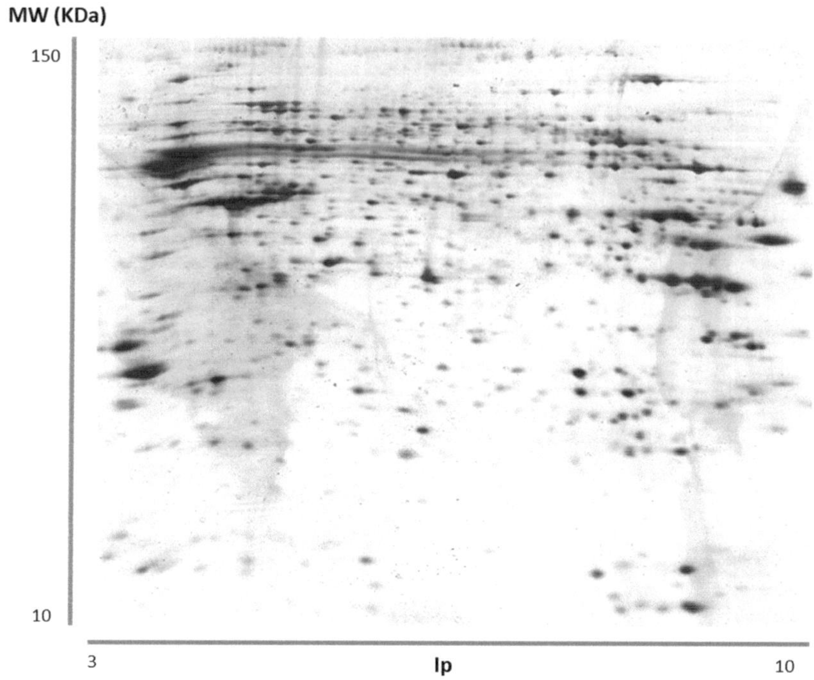

Figure 10.3 Two dimensional gel obtained from mice brain. MW: molecular weight; Ip: isoelectric point

dimension, proteins are separated according to their molecular weight by electrophoresis using SDS polyacrylamide gels (SDS-PAGE). After adequate staining, the result is a 2D gel in which each spot usually corresponds to an unique protein (Figure 10.3). This technique can separate several thousands of intact proteins in a single gel. The 2D gels obtained from different samples may be compared to detect changes in spot expression (increases or decreases in spot intensity, or appearance or disappearance of spots). Spots showing significant changes are then cut out from the gels to identify the proteins they contain by MS. For MS analysis, proteins are first cleaved to peptides with the enzyme trypsin, which cuts peptide bonds at specific sites. Then the obtained peptides are ionized, separated according to their mass-to-charge ratio, and quantified. The mass spectrometers most commonly used for proteomic studies combine matrix-assisted laser desorption/ionization (MALDI) with a time of flight (TOF) mass analyzer. Peptide mass patterns are specific for each protein, thus the mass patterns obtained by MS are searched against a comprehensive protein database (such as the ones included in UniProt) to retrieve the identity of the proteins. The 2-DE/MS approach is one of the most widely used in toxicoproteomics (Rabilloud and Lescuyer 2015), although it has some limitations that have pushed the development of other gel-based methods (such as differential gel electrophoresis [DIGE]) and, especially, non-gel-based approaches. In non-gel-based proteomic protocols, such as "shotgun" proteomics, proteins are extracted from the biological sample and digested, and the resulting peptides are separated by liquid chromatography (LC) prior to being analyzed by tandem MS (MS/MS). This LC-MS/MS approach overcomes the limitations of 2-DE/MS to identify and separate low-abundance proteins, highly acidic or basic proteins, and membrane proteins. Finally, some proteomic protocols are also implemented to improve protein quantification and include stable isotope labelling of the samples by different strategies such as stable isotope labelling by amino acids in cell culture (SILAC), isobaric tags for relative and absolute quantitation (iTRAQ), isotope coded affinity tagging (ICAT) or tandem mass tag (TMT) (Bąchor et al. 2019).

10.3.3 Applications

Proteomic approaches have been extensively applied to the study of the effects of a great variety of toxic compounds including therapeutic drugs, drugs of abuse, environmental pollutants, biological toxins and even nanomaterials, both *in vitro* and *in vivo*, and using different animal models and humans (Miller et al. 2014; Titz et al. 2014; Rietjens et al. 2018). Regarding humans, the Human Proteome Organization (HUPO) also includes a toxicoproteomics interest group (HTP) focused on the study of drug-induced organ injury and the indirect detection of such events in body fluids as a means to identify safety biomarkers (HUPO 2018).

Toxicoproteomics can help to uncover the molecular mechanisms involved in ADR of therapeutic drugs (Rabilloud and Lescuyer 2015). For example, overdose of the over-the-counter analgesic and antipyretic drug, acetaminophen, is a leading cause of acute liver failure and one of the major reasons for liver transplantation in the United States. In a recent study, a quantitative proteomic analysis, based on stable isotope labelling and 2D-LC-MS/MS, was performed in the livers of rats treated with low and high doses of acetaminophen; a total of 31 proteins were identified whose expression was uniquely correlated to liver damage. These proteins were mainly involved in regulation of glutathione metabolism and NF-E2-related transcription factor 2 (NRF2) regulation of oxidative stress. Among them, heme oxygenase 1 also showed a dramatic increase in plasma, thus this protein could be a good plasma biomarker of liver injury (Gao et al. 2017). Another application of toxicoproteomics related to drug therapy is precision medicine. This application is of special interest in the case of anticancer drugs. As an example, a proteomic study, using LC-MS/MS, demonstrated that combination of sulbactam and doxorubicin enhances the cytotoxicity of doxorubicin in the breast cancer cells by inhibiting the expression of ATP-binding cassette transporter proteins and proteins associated with the process of transcription and initiation of translation; moreover, this cotreatment also reduces the ADR of doxorubicin (Wen et al. 2018).

Drugs of abuse toxicity has also been extensively studied by using proteomics. An interesting approach in this field is found in the study of Vicente-Rodríguez et al. (2015), which is focused on the effect of cocaine on phosphorylated proteins. Phosphorylation is a crucial posttranslational modification, because many enzymes and receptors are activated/deactivated by phosphorylation and dephosphorylation events. In this study, a phosphoprotein enrichment (using immobilized metal affinity chromatography) was applied to brain striatum samples obtained from mice chronically treated with cocaine prior to their analysis by 2-DE and MALDI-TOF MS. Most of the phosphoproteins identified in this study are related to neurodegenerative disorders, such as Parkinson's disease or Alzheimer's disease, and these results may help to provide the molecular basis for the link between drug addiction and neurodegenerative diseases that has been observed in epidemiological studies (Lappin et al. 2018).

Various classes of environmental pollutants have been also studied by proteomic approaches but with more focus on substances that are known to cause major public health problems, such as polycyclic aromatic hydrocarbons or endocrine disruptors (Rabilloud and Lescuyer 2015). Among the first ones, benzo(a)pyrene is one of the most investigated, particularly because of its cytotoxic, mutagenic, and carcinogenic properties. In urban areas, benzo(a)pyrene is adsorbed by airborne carbon nanoparticles (mainly derived from biomass burning) which modify its toxicity. This has been demonstrated by proteomic techniques that revealed that the toxicity produced by carbon nanopowder-benzo(a)pyrene complex is not derived from an additive or synergistic effect of the individual compounds but from the alteration of different toxicological pathways. These pathways are mainly related to energy shift, stress response, apoptosis, and cellular trafficking (Binelli et al. 2018).

Protein toxins, derived from bacteria and plants, are among the most powerful poisons known, and some of them are used as bioterrorism agents. Bioassays are widely used for the detection of protein toxins, but they require purification of the toxins prior to testing and the use of live animals or cell culture to determine the biological activity of toxins. However, the response to bioterrorism attacks requires methods that can quickly detect and identify the toxins used, enabling effective treatment and decontamination. Proteomic methods can give results much faster and are simpler to perform than bioassays. Thus, LC-MS methods are being used to detect toxins from *Staphylococcus aureus*, *Clostridium botulinum* and *Bacillus anthracis*, among others species. These proteomic approaches enable an unequivocal detection of potential sources of contamination and the protein toxin itself (Duracova et al. 2018).

The study of nanomaterials toxicity is one of the newest applications of proteomics. Nanomaterials are made of ultrafine particles of different minerals (such as carbon, silver, titanium, zinc and silicon), some of which can cross the blood-brain barrier or be endocytosed by cells; therefore they are widely used in medicine as drug delivery systems, as biosensors or for tissue engineering, among other applications. However, some studies suggest that nanomaterials could also damage human health, such as by inflammation of the lungs, asthma or cancer

(Pietroiusti et al. 2018). Therefore, it would be interesting to decipher the mechanisms involved in nanomaterials-mediated toxicity. Proteomic studies are helping to decipher the biochemical processes altered by nanomaterials, which include cell morphology, cell differentiation, cell metabolism, cell mobility, and cellular immunity or redox regulation (Nath Roy et al. 2017). This information is highly valuable for assessing and managing the potential toxicity derived from existing and future uses of nanomaterials.

10.4 METABOLOMICS

10.4.1 Definition

At its simplest, the metabolome can be defined as the complete set of metabolites within the biological system under study (e.g. culture cells, biofluid, tissue, organ, organism). These are small molecules (typically, less than 1500 Da), the products and intermediates of anabolism and catabolism. Therefore, although several terms can be considered sometimes equivalent or included within metabolomics (see Table 10.3), we could merely consider it the study of the metabolome.

10.4.2 Metabolomic Techniques

There is not a specific methodology, instrumental technique or data management process fully characteristic of this discipline. Instead, it is the whole workflow for untargeted metabolomics (see Figure 10.4) that has become characteristic. The special requirements, in terms of reproducibility and the fact that most of the metabolites remain unidentified in databases, implies the need for specific combinations of procedures to achieve the widest metabolite coverage with the minimum contamination.

Due to the large variability in the physicochemical properties of analytes, together with the enormous differences in concentrations, there is no single analytical technique that can fulfil all the requirements to provide an adequate signal for all (Alonso 2015). Metabolome analysis is generally conducted through two detection techniques: nuclear magnetic resonance (NMR) and mass spectrometry (MS).

The NMR profile can contain qualitative and quantitative information on hundreds of different small molecules present in the sample. Although sensitivity is poorer in NMR than in MS, the elucidation capabilities are unquestionable: each compound exhibits a unique pattern of chemical

Table 10.3 Metabolomics-Related Definitions

Term	Definition
Metabolites	Small molecules, low-molecular weight (<1500 Da) intermediates, building blocks for other biological components, essential for the regulation, growth and normal function of the cell
Metabolome	The final downstream product of the genome, the complete set of all low molecular-weight metabolites present in a cell or organism. The metabolome is divided into exometabolomes (metabolites from outside the system) and endometabolomes (generated by the system).
Metabotype	The metabolic phenotype
Metabolomics	The study of the complete set of metabolites varying according to the physiology and developmental or pathological state of the cell, tissue, organ or organism
Metabolic fingerprinting	Unbiased, global, high-throughput approach to classifying samples based on metabolite patterns, or "fingerprints", which change in response to disease, environmental or genetic perturbations with the ultimate goal of identifying discriminating metabolites
Metabolic profiling	Identification and quantification or semiquantification of specific groups of metabolites, which are generally related to a specific metabolic pathway
Metabolic footprinting	Analysis of the metabolites secreted/excreted by an organism under controlled conditions
Untargeted metabolomics	Global in scope, hypothesis free, with the aim of simultaneously measuring as many metabolites as possible from biological samples
Targeted metabolomics	Involves using one particular analytical method for the determination and quantification of a small set of known metabolites
Metabolic pathway	A set of chemical reactions inside a cell

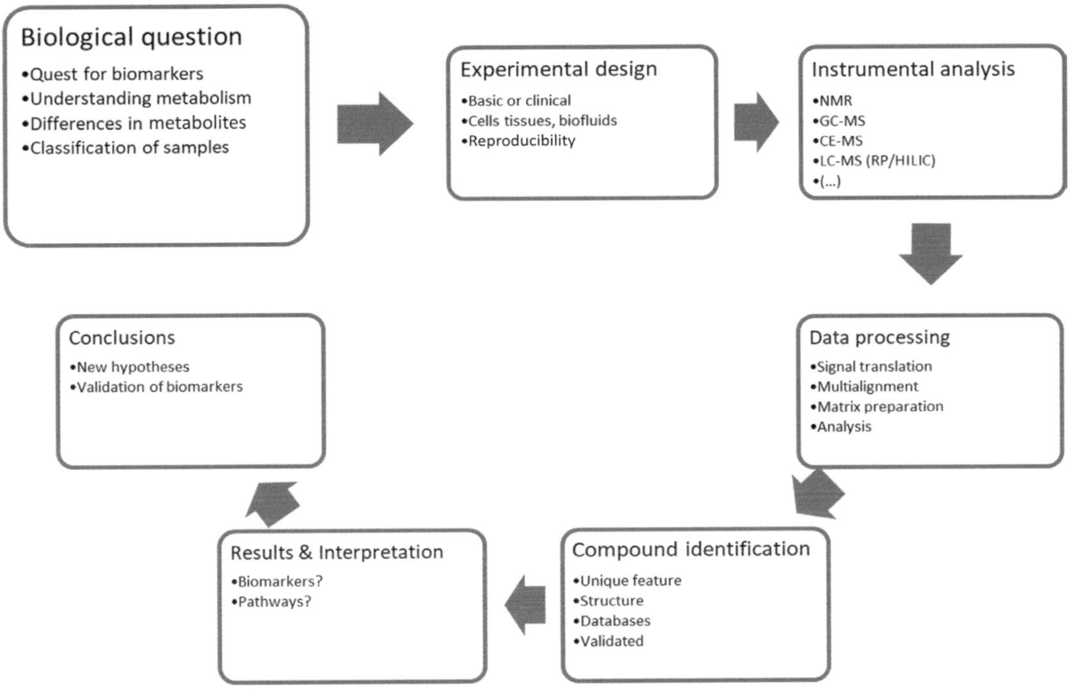

Figure 10.4 Workflow for untargeted metabolomics (metabolic profiling)

shifts (a fingerprint), and therefore the compounds present in a complex sample can be undoubtedly identified. Moreover, as the intensity (integrated signal area) of each signal is directly proportional to the concentration of the corresponding compound in the sample, the comparison between samples to find differential compounds is straightforward.

Metabolomics was first developed using NMR and was the topic for a special issue of the *Journal of Biomolecular NMR* in April 2011 ("NMR Investigations of Metabolomics"). In recent years, MS has become the most commonly employed technique in metabolomics (Armitage et al. 2013). Two approaches regarding MS can be considered, with different subtypes: MS can be hyphenated to a separation technique (gas chromatography [GC], liquid chromatography [LC], capillary electrophoresis [CE] and supercritical fluid chromatography [SFE]) or not (direct MS [DMS]). MS benefits from detection permitted at high sensitivity and also structural elucidation based on spectral libraries and tandem MS, even in complex biological samples.

GC-MS is very well suited for the analysis of energy metabolism, as it is highly appropriate for volatile derivatives of metabolites; amino acids, monosaccharides, fatty acids, disaccharides and cholesterol are easily identified in the chromatogram (Koek et al. 2011). Such metabolites are mostly hydrophilic and therefore freely soluble in typical biofluids such as serum/plasma or urine. CE-MS can be applied to the analysis of similar compounds to a certain extent, benefiting from reduced sample treatment compared to GC-MS (Ramautar et al. 2017), but it has different strengths and drawbacks (Table 10.4).

The most widely employed technique for metabolic fingerprinting is LC-MS, which involves the minimum requirement for sample treatment and alteration or hydrolysis of the metabolites during the analysis among the hyphenated techniques (Roux et al. 2011). The most common separation mode for LC-MS-based metabolomic studies involves reversed-phase chromatographic separation of analytes. This is especially suited to metabolites with medium-low polarity, although analysis is limited because polar metabolites including sugars or amino acids that CE-MS and GC-MS typically analyze are not detected. Such metabolites can be analyzed by LC-MS, however it requires the application of hydrophilic liquid chromatography (HILIC; Rojo et al. 2012).

Once the signals from every sample have been obtained, they have to be processed in order to convert them into a file format suitable for further analysis. The simplest treatment will give us a plain text file with information about the retention time, the mass-to-charge (MS) or drift (NMR),

Table 10.4 Advantages and Drawbacks of Each Technique for Metabolomics, and Main Metabolites That Can Be Studied with Each Technique

Technique	Advantages	Drawbacks	Main metabolites
LC-MS	High throughput Wide metabolite coverage	Signal processing	Lipids (in reverse phase mode) Other polar (in HILIC mode)
GC-MS	Identification with databases	Sample treatment	Amino acids, sugars and organic acids
CE-MS	Little sample treatment, ideal for polar metabolites	Robustness Sensitivity	Amino acids, nucleotides and related metabolites
DMS	Fast	Metabolite identification and sample comparison difficult	Lipids

and the intensity. The subsequent multialignment of all the signals from all samples will provide a matrix suitable for statistical analysis. Multialignment for multidimensional signals was a challenge for metabolomics, and the experience from multiple sequence alignment (MSA) for DNA analysis was employed to develop several algorithms and tools for it. Nowadays, many tools are available (for a comprehensive list, visit https://fiehnlab.ucdavis.edu/staff/kind/Metabolomics/Peak-Alignment). The differences between them exceed the purpose of this chapter, but it is worth mentioning that issues such as peak finding, feature definition, noise determination, drift tolerance and many more are very important for the final result.

It is assumed that, currently, the main bottleneck in metabolomics is the identification of the significant metabolites when they were found in LC-MS, CE-MS or DMS measurements. The identification process starts with the quest for information in publicly available databases (most free of charge). The input is the exact mass, and the output is a chemical identity. But databases for metabolomics are not fully standardized. There are many, different in the number of records (compounds), in the fields for each record (compound properties), as well as the searchable fields (mass, m/z, MS/MS spectrum, name, etc.). Probably the most important databases nowadays are the Human Metabolome Database (HMDB; http://hmdb.ca), METLIN (https://metlin.scripps.edu/) and LIPID MAPS Structure Database (LMSD; www.lipidmaps.org/), which provide chemical properties and biochemical information as well as spectral information for 10^5, 10^6, and 10^4 compounds, respectively. Compounds in HMDB can be searched through more than 100 different fields, but METLIN includes different possibilities to find compounds that match different MS experimental results. In addition, LIPID MAPS is a consortium which, among other activities, is devoted to unifying nomenclature and classification of lipids. To facilitate the simultaneous queries in different databases, tools as CEU Mass Mediator (http://ceumass.eps.uspceu.es/) permit obtaining the information from the previously mentioned databases, together with other utilities.

Nevertheless, current databases do not collect many of the compounds that can be detected with the electrospray source MS techniques (LC-MS, CE-MS, DMS), whereas for NMR and GC-MS, virtually all metabolites than can be analyzed can be identified, because there are fewer compounds that can be detected (although with very high metabolic relevance), and the spectral results are fully reproducible from instrument to instrument. In this case, compounds are identified before with statistical analysis, which is used to point at the most significant compounds that discriminate between the groups/conditions.

In the case of LC-MS, DMS and CE-MS, in some cases (if there is the possibility of performing tandem MS in the equipment) additional analysis are performed to obtain the MS/MS fragmentation spectrum to proceed to structural elucidation. Moreover, it is also possible to compare experimental MS/MS data with *in silico* predicted spectrum (Gil de la Fuente et al. 2017). This elucidation is not automatized yet, and therefore it is not applied to the complete set of detected compounds, only to those ions that might have significant results.

10.4.3 Applications

Metabolomics is increasingly used as a tool in toxicology because it offers the possibility of studying all the biochemical perturbations caused by a potentially hazardous compound, including its mode of action. *In vitro* approaches are increasingly recognized in toxicology, and the term "toxicometabolomics" is becoming increasingly used.

In 2007, the US National Research Council (NRC) published *Toxicity Testing in the 21st Century: A Vision and a Strategy*. This report proposed a paradigm shift in toxicology, in which current animal tests should be replaced by *in vitro* cell systems combined with new, promising techniques such as metabolomics. A major goal of the field is the use of metabolomics information in comparison to known standards to predict actions of unknown chemicals in biological systems (Rusyn et al. 2012). To develop such models, metabolite patterns related to well-known "training" compounds would be used to develop classification schemes. These would then be applied to the metabolome patterns triggered by unknown compounds in order to predict their toxicological hazard. For example, HL-1 cardiomyocytes after exposure to drugs from different classes (analgesics, antimetabolites, anthracyclines, antihistamines, channel blockers) were analyzed to determine specific metabolic footprints in response to the tested drugs (Strigun et al. 2011), and simple metabolome analysis was shown to be useful to classify drugs into mechanism of action classes, according to their classification into multivariate analysis models. Moreover, 18 metabolites were shown to be the most influencers, and the metabolic response could be associated to the variation in these metabolites.

In 2005, the Consortium for Metabolomic Toxicology (COMET) was formed between five pharmaceutical companies and the Imperial College of Science, Technology and Medicine (IC) in London (Lindon et al. 2005). The consortium performed metabolomic studies to predict liver and kidney toxicity using urine and serum samples from rodents after 7 days of exposure to 147 compounds (individually tested). The generated database still serves as a useful tool in the area of drug discovery and development within these companies (Spagou et al. 2011).

The chemical industry has also recognized metabolomics as a useful tool for toxicity testing. In 2012, Metanomics Health, a BASF Group company, launched their new MetaMap Tox database (van Ravenzwaay et al. 2012). The database has been developed over several years, testing more than 500 chemicals (28-day rat studies) to generate metabolomic patterns for different toxicological targets (e.g. liver, kidney, thyroid, testes, blood, nervous system, endocrine system). The metabolomic approach suggests a toxicological mode of actions at an early state and can therefore, speed up safety decisions and lower the cost through reduced numbers of animal studies.

Blood, serum and urine are the most commonly used samples for *in vivo* metabolomics studies, but the technique can be applied to other biofluids or tissue samples. One example is a metabolomic study of liver samples from mice showing that polychlorinated biphenyls (PCBs) induce nonalcoholic fatty liver disease when they are administered together with a high-fat diet (Shi et al. 2012). The metabolomic approach was useful to show the association of the depletion of antioxidants (glutathione) to the development of the disease.

In addition to the mechanism of action, metabolomics can be used for discriminating the critical pathways that lead to toxicity, as opposed to other chemical-induced changes that are only adaptations, counterregulations or epiphenomena. For instance, plasma and urine NMR profiles from mice with chemically induced hepatic necrosis were studied together with transcriptomic data, and the resulting alterations in amino acid and glutathione metabolism enabled the generation of new hypotheses on the cellular mechanism involved in hepatotoxicity (Heijne et al. 2005).

In vitro assays and metabolomic approaches are still not intensively applied together in toxicology, but they are increasingly used and have shown to be very promising. One advantage of cell-based metabolomics is that both extracellular and intracellular metabolites can be measured. Metabolites detected in extracellular samples (i.e. metabolic footprinting; see Table 10.3) can be associated to *in vivo* identified biomarkers from biofluids, whereas the intracellular samples (cell lysate) can provide information about the toxic mechanisms at the cellular level. The main studies are still performing measurements of cell culture media, but the assessment of intracellular metabolites is constantly increasing.

10.5 INTEGRATION OF GENOMICS, PROTEOMICS AND METABOLOMICS FOR TOXICOLOGY STUDIES: Systems Toxicology

Research findings from the different 'omics fields of study (i.e. genomics, proteomics and metabolomics) should not be taken individually but instead should inform and complement one another. This integrated approach is the basis of systems biology, or systems toxicology when specifically focused on toxicant effects. In systems toxicology, the data obtained by 'omics techniques are analyzed and integrated by advanced computational tools in order to elucidate toxicity pathways and to create mathematical models to predict adverse outcomes derived from toxicant exposure (Hartung et al. 2017). Systems toxicology is thus a highly valuable means not only to design effective diagnosis and therapies for diseases related to toxic agents (which is the basis of precision medicine) but also to improve the process of risk assessment (Figure 10.5).

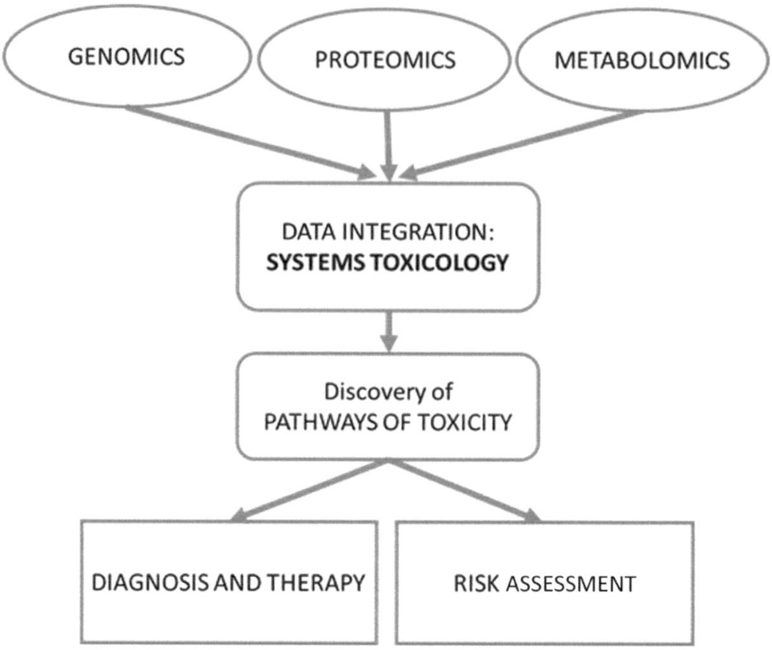

Figure 10.5 Integration of 'omics technologies for systems toxicology and their applications.

System toxicology studies still have some limitations, such as the uneven maturity of the different 'omics technologies and knowledge (e.g. mapping the human proteome and metabolome is still ongoing) or the difficulty for a certain translation of molecular perturbations into phenotypic toxicity. The collaborative efforts of academic, clinical, industrial and regulatory bodies are key to transforming these limitations into promising applications.

10.6 CONTRIBUTION OF THE 'OMICS TO THE KNOWLEDGE OF THE EXPOSOME: Role of Emerging 'Omics

It is widely accepted that cumulative exposure to a wide variety of environmental agents negatively affects human health. The evaluation of several exposures simultaneously can provide a more accurate analysis of the impact of the environment on human health. In this context, the concept of exposome arose. It was established in 2005 by Christopher Wild and refers to the totality of environmental exposures during an individual lifetime, from conception to death (Wild 2005). We can distinguish three domains in the exposome: the internal, the specific external and the general external factors. Internal factors are those internal to the body such as metabolism, endogenous circulating hormones, gut microflora, inflammation, lipid peroxidation, oxidative stress and ageing. Specific external exposures include diet, lifestyle, environmental and occupational exposures. General external factors include the wider social, economic and psychological influences on the individual (Rappaport 2011; Wild 2012).

Exposomics is the study of the exposome and represents a significant challenge due to the wide scope and the dynamic nature of the exposome. However, 'omics can greatly contribute to the knowledge of the internal factors (Wild 2012). For instance, the use of transcriptomic and proteomic techniques have revealed the molecular changes induced by occupational exposure to benzene in Chinese workers (Zhang et al. 2010). Another example is the PISCINA study (part of a large project called EXPOsOMICS), in which metabolomic techniques have helped to evaluate the effect of water disinfection by-products in the blood metabolome of swimmers (Vineis et al. 2017). In addition to the classical 'omics, other emerging 'omics are also being applied to the study of the exposome, such as lipidomics (a large-scale study of cellular lipids), immunomics (immune-related genomics and proteomics) and, especially, adductomics (Wild 2012). Adductomics refers to the comprehensive analysis of DNA and protein adducts. Toxic compounds are frequently metabolized to reactive species that covalently bind with DNA or proteins to produce adducts. Once formed, these adducts may produce direct cell toxicity, activate an immune response or initiate mutagenic/carcinogenic processes. Therefore adductomics is a powerful tool to link exposure to toxicants with the occurrence of different pathological processes (Pereira and Antunes 2019). Currently, the most utilized technique for the untargeted screening of adducts is LC-MS/MS (Shibata and Uchida 2019). This technique has been employed in protein adductomics to detect adducts of blood proteins, such as albumin or hemoglobin. That is the case of the study of Carlsson et al. (2014), which compared the hemoglobin adductome of smokers and nonsmokers and detected 19 previously unknown human hemoglobin adducts. DNA adductomics is also being widely applied to the screening of DNA damage in the human genome by using different human samples (Guo and Turesky 2019); however, it requires an invasive sampling of tissue to obtain sufficient DNA for sensitive analysis. To circumvent this, DNA adductomics has been applied to urine. Accordingly, Cooke et al. (2018) compared the adductome map of urine from a patient under mechanical ventilation in an intensive care unit to one from a healthy subject. More putative adducts were detected and at a higher intensity in the urine of the patient compared to the healthy subject. This noninvasive approach facilitates the exposure assessment in human individuals and populations, across the life course, which supports the aims of exposomics.

BIBLIOGRAPHY

Alonso, A., S. Marsal, and A. Julià. 2015. Analytical methods in untargeted metabolomics: State of the art in 2015. *Front Bioeng Biotechnol* 3: 23.

Armitage, E. G., F. J. Rupérez, and C. Barbas. 2013. Metabolomics of diet-related diseases using mass spectrometry. *TrAC Trends Analytical Chem* 52: 61–73.

Ates, G., B. Mertens, A. Heymans, et al. 2018. A novel genotoxin-specific qPCR array based on the metabolically competent human HepaRG™ cell line as a rapid and reliable tool for improved in vitro hazard assessment. *Arch Toxicol* 92(4): 1593–608.

Bąchor, R., M. Waliczek, P. Stefanowicz, and Z. Szewczuk. 2019. Trends in the design of new isobaric labeling reagents for quantitative proteomics. *Molecules* 24(4). pii: E701.

Berger, S. L., T. Kouzarides, R. Shiekhattar, and A. Shilatifard. 2009. An operational definition of epigenetics". *Genes Dev.* 23: 781–83.

Binelli, A., S. Magni, C. La Porta, et al. 2018. Cellular pathways affected by carbon nanopowder-benzo(α)pyrene complex in human skin fibroblasts identified by proteomics. *Ecotoxicol Environ Saf* 60: 144–53.

Carlsson, H., H. von Stedingk, U. Nilsson, and M. Törnqvist. 2014. LC-MS/MS screening strategy for unknown adducts to N-terminal valine in hemoglobin applied to smokers and nonsmokers. *Chem Res Toxicol* 27(12): 2062–70.

Chambliss, A. B., and D. W. Chan. 2016. Precision medicine: From pharmacogenomics to pharmacoproteomics. *Clin Proteom* 13: 25–33.

Cooke, M. S., C. W. Hu, Y. J. Chang, and M. R. Chao. 2018. Urinary DNA adductomics: A novel approach for exposomics. *Environ Int.* 121: 1033–8.

Corvi, R. 2002. Genomics: An in vitro toxicology point of view. *Atla* 30(Suppl 2): 129–31.

Csizmok, V., and J. D. Forman-Kay. 2018. Complex regulatory mechanisms mediated by the interplay of multiple post-translational modifications. *Curr Opin Struct Biol* 58–67.

Duracova, M., J. Klimentova, A. Fucikova, and J. Dresler. 2018. Proteomic methods of detection and quantification of protein toxins. *Toxins (Basel)* 10(3): pii: E99.

Gao, Y., Z. Cao, X. Yang, et al. 2017. Proteomic analysis of acetaminophen-induced hepatotoxicity and identification of heme oxygenase 1 as a potential plasma biomarker of liver injury. *Proteomics Clin Appl* 11(1–2): 1–2. https://doi.org/10.1002/prca.201600123.

Gil de la Fuente, A., E. G. Armitage, A. Otero, C. Barbas, and J. Godzien. 2017. Differentiating signals to make biological sense: A guide through databases for MS-based non-targeted metabolomics. *Electrophoresis* 38(18): 2242–56.

Guo, J. and Turesky, R,J. 2019. Emerging Technologies in Mass Spectrometry-Based DNA Adductomics. *High Throughput.* 8(2):13.

Hartung, T., R. E. FitzGerald, P. Jennings, et al. 2017. Systems toxicology: Real world applications and opportunities. *Chem Res Toxicol* 4: 870–82.

Heijne, W, H., R. J. Lamers, P. J. van Bladeren, J. P. Groten, J. H. van Nesselrooij, and B. van Ommen. 2005. Profiles of metabolites and gene expression in rats with chemically induced hepatic necrosis. *Toxicol Pathol* 33(4): 425–33.

Hicks, J. K., J. R. Bishop, K. Sangkuhl, et al. 2015. Clinical Pharmacogenetics Implementation Consortium (CPIC) Guideline for CYP2D6 and CYP2C19 genotypes and dosing of selective serotonin reuptake inhibitors. *Clin Pharmacol Ther* 98(2): 127–34.

Human Proteome Organization (HUPO). *Toxicoproteomics.* www.hupo.org/Toxicoproteomics (accessed October 29, 2020).

Kenyon, G. L., D. M. DeMarini, E. Fuchs, et al. 2002. National Research Council Steering Committee. Defining the mandate of proteomics in the post-genomics era: Workshop report. *Mol Cell Proteomics* 10: 763–80.

Koek, M. M., R. H. Jellema, J. van der Greef, A. C. Tas, and T. Hankemeier. 2011. Quantitative metabolomics based on gas chromatography mass spectrometry: Status and perspectives. *Metabolomics.* 7: 307–28.

Laine, J. E., and R. C. Fry. 2016. A systems toxicology-based approach reveals biological pathways dysregulated by prenatal arsenic exposure. *Ann Glob Health* 2016 Jan-Feb;82(1): 189–96.

Lappin, J. M., S. Darke, and M. Farrell. 2018. Methamphetamine use and future risk for Parkinson's disease: Evidence and clinical implications. *Drug Alcohol Depend* 187: 134–40.

Lindon, J. C., H. C. Keun, T. M. Ebbels, J. M. Pearce, E. Holmes, and J. K. Nicholson. 2005. The Consortium for Metabonomic Toxicology (COMET): Aims, activities and achievements. *Pharmacogenomics* 6: 691–9.

Miller, I., T. Serchi, A. J. Murk, and A. C. Gutleb. 2014. The added value of proteomics for toxicological studies. *J Toxicol Environ Health B Crit Rev* 17(4): 225–46.

Nath Roy, D., R. Goswami, and A. Pal. 2017. Nanomaterial and toxicity: What can proteomics tell us about the nanoToxicology? *Xenobiotica* 47(7): 632–43.

Naujokas, M. F., B. Anderson, H. Ahsan, et al. 2013. The broad scope of health effects from chronic arsenic exposure: Update on a worldwide public health problem. *Environ Health Perspect* 121: 295–302.

Pereira, S. A., and A. M. M. Antunes. 2019. Special Issue "Adductomics: Elucidating the environmental causes of disease". *High-Throughput* 8: 17.

Pietroiusti, A., H. Stockmann-Juvala, F. Lucaroni, and K. Savolainen. 2018. Nanomaterial exposure, toxicity, and impact on human health. *Wiley Interdiscip Rev Nanomed Nanobiotechnol.* https://doi.org/10.1002/wnan.1513.

Rabilloud, T., and P. Lescuyer. 2015. Proteomics in mechanistic toxicology: History, concepts, achievements, caveats, and potential. *Proteomics* 5–6: 1051–74.

Ramautar, R., G. W. Somsen, and G. J. de Jong. 2017. CE-MS for metabolomics: Developments and applications in the period 2014–2016. *Electrophoresis* 38(1): 190–202.

Rappaport, S. M. 2011. Implications of the exposome for exposure science. *J Expo Sci Environ Epidemiol* 21(1): 5–9.

Rietjens, I. M. C. M., J. Vervoort, A. Maslowska-Górnicz, N. Van den Brink, and K. Beekmann. 2018. Use of proteomics to detect sex-related differences in effects of toxicants: Implications for using proteomics in toxicology. *Crit Rev Toxicol* 26: 1–16.

Rojo, D., C. Barbas, and F. J. Rupérez. 2012. LC-MS metabolomics of polar compounds. *Bioanalysis* 4: 1235–43.

Roux, A., D. Lison, C. Junot, and J.-F. Heilier. 2011. Applications of liquid chromatography coupled to mass spectrometry-based metabolomics in clinical chemistry and toxicology: A review. *Clin Biochem* 44: 119–35.

Rusyn, I., A. Sedykh, Y. Low, K. Z. Guyton, and A. Tropsha. 2012. Predictive modeling of chemical hazard by integrating numerical descriptors of chemical structures and short-term toxicity assay data. *Toxicol Sci.* 127(1): 1–9.

Shi, X., B. Wahlang, and X. Wei. 2012. Metabolomic analysis of the effects of polychlorinated biphenyls in nonalcoholic fatty liver disease. *J Proteome Res* 11(7): 3805–15.

Shibata, T., and K. Uchida. 2019. Protein adductomics: A comprehensive analysis of protein modifications by electrophiles. *Free Radic Biol Med* pii: S0891–5849(18)32648–0.

Spagou, K., I. D. Wilson, and P. Masson. 2011. HILIC-UPLC-MS for exploratory urinary metabolic profiling in toxicological studies. *Anal Chem* 83: 382–90.

Strigun, A., J. Wahrheit, S. Beckers, E. Heinzle, and F. Noor. 2011. Metabolic profiling using HPLC allows classification of drugs according to their mechanisms of action in HL-1 cardiomyocytes. *Toxicol Appl Pharmacol* 252(2): 183–91.

Titz, B., A. Elamin, F. Martin, et al. 2014. Proteomics for systems Toxicology. *Comput Struct Biotechnol J* 18: 73–90.

US EPA. 2018. *About the TSCA Chemical Substance Inventory.* www.epa.gov/tsca-inventory/about-tsca-chemical-substance-inventory (accessed October 29, 2020).

van Ravenzwaay, B., M. Herold, H. Kamp, et al. 2012. Metabolomics: A tool for early detection of toxicological effects and an opportunity for biology based grouping of chemicals-from QSAR to QBAR. *Mutat Res* 746: 144–50.

Venter, J. C. et al. 2011. The sequence of the human genome. *Science* 291(5507): 1304–51.

Verbist, B., G. Klambauer, L. Vervoort, et al. 2015. Using transcriptomics to guide lead optimization in drug discovery projects: Lessons learned from the QSTAR project. *Drug Discov Today* 20(5): 505–13.

Vicente-Rodríguez, M., G. Herradón, M. Ferrer-Alcón, M. Uribarri, and C. Pérez-García. 2015. Chronic cocaine use causes changes in the striatal proteome depending on the endogenous expression of pleiotrophin. *Chem Res Toxicol* 28(7): 1443–54.

Vineis, P., M. Chadeau-Hyam, H. Gmuender, et al. 2017. EXPOsOMICS Consortium. The exposome in practice: Design of the EXPOsOMICS project. *Int J Hyg Environ Health* 220(2 Pt A): 142–51.

Wang, Z., M. Gerstein, and M. Snyder. 2009. RNA-Seq: A revolutionary tool for transcriptomics. *Nat Rev Genet* 10(1): 57–63.

Wen, S. H., S. C. Su, B. H. Liou, C. H. Lin, and K. R. Lee. 2018. Sulbactam-enhanced cytotoxicity of doxorubicin in breast cancer cells. *Cancer Cell Int* 18: 128.

Wetmore, B. A., and B. A. Merrick. 2004. Toxicoproteomics: Proteomics applied to Toxicology and pathology. *Toxicol Pathol* (6): 619–42.

Wild, C. P. 2005. Complementing the genome with an "exposome": The outstanding challenge of environmental exposure measurement in molecular epidemiology. *Cancer Epidemiol Biomarkers Prev* 14(8): 1847–50.

Wild, C. P. 2012. The exposome: From concept to utility. *Int J Epidemiol* 41(1): 24–32.

Wilkins, M. R., J. C. Sanchez, A. A. Gooley, et al. 1996. Progress with proteome projects: Why all proteins expressed by a genome should be identified and how to do it. *Biotechnol Genet Eng Rev* 13: 19–50.

Zhang, L., C. M. McHale, N. Rothman, et al. 2010. Systems biology of human benzene exposure. *Chem Biol Interact* 184(1–2): 86–93.

11 Toxic Effects of Pharmaceuticals and Cosmetics

María de los Ángeles, Jesús Molpeceres, María del Rosario Aberturas and Antonio Peña-Fernández

CONTENTS

11.1 INTRODUCTION

Drug toxicity refers to the level of harm that a compound can cause to an organism. It has been estimated that toxicity is responsible for almost one-third of drug candidates not continuing their development and contributes greatly to the high cost of drug development overall. It is important to establish adverse reactions, which are defined as "an appreciably harmful or unpleasant reaction, resulting from an intervention related to the use of a medicinal product, which predicts hazard from future administration and warrants prevention or specific treatment, or alteration of the dosage regimen, or withdrawal of the product" (Edwards and Aronson 2000).

In this chapter, a summary of the main hazardous ingredients of pharmaceuticals and cosmetics will be presented. The growing demand for cosmetics and personal care products throughout the world has led to increased awareness of their safety and/or risk. This chapter is dedicated to addressing the special multidisciplinary challenges of chemical ingredients in personal care products and deals with exposure to the environment. In support of this hypothesis, we supply cases of target studies and environmental behaviour for classes of compounds that are listed in the International Nomenclature of Cosmetics Ingredients (INCI) and have been classified as potentially harmful.

11.2 HAZARDOUS INGREDIENTS IN PHARMACEUTICALS

Excipients are pharmaceutical ingredients which are pharmacologically inert and are used in drug formulations for fulfilling various roles, for instance to improve dosage form volume, disintegration of solid dosage forms, binding of particulates, lubrication during processing, taste masking or modifying drug release. Ideal excipients are considered safe, inert and multifunctional. The safety of excipients needs to be well established for their correct usage in the diverse pharmaceutical products and to be evaluated systematically for possible risk in humans or animals. Therefore, careful consideration should be given to the selection and variety the excipients. Depending on data available and precedent, excipients may have to be assessed experimentally for genotoxicity, general toxicity or carcinogenicity.

 DOI: 10.1201/9780203730584-11

It is known that many excipients are potentially toxic at high doses, although in general they are safe in humans at therapeutic doses. Some are cyclodextrins, dextrans and polyethylene glycol, surfactants and preservatives; other pharmaceutical excipients can even modify the cytotoxic attributes of preservatives, such as the surfactants mentioned (Haschek et al. 2013; Herman 2018).

11.2.1 Cyclodextrins

Cyclodextrins (CDs) are macrocyclic oligosaccharides composed of α(1,4)-linked glucopyranose subunits, with lipophilic internal cavities type cage cavity-type cages and hydrophilic external surfaces (Martin 2004; Zhang et al. 2012; Crini 2014; García et al. 2019). The natural cyclodextrins α-, β-, γ- and δ-CD, with a corresponding number of glucose units ($\alpha = 6$, $b = 7$, $\gamma = 8$ and $\delta = 9$), and their synthetic derivatives are divided into three groups: hydrophilic, such as 2-hydroxypropyl-β-CD (HP-β-CD); hydrophobic, such as 2,6-di-O-ethyl-β-CD; and ionizable, such as sulfobutylether-β-CD (SBE-β-CD; Carneiro et al. 2019). The CDs can solubilize and increase the absorption rate as well as stabilize drugs by forming noncovalent inclusion complexes with various inorganic, organic or organometallic, cationic, anionic or neutral molecules. The toxicity profile of CDs can differ depending on the route of administration and the type of CD used. CDs are poorly absorbed orally so they have limited toxicity, although high doses can be harmful and lead to irreversible damage and renal dysfunction. Therefore, CDs if used in safe concentration ranges are nontoxic and considered safe for drug complexion (Carneiro et al. 2019). Animal toxicity is known: if β-CD is administered at higher doses in animals parenterally, it will increase liver enzymes; decrease erythrocyte parameters, epithelial vacuolation of proximal renal tubules and urinary bladder; and cause accumulation of foamy macrophages in the lung and Kupffer cells in the liver (Haschek et al. 2013). However, the toxicity of CD in humans depends on the route of administration and the type used; when given orally, CDs are mostly nontoxic at therapeutic doses because they are negligibly absorbed from the GI tract (Carneiro et al. 2019).

11.2.2 Dextrans

Dextrans are complex straight (α-1,6 linkages) or branched (α-1,3 linkages) polysaccharide polymers made of glucose molecules. Dextrans of low molecular weight have been used clinically as antithrombotics, anticoagulants and blood volume expanders whose inclusion can improve blood flow. Dextran is relatively clinically safe to use, although anaphylaxis, volume overload, pulmonary and cerebral oedema, platelet dysfunction and acute renal failure have been observed (Haschek et al. 2013; Kerlin and Li 2013).

11.2.3 Polyethylene Glycol

Polyethylene glycol is a polymer of ethylene oxide; it is also known as polyethylene oxide or polyoxyethylene depending on the molecular weight or mass. Polyethylene glycol may contain potentially toxic impurities such as ethylene oxide and 1,4-dioxane. Polyethylene glycol has been widely used as a laxative in medicine and an excipient in pharmaceutical products. Polyethylene glycol, when conjugated with peptides or proteins, induces renal tubular epithelial cell vacuolation in animals at high doses (Kerlin and Li 2013).

11.2.4 Surfactants

Surfactants are usually organic compounds that decrease the surface tension between two phases. They are amphiphilic structures, containing both hydrophobic groups (their tails) and hydrophilic groups (their heads). Surfactants form aggregates, called micelles, where the hydrophobic tails form the core of the aggregate and the hydrophilic heads are in contact with the surrounding liquid. Surfactants cause diverse effects on the skin and into the body: they damage enzyme activity and thus disrupt the body's normal physiological functions. In general, nonionic surfactants have minimal irritation to the skin and hypersensivity induced by nonionic detergents. Cationic surfactants have the greatest toxicity, and the toxicity of anionic surfactants is between nonionic and cationic surfactants. Some surfactants, as sodium dodecyl benzenesulfonate, are absorbed through the skin, damaging the liver and causing teratogenic and carcinogenic effects (Maithili et al. 2015). Cremophor-EL and polysorbate-80 produce side effects such as acute hypersensitivity and systemic inmunostimulation, due to complement activation of derivate inflammatory mediators. From the above, it is evident that many surfactants are not inert vehicles, which gives them potentially important clinical implications (Gelderblom et al. 2001; Weiszhar et al. 2012).

One of the missions of the US Food and Drug Administration (FDA) and its Center for Drug Evaluation and Research (CDER) and Center for Biological Evaluation and Research (CBER) is to determine whether a drug/biological product contains impurities. The subcommittees' focus involves evaluating the toxicities of impurities such as degradants, extractables, leachables, excipients, flavours, solvents and other contaminants in a consistent manner. Impurities in drug products have become subject to greater scrutiny, and various international and national guidelines, guidance, and regulations have been proposed and accepted for use.

The International Council for Harmonisation of Technical Requirements for Pharmaceuticals for Human Use (ICH) conveys no legal authority. These guidelines are submitted to the appropriate national (or regional) regulatory authorities for official adoption and publication. The ICH Steering Committee establishes four categories of guidelines, symbolized alphanumerically by safety (S), efficacy (E), quality (Q) and combined topics (M).

11.3 HAZARDOUS INGREDIENTS IN COSMETICS

The US Federal Food, Drug, and Cosmetic Act (FD&C Act) defines cosmetics as "articles intended to be rubbed, poured, sprinkled, or sprayed on, introduced into, or otherwise applied to the human body . . . for cleansing, beautifying, promoting attractiveness, or altering the appearance" (FD&C Act, sec. 201). The FD&C Act is a set of laws giving authority to the US FDA to oversee the safety of food, drugs and cosmetics. Preservatives and enhancers of the sensory characteristics are added into cosmetic products to preserve them, to improve their organoleptic characteristics and to increase their shelf-life. Organoleptic properties refer to physical characteristics of matter that senses can perceive, such as colour, taste, scent and texture. In view of the fact that there exists elevated demand for these products around the world, it is important that significant knowledge exists concerning is the safety of cosmetics and their constituents (see below). The presence of some of these ingredients is associated with cancer, reproductive disorders and other damage. A few of these products have been banned, such as bisphenol A in children's products in Europe, United States and Canada, and some parabens in cosmetics in those countries and Japan (Casassus 2018). This section aims to raise awareness about the dangerous effects of preservatives and provide recommendations for their use in order to improve their safety and efficacy.

Primarily, cosmetic absorption routes are the oral, respiratory and dermal routes, which can cause harmful effects on health. Dermal exposure is expected to be the most significant route because most cosmetic products are directly applied to the skin. Oral and respiratory exposure can occur from wearing of cosmetics products around the mouth and from hand-to-mouth contact (Mangelsdorf et al. 2014).

A frequent assessment of health risks and control of consumer products are considered of vital importance to reduce serious toxic effects, such as respiratory and dermal health effects in relation to air fresheners (Joo-Hyon et al. 2018). The following discusses some of the main ingredients that may be able to cause damage to health, for example, preservatives, fragrances and heavy metal impurities (Siti Zulaika et al. 2015), and UV filters.

11.3.1 Preservatives

Preservatives are added to cosmetics to prevent decomposition by microbial growth or by unwanted chemical changes to maintain their microbiological purity during manufacture, packing and storage, but especially during the entire period of use, and thereby ensure the safety of all users (Herman 2018). They are habitually used in combination with other types of preservatives to better protect against a broad diversity of microorganisms. However, some preservatives can cause serious health hazards. For example, sulphites may cause allergic reactions and exacerbation of asthma or hypersensitivity (Anand and Sati 2013), neurological damage or cancer; numerous investigations have reported on artificial preservatives which are commonly used in the manufacture of cosmetics and personal care products. Allergy is a disorder very characteristic of these undesirable effects. The main cosmetics responsible to cause the symptoms of allergy are facial care products, body care products, perfumes, eye care products or nail cosmetics. Furthermore, it is very troubling that some preservative-resistant bacterial strains, particularly those producing extended-spectrum β-lactamases, isolated from cosmetic products show a degree of cross-resistance with antibiotics (Orús et al. 2015).

One approach to minimizing these undesirable effects is the use of preservative boosters, or to create self-preserving (preservative-free) cosmetic formulations (Herman 2018). Preservatives boosters are defined as cosmetic ingredients with antimicrobial properties, which can significantly reduce the required concentration of synthetic preservatives used in cosmetic products. Moreover, preservative boosters show not only antimicrobial activity, but many of them showed other desirable properties useful in cosmetic products (moisturizing, antioxidant, etc.). Some of these synthetic preservative boosters are titanium dioxide nanoparticles in emulsion (Papageorgiou et al. 2010), caffeic acid in emulsion (Aerts et al. 2016), and phenolic compounds, as caffeic acid. Synthetic preservatives affect the cytoplasmic membrane, altering its structure and function, changing active transport processes and leading to coagulation of the cell contents. Some agents in combination with preservatives are used as a potentiating agent to obtain better protection against microbial growth. These agents are ethylhexyl glycerin (Langsrud et al. 2016) or ethylhexyl glycerin plus 1,2-pentanediol (Kunicka-Styczyńska et al. 2015). The antimicrobial mechanism of action of ethylhexyl glycerin is based on its surfactant-like structure. Due to this specific organization, ethylhexyl glycerin affects the interfacial tension at the cell membrane of microorganisms that damages the cell membrane integrity and allows the preservative to penetrate more effectively improving the preservative activity.

There is a wide variety of preservatives such as nitrates, benzoates, sulphites, parabens, formaldehyde, butylhydroxytoluene and butylhydroxyanisole. Table 11.1 lists the main preservatives detected in cosmetic products (Siti Zulaikha et al. 2015).

Table 11.1 Preservatives in Cosmetic Products

Chemical name	Glossary
Salts and esters of benzoic acid	Ammonium benzoate, butyl benzoate, calcium benzoate, ethyl benzoate, isobutyl benzoate, isopropyl benzoate, magnesium benzoate, mea-benzoate, methyl benzoate, phenyl benzoate, potassium benzoate, propyl benzoate
Benzoic acid and its sodium salt	Benzoic acid, sodium benzoate
Propionic acid and its salts	Propionic acid, ammonium propionate, calcium propionate, magnesium propionate, potassium propionate, sodium propionate
Salicylic acid and its salts	Salicylic acid, calcium salicylate, magnesium salicylate, mea-salicylate, sodium salicylate, potassium salicylate, tea-salicylate
Hexa-2,4-dienoic acid and its salts	Sorbic acid, calcium sorbate, sodium sorbate, potassium sorbate
Formaldehyde, paraformaldehyde	Formaldehyde, paraformaldehyde
Biphenyl-2-ol and its salts	O-phenylphenol, mea o-phenylphenate, potassium o-phenylphenate, sodium o-phenylphenate
Pyrithione zinc	Zinc pyrithione
Inorganic sulphites and hydrogen sulphites	Sodium sulfite, ammonium bisulfite, ammonium sulfite, potassium sulfite, potassium hydrogen sulfite, sodium bisulfite, sodium metabisulfite, potassium metabisulfite
Chlorobutanol	Chlorobutanol
Butyl 4-hydroxybenzoate and its salts, propyl 4-hydroxybenzoate and its salts	Butylparaben, propylparaben, sodium propylparaben, sodium butylparaben, potassium butylparaben, potassium propylparaben
4-Hydroxybenzoic acid its salts and esters, other than isopropyl, isobutyl, phenyl, benzyl and pentyl esters	4-Hydroxybenzoic acid, methylparaben, potassium ethylparaben, potassium paraben, sodium methylparaben, sodium ethylparaben, ethylparaben, sodium paraben, potassium methylparaben, calcium paraben
3-Acetyl-6-methylpyran-2,4(3H)-dione and its salts	Dehydroacetic acid, sodium dehydroacetate
Formic acid and its sodium salt	Formic acid, sodium formate
3,3′-Dibromo-4,4′-hexamethylenedioxydibenzamidine and its salts (including isethionate)	Dibromohexamidine isethionate
Thiomersal	Thimerosal
Phenylmercuric salts (including borate)	Phenyl mercuric acetate, phenyl mercuric benzoate
Undec-10-enoic acid and its salts	Undecylenic acid, potassium undecylenate, sodium undecylenate, calcium undecylenate, mea-undecylenate, tea-undecylenate
5-Pyrimidinamine, 1,3-bis(2-ethylhexyl)hexahydro-5-methyl-	Hexetidine
5-Bromo-5-nitro-1,3-dioxane	5-bromo-5-nitro-1,3-dioxane

(Continued)

Table 11.1 (Continued)

Chemical name	Glossary
Bronopol	2-bromo-2-nitropropane-1,3-diol
2,4-Dichlorobenzyl alcohol	Dichlorobenzyl alcohol
1-(4-Chlorophenyl)-3-(3,4-dichlorophenyl)urea	Triclocarban
Chlorocresol	P-chloro-m-cresol
5-Chloro-2-(2,4-dichlorophenoxy)phenol	Triclosan
Chloroxylenol	Chloroxylenol
N,N''-Methylenebis[N'-[3-(hydroxymethyl)-2,5-dioxoimidazolidin-4-yl]urea]	Imidazolidinyl urea
Poly(hexamethylenebiguanide) hydrochloride, poly(iminoimidocarbonyl) iminohexamethylenehydrochloride, poly(iminocarbon imidoyliminocarbonimidoylimino-1,6-hexanediyl)	Polyaminopropyl biguanide
2-Phenoxyethanol	Phenoxyethanol
Methenamine	Methenamine
Methenamine 3-chloroallylochloride	Quaternium-15
1-(4-Chlorophenoxy)-1-(imidazol-1-yl)-3,3-dimethylbutan-2-one	Climbazole
1,3-Bis(hydroxymethyl)-5,5-dimethylimidazolidine-2,4-dione	DMDM hydantoin
Benzyl alcohol	Benzyl alcohol
1-Hydroxy-4-methyl-6-(2,4,4-trimethylpentyl)-2 pyridon and its monoethanolamine salt	1-hydroxy-4-methyl-6-(2,4,4-trimethylpentyl)-2 pyridon, piroctone olamine
2,2'-Methylenebis(6-bromo-4-chlorophenol)	Bromochlorophene
4-Isopropyl-m-cresol	O-cymen-5-ol
Mixture of 5-chloro-2-methyl-isothiazol-3(2H)-one and 2-methylisothiazol-3(2H)-one with magnesium chloride and magnesium nitrate	Methylchloroisothiazolinone and methylisothiazolinone
2-Benzyl-4-chlorophenol	Chlorophene
2-Chloroacetamide	Chloroacetamide
N,N'-bis(4-chlorophenyl)-3,12-diimino-2,4,11,13-tetraazatetradecanediamidine and its digluconate, diacetate and dihydrochloride	Chlorhexidine, chlorhexidine diacetate, chlorhexidine digluconate, chlorhexidine dihydrochloride
1-Phenoxypropan-2-ol	Phenoxyisopropanol
Alkyl (C12-C22) trimethyl ammonium bromide and chloride	Behentrimonium chloride (1), cetrimonium bromide, cetrimonium chloride (2), laurtrimonium bromide, laurtrimonium chloride, steartrimonium bromide, steartrimonium chloride (2)
4,4-Dimethyl-1,3-oxazolidine	Dimethyl oxazolidine
N-(Hydroxymethyl)-N-(dihydroxymethyl-1,3-dioxo-2,5-imidazolidinyl-4)-N'-(hydroxymethyl)urea	Diazolidinyl urea
Benzenecarboximidamide, 4,4'-(1,6-hexanediylbis(oxy)) bis-, and its salts (including isothionate and p-hydroxybenzoate)	Hexamidine, hexamidine diisethionate, hexamidine diparaben, hexamidine paraben
Glutaraldehyde (pentane-1,5-dial)	Glutaral
5-Ethyl-3,7-dioxa-1-azabicyclo[3.3.0] octane	7-ethylbicyclooxazolidine
3-(p-Chlorophenoxy)-propane-1,2-diol	Chlorphenesin
Sodiumhydroxymethylaminoacetate	Sodium hydroxymethylglycinate
Silver chloride deposited on titanium dioxide	Silver chloride
Benzenemethanaminium, N,N-dimethyl-N-[2-[2-[4-(1,1,3,3,-tetramethylbutyl)phenoxy]ethoxy]ethyl]-, chloride	Benzethonium chloride
Benzalkonium chloride, bromide and saccharinate	Benzalkonium chloride, benzalkonium bromide, benzalkonium saccharinate
Methanol (phenylmethoxy)-	Benzylhemiformal
3-Iodo-2-propynylbutylcarbamate	Iodopropynyl butylcarbamate
2-Methyl-2H-isothiazol-3-one	Methylisothiazolinone
Ethyl lauroyl arginate HCl (preservative)	
1,2,3-Propanetricarboxylic acid, 2-hydroxy-, monohydrate and 1,2,3-propanetricarboxylic acid, 2-hydroxy-silver salt, monohydrate	Citric acid and silver citrate

11.3.1.1 Parabens

One group of preservatives are parabens, which are antimicrobial and antifungal compounds, which act against an extremely broad spectrum of microorganism. Chemically, they are a homologous series of p-hydroxybenzoic acid esters that differ in the ester group (4-hydroxybenzoic acid). The parabens used most commonly in cosmetics are methylparaben, ethylparaben, propylparaben, butylparaben, heptylparaben, isobutylparaben, isopropylparaben, and benzylparaben, and their sodium salts; methylparaben and propylparaben may also occur in combination with phthalates. The European Union permits the use of parabens in cosmetic products with a maximum concentration of 0.4% (w/v) and a total maximum concentration of 0.8% (w/v) (EU Cosmetics Directive 76/768/EEC). The chain length of the alkyl group on the paraben is directly proportional to their antimicrobial activity; for example, butylparaben has 4-fold greater ability to inhibit microbial growth than ethylparaben, and inversely related to their solubility in water, as the value of octanol-water partition coefficient rises (Taxvig et al. 2008; Błędzka et al. 2014). Table 11.2 summarizes the physicochemical characteristics of parabens. Compared to other preservatives, parabens have great advantages as regards their physical and chemical characteristics: they have relatively low toxicity, good stability, nonvolatility, little cost, no perceptible odour or taste, are practically pH neutral, do not produce discoloration or hardening in cosmetic formulations and are generally well tolerated by the skin (Błędzka et al. 2014). Parabens have been detected in human urine, serum, adipose tissue, milk, amniotic fluid and placental tissue by reverse-phase high performance liquid chromatography (HPLC), and detected by atmospheric pressure chemical ionization or atmospheric pressure photoionization–isotope dilution–tandem mass spectrometry (MS/MS) (Hines et al. 2015) and in human breast cancer tissues using the analytical chemistry technique, HPLC-MS, that combines the physical separation capabilities of liquid chromatography with mass spectrometry (MS) for more robust compound identification (Barr et al. 2012).

Parabens are a class of preservatives widely used in consumer products, mainly in personal care products, like shampoos, cosmetics, shaving gels, body lotions, deodorant or antiperspirants, and sunscreens. In addition, it should also be noted they have a wide use in other areas, such as food additives and drugs.

Despite the restrictions (Regulation (EU) No 358/2014), the extensive use of parabens has led to continuous exposure levels, and it is recognized that the absorbed amounts can cause problems for human health. In recent years, the consequences of continued exposure has become a focus of debate over the potential role of parabens in some diseases, such as endocrine disruption and carcinogenesis, as well as reproductive problems (Błędzka et al. 2014; Goyal et al. 2014). The definition of endocrine disruption proposed by the World Health Organization (WHO) in 2002 is widely accepted scientifically, an endocrine disrupter is an exogenous substance or mixture that alters function of the endocrine system and consequently causes adverse health effects in an intact organism, or its progeny or (sub)populations (WHO/IPCS 2002). The acronym IPCS stands for the International Programme for Chemical Safety (which involves WHO; the United Nations Environment Programme [UNEP] and the International Labour Organization [ILO]). Endocrine-disrupting chemicals have been linked to a variety of medical conditions; one of the most troubling is their association with breast cancer (Vandenberg et al. 2012; Zoeller et al. 2012; Pan et al. 2016). The explanation of why parabens might be able to cause cancer is that they are considered estrogenic because they activate the same estrogen receptor as the natural hormone estradiol. There are many studies that address the estrogenicity of parabens and the role in the possible incidence of breast cancer (Darbre and Harvey 2008; Witorsch and Thomas 2010; Goyal et al. 2014) due to interferences with estrogen hormone action. It is a datum confirmed that the usage of parabens in deodorants and antiperspirants constitute is a risk factor in the rising incidence of breast cancer.

Table 11.2 Physical and Chemical Characteristics of Parabens

Chemical name	Chemical formula	Molecular weight	Log octanol-water coefficients
Methylparaben	$C_8H_8O_3$	152.16	1.66
Ethylparaben	$C9H_{10}O_3$	166.18	2.19
Propylparaben	$C_{10}H_{12}O_3$	180.21	2.71
Butylparaben	$C_{11}H_{14}O_3$	194.23	3.24

Cellular actions of estrogens are mediated through binding to specific receptor proteins. There are two types of nuclear estrogen receptors, α and β (ERα or ERβ), which exist as multiple isoforms and which function as ligand-activated transcription factors to influence patterns of gene expression. Final outcomes are dependent on the affinity of ligand binding to receptor, the concentrations of receptors in the target cells and the presence of coacting factors/transcription factors (Darbre 2019). Many studies have implicated the estrogenicity of parabens with very weak estrogenic activity both *in vitro* and *in vivo* (Witorsch and Thomas 2010; Goyal et al. 2014). Emerging experimental studies suggest that parabens could affect metabolism by altering the microbiome or indicator pathways involved in adipocyte differentiation (Quirós-Alcalá et al. 2018).

11.3.1.2 Phthalates

Phthalates are dialkyl or aryl alkyl esters of 1,2 benzenedicarboxylic acids, and are a group of chemicals used in hundreds of everyday products as "plasticizers"—enhancing the flexibility of plastics, and as lubricating oils, food packaging, pharmaceuticals, and personal care products, such as nail polish, hair sprays, aftershave lotions, soaps, shampoos, perfumes and other fragrance preparations.

Phthalates are omnipresent environmental endocrine disrupting chemicals that are suspected to be a risk factor for disorders, such as attention deficit hyperactivity disorder or autism spectrum disorder (Hu et al. 2017), and to interfere with the androgenic action of development and produce adverse effects on male reproductive function (Fisher 2004; Witorsch and Thomas 2010). Consequently, phthalate use has been restricted because of these potential adverse effects (Rowdhwal and Chen 2018; Ito et al. 2019).

It is very worrying that this compound appears in baby care products, such as infant lotion or baby powder (Fisher et al. 2019). Likewise, in different studies, prenatal and postnatal exposure may contribute to many male reproductive disorders, including the incidence of cryptorchidism and hypospadias and reduced testicular volume, testosterone and anogenital distance in newborn boys (Radke et al. 2018). These collective manifestations of what is referred to as testicular dysgenesis syndrome and which appear idiopathic, have sparked interest in the possible role of the health impacts of *in utero* exposure to phthalates (Welsh et al. 2008). Nevertheless, in 2018 there are no data available, or weak or uncertain data which link antenatal exposure to phthalates to long-term health effects in men (Rodprasert et al. 2019). It is also important to consider that the influence of the phthalate exposure, if significant, may vary according to the timing of the pregnancy exposure (Martino-Andrade et al. 2016), and be dose dependent. To date, although there have been a significantly increasing number of studies on the relationship between phthalate exposure during pregnancy and children, the results are inconsistent.

Nonetheless, via the Consumer Product Safety Improvement Act (CPSIA), in 2018 the US Congress prohibited children's toys or child care articles to contain concentrations of more than 0.1% of each type of phthalate: di-(2-ethylhexyl) phthalate; dibutyl phthalate; or, benzyl butyl phthalate, diisononyl phthalate, diisobutyl phthalate, di-n-pentyl phthalate, di-n-hexyl phthalate and dicyclohexyl phthalate.

11.3.1.3 Triclosan

Triclosan is an effective antimicrobial agent (5-chloro-2-(2,4-dichlorophenoxy) phenol) against most gram-negative and gram-positive bacteria, and has been in use for over 40 years (Yueh and Tukey 2016). Nowadays is used in many contemporary consumer and personal healthcare products, such as hand soaps, shampoos, detergents, toothpastes, sunscreen, and deodorants together with household items, including plastics and textiles. Due to its extensive use, triclosan has been detected in human tissues and the environment, using HPLC-MS/MS to quantify. HPLC-UV-MS detection are among the most important analytical methodologies to measure trace levels of triclosan. Triclosan is readily absorbed into human skin and oral mucosa, so it has been found in a variety of human body tissues and fluids; epidemiological studies have highlighted its presence in a wide range of concentrations in fluids such as urine, blood, and breast milk (Yueh et al. 2016). Various studies have revealed widespread use of triclosan may represent a potential public health risk regarding development of concomitant resistance to clinically important antimicrobials against bacteria such as *Escherichia coli* and *Salmonella enterica* (Suller and Russell 2000; Yueh and Tukey 2016; Yazdankhah et al. 2006).

The mechanisms underlying triclosan mediated toxicity include hormone dyshomeostasis, induction of oxidative stress, apoptosis and inflammation (Witorsch 2014; Spanier et al. 2014; Ruszkiewicz et al. 2017); notably, several of these effects were also related to neuronal functions.

One study points to the possibility that triclosan induced microRNAs regulation, a class of short noncoding RNAs (18–22 nucleotides) that regulate translation and stability of a great deal of mRNAs influences the central nervous system, which offers hints to the molecular mechanisms of triclosan toxicity (Wang et al. 2019).

Witorsch and Thomas (2010) give details about mechanism of action of triclosan exhibiting different interactions with the endocrine system that involve interactions with aryl hydrocarbon showing agonist activity, sex hormone receptors (estrogen and androgen) evidencing its ability to displace estradiol from cytosolic estrogen receptor of estrogensensitive cells (such as MCF-7 human breast cancer cells), and effects on the hypothalamus pituitary thyroid axis. Structural similarities between triclosan and the thyroid hormones T3 and T4 have recently generated interest as to whether this compound might affect the hypothalamus-pituitary-thyroid axis. However, *in vitro* and *in vivo* findings have not been consistent, leaving room for uncertainty in predicting health outcomes (Goodman et al. 2018). There are studies suggesting that reproductive function is significantly reduced after chronic exposure to triclosan (Paul et al. 2012; Lin et al. 2014). Weatherly and Gosse (2017) described routes and metabolism and transformation processes, as well as reproductive problems. Triclosan is readily absorbed by human skin and oral mucosa due to its lipophilic properties and found in various human tissues and fluids, with a half-life of 21 hours, such that brief use cause a prolonged exposure of several hours. Due to the ability of triclosan to penetrate and remain in the tissues, the concentration will be high enough to induce harmful effects in humans. Triclosan is not rapidly metabolized dermally and the majority remains for at least 24 hours. Main route of elimination in humans is via the urine, and fecal elimination is the secondary route. However, *in vitro* and *in vivo* findings have not been consistent, leaving room for uncertainty in predicting health outcomes (Witorsch 2014; Goodman et al. 2018). Another antimicrobial agent, similar to triclosan, triclocarban (3,4,4′-trichlorocarbanilide) has also been linked to numerous human health problems due to its extensive use in numerous personal care products, for example soap, toothpaste or medical disinfectants (Halden et al. 2017).

In summary, given the wide use as an antiseptic of triclosan, its implications for human health suggest that this compound should be considered a priority contaminant (Dhillon et al. 2015), also for this reason, the FDA issued recently a directive prohibiting its use in soaps, justifying this measure on the lack of data on its effectiveness and safety and indicating the need for further studies (Ruszkiewicz et al. 2017).

11.3.2 Fragrances

Fragrances are organic compounds with pleasant odours that are used in each aspect of our daily lives such as in perfumes, deodorants, shampoos and conditioners, laundry products, cleaning products and air fresheners (Abedia et al. 2018). Continued exposure to fragrances and other perfumed products has been associated with a variety of adverse human health effects, including migraine headaches, contact dermatitis, allergies, respiratory difficulties and mucosal symptoms (Steinemann et al. 2011; Basketter et al. 2015; Weinberg et al. 2017; Steinemann 2017). Perfume can worsen asthma and perhaps even contribute to its development in children. It is ranked the second most common cause of allergy in some patients (Okereke et al. 2015). Primary sources of these pollutants are mentioned in Table 11.3 (Chen et al. 2017). Pollutants are categorized into two types: primary pollutants are directly emitted from the source, and secondary pollutants form when primary pollutants chemically react in the atmosphere. Then fragranced consumer products produce different volatile compounds, including terpenes. One can find monoterpene hydrocarbons (limonene), sesquiterpene hydrocarbons (α-farnesene), alcohols (cis-3-hexenol), monoterpene alcohols (linalool), sesquiterpene alcohols (farnesol), phenols (eugenol), aldehydes (2,6-nonadienal), terpene aldehydes (citral), ketones (cyclohexanone), terpene ketones (β-ionone), lactones (γ-undecalactone), esters (methylsalicylate), terpene esters (linalyl acetate), oxides (eucalyptol), which are primary pollutants and react with ozone to generate secondary pollutants such as formaldehyde and acetaldehyde (Wiecka et al. 2018).

Chisvert et al. (2018) describe new methods to control those fragrance chemicals that have been shown to pose a safety hazard. The use of spectroscopic analytical techniques, like ultraviolet/visible spectrometry, infrared spectrometry and nuclear magnetic resonance, also provides valuable information about quality. So, more sophisticated analytical techniques are necessary, especially considering that a perfume could be made up of hundreds of fragrance ingredients. Separation techniques, like chromatography, are the most suitable analytical techniques for these purposes with MS detection (Begnaud and Chaintreau 2016).

Table 11.3 Fragrances in Cosmetic Products

Chemical name/IUPAC name	Chemical structure	Chemical formula	Molecular weight (g/mol)
Limonene (1-methyl-4-(1-methylethenyl)-cyclohexene)	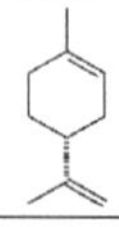	$C_{10}H_{16}$	136.24 g/mol
Alpha-pinene ((1*S*,5*S*)-2,6,6-trimethylbicyclo[3.1.1]hept-2-ene ((−)-α-pinene))	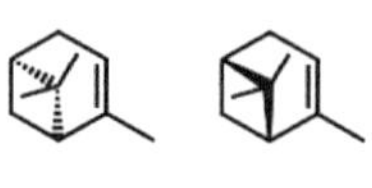	$C_{10}H_{16}$	136.23 g/mol
Beta-pinene (6,6-dimethyl-2-methylidenebicyclo[3.1.1]heptane)	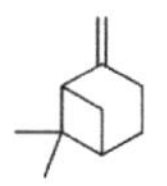	$C_{10}H_{16}$	136.23 g/mol
Linalool (2,6-dimetil-2,7-octadien-6-ol)	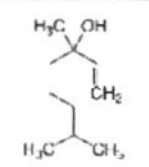	$C_{10}H_{18}O$	154.25 g/mol
Hexyl cinnamal (2-benzylideneoctanal)	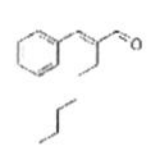	$C_{15}H_{20}O$	216.319 g/mol
Benzyl salicylate (benzyl 2-hydroxybenzoate)	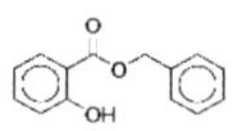	$C_{14}H_{12}O_3$	228.24 g/mol
Butylphenyl methylpropional (3-(4-*tert*-butylphenyl)-2-methylpropanal)	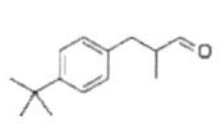	$C_{14}H_{20}O$	204.31 g/mol
Citronellol (3,7-dimetiloct-6-en-1-ol)	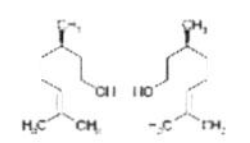	$C_{10}H_{20}O$	156.27 g/mol

International organizations, such as the EC Regulation, International Fragrance Association (IFA) and the Research Institute for Fragrance Materials (RIFM), prohibit or limit the content of some of these compounds in cosmetic products. A high percentage of the adult European population suffers from contact allergy, with fragrance substances recognized as one of the main causes (Bennike et al. 2018). More than 120 chemicals used for their scent in consumer products have been classified as possessing skin sensitizing properties (Uter et al. 2013). Fragranced consumer products produce different volatile compounds and may include 10 to 100 individual ingredients, including terpenes, limonene, alpha-pinene and beta-pinene, which are primary pollutants and can react with ozone to generate secondary pollutants such as formaldehyde and acetaldehyde (Wiecka et al. 2018). Different methods have been described to control chemical fragrances that have proven to be hazardous (Villatoroa et al. 2016; Chisvert et al. 2018), including gas chromatography in combination with MS and GC-olfactometry or "GC-sniffing" techniques coupled with condensed phase Fourier-transform infrared spectroscopy or Time of flight–MS (ToFMS).

11.3.2.1 Limonene

Limonene (1-methyl-4-(1-methylethenyl) cyclohexene) is a cyclic monoterpene used as a flavouring agent in food manufacturing. It is a chiral molecule, and biological sources produce one enantiomer: the principal industrial source, citrus fruit, contains D-limonene ((+)-limonene), which is the (R)-enantiomer, is a monoterpene widely used as a fragrance additive in cosmetics, foods, and industrial solvents as it possesses a pleasant lemon-like odour.

D-limonene has been designated as a substance with low toxicity. Conversely, the potential for irritation or sensitization of the skin after the widespread use of this agent in various consumer products has been reported (Kim et al. 2013; Panico et al. 2019; Natsch et al. 2019; de Groot 2020). Oxidized derivatives, such as peroxides, are responsible for the allergenic activity, as in the case of limonene or linalool. Both terpenes are prehaptens, meaning that they are pure fragrance chemicals, with low sensitizing potency, but are modified outside the skin by air oxidation resulting in a compound with increased allergenic potential. For both linalool and limonene, specific hydroperoxides are formed during oxidation, which have been identified as the primary haptens causing skin sensitization (Sköld et al. 2004). It is important to mention that numerous studies have suggested that D-limonene does not appear to present a serious health risk following human exposure and that the possible appearance of skin irritation requires the regulation of this chemical as an ingredient in cosmetics. In conclusion, the use of D-limonene in cosmetics is safe according to the current regulatory guidelines for cosmetics (European Union 2009). The FDA includes limonene on its list of substances considered generally recognized as safe (GRAS) as a flavouring substance.

11.3.3 Heavy Metals

Metals in cosmetics may suffer retention and act directly on the skin or be absorbed through the skin into the blood, accumulate in the body and exert toxic effects in various organs. Borowska and Brzóska (2015) explain the implications of metals in cosmetic products for the human health and its relationship for causing serious diseases. They have described some cases of topical effects, mainly allergic contact dermatitis; and systemic effects, generally cancer, reproductive and developmental disorders, neurological, cardiovascular, skeletal or immunological problems; kidney damage; headaches; vomiting, nausea and diarrhoea, or intellectual retardation (Ayenimo et al. 2010). These heavy metals are found in shampoo, lipstick, cream, eye shadow and powder. Heavy metals most commonly detected in cosmetic and personal care products are cadmium, lead, nickel, mercury or others: chromium, potassium, dichromate, cobalt or aluminium compounds have been used commercially in antiperspirants since 1903 (Okereke et al. 2015; Fischer et al. 2017; Ullah et al. 2017).

Kuan-Ling et al. (2017) clarify techniques to identify metals in cosmetics; these authors have determined lead, cadmium and mercury present in cosmetics using "sweeping" via dynamic chelation by capillary electrophoresis. Kazi et al. (2019) developed an innovative ultrasonic assisted microextraction method using a deep eutectic solvent for entrapping the organ metallic complexes to obtain traces of cadmium and lead in lipsticks and eye shadows. The heavy metal ions, when they meet the human body, get absorbed and form complexes with several groups, carboxylic acid (-COOH), amine ($-NH_2$), and thiol (-SH), of proteins resulting in malfunctioning or death of the cells. The treatment of metal intoxication is achieved by using a chelating agent (chelation therapy) that binds with the metal ions, and the complexes are then eliminated from the body (Kim et al. 2019). For example, nickel, a major skin allergen, nitrilotriacetic acid is used to chelate metal ions of Ni^{2+} or 2,3-dimercaprol developed as a chelating agent for cadmium and mercury, and calcium disodium ethylenediaminetetraacetic acid is an FDA-approved synthetic polyamino-polycarboxylic acid used to treat lead poisoning (Mahato et al. 2018).

11.3.3.1 Cadmium

Cadmium is toxic at extremely low levels. Acute ingestion of high levels of cadmium (as 10 mg of inorganic cadmium) can lead to severe pulmonary and stomach irritation, causing vomiting and diarrhoea and pulmonary, hepatic or renal injury depending on the route of poisoning; whereas exposure to lower levels for a long time can lead to kidney damage, bone deformity, and the ability of bones to break easily. Ingestion of >100 mg of its soluble salts can be lethal (Hung and Chung 2004; Kazi et al. 2019). Acute or chronic ingestion of cadmium salts is rare but it may lead to death.

11.3.3.2 Lead

Lead poisoning occurs in humans when exposed to high doses or chronic exposure. Lead exposure has also been linked to miscarriage, hormonal changes, reduced fertility in men and women, menstrual irregularities, delayed onset of puberty in girls, viability, integrity, elevation in morphological abnormalities, and sperm DNA integrity (Kumar 2018). Lead poisoning can also cause nausea, vomiting and abdominal pain, renal damage or encephalopathy. Lead and inorganic lead compounds have been classified as a suspected carcinogen in humans; subsequently the FDA has published draft guidance for industry that recommends a maximum level of 10 ppm for lead as an impurity in cosmetics.

11.3.3.3 Nickel

Allergy to nickel is also common, and it can cause severe contact dermatitis. This is one of the principal elements responsible for causing allergic dermatitis, together with cobalt, copper and chromium. Considering the prolonged contact time of cosmetic products with the skin, the risk of allergic dermatitis contact might be increased (Volpe et al. 2012).

11.3.3.4 Mercury

Mercury salts inhibit the formation of melanin, resulting in a lighter skin tone. Mercury in skin-lightening products are capable of also causing skin rashes, skin discoloration and scarring, as well as a reduction in the skin's resistance to bacterial and fungal infections. Mercury salts inhibit the formation of melanin, resulting in a lighter skin tone. Skin lightening products are used worldwide, but their use is particularly widespread in many African, Asian and Caribbean countries Other effects include anxiety, fatigue, nervousness, irritability, depression or psychosis and peripheral neuropathy. Mercury compounds are allowed in cosmetics only as preservatives in eye area products. It must not be more than 1 ppm in the finished product, but not have been found any eye area cosmetic products currently on the market. The European Union (2019) and many other countries have regulations banning mercury-containing cosmetics. FDA limits the amount of mercury in cosmetics to trace amounts under 1 ppm (Sun et al. 2017). The European Union (European Commission 2019) and many other countries have regulations banning mercury-containing cosmetics. These include Canada (Health Canada 2012) and the United States (US FDA 2019).

11.3.4 Other Compounds

It is well known that sunbathing as a leisure activity has progressively increased; similarly, growth in the number of disorders caused by excess UV radiation have increased, including skin cancers or photoaging (D'Orazio et al. 2013; Kiriiri et al. 2019). Filters (or chemical filters) are chemical substances present in photo-protector formulas that interfere directly with solar radiation through its absorption, dispersion or reflection. They are widely used in a great selection of personal care products to protect the skin from UV-induced damage and sometimes to stabilize the colour and the fragrance of cosmetic formulations. Ultraviolet filters can be organic (chemical filters) and inorganic (physical filters) based on their mechanism of action; and they will be discussed separately below because they raise distinct environmental and health issues.

The wavelength ultraviolet radiation that can reach the Earth's surface comprises three categories based on the range of protection: 280–315 nm for shorter UVB radiation, 315–400 nm for longer UVA, and 100–280 nm for UVC (Gabard 2009). UVB radiation causes inflammation of the skin and erythema. UVA radiation is the main cause of premature aging of the skin system. But the worst manifestation of excessive exposure to UV radiation is the development of skin cancer. A direct relationship between carcinogenic action and solar radiation has been confirmed, and the International Agency for Research on Cancer (IARC) has classified solar radiation, UVB rays and UVA rays as human carcinogens (Fraumeni and Schottenfeld 2006).

On other hand, UV filters, their metabolites and/or their degradation products have been reported to be potentially bioaccumulative, leading to endocrine-disrupting effects (Krause et al. 2012). It is known that some of them can be absorbed through the skin (León-González et al. 2013), and even the presence of UV filters in human milk is evidenced (Schlumpf et al. 2010).

11.3.4.1 Organic Filters

Organic filters are molecules with an aromatic structure having a carboxyl group that undergoes isomerization under the influence of energy absorbed from radiation. Their efficacy is mainly due to physicochemical properties, such as absorption coefficient and absorption spectra (Skotarczak et al. 2015). A list of the most recently investigated UV filters, classified by the type of ultraviolet radiation they absorb, is reported in Table 11.4, together with their structure and some physicochemical characteristics, including benzophenone derivatives, p-aminobenzoic acid derivatives, and others.

There are indications that some UV filters can penetrate the skin, resulting in systemic exposure to their components; they may degrade under sunlight resulting in toxic degradation products and could be potential endocrine disruptors. Exposure to organic UV filters may alter human immune system function (Ao et al. 2018). The studies concerning these probable negative effects of UV filters used different methods, so the results are very controversial. Consequently, validated

Table 11.4 Organic Filters in Cosmetic Products

Name Absorption spectrum range	Chemical structure	Molecular weight (g/mol)	Water solubility (mg/L)
Diethylamiohydroxybenzoy lhexylbenzoate UVA	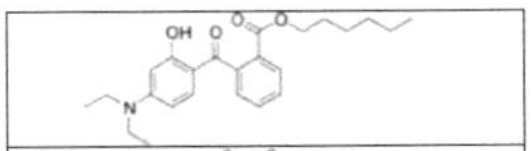	397.515	<0.01 (at 20°C)
Butyl methoxydibenzoylmethane (avobenzone) UVA	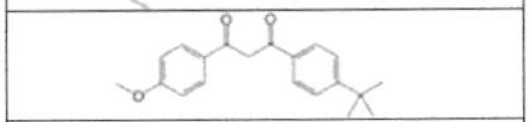	310.393	2.2 (at 25°C)
4-methylbenzylidene camphor (enzacamene) UVB	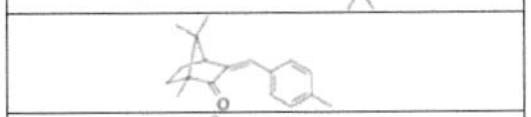	258.397	1.3 (at 20°C)
Octocrylene (octocrilene) UVB	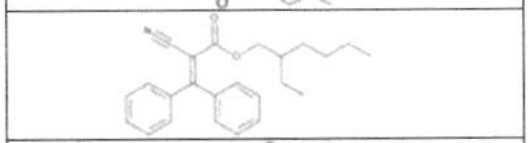	361.485	0.0038
Isoamyl p-methoxycinnamate (amiloxate) UVB	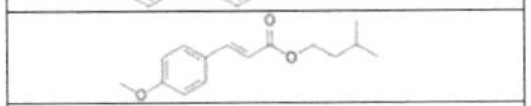	248.322	4.9 (at 25°C)
Ethylhexyltriazone UVB	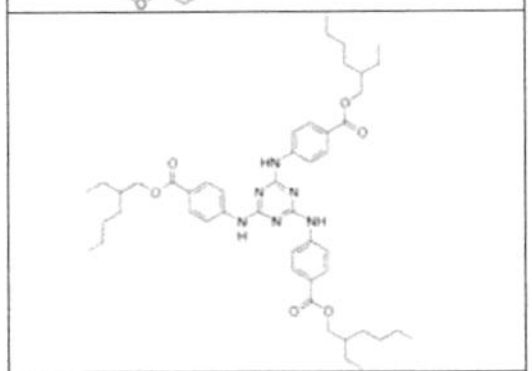	823.092	<0.001 (at 20°C)
Ethylhexylmethoxycinnamate (octinoxate) UVB	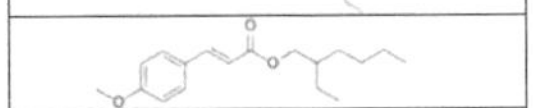	290.403	0.041 (at 24°C and pH 7.1)
Ethylhexyldimethyl PABA (padimate-O) UVB	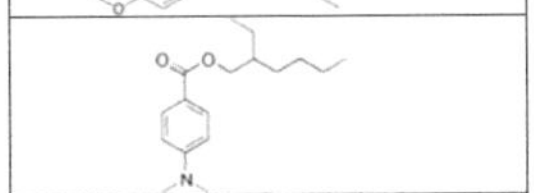	277.408	0.54 (at 25°C)
Benzophenone-3 (oxybenzone) UVA + UVB	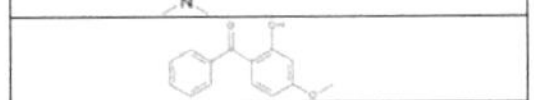	228.247	3.7 (at 20°C)
Bis-ethylhexyloxyphenol methoxyphenol triazine (bemotrizinol) UVA + UVB	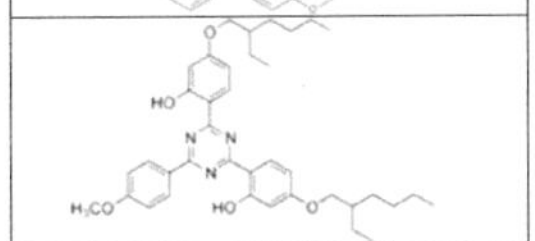	627.826	$<10^{-4}$
Phenylbenzimidazole sulfonic acid (ensulizole) UVA + UVB	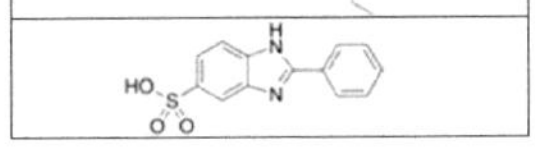	274.294	>30% (sodium or triethanol-ammonium salt at 20°C)

Source: Based on Tampucci et al. (2018).

methods are needed to be sure of sunscreen safety. The method employs solid-phase extraction and HPLC tandem MS, using electrospray ionization (León-González et al. 2013).

11.3.4.2 Inorganic Filters

Inorganic filters are inert and nonirritant substances able to protect the skin from the incident solar radiation due to physical phenomena, such as scattering and reflection, by forming a layer over the skin that works as mechanical barrier. Notwithstanding that the number of physical filters on the market are very few, they possess many advantages, such as high photo-protection level in the longer UVA range, photostability and low photoallergic potential. The most investigated inorganic UV filters are metallic oxides, zinc oxide (ZnO) and titanium dioxide (TiO_2). Both have been used for years in their micronized forms, but nowadays the nano-sized form is the form of choice. Recent advances in nanotechnology have led to the production of nanoparticles of metallic oxides, ZnO and TiO_2 nanoparticles, whose ability to absorb UV radiation is increased with respect to micronized ones (Tampucci et al. 2018). Unfortunately, these nanoparticles, in addition to being effective sunblock and eliminating the anaesthetic formation of an opaque film on the skin (due to visible light scattering), seem to possess a photocatalytic activity associated with oxidative stress and genotoxicity (see Chapter 5; Gulson et al. 2015; Cooke 2018).

11.4 ENVIRONMENTAL IMPACT OF PHARMACEUTICALS AND PERSONAL CARE PRODUCTS

Pharmaceuticals and personal care products could produce an environmental harmful effect; numerous studies have allowed us to evaluate the risks of toxicity, persistence and bioaccumulation that these chemicals cause and to distinguish their impacts on the environment. The bioaccumulation of harmful chemicals has been associated with cancer, reproductive and developmental disorders, contact dermatitis, lung injury, aging, skin disease and reaction, allergies, or damage to nails and hair. Chemicals such as trace levels of triclosan (5-chloro-2-(2,4-dichlorophenoxy) are detected in many nonhuman organisms, and it has been shown to be particularly toxic to aquatic species (Dhillon et al. 2015). Triclosan is an endocrine-disrupting compound, as noted above, and as with many other consumer products it can be absorbed by plants irrigated with wastewater or can accumulate in the fatty tissues of aquatic organisms. It is worrisome that growing presence of triclosan in water resources presents a potential danger for human and environmental health (Olaniyan et al. 2016).

UV filters have been detected in surface waters such as seas, lakes, rivers and soils due to its presence in sunscreens. The filters most frequently identified in sediment are ethylhexyl methoxycinnamate, octocrylene, butyl methoxydibenzoylmethane, OD-PABA (octyldimethyl-p-aminobenzoic acid) and benzophenone derivatives. The toxic effects of filters on the environment are known, and the toxicity of organic UV filters was evaluated in several aquatic organisms; some frequently used UV filters have proven to be toxic for phytoplankton species, microalgae, protozoa and crustaceans. UV filters inhibit growth in marine phytoplankton and tend to bioaccumulate in food webs. Danovaro et al. (2008) demonstrated that some organic UV filters can cause coral bleaching (i.e. the release of coral symbiotic zooxanthellae) at very low concentrations; the results of their experiments, jointly with the estimated release of sunscreens in reef areas, strongly suggest that at least 10% of the reefs are likely to meet coral bleaching due to UV filter pollution (Cooke 2018). There are many environmental impact studies with the continued use of inorganic UV filters; very similar conclusions are obtained regarding the environmental damage that they produce. Zinc and titanium dioxides are used widely due to their broad UV protection spectrum and their limited penetration into the skin; this circumstance has an important impact on bleaching.

In Chapter 22, pharmaceuticals and cosmetics, as emerging environmental contaminants, will be addressed in more depth.

BIBLIOGRAPHY

Abedia, G., Z. Talebpoura, and F. Jamechenarboo. 2018. The survey of analytical methods for sample preparation and analysis of fragrances in cosmetics and personal care products. *TrAC Trends Anal Chem* 102: 41–59.

Aerts, O., L. Verhulst, and A. Goossens. 2016. Ethylhexylglycerin: A low-risk, but highly relevant, sensitizer in 'hypo-allergenic' cosmetics. *Contact Dermat* 74: 281–8.

Anand, S. P., and N. Sati. 2013. Artificial preservatives and their harmful effects: Looking toward nature for safer alternatives. *Int J Pharm Sci Res* 4(7): 2496–501.

Ao, J., T. Yuan, L. Gao, et al. 2018. Organic UV filters exposure induces the production of inflammatory cytokines in human macrophages. *Sci Total Environ* 1(635): 926–35.

Ayenimo, J. G., A. M. Yusuf, A. S. Adekunle, and O. W. Makinde. 2010. Heavy metal exposure from personal care products. *Bull Environ Contam Toxicol* 84: 8–14.

Barr, L., G. Metaxas, C. A. J. Harbach, L. A. Savoy, and P. D. Darbre. 2012. Measurement of paraben concentrations in human breast tissue at serial locations across the breast from axilla to sternum. *J Appl Toxicol* 32: 219–32.

Basketter, D. A., S. Lemoine, and J. P. McFadden. 2015. Skin sensitization to fragrance ingredients: Is there a role for household cleaning/maintenance products? *Eur J Dermatol* 25: 7–13.

Begnaud, F. A., and A. Chaintreau. 2016. Good quantification practices of flavours and fragrances by mass spectrometry. *Philos Trans R Soc A* 374: 20150365.

Bennike, N. H., N. B. Oturai, S. Müller, et al. 2018. Fragrance contact allergens in 5588 cosmetic products identified through a novel smartphone application. *JEADV* 32: 79–85.

Błędzka, D., J. Gromadzińska, and W. Wąsowicz. 2014. Parabens. From environmental studies to human health. *Environ Int* 67: 27–42.

Bornehag, C. G., F. Carlstedt, B. A. Jönsson, et al. 2014. Prenatal phthalate exposures and anogenital distance in Swedish boys. *Environ Health Perspect* 123(1): 101–7.

Borowska, M., and M. M. Brzóska. 2015. Metals in cosmetics: Implications for human health. *J Applied Toxicol* 35(6): 551–72.

Carneiro, S. B., F. I. Costa, L. Heimfarth, et al. 2019. Cyclodextrin—drug inclusion complexes: In vivo and in vitro approaches. *Int J Mol Sci* 20(3): 642–55.

Casassus, B. 2018. Hormone disrupting chemicals: Slow progress to regulation. *BMJ* 30: 361: k1876.
Chen, K., S. Jiang, and Y. Chen. 2017. Determining lead, cadmium and mercury in cosmetics using sweeping via dynamic chelation by capillary electrophoresis. *Anal Bioanal Chem* 409(9): 2461–9.
Cheng, M., I. E. Galbally, S. B. Molloy, et al. 2015. Factors controlling volatile organic compounds in dwellings in Melbourne, Australia. *Indoor Air* 26: 219–30.
Chisvert, A., M. López-Nogueroles, P. Miralles, and A. Salvador. 2018. Perfumes in cosmetics: Regulatory aspects and analytical methods. In *Analysis of Cosmetic Products*, 2nd ed., ed. A. Salvador and A. Chisvert, 225–48. Elsevier.
Cooke, M. S. 2018. Mycosporine-like amino acids: Does Nature make a better sunscreen? *B J Derm* 178(6): 1239–40.
Crini, G. 2014. Review: A history of cyclodextrins. *Chem Rev* 114(21): 10940–75.
Danovaro, R., L. Bongiorni, C. Corinaldesi, et al. 2008. Los protectores solares causan blanqueamiento de corales promoviendo infecciones virales. *Reinar Perspectiva de salud* 116: 441–7.
Darbre, P. D. 2019. The potential for estrogen disrupting chemicals to contribute to migration, invasion and metastasis of human breast cancer cells. *J Cancer Metastasis Treat* 5: 58–71.
Darbre, P. D., and P. W. Harvey. 2008. Paraben esters: Review of recent studies of endocrine toxicity, absorption, esterase and human exposure, and discussion of potential human health risks. *J Appl Toxicol* 28(5): 561–78.
de Groot, A. 2020. Fragrances. Contact allergy and other adverse effects. *Dermatitis* 31(1): 13–35.
Dhillon, G. S., S. Kaur, R. Pulicharla, et al. 2015. Triclosan: Current status, occurrence, environmental risks and bioaccumulation potential. *Int J Environ Res Public Health* 12: 5657–84.
D'Orazio, J., S. Jarrett, A. Amaro-Ortiz, and T. Scott. 2013. La radiación UV y la piel. *Int J Mol Sci* 14(6): 12222–48.
Edwards, I. R., and J. K. Aronson. 2000. Adverse drug reactions: Definitions, diagnosis, and management. *Lancet* 7: 1255–9.
European Commission. 2019. Annex V. List of preservatives allowed in cosmetic products. In: Regulation (EC) No. 1223/2009 of the European Parliament and of the Council. Updated July 11, 2019. Brussels, The European Commission.
European Union. 2009. Regulation (EC) No 1223/2009 of the European Parliament and of the Council of 30 November 2009 on Cosmetic Products.
Fischer, A., B. Brodziak-Dopierała, K. Loska, and J. Stojko. 2017. The assessment of toxic metals in plants used in cosmetics and cosmetology. *Int J Environ Res Public Health* 24: 14(1280).
Fisher, J. S. 2004. Environmental anti-androgens and male reproductive health: Focus on phthalates and testicular dysgenesis syndrome. *Reproduction* 127: 305–15.
Fisher, M., T. E. Arbuckle, S. MacPherson, J. M. Braun, M. Feeley, and E. Gaudreau. 2019. Phthalate and BPA exposure in women and newborns through personal care product use and food packaging. *Environ Sci Technol* 53(18): 10813–26.
Fraumeni, F., and D. Schottenfeld. 2006. *Cancer Epidemiology and Prevention*. New York & Oxford: Oxford University Press.
Gabard, B. 2009. Sun protection and sunscreens. In *Handbook of Cosmetic Science and Technology*, 3rd ed., ed. A. O. Barel, M. Paye, and H. I. Maibach, 323–32. Boca Raton, FL: CRC Press.
García, P., V. Codina, and H. Quiroga. 2019. Synthesis and characterization of a new cyclodextrin derivative with improved properties to design oral dosage forms. *Drug Deliv Transl Res* 9(1): 273–83.
Gelderblom, H., J. Verweij, K. Nooter, and A. Sparreboom. 2001. Cremophor EL: The drawbacks and advantages of vehicle selection for drug formulation. *Eur J Cancer* 37: 1590–8.
Goodman, M., D. Q. Naiman, and J. S. Kind. 2018. Systematic review of the literature on triclosan and health outcomes in humans. *J Critical Rev Toxicol* 48: 1–51.
Goyal, S., S. K. Amar, H. N. Kushwaha, et al. 2014. Toxicological potential of parabens a widely used preservative. *Global J Multidiscipl Stud* 4(01).
Gulson, B., M. J. McCall, D. M. Bowman, and T. Pinheiro. 2015. A review of critical factors for assessing the dermal absorption of metal oxide nanoparticles from sunscreens applied to humans, and a research strategy to address current deficiencies. *Arch Toxicol* 89: 1909–30.
Halden, R. U., A. E. Lindeman, A. E. Aiello, et al. 2017. The Florence Statement on Triclosan and Triclocarban. *Environ Health Perspect* 126(6): 064501.
Haschek, W. M., C. G. Rousseaux, M. A. Wallig, B. Bolon, and R. Ochoa. 2013. *Haschek and Rousseaux's Handbook of Toxicologic Pathology*. Cambridge, MA: Academic Press.

Health Product Vigilance Framework, 2012. Health Canadá.
Herman, A. 2018. Antimicrobial ingredients as preservative booster and components of self-preserving cosmetic products. *Curr Microbiol* 76(6): 744–54.
Hines, E. P., P. Mendola, O. S. von Ehrenstein, X. Ye, A. M. Calafat, and S. E. Fenton. 2015. Concentrations of environmental phenols and parabens in milk, urine and serum of lactating North Carolina women. *Reprod Toxicol* 54: 120–8.
Hu, D., Y. X. Wang, W. J. Chen, Y. Zhang, H. H. Li, and J. Xiong. 2017. Associations of phthalates exposure with attention deficits hyperactivity disorder: A case-control study among Chinese children. *Environ Pollut* 229: 375–85.
Hung, Y., and H. Chung. 2004. Acute self-poisoning by ingestion of cadmium and barium. *Neph Dialysis Transp* 19(5): 1308–9.
Ito, Y., M. Kamijima, and T. Nakajima. 2019. Di(2-ethylhexyl) phthalate-induced toxicity and peroxisome proliferator-activated receptor alpha: A review. *Environ Health Prev Med* 24(47): 1–9.
Joo-Hyon, K., Daeyeop, L., Hyunwoo, L., Taksoo, K., Kwangsul, S., and Jungkwan, S. 2018. Risk assessment to human health: Consumer exposure to ingredients in air fresheners. *Regul Toxicol Pharmacol* 98: 31–40.
Kazi, T. G., H. I. Afridi, M. Bhatti, and A. Akhtar. 2019. A rapid ultrasonic energy assisted preconcentration method for simultaneous extraction of lead and cadmium in various cosmetic brands using deep eutectic solvent: A multivariate study. *Ultrason Sonochem* 51: 40–8.
Kerlin, R. L., and X. Li. 2013. Pathology in non-clinical drug safety assessment. In *Haschek and Rousseaux's Handbook of Toxicologic Pathology*, 3rd ed., ed. W. Haschek, C. Rousseaux, and M. Wallig, 725–50. Cambridge, MA: Academic Press.
Kim, J. H., D. Lee, H. Lim, T. Kim, K. Suk, and J. Seo. 2018. Risk assessment to human health: Consumer exposure to ingredients in air fresheners. *Regul Toxicol Pharmacol* 29(98): 31–40.
Kim, J. J., Y. Kim, and V. Kumar. 2019. Heavy metal toxicity: An update of chelating therapeutic strategies. *J Trace Elem Med Bio* 54: 226–31.
Kim, Y. W., M. J. Kim, B. Y. Chung, et al. 2013. Safety evaluation and risk assessment of d-Limonene. *J Toxicol Environ Health B Crit Rev* 16(1): 17–38.
Kiriiri, G., A. N. Mwangi, and S. M. Maru. 2019. Sunscreen products: Rationale for use, formulation development and regulatory considerations. *Saudi Pharm J* 27(7): 1009–18.
Krause, M., A. Klit, M. Blomberg, et al. 2012. Sunscreens: Are they beneficial for health? An overview of endocrine disrupting properties of UV-filters. *Int J Androl* 35: 424–36.
Kuan-Ling, C., J. Shiuh-Jen, and C. Yen-Ling. 2017. Determining lead, cadmium and mercury in cosmetics using sweeping via dynamic chelation by capillary electrophoresis. *Anal Bioanal Chem* 409(9): 2461–9.
Kumar, S. 2018. Occupational and environmental exposure to lead and reproductive health impairment: An overview. *Indian J Occup Environ Med* 22(3): 128–37.
Kunicka-Styczyńska, A., K. Śmigielski, R. Prusinowska, K. Rajkowska, B. Kuśmider, and M. Sikora. 2015. Preservative activity of lavender hydrosols in moisturizing body gels. *Lett Appl Microbiol* 60: 27–32.
Langsrud, S., K. Steinhauer, S. Lüthje, K. Weber, P. Goroncy-Bermes, and A. L. Holck. 2016. Ethylhexylglycerin impairs membrane integrity and enhances the lethal effect of phenoxyethanol. *PLoS One* 11(10): e0165228.
León-González, Z., A. Chisvert, I. Fernández, and A. Salvador. 2013. A solid-phase extraction liquid chromatography-tandem mass spectrometry method for the percutaneous absorption assessment of 3-(4′-methylbenzylidene)camphor via human urine analysis. *Anal Methods* 5: 367–71.
Lin, D., Y. Li, Q. Zhou, Y. Xu, and D. Wang. 2014. Effect of triclosan on reproduction, DNA damage and heat shock protein gene expression of the earthworm Eisenia fetida. *Ecotoxicology* 23: 1826–32.
Mahato, M., M. E. Sherman, N. K. Mudnakudu, N. Joshi, E. Briand, and J. M. Karp. 2018. Prevention of metal exposure: Chelating agents and barrier creams. In *Metal Allergy*, ed. J. Chen and J. Thyssen. Cham: Springer.
Maithili, A., S. Anuradha, and M. Nupur. 2015. Toxic effects of cosmetics on the users: A review. *J Environ Res Dev* 9(3A): 900–5.
Mangelsdorf, I., N. Costa, K. Schroeder, J. Cherrie, and I. Dobrev. 2014. Environmental health criteria 242: Dermal exposure. *Environ Health Criteria*. 242.
Martin, L. M. 2004. Cyclodextrins and their uses: A review. *Process Biochem* 39(9): 1033–46.
Martino-Andrade, A. J., F. Liu, S. Sathyanarayana, E. S. Barrett, J. B. Redmon, and R. H. Nguyen. 2016. Timing of prenatal phthalate exposure in relation to genital endpoints in male newborns. *Andrology* 4: 585–93.

Natsch, A., M. Nägelin, H. Leijs, M. van Strien, E. Giménez-Arnau, M. Vey, and C. González. 2019. Exposure source for skin sensitizing hydroperoxides of limonene and linalool remains elusive: An analytical market surveillance. *Food Chem Toxicol* 127: 156–62.

Okereke, J. N., A. C. Udebuani, E. U. Ezeji, K. O. Obasi, and M. C. Nnoli. 2015. Possible health implications associated with cosmetics: A review. *Sci J Public Health* 3(5–1): 58–63.

Olaniyan, L. W., N. Mkwetshana, and A. I. Okoh. 2016. Triclosan in water, implications for human and environmental health. *Springerplus* 5(1): 1639.

Orús, P., L. Gomez-Perez, S. Leranoz, and M. Berlanga. 2015. Increasing antibiotic resistance in preservative-tolerant bacterial strains isolated from cosmetic products. *Int Microbiol* 18: 51–9.

Pan, S., C. Yuan, A. Tagmount, et al. 2016. Paraben sand human epidermal growth factor receptor ligand cross-talk in breast cancer cells. *Environ Health Perspect* 124: 563–9.

Panico, A., F. Serio, F. Bagordo, et al. 2019. Skin safety and health prevention: An overview of chemicals in cosmetic products. *J Prev Med Hyg* 60(1): E50–E57.

Papageorgiou, S., A. Varvaresou, E. Tsirivas, and C. Demetzos. 2010. New alternatives to cosmetics preservation. *J Cosmet Sci* 61: 107–23.

Paul, K. B., J. M. Hedge, R. Bansal, et al. 2012. Developmental triclosan exposure decreases maternal, fetal, and early neonatal thyroxine: A dynamic and kinetic evaluation of a putative mode-of-action. *Toxicol* 300: 31–45.

Quirós-Alcalá, L., J. P. Buckley, and M. Boyle. 2018. Parabens and measures of adiposity among adults and children from the U.S. general population: NHANES 2007–2014. *Int J Hygiene Environ Health* 221(4): 652–60.

Radke, E. G., J. M. Braun, J. D. Meeker, and G. S. Cooper. 2018. Phthalate exposure and male reproductive outcomes: A systematic review of the human epidemiological evidence. *Environ Int* 121: 764–93.

Rodprasert, W., K. M. Main, J. Toppari, and H. E. Virtane. 2019. Associations between male reproductive health and exposure to endocrine-disrupting chemicals. *Curr Opinion Endocr Metab Res* 7: 49–61.

Rowdhwal, S. S. S., and J. Chen. 2018. Toxic effects of di-2-ethylhexyl phthalate: An overview. *Biomed Res Int* 2018: 1–10.

Ruszkiewicz, J. A., S. Li, M. B. Rodriguez, and M. Aschner. 2017. Is Triclosan a neurotoxic agent? *J Toxicol Environ Health Crit Rev Part B* 20(2): 104–17.

Schlumpf, M., K. Kypke, M. Wittassek, et al. 2010. Exposure patterns of UV filters, fragrances, parabens, phthalates, organochlor pesticides, PBDEs, and PCBs in human milk: Correlation of UV filters with use of cosmetics. *Chemosphere* 81: 1171–83.

Siti Zulaikha, R., S. I. Sharifah, and S. M. Praveena. 2015. Hazardous ingredients in cosmetics and personal care products and health concern: A review. *Public Health Res* 5(1): 7–15.

Sköld, M., A. Börje, E. Harambasic, and A. T. Karlberg. 2004. Contact allergens formed on air exposure of linalool. Identification and quantification of primary and secondary oxidation products and the effect on skin sensitization. *Chem Res Toxicol* 17(12): 1697–705.

Skotarczak, K., A. Osmola-Man Kowska, M. Lodyga, A. Polanska, M. Mazur, and Z. Adamski. 2015. Photoprotection: Facts and controversies. *Eur Rev Med Pharmacol Sci* 19: 98–112.

Spanier, A. J., T. Fausnight, F. Camacho, and J. M. Braun. 2014. The associations of triclosan and paraben exposure with allergen sensitization and wheeze in children. *Allergy Asthma Proc* 35: 475–81.

Steinemann, A. 2017. Health and societal effects from exposure to fragranced consumer products. *Prev Med Rep* 5: 45–7.

Steinemann, A. C., I. C. MacGregor, S. M. Gordon, et al. 2011. Fragranced consumer products: Chemicals emitted ingredients unlisted. *Environ Impact Assess Rev* 31: 328–33.

Suller, M. T., and A. D. Russell. 2000. Triclosan and antibiotic resistance in Staphylococcus aureus. *J Antimicrob Chemother* 46: 11–18.

Sun, G. F., W. T. Hu, Z. H. Yuan, B. A. Zhang, and H. Lu. 2017. Characteristics of mercury intoxication induced by skin-lightening products. *Chin Med J (Engl)* 130(24): 3003–4.

Tampucci, S., S. Burgalassi, P. Chetoni, and D. Monti. 2018. Cutaneous Permeation and penetration of sunscreens: Formulation strategies and in vitro methods. *Cosmetics* 5(1): 1.

Taxvig, C., A. M. Vinggaard, U. Hass, et al. 2008. Do parabens have the ability to interfere with steroidogenesis? *Toxicol Sci* 106: 206–13.

U.S. FDA. 2019. *CFR—Code of Federal Regulations Title 21*. Chapter I: Food and Drug Administration. Subchapter G: Cosmetics. Silver Spring: United States Department Food and Drug Administration.

Ullah, H., S. Noreena, F. A. Rehmana, A. Waseem, S. Zubair, and M. Adnan. 2017. Comparative study of heavy metals content in cosmetic products of different countries marketed in Khyber Pakhtunkhwa, Pakistan. *Arab J Chem* 10(1): 10–18.

Uter, W., J. D. Johansen, and A. Börje. 2013. Categorization of fragrance contact allergens for prioritization of preventive measures: Clinical and experimental data and consideration of structure-activity relationships. *Contact Dermatitis* 69(4): 196–230.

Vandenberg, L. N., T. Colborn, T. B. Hayes, J. J. Heindel, D. R. Jr. Jacobs, and D. H. Lee. 2012. Hormones and endocrine-disrupting chemicals: Low-dose effects and nonmonotonic dose responses. *Endocr Rev* 33: 378–455.

Villatoroa, C., L. Vera, and H. Gygax. 2016. Comparative study of odours present in twin fragrances by GC-sniffing-ToFMS. *Chem Eng Trans* 54: 133–8.

Volpe, M. G., M. Nazzaro, R. Coppola, F. Rapuano, and R. P. Aquino. 2012. Determination and assessments of selected heavy metals in eye shadow cosmetics from China, Italy, and USA. *Microchemical J* 101: 65–9.

Wang, F., F. Liu, and C. Wanguang. 2019. Exposure to triclosan changes the expression of microRNA in male juvenile zebrafish. *Chemosphere* 214: 651–8.

Weatherly, L. M., and J. A. Gosse. 2017. Triclosan exposure, transformation, and human health effects. *J Toxicol Environ Health, Part B* 20(8): 447–69.

Weinberg, J. L., J. Flattery, and R. Harrison. 2017. Fragrances and work-related asthma-California surveillance data, 1993–2012. *J Asthma* 54(10): 1041–50.

Weiszhar, Z., J. Czucz, C. Revesz, L. Rosivall, J. Szebeni, and Z. Roznyay. 2012. Complement activation by polyethoxylated pharmaceutical surfactants: Cremohor-EL, Tween 80 and Tween 20. *Eur J Pharm Sci* 45: 492–8.

Welsh, M., P. T. K. Saunder, M. Fisken, et al. 2008. Identification in rats of a programming window for reproductive tract masculinization, disruption of which leads to hypospadias and cryptorchidism. *J Clin Invest* 118: 1479–90.

WHO/IPCS. 2002. *Global Assessment of the State-of-the-Science of Endocrine Disruptors*. www.who.int/ipcs/publications/new_issues/endocrine_disruptors/en/ (accessed November 20, 2020).

Wiecka, S., O. Olssona, K. Kümmerera, and U. Klaschkab. 2018. Fragrance allergens in household detergents. *Regul Toxicol Pharm* 97: 163–9.

Witorsch, R. J. 2014. Critical analysis of endocrine disruptive activity of triclosan and its relevance to human exposure through the use of personal care products. *Crit Rev Toxicol* 44: 535–55.

Witorsch, R. J., and J. A. Thomas. 2010. Personal care products and endocrine disruption: Acritical review of the literature. *Critical Rev Toxicol* 40(S3): 1–30.

Yazdankhah, S. P., A. Scheie, E. A. Høiby, et al. 2006. Triclosan and antimicrobial resistance in bacteria: An overview. *Microb Drug Resist* 12(2): 83–90.

Yueh, M. F., and R. H. Tukey. 2016. Triclosan: A widespread environmental toxicant with many biological effects. *Annu Rev Pharmacol Toxicol* 56: 251–72.

Zhang, H., T. Tan, W. Feng, and D. van der Spoel. 2012. Molecular recognition in different environments: β-cyclodextrin dimer formation in organic solvents. *J Phys Chem B* 116(42): 12684–93.

Zoeller, R. T., T. R. Brown, L. L. Doan, A. C. Gore, N. E. Skakkebaek, and A. M. Soto. 2012. Endocrine-disrupting chemicals and public health protection: A statement of principles from The Endocrine Society. *Endocrinology* 153: 4097–110.

12 Toxicology of Herbal Medicines

Hilmi Orhan, Ege Arzuk, Fuat Karakuş and Ali Ergüç

CONTENTS

12.1 HISTORY OF HERBAL MEDICINES IN VARIOUS CULTURES

There are several terms used to refer to herbal-based products that are used as medicine against diseases: herbal medicine, herbal product, herbal medicinal product and so forth. The use of plants in healing as well as hunting most probably dates back to the first appearance of hominids on Earth. One of the oldest tangible findings are 4000-year-old antique Egyptian arrowheads that showed probable herbal poisons (Clark et al. 1974). A much older example is a wooden stick found in Border Cave, South Africa, from 24,500 BP (D'Errico et al. 2012a, 2012b). This stick was found to contain an unusual poison: ricinoleic acid (D'Errico et al. 2012a, 2012b; Evans 2012). In order to characterise the use of poisons in ancient weapons, the first nonanalytical study in the 1970s involved preparing an aqueous extract of the dark stains on an Egyptian arrowhead and injecting the extract into two mice. Although the mice survived, the material on the arrowhead appeared to have sedative and hypnotic effects, together with loss of equilibrium, infrequent spasms and extension of the rear legs at a high dose (Clark et al. 1974). Although the analytical proof is not definitive, the residue on the arrowhead was concluded to contain ouabain, a cardiotoxic alkaloid from the genus *Acokanthera*. This plant is known to have grown in ancient Egypt. Figure 12.1 shows *Acokanthera schimperi* (Botanicus 2020) and the chemical structure of ouabain.

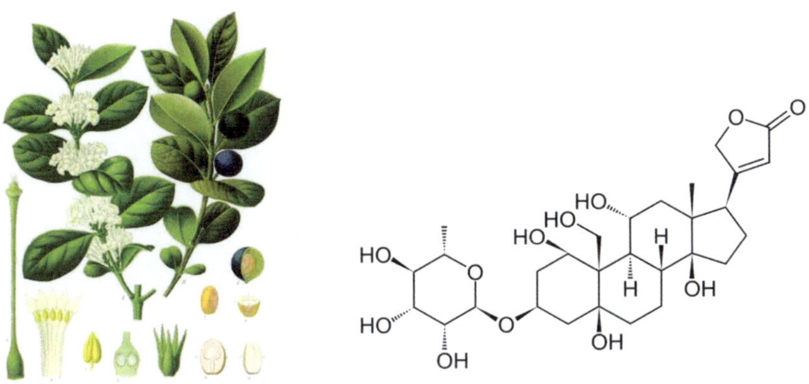

Figure 12.1 A drawing of the poisonous plant *Acokanthera schimperi*, the source of oubain, a compound that was probably present on arrowheads used in ancient Egypt. The chemical structure of ouabain is also shown.

DOI: 10.1201/9780203730584-12

In another example from the Asian island of Borneo, aconitine was recently identified in a dart and arrow poison sample; this poisonous alkaloid is from aconitum (monkshood; Borgia et al. 2017). In addition to poisoned arrowheads, Egyptians used diverse herbal preparations for mummification, and it is not irrational to deduce that they used similar preparations for medical purposes. They had relatively advanced medical knowledge for their time; the ancient Greek historian Herodotus described an advanced specialisation system among physicians and great knowledge of anatomy upon his journey to Egypt (Halioua et al. 2005). A number of surviving papyri, named after the discoverers, are about these medical and pharmaceutical records (Nunn 1996). Among these, the Ebers Papyrus, Kahun Papyri and Hearst Papyrus contain compilations of a large number of prescriptions and recipes, including various herbs and plants (Bryan 1930; Ghalioungui 1975; Hamed 2009). Figure 12.2 shows a piece of the Ebers Papyrus that describes gastrointestinal disorders and treatments. In other columns there is detailed information about cures for various external or systemic diseases by different means, including herbal preparations and cosmetic applications. For example, column 61 describes spreading an herbal mixture on the eyelids in order to improve vision. In total, the Ebers Papyrus includes 900 diagnoses and medical prescriptions. It dates to the 18th dynasty (ca. 1550 BC) during the reign of Amenhotep III.

Due to the prerationalism characteristics of that era, not all prescriptions and recipes were effective; most were rooted in basic trial-and-error experiences, and others had religious and

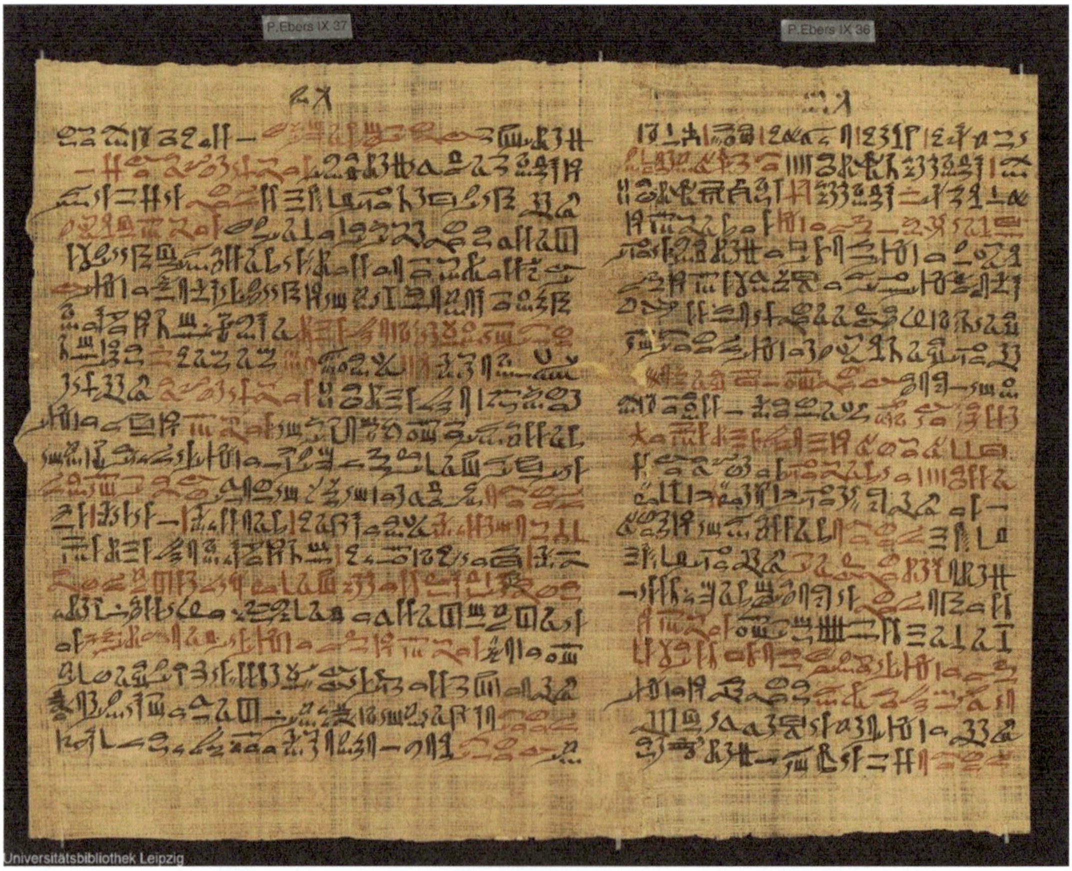

Figure 12.2 Columns 36 and 37 of the Ebers Papyrus, which describes treatments for gastrointestinal disorders. The red ink indicates the headings and quantities and the black ink represents the text; the script is written from right to left without periods, commas or spaces between words.

Source: Photograph via Public Domain Mark 1.0, with the permission of Leipzig University Library, Papyrus Ebers cols. 36 and 37.

mythological motivation and inspiration. Examples of strange formulae include crocodile dung, honey or sour milk for contraception (David 2008). On the other hand, there are a substantial number of successful examples of traditional herbal applications/medicines: bark from the cinchona tree contains quinine, which is still in use as a drug for malaria; willow tree bark contains salicylic acid (a precursor of aspirin), a pain reliever and antipyretic. In contrast to the burst of synthetic and more effective drugs over more than a century, 90% of the world population still relies upon herbal medicines for their healthcare needs. Notably, the same molecules once used as poisons have formed the basis of modern medical treatments.

When we examine the history of herbal medicines from the Far East, Chinese herbal remedies date back to their ancient Egyptian counterparts; however, there are fewer written records about Chinese remedies. The earliest known record of Chinese herbal medicines is *Huangdi neijing* (The yellow emperor's inner classic) from the 3rd century BC; the original is believed to have been written by the emperor himself at a much earlier date (Britannica 2020). Emperor Huangdi and with the two predecessor emperors, Fu Xi and Shennong, were deeply interested in medicine. Among them, Shennong is recognised as the founder of Chinese medicine; he is also known as the 'Divine Husbandman'. His death supposedly occurred by tasting numerous plants and herbs that probably included poisonous ones. According to the Chinese myth, he allegedly ate a yellow flower of a weed (aconitum?!) that caused his intestines to rupture before he had time to drink the antidote tea (Pan et al. 2014).

Together with acupuncture, Chinese herbal medicines have been using to achieve harmony between the two forces of yin (passive) and yang (active), which they believed to exist inherently in every human being. In 1578, Li Shizhen compiled the *Bencao gangmu* (Compendium of *materia medica*). This compendium lists 1,892 medical herbs and 11,000 prescriptions from the Ming dynasty (1368–1644). Furthermore, there is a recent scientifically established example: Youyou Tu was awarded the 2015 Nobel Prize in Physiology or Medicine because of her discovery of the key antimalarial drug artemisinin, which has its roots in traditional Chinese medicine (Tu 2016).

Herbal medicine in Japan has its roots in Chinese medicine. The generic Japanese name 'Kampo' is a combination of 'Kan' and 'Ho': the former is the name of the Han dynasty (206 BC–AD 220) in China, and the latter means 'way of treatment' (Keio University 2020). In addition to plants, minerals and animal parts are also utilised as Kampo. The fundamental knowledge came to Japan by Buddhist monks in the 6th and 7th centuries BC; it was transformed consistently and coherently by Japanese culture (Keio University 2020). In addition to Kampo, 'western herbs' have also gained popularity for the last several decades in Japan, although the exact details are not sufficiently clear (Teng et al. 2016).

In India, the history of herbal remedies and traditional applications goes back to the 5th millennium BC, around the same time as the Egyptian papyri. Indian records in ancient herbal medicine literature are noted in the *Rigveda, Yajurveda* and *Atharva Veda* between 1700 and 1000 BC (Adhikari and Paul 2018). The *Atharva Veda* is also the first record of the use of poisons on arrows. Specifically, the use of aconitine to poison arrows for use in war is mentioned (Nougayrol 1952; Bisset 1989).

The application-oriented uses of plants and multiherbal formulations have been summarised in several manuscripts: two old (*Charaka Samhita* from 990 BC and *Sushruta Samhita* from 660 BC) and one relatively recent (*Dhanwantari Nighantu* from 1800; Kumar et al. 2007). Ayurvedic ('knowledge of life') medicine from ancient Indian culture, similar to its Chinese and Egyptian counterparts, is suffused with beliefs and myths, as well as tangible outcomes from the trial-and-error approach. Among the classical texts of Ayurveda, *Bhava Prakasa* focuses on additional information related to plants and diet, and *Agada Tantra* focuses on toxicology. The Ayurvedic medical system includes diverse herbal medicines and therapeutic baths. This philosophy is still practised in India as well as Bangladesh, Nepal, Pakistan and Sri Lanka (WHO 2001).

Moving westward, Azerbaijan includes a combination of different herbal medicine traditions from the region. This country has been at the crossroads of many cultures and has been influenced by various healing systems such as Turkish, Iranian, Semitic and Greek medicine. Between the 4th and the 1st millennia BC, Guti tribes in Azerbaijan used medical and aromatic plants and herbs such as lavender, laden, myrrh, sesame, dates, saffron, onion and garlic, some of which were imported from ancient Egypt, Syria and Palestine. The overall healing practises were influenced by Sumerian, Acadian, Babylonian and Assyrian medicine after Gutis conquered Mesopotamia in 2200 BC (Alakbarli 2006). Although Assyrian medicine had the greatest influence on the healing system practises in the region, Gutis and Lullubis developed their own folk treatment methods by using herbs, milk products and spices soaked in wine, especially in the Azerbaijan highlands.

Contemporary herb wines were used in the treatment of diarrhoea and weakness, among other conditions. In subsequent centuries, medicine in the country incorporated Arian traditions with the migration of Iranian tribes to the region and mixing with local people during the 1st millennium BC (Parpola 1988). The fundamental role of herbs in Azerbaijan medicine was emphasised as 'word [magic], herbs and knife [surgery]' in the ancient religious book of Zoroastrians, the *Avesta* (Darmesteter 1898).

In the north, Russia was relatively closed and unknown to the other regions of the world with regard to folk medicine, and particularly traditional herbal treatments. Therefore, the history of herbal medicines during ancient times in this country is not known. In 1005, the first pharmacy was opened by a monk from Athos, and the Greek influence was strengthened in the country. It is stunning to note that Russian herbalists were able to cure infected wounds with 'banya mold' seven centuries before the discovery of penicillin in England (Solovieva 2005; Shikov et al. 2014). Between 1236 and 1480, Russia was occupied by Tatars and Mongolians, who brought their traditions on herbal medicines. These novel traditions were mixed with and enriched Russian herbal practice. Russian flora was systematically investigated and documented by the German botanist Johann Georg Gmelin between 1732 and 1743 (Shikov et al. 2014), soon after the establishment of the State Academy of Science in the country. This event was followed by other efforts to define and utilise local flora for medicinal purposes (Muraviova 1991; Shikov et al. 2014). Soon after, the Russian pharmacopoeia was compiled, including 316 texts on herbal medicinal preparations. The 19th century is highlighted as the beginning of the search for the chemical composition of medicinal plants. In 1833, atropine was the first herbal compound that was isolated and characterised. This century was also the beginning of the awareness that the biological effects of plants originate from their specific constituent(s) (Heinrich et al. 2012). There was another great contribution from Russia to such studies on herbal medicines: Mikhail Tswett established the principles of chromatography and performed the first applications with ink (a mixture of pigments). These experiments provided great advancement in the isolation of mixtures of complex chemicals, including plant extracts (Tswett 1905; Ettre and Sakodynskii 1993). Today, advanced analytical techniques that are fundamentally based on chromatography are the most powerful tools not only for identification of herbal constituents but also for characterisation and quality control measures of herbal medicines.

In Oceania, the history of Malaysian herbal medicines has some rich characteristics. More than 1300 medicinal plants have been recorded in this country. The traditional medicine in Malaysia, *Unani*, has its roots in the Hellenic system. It was significantly influenced by Indonesian, Chinese, Indian and indigenous Orang Asli (aboriginal peoples) traditional practices of herbal medicine (Zakaria and Mohd 1994; Alsarhan et al. 2014).

The Indonesian tradition of herbal medicine is called *Jamu* and dates back to the 8th century. Jamu prescriptions were collected in ancient documents written on palm tree leaves by Javanese tribes (Soedibyo 1989, 1990). Another ancient manuscript was published in 1831 (Sutarjadi 1990). Jamu is still in use today to express all kinds of traditional medicines (Riswan and Sangat-Remantyo 2002).

The first known recordings on traditional use of herbal medicines in the New World, America, date back to 1552 in Mexico; the Badianus Manuscript compiled numerous coloured illustrations of medicinal plants. This text was produced after the Spanish conquerors were very impressed by the medical knowledge of indigenous Aztecs. The Badianus Manuscript was written by Martinus de la Cruz in 1552 and translated by Juannes Badianus into Latin. In 1615, the following script, *Historia Natural de la Nueva Espana* (Natural history of New Spain; Figure 12.3) gathered detailed information about medicinal aspects of animals and minerals as well as medicinal plants.

Francisco Hernández de Toledo was a court physician who was appointed as a researcher by King Philip II of Spain in 1570 to investigate and document medicinal plants of New Spain, which later became 'Mexico'. He completed the mission and returned to Spain in 1577 with comprehensive notes comprising numerous descriptions of unknown local plants. He prepared a brief introduction that was published; however, he died in 1587 without seeing his complete work published. Following several attempts by different people, Hernández's work was finally published in Mexico in 1615 by the Dominican priest Francisco Ximénez.

In present-day Mexico, as a result of ethnopharmacy and ethnobotany studies that began in the 1990s, Aguilar et al. (1994) established the first Mexican herbarium. Currently, digital libraries incorporate 1045 monographs about medicinal plants in Mexico (Valdivia-Correa et al. 2016).

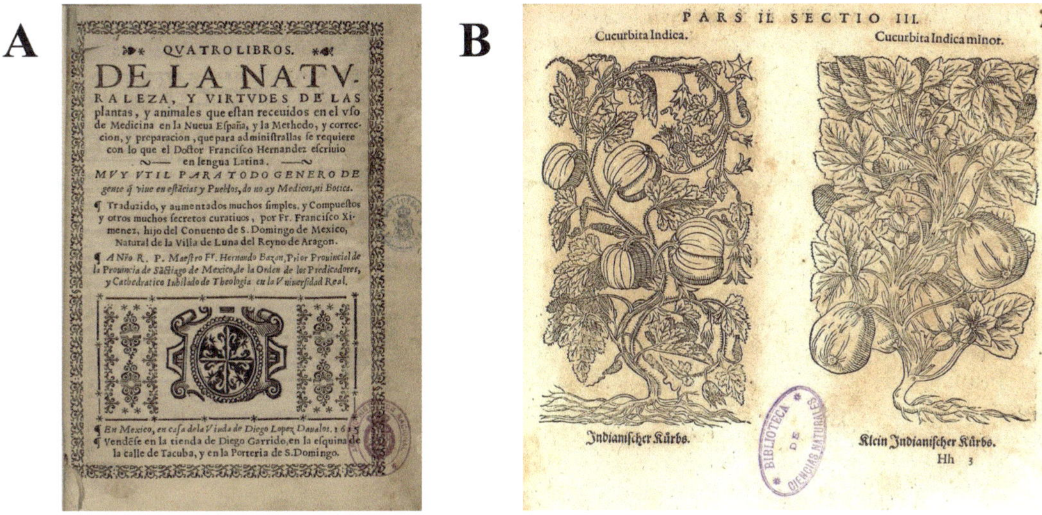

Figure 12.3 The earliest scripts on medical herbs in the Americas: (A) the cover of *Historia Natural de la Nueva Espana* by Francisco Hernández de Toledo and published by Francisco Ximénez in Mexico in 1615; (B) one of its pages representing a description of a plant.

Source: Creative Commons Licence 4.0; Digital Library, Royal Botanic Garden Madrid, RJB-CSIC; https://bibdigital.rjb.csic.es.

12.2 REGULATION OF HERBAL MEDICINES

The major difference between officially approved pharmaceuticals and herbal medicines is the standard chemical content of the former, whereas the chemical content of the latter, including active substance(s), is subjected to marked variation due to geographic localisation, seasonal changes, good agricultural and collection practises, time of harvesting and applied process towards the final product. Due to this major difference, authoritative regulation on herbal products is also different compared to pharmaceuticals; herbal medicines are regulated under different directives around the world.

In Europe, herbal products as medicines are regulated by Directives 65/65/EEC (1965) (European Commission 1965) and 2001/83/EC (European Commission 2001). The three main regulatory pathways for approval of an herbal medicinal product in EU are (1) *standalone or mixed application*, established in article 8–3; (2) *traditional use*, described in articles 16–1; and (3) *well-established use* marketing authorisation, instituted in article 10a. In the *standalone or mixed application*, a company's own development or a combination of their own studies and literature data on safety and efficacy are the main requirements. In the *traditional use* registration, there is no need for clinical tests and trials as long as sufficient safety and efficacy data are provided. In the *well-established use* marketing authorisation, scientific literature should establish the efficacy and acceptable level of safety of the active substance(s) of herbal medicine that has been in medicinal use within the EU for at least 10 years (European Commission 2001).

Herbal products are marketed with a special status as dietary supplements in the United States and regulated by Dietary Supplement Health and Education Act of 1994. A dietary supplement is defined as a product (other than tobacco) intended to supplement the diet, such as vitamins, minerals, herbs or other botanicals, and amino acids. This product can be a concentrate constituent, metabolite, extract or combination of the above-mentioned ingredients (Liu and Salmon 2010; Alostad et al. 2018). Dietary supplements do not require registration in the United States. However, manufacturers or distributors must notify the Food and Drug Administration (FDA) if the supplement contains a new dietary ingredient (NDI), and the manufacturer must prove that this NDI is safe. An NDI is a dietary ingredient that was not sold in the United States before 1994. All other ingredients marketed before 1994 are considered as safe (Alostad et al. 2018).

Given that herbal products are not accepted as drugs, they are not evaluated by the FDA and therefore, are not intended to diagnose, treat, cure or prevent any disease. The label and related media should explicitly state that a product is a dietary supplement. Dietary supplements are only allowed in oral preparations; all others intended for different routes of administration are termed *botanical drugs*. Botanical drugs have the strictest registration requirements (Alostad et al. 2018).

Traditional Chinese herbal medicines (TCHM) are regulated by the China Food and Drug Administration similar to pharmaceutical drugs (SFDA 2007) under the provisions of drug registration (Teng et al. 2016) that came into force in 2007. One year later, supplementary rules for registration of TCHM were entered in order to regulate registration issues for Chinese herbal crude drugs and Chinese herbal preparations. Similar to the approaches of the European Union (EU) and the United States, traditional formulations with a long history of use can be registered without clinical studies (Teng et al. 2016).

12.3 MAJOR TOXIC EPIDEMICS OF HERBAL PREPARATIONS AND MEDICINES

There has been a general global persuasion/sense that natural is safer than synthetic. This preconception ignores the apparent examples of plant-based toxins that have existed during human history. By perpetuating this preconception, as well as other reasons, adverse effects of herbal preparations and medicines have been and continue to be a problem around the world. This section presents a compilation of these issues.

In 1993, a group of young female patients ($n = 9$) were diagnosed with a rapidly progressive renal disease that is associated with the consumption of a Chinese herbal preparation. The renal disease was characterised as interstitial renal fibrosis, tubular atrophy and low numbers of chronic inflammatory cells (Gökmen et al. 2013; Vanherweghem et al. 1993). These women were treated with a combination of Chinese herbs in a weight loss clinic, one of which contained the causative agent aristolochic acid (AA), a nephrotoxin. The initial herb in the regimen, *Stephania tetrandra*, had been replaced by *Aristolochia fangchi*. Following the epidemic, the presence of AA was confirmed by phytochemical analysis (Vanhaelen et al. 1994). Therefore, the disease was termed aristolochic acid nephropathy (AAN). These and other reported patients in the literature had renal insufficiency, anaemia and mild proteinuria. AAN is notable for its rapid progression to end-stage kidney disease even after withdrawal of AA-containing products (Gökmen et al. 2013). The patient can only survive if she or he receives a kidney transplant. The victim also develops urothelial tumours within a few years later. AA-DNA adducts have been detected in renal tissue of poisoned patients (Nortier et al. 2000; Schmeiser et al. 1996). A recent study showed the molecular mechanism of AA-induced urothelial cancer: the formation of aristolactam-DNA adducts that result in A:T → T:A transversions in the *TP53* gene in tissue samples (Chen et al. 2012). Consequently, the IARC has classified AA as human carcinogen (Group 1; IARC 2012). Up to that time, it was known that AA exposure was also responsible for Balkan endemic nephropathy (BEN), which caused kidney toxicity and, subsequently, urothelial cancer in at least 25,000 victims in 1950s (Tatu et al. 1998). AA was first suggested as a cause of BEN in 1969, when wheat flour that victims consumed was found contaminated by the seeds of *A. clematitis*, an abundant weed in wheat fields of Bosnia, Bulgaria, Croatia, Romania and Serbia (Ivic 1969). However, BEN has also been attributed to a mycotoxin, ochratoxin A, produced by fungi of the genera *Penicillium* and *Aspergillus* (Godin et al. 1997; Ueno 1985).

Another problematic herbal compound is usnic acid, used for weight loss. This compound is a lichen metabolite that possesses antimicrobial (Lauterwein et al. 1995), antiviral (Scirpa 1999) and anti-inflammatory (Vijayakumar et al. 2000) properties, due to its uncoupling action on mitochondria (Abo-Khatwa et al. 1996; Ingólfsdóttir 2002). The same properties are the basis for the use of usnic acid for weight loss. However, excessive doses of usnic acid are required to achieve this effect, a factor that markedly increases the risk of severe liver injury (Favreau et al. 2002). Due to this high risk, usnic acid is no longer sold as a dietary supplement in the United States. However, it can be purchased from the internet in many countries.

Aconitum is related to a multicultural infamous poisonous plant mentioned in the first section. It has been used as arrow poison in many cultures from the Far East to the New World and was the poison given to execute Socrates in ancient times. It is still employed in homeopathy to treat several disorders such as anxiety and neuralgias as well as cold and fever (Borgia et al. 2017). On the other hand, given that the therapeutic window of this compound is very narrow, slight overdoses can easily cause adverse effects, including fatalities. Seven patients died in China between 1999 and 2008 after the ingestion of traditional Chinese medications containing aconitine (Liu et al. 2011).

The worldwide occurrence of poisonings due to the contamination of herbal teas and herbs can pose a serious health hazard to society. Such contamination most likely occurs during harvest and/or processing. Among various measures to prevent herbal poisonings, good agricultural and collection practises, as well as good manufacturing practises, have been emphasised by the World Health Organization (WHO 2003). Medicinal plants collected in the wild may be contaminated by other species through misidentification, accidental contamination or intentional adulteration (WHO 2004). Important examples of contaminating plants/herbs in terms of toxic potential include digitalis, podophyllum and aconitum species. Uncontrolled/unauthorised collection of plants in the wild poses a risk to human health as well as the existence of such plants themselves. Wild type ginseng (*Panax ginseng*), goldenseal, echinacea, black cohosh, slippery elm and kava kava have been at the top of the endangered species list (WHO 2004) because the demand for these plants is greater than the available natural supply.

There were 14 reports of poisoning due to the contamination of herbal teas between 1978 and 2018 (Bryson et al. 1978; Rhoads et al. 1984; Galizia 1983; Routledge and Spriggs 1989; Awang and Kindack 1989; Oerlemans et al. 2017; Van Riel et al. 2014; Baca-García et al. 2007; Hsu et al. 1995; Scholz et al. 1980; Cagin et al. 2017; Mousa et al. 2018; Reddy et al. 2016; Zheng et al. 2016). Most of them were due to contamination by tropane alkaloids: atropine, hyoscyamine and scopolamine. Anticholinergic symptoms were observed in most of those patients.

12.4 ADVERSE HEALTH EFFECTS ASSOCIATED WITH THE USE OF HERBAL MEDICINES

Adverse health effects associated with the use of herbal medicines can be classified under four main categories: (1) adversity due to the *original constituents* of the herb/plant; (2) adversity because of an *interaction* between plant constituents and other simultaneously used herbal medicines or pharmaceuticals; (3) adversity due to *contaminants*, either in the form of environmental pollutants, such as metals and halogenated compounds, or in the form of microorganisms and other plants; and (4) adversity due to *intentional adulterants* in the form of synthetic pharmaceuticals or different plants that have a similar appearance and lower cost. As emphasised in the introduction to this chapter, there are two major problems with regard to herbal medicines: the lack of standardisation of active ingredient(s) and the lack of sufficient data obtained that are scientifically sound, controlled experiments in order to make reliable human health risk assessments.

A recent review series summarised, in a tabular form, the major organ toxicities caused by herbs and herbal medicines for the liver, kidney and heart, as well as the potential of herbs and dietary supplements of causing cancer (Brown 2017b, 2017c, 2018a, 2018b). The same author published regulatory issues of herbs and dietary supplements in the United States in the fifth article from this review series (Brown 2017a). According to these articles, the largest number of publications on the adverse effects of herbs and herbal medicines on the target organs have been on germander, black cohosh, kava extract and green tea extract for hepatotoxicity; the aristolochic acid–containing herbs guang fang ji and chocolate vine or mu tong for nephrotoxicity; and Don quai Jin bu huan, Thundergod vein, lei gong teng, Ting kung teng, Aswagandha, blue cohosh and Yohimbe for cardiotoxicity (Brown 2017b, 2017c, 2018b). A comprehensive search of literature from the past 50 years, in terms of increased risk of cancer due to herb use, highlighted three herbs: guang fang ji and dietary supplements containing guang fang ji or aristolochic acid and two others used as food: bracken fern and *Ilex paraguariensis* (consumed as herbal infusion under the name 'hot mate' in some regions of South America). Although aloe vera, coltsfoot, comfrey, *Ginkgo biloba*, goldenseal (gordolobo yerba), kava, mader root, soy and sassafras have been implicated in cancer by various mechanistic hypotheses, experimental results and epidemiological analyses represent contradicting results (Brown 2018a).

Due to their widespread global use and priority in terms of posing risk to human health, eight herbal medicines are given special emphasis in this chapter: ginseng, green tea, echinacea, *G. biloba*, saw palmetto, kava kava, St. John's wort and garlic. Several other herbal preparations that have greater toxicity potential are not included because their trade on the internet has been seriously restricted. Those preparations include germander, black cohosh and usnic acid in some countries (Brown 2017a); this restriction has dramatically diminished public exposure.

12.4.1 Ginseng (*Panax*)

Ginseng has many species due to its widespread geographical distribution in nature with the most popular being *P. ginseng* (Asian, Chinese or Korean ginseng) and *P. quinquefolius* (American and Canadian ginseng). Others include *P. japonicus* (Japanese ginseng), *P. notoginseng* (sanqui or tienqi ginseng), *P. elegantior* (pearl ginseng), *P. pseudoginseng* (Himalayan ginseng) and *P. zingiberensis* (ginger ginseng; Ocollura 1997). The root colour depends on the drying process; peeling and drying in the sun generates white root ginseng, whereas drying unpeeled root by steam generates red root ginseng. As an herbal medicine, ginseng root has many different constituents with diverse biological activities. These include triterpene saponins, polyacetylenes, sesquiterpenes, polysaccharides, peptidoglycans, fatty acids, carbohydrates and phenolic compounds (Sticher 1998). The major components are ginsenosides, which are claimed to have anti-inflammatory, antioxidant and anticancer effects. There have also been clinical studies that have suggested ginseng may improve psychological function, immune function and diabetic conditions (Kiefer and Pantuso 2003). Various ginseng products, such as root, powder, tablets, capsules, liquid extract and tea, are marketed in Korea (Yun 2001). Standardised *P. ginseng*, which contains 4% ginsenosides, and *P. quinquefolius* are available. Thereby, the efficacy and safety of ginseng products have been established (Ebo et al. 2008).

Ginseng products are assumed to be generally safe, although mild toxicity and a few adverse events have been reported in long-term use at high doses (Kiefer and Pantuso 2003). A clinical study designed with 2-year follow-up revealed that *P. ginseng* ingestion was associated with morning diarrhoea (35%), skin eruption (25%), nervousness (25%), sleeplessness (20%), hypertension (17%), oedema, decreased appetite, depression and hypotension (10%) in 133 individuals (Siegel 1979). A systematic literature review also suggested that *P. ginseng* has a low potential to interact with clinical drugs (Coon and Ernst 2002; Lee et al. 2008; Izzo and Ernst 2009). On the other hand, adverse reactions associated with the abuse of ginseng have been reported in literature (Paik and Lee 2015). Between 1970 and 2012, hypertension (Martinez-Mir et al. 2004); gynaecomastia and genital organ bleeding (Palop-Larrea et al. 2000; Kabalak et al. 2004; Kakisaka et al. 2012); hepatic, cardiovascular and renal toxicities (Caron et al. 2002; Liao et al. 2010; Bilgi et al. 2010; Torbey et al. 2011); allergy (Wiwanitkit and Taungjaruwinai 2004; Lee et al. 2006; Kim et al. 2008; Lee et al. 2012); and affective disorders (Engelberg et al. 2001; Vázquez and Aguera-Ortiz 2002) have been described in case reports. American ginseng reportedly interacts with anticoagulants (Yuan et al. 2004).

The mechanisms of some adverse reactions associated with ginseng ingestion have been studied. Allergic reactions occur via both immunoglobulin E (IgE)-dependent and independent manners in individuals likely predisposed or presensitised to ginseng (Lee et al. 2006; Kim et al. 2008). Endocrine disruption associated with exposure to ginseng is based on oestrogen receptor activation by ginseng phytoestrogens (ginsenosides and panaxosides) in a ligand-independent manner (Cho et al. 2004; Lau et al. 2008).

Due to a wide variety of diverse constituents, ginseng species can be expected to interact with clinical drugs that a patient is concurrently using. Ginseng inhibits cytochrome P450 (CYP)3A4; it was found that plasma concentration of anticancer imatinib rose to toxic levels in a patient who was also using ginseng simultaneously (Bilgi et al. 2010). The same inhibitory effect of ginseng on CYP3A4 activity resulted in a significant elevation of plasma concentrations of cardiovascular agents such as digoxin and nifedipine, as well as the risk of adverse reactions (Bressler 2005).

Ginseng may interact with the blood coagulation process by two means: it inhibits platelet aggregation and prolongs the blood coagulation time *in vitro* and in experimental animals (Kimura et al. 1988; Park et al. 1996; Lau et al. 2008). Thus, it may enhance the anticoagulant effects of warfarin. On the other hand, ginseng increases CYP2C9 activity and thereby decreases the efficiency of warfarin and predisposes patients to thrombotic events (Yuan et al. 2004). Several case reports on this issue showed that ginseng significantly reduces the international normalised ratio (INR; measure of blood clotting) and peak warfarin plasma level. The net result is a significant decrease in the anticoagulant effect upon American ginseng consumption (Jiang et al. 2004; Zhu et al. 1999; Janetzky and Morreale 1997; Yuan et al. 2004). Ginseng interacts with two pharmacological groups of drugs, but the interaction mechanisms remain unknown (Coon and Ernst 2002). The first group includes monoamine oxidase inhibitor antidepressants such as isocarboxazid, tranylcypromine and phenelzine. Interaction of ginseng with these drugs increases the risk of central nervous system (CNS) effects (e.g. mania-like symptoms, headache, insomnia and tremulousness; Coon and Ernst 2002). The second group of drugs includes diuretics such as ethacrynic acid, furosemide and torsemide. Ginseng significantly decreases the efficacy of these drugs and increases the risk of heart failure due to fluid overload (Charrois et al. 2006).

The lack of global standardisation of safety and efficacy requirements means that herbal medicines or crude herbs are subjected to circumstantial analyses for various contaminants. *P. ginseng* and *P. quinquefolius* species in 47 different samples from 20 suppliers were analysed for various metal content including arsenic, aluminium, molybdenum, chromium, copper, magnesium, zinc, cadmium, mercury and lead (Durgnat et al. 2005). The concentrations of the metals of most concern (i.e. cadmium, mercury, lead and arsenic) have been found to be below the maximum residue level (MRL). Among the other elements, aluminium was the most abundant (Durgnat et al. 2005). The same research group analysed contamination levels of various pesticides in 30 ginseng samples from 17 suppliers; 80% of the same samples contained pesticides above the limit of detection. The concentrations of quintozene, hexachlorobenzene, total hexachlorocyclohexane, lindane, total heptachlor, chlorpyrifos and folpet exceeded the MRL in 43% of the samples, with levels up to 55- and 30-fold higher than the MRL.

Another (usually) intentional adulteration of ginseng is misbranding different plant materials both inside and outside the same Araliaceae family as ginseng. Examples are the plants called 'seng' or 'shen' in traditional Chinese medicine that are regarded as 'poor man's ginseng' (Foster 2016). Relatively low-priced American ginseng is sometimes intentionally labelled and marketed as Asian ginseng. Intentional adulteration with nonginseng plants frequently occurs in the powder form, as it may be more difficult to recognise with the naked eye. Ginseng leaf extracts are sometimes added to root extracts in order to increase the level of ginsenosides, which are considered to be the most important ginseng components responsible for efficacy. However, the profile and, hence, the biological activity of the leaf may be different than the root constituents. It has also been emphasised that the chemistry of various plant parts in the genus *Panax* is also species dependent (Foster 2016).

12.4.2 Green Tea (*Camellia sinensis*)

Tea is one of the most consumed beverages all over the world; it is prepared by infusion from leaves, a technique called brewing. There are two main types of tea, depending on the degree of postharvest fermentation (enzymatic oxidation): green (unfermented) and black (fully fermented). These types differ in terms of flavour, colour and composition of both the leaves and the beverage (Dias et al. 2014; Wu et al. 2014; Tenore et al. 2015). The major components of green tea (unfermented) are catechins, with 5%–12% epigallocatechin gallate (EGCG) and epicatechin gallate (ECG; Bruneton 1999). These polyphenols are considered responsible for the potential health benefits of tea consumption; EGCG is more efficient in terms of antioxidant capacity (Mazzanti et al. 2009). The strong antioxidant properties of these polyphenols have been claimed to provide benefits to obesity (attributed to the fat-burning effect of catechins), diabetes mellitus, cardiovascular disorders and even cancer, although such benefits have not been clinically confirmed (Wierzejska 2014; Wang et al. 2014; Keske et al. 2015; Butt et al. 2015; Isomura et al. 2016). In line with these findings, unfermented green tea is supposedly better compared to fully fermented black tea.

In contrast to the attributed health benefits (Fujiki et al. 2015; Legeay et al. 2015), EGCG has also been implicated in liver injuries (Sarma et al. 2008; García-Cortés et al. 2016; Harrison-Dunn 2016). Animal studies with EGCG have suggested that hepatic effects are, at least partially if not fully, governed by genetics. A small fraction (16%) of genetically heterogenous mice experienced severe hepatotoxicity. This observation was similar to humans (Church et al. 2015). The hepatotoxic potential is associated with the consumed amount of green tea. It has been used at much higher concentrations in the form of green tea extract for losing weight. Due to 13 liver injury cases in 2013, one commercial extract was withdrawn from the market in France and Spain (Gloro et al. 2005).

Green tea can be a rich source of vitamin K, a cofactor that plays an essential role in coagulation. Therefore, green tea consumption can theoretically antagonise the effect of the anticoagulant drug warfarin. Indeed, there is a case report about this interaction (Taylor and Wilt 1999) that has been cited by other researchers (Cheng 2007; Holbrook et al. 2005; Nutescu et al. 2006). On the other hand, there have been no further studies in the literature to support this observation. Green tea has been found to inhibit solute carrier organic anion transporter family member 1A2 (OATP1A2)-mediated uptake of nadolol in the intestine, an eventuality that may in turn result in a decrease in plasma concentration of the drug and antagonise its antihypertensive effect (Misaka et al. 2014).

Similar to the competitive interaction between folate and methotrexate during intestinal uptake, green tea catechins may also diminish folic acid absorption in a dose-dependent manner (Alemdaroglu et al. 2008). As green tea catechins inhibit OATP1A2 as well as OATP2B1, consumption of green tea and/or green tea extract decreases the bioavailability of rosuvastatin (Hegazy 2014; Kim et al. 2017; Misaka et al. 2014). The interaction between catechins and sildenafil or tacrolimus involves CYP3A inhibition, with increased drug concentration and both CYP3A and p-glycoprotein inhibition and an increased risk of toxicity from both drugs (Hegazy 2014; Kim et al. 2017; Misaka et al. 2014). Notably, there is a high interindividual variability with regard to the effect of green tea on drug exposure (Werba et al. 2008).

In terms of contamination, the same favourable subtropical climate for tea cultivation is also ideal for proliferation of toxicologically concerned moulds such as *Aspergillus flavus* and *A.*

ochraceus. These moulds generate aflatoxins and ochratoxins, which are highly potent toxins with mutagenic nephrotoxic and carcinogenic effects, respectively. China and India are the major tea-producing countries, as well as the countries most affected by aflatoxins (Klingelhöfer et al. 2018). The percentage of mycotoxin transfer from raw tea into a brewed beverage has been well studied for ochratoxin A (Sedova et al. 2018). The same study estimated exposures to aflatoxin B1 and ochratoxin A based on an average contamination level; other studies estimated 3%–40% exposure to aflatoxin B1 and 5%–24% exposure to ochratoxin A from different cluster diets (Sedova et al. 2018; Carraturo et al. 2018; Chaly et al. 2018; Haas et al. 2013; Pouretedal and Mazaheri 2013; Weidenbörner 2013; Wu et al. 2014; Li et al. 2015; Benford et al. 2010).

12.4.3 Echinacea

Echinacea purpurea, E. angustifolia and *E. pallida* are three major *Echinacea* species that have been extensively studied in terms of biological activities (Rininger et al. 2000; Nieri et al. 2003). Echinacea is generally accepted as an immunostimulant; therefore, it is one of the best-selling herbal medicines in Europe and the United States as a cold and flu remedy (Starvaggi et al. 2008). It has also been indicated as an anti-infective plant and has been used widely for this purpose (Bradley 1992; Tyler 1993; Williamson and Wren 2003). Rhizomes and/or roots of the echinacea plant are used to produce herbal medicine from all three species, and purpurea flowers are used either in dried powder or in fresh juice form for these indications (ESCOP 2003; Wichtl 2004). Available pharmaceutical forms are tablets, capsules, teas and tinctures; there are also preparations for parenteral use (Barnes et al. 2005). The immunomodulatory effects have been attributed to constituents such as polysaccharides, glycoproteins, alkamides and caffeic acid derivatives (Bauer et al. 1988). Given that pharmacokinetics of active echinacea components have revealed that alkamides—but not caffeic acid derivatives—are absorbed readily and substantially, researchers have suggested that the latter may not be responsible for the immunomodulatory effect of echinacea (Matthias et al. 2004).

In 2004, WHO's Pharmacovigilance Centre in Uppsala received 259 reports regarding echinacea, including 537 adverse reactions associated with the use of *E. purpurea*. However, most of them were mild reactions (the list includes those with a minimum of 10 or more instances): erythematosus (23 cases), urticaria (23), rash (18), dyspnoea (18), pruritus (17), nausea (14), angioedema (10) and abdominal pain (10; Barnes et al. 2005). On the other hand, more serious toxic reactions have also been published in a case report format, such as thrombocytopaenia (Liatsos et al. 2006), leucopenia (Kemp and Franco 2002), eosinophilia (Maskatia and Baker 2010), Sjögren syndrome (Logan and Ahmed 2003), erythema nodosum (Lee Soon and Crawford 2001), exanthema (Lee and Werth 2004), anaphylaxis and other allergic reactions (Mullin 1998; Mullins and Robert 2002).

Children are not among the sensitive populations for the serious risks posed by echinacea. In a study performed to assess the efficiency of *E. purpurea* juice on upper respiratory tract infection in children aged between 2 and 11 years, the preparation was ineffective. Hence, its use was associated with an increased risk of rash (Taylor et al. 2003). After an assessment of the available data, herbal products that contain echinacea appears to be safe including during pregnancy and lactation (Gallo et al. 2000; Gallo and Koren 2001; Nordeng and Havnen 2004; Perri et al. 2006).

The data about *in vivo* interaction of *Echinacea* species with CYPs are controversial: there have been reports of inhibition and no effects for CYP1A2 and CYP3A4 activities (Gorski et al. 2004; Gurley et al. 2004; Gurley et al. 2008; Freeman and Spelman 2008). These controversies have been attributed to differences in the species and in the utilised plant parts, as well as dose regimen and duration of treatment (Izzo and Ernst 2009). There is still no clinically significant report on the interaction between *Echinacea* species and drugs, although a decrease in the effect of immunosuppressants could theoretically be expected (Izzo and Ernst 2009).

Echinacea species have long been intentionally adulterated and sold on the market, mostly with *Parthenium integrifolium*. The roots of *P. integrifolium* are heavier and can be unearthed much easier than those from *Echinacea* species. Furthermore, *E. purpurea* is rare in nature (Hobbs 1994). Monitoring of the US markets in the past showed that most commercially available *E. purpurea* was in fact *P. integrifolium* (Hobbs 1994; Bauer and Wagner 1987).

12.4.4 *Ginkgo biloba*

G. biloba is a tree originally from China that has been cultivated in other geographical regions all over the world throughout history (Chan et al. 2007). The seeds have been used in traditional Chinese medicine against various disorders such as cough, asthma, enuresis, skin infections and intestinal worm infections (Van Beek and Montoro 2009). On the other hand, the leaf extract of the tree is mainly used nowadays. This extract represents a standardised and concentrated form in European and other markets. The biological activity of ginkgo comes from its terpenoids such as ginkgolides (A, B, C and J types) and bilobalides, as well as flavonoid glycosides such as quercetin, kaempferol, isorhamnetin and a number of organic acids (Vasseur et al. 1994; Salvador 1995; Brown 1996). Ginkgolides and bilobalides possess antiplatelet activity because they antagonise platelet-activating factor (Diamond et al. 2000). Owing to these effects, it is especially used to improve cerebral circulation in memory loss, Alzheimer's disease and other cognitive disorders, as well as in circulation related to other diseases such as arrhythmias and ischaemic heart disease, cancer, diabetes and thromboses (Diamond et al. 2000; Fugh-berman 2000). In contrast to these beneficial components, ginkgo contains potentially toxic compounds as well such as gingkolic acids, alkylphenol and alkylbenzoic acid derivatives, which have allergic, immunotoxic and other undesirable properties (Jaggy and Koch 1997). These compounds have been limited from the commercial extracts, whereas the other beneficial flavonol glycosides and terpene lactones have been enriched.

G. biloba has the fourth highest number of associated adverse reactions of all herbal preparations. Among the reports in the literature, 28 papers have referred to adverse reactions and 14 papers have referred to drug interactions (Di Lorenzo et al. 2015). The symptoms of poisoning are convulsion, vomiting and loss of consciousness (Miwa et al. 2001; Hori et al. 2004; Wada 2005; Hasegawa et al. 2006). The responsible constituents in ginkgo seeds are neurotoxic 4′-*O*-methylpyridoxine (MPN), known as ginkgotoxin (Wada et al. 1985; Wada et al. 1988), which is a glucoside derivative (Yoshimura et al. 2006; Kobayashi et al. 2015). MPN has structural similarity to vitamin B6 and interacts with it at various levels, including biosynthesis, biotransformation and its mode of action. Due to the interference with vitamin B6 biosynthesis (Mizuno et al. 1980; Kästner et al. 2007; Leistner and Drewke 2010; Kobayashi et al. 2015), ginkgo poisoning results in vitamin B6 deficiency and leads to diminished γ-aminobutyric acid (GABA) synthesis. Vitamin B6 is the major agent in the treatment of ginkgo seed poisoning (Hori et al. 2004; Kajiyama et al. 2001; Ebo et al. 2008), and most of the cases involve reversible toxicity. Furthermore, ginkgo leaves contain small amount of MPN (Arenz et al. 1996). The US FDA database Special Nutritionals Adverse Event Monitoring System reported seven cases of seizures associated with ingestion of ginkgo biloba leaf extract in 2001 (Gregory 2001). Three possible perpetrators have been suggested for ginkgo-induced seizures: MPN, antagonism of platelet activating factor and CYP2C19 induction (Samuels et al. 2008). In line with the intended use of ginkgo extract, spontaneous bleeding has also been observed, although there have been controversial reports. A substantial amount of clinical research has shown the potential of ginkgo for inducing cardiovascular adverse events such as ventricular arrhythmia (Cianfrocca et al. 2002), ischaemic stroke and transient ischaemic attack (Dodge et al. 2008), frequent nocturnal palpitations (Russo et al. 2011), as well as dermatological problems such as generalised exanthematous pustulosis (Pennisi 2006) and contact dermatitis (Hotta et al. 2013). Again, because of antiplatelet efficiency, it has been suggested that ginkgo products should be used carefully by pregnant women and during lactation (Dugoua et al. 2006). In 2013, the IARC classified the leaf extract as a possible human carcinogen (Group 2B;IARC 2012; Grosse et al. 2013).

Ginkgo products should not be used concurrently with drugs that are effective with regard to platelet function and/or coagulation, such as warfarin or nonsteroidal inflammatory drugs (NSAIDs), in order to avoid spontaneous bleeding, as they will interact additively (Rosenblatt and Mindel 1997; Shaw et al. 1997; Meisel et al. 2003). This adverse reaction is particularly severe with acetylsalicylic acid or aspirin (spontaneous hyphema; Rosenblatt and Mindel 1997); warfarin (intracerebral haemorrhage; Matthews et al. 1998); and ibuprofen (comatose state with an intracerebral mass bleeding of which the patient died; Meisel et al. 2003). On the other hand, subsequent trials did not confirm these reactions (Aruna and Naidu 2007; Mohutsky et al. 2006; Greenblatt et al. 2006; Gardner et al. 2007). Two conclusions from 2005 emphasised that 'the causality between ginkgo intake and bleeding is unlikely' (Izzo et al. 2005) and 'available evidence does not demonstrate that extract of *G. biloba* causes significant changes in blood coagulation parameters' (Savovic et al. 2005).

Using ginkgo does not alter CYP1A2, CYP2E1, CYP2D6 or CYP2C9 activity, but it induces CYP2C19 and partially inhibits CYP3A4 (Gurley et al. 2002, 2005; Aruna and Naidu 2007; Mauro et al. 2003; Robertson et al. 2008; Uchida et al. 2006; Yin et al. 2004). Concurrent use of ginkgo with CYP2C19 substrates, such as omeprazole, may significantly reduce the efficiency of drugs (Yin et al. 2004).

G. biloba extract is usually adulterated with pure flavonols (rutin, quercetin and kaempferol) in order to increase the amount that violates the standard content. In order to decrease the cost, some producer companies use 3% hydrochloric acid instead of expensive organic solvents during extraction that eventually increases free quercetin, kaempferol and isorhamnetin (Gafner 2018).

12.4.5 Saw Palmetto (*Serenoa repens*)

Saw palmetto (SP) from the Arecaceae family is a low-growing palm tree endemic to the West Indies and Southeastern America (Murray and Pizzorno 1994). Like its name, the leaves resemble a saw, and this small tree-like plant is also known as *S. repens*, scrub palmetto, American dwarf palm tree, cabbage palm, *Sabal serrulata* and *Serenoa serrulata*. The medicinal part is its small, yellowish berries. These berries contain fatty acids and phytosterol derivatives such as β-sitosterol, stigmasterol, cycloartenol, lupeol, lupenon and methylcloartenol (Murray and Pizzorno 1994). Sitosterol and other steroidal components have been considered responsible for the biological/pharmacological activities. Saw palmetto has reported anti-oestrogenic activity (Di Silverio et al. 1992) and provides relief in prostatic hyperplasia-associated urinary disorders (Champault et al. 1984; Boyle et al. 2000). Traditional use of SP has also been in line with these pharmacological actions; it has been used to treat genitourinary problems, prostate swelling, erectile dysfunction, enhance sperm production, increase breast size and testicular atrophy and as a mild diuretic (Lowe and Ku 1996; Ernst 2002). Using SP against prostate swelling was first confirmed by 19th-century medical experts (Lowe and Ku 1996). The metabolism and excretion of dihydrotestosterone by SP is increased through inhibition of cellular and nuclear receptor binding (Marks 2004).

Androgenic alopecia and several more hair disorders have been reported to be treatable with SP extract (Murugusundram 2009). Diuretic, antidiarrhoeic and stomach ache relieving effects of SP have also been reported (Wilt et al. 1998).

There have been few adverse effects associated with the use of SP extract: headache, fatigue, nausea, vomiting and diarrhoea have been observed in clinical trials (Gerber et al. 2001; Agbabiaka et al. 2009; Ju et al. 2015; Vela-Navarrete et al. 2018). Although rare, more serious toxicities have also reported, including hyper- and polymenorrhea, menarche (Miroddi et al. 2012; Morabito et al. 2015), hepatic and pancreatic toxicity (Jibrin et al. 2006; Wargo et al. 2010; Lapi et al. 2010; Bruminhent et al. 2011), intraoperative haemorrhage (Cheema et al. 2001), and coagulopathy (Villanueva and González 2009).

SP has minimal drug interactions in humans. In a clinical study with alprazolam, midazolam, caffeine, chlorzoxazone, debrisoquine and dextromethorphan as probe substrates, there was no significant effect on CYP1A2, CYP2D6, CYP2E1 or CYP3A4 activity in healthy volunteers (Gurley et al. 2004; Markowitz et al. 2003). In another study, however, the use of the plant extract increased the activity of anticoagulant drug rivaroxaban, probably by inhibiting both P-glycoprotein and CYPP3A4 activity (Shivamurthy et al. 2014).

SP extracts and dietary supplements have been adulterated with fruit from other related palm species, mixing products with exhaustively extracted berry powder, the addition of vegetable oils such as canola, coconut, olive palm, peanut and sunflower, as well as completely replacing SP extract with one of these vegetable oils (Gafner 2018).

12.4.6 Kava (*Piper methysticum*)

Kava is a beverage prepared from extract from the roots of *P. methysticum*, with psychoactive activity. It is used by the people of many Pacific islands, such as Hawaii, Fiji and Polynesia for social gatherings and ceremonies (Singh 1992; Norton 1994). The products or the plant itself has many names in the United States and other countries, such as ava, ava pepper, awa, intoxicating pepper, kava, kava kava, kava pepper, kava root, kava-kava, kawa, kawa kawa, kawa-kawa, kew, rauschpfeffer, Sakau, tonga, wurzelstock and yangona (FDA 2002).

Fresh or dried kava rhizomes are extracted by grinding and soaking them in solvents such as water, ethanol or acetone (Singh 1992; Norton 1994; Lebot et al. 1997; Denham et al. 2002). Three chemical classes appear as arylethylene-α-pyrones, chalcones and other flavanones and conjugated diene ketones. Kavalactones (substituted 4-methoxy-5,6-dihydro-α-pyrones or kava pyrones) have the highest purported pharmacological activities (Fu et al. 2008; Martin et al. 2014; Smith and Leiras 2018). The active principles of kava are mostly contained in the lipid-soluble resin (Singh 2005).

Clinical trials have analysed the effects of the drug on generalised anxiety disorder symptoms, including excessive worrying, insomnia (Newman et al. 2013), headaches, fatigue, muscle strain and tension (Pluess et al. 2009), due to the anxiolytic and muscle relaxing properties of kavalactones (Sarris et al. 2011). Kava is an effective herbal anxiolytic (Pittler and Ernst 2003). Kava might have additive effects with benzodiazepines; both act on the same receptors and on the same areas of the CNS, with increased GABA receptors (Jussofie et al. 1994). Alcoholic extraction yields higher kavalactone content (Martin et al. 2014). These higher quantities result in a greater feeling of calmness and provide greater efficacy in treating excess apprehension found in basic anxiety or anxiety disorders (Pittler and Ernst 2000; Sarris et al. 2011).

The modern pharmacologic usage of kava in Europe started around 1900 as diuretics and for treatment of gonorrhoea and nervous disorders (Norton 1994; Bilia et al. 2002). In the United States, kava is promoted as a 'natural' alternative to anxiety drugs such as Xanax and Valium (Ernst 2006).

The consumption of aqueous kava extracts is commonly well tolerated in the absence of major side effects (Denham et al. 2002; Currie and Clough 2003; Moulds and Malani 2003), with increased serum activities of γ-glutamyl transpeptidase (Mathews et al. 1988; Clough et al. 2003; Russmann et al. 2003), alkaline phosphatase (Clough et al. 2003) and alanine aminotransferase (Russmann et al. 2003) observed in some patients. In Western countries, kava is mainly used as ethanolic or acetonic rather than aqueous extracts for treatment of anxiety disorders (Stevinson et al. 2002; Pittler and Ernst 2003; Ernst 2004), and discussions have emerged regarding their possible hepatotoxic properties (BfArM 2002; Loew and Franz 2003; Schulze et al. 2003; Stickel et al. 2003; Teschke et al. 2003; Block et al. 2004; Clouatre 2004; Teschke et al. 2008a, 2008b). Unfortunately, higher portions of kavalactones are also associated with kava kava–induced liver failure (Martin et al. 2014). In 2003, when the Pacific kava paradox was first proposed (Currie and Clough 2003; Moulds and Malani 2003), the first case reports appeared that demonstrated hepatotoxicity due to kava use prepared traditionally as water extracts rather than by organic solvents. In 2007, the WHO kava hepatotoxicity study listed five cases due to aqueous kava extracts. Three cases were coded for kava with a possible causality and two had a probable causality; two of those cases were from traditionally prepared kava. The WHO report also referred to other cases of patients with liver disease, but a relation to the use of traditional aqueous kava extracts was not established using the WHO scale (Coulter et al. 2007). This report essentially confirmed that traditional aqueous kava extracts may exert rare potential hepatotoxicity that is similar to acetonic and ethanolic extracts. It is unclear at what doses were used and the period of use associated with the risk of liver damage. Both chronic and heavy use has been associated with neurotoxicity, pulmonary hypertension and dermatologic changes. The use of kava at frequent high doses causes kava dermopathy (Ruze 1990; Norton 1998).

There are still no specific dosages or extended user guidelines defined by the FDA for kava kava (FDA 2013). Decreased cognitive function (reduced visual attention accuracy) and awareness are frequent side effects of kava kava at higher doses, but it has also been shown to increase visual processing and working memory (Sarris et al. 2011). Previous trials have also demonstrated that kava has an absence of severe side effects if the dose remains under 400 mg of kavalactones per day (Lakhan and Vieira 2010), and it has increased effectiveness when taken by females and/or younger adults (Witte et al. 2005).

The mechanism of hepatotoxicity has not been clearly elucidated; however, several potential hepatotoxic constituents (i.e. pipermethystine, flavokavain B and mould hepatotoxins) and cofactors (i.e. alterations in hepatic microsomal cytochrome P450, cyclooxygenase inhibition, P-glycoprotein and glutathione) have been extensively reviewed (Teschke 2010a; Teschke et al. 2011). The author suggested that kava hepatotoxicity is partially preventable by quality control, prescription adherence and avoidance of comedications, because it occurs primarily due to daily overdose (exceeding 250 mg of kava lactones), prolonged treatment and the use of the kava plant's aerial parts, which may contain the hepatotoxic alkaloid piper methysticin and be contaminated kava raw material (Teschke and Schulze 2010; Teschke 2010b; Teschke et al. 2011). The US FDA began advising consumers on the potential risk of severe liver injury associated with the use of kava-containing dietary supplements in 2002, and kava-containing products have been banned from the markets of some countries in Europe, although they are still available in the United States, Canada, Australia, New Zealand and South Pacific Islands, as well as via the internet (Teschke and Schulze 2010).

The active components of kava kava, kavalactones have been shown to inhibit several CYPs, namely, CYP1A2, 2C9, 2C19, 2D6, 3A4 and 4A9/11 (Mathews et al. 2002; Zou et al. 2002; Unger et al. 2002). Hence, kava has a high potential to cause pharmacokinetic drug interactions with other medications (either herbal or not) that are metabolised by CYPs. Furthermore, there is also a possibility for pharmacodynamic interactions. An interaction with alprazolam reportedly caused a serious outcome in one patient (Almeida and Grimsley 1996). Kava possesses dopamine antagonistic properties (Schelosky et al. 1995), and cases of patients developing clinical signs suggestive of central dopaminergic antagonism have been described. The dopamine antagonistic properties of kava could explain the increase in the duration and number of 'off' periods in a patient with Parkinson's disease treated concomitantly with levodopa (Jamieson and Duffield 1990).

There is reportedly a synergistic interaction between alcohol and kava in terms of the hypnotic effect when they are administered together (Jamieson and Duffield 1990). It is not recommended to use kava with alcohol, although contrary data was also published several years later (Herberg 1993). Due to the above-mentioned inhibition of major drug-metabolising CYPs by kava kava, there may be interactions with critical drugs such as anticoagulants (e.g. aspirin, dipyridamole, warfarin, phenprocoumon). However, there has yet to be such a study (Mathews et al. 2002).

Because kava kava is frequently adulterated with *P. wichmannii* or *Macropiper latifolium*, the wild species, these may be present in the exported kava products (Coulter et al. 2007).

12.4.7 St. John's Wort (*Hypericum perforatum*)

St. John's wort (*H. perforatum*; SJW) has been traditionally used by many cultures; the first recordings date back to the ancient Greeks (Hammerness et al. 2003). Traditionally, SJW has been used to treat both internal and external illnesses in the form of oils and tinctures; it has remained a favourable treatment for anxiety, depression, wounds, burns, sunburns, ulcers, varicose veins, haemorrhoids, myalgia and sciatica, among others (WHO 2002). In terms of safety, SJW is superior over conventional antidepressants because it has fewer adverse effects (Linde et al. 1996; Di Carlo et al. 2001).

Chemical composition analyses of SJW yielded phloroglucinols (hyperforin etc.), naphthodianthrones (hypericin, etc.), xanthones in trace amounts, essential oil with 2-methyloctane, α-pinene and others terpenes, biflavones (biapigenin and amentoflavone), flavonoids (quercitrin, etc.), procyanidines, small amounts of chlorogenic acid and related substances. As has been emphasised for many plants, the actual concentrations of these components in preparations depend on various factors such as geographic origin of the plant, harvest of wild cultivated plants, harvest time, processing of fresh or dried plant material, lipophilicity and temperature of the extraction solvent and final formulation, exposure to sunlight, time and conditions of storage (Wölfle et al. 2014). In addition to a better safety profile as an alternative to conventional antidepressants, SJW has either no (Hubner et al. 1994; Brockmöller et al. 1997; Klier et al. 2002; Lecrubier et al. 2002) or few adverse effects in other uses such as gastrointestinal irritations, allergic reactions, tiredness, restlessness and sleep disturbance (Woelk et al. 1994; Sommer and Harrer 1994; Hänsgen et al. 1994; Ernst et al. 1998; Schempp et al. 2003a, 2003b). Among these, photosensitivity is a potentially serious adverse effect (Golsch et al. 1997; Bove 1998; Schempp et al. 2000), although it is rarely seen (Ernst 1999). Studies performed on animals and humans have suggested that doses much higher than the recommended daily dose is required for a serious phototoxic reaction in humans (Tyler et al. 1999; Bernd et al. 1999). In an *in vitro* analysis of the constituents, hypericin, pseudohypericin and hyperforin appeared to be responsible for phototoxic effects (Onoue et al. 2011).

SJW is an effective herbal medicine for mild to major depression (Linde et al. 2008). As mentioned above, SJW appears to be relatively safer compared to clinical antidepressants (Di Carlo et al. 2001; Linde et al. 2008). On the other hand, SJW is not free from pharmacokinetic interactions; the major CYPs in terms of drug metabolising capacity—CYP3A4, CYP2E1 and CYP2C19 and/or P-glycoprotein—are induced by the use of SJW (Zou et al. 2002; Izzo and Ernst 2009). Among the constituents, is responsible for the induction of both types of proteins. Due to this inductive effect, SJW clinically interacts with numerous drugs, including immunosuppressants, contraceptives, cardiovascular, antiretroviral and anticancer drugs, drugs acting on the CNS (such as antidepressants, anxiolytics, antiepileptics and muscle-relaxing agents), gastrointestinal tract and respiratory system, hypoglycaemics, anti-inflammatories, antimicrobials and antimigraine medicines (Izzo and Ernst 2009). Among these, interaction with cyclosporine may have serious clinical consequences. Due to CYP3A4 and P-glycoprotein induction, SJW decreases blood concentrations of cyclosporine and tacrolimus which, in some cases, may be associated with rejection of a transplanted organ (Dresser et al. 2003; Mai et al. 2004). The use of SJW also interacts with warfarin or phenprocoumon and results in decreased INR values (Ernst 1999; Kaminsky and Zhang 1997). It is likely that SJW reduces plasma warfarin concentrations through induction of CYP3A4 and possibly other CYP isoforms. Such interactions have been confirmed in clinical trials showing decreased plasma concentrations of warfarin and phenprocoumon (Jiang et al. 2004).

Another clinically important pharmacological groups of drugs that pharmacokinetically interact with SJW and result in decreased plasma concentrations are the P-glycoprotein substrate cardiotonic digoxin (Johne et al. 1999; Mueller et al. 2004; Gurley et al. 2008); CYP3A4-metabolised antiarrhythmic ivabradine (Portolés et al. 2006); the CYP3A4-metabolised calcium channel blockers nifedipine and verapamil (Wang et al. 2009; Tannergren et al. 2004); the P-glycoprotein substrate β-adrenoreceptor antagonist talinolol; the CYP3A4-metabolised and P-glycoprotein substrate antihyperlipidemic drugs simvastatin (Sugimoto et al. 2001) and atorvastatin (Andrén et al. 2007) and anticancer drug imatinib (Mathijssen 2002); the CYP3A4-metabolised benzodiazepines alprazolam, midazolam (Gurley et al. 2005; Gurley et al. 2002; Dresser et al. 2003; Hall et al. 2003; Mueller et al. 2006) and quazepam (Kawaguchi et al. 2004); CYP3A4-metabolised theophylline (also by CYP1A2 and CYP2E1; Morimoto et al. 2004); and CYP3A4-metabolised oral contraceptive (also metabolised by CYP1A2; Pfrunder et al. 2003; Murphy et al. 2005). SJW pharmacodynamically interacts with drugs that increase serotonin response in the brain. Therefore, SJW causes serotonin syndrome when taken simultaneously with serotonin reuptake inhibitors and/or serotonin receptor antagonists (Barbenel et al. 2000).

SJW is frequently adulterated with other *Hypericum* species such as *H. barbatum, H. hirsutum, H. maculatum, H. montanum, H. tetrapterum, H. patulum* and *H. crux-andreae*. Another frequent adulteration is adding bright colours to the SJW products to mislead customers; these include E123-amaranth, E133-brilliant blue, E110-sunset yellow, and E102-tartrazine (Booker et al. 2018).

12.4.8 Garlic (*Allium sativum* L.)

Garlic (*A. sativum*) is one of the oldest plants that has been used both as a remedy and as food and spice by many cultures throughout history. Although it originated from Asia, garlic has been spread all over the world and is currently cultivated in many countries. Crushed cloves are mostly used, whereas fresh tillers are also used by various means, especially as food. Pharmacological actions of garlic can be classified under five subheadings: antimicrobial (antibacterial, antifungal, antiparasitic and antiviral), cardiovascular (antihypertensive, anti-atherosclerotic, antithrombotic and action on blood factors), anticancer (antimutagenic), anti-inflammatory and immunomodulatory (Mikaili et al. 2013). The pungent and penetrating odour is due to the volatile sulphur-containing components, which are also responsible for the plant's pharmacological and biological actions. These components are allyl cysteine sulphoxide (alliin), which is instantly converted to diallyl dithiosulphinate (allicin) by the enzyme alliinase when garlic is crushed or chopped (Iberl et al. 1990; Londhe et al. 2011); diallyl disulphide (Avato et al. 2000); S-allylcysteine (Cruz et al. 2007); diallyl trisulphide (Avato et al. 2000); and ajoene (Weber et al. 1992).

Garlic has been used orally and topically throughout history, mostly to prevent and treat infections. Currently it is more often used to treat hypercholesterolaemia and hypertension and to prevent arteriosclerosis (Rotblatt and Ziment 2002; Barnes et al. 2002; Capasso et al. 2003; Schulz et al. 2004; Ernst et al. 2006). Although the unpleasant garlic breath and body odour associated with garlic intake limit its use (Stevinson 2000; Beck and Grunwald 1993; Schmidt and Schenk 1992), the popularity of garlic as a dietary supplement has increased over the last several decades.

The fresh form of garlic is a potent irritant; it can cause chemical burns (Dietz et al. 2004), as well as *in vivo* irritation in the alimentary tract after consumption of raw garlic, even after intake of garlic tablets (Barnes and Anderson 2002). Several clinical studies have confirmed gastrointestinal irritation after therapeutic doses of garlic (Sanchez-Hernandez et al. 2000; Ackermann et al. 2001). The responsible components appear to be heat sensitive, as topical or gastrointestinal irritation is only observed with raw garlic (Eming et al. 1999). Due to its topical or oral use, allergic contact dermatitis (Bojs and Svensson 1988; Burden et al. 1994), urticaria and angioedema (Asero et al. 1998; Sanchez-Hernandez et al. 2000; Pires et al. 2002), and anaphylaxis (Pérez-Pimiento et al. 1999) have been reported. Although relatively rare, occupational allergic reactions have also been observed, especially among cooks, practitioners of phytotherapy, workers in the food industry and domestic users. Both garlic and onion can provoke a type IV hypersensitivity reaction in sensitised individuals; 33 such food handlers developed occupational dermatoses (Hjorth and Roed-Petersen 1976). In one study, 44 of the 53 participants, mainly women, who had a history of contact dermatitis responded positively to garlic in a patch test (Sinha 1977). Similarly, 47 of 50 caterers who regularly handled garlic developed eczema (Cronin 1987). Garlic causes allergy by other routes as well; 7 out of 12 garlic workers showed signs of allergic bronchial asthma after exposed to garlic dust. Garlic sensitisation has been demonstrated by bronchial challenge and IgE tests (Añibarro et al. 1997). Garlic constituents responsible for allergies are mono-, di- and trisulphides, with diallyl disulphide having the highest activity (Garty 1993; Farrell and Staughton 1996).

Other than asthma, dyspnoea, cough and rhinitis due to garlic contact or inhalation have also been reported (Henson 1940; Falleroni et al. 1981; Lybarger et al. 1982; Couturier and Bousquet 1982; Seuri et al. 1993; Lemière et al. 1996; Armentia and Vega 1996; Alvarez et al. 2003).

An interesting allergic reaction was reported in a 56-year-old woman who had eczema on her lips triggered by sunlight. After a patch test with 68 ingredients of lip products and 13 additional common allergens, diallyl disulphide appeared as the causative agent (Scheman and Gupta 2001). This photoallergic reaction of diallyl disulphide was confirmed in a cohort several years later (Alvarez et al. 2003).

Garlic has antimicrobial activity against a wide range of gram-negative and gram-positive bacteria, including *Staphylococcus, Salmonella, Vibrio, Mycobacteria* and *Proteus* species (Abdullah et al. 1988). It also exhibits antifungal activity against *Candida albicans*, antiparasitic activity against *Schistosoma mansoni, Plasmodium falciparum* and *Trypanosoma brucei brucei* (Waag et al. 2010), *Entamoeba histolytica* and *Giardia lamblia* (Ankri and Mirelman 1999), and antiviral activity against cytomegalovirus (HCMV), influenza B virus, herpes simplex virus type 1 and 2, parainfluenza virus type 3, vaccinia virus, vesicular stomatitis virus and human rhinovirus type 2 (Ankri and Mirelman 1999). The antiviral effect has mostly been attributed to ajoene in garlic (Weber et al. 1992), whereas the others have been attributed to allicin.

The recent use of garlic is mainly for its actions on the cardiovascular system such as anti-hypertensive, antiplatelet, antithrombotic and fibrinolytic activities (Capasso et al. 2003; Ariga and Seki 2006; Rahman and Lowe 2006; Ernst et al. 2006). A direct decrease in blood pressure without the cholinergic pathway is suggested to involve a peripheral mechanism (Nwokocha et al. 2011). Similar data were obtained both in humans and in animals (Al-Qattan et al. 2006; Harauma and Moriguchi 2006). Excess amounts of garlic, both as food or tablets, resulted in alterations with coagulative process (Rose et al. 1990; Burnham 1995; German et al. 1995; Ackermann et al. 2001). These effects are due to decreased platelet aggregation and prolonged bleeding episodes (Bordia 1978). One of the mechanisms of platelet aggregation was reported as inhibition of calcium uptake by platelets (Srivastava 1986). Although it has been hypothesised that simultaneous garlic and warfarin use may pose serious haemorrhagic risks, two trials have reported contrary results (Macan et al. 2006; Mohammed Abdul et al. 2008). Another adverse interaction of garlic has been reported with the antiretroviral drug saquinavir: there was a significant decline in the plasma concentrations of this drug, with a consequent reduction in the drug's anti-AIDS efficacy, in healthy volunteers administered garlic for 3 weeks (Piscitelli et al. 2002). The mechanism of this interaction has been suggested to be the induction of intestinal CYP3A4, although experimental studies in humans did not confirm this hypothesis (Gurley et al. 2005; Cox et al. 2006). Saquinavir is also a P-glycoprotein substrate; however, this transporter is unlikely to be induced by garlic (Borrelli et al. 2007). Accumulating clinical data suggest that excess amounts of garlic should be avoid during pregnancy and lactation (Mennella and Beauchamp 1993; Ziaei et al. 2001).

BIBLIOGRAPHY

Abdullah, T. H., O. Kandil, A. Elkadi, and J. Carter. 1988. Garlic revisited: Therapeutic for the major diseases of our times?. *J Natl Med Assoc* 80(4): 439–45.

Abo-Khatwa, A., A. A. al-Robai, and D. A. al-Jawhari. 1996. Lichen acids as uncouplers of oxidative phosphorylation of mouse-liver mitochondria. *Nat Toxins* 4(2): 96–102.

Ackermann, R. T., C. D. Mulrow, G. Ramirez, C. D. Gardner, L. Morbidoni, and V. A. Lawrence. 2001. Garlic shows promise for improving some cardiovascular risk factors. *Arch Intern Med* 161(6): 813–24.

Adhikari, P. P., and S. B. Paul. 2018. History of Indian traditional medicine: A medical inheritance. *Asian J Pharm Clin Res* 11(1): 421.

Agbabiaka, T. B., M. H. Pittler, B. Wider, and E. Ernst. 2009. Serenoa Repens (Saw Palmetto): A systematic review of adverse events. *Drug Saf* 32(19): 637–47.

Aguilar, A., J. R. Camacho, S. Chino, P. Jácquez, and M. E. López. 1994. *Herbario Medicinal del Instituto Mexicano del Seguro Social. Información Etnobotanica* [Medicinal Herbarium of the Mexican Institute of Social Security. Ethnobotanical Information], 1st ed. México: Instituto Mexicano del Seguro Social.

Alakbarli, F. 2006. *Azerbaijan: Medieval Manuscripts. History of Medicine. Medicinal Plants.* Baku: Nurlan Press.

Alemdaroglu, N. C., U. Dietz, S. Wolffram, H. Spahn-Langguth, and P. Langguth. 2008. Influence of green and black tea on folic acid pharmacokinetics in healthy volunteers: Potential risk of diminished folic acid bioavailability. *Biopharm Drug Dispos* 29(6): 335–48.

Almeida, J. C., and E. W. E. Grimsley. 1996. Coma from the health food store: Interaction between Kava and Alprazolam. *Ann Intern Med* 125(11): 937–42.

Alostad, A. H., D. T. Steinke, and E. I. Schafheutle. 2018. International comparison of five herbal medicine registration systems to inform regulation development: United Kingdom, Germany, United States of America, United Arab Emirates and Kingdom of Bahrain. *Pharmaceut Med* 32(1): 39–49.

Al-Qattan, K. K., M. Thomson, S. Al-Mutawa'a, D. Al-Hajeri, H. Drobiova, and M. Ali. 2006. Nitric oxide mediates the blood-pressure lowering effect of garlic in the rat two-kidney, one-clip model of hypertension. *J Nutr* 136(3 Suppl): 774S-776S.

Alsarhan, A., N. Sultana, A. Al-Khatib, and M. R. Abdul Kadir. 2014. Review on some Malaysian traditional medicinal plants with therapeutic properties. *J Basic Appl Sci* 10: 149–59.

Alvarez, M. S., S. Jacobs, S. B. Jiang, R. R. Brancaccio, N. A. Soter, and D. E. Cohen. 2003. Photocontact allergy to diallyl disulfide. *Am J Contact Dermatitis* 14(3): 161–5.

Andrén, L., Å. Andreasson, and R. Eggertsen. 2007. Interaction between a commercially available St. John's Wort Product (Movina) and Atorvastatin in patients with hypercholesterolemia. *Eur J Clin Pharmacol* 63(10): 913–16.

Añibarro, B., J. L. Fontela, and F. De La Hoz. 1997. Occupational asthma induced by garlic dust. *J Allergy Clin Immunol* 100(6 Pt 1): 734–8.

Ankri, S., and D. Mirelman. 1999. Antimicrobial properties of allicin from garlic. *Microbes Infect* 1(2): 125–9.

Arenz, A., M. Klein, K. Fiehe, et al. 1996. Occurrence of neurotoxic 4'-O-methylpyridoxine in Ginkgo Biloba leaves, ginkgo medications and Japanese ginkgo food. *Planta Med* 62(6): 548–51.

Ariga, T., and T. Seki. 2006. Antithrombotic and anticancer effects of garlic-derived sulfur compounds: A review. *Biofactors* 26(2): 93–103.

Armentia, A., and J. M. Vega. 1996. Can inhalation of garlic dust cause asthma? *Allergy* 51(2): 137–8.

Aruna, D., and M. U. R. Naidu. 2007. Pharmacodynamic interaction studies of Ginkgo Biloba with Cilostazol and Clopidogrel in healthy human subjects. *Br J Clin Pharmacol* 63(3): 333–8.

Asero, R., G. Mistrello, D. Roncarolo, P. L. Antoniotti, and P. Falagiani. 1998. A case of garlic allergy. *J Allergy Clin Immunol* 101(3): 427–28.

Avato, P., E. Tursi, C. Vitali, V. Miccolis, and V. Candido. 2000. Allylsulfide constituents of garlic volatile oil as antimicrobial agents. *Phytomedicine* 7(3): 239–43.

Awang, D. V. C., and D. G. Kindack. 1989. Atropine as possible contaminant of comfrey tea. *Lancet* 2(8653): 44.

Baca-García, E., H. Blasco-Fontecilla, C. Blanco, C. Díaz-Sastre, M. M. Pérez-Rodríguez, and J. Sáiz-Ruiz. 2007. Acute atropine intoxication with psychiatric symptoms by herbal infusion of Pulmonaria Officinalis (Lungwort). *Eur J Psychiatr* 21(2): 93–7.

Barbenel, D. M., B. Yusufi, D. O'Shea, and C. J. Bench. 2000. Mania in a patient receiving testosterone replacement post-orchidectomy taking St John's Wort and sertraline. *J Psychopharmacol* 14(1): 84–6.

Barnes, J., L. A. Anderson, S. Gibbons, and J. D. Phillipson. 2005. Echinacea species (Echinacea angustifolia (DC.) Hell., Echinacea pallida (Nutt.) Nutt. Echinacea purpurea (L.) Moench): A review of their chemistry, pharmacology and clinical properties. *J Pharm Pharmacol* 57(8): 929–54.

Barnes, J., L. Anderson, and J. Phillipson. 2002. *Herbal Medicines*. London: Pharmaceutical Press.

Bauer, R., and H. Wagner. 1987. Neue Ergebnisse zur Analytik von Echinacea Wurzeln. *Sci Pharm* 55: 159–61.

Bauer, V. R., K. Jurcic, J. Puhlmann, and H. Wagner. 1988. Immunologic in vivo and in vitro studies on echinacea extracts. *Arzneimittelforschung* 38(2): 276–81.

Beck, E., and J. Grunwald. 1993. Allium Sativum in Der Stufentherapie Der Hyperlipidaemie. *Med Welt* 44: 516–20.

Benford, D., P. M. Bolger, P. Carthew, et al. 2010. Application of the Margin of Exposure (MoE) approach to substances in food that are genotoxic and carcinogenic. *Food Chem Toxicol* 48(Suppl 1): S2–24.

Bernd, A., S. Simon, A. Ramirez Bosca, et al. 1999. Phototoxic effects of Hypericum extract in cultures of human keratinocytes compared with those of psoralen. *Photochem Photobiol* 69(2): 218–21.

Bilgi, N., K. Bell, A. N. Ananthakrishnan, and E. Atallah. 2010. Imatinib and Panax Ginseng: A potential interaction resulting in liver toxicity. *Ann Pharmacother* 44(5): 926–28.

Bilia, A. R., S. Gallon, and F. F. Vincieri. 2002. Kava-Kava and anxiety: Growing knowledge about the efficacy and safety. *Life Sci* 70(22): 2581–97.

Bisset, N. G. 1989. Arrow and dart poisons. *J Ethnopharmacol* 25(1): 1–41.

Block, K. I., C. Gyllenhaal, and M. N. Mead. 2004. Safety and efficacy of herbal sedatives in cancer care. *Integr Cancer Ther* 3(2): 128–48.

Bojs, G., and A. Svensson. 1988. Contact allergy to garlic used for wound healing. *Contact Derm* 18(3): 179–81.

Booker, A., A. Agapouda, D. A. Frommenwiler, F. Scotti, E. Reich, and M. Heinrich. 2018. St John's Wort (Hypericum Perforatum) products—An assessment of their authenticity and quality. *Phytomedicine* 40: 158–64.

Bordia, A. 1978. Effect of garlic on human platelet aggregation in vitro. *Atherosclerosis* 30(4): 355–60.

Borgia, V., M. G. Carlin, and J. Crezzini. 2017. Poison, plants and palaeolithic hunters. An analytical method to investigate the presence of plant poison on archaeological artefacts. *Quatern Int* 427(Part B): 94–103.

Borrelli, F., R. Capasso, and A. A. Izzo. 2007. Garlic (Allium Sativum L.): Adverse effects and drug interactions in humans. *Mol Nutr Food Res* 51(11): 1386–97.

Botanicus. 2020. *Botanicus.org: Information about 'Köhler's Medizinal-Pflanzen in naturgetreuen Abbildungen mit kurz erläuterndem Texte: Atlas zur Pharmacopoea germanica, austriaca, belgica, danica, helvetica, hungarica, rossica, suecica, Neerlandica, British pharmacopoeia, zum Codex medicamentarius, sowie zur Pharmacopoeia of the United States of America /herausgegeben von G. Pabst.* www.botanicus.org/bibliography/b12428048 (accessed November 4, 2020).

Bove, G. M. 1998. Acute neuropathy after exposure to sun in a patient treated with St John's Wort. *Lancet* 352(9134): 1121–2.

Boyle, P., C. Robertson, F. Lowe, and C. Roehrborn. 2000. Meta-analysis of clinical trials of permixon in the treatment of symptomatic benign prostatic hyperplasia. *Urology* 55(4): 533–9.

Bradley, P. R. 1992. *A Handbook of Scientific Information on Widely Used Plant Drugs: Companion to Vol. 1 of the British Herbal Pharmacopoeia.* Bournemouth, Dorset: British Herbal Medicine Association.

Bressler, R. 2005. Herb-drug interactions: Interactions between ginseng and prescription medications. *Geriatrics* 60(8): 16–117.

Britannica. 2020. The Editors of Encyclopaedia Britannica. Traditional Chinese Medicine. In *Encyclopædia Britannica.* Encyclopedia Britannica, Inc., October 1, 2018. www.britannica.com/science/traditional-Chinese-medicine (accessed November 4, 2020).

Brockmöller, J., T. Reum, S. Bauer, R. Kerb, W. D. Hübner, and I. Roots. 1997. Hypericin and pseudohypericin: Pharmacokinetics and effects on photosensitivity in humans. *Pharmacopsychiatry* 30(2): 94–101.

Brown, A. C. 2017a. An overview of herb and dietary supplement efficacy, safety and government regulations in the United States with suggested improvements. Part 1 of 5 series. *Food Chem Toxicol* 107: 449–71.

Brown, A. C. 2017b. Kidney toxicity related to herbs and dietary supplements: Online table of case reports. Part 3 of 5 series. *Food Chem Toxicol* 107: 502–19.

Brown, A. C. 2017c. Liver toxicity related to herbs and dietary supplements: Online table of case reports. Part 2 of 5 series. *Food Chem Toxicol* 107: 472–501.

Brown, A. C. 2018a. Cancer related to herbs and dietary supplements: Online table of case reports. Part 5 of 5 series. *J Diet Suppl* 15(4): 556–81.

Brown, A. C. 2018b. Heart toxicity related to herbs and dietary supplements: Online table of case reports. Part 4 of 5 series. *J Diet Suppl* 15(4): 516–55.

Brown, D. J. 1996. Phytotherapy-herbal medicine meets clinical science. *NARD* 118: 41–52.

Bruminhent, J., P. Carrera, Z. Li, R. Amankona, and M. I. Roberts. 2011. Acute pancreatitis with Saw Palmetto use: A case report. *J Med Case Rep* 5: 414.

Bruneton, J. 1999. *Pharmacognosy, Phytochemistry, Medicinal Plants,* 2nd ed. Paris: Lavoisier Tec & Doc.

Bryan, C. P. 1930. *The Papyrus Ebers.* London: Geoffrey Bless.

Bryson, P. D., A. S. Watanabe, B. H. Rumack, and R. C. Murphy. 1978. Burdock root tea poisoning. Case report involving a commercial preparation. *JAMA* 239(20): 2157.

Bundesinstitut für Arzneimittel und Medizinprodukte (BfArM). 2002. Bonn: Federal Institute for Drugs and Medicinal Products in Germany. www.bfarm.de/de/arzneimittel/am sicher/stufenpl/Besch-kava-Final.pdf (accessed November 4, 2020).

Burden, A. D., S. M. Wilkinson, M. H. Beck, and R. J. Chalmers. 1994. Garlic-induced systemic contact dermatitis. *Contact Derm* 30(5): 299–300.

Burnham, B. E. 1995. Garlic as a possible risk for postoperative bleeding. *Plast Reconstr Surg* 95(1): 213.

Butt, M. S., R. S. Ahmad, M. T. Sultan, M. M. Qayyum, and A. Naz. 2015. Green tea and anticancer perspectives: Updates from last decade. *Crit Rev Food Sci Nutr* 55(6): 792–805.

Cagin, Y. F., Y. Seckin, F. Firat, and E. Samdanci. 2017. Acute toxic hepatitis induced by a herbal medicine: Anchusa Boraginaceae. *Acta Gastroenterol Belg* 80(4): 533–6.

Capasso, F., T. S. Gaginella, G. Grandolini, and A. A. Izzo. 2003. *Phytotherapy: Quick Reference to Herbal Medicine*. Berlin: Springer.

Caron, M. F., A. L. Hotsko, S. Robertson, L. Mandybur, J. Kluger, and C. M. White. 2002. Electrocardiographic and hemodynamic effects of Panax Ginseng. *Ann Pharmacother* 36(5): 758–63.

Carraturo, F., O. D. Castro, J. Troisi, et al. 2018. Comparative assessment of the quality of commercial black and green tea using microbiology analyses. *BMC Microbiol* 18(1): 4.

Chaly, Z. A., I. B. Sedova, and M. G. Kiseleva. 2018. Occurrence of mycotoxins in tea. *Prob Nutr* 87: 199–200.

Champault, G., A. M. Bonnard, J. Cauquil, and J. C. Patel. 1984. Medical treatment of prostatic adenoma. Controlled trial: PA 109 vs placebo in 110 patients. *Ann Urol* 18(6): 407–10.

Chan, P. C., Q. Xia, and P. P. Fu. 2007. Ginkgo Biloba leave extract: Biological, medicinal, and toxicological effects. *J Environ Sci Heal C* 25(3): 211–44.

Charrois, T. L., J. Hrudey, and S. Vohra. 2006. Ginseng: Practical management of adverse effects and drug interactions. *Can Pharm J* 139(2): 44–6.

Cheema, P., O. El-Mefty, and A. R. Jazieh. 2001. Intraoperative haemorrhage associated with the use of extract of Saw Palmetto herb: A case report and review of literature. *J Int Med* 250(2): 167–69.

Chen, C. H., K. G. Dickman, M. Moriya, et al. 2012. aristolochic acid-associated urothelial cancer in Taiwan. *Proc Natl Acad Sci USA* 109(21): 8241–46.

Cheng, T. O. 2007. Green tea may inhibit warfarin. *Int J Cardiol* 115(2): 236.

Cho, J., W. Park, S. Lee, W. Ahn, and Y. Lee. 2004. Ginsenoside-Rb1 from Panax Ginseng C.A. Meyer activates estrogen receptor-α and -β, independent of ligand binding. *J Clin Endocrinol Metab* 89(7): 3510–5.

Church, R. J., D. M. Gatti, T. J. Urban, et al. 2015. Sensitivity to hepatotoxicity due to Epigallocatechin Gallate is affected by genetic background in diversity outbred mice. *Food Chem Toxicol* 76: 19–26.

Cianfrocca, C., F. Pelliccia, A. Auriti, and M. Santini. 2002. Ginkgo Biloba-induced frequent ventricular arrhythmia. *Ital Heart J* 3(11): 689–91.

Clark, J. D., J. L. Phillips, and P. S. Staley. 1974. Interpretations of prehistoric technology from Ancient Egyptian and other sources. Part I: Ancient Egyptian bows and arrows and their relevance for African prehistory. *Paléorient* 2(2): 323–88.

Clouatre, D. L. 2004. Kava Kava: Examining new reports of toxicity. *Toxicol Lett* 150(1): 85–96.

Clough, A. R., R. S. Bailie, and B. Currie. 2003. Liver function test abnormalities in users of aqueous Kava extracts. *J Toxicol—Clin Toxicol* 41(6): 821–29.

Coon, J. T., and E. Ernst. 2002. Panax Ginseng: A systematic review of adverse effects and drug interactions. *Drug Saf* 25(5): 323–44.

Coulter, D., C. Tamayo, S. Sotheeswaran, and C. Ulbricht. 2007. *Assessment of the Risk of Hepatotoxicity with Kava Products*. WHO Library Cataloguing-in-Publication Data, 1–82. https://apps.who.int/iris/bitstream/handle/10665/43630/9789241595261_eng.pdf?sequence=1&isAllowed=y (accessed November 4, 2020).

Couturier, P., and J. Bousquet. 1982. Occupational allergy secondary inhalation of garlic dust. *J Allergy Clin Immunol* 70(2): 145.

Cox, M. C., J. Low, J. Lee, et al. 2006. Influence of garlic (Allium Sativum) on the pharmacokinetics of docetaxel. *Clin Cancer Res* 12(15): 4636–40.

Cronin, E. 1987. Dermatitis of the hands in caterers. *Contact Derm* 17(5): 265–9.

Cruz, C., R. Correa-Rotter, D. J. Sánchez-González, et al. 2007. Renoprotective and antihypertensive effects of S-Allylcysteine in 5/6 nephrectomized rats. *Am J Physiol Renal* 293(5): F1691–98.

Currie, B. J., and A. R. Clough. 2003. Kava hepatotoxicity with Western herbal products: Does It occur with traditional Kava use? *Med J Austr* 178(9): 421–22.

D'Errico, F., L. Backwell, P. Villa, et al. 2012a. Reply to Evans: Use of poison remains the most parsimonious explanation for Border Cave castor bean extract. *Proc Natl Acad Sci USA* 109(48): 3291–2.

D'Errico, F., L. Backwell, P. Villa, et al. 2012b. Early evidence of san material culture represented by organic artifacts from Border Cave, South Africa. *Proc Natl Acad Sci USA* 109(33): 13214–9.

Darmesteter, J. 1898. *Sacred Books of the East*. New York: The Christian Literature Company.

David, A. R. 2008. *Egyptian Mummies and Modern Science*. Cambridge: Cambridge University Press.

Denham, A., M. McIntyre, and J. Whitehouse. 2002. Kava: The unfolding story: Report on a work-in-progress. *J Altern Complement Med* 8(3): 237–63.

Di Carlo, G., F. Borrelli, E. Ernst, and A. A. Izzo. 2001. Prozac from the plant kingdom. *Trends Pharmacol Sci* 22(6): 292–7.

Di Lorenzo, C., A. Ceschi, H. Kupferschmidt, et al. 2015. Adverse effects of plant food supplements and botanical preparations: A systematic review with critical evaluation of causality. *Br J Clin Pharmacol* 79(4): 578–92.

Di Silverio, F., G. D'Eramo, C. Lubrano, et al. 1992. Evidence that Serenoa Repens extract displays an antiestrogenic activity in prostatic tissue of benign prostatic hypertrophy patients. *Eur Urol* 21(4): 309–14.

Diamond, B. J., S. C. Shiflett, N. Feiwel, et al. 2000. Ginkgo Biloba extract: Mechanisms and clinical indications. *Arch Phys Med Rehab* 81(5): 668–78.

Dias, T. R., M. G. Alves, G. D. Tomás, S. Socorro, B. M. Silva, and P. F. Oliveira. 2014. White tea as a promising antioxidant medium additive for sperm storage at room temperature: A Comparative study with green tea. *J Agric Food Chem* 62(3): 608–17.

Dietz, D. M., J. R. Varcelotti, and K. R. Stahlfeld. 2004. Garlic burns: A not-so-rare complication of a naturopathic remedy?. *Burns* 30(6): 612–3.

Dodge, H. H., T. Zitzelberger, B. S. Oken, D. Howieson, and J. Kaye. 2008. A randomized placebo-controlled trial of Ginkgo Biloba for the prevention of cognitive decline. *Neurology* 70(19 Pt 2): 1809–17.

Dresser, G. K., U. I. Schwarz, G. R. Wilkinson, and R. B. Kim. 2003. coordinate induction of both cytochrome P4503A and MDR1 by St John's Wort in healthy subjects. *Clin Pharmacol Ther* 73(1): 41–50.

Dugoua, J. J., E. Mills, D. Perri, and G. Koren. 2006. Safety and efficacy of Ginkgo (Ginkgo Biloba) during pregnancy and lactation. *Can J Clin Pharmacol* 13(3): e277–84.

Durgnat, J. M., J. Heuser, D. Andrey, and C. Perrin. 2005. Quality and safety assessment of Ginseng extracts by determination of the contents of pesticides and metals. *Food Addit Contam* 22(12): 1224–30.

Ebo, D. G., C. H. Bridts, M. M. Hagendorens, N. E. Aerts, L. S. De Clerck, and W. J. Stevens. 2008. Basophil activation test by flow cytometry: Present and future applications in allergology. *Cytometry B Clin Cytom* 74(4): 201–10.

Eming, S. A., J. O. Piontek, N. Hunzelmann, H. Rasokat, and K. Scharffetter-Kachanek. 1999. Severe toxic contact dermatitis caused by garlic. *Br J Dermatol* 141(2): 391–2.

Engelberg, D., A. McCutcheon, and S. Wiseman. 2001. A case of ginseng-induced mania. *J Clin Psychopharmacol* 21(5): 535–7.

Ernst, E. 1999. Second thoughts about safety of St. John's Wort. *Lancet* 354(9195): 2014–16.

Ernst, E. 2002. The risk-benefit profile of commonly used herbal therapies: Ginkgo, St. John's Wort, Ginseng, Echinacea, Saw Palmetto, and Kava. *Ann Int Med* 136(1): 42–53.

Ernst, E. 2004. Kava update: A European perspective. *N Z Med J* 117(1205): U1143.

Ernst, E. 2006. Herbal remedies for anxiety—A systematic review of controlled clinical trials. *Phytomedicine* 13(3): 205–8.

Ernst, E., M. H. Pittler, and B. Wider. 2006. *The Desktop Guide to Complementary and Alternative Medicine: An Evidence-Based Approach*, 2nd ed. London: Mosby Elsevier.

Ernst, E., J. I. Rand, J. Barnes, and C. Stevinson. 1998. Adverse effects profile of the herbal antidepressant St. John's Wort (Hypericum Perforatum L.). *Eur J Clin Pharmacol* 54(8): 589–94.

Ettre, L. S., and K. I. Sakodynskii. 1993. Tswett and the discovery of chromatography II: Completion of the development of chromatography (1903–1910). *Chromatographia* 35: 329–38.

European Commission. 1965. Council Directive 65/66/EEC. *Official Journal of the European Communities* 369: 20–24. http://eur-lex.europa.eu/legal-content/EN/TXT/PDF/?uri=CELEX:31965L0065&from=EN (accessed November 4, 2020).

European Commission. 2001. Informations and notices. *Official Journal of the European Union* 44. https://eur-lex.europa.eu/legal-content/EN/TXT/PDF/?uri=OJ:C:2001:328:FULL&from=EN (accessed November 4, 2020).

European Scientific Cooperative on Phytotherapy (ESCOP). 2003. *ESCOP Monographs The Scientific Foundation for Herbal Medicinal Products*, 2nd ed. Exeter, UK: ESCOP, Stuttgart: Georg Thieme Verlag, New York: Thieme New York.

Evans, A. A. 2012. Arrow poisons in the palaeolithic? *Proc Natl Acad Sci USA* 109(48): 3290.

Falleroni, A. E., C. R. Zeiss, and D. Levitz. 1981. Occupational asthma secondary to inhalation of garlic dust. *J Allergy Clin Immunol* 68(2): 156–60.

Farrell, A. M., and R. C. Staughton. 1996. Garlic burns mimicking herpes zoster. *Lancet* 347(9009): 1195.

Favreau, J. T., M. L. Ryu, G. Braunstein, et al. 2002. Severe hepatotoxicity associated with the dietary supplement LipoKinetix. *Ann Intern Med* 136(8): 590–5.

FDA, U.S. Food and Drug Administration. 2002. *Letter to Health-Care Professionals: FDA Issues Consumer Advisory that Kava Products May be Associated with Severe Liver Injury*. Rockville, MD: U.S. Department of Health and Human Services. https://ods.od.nih.gov/HealthInformation/kava.aspx.

FDA, U.S. Food and Drug Administration. 2013. *Safety Alerts for Human Medical Products: Kava (piper methysticum)*. Silver Spring, MD: U.S. Department of Health and Human Services.

Foster, S. 2016. Towards an understanding of ginseng adulteration: The tangled web of names, history, trade, and perception. HerbalGram. *American Botanical Council* 111: 36–57.

Freeman, C., and K. Spelman. 2008. A critical evaluation of drug interactions with Echinacea Spp. *Mol Nutr Food Res* 52(7): 789–98.

Fu, P. P., Q. Xia, L. Guo, H. Yu, and P. C. Chan. 2008. Toxicity of Kava Kava. *J Environ Sci Heal C* 26(1): 89–112.

Fugh-Berman, A. 2000. Herb-drug interactions. *Lancet* 355: 134–8.

Fujiki, H., E. Sueoka, T. Watanabe, and M. Suganuma. 2015. Primary cancer prevention by green tea, and tertiary cancer prevention by the combination of green tea catechins and anticancer compounds. *J Cancer Prev* 20(1): 1–4.

Gafner, S. 2018. Adulteration of Ginkgo Biloba leaf extract. *Botanical Adulterants Bulletin* 1–8.

Galizia, E. J. 1983. Clinical curio: Hallucinations in elderly tea drinkers. *Br Med J* 287: 979.

Gallo, M., and G. Koren. 2001. Can herbal products be used safely during pregnancy? Focus on Echinacea. *Can Fam Phys* 47: 1727–8.

Gallo, M., M. Sarkar, W. Au, et al. 2000. Pregnancy outcome following gestational exposure to echinacea: A prospective controlled study. *Arch Int Med* 160(20): 3141–3.

García-Cortés, M., M. Robles-Díaz, A. Ortega-Alonso, I. Medina-Caliz, and R. J. Andrade. 2016. Hepatotoxicity by dietary supplements: A tabular listing and clinical characteristics. *Int J Mol Sci* 17(4): 537.

Gardner, C. D., J. L. Zehnder, A. J. Rigby, J. R. Nicholus, and J. W. Farquhar. 2007. Effect of Ginkgo Biloba (EGb 761) and Aspirin on platelet aggregation and platelet function analysis among older adults at risk of cardiovascular disease: A randomized clinical trial. *Blood Coagul Fibrinolysis* 18(8): 787–93.

Garty, B. Z. 1993. Garlic burns. *Pediatrics* 91(3): 658–9.

Gerber, G. S., D. Kuznetsov, B. C. Johnson, and J. D. Burstein. 2001. Randomized, double-blind, placebo-controlled trial of Saw Palmetto in men with lower urinary tract symptoms. *Urology* 58(6): 960–4.

German, K., U. Kumar, and H. N. Blackford. 1995. Garlic and the risk of TURP bleeding. *Br J Urol* 76(4): 518.

Ghalioungui, P. 1975. Les plus anciennes femmes-médicins de l'histoire [The oldest female doctors in history]. *Bull Inst Fr Archeol Orient* 75: 159–64.

Gloro, R., I. Hourmand-Ollivier, B. Mosquet, et al. 2005. Fulminant hepatitis during self-medication with hydroalcoholic extract of green tea. *Eur J Gastroenterol Hepatol* 17(10): 1135–7.

Godin M., J.-P. Fillastre, P. Simon, A. Francois, F. Le Roy, and J.-P. Morin. 1997. Is ochratoxin A a nephrotoxic in human beings? *Adv Nephrol Necker Hosp* 26: 181–206.
Gökmen, M. R., J. P. Cosyns, V. M. Arlt, et al. 2013. The epidemiology, diagnosis, and management of aristolochic. *Ann Int Med Rev* 158(6): 469–77.
Golsch, S., E. Vocks, J. Rakoski, K. Brockow, and J. Ring. 1997. Reversible Erhöhung der Photosensitivität Im UV-B-Bereich Durch Johanniskrautextrakt-Präparate. *Der Hautarzt* 48: 249–52.
Gorski, J. C., S. M. Huang, A. Pinto, et al. 2004. The effect of echinacea (Echinacea Purpurea Root) on cytochrome P450 activity in vivo. *Clin Pharmacol Ther* 75(1): 89–100.
Greenblatt, D. J., L. L. Von Moltke, Y. Luo, et al. 2006. Ginkgo Biloba does not alter clearance of flurbiprofen, a cytochrome P450–2C9 substrate. *J Clin Pharmacol* 46(2): 214–21.
Gregory, P. J. 2001. Seizure associated with Ginkgo Biloba? *Ann Int Med* 134(4): 342.
Grosse, Y., D. Loomis, B. Lauby-Secretan, et al. 2013. Carcinogenicity of some drugs and herbal products. *Lancet Oncol* 14(9): 807–8.
Gurley, B. J., S. F. Gardner, M. A. Hubbard, et al. 2002. Cytochrome P450 Phenotypic ratios for predicting herb-drug interactions in humans. *Clin Pharmacol Ther* 72(3): 276–87.
Gurley, B. J., S. F. Gardner, M. A. Hubbard, et al. 2004. In vivo assessment of botanical supplementation on human Cytochrome P450 Phenotypes: Citrus Aurantium, Echinacea Purpurea, Milk Thistle, and Saw Palmetto. *Clin Pharmacol Ther* 76(5): 428–40.
Gurley, B. J., S. F. Gardner, M. A. Hubbard, et al. 2005. Clinical assessment of effects of botanical supplementation on Cytochrome P450 Phenotypes in the elderly. *Drugs Aging* 22(6): 525–39.
Gurley, B. J., A. Swain, M. A. Hubbard, et al. 2008. Clinical assessment of cyp2d6-mediated herb-drug interactions in humans: Effects of milk thistle, Black Cohosh, Goldenseal, Kava Kava, St. John's Wort, and Echinacea. *Mol Nutr Food Res* 52(7): 755–63.
Haas, D., B. Pfeifer, C. Reiterich, R. Partenheimer, B. Reck, and W. Buzina. 2013. Identification and quantification of fungi and mycotoxins from Pu-Erh Tea. *Int J Food Microbiol* 166(2): 316–22.
Halioua, B. 2005. *Medicine in the Days of the Pharaohs*. Cambridge, MA: Belknap Press of Harvard University Press.
Hall, S. D., Z. Wang, S. M. Huang, et al. 2003. The interaction between St John's Wort and an oral contraceptive. *Clin Pharmacol Ther* 74(6): 525–35.
Hamed, A. E. 2009. *Medicine in Old Egypt*. Edited and prepared by Prof. Hamed A. Ead. www.levity.com/alchemy/islam22.html (accessed November 4, 2020).
Hammerness, P., E. Basch, C. Ulbricht, et al. 2003. St. John's Wort: A systematic review of adverse effects and drug interactions for the consultation psychiatrist. *Psychosomatics* 44(4): 271–82.
Hänsgen, K. D., J. Vesper, and M. Ploch. 1994. Multicenter double-blind study examining the antidepressant effectiveness of the Hypericum Extract LI 160. *J Geriatr Psychiatr Neurol* 7(Suppl 1): S15–8.
Harauma, A., and T. Moriguchi. 2006. Aged garlic extract improves blood pressure in spontaneously hypertensive rats more safely than raw garlic. *J Nutr* 136(Suppl 3): 769S-73S.
Harrison-Dunn, A. 2016. *Green Tea Extracts May Cause Liver Damage, Norway Warns*. www.nutraingredients.com/Article/2016/03/18/Green-tea-extracts-may-cause-liver-damage-Norway-warns# (accessed November 4, 2020).
Hasegawa, S., Y. Oda, T. Ichiyama, Y. Hori, and S. Furukawa. 2006. Ginkgo nut intoxication in a 2-year-old male. *Pediatr Neurol* 35(4): 275–6.
Hegazy, S. 2014. The effect of green tea on Sildenafil pharmacokinetics in Egyptian healthy volunteers. *Br J Pharm Res* 4(3): 289–300.
Heinrich, M., J. Barnes, S. Gibbons, and M. E. Williamson. 2012. *Fundamentals of Pharmacognosy and Phytotherapy*, 2nd ed. London: Churchill Livingstone.
Henson, G. E. 1940. Garlic: An occupational factor in the etiology of bronchial asthma. *JFMA* 27: 86.
Herberg, K. W. 1993. Effect of Kava-special extract WS 1490 combined with ethyl alcohol on safety-relevant performance parameters. *Blutalkohol* 30(2): 96–105.
Hjorth, N., and J. Roed-Petersen. 1976. Occupational protein contact dermatitis in food handlers. *Contact Derm* 2(1): 28–42.
Hobbs, C. 1994. Echinacea: A literature review; botany, history, chemistry, pharmacology, toxicology, and clinical uses. The Journal of the American Botanical Council. *HerbalGram* 30: 33.
Holbrook, A. M., J. A. Pereira, R. Labiris, H. McDonald, J. D. Douketis, and M. Crowther. 2005. Systematic overview of warfarin and its drug and food interactions. *Arch Int Med* 165(10): 1095–106.

Hori, Y., M. Fujisawa, K. Shimada, A. Oda, S. Katsuyama, and K. Wada. 2004. Rapid analysis of 4-O-methylpyridoxine in the serum of patients with Ginkgo Biloba seed poisoning by ion-pair high-performance liquid chromatography. *Biol Pharm Bull* 27(4): 486–91.

Hotta, E., R. Tamagawa-Mineoka, and N. Katoh. 2013. Allergic contact dermatitis due to ginkgo tree fruit and leaf. *Eur J Dermatol* 23(4): 548–9.

Hsu, C. K., P. Leo, D. Shastry, W. Meggs, R. Weisman, and R. S. Hoffman. 1995. Anticholinergic poisoning associated with herbal tea. *Arch Int Med* 155: 2245–8.

Hubner, W. D., S. Lande, and H. Podzuweit. 1994. Hypericum treatment of mild depressions with somatic symptoms. *J Geriatr Psychiatr Neurol* 7(Suppl 1): 12–14.

IARC (International Agency for Research on Cancer). 2012. *Monographs on the Evaluation of Carcinogenic Risks to Humans*. Lyon, France: International Agency for Research on Cancer. https://publications.iarc.fr/Book-And-Report-Series/Iarc-Monographs-On-The-Identification-Of-Carcinogenic-Hazards-To-Humans/Biological-Agents-2012 (accessed November 4, 2020).

Iberl, B., G. Winkler, B. Muller, and K. Knobloch. 1990. Quantitative determination of allicin and alliin from garlic by HPLC. *Planta Med* 56(3): 320–6.

Ingólfsdóttir, K. 2002. Usnic acid. *Phytochemistry* 61(7): 729–36.

Isomura, T., S. Suzuki, H. Origasa, et al. 2016. Liver-related safety assessment of green tea extracts in humans: A systematic review of randomized controlled trials. *Eur J Clin Nutr* 1221–9.

Ivic, M. 1969. Etiology of endemic nephropathy. *Lijec Vjesn* 91: 1273–81.

Izzo, A. A., G. Di Carlo, F. Borrelli, and E. Ernst. 2005. Cardiovascular pharmacotherapy and herbal medicines: The risk of drug interaction. *Int J Cardiol* 98(1): 1–14.

Izzo, A. A., and E. Ernst. 2009. Interactions between herbal medicines and prescribed drugs: An Updated systematic review. *Drugs* 69(13): 1777–98.

Jaggy, H., and E. Koch. 1997. Chemistry and biology of alkylphenols from ginkgo biloba L. *Pharmazie* 52(10): 735–8.

Jamieson, D. D., and P. H. Duffield. 1990. Positive interaction of ethanol and Kava resin in mice. *Clin Exp Pharmacol Physiol* 17(7): 509–14.

Janetzky, K., and A. P. Morreale. 1997. Probable interaction between warfarin and ginseng. *Am J Health Syst Pharm* 54(119): 692–3.

Jiang, X., K. M. Williams, W. S. Liauw, et al. 2004. Erratum: Effect of St John's Wort and ginseng on the pharmacokinetics and pharmacodynamics of warfarin in healthy subjects. *Br J Clin Pharmacol* 58(1): 102.

Jibrin, I., A. Erinle, A. Saidi, and Z. Y. Aliyu. 2006. Saw Palmetto-induced pancreatitis. *South Med J* 99(6): 611–2.

Ju, X. B., X. J. Gu, Z. Y. Zhang, et al. 2015. Efficacy and safety of Saw Palmetto extract capsules in the treatment of benign prostatic hyperplasia. *Zhonghua Nan Ke Xue* 21(12): 1098–101.

Jussofie, A., A. Schmiz, and C. Hiemke. 1994. Kavapyrone enriched extract from piper methysticum as modulator of the GABA binding site in different regions of rat brain. *Psychopharmacology* 116(4): 469–74.

Kabalak, A. A., O. B. Soyal, A. Urfalioglu, F. Saracoglu, and N. Gogus. 2004. Menometrorrhagia and tachyarrhythmia after using oral and topical ginseng. *J Womens Health* 13(7): 830–3.

Kajiyama, Y., K. Fujii, H. Takeuchi, and Y. Manabe. 2001. Ginkgo seed poisoning. *Pediatrics* 109(2): 325–27.

Kakisaka, Y., T. Ohara, H. Tozawa, et al. 2012. Panax Ginseng: A newly identified cause of gynecomastia. *Tohoku J Exp Med* 228(2): 143–5.

Kaminsky, L. S., and Z. Y. Zhang. 1997. Human P450 metabolism of warfarin. *Pharmacol Ther* 73(1): 67–74.

Kästner, U., C. Hallmen, M. Wiese, E. Leistner, and C. Drewke. 2007. The human pyridoxal kinase, a plausible target for ginkgotoxin from Ginkgo biloba. *FEBS J* 274(4): 1036–45.

Kawaguchi, A., M. Ohmori, S. I. Tsuruoka, et al. 2004. Drug interaction between St John's Wort and Quazepam. *Br J Clin Pharmacol* 58(4): 403–10.

Keio University. 2020. *Kampo Virtual Class: Japanese Kampo Medicine*. Kampo Igaku Center, Keio University Medical School. www.keio-kampo.jp/vc/education1_1.html (accessed November 4, 2020).

Kemp, D. E., and K. N. Franco. 2002. Possible leukopenia associated with long-term use of echinacea. *J Am Board Fam Med* 15(5): 417–9.

Keske, M. A., H. L. Ng, D. Premilovac, et al. 2015. Vascular and metabolic actions of the green tea polyphenol Epigallocatechin Gallate. *Curr Med Chem* 22(1): 59–69.

Kiefer, D., and T. Pantuso. 2003. Panax Ginseng. *Am Fam Physician* 68(8): 1539–42.
Kim, K. M., H. S. Kwon, S. G. Jeon, et al. 2008. Korean Ginseng-induced occupational asthma and determination of IgE Binding components. *J Korean Med Sci* 23(2): 232–5.
Kim, T. E., N. Ha, Y. Kim, et al. 2017. Effect of epigallocatechin-3-gallate, major ingredient of green tea, on the pharmacokinetics of rosuvastatin in healthy volunteers. *Drug Des Devel Ther* 11: 1409–16.
Kimura, Y., H. I. Okuda, and S. Arichi. 1988. Effects of various ginseng saponins on 5-hydroxytryptamine release and aggregation in human platelets. *J Pharm Pharmacol* 40(12): 838–43.
Klier, C. M., M. R. Schäfer, B. Schmid-Siegel, G. Lenz, and M. Mannel. 2002. St. John's Wort (Hypericum Perforatum) is it safe during breastfeeding? *Pharmacopsychiatry* 35(1): 29–30.
Klingelhöfer, D., Y. Zhu, M. Braun, M. H. K. Bendels, D. Brüggmann, and D. A. Groneberg. 2018. Aflatoxin—Publication analysis of a global health threat. *Food Control* 89: 280–90.
Kobayashi, D., T. Yoshimura, A. Johno, M. Ishikawa, K. Sasaki, and K. Wada. 2015. Decrease in pyridoxal-5′-phosphate concentration and increase in pyridoxal concentration in rat plasma by 4′-O-methylpyridoxine administration. *Nutr Res* 35(7): 637–42.
Kumar, B., M. Vijayakumar, R. Govindarajan, and P. Pushpangadan. 2007. Ethnopharmacological approaches to wound healing-exploring medicinal plants of India. *J Ethnopharmacol* 114(2): 103–13.
Lakhan, S. E., and K. F. Vieira. 2010. Nutritional and herbal supplements for anxiety and anxiety-related disorders: Systematic review. *Nutr J* 9(1): 1–14.
Lapi, F., E. Gallo, E. Giocaliere, et al. 2010. Acute liver damage due to Serenoa Repens: A case report. *Br J Clin Pharmacol* 69(5): 558–60.
Lau, W. S., R. Yat-Kan-Chan, D. A. Guo, and M. S. Wong. 2008. Ginsenoside Rg1 exerts estrogen-like activities via ligand-independent activation of ERα pathway. *J Steroid Biochem Mol Biol* 108(1–2): 64–71.
Lauterwein, M., M. Oethinger, K. Belsner, T. Peters, and R. Marre. 1995. In vitro activities of the lichen secondary metabolites Vulpinic Acid, (+)-Usnic Acid, and (-)-Usnic Acid against aerobic and anaerobic microorganisms. *Antimicrob Agents Chemother* 39(11): 2541–3.
Lebot, V., M. D. Merlin, and L. Lindstrom. 1997. *Kava-the Pacific Elixir: The Definitive Guide to Its Ethnobotany, History, and Chemistry*. Rochester, VT: Healing Arts Press.
Lecrubier, Y., G. Clerc, R. Didi, and M. Kieser. 2002. Efficacy of St. John's Wort extract WS 5570 in major depression: A double-blind, placebo-controlled trial. *Am J Psychiatr* 159(8): 1361–6.
Lee, A. N., and V. P. Werth. 2004. Activation of autoimmunity following use of immunostimulatory herbal supplements. *Arch Dermatol* 140(6): 723–7.
Lee, J. Y., H. J. Jin, J. W. Park, S. K. Jung, J. Y. Jang, and H. S. Park. 2012. A case of Korean Ginseng-induced anaphylaxis confirmed by open oral challenge and basophil activation test. *Allergy Asthma Immunol Res* 4(2): 110–1.
Lee, J. Y., Y. D. Lee, J. W. Bahn, and H. S. Park. 2006. A case of occupational asthma and rhinitis caused by Sanyak and Korean Ginseng dusts. *Eur J Allergy Clin Immunol* 61(3): 392–3.
Lee, S. H., Y. M. Ahn, S. Y. Ahn, H. K. Doo, and B. C. Lee. 2008. Interaction between Warfarin and Panax Ginseng in ischemic stroke patients. *J Altern Complement Med* 14(6): 715–21.
Legeay, S., M. Rodier, L. Fillon, S. Faure, and N. Clere. 2015. Epigallocatechin Gallate: A review of its beneficial properties to prevent metabolic syndrome. *Nutrients* 7(7): 5443–68.
Leistner, E., and C. Drewke. 2010. Reviews Ginkgo Biloba and ginkgotoxin. *J Nat Prod* 73: 86–92.
Lemière, C., A. Cartier, S. B. Lehrer, and J. L. Malo. 1996. Occupational asthma caused by aromatic herbs. *Eur J Allergy Clin Immunol* 51(9): 647–9.
Li, W., K. Xu, R. Xiao, G. Yin, and W. Liu. 2015. Development of an HPLC-based method for the detection of aflatoxins in Pu-Erh tea. *Int J Food Prop* 18(4): 842–8.
Liao, W. I., Y. Y. Lin, S. J. Chu, C. W. Hsu, and S. H. Tsai. 2010. Bradyarrhythmia caused by ginseng in a patient with chronic kidney disease. *Am J Emerg Med* 538: e5–538.e6.
Liatsos, G., I. Elefsiniotis, R. Todorova, and A. Moulakakis. 2006. Severe Thrombotic Thrombocytopenic Purpura (TTP) induced or exacerbated by the immunostimulatory herb echinacea. *Am J Hematol* 81(3): 224.
Linde, K., G. Ramirez, C. D. Mulrow, A. Pauls, W. Weidenhammer, and D. Melchart. 1996. St John's Wort for depression: An overview and meta-analysis of randomised clinical trials. *Br Med J* 313(7052): 253–8.
Linde, K., M. M. Berner, and L. Kriston. 2008. St. John's Wort for major depressive disorder: A systematic review. *Cochrane Database Syst Rev* 8(4): 1–56.

Liu, F. X., and J. W. Salmon. 2010. Comparison of herbal medicines regulation between China, Germany, and the United States. *Integr Med* 9: 54–61.

Liu, Q., L. Zhuo, L. Liu, et al. 2011. Seven cases of fatal aconite poisoning: Forensic experience in China. *Forensic Sci Int* 212(1–3): e5–9.

Loew, D., and G. Franz. 2003. Quality aspects of traditional and industrial Kava-extracts. *Phytomedicine* 10(6–7): 610–2.

Logan, J. L., and J. Ahmed. 2003. Critical hypokalemic renal tubular acidosis due to Sjögren's syndrome: Association with the purported immune stimulant echinacea. *Clin Rheum* 22(2): 158–9.

Londhe, V. P., A. T. Gavasane, S. S. Nipate, D. D. Bandawane, and P. D. Chaudhari. 2011. Role of Garlic (Allium Sativum) in various diseases: An overview. *J Pharm Res* 4(1): 129–34.

Lowe, F. C., and J. C. Ku. 1996. Phytotherapy in treatment of benign prostatic hyperplasia: A critical review. *Urology* 48(1): 12–20.

Lybarger, J. A., J. S. Gallagher, D. W. Pulver, A. Litwin, S. Brooks, and I. L. Bernstein. 1982. Occupational asthma induced by inhalation and ingestion of garlic. *J Allergy Clin Immunol* 69(5): 448–54.

Macan, H., R. Uykimpang, M. Alconcel, et al. 2006. Aged garlic extract may be safe for patients on warfarin therapy. *J Nutr* 136(3): 793S-5S.

Mai, I., S. Bauer, E. S. Perloff, et al. 2004. Hyperforin content determines the magnitude of the St John's Wort-cyclosporine drug interaction. *Clin Pharmacol Ther* 76(4): 330–40.

Markowitz, J. S., J. L. Donovan, C. L. DeVane, R. M. Taylor, Y. Ruan, J. S. Wang, et al. 2003. Multiple doses of Saw Palmetto (Serenoa Repens) did not alter cytochrome P450 2D6 and 3A4 activity in normal volunteers. *Clin Pharmacol Ther* 74(6): 536–42.

Marks, L. S. 2004. 5alpha-reductase: History and clinical importance. *Rev Urol* 6(9): S11–21.

Martin, A. C., E. Johnston, C. Xing, and A. D. Hegeman. 2014. Measuring the chemical and cytotoxic variability of commercially available Kava (Piper Methysticum G. Forster). *PLoS ONE* 9(11): e111572.

Martinez-Mir, I., E. Rubio, F. J. Morales-Olivas, and V. Palop-Larrea. 2004. Transient ischemic attack secondary to hypertensive crisis related to Panax Ginseng. *Ann Pharmacother* 38(11): 1969–70.

Maskatia, Z. K., and K. Baker. 2010. Hypereosinophilia associated with echinacea use. *South Med J* 103(11): 1173–4.

Mathews, J. D., M. D. Riley, L. Fejo, et al. 1988. Effects of the heavy usage of Kava on physical health: Summary of a pilot survey in an Aboriginal community. *Med J Austr* 148(11): 548–55.

Mathews, J. M., A. S. Etheridge, and S. R. Black. 2002. Inhibition of human cytochrome P450 activities by Kava extract and kavalactones. *Drug Metab Dispos* 30(11): 1153–7.

Mathijssen, R. H., J. Verweij, P. de Bruijn, W. J. Loos, and A. Sparreboom. 2002. Effects of St. John's wort on irinotecan metabolism. *J Natl Cancer Inst* 21;94(16): 1247–9.

Matthews, J. R., S. L. Lewis, and J. Rowin. 1998. Association of Ginkgo Biloba with intracerebral hemorrhage. *Neurology* 50(6): 1933–34.

Matthias, A., K. G. Penman, K. M. Bone, and R. P. Lehmann. 2004. *Echinacea—What Constituents are Therapeutically Important*? International Congress on Natural Product Research, Phoenix, AZ, 43 (abstract).

Mauro, V. F., L. S. Mauro, J. F. Kleshinski, S. A. Khuder, Y. Wang, and P. W. Erhardt. 2003. Impact of Ginkgo Biloba on the pharmacokinetics of digoxin. *Am J Ther* 10(4): 247–51.

Mazzanti, G., F. Menniti-Ippolito, P. A. Moro, et al. 2009. Hepatotoxicity from green tea: A review of the literature and two unpublished cases. *Eur J Clin Pharmacol* 65(4): 331–41.

Meisel, C., A. Johne, and I. Roots. 2003. Fatal intracerebral mass bleeding associated with Ginkgo Biloba and Ibuprofen. *Atherosclerosis* 167(2): 367.

Mennella, J. A., and G. K. Beauchamp. 1993. The effects of repeated exposure to garlic-flavored milk on the nursling's behaviour. *Pediatr Res* 34(6): 805–8.

Mikaili, P., S. Maadirad, M. Moloudizargari, S. Aghajanshakeri, and S. Sarahroodi. 2013. Therapeutic uses and pharmacological properties of garlic, shallot, and their biologically active compounds. *Iran J Basic Med Sci* 16: 1031–48.

Miroddi, M., A. Carnì, C. Mannucci, M. Moleti, M. Navarra, and G. Calapai. 2012. Hot flashes in a young girl: A wake-up call concerning serenoa repens use in children. *Pediatrics* 130(5): e1374–6.

Misaka, S., J. Yatabe, F. Müller, et al. 2014. Green tea ingestion greatly reduces plasma concentrations of nadolol in healthy subjects. *Clin Pharmacol Ther* 95(4): 432–8.

Miwa, H., M. Iijima, S. Tanaka, and Y. Mizuno. 2001. Generalized convulsions after consuming a large amount of gingko nuts. *Epilepsia* 42(2): 280–1.
Mizuno, N., K. Kawakami, and E. Morita. 1980. Competitive Inhibition between 4'-substituted pyridoxine analogues and pyridoxal for pyridoxal kinase from mouse brain. *J Nutr Sci Vitaminol* 26(6): 535–43.
Mohammed Abdul, M. I., X. Jiang, K. M. Williams, et al. 2008. Pharmacodynamic interaction of warfarin with cranberry but not with garlic in healthy subjects. *Br J Pharmacol* 154(8): 1691–700.
Mohutsky, M. A., G. D. Anderson, J. W. Miller, and G. W. Elmer. 2006. Ginkgo Biloba: Evaluation of CYP2C9 drug interactions in vitro and in vivo. *Am J Ther* 13(1): 24–31.
Morabito, P., M. Miroddi, S. Giovinazzo, E. Spina, and G. Calapai. 2015. Serenoa Repens as an endocrine disruptor in a 10-year-old young girl: A new case report. *Pharmacology* 96(1–2): 41–3.
Morimoto, T., T. Kotegawa, K. Tsutsumi, Y. Ohtani, H. Imai, and S. Nakano. 2004. Effect of St. John's Wort on the pharmacokinetics of theophylline in healthy volunteers. *J Clin Pharmacol* 44(1): 95–101.
Moulds, R. F. W., and J. Malani. 2003. Kava: Herbal panacea or liver poison? *Med J Aust* 178(9): 451–3.
Mousa, M. S., A. Sephien, J. Gutierrez, and C. O'Leary. 2018. N-acetylcysteine for acute hepatitis induced by Kratom herbal tea. *Am J Ther* 25(5): e550–e551.
Mueller, S. C., J. Majcher-Peszynska, B. Uehleke, et al. 2006. The extent of induction of CYP3A by St. John's Wort varies among products and is linked to hyperforin dose. *Eur J Clin Pharmacol* 62(1): 29–36.
Mueller, S. C., B. Uehleke, H. Woehling, et al. 2004. Effect of St John's Wort dose and preparations on the pharmacokinetics of digoxin. *Clin Pharmacol Ther* 75(6): 546–57.
Mullins, R. J. 1998. Echinacea-associated anaphylaxis. *Med J Aust* 168(4): 170–71.
Mullins, R. J., and H. Robert. 2002. Adverse reactions associated with echinacea: The Australian experience. *Ann Allergy Asthma Immunol* 88(1): 42–51.
Muraviova, D. A. 1978. *Farmakognoziya*. Moscow: Meditzina.
Murphy, P. A., S. E. Kern, F. Z. Stanczyk, and C. L. Westhoff. 2005. Interaction of St. John's Wort with oral contraceptives: Effects on the pharmacokinetics of norethindrone and ethinyl estradiol, ovarian activity and breakthrough bleeding. *Contraception* 71(6): 402–8.
Murray, M., and J. Pizzorno. 1994. *Encyclopedia of Natural Medicine*. Seattle, WA: John Bastyr University Publishing.
Murugusundram, S. 2009. Serenoa Repens: Does it have any role in the management of androgenetic alopecia? *J Cutan Aesthet Surg* 2(1): 31–2.
Newman, M. G., S. J. Llera, T. M. Erickson, A. Przeworski, and L. G. Castonguay. 2013. Worry and generalized anxiety disorder: A review and theoretical synthesis of evidence on nature, etiology, mechanisms, and treatment. *Annu Rev Clin Psychol* 9(1): 275–97.
Nieri, P., B. Adinolfi, I. Morelli, M. C. Breschi, G. Simoni, and E. Martinotti. 2003. Genetic Characterization of the three medicinal echinacea species using RAPD analysis. *Planta Med* 69(7): 685–6.
Nordeng, H., and G. C. Havnen. 2004. Use of herbal drugs in pregnancy: A survey among 400 Norwegian women. *Pharmacoepidemiol Drug Saf* 13(6): 371–80.
Nortier, J. L., M. C. Muniz Martinez, H. H. Schmeiser, et al. 2000. Urothelial carcinoma associated with the use of a Chinese herb (Aristolochia Fangchi). *N Engl J Med* 342(23): 1686–92.
Norton, S. A. 1998. Herbal medicines in Hawaii from tradition to convention. *Hawaii Med J* 57(1): 382–6.
Norton, S. A., and P. Ruze. 1994. Kava dermopathy. *J Am Acad Dermatol* 31(1): 89–97.
Nougayrol, J. 1952. Ningirsu Vainqueur De Zû. [Ningirsu Winner of Zû]. *Revue d'Assyriologie et d'archéologie orientale* 46(2): 87–97.
Nunn, J. F. 1996. *Ancient Egyptian Medicine*. London: British Museum Press.
Nutescu, E. A., N. L. Shapiro, S. Ibrahim, and P. West. 2006. Warfarin and its interactions with foods, herbs and other dietary supplements. *Expert Opin Drug Saf* 5(3): 433–51.
Nwokocha, C. R., R. I. Ozolua, D. U. Owu, M. I. Nwokocha, and A. C. Ugwu. 2011. Antihypertensive properties of Allium Sativum (Garlic) on normotensive and two kidney one clip hypertensive rats. *Niger J Physiol Sci* 26(2): 213–8.
Ocollura, J. 1997. Ginseng tonic of life. *Vegetarian Times* 235: 94.

Oerlemans, C., I. de Vries, and A. J. H. P. van Riel. 2017. Anticholinergic syndrome caused by contaminated herbal tea; acting swiftly to identify the source. *Ned Tijdschr Geneeskd* 161: D1261 (In Dutch).

Onoue, S., Y. Seto, M. Ochi, et al. 2011. In vitro photochemical and phototoxicological characterization of major constituents in St. John's Wort (Hypericum Perforatum) extracts. *Phytochemistry* 72(14–15): 1814–20.

Paik, D. J., and C. H. Lee. 2015. Review of cases of patient risk associated with ginseng abuse and misuse. *J Ginseng Res* 39(2): 89–93.

Palop-Larrea, V., J. L. Gonzálvez-Perales, C. Catalán-Oliver, A. Belenguer-Varea, and I. Martínez-Mir. 2000. Metrorrhagia and ginseng. *Ann Pharmacother* 34(11): 1347–8.

Pan, S. Y., G. Litscher, S. H. Gao, et al. 2014. Historical perspective of traditional indigenous medical practices: The current renaissance and conservation of herbal resources. *Evid Based Complement Alternat Med* 525340.

Park, H. J., J. H. Lee, Y. B. Song, and K. H. Park. 1996. Effects of dietary supplementation of lipophilic fraction from Panax Ginseng on CGMP and CAMP in rat platelets and on blood coagulation. *Biol Pharm Bull* 19(11): 1434–9.

Parpola, A. 1988. The coming of the Aryans to Iran and India and the cultural and ethnic identity of the Dasas. *Studia Orientalia Electronica* 64: 195–302.

Pennisi, R. S. 2006. Acute generalised exanthematous pustulosis induced by the herbal remedy Ginkgo Biloba. *Med J Aust* 184(11): 583–4.

Pérez-Pimiento, A. J., I. Moneo, M. Santaolalla, S. de Paz, B. Fernández-Parra, and A. R. Domínguez-Lázaro. 1999. Anaphylactic reaction to young garlic allergy. *Eur J Allergy Clin Immunol* 54(6): 626–9.

Perri, D., J. J. Dugoua, E. Mills, and G. Koren. 2006. Safety and efficacy of echinacea (Echinacea Angustafolia, e. Purpurea and e. Pallida) during pregnancy and lactation. *J Clin Pharmacol* 13(3): e262–7.

Pfrunder, A., M. Schiesser, S. Gerber, M. Haschke, J. Bitzer, and J. Drewe. 2003. Interaction of St John's Wort with low-dose oral contraceptive therapy: A randomized controlled trial. *Br J Clin Pharmacol* 56(6): 683–90.

Pires, G., E. Pargana, V. Loureiro, M. M. Almeida, and J. R. Pinto. 2002. Allergy to garlic. *Allergy* 57(10): 957–68.

Piscitelli, S. C., A. H. Burstein, N. Welden, K. D. Gallicano, and J. Falloon. 2002. The effect of garlic supplements on the pharmacokinetics of Saquinavir. *Clin Infect Dis* 34(2): 234–8.

Pittler, M. H., and E. Ernst. 2000. Efficacy of Kava extract for treating anxiety: Systematic review and meta-analysis. *J Clin Psychopharmacol* 20(1): 84–9.

Pittler, M. H., and E. Ernst. 2003. Kava extract for treating anxiety. *Cochrane Database Sys Rev* (1): CD003383.

Pluess, M., A. Conrad, and F. H. Wilhelm. 2009. Muscle tension in generalized anxiety disorder: A critical review of the literature. *J Anxiety Disord* 23(1): 1–11.

Portolés, A., A. Terleira, A. Calvo, I. Martínez, and G. Resplandy. 2006. Effects of hypericum perforation on ivabradine pharmacokinetics in healthy volunteers: An open-label, pharmacokinetic interaction clinical trial. *J Clin Pharmacol* 46(10): 1188–94.

Pouretedal, Z., and M. Mazaheri. 2013. Aflatoxins in black tea in Iran. *Food Addit Contam Part B* 127–9.

Rahman, K., and G. M. Lowe. 2006. Garlic and cardiovascular disease: A critical review. *J Nutr* 136(3): 736S-40S.

Reddy, S., P. Mishra, S. Qureshi, S. Nair, and T. Straker. 2016. Hepatotoxicity due to red bush tea consumption: A case report. *J Clin Anesth* 35: 96–8.

Rhoads, P. M., T. G. Tong, W. Banner, and R. Anderson. 1984. Anticholinergic poisonings associated with commercial Burdock root tea. *J Toxicol Clin Toxicol* 22(6): 581–4.

Rininger, J. A., S. Kickner, P. Chigurupati, A. McLean, and Z. Franck. 2000. Immunopharmacological activity of echinacea preparations following simulated digestion on murine macrophages and human peripheral blood mononuclear cells. *J Leukoc Biol* 68(4): 503–10.

Riswan, S., and H. Sangat-Remantyo. 2002. Jamu as traditional medicine in Java, Indonesia. *South Pacific Study* 23(1): 1–10.

Robertson, S. M., R. T. Davey, J. Voell, E. Formentini, R. M. Alfaro, and S. R. Penzak. 2008. Effect of Ginkgo Biloba extract on Lopinavir, Midazolam and Fexofenadine pharmacokinetics in healthy subjects. *Curr Med Res Opin* 24(2): 591–9.

Rose, K. D., P. D. Croissant, C. F. Parliament, and M. B. Levin. 1990. Spontaneous spinal epidural hematoma with associated platelet dysfunction from excessive garlic ingestion: A case report. *Neurosurgery* 26(5): 880–2.
Rosenblatt, M., and J. Mindel. 1997. Spontaneous hyphema associated with ingestion of Ginkgo Biloba extract. *N Eng J Med* 336(15): 1101–9.
Rotblatt, M., and I. Ziment. 2002. *Evidence Based Herbal Medicine*. Philadelphia: Hanley & Belfus.
Routledge, P. A., and T. L. B. Spriggs. 1989. Atropine as possible contaminant of Comfrey tea. *Lancet* 1(8644): 963–4.
Russmann, S., Y. Barguil, P. Cabalion, M. Kritsanida, D. Duhet, and B. H. Lauterburg. 2003. Hepatic injury due to traditional aqueous extracts of Kava root in New Caledonia. *Eur J Gastroenterol Hepatol* 15(9): 1033–6.
Russo, V., A. Rago, G. M. Russo, R. Calabrò, and G. Nigro. 2011. Ginkgo Biloba: An ancient tree with new arrhythmic side effects. *J Postgrad Med* 57(3): 221.
Ruze, P. 1990. Kava-induced dermopathy: A niacin deficiency? *Lancet* 335(8703): 1442–5.
Salvador, R. L. 1995. Herbal medicine—Ginkgo. *Can Pharm J* 52: 39–41.
Samuels, N., Y. Finkelstein, S. R. Singer, and M. Oberbaum. 2008. Herbal medicine and epilepsy: Proconvulsive effects and interactions with antiepileptic drugs. *Epilepsia* 49(3): 373–80.
Sanchez-Hernandez, M. C., M. Hernandez, J. Delgado, et al. 2000. Allergenic cross-reactivity in the Liliaceae family. *Eur J Allergy Clin Immunol* 55(3): 297–9.
Sarma, D. N., M. L. Barrett, M. L. Chavez, et al. 2008. Safety of green tea extracts: A systematic review by the US pharmacopeia. *Drug Saf* 31(6): 469–84.
Sarris, J., A. Panossian, I. Schweitzer, C. Stough, and A. Scholey. 2011. Herbal medicine for depression, anxiety and insomnia: A review of psychopharmacology and clinical evidence. *Eur Neuropsychopharmacol* 21(12): 841–60.
Sarris, J., E. Laporte, and I. Schweitzer. 2011. Kava: A comprehensive review of efficacy, safety, and psychopharmacology. *Aust N Z J Psychiatry* 45(1): 27–35.
Savovic, J., B. Wider, and E. Ernst. 2005. Effects of Ginkgo Biloba on blood coagulation parameters. *Evid Based Complement Alternat Med* 2(3): 167–76.
Schelosky, L., C. Raffauf, K. Jendroska, and W. Poewe. 1995. Kava and dopamine antagonism. *J Neurol Neurosurg Psychiatry* 58(5): 639–40.
Scheman, A. J., and S. Gupta. 2001. Photoallergic contact dermatitis from Diallyl Disulfide. *Contact Derm* 45(3): 179.
Schempp, C. M., R. Lüdtke, B. Winghofer, and J. C. Simon. 2000. Effect of topical application of Hypericum Perforatum extract (St. John's Wort) on skin sensitivity to solar simulated radiation. *Photodermatol Photoimmunol Photomed* 16(3): 125–8.
Schempp, C. M., B. Winghofer, K. Müller, et al. 2003b. Effect of oral administration of Hypericum Perforatum extract (St. John's Wort) on skin erythema and pigmentation induced by UVB, UVA, visible light and solar simulated radiation. *Phytother Res* 17(2): 141–6.
Schempp, C. M., T. Windeck, S. Hezel, and J. C. Simon. 2003a. Topical treatment of atopic dermatitis with St. John's Wort cream: A randomized, placebo controlled, double blind half-side comparison. *Phytomedicine* 10(4): 31–7.
Schmeiser, H. H., C. A. Bieler, M. Wiessler, C. van Ypersele de Strihou, and J. P. Cosyns. 1996. Detection of DNA adducts formed by aristolochic acid in renal tissue from patients with Chinese herbs nephropathy. *Cancer Res* 56: 2025–8.
Schmidt, U., and N. Schenk, 1992. *Geruchsbildung bei repetiert Einnahme von standardisierten Knoblauchpulver-Dragees Kwai (LI 111) in Abbhangigkeit von der Tagesdosis*. Wissenschaftlicher Bericht.
Scholz, H., S. Kascha, and H. Zingerle. 1980. Atropine poisoning from "health tea". *Fortschr Med* 98: 1525–6.
Schulz, V., R. Hänsel, M. Blumenthal, and V. E. Tyler. 2004. Medicinal plants, phytomedicines, and phytotherapy. In *Rational Phytotherapy*. Berlin: Springer.
Schulze, J., W. Raasch, and C. P. Siegers. 2003. Toxicity of Kava pyrones, drug safety and precautions—A case study. *Phytomedicine* 10(4): 68–73.
Scirpa, P., G. Scambia, V. Masciullo, et al. 1999. A zinc sulfate and usnic acid preparation used as post-surgical adjuvant therapy in genital lesions by Human Papillomavirus. *Minerva Ginecol* 51(6): 255–60.
Sedova, I., M. Kiseleva, and V. Tutelyan. 2018. Mycotoxins in tea: Occurrence, methods of determination and risk evaluation. *Toxins* 10(11). pii: E444.

Seuri, M., A. Taivanen, P. Ruoppi, and H. Tukiainen. 1993. Three cases of occupational asthma and rhinitis caused by garlic. *Clin Exp Allergy* 23(12): 1011–4.

SFDA. 2007. *Provisions for Drug Registration*. Normative Documents, 1–41. http://english.nmpa.gov.cn/2019-07/25/c_390615.htm (accessed November 4, 2020).

Shaw, D., C. Leon, S. Kolev, and V. Murray. 1997. Traditional remedies and food supplements. *Drug Saf* 17(5): 342–56.

Shikov, A. N., O. N. Pozharitskaya, V. G. Makarov, H. Wagner, R. Verpoorte, and M. Heinrich. 2014. Medicinal plants of the Russian pharmacopoeia; Their history and applications. *J Ethnopharmacol* 154(3): 481–536.

Shivamurthy, P., N. Brar, and M. L. Therrien. 2014. Isolated hemopericardium associated with rivaroxaban: First case report. *Pharmacotherapy* 34(9): e169–72.

Siegel, R. K. 1979. Ginseng abuse syndrome: Problems with the panacea. *J Am Med Assoc* 241(15): 1614–5.

Singh, Y. N. 1992. Kava: An overview. *J Ethnopharmacol* 37(1): 13–45.

Singh, Y. N. 2005. Potential for interaction of Kava and St. John's Wort with drugs. *J Ethnopharmacol* 100(1–2): 108–13.

Sinha, S. M. 1977. Vegetables responsible for contact dermatitis of the hands. *Arch Dermatol* 113(6): 776–9.

Smith, K., and C. Leiras. 2018. The effectiveness and safety of kava kava for treating anxiety symptoms: A systematic review and analysis of randomized clinical trials. *Complement Ther Clin Pract* 33: 107–17.

Soedibyo, M. 1989. *Falsafah Jamu sebagai Titik Pengembangan Obat Tradisional* [Herbal Philosophy as a Traditional Drug Development Point]. Paper presented on Simposium Kosmetika dan Obat Tradisional II, Departemen Kesehatan, Jakana.

Soedibyo, M. 1990. *Javanese Traditional Medicine*. Paper presented on the International Congress Traditional Medicine and Medicinal Plants, Denpasar, Bali, October 15–17, Indonesia.

Solovieva, V. A. 2005. *Traditional Methods of Health Promotion* [Narodnye metody ukreplenija zdorovja], 352p. St-Petersburg: Izdatelsky dom "Neva".

Sommer, H., and G. Harrer. 1994. Placebo-controlled double-blind study examining the effectiveness of an hypericum preparation in 105 mildly depressed patients. *J Geriatr Psych Neurol* 7(1): 9–11.

Soon, S. L., and R. I. Crawford. 2001. Recurrent erythema nodosum associated with echinacea herbal therapy. *J Am Acad Dermatol* 44(2): 298–9.

Srivastava, K. C. 1986. Evidence for the mechanism by which garlic inhibits platelet aggregation. *Prostaglandins Leukot Med* 22(3): 313–21.

Starvaggi, C. L., M. Motta, P. Accornero, and M. Baratta. 2008. Effect of Echinacea Augustifolia extract on cell viability and differentiation in mammary epithelial cells. *Phytomedicine* 15(8): 555–62.

Stevinson, C. 2000. Garlic for treating hypercholesterolemia. *Ann Int Med* 133(6): 420.

Stevinson, C., A. Huntley, and E. Ernst. 2002. A systematic review of the safety of Kava extract in the treatment of anxiety. *Drug Saf* 25(4): 251–61.

Sticher, O. 1998. Getting to the root of ginseng. *Chemtech* 28: 26–32.

Stickel, F., H. M. Baumüller, K. Seitz, et al. 2003. Hepatitis induced by Kava (Piper Methysticum Rhizoma). *J Hepatol* 39(1): 62–7.

Sugimoto, K. I., M. Ohmori, S. Tsuruoka, et al. 2001. Different effects of St John's Wort on the pharmacokinetics of Simvastatin and Pravastatin. *Clin Pharmacol Ther* 70(6): 518–24.

Sutarjadi. 1990. *Pengkajian dan Pengembangan Ilmu tentang Jamu Indonesia* [The Study and Development of Knowledge on Herbal Herbs in Indonesia]. Paper presented on Simposium Meningkatkan Citra dan Manfaat Jamu dan Obat traditional serta Keamanannya terhadap KesehatanTubuh Manusia. Universitas Airlangga, Surabaya.

Tannergren, C., H. Engman, L. Knutson, M. Hedeland, U. Bondesson, and H. Lennernäs. 2004. St John's Wort decreases the bioavailability of R- and S-Verapamil through induction of the first-pass metabolism. *Clin Pharmacol Ther* 75(4): 298–309.

Tatu, C. A., W. H. Orem, R. B. Finkelman, and G. L. Feder. 1998. The etiology of Balkan endemic nephropathy: Still more questions than answers. *Environ Health Perspect* 106(11): 689–700.

Taylor, J. A., W. Weber, L. Standish, et al. 2003. Efficacy and Safety of echinacea in treating upper respiratory tract infections in children: A randomized controlled trial. *J Am Med Assoc* 290(21): 2824–30.

Taylor, J. R., and V. M. Wilt. 1999. Probable antagonism of warfarin by green tea. *Ann Pharmacother* 33(4): 426–8.
Teng, L., Q. Zu, G. Li, et al. 2016. Herbal medicines: Challenges in the modern world. Part 3. China and Japan. *Exp Rev Clin Pharmacol* 9(9): 1225–33.
Tenore, G. C., M. Daglia, R. Ciampaglia, and E. Novellino. 2015. Exploring the nutraceutical potential of polyphenols from black, green and white tea infusions—An overview. *Curr Pharmacol Biotechnol* 16(3): 265–71.
Teschke, R. 2010a. Kava hepatotoxicity: Pathogenetic aspects and prospective considerations. *Liver Int* 30(9): 1270–9.
Teschke, R. 2010b. Kava hepatotoxicity. A clinical review. *Ann Hepatol* 9(3): 251–65.
Teschke, R., A. Schwarzenboeck, and A. Akinci. 2008a. Kava hepatotoxicity: A European view. *N Z Med J* 121(1283): 90–8.
Teschke, R., A. Schwarzenboeck, and K. H. Hennermann. 2008b. Kava hepatotoxicity: A clinical survey and critical analysis of 26 suspected cases. *Eur J Gastroenterol Hepatol* 20(12): 1182–93.
Teschke, R., A. Schwarzenboeck, W. Schmidt-Taenzer, A. Wolff, and K. Heinz Hennermann. 2011. Herb induced liver injury presumably caused by Black Cohosh: A survey of initially purported cases and herbal quality specifications. *Ann Hepatol* 10(3): 249–59.
Teschke, R., W. Gaus, and D. Loew. 2003. Kava extracts: Safety and risks including rare hepatotoxicity. *Phytomedicine* 10(5): 440–6.
Teschke, R., and J. Schulze. 2010. Risk of Kava hepatotoxicity and the FDA consumer advisory. *J Am Med Assoc* 304(19): 2174–5.
Torbey, E., N. A. Rafeh, G. Khoueiry, M. Kowalski, and S. Bekheit. 2011. Ginseng: A potential cause of long QT. *J Electrocardiol* 44(3): 357–8.
Tswett, M. S. 1905. On a new category of adsorption phenomena and on its application to biochemical analysis (Polish). *Proc Warsaw Soc Natur* 14(6): 20–39.
Tu, Y. 2016. Artemisinin: A gift from traditional Chinese medicine to the world (Nobel Lecture). Angew *Chem Int Ed Engl* 55(35): 10210–26.
Tyler, V. E. 1993. *The Honest Herbal: A Sensible Guide to the Use of Herbs and Related Remedies.* New York: Pharmaceutical Products Pr.
Tyler, V. E., M. Blumenthal, and W. R. Busse. 1999. *The Complete German Commission e Monographs: Therapeutic Guide to Herbal Medicines.* Austin, TX: American Botanical Council.
Uchida, S., H. Yamada, D. L. Xiao, et al. 2006. Effects of Ginkgo Biloba extract on pharmacokinetics and pharmacodynamics of Tolbutamide and Midazolam in healthy volunteers. *J Clin Pharmacol* 46(11): 1290–8.
Unger, M., U. Holzgrabe, W. Jacobsen, C. Cummins, and L. Z. Benet. 2002. Inhibition of cytochrome P450 3A4 by extracts and kavalactones of Piper Methysticum (Kava-Kava). *Planta Med* 68(12): 1055–8.
Valdivia-Correa, B., C. Gómez-Gutiérrez, M. Uribe, and N. Méndez-Sánchez. 2016. Herbal medicine in Mexico: A cause of hepatotoxicity. A critical review. *Int J Mol Sci* 17(2): 1–10.
Van Beek, T. A., and P. Montoro. 2009. Chemical analysis and quality control of Ginkgo Biloba leaves, extracts, and phytopharmaceuticals. *J Chromatogr A* 1216(11): 2002–32.
Van Riel, A. J. H. P., J. Meulenbelt, and I. de Vries. 2014. Anticholinergic poisoning after drinking contaminated herbal tea. *Clin Toxicol* 52: 383.
Vanhaelen, M., R. Vanhaelen-Fastre, P. But, and J. L. Vanherweghem. 1994. Identification of aristolochic acid in Chinese herbs. *Lancet* 343(8890): 174.
Vanherweghem, J. L., M. Depierreux, C. Tielemans, et al. 1993. Rapidly progressive interstitial renal fibrosis in young women: Association with slimming regimen including Chinese herbs. *Lancet* 341(8842): 387–91.
Vasseur, M., J. Thierry, F. V. Defeudis, and K. Drieu. 1994. Effects of repeated treatments with an extract of Ginkgo Biloba (EGb 761), Bilobalide and Ginkgolide B on the electrical activity of pancreatic β Cells of normal or alloxan-diabetic mice: An ex vivo study with intracellular microelectrodes. *Gen Pharmacol* 25(1): 31–46.
Vázquez, I., and L. F. Aguera-Ortiz. 2002. Herbal products and serious side effects: A case of ginseng-induced manic episode. Acta *Psychiatr Scand* 105(1): 76–7.
Vela-Navarrete, R., A. Alcaraz, A. Rodríguez-Antolín, et al. 2018. Efficacy and safety of a hexanic extract of Serenoa Repens (Permixon®) for the Treatment of Lower Urinary Tract Symptoms Associated with Benign Prostatic Hyperplasia (LUTS/BPH): Systematic review and meta-analysis of randomised controlled trials and observations. *BJU Int* 122(6): 1049–65.

Vijayakumar, C. S., S. Viswanathan, M. Kannappa Reddy, S. Parvathavarthini, A. B. Kundu, and E. Sukumar. 2000. Anti-inflammatory activity of (+)-Usnic Acid. *Fitoterapia* 71(5): 564–6.

Villanueva, S., and J. González. 2009. Coagulopathy induced by Saw Palmetto: A case report. *Bol Asoc Med* P R 101(3): 48–50.

Waag, T., C. G. J. Rath, A. Stich, M. Leippe, and T. Schirmeister. 2010. Allicin and derivates are cysteine protease inhibitors with antiparasitic activity. *Bioorg Med Chem Lett* 20(18): 5541–3.

Wada, K. 2005. Ginkgo seed food poisoning. *Chudoku Kenkyu* 18(1): 11–6.

Wada, K., S. Ishigaki, K. Ueda, et al. 1988. Studies on the constitution of edible and medicinal plants. I. Isolation and identification of 4-O-methylpyridoxine, toxic principle from the seed of Ginkgo Biloba L. *Chem Pharm Bull* 36(5): 1779–82.

Wada, K., S. Ishigaki, K. Ueda, M. Sakata, and M. Haga. 1985. An antivitamin B6, 4′-methoxypyridoxine, from the seed of Ginkgo Biloba L. *Chem Pharm Bull* 33(8): 3555–7.

Wang, L., X. Zhang, J. Liu, L. Shen, and Z. Li. 2014. Tea consumption and lung cancer risk: A meta-analysis of case-control and cohort studies. *Nutrition* 30(10): 1122–7.

Wargo, K. A., E. Allman, and F. Ibrahim. 2010. A possible case of Saw Palmetto-induced pancreatitis. *South Med J* 103(7): 683–5.

Weber, N. D., D. O. Andersen, J. A. North, B. K. Murray, L. D. Lawson, and B. G. Hughes. 1992. In vitro virucidal effects of Allium Sativum (Garlic) extract and compounds. *Planta Med* 58(5): 417–23.

Weidenbörner, M. 2013. *Mycotoxins in Foodstuffs*, 2nd ed. New York: Springer Science + Business Media.

Werba, J. P., M. Giroli, V. Cavalca, M. C. Nava, E. Tremoli, and L. Dal Bo. 2008. The effect of green tea on simvastatin tolerability. *Ann Int Med* 149(4): 286–7.

WHO (World Health Organization). 2001. *Legal Status of Traditional Medicine and Complementary/Alternative Medicine: A Worldwide Review*. Switzerland: Geneva.

WHO (World Health Organization). 2002. *Trade and Production of Herbal Medicines and Natural Health Products*. Switzerland: Geneva.

WHO (World Health Organization). 2003. *WHO Guidelines on Good Agricultural and Collection Practices (GACP) for Medicinal Plants*. https://apps.who.int/iris/bitstream/handle/10665/42783/9241546271.pdf?sequence=1 (accessed November 4, 2020).

WHO (World Health Organization). 2004. *Drug Information 2004*, 18(1): 27–30.

Wichtl, M. 2004. *Herbal Drugs and Phytopharmaceuticals: A Handbook for Practice on a Scientific Basis*. Marburg, Germany.

Wierzejska, R. 2014. Tea and health: A review of the current state of knowledge. *Przegl Epidemiol* 68(3): 501–6.

Williamson, E. W., and R. C. Wren. 2003. *Herbal Cyclopaedia the Authoritative Reference Work on Plants with a Known Medical Use*. Saffron Walden: Daniel.

Wilt, T. J., A. G. Stark, R. MacDonald, J. Lau, and C. Mulrow. 1998. Saw Palmetto extracts for treatment of benign prostatic hyperplasia: A systematic review. *J Am Med Assoc* 280(18): 1604–9.

Witte, S., D. Loew, and W. Gaus. 2005. Meta-analysis of the efficacy of the acetonic Kava-Kava extract WS1490 in patients with non-psychotic anxiety disorders. *Phytother Res* 19(3): 183–8.

Wiwanitkit, V., and W. Taungjaruwinai. 2004. A case report of suspect ginseng allergy. *Med Gen Med* 6(3): 9.

Woelk, H., G. Burkard, and J. Grunwald. 1994. Benefits and risks of the hypericum extract LI 160: Drug monitoring study with 3250 patients. *J Geriatr Psychiatr Neurol* 7(1): 34–8.

Wölfle, U., G. Seelinger, and C. M. Schempp. 2014. Topical application of St John's Wort (Hypericum Perforatum). *Planta Med* 80(2–3): 109–20.

Wu, Q. J., Qing H. D., W. J. Sun, Y. H., Q. Wang, and W. L. Zhou. 2014. Discrimination of Chinese teas with different fermentation degrees by Stepwise Linear Discriminant Analysis (S-LDA) of the chemical compounds. *J Agric Food Chem* 62(38): 9336–44.

Yin, Q. P., B. Tomlinson, M. Y. Waye, H. L. Chow, and M. S. S. Chow. 2004. Pharmacogenetics and herb-drug interactions: Experience with Ginkgo Biloba and Omeprazole. *Pharmacogenetics* 14(12): 841–50.

Yoshimura, T., N. Udaka, J. Morita, et al. 2006. High performance liquid chromatographic determination of ginkgotoxin and ginkgotoxin-5′-glucoside in Ginkgo Biloba seeds. *J Liq Chromatogr R T* 29(4): 605–16.

Yuan, C. S., G. Wei, L. Dey, et al. 2004. Brief communication: American ginseng reduces warfarin's effect in healthy patients: A randomized, controlled trial. *Ann Int Med* 141(1): 23–7.

Yun, T. K. 2001. Brief introduction of Panax Ginseng C.A. Meyer. *J Korean Med Sci* 16(Suppl 4): 16–8.
Zakaria, M. B., and M. A. Mohd. 1994. *Traditional Malay Medicinal Plants*. Kuala Lumpur: Penerbit Fajar Bakti.
Zheng, E. X., S. Rossi, R. J. Fontana, et al. 2016. Risk of liver injury associated with green tea extract in SLIMQUICK® weight loss products: Results from the DILIN prospective study. *Drug Saf* 39(8): 749–54.
Zhu, M., K. W. Chan, L. S. Ng, Q. Chang, S. Chang, and R. C. Li. 1999. Possible influences of ginseng on the pharmacokinetics and pharmacodynamics of warfarin in rats. *J Pharm Pharmacol* 51(2): 175–80.
Ziaei, S., S. Hantoshzadeh, P. Rezasoltani, and M. Lamyian. 2001. The effect of garlic tablet on plasma lipids and platelet aggregation in nulliparous pregnants at high risk of preeclampsia. *Eur J Obstet Gynecol Reprod Biol* 99(2): 201–6.
Zou, L., M. R. Harkey, and G. L. Henderson. 2002. Effects of herbal components on CDNA-expressed cytochrome P450 enzyme catalytic activity. *Life Sci* 71(13): 1579–89.

13 Toxicology of Industrial Chemicals

Antonio Peña-Fernández, Mark D. Evans and María de los Ángeles Peña

CONTENTS

13.1 INTRODUCTION

Chemicals are key for human development and have a significant role in modern societies, being essential in our lives for myriad applications, including industry (e.g. metals, plastics, technology); transportation (e.g. vehicles, fuels); agriculture, food production and preservation (e.g. soil chemical amendments, herbicides, pesticides); and pharmacy and health consumer products (e.g. drugs, pharmaceuticals, lotions, creams). Currently, the amount of chemicals available in the market is very extensive and the production of emerging chemical substances with unknown toxicological effects is increasing every day. Thus, the Chemical Abstract Service (CAS), a division of the American Chemical Society for the provision of reliable and accurate information on chemicals published in scientific documents, has registered more than 175 million unique organic and inorganic chemical substances to date (CAS REGISTRY 2021).[1] Moreover, the United Nations Environment Programme (OECD 2019) has recorded that chemical production is rapidly growing globally and is projected to exceed that of population growth until 2030, which highlights the importance of not only promoting the evaluation of risks and information about the chemicals but also their sustainable consumption and production.

The overwhelming amount of chemical substances in modern societies therefore highlights the importance of understanding the potential impact that these substances, including emerging chemicals, may have in our health and in the environment, in both the short and long terms. Appropriate regulations and policies should be therefore be in place, as the increasing ubiquity and presence of different industrial chemicals in our environment following economic and technological development is increasing and enhancing human exposure. To guarantee a high level of protection of human health and the environment on chemical substances, the European Union created an ambitious legislative framework, the Regulation concerning the Registration, Evaluation, Authorization and Restriction of Chemicals (REACH; Regulation No. 1907/2006), which is also aiming to enhance the competitiveness of the EU chemical industry. This regulation introduces the obligation to create a registry of all chemical substances that are marketed within the EU and

DOI: 10.1201/9780203730584-13

not to be able to market any substance that is not appropriately registered and evaluated. REACH also facilitates the progressive substitution of the most hazardous chemicals for suitable alternatives when available. From June 1, 2008, to December 31, 2020, 23,316 substances have been registered according to the European Chemicals Agency (ECHA).[2] The chemical industry is therefore transitioning to a more environmentally friendly position in which there is a need to identify and implement comprehensive risk assessment studies to understand the hazards that manufactured chemicals may have to humans and the environment. A similar policy initiative has been recently launched in the United States, the Frank R. Lautenberg Chemical Safety for the 21st Century Act (Lautenberg Chemical Safety Act 2016).[3]

Moreover, ECHA has launched the 'infocards' system to facilitate the quick dissemination of safe handling of chemicals to the public and their risks on human health and the environment (ECHA 2018). Infocards are accessible at the ECHA's Information of Chemicals website.[4]

13.2 CHEMICAL INCIDENTS

Besides chronic exposure to industrial chemicals from ongoing and past industrial activity, sudden exposure of selected populations may also occur from industrial accidents. Such accidents are called hazardous material (HazMat) incidents to distinguish from deliberate release of hazardous Chemical, Biological, Radiological, Nuclear and Explosive (CBRNE) agents to the environment and/or attacks with these substances (Calder and Bland 2015).

Although rare, major incidents involving the release of industrial chemicals can have devastating impacts on human health, including long-term deleterious effects such as cancers and birth defects. Examples of accidental exposure to industrial chemicals affecting thousands of individuals include the Seveso disaster, in which large amounts of the dioxin 2,3,7,8-tetrachlorodibenzo-p-dioxin (TCDD) were released into the air in Italy in 1976 following an explosion at a chemical factory (Warner et al. 2020); or the accidental leakage of tonnes of methyl isocyanate gas and different toxic industrial chemicals from a storage tank of the Union Carbide India Limited (UCIL) company in Bhopal, India, in 1984 (Ganguly et al. 2018). Such sudden and unexpected exposures to hazardous chemicals in the aftermath of an incident require a coordinated, multiagency response to deal with the immediate health/environmental impact and the recovery phases, which include prevention, preparedness, detection and alert, response and recovery. Appropriate communication to the public should be considered in each stage.

Governments should also pay special attention to preparedness and planning, to appropriately respond to chemical incidents and minimise the deleterious effects that these events have on humans and the environment (Griffiths 2019). Preparedness is increasingly important as, although major HazMat incidents are rare, industrial chemical spillages and release into the environment are frequent and can affect large populations (e.g. tens of thousands people annually in Europe). During 2005, 1,040 and 8,603 chemical incidents were reported in the UK and United States, respectively, producing thousands of casualties and several deaths (Bos et al. 2013).

13.2.1 Mass Casualty Decontamination

Evacuation and disrobing are immediate actions following large HazMat incidents, measures which should be immediately considered by the risk assessor and should form part of the planning and preparedness, especially in areas with large industrial chemical settings close to residential areas and cities. When performed immediately in the aftermath of a HazMat incident and performed appropriately to avoid contact and transfer of the chemicals to the skin, disrobing is an effective and fast process to remove contaminants from clothes and reduce human exposure. To effectively undertake disrobing countermeasures, UK fire and rescue services have 'disrobe' packs, so the affected population can re-robe (i.e. have new 'clothing') to avoid hypothermia and ethical or religious issues (Goode et al. 2019).

The Optimisation through Research of Chemical Incident Decontamination Systems (ORCHIDS) programme in Europe has developed initial and specialist operational response processes to optimise a Model Response, which will deliver a group of trained first responders to the impacted environment to aid with the initial response. The Model Response, which sets out the operational parameters to respond to CBRN and HazMat incidents in the UK, involves a team of experts with bespoke resources and procedures to effectively implement disrobing and decontamination facilities in conjunction with medical countermeasures. This team also have relevant protection and detection capabilities to aid with the response (Chilcott et al. 2019a). These authors have recently optimised the initial operational response process with capabilities for undertaking rapid disrobing and dry decontamination processes to enable an initial chemical response at the scene before more specialised resources arrive, as dry absorbents have shown similar but safer effectiveness in mass decontamination of liquid noncorrosive contaminants compared to wet decontamination processes (Kassouf et al. 2017). This optimised approach has informed the US federal government guidance (Primary Response Incident Scene Management [PRISM]) to respond to industrial chemical accidents and attacks involving the deliberate release of chemical agents (Chilcott et al. 2019b). In conjunction with the evidence-based guidelines, the United States has developed the web-based resource, ASPIRE[5] (Algorithm Suggesting Proportionate Incident Response Engagement), to aid first responders to identify the most effective decontamination approach for the chemical involved. ASPIRE also provides indication of how long the chemical would evaporate from artificial skin since time of exposure, which will provide key information about the time frame for implementing mass decontamination (Butler 2019).

13.2.2 Recovery and Remediation

Additionally, countries should have appropriate guidelines, preparation and capabilities in place to undertake the last phase of any chemical incident response (i.e. the recovery stage). Figure 13.1 is an overview of the incident response and the resources needed to manage different stages, which depends on the site and the scale of the event (Duarte-Davidson et al. 2014).

The recovery and remediation stage is crucial not only to decontaminate and return the affected environment to 'normal', but also to minimise the spread of the chemicals involved (which could affect other areas and subenvironments) and prevent long-term effects due to potential future exposures to the chemical(s) released (Peña-Fernández et al. 2014; Wyke et al. 2020). These plans should also consider the long-term follow-up of the residual effects of the chemical incident (including environmental monitoring and human biomonitoring for appropriate risk assessment characterisation). The UK has developed appropriate guidelines and advice (named the UK Recovery Handbooks) to guide the appropriate recovery and remediation of the environment following a chemical (UKRHCI; Wyke et al. 2020), biological (UKRHBI; Pottage et al. 2015) and Radiation (UKRHRI; Nisbet et al. 2015), which are freely accessible from the Public Health England (PHE) website.[6] The UK Recovery Handbooks have covered a gap in the literature and have been designed to be user-friendly technical guidance documents with innovative decision frameworks for identifying and evaluating effective up-to-date recovery strategies. However, most countries lack similar and standardised recovery guidelines and preparedness for remediating the affected environment in the aftermath of events involving the release of chemical hazards (Duarte-Davidson et al. 2014).

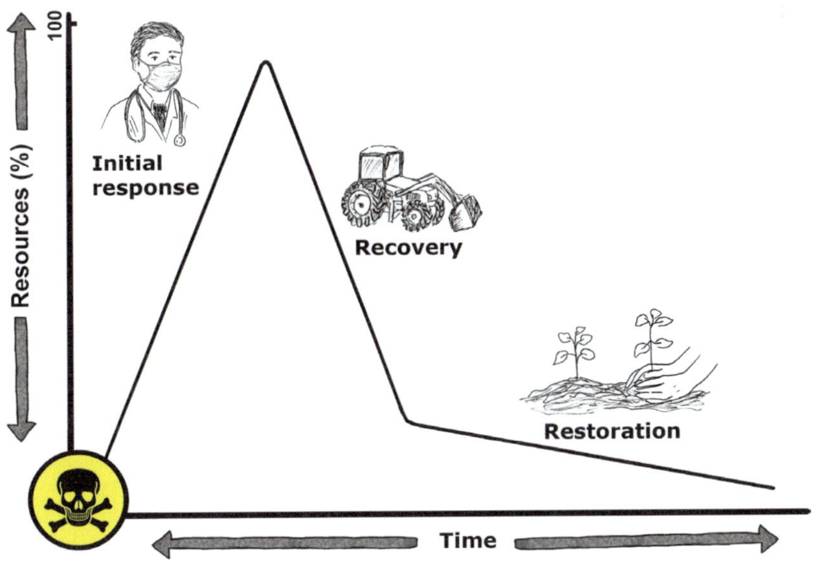

Figure 13.1 Overview of the chemical incident response and the amount of resources needed to manage the different phases.

Source: Adapted from Wyke et al. (2020).

As each chemical incident is unique, the recovery strategy will need to be considered on a site- and incident-specific basis. Thus, multiple factors need to be considered when implementing the response to a HazMat; these factors are separated into four key groups: temporal and spatial, technical, social and ethical (Peña-Fernández et al. 2014). Moreover, the physicochemical nature and toxicological properties of the chemical or mixture of chemicals involved (and its reaction or degradation by-products) will influence the remedial response following a chemical incident and will need to be assessed on a site- and incident-specific basis. Furthermore, the characteristics of the contaminated environment(s), such as the location (urban or rural), size and scale of the incident, local population demographics and weather conditions (Wyke et al. 2014, 2020), are all aspects that decision makers need to evaluate prior to undertaking remediation. Such physicochemical properties include the physical state of the substance (solid, gas or liquid), including volatility of the latter; its relative water or lipid solubility and its absorptive properties on surfaces, including soil. Thus, some industrial chemicals although highly toxic may also be highly volatile, and depending on the affected environment (e.g. in the countryside where there are fewer people likely to be affected) may not require an action. Contrarily, events involving persistent chemicals are more likely to require rapid action and active remediation. Additionally, solids may exist as dusts, for example, that can easily spread in air. Similarly, the physical nature of the affected environment is also important, as this includes water or soil contamination, for example, and ambient temperature which may facilitate or impede recovery strategies (Peña-Fernández et al. 2014; Wyke et al. 2014). The characteristics of the site (e.g. the presence of organic matter in soil, microbial degradation, exposure to sunlight, temperature and pH) also need to be considered and could also influence the half-life of a chemical and, therefore, the remedial response needed.

High lipid solubility leads to relative accumulation of substances in fatty tissues and contributes to bioaccumulation. Speciation, the chemical form of a toxic substance, also influences toxicity and can be affected by the contaminated environment. Human toxicity of substances may diminish naturally in certain environments by the action of light or other degradative processes in addition to biotransformation by other organisms. However, the converse may also be true depending on the substance involved; that is, environmental processes enhance toxicity by the production of more toxic by-products or metabolites of the parent compound. Bioaccessibility describes the amount of contaminant bioavailable to an organism, and thus limiting bioaccessibility can directly impact bioavailability and hence toxicity. Bioaccessibility can be modified in remediation strategies to ultimately limit the toxicity of the chemical agents (Peña-Fernández et al. 2014).

Chapter 28 provides a detailed description of the UKRHCI and the different PHE recovery tools created to facilitate the implementation and use of the novel multistage methodology, and also instructions on how to use these guidance and tools. Chapter 28 also provides insight into how protection and recovery options (actions or techniques to return the impacted environment back to 'normal') are appropriately selected based on the physicochemical characteristics of the chemical(s) involved and the characteristics of the site.

13.3 ROUTES OF EXPOSURE OF INDUSTRIAL CHEMICALS

The development of appropriate networks with toxicologists and public and environmental health professionals able to perform risk assessments to chemical substances, and to provide advice and support to prevent and respond to HazMat incidents, are therefore key to facing and minimising the threats that industrial chemicals can pose in our societies (Duarte-Davidson et al. 2014). The European Chemical Emergency Network (ECHEMNET) project has created a Rapid Risk Assessment (RRA) methodology and guidance which, although not perfect because of the limited information available in the aftermath of a HazMat incident, will provide vital information to quickly assess the risks and associated impacts of these events and to appropriately respond to them in a short time frame (Goode et al. 2019). This RRA methodology is currently strengthening within the European Multiple Environmental Threats Emergency Network (EMETNET)[7] to support the European Commission, which also provides a network of risk assessors with expertise in emerging environmental threats. This strengthened RRA methodology also identifies a signpost to useful web-based tools, such as the Flash Environmental Assessment Tool (FEAT)[8] to assess potential acute risks in the aftermath of natural disasters, developed by The United Nations Environment Programme Office for the Coordination of Humanitarian Affairs (OCHA) Joint Environment Unit.

Industrial chemicals can principally enter the human body three ways: (1) through the skin or mucous membranes (dermal route); (2) through the mouth (oral route); and (3) by breathing (inhalation route). The exposure can be direct: the highest intake of chemical agents released in air will be through direct inhalation or through eye contact. It can also be indirect: chemical agents can be transferred from air to soil and then deposited onto vegetation and transferred to foods.

The dermal or topical route uses the skin and mucous membranes for the absorption of a specific compound. This includes the conjunctiva and the buccal, sublingual, nasal, optical and urogenital mucosa. Topical absorption occurs directly after an industrial chemical contacts skin or mucous membranes. Absorption will continue if the toxic agent remains in contact with a surface. In general, solid forms of pesticides (e.g. powders, dusts and granules) are not as readily absorbed through the skin and other tissues as liquid formulations. The topical absorption rate depends on many factors, that can be grouped into three areas, related to the toxic agent, i.e., concentration, type of vehicle or presence of other substances; anatomical and physiological characteristics of the skin type, i.e., content of water, pH and temperature, local blood flow, perspiration, surface of contaminated skin, thickness of the stratum corneum, the sebum composition, and the distance of blood vessels to the surface of the skin characteristics (e.g. sex, age and genetic variation), differences between ethnic groups/races, preexisting disease and occupation (e.g. people working with pesticides are more likely to be exposed). Contact dermatitis is one of the most common types of skin disorders. Pesticides, solvents, and organic contaminants, such as polycyclic aromatic hydrocarbons (PAHs) and polychlorinated dibenzo-p-dioxins and dibenzofurans (PCDD/Fs), are some of the main chemical groups that have been recognised as posing health problems by dermal absorption (Tielemans et al. 1999; Macfarlane et al. 2013; Coalova et al. 2014).

Oral exposure is the main route of absorption in most acute toxic cases. Chemicals can be ingested by accident, carelessly or intentionally. There are many factors that influence the absorption rate of chemicals in the gastrointestinal tract, such as the physical and chemical characteristics of the substance (partition coefficient, size or solubility); amount present in food and water, surface area of absorption, local pH, peristalsis and gastric secretions, or the action of microflora that can biologically transform toxins. Ingestion is the most significant route of exposure to metals and metalloids (Jayarathne et al. 2018; Khademi et al. 2019).

Respiratory exposure can be particularly significant for industrial chemicals, because the lungs can rapidly absorb these substances that then directly enter the bloodstream. The deposition of particles and aerosols in the respiratory tract depends on physical and physiological factors, as well as the size and density of the particles, or the breathing conditions, the geometry of the airways and the mucociliary clearance mechanism. The low and relatively constant retention of particles in the lungs of very exposed people suggests the existence of a highly effective system of particle removal (Fedorovitch 2019). In the upper respiratory tree, the elimination oversees a defence system called mucociliary clearance. Compounds like isocyanates are a leading cause of occupational asthma (Bello et al. 2007; Daniels 2018). Isocyanates are used in a myriad of industrial products, including paints, coatings, polyurethane products, foams, adhesives, and sealants (Wisnewski 2007).

Information about how to quantify environmental exposures to chemicals using the guidance developed by the US Environmental Protection Agency (EPA) to perform risk assessments is provided in a practical and descripted manner in Chapter 27.

13.4 PESTICIDES

Pesticides are chemical compounds that are used to prevent, repel, mitigate or destroy any organisms that are considered pests (Settimi et al. 2016); such as insects (insecticides), rodents (rodenticides), unwanted plants (herbicides) and microorganisms (algicides, fungicides, bactericides, acaricides) in agriculture and the home. By their nature, as they are biologically active, pesticides can potentially cause harm to humans and the environment, and therefore they need to be used and disposed safely and properly. Owing to the extension of uses and increasing production, the Food and Agricultural Organisation of the United Nations (FAO) adopted the International Code of Conduct on Pesticides Management (FAO and WHO 2014). Successful implementation of these conventions requires information about the frequency and circumstances of pesticide exposures and about the health impact of exposure, although in many cases such information is still lacking. Acute intoxications involving pesticides is required to be reported in the EU (Settimi et al. 2016).

For regulation purposes, the EU distinguishes between biocidal pesticides (those that act against pests, excluding plants) and plant protection pesticides (agents that target pests affecting plants and/or their products; Settimi et al. 2016). The European Commission approves the use of plant protection pesticides within the EU; the European Food Safety Authority (EFSA) is responsible for performing the necessary risk assessments and provide scientific advice, so the European

Commission can make evidence-based decisions to approve plant protection pesticides. In the United States, the EPA is responsible for evaluating the potential adverse effects of pesticides against federal safety standards (Federal Insecticide, Fungicide, and Rodenticide Act [FIFRA]).[9]

The sustainability of agricultural crop systems is a fundamental activity on which present and future populations rely (Mesnage and Séralini 2018). Some countries apply restrictions in using certain pesticides or their use is completely banned because of their serious adverse effects. This is the example for chlorpyrifos in Europe (Pesticide Action Network [PAN]),[10] which is used in agriculture or farms to control ticks on crop pests and cattle. The European Commission has established a framework to achieve the sustainable use of plant protection pesticides, which requires the adoption of national action plans to reduce not only the risks of these substances but also encourage them to adopt strategies to reduce their use.

According to the EPA,[11] pesticides are composed of 'active ingredients' (the active chemicals that will act against the pests) and 'inert ingredients' (substances that are present in the formulation of the pesticide but are not active ingredients). Inert ingredients, which are necessary for the performance and effectiveness of the pesticides, must also be assessed for safety, as they can have hazardous effects in humans and the environment.

Some pesticide inert ingredients are surfactants; these are organic amphiphilic compounds that are active on the surface or at the interface between an organic phase and aqueous phase. Surfactants allow solubilisation, suspension or dispersion of the active pesticide ingredient in aqueous solution and can improve the biological activity by modifying spray droplet size, retention and spreading on some surfaces. The most common compounds are nonionic substances, such as ethoxylated alkylphenols, necessary to improve the absorption and stability of the product and to promote the action of pesticides (Travlos et al. 2017; Mesnage and Antoniou 2018). The adjuvants used in pesticide formulations also include spreaders, adhesives, antifoaming agents, colorants and retardants that modify the physicochemical properties of the mixture.

The WHO uses the acute toxicity hazard categories from the Globally Harmonized System of Classification and Labelling of Chemicals (GHS) as a starting point for its classification of chemical hazards as in shown in Table 13.1. The GHS is a worldwide initiative to communicate hazards by using pictograms, now being widely used for the classification and labelling of chemicals worldwide according to their health, physical and environmental hazardous effects. The EU implemented the GHS in 2008 as part of the classification and labelling of hazardous substances;[12] ECHA manages the law on classification and labelling of pesticides. Contrarily, GHS has not been adopted in the United States for pesticides yet.[13]

The toxicity of a pesticide can be determined by laboratory testing on animals such as rats, mice and rabbits. The measurement of LD_{50} (lethal dose) describes the dose of a pesticide that will kill half of a group of test animals from a single exposure (dose) by either the dermal, oral or inhalation routes. Thus, a pesticide with a lower LD_{50} is more toxic than a pesticide with a higher value because it takes less of the pesticide to kill half of the test animals. Unlikely to present an acute hazard corresponds to classification U; acute toxicity of a pesticide refers to the effects from a single exposure or repeated exposure over a short time. As described previously, various signs and symptoms are associated with acute poisoning, such as central nervous system (CNS) involvement, blurred vision, confusion, sweating, lacrimation, bradycardia, hypothermia, diarrhoea,

Table 13.1 Recommended Classification of Pesticides by Hazard as WHO Class

WHO class	Hazardous	LD_{50} for the rat* (mg/kg body weight)	
		Oral	Dermal
Ia	Extremely hazardous	<5	<50
Ib	Highly hazardous	5–50	50–200
II	Moderately hazardous	50–2000	200–2000
III	Slightly hazardous	Over 2000	Over 2000
U	Unlikely to present acute hazard	5000 or higher	

* Median lethal dose (dose required to kill half the members of a tested population after specified test duration).

Source: Adapted from WHO (2019).

intestinal colic or bronchospasm (Sánchez et al. 2019). A pesticide with a high acute toxicity can be deadly even if a small amount is absorbed.

On the other hand, chronic toxicity refers to the effects of long-term or repeated lower-level exposures to a toxic substance. The effects of chronic exposure do not appear immediately after first exposure and may take years to produce signs and symptoms. Effects associated with chronic exposure may include carcinogenicity, mutagenicity, teratogenicity, oncogenicity, liver damage, reproductive disorders, nerve damage or allergenic sensitisation. The effects of chronic toxicity, as with acute toxicity, are dose-related. In other words, low-level exposure to chemicals that have potential to cause long-term effects may not cause immediate deleterious effects, but repeated exposures through careless handling or misuse can greatly increase the risk of chronic adverse effects.

The EPA manual *Recognition and Management of Pesticide Poisonings* (Roberts and Reigart 2013) provides clinicians and healthcare providers with information to identify, manage and treat poisonings due to pesticides.

As explained above, the term 'pesticide' applies to a wide range of chemicals that include chemical pesticides (insecticides, herbicides, rodenticides and fungicides) and biopesticides. The classification of both types is shown in Table 13.2 (Sharma 2006; Tadeo et al. 2019).

Pesticides can also show bioaccumulation, by which certain compounds can accumulate and keep on accumulating in living organisms, posing a threat to health, life and the environment.

Table 13.2 Types of Pesticides

Class	Classification	Families	Examples
Chemical pesticides	Insecticides	Organophosphates	Parathion, chlorpyrifos, diazinon, fenthion, malathion
		Carbamates	Aldicarb, carbofuran, carbaryl, ethinenocarb, fenobucarb, oxamyl, methomyl, pirimicarb, propoxur
		Organochlorines	Dicofol, lindane, DDT and its analogues, cyclodienes (aldrin, dieldrin, endrin, heptachlor, chlordane, endosulfan)
		Pyrethrins/Pyrethroids	Type I Type II
		Others	Neonicotinoids (imidacloprid, thiamethoxam) Ryanoids (chlorantraniliprole, cyantraniliprole, flubendiamide) Insect growth regulator (IGR) (diflubenzuron)
	Herbicides	Nitriles	Bromoxynil
		Aryloxyalkylcarboxylic acids	(2,4-dichlorophenoxy)acetic acid
		Oxyacetanilides	Flufenacet
		Semicarbazones	Diflufenzopyr
		Cyclohexanediones	Cycloxydim
		Pyridazine derivatives	Flurtamone
		Organophosphorus herbicides	Glyphosate
		Triazinones	Metamitron
	Rodenticides	Anticoagulants	Brodifacoum, bromadiolone, difenacoum, pindone, diphacinone, chlorophacinone
		Non-anticoagulants	Strychnine, bromethalin, cholecalciferol, zinc phosphide
	Fungicides	Azoles/benzimidazoles	Cyproconazole, flusilazole Benomyl, carbendazim
		Dithiocarbamates	Benomyl
		Morpholines	Macozeb

It is be considered is the sum of two processes: bioconcentration and biomagnification. Bioaccumulation factors (BAFs) are widely used in scientific and regulatory programs to assess chemical hazards (Mackay et al. 2018). Trophic magnification factors (TMFs) can provide valuable insights into the bioaccumulation behaviour of chemicals in food webs. However, these bioaccumulation factors may be subject to uncertainty because of systematic bias and the effect of different confounding variables (Kim et al. 2016). Explained below are some details of the types of pesticides summarised in Table 13.2.

13.4.1 Insecticides

Insecticides can be of chemical and biological origin and are used to kill or control insects, such as mosquitoes and ticks, which can act as vectors of important pathogens. They can be classified according to their chemical structure, as shown in Table 13.3. The insecticides can bind to different enzymes, receptors, and other proteins, acting on the insect's nervous system or as a growth regulator; the necessary sites and adducts, along with residue of insecticides and their metabolites, can be used as biomarkers of exposure and effects (Gupta et al. 2019). The EU, in its Good Agricultural Practice Regulations, carefully regulates pesticide residues or traces of the pesticides that are left in the treated product, which must be lower than the maximum residue levels (MRLs) fixed for all foods and animal feed. MRLs currently in use can be found at the European Commission dedicated web page.[14] Residue of pesticides are regularly monitored in European countries for safety, including their chronic cumulative exposure (e.g. for organophosphorus, carbamates and pyrethroid and pyrethrin insecticides in fruit and vegetables; Quijano et al. 2016).

Among their many applications, insecticides are commonly used in agriculture and industry as well as in public health and the household. The most commonly used are organophosphates, pyrethroids and carbamates due to their best price, easy availability, low persistence in the environment (shorter half-life) and broad spectrums of utilisation, which are briefly described below.

13.4.1.1 Organophosphates (OPs)

Organophosphates (OPs) are synthesised by the esterification of phosphoric acid and alcohol. These substances can be found and used as insecticides and also fungicides and herbicides (Adeyinka et al. 2020). Currently, OPs are the most commercially favoured group of pesticides, with large application areas all over the world, as they are effective even at low concentrations and will hydrolyse rapidly in the environment (Martín Reina et al. 2017). However, owing to the fact that all OPs can potentially cause acute and subacute toxicity, damage the nervous system and even cause death, governments are banning or reducing their uses, such as parathion and chlorpyrifos. Thus, the use of chlorpyrifos, which has been linked with cancer (Engel et al. 2017), has been recently banned in the EU and for home use in the United States (Roberts and Reigart 2013; Hites 2021). As all OPs can inhibit cholinesterase by the phosphorylation of the acetylcholinesterase

Table 13.3 Types of Insecticides

Compound	Chemical structure	Target	Type of interaction
Organophosphates	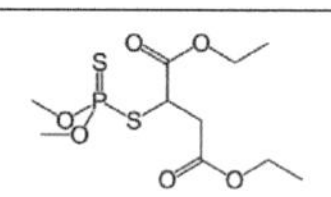	Acetylcholinesterase	Inhibition
Carbamates	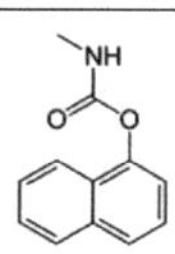	Acetylcholinesterase	Inhibition
Organochlorines	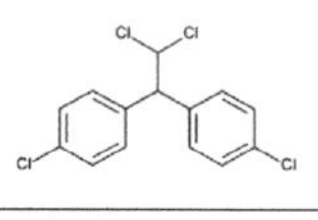	Sodium channels	Activation
Pyrethroids	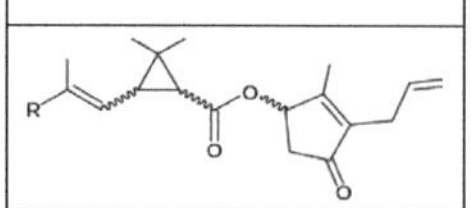	Sodium channels (all) GABA-receptor-gated chloride channels (type II)	Activation (all) Inhibition (type II)

enzyme (required for the normal control of nerve impulse transmission) at nerve endings, the toxicity resulting from multiple exposure to different OPs or multiple exposure from different routes to the same OP can be higher due to an additive or cumulative effect (Roberts and Reigart 2013; Benbrook and Davis 2020). Moreover, some OPs, such as diazinon and fenthion, are highly lipophilic, being able to accumulate in fatty tissues that may lead to delayed toxic effects due to late release. Humans can be efficiently exposed to these substances through ingestion, inhalation and dermal penetration depending on the agent. Chronic exposures to low concentrations of OPs has been linked with neurological effects in humans; meanwhile, an acute exposure to a high concentration of these pesticides can cause long-lasting behavioural effects in the person exposed (Bjørling-Poulsen et al. 2008). OPs can also mimic oestrogen function (endocrine disruptor), resulting in reproductive and developmental adverse effects (Martín Reina et al. 2017).

13.4.1.2 Carbamates

Carbamates are N-methyl esters of carbamic acid, which are commonly used in agriculture to protect crops against insects and other agricultural and household pests (Zhang et al. 2015). Derivatives of carbamic acid can be also used as herbicides and fungicides (Gupta 2014). Carbamate pesticides are hazardous to humans causing similar symptoms to those described previously for organophosphates; that is, they facilitate the carbamylation of the acetylcholinesterase at neuronal synapses and neuromuscular junctions, which may lead to respiratory depression, sensory and behavioural disturbances, loss of coordination, lethargy and coma (Roberts and Reigart 2013; Silberman and Taylor 2020). *In vivo* experiments with rats chronically exposed to carbofuran during gestation have shown cognitive and hippocampus impairments together with decreased neurogenesis in the offspring (Martín Reina et al. 2017). However, the phosphorylation of acetylcholinesterase into carbamyl-acetylcholinesterase dissociates more readily than the phosphoryl-acetylcholinesterase complex produced when exposed to OPs, as the phosphorylation due to carbamates is reversible with a duration of toxicity that is typically less than 24 hours. Moreover, carbamates are rapidly metabolised (Martín Reina et al. 2017). Thus, from all the illnesses related to acute occupational exposures to pesticides in the United States from 2005 to 2009, 10.4% were related to organophosphorous compounds, whereas only 1% were related to exposure to thiocarbamates/dithiocarbamates (Roberts and Reigart 2013). If we compare the median lethal doses, carbofuran (carbamate) has a rat oral LD_{50} of 5 mg/kg, whereas parathion (OP) has a rat oral LD_{50} between 3–8 mg/kg. Carbamates can be absorbed orally, dermally and via inhalation, although absorption and toxic effects are higher when ingested (Roberts and Reigart 2013). In humans, toxic exposures are usually intentional and have been described for different carbamates, including aldicarb, carbofuran, carbaryl, ethinenocarb, fenobucarb, oxamyl, methomyl, pirimicarb and propoxur (Silberman and Taylor 2020).

13.4.1.3 Organochlorines (OCs)

Organochlorine (OC) insecticides are a group of synthetic chlorinated compounds, produced from hydrocarbons, which may also contain other elements, such as oxygen and sulphur (Jayaraj et al. 2016), and belong to the class of persistent organic pollutants (POPs). Owing to their high persistence in the environment and despite their success in controlling malaria and typhus, OCs have been nearly banned in developed countries or their used is limited. For example, the EPA has only dicofol and endosulfan as registered OCs. Lindane is still used for treatment of lice and scabies, although paediatricians are no longer recommending their use as it was linked with acute neurological toxicity cases (Roberts and Reigart 2013). However, significant concentrations of OCs are continuing to be reported globally, as is the case for the organochlorine insecticide dichlorodiphenyltrichloroethane (DDT), despite it being banned in the 1980s (Buah-Kwofie and Humphries 2021).

The OCs generally have lower acute poisoning effects than OPs. However, as these substances decompose slowly and are lipophilic, they can accumulate in fatty tissues and biomagnify in the trophic chain. Thus, OCs can be very persistent with an estimated half-life of about 10–30 years. The OCs includes the DDT-type compounds and the chlorinated alicyclics (aldrin, dieldrin, endrin, heptachlor, chlordane and endosulfan). The routes of exposure differ between compounds, although they can be absorbed orally, dermally and by inhalation. Acute exposure to OCs produce toxic effects in the CNS by hyperexcitation of the neural tissues, leading to convulsion and coma. These compounds interact with sodium channels, keeping them open longer, which results in an increase in the likelihood of action potentials developing, thus creating a condition of hyperexcitability, whose main clinical sign is tremors. Infants and children are therefore particularly susceptible to exposure to OCs, which also have potential carcinogenic effects (Roberts and Reigart 2013). Moreover, prenatal exposure to OC pesticides has been associated with increased risk for neural tube defects in humans, which is a relatively common and disabling congenital defect in the foetus (Yin et al. 2021).

13.4.1.4 Pyrethrins and Pyrethroids

Pyrethrins and pyrethroids are among the most frequently used agricultural and household insecticides as they have generally been regarded as having low toxicity in mammals (Martín Reina et al. 2017). Pyrethrins are a modification of natural pyrethrum, which is a oleoresin with insecticide properties extracted from chrysanthemum flowers (Konopelko et al. 2018). Pyrethrins exhibit high selective insecticidal activity, nonpersistence in the environment, and low bioavailability and toxicity in mammals (Ye et al. 2006). Pyrethrum is a dermal and respiratory allergen that can produce anaphylactic reactions in humans (Roberts and Reigart 2013).

On the other hand, pyrethroids were chemically synthesised to enhance their environmental stability. The average reported half-lives for pyrethrins and pyrethroids in plants range from 1.1 to 5 days (Fantke and Juraske 2013). Pyrethroids are divided into two groups according to their chemical structure: type I pyrethroids present only the basic cyclopropane carboxylic ester structure (permethrin, d-phenothrin, bifenthrin), which mostly affects peripheral nerves and can cause rapid onset of aggressive behaviour, ataxia, progressive paralysis and convulsions, coma and death; and type II, which are enhanced synthetic pyrethroids with an alpha-cyano group (deltamethrin, cyfluthrin, fenvalerate, lambda-cyhalothrin cypermethrin), which can affect the CNS and provokes salivation and choreoathetosis (sinuous writhing) and is usually characterised by pawing and burrowing behaviour (Martín Reina et al. 2017). Humans are mostly exposed to pyrethroids through oral ingestion, as they present low dermal and inhalation absorption rates (Roberts and Reigart 2013).

13.4.2 Herbicides

The intensive use of herbicides, key in modern agriculture, especially with the use of genetically modified crops, has led to widespread environmental contamination because they are found in surface water or groundwater sources, soils, turfs and other vegetations all over the world. They can volatilise, leach and runoff, accumulate in soils, water and organisms, which can cause long-lasting toxic effects on humans and the environment (Wilms et al. 2020). Herbicides can interact and interfere with the metabolic processes and other biochemical pathways of weeds and cause irreversible damage, tissue injury, leading to the eventual death and elimination of the weeds. Herbicides may vary based upon their complex chemical structures, characteristics, and properties with other members of their chemical 'family' and are grouped according to their mode of action and target specificity. The mode of action of herbicides includes inhibition, interruption, disruption or mitigation of the regular plant growth. Herbicides are grouped based upon their chemical structures, which consist of a common molecular structure with varying side chain substituents, as summarised in Table 13.4.

13.4.3 Rodenticides

Rodenticides are chemical substances meant to control and kill rodents, commonly rats and mice. Although all rodenticides can be toxic when ingested, humans can also be exposed through inhalation and skin contact. Toxicity can be induced even by 1-minute exposure, meaning that a high concentration will be required to achieve a lethal effect. Rodenticides must always be stored away from children and pets, as they may find them attractive.

The use of rodenticides has evolved with the ages from metals to anticoagulants and non-anticoagulants. Anticoagulants, which are the most commonly used currently, can also be classified into first (developed before 1970) and second generation (developed in the 1970s). They act via the blockade of the vitamin K cycle leading to the nonavailability of essential clotting factors such as the factors II and VII, causing death due to fatal bleeding. Although first-generation anticoagulants are more toxic, second-generation substances are more likely to kill rodents after a day's exposure and they present higher half-lives remaining longer in body tissues, hence presenting higher risks to nontarget mammals including humans. As a result, the United States has banned the use of second-generation anticoagulants in products for human consumption. Examples of active, EU-approved ingredients include brodifacoum, bromadiolone, difenacoum, diphacinone, chlorophacinone and warfarin (Rial-Berriel et al. 2020). These substances can be highly resistant to degradation and have long elimination half-lives in the liver of almost a year following a single

Table 13.4 Classes of Herbicides

Mechanism of action	Site of action	Chemical family
Lipid-synthesis inhibitors	ACCase inhibitor	Aryloxyphenoxypropionate, cyclohexanedione, phenylpyrazolin
Amino-acid synthesis inhibitors	ALS inhibitors	Imidazolinones, pyrimidinylthiobenzoates, sulfonylureas, triazolopyrimidines
Root-growth inhibitors	Microtubule inhibitors	Benzamide, benzoic acid (DCPA), dinitroaniline, phosphoramidate, pyridine
Plant-growth inhibitors	Site of action unknown	Benzoic acid, phenoxycarboxylic acid, pyridine carboxylic acid, quinoline carboxylic acid
Photosynthesis inhibitors	Photosystem II inhibitors	Triazine, triazinone, phenylcarbamates, pyridazinones
Shoot-growth inhibitors	Lipid-synthesis inhibitors	Phosphorodithioates, thiocarbamates
Amino-acid synthesis inhibitors	EPSP synthase inhibitors	Phosphonates
Pigment-synthesis inhibitors	HPPD inhibitors	Pyridazine derivatives
Cell-membrane disrupters	PPO inhibitors	Diphenylether, aryl triazolinone, *N*-phenylphthalimides, oxadiazoles, pyrimidindiones thiadiazoles
Shoot-growth inhibitors	Very-long-chain fatty acid inhibitors	Chloroacetamide, acetamide, oxyacetamide tetrazolinone

Abbreviations: ACCase = acetyl CoA coarboxylase; ALS = acetolactate synthase; EPSP synthase = 5-enolpyruvyl-shikimate-3-phosphate synthase; HPPD = 4-hydroxyphenyl-pyruvate-dioxygenase; PPO = protoporphyrinogen oxidase.

exposure in nontarget mammals (Nakayama et al. 2019). Bromadiolone has also shown to induce acute neurotoxicity and encephalopathy in humans (Li et al. 2021).

Non-anticoagulant rodenticides include strychnine, bromethalin, cholecalciferol and zinc phosphide, which produce toxicity through different mechanisms of action (Gupta 2018). Some active ingredients can be extremely toxic for humans, such as zinc phosphide, which can transform onto phosphine gas in the presence of water and acid. In contrast, cholecalciferol (vitamin D_3) will be rarely toxic in humans. Although the use of strychnine is banned in the EU since 2006, there are still reports of human intoxication in European countries (Boucaud-Maitre et al. 2019).

13.4.4 Fungicides

Fungicides are biocidal chemical compounds or biological organisms used to treat diseases derived from fungal pathogens, such as root rot, loose smut, powdery mildew and leaf rust (Gao et al. 2012; Han et al. 2014). Fungicides are widely applied to control postharvest disease in fruits and vegetables. Moreover, they have a myriad of applications in various sectors, although they present a risk of poisoning in food and agricultural (occupational) security production owing to their residues (Hamdi et al. 2021), which are carefully regulated in the EU as described previously.

Fungicides can be classified according to their origins in two major groups: biofungicides, which contain living microorganisms (e.g. bacteria, fungi) as active ingredients that are antagonistic to the pathogens that cause disease, and chemical fungicides, which are synthesised from organic and inorganic chemicals (Jung et al. 2020; Dukare et al. 2020).

Generally, newer fungicides have low to moderate toxicity, those being hazardous for human health been banned in the EU and United States, as in the case of chlorothalonil and mancozeb, which have been recently banned in the EU. Fungicides can be harmful for both humans and the environment. However, fungicides can have different deleterious effects that will vary among them, but some can behave as endocrine disruptors and affect the reproductive system, or have teratogenic, mutagenic or carcinogenic effects (Gupta 2018). Novel nanoparticle formulations have shown an enhancement in the properties of fungicides meanwhile lowering their toxicity, although their effects on human health and the environment have been poorly studied due to their novelty (Tleuova et al. 2020).

13.4.5 Biopesticides

Biopesticides are types of pesticides derived from natural materials, including animals, plants, bacteria and certain natural elements such as copper (Cu) and sulphur. These substances are gaining increasing importance, as are less toxic alternatives to synthetic chemical pesticides, can be target specific (i.e. will affect mostly the pest rather than different organisms), can be used in lower concentrations and decompose quickly, and can be a major component of many pest control programs (Wani et al. 2020; Fletcher et al. 2020). Despite the multiple benefits of biopesticides for human health and the environment, biopesticide active substances registered in the EU is lower than in the United States owing to the complexity of the regulations in the EU (Balog et al. 2017). The European Commission is planning to strengthen the risk assessment process of pesticides and facilitate the approval of biopesticides.

According the US EPA, biopesticides can be divided into three main classes:[15]

- *Microbial pesticides*, which consist of microorganisms (e.g. entomopathogenic fungi, viruses and nematode-associated bacteria) acting against different pests such as invertebrate pests. Examples include the insect pathogenic bacterium of *Bacillus thuringiensis*, which produces endotoxins during bacterial spore formation that causes lysis of gut cells in susceptible insects.
- *Biochemical pesticides*, which are naturally occurring substances that can control (or modify, as in the case of insect semiochemicals or pheromones) pests by mechanisms that are not toxic, such as interfering in the mating of insects. They include, for example, pyrethrins, which are fast-acting insecticidal compounds produced by *Chrysanthemum cinerariaefolium*.
- *Plant-incorporated protectants (PIPs)*, which have been synthesised by plants by genetic modification. Their use is controversial, especially in many European countries.

13.5 METALS

Metals and metalloids are naturally present in the composition of different rocks and minerals in the Earth's crust. The term metalloid is used for elements in the periodic table that exhibit some of the properties of both nonmetals and metals (Deng et al. 2020). However, anthropogenic activities, particularly in the last few centuries, are contributing to a considerable increase of the presence of these elements in our environments, mainly in topsoil and water (Martínez-Alcalá and Bernal 2020), which in turn has dramatically increased human exposure. Thus, metal(loid)s are released in a wide variety of human activities including industrial, urban, agriculture, hunting and fishing, mining and smelting activities, as well as combustion of fossil fuels. The presence of rarer elements in the environment has also been increased by the exponential electronic and technological industry development. However, although the presence of rare earth elements (REEs) in the environment is being enhanced due to their increasing uses in a diverse and expanding array of high-technology applications (e.g. electric vehicles, energy-efficient lighting, web-enabled devices), the scientific community has poor knowledge of their acute and/or chronic impact on human health (Henríquez-Hernández et al. 2020).

Metals are environmentally persistent and ubiquitous. Exposure to metal(loid) environmental pollutants occurs through the ingestion of contaminated biotic or abiotic matter, absorption through the skin, inhalation of volatile pollutants or maternal transfer. The presence of these pollutants in soils and their availability to living organisms including humans will depend on different factors, including the physicochemical properties of the soils (pH, organic matter content, electric conductivity) and their texture (García-Carmona et al. 2019). Thus, metal(loid)s are more mobile in acidic soils (i.e. with low pH values; Madejón et al. 2006). Meanwhile, the content of organic matter plays an important role for their retention in soils, reducing solubility and bioavailability (a fraction of the pollutant that is available to be taken up by organisms and will be related with the hazardous capability) of metal(loids) (Madejón et al. 2006; Moreno-Jiménez et al. 2012; García-Carmona et al. 2019), thus decreasing their entry into the food chain and potential human risks (e.g. in vegetables and fruits). Jardine et al. (2013) have observed that the bioaccessibility of chromium(IV) decreased with increasing soil organic matter and pH. Therefore, the interaction between the physicochemical properties of the soil and their texture at the local level will be key information when developing soil guideline values, regulatory risk limits and human risk assessments for metal(loid)s present in soils (García-Carmona et al. 2019).

As a result, the solubility and bioavailability of metal(loid)s in contaminated soils could be more important in remediation studies than their total concentration (Wyke et al. 2014). In this way, it has been observed that the application of phosphate amendments in soils contaminated with lead (Pb) can immobilise this toxic metal through the formation of pyromorphite (insoluble lead phosphate), a technique that can reduce Pb bioavailability in the soil (Hettiarachchi et al. 2003). In calcareous soils, the application of phosphoric acid significantly reduces Pb bioaccessibility in these types of soils (Yang and Mosby 2006). The application of amendments with a high carbonate content can also stabilise metal(loid)s in soils (Garrido 2008), because carbonates can enhance metal retention in soils (Sastre et al. 2007). However, careful consideration should be given to the application of amendments depending on the metal pollutant, as organic matter-based amendments can stimulate microbial processes that could facilitate the production of their organic elemental forms, such as methylmercury, which is more toxic and bioaccumulative than inorganic mercury (Hg) (Eckley et al. 2021).

13.5.1 Metal(loid) Toxicity

Metal(loid)s play an important role in living beings, as some of them are a fundamental part of their physiological functions and the maintenance of their biochemical systems, such as manganese, chromium, zinc (Zn) or Cu. Inadequate supply of these essential elements is related to the development of deficiency syndromes and pathologies (Tchounwou et al. 2012). In contrast, other metal(loid)s can be highly deleterious to living organisms even at very low doses, such as arsenic (As) and Pb. Moreover, essential metal(loid)s for human life such as iron (Fe) or Cu can become hazardous above certain concentrations in the environment (e.g. when produced industrially). Currently, the toxic effects of most metal(loid)s are well described, but not their essentiality; some authors have pointed out that there is evidence that aluminium, manganese, nickel and vanadium could serve as essential nutrients in some biological functions (Trumbo et al. 2001; Rehder 2013). Thus, deficiency of, or excess exposure to, metal(loid)s can contribute significantly to the global burden of disease (Mamtani et al. 2011; Rehman et al. 2018). Humans are exposed to toxic/carcinogenic metal(loid)s through a variety of processes due to industrial activity, with selected populations exposed to much higher levels (e.g. as an occupational exposure or as result of living in particular areas of the world). Public appreciation of the toxicity and carcinogenic effects of metal(loid)s is possibly focused on Pb, Hg and perhaps As, to which large numbers of people are exposed to in Bangladesh, for example, arising from inherently high levels of this element in the rocks and soil in that and other regions of the world. Metal(loid)s are thought to exert their toxicity at or above critical concentrations in cells and tissues, and target tissues and cells are targets because of their preferential accumulation of particular metal(loids). Because metal(loid) s are not catabolised to other metabolites, undergoing redox reactions or changes in speciation, they can persist and exert their effects chronically and also accumulate in areas such as the CNS. Metal(loid) poisoning manifests in several pathological states, with some toxic exposures leading to characteristic pathology and tissue injury (e.g. the induction of bone disease, anaemia and kidney injury from excessive exposure to cadmium [Cd] and more specific pathologies such as lung disease arising from exposure to beryllium [Be]). The toxic effects of selected metals may also be more prominent in sensitive populations, such as the effect of Pb on the developing nervous system. Neurotoxicity, cardiovascular disease, kidney dysfunction, carcinogenesis and adverse reproductive effects are some of the toxic effects of metals. Mechanisms of toxicity for selected metals are discussed in material below. Often the toxic effects arise because of interference by toxic metals in normal processes, where the toxic metal may replace the normally functioning metal in a protein. Also, metal(loid) binding can modify protein structure to alter normal metabolism or cell signalling processes; for certain elements engagement in redox cycling, those are futile reduction oxidation processes that generate oxidants leading to oxidative injury to cells and subsequent dysfunction.

A brief discussion of the industrial metals Cu, Cd, Pb and Hg is presented below. These metals were selected owing to their significant global production and their toxicological relevance on human health. The nonferrous metals Cu and Pb were also selected because industry generates large amount of wastes (Pietrzykowski and Krzaklewski 2018). Cd, Pb and Hg are also important foodborne metals that contribute substantially to the burden of diseases worldwide: the ingestion of As, methylmercury, Pb and Cd resulted in more than 1 million illnesses in 2015 (Gibb et al. 2019). Comprehensive information about As and Be, which are potent carcinogenic elements, appears in Chapter 27 as part of a risk assessment study.

13.5.2 Copper

Copper (Cu) is an essential element for the healthy functioning of living organisms. The main symptoms of Cu deficiency are anaemia (microcytic, normocytic or macrocytic) and neutropenia, skeletal demineralisation, degeneration of vascular tissue with arterial aneurysms, depigmentation of hair and skin, cerebral and cerebellar degeneration, hypotonia and hypothermia (Wazir and Ghobrial 2017; Altarelli et al. 2019).

Cu deficiency is generally caused by poor availability in the diet or in individuals who are subject to parenteral nutrition. Deficiencies may also appear due to increased excretion (e.g. in patients undergoing dialysis). Regarding its toxicity, Cu is considered a nontoxic element, although excessive exposure to this metal can cause serious effects on human health as it could contribute to oxidative modification of DNA (Yaman and Akdeniz 2004; Taylor et al. 2020). Its toxic effects are caused both by the cation itself and by interfering with the absorption and distribution of Fe and Zn. In acute poisonings due to ingestion of soluble Cu salts, haemolytic symptoms appear accompanied by lesions in the liver and brain, with necrosis that can lead to death (Franchitto et al. 2008). In children, chronic exposure through food preserved in containers with Cu has been associated with the appearance of liver lesions. Wilson disease and Menkes disease are inherited disorders of Cu transport and metabolism (Kaler 2020).

The EPA has stated that there is no evidence of carcinogenic effects for Cu or its compounds by any route of exposure for humans (ATSDR 2004). The International Agency for Research on Cancer (IARC) has classified Cu 8-hydroxyquinoline as Group 3 (i.e. not classifiable as to its carcinogenicity to humans; more information about the IARC classification can be found at their website).[16] The Occupational Safety and Health Administration (OSHA) has established that levels of Cu in the air in workplaces should not exceed 0.1 mg/m^3.

13.5.3 Cadmium

Cadmium (Cd) is released from burning fossil fuels, mining and waste incinerators. Leafy vegetables and grains can take up Cd from soils, which represents a significant route of human exposure. Animal studies show that Cd is potentially more toxic to an infant than to a foetus, apparently because it induces anorexia in pregnant animals that ingest it, resulting in low uptake (Satarug 2019). Low-level exposure to Cd may lead to damage to the kidneys, liver, skeletal system and cardiovascular system, as well as to deterioration of sight and hearing (Mezynska and Brzóska 2018) and even dermal conditions (Wacewicz-Muczyńska et al. 2021).

The toxic effects of Cd depend on the concentration and duration of exposure as well as the route of entry into the body and its physical and chemical form, being those compounds that are more soluble in water generally more toxic (Rafati et al. 2017; Genchi et al. 2020). Cd has a long biological half-life and can therefore bioaccumulate in the human body over time due to its low rate of excretion (Wang and Du 2013).

In the kidneys, Cd affects tubular function, producing proteinuria, aminoaciduria, glycosuria and kidney stones. At the pulmonary level, the main disease it produces is lung cancer, although there is some controversy about this. Other effects are chronic obstructive pulmonary disease and emphysema. At the cardiovascular level, Cd has been linked to an increase in blood pressure. Chronic exposures cause 'itai-itai' disease ('it hurts-it hurts' in Japanese), a disease that was recognised in Japan following a mass Cd poisoning related to industrialisation. Cd poisoning is characterised by osteomalaecia with severe bone pain resulting from proximal tubular injury (Nishijo et al. 2017; Imura et al. 2019).

In addition, Cd is highly neurotoxic, because it can increase blood-brain barrier permeability and can accumulate in different brain regions and peripheral neuronal systems (Wang and Du 2013).

The IARC has classified Cd and its compounds in Group 1 (as human carcinogens; IARC 2009). The EPA classifies them in Group B1 (as probable human carcinogens; ATSDR 2012). Cd can induce oxidative stress, DNA damage and cell death processes, resulting in increase the risk of cancer and cancer-related diseases. OSHA has established a limit for workers' exposure of less than 5 μg/m^3 day.

13.5.4 Lead

Lead (Pb) is one of the most important environmental contaminants; it is toxic even at very low concentrations. Pb has been used since antiquity, and it is ubiquitous and persistent in the environment, presenting rapid bioaccumulation in tissues and a long biological half-life. Burning fossil fuels, mining and manufacturing have been the major sources of this metal to the environment (Cheema et al. 2020). Governments are banning or reducing the use of this metal in gasoline, paints and ceramic products, caulking and pipe solder owing to its deleterious effects on human health. The use of Pb as an additive in gasoline was banned in 1996 in the United States. It is a cumulative toxicant that affects multiple body systems, including the neurological, haematological, gastrointestinal, cardiovascular, renal, skeletal and endocrine systems (Boskabady et al. 2018).

Pb affects the nervous system and causes slow nerve conduction, fatigue, mood swings, drowsiness, concentration disorders, headaches and coma. In the circulatory system it produces increasing blood pressure, peripheral vascular resistance, cardiac output and, in severe cases, encephalopathy (Briffa et al. 2020). Gastrointestinal disorders from Pb exposure include colic/pain, nausea, vomiting, diarrhoea and constipation. Pb also causes fertility disorders due to its effects on reproductive hormones (Charkiewicz and Backstrand 2020).

Furthermore, because of its accumulation in the bone system, Pb has been related to different bone pathologies, mainly Paget disease (Singer 2020). It will also affect the renal system, being related to various kidney disorders, such as lead nephropathy or renal adenocarcinoma.

Infants and children could be at greater risk of exposure to Pb because of their behavioural traits, such as putting objects in their mouths and increased time spent outdoors, which potentially increases the likelihood of children coming into contact with Pb present in the soil and house dust (Donangelo et al. 2021). Thus, Pb poisoning is more common in children than adults. The exposure to this element is especially dangerous for the foetus; high exposure to this metal in pregnant women can lead to premature births, children with low birth weight and even miscarriages. Children are a population particularly sensitive to Pb contamination (Yamada et al. 2020), especially under the age of 6, most likely because of the rapid brain growth and development with associated periods of heightened vulnerability, and because of high demand for nutrients. Pb can interfere with the ability to learn, impairs memory and interferes with growth and development. Pb has effects on speech, hearing, hyperactivity and nerve conduction; it can cause intestinal discomfort, constipation, vomiting, weight loss and muscle aches. Various authors have observed damage to cognitive function, learning difficulties and reduced intelligence quotient (IQ). The US Centers for Disease Control and Prevention (CDC) has established a blood lead reference value of 5 μg/dL to identify children who have been exposed to this metal.[17]

The EPA has classified Pb as a probable carcinogenic in humans (ATSDR 2020). The IARC has determined that inorganic Pb is probably carcinogenic to humans (Group 2A), whereas there is insufficient information regarding its organic compounds (Group 3) (IARC 2006).

13.5.6 Mercury

Mercury (Hg) is a ubiquitous pollutant: about 50% of this element found in the environment originates from natural sources, and the remainder emanates from industrial processes, including paper and pulp production and coal-burning furnaces. Metallic Hg can travel widely in air and water and is converted in water by bacterial methylation action to the more readily absorbed organic compounds (primarily methylmercury), which are then easily bioconcentrated through the aquatic food chain (Santos-Sacramento et al. 2021). Populations for whom fish and rice are a main part of their diet can be highly exposed to Hg (Dórea 2020). Chronic exposure to Hg vapour in workers is of significant concern because it can cross the blood-brain barrier as it is very fat soluble (thus affecting the CNS; Roda et al. 2016).

Exposure to Hg is mostly characterised by a neurological-psychiatric (neurotoxic) and renal (nephrotoxic) syndrome; it can also affect the cardiovascular system and act as a genotoxic agent. The main target is the CNS; neurological disorders may initially resemble amyotrophic lateral sclerosis. Chronic exposures to low concentrations of Hg are more common than acute exposures.

Exposure to metallic Hg orally is not significant, because only 0.1% is absorbed. In contrast, organic Hg compounds can be absorbed more than 95%, mainly by consumption of contaminated fish, as indicated previously. As a result, it has been recommended that dietary exposure to Hg be reduced: pregnant women and children particularly should avoid consuming large quantities of fish, especially predatory fish species such as shark, swordfish and tuna (Murillo-Cisneros et al. 2021). Hg can also pass into breast milk (ATSDR 1999). Furthermore, several studies have indicated that exposure to Hg would affect the hormonal system as an endocrine disruptor (e.g. by inducing the secretion of female hormones in both sexes; Henriques et al. 2019).

The IARC (1993) classifies methylmercury compounds in Group 2B (possibly carcinogenic), whereas elemental Hg and inorganic Hg compounds are classified in group 3 (no evidence of carcinogenic activity). The EPA has determined that mercuric chloride and methylmercury are possibly carcinogenic to humans (ATSDR 1999).

13.6 CONCLUSIONS

Exponential industrial and technological developments, together with the increase in human demographics and subsequent impact on the agriculture and pharmaceutical industries, are facilitating a global overexposure to thousands of industrial chemicals and generating emerging (and unknown) environmental threats for humans. Appropriate tools and guidance are needed to be in place to tackle these threats to industrial chemicals and to respond to HazMat incidents.

Environmental and toxicology experts are needed to implement toxicological evaluation, monitoring and characterisation programmes necessary to be able to appropriately manage and regulate the industrial chemical substances that are marketed. Future professionals will need to identify novel and less hazardous industrial chemicals to protect humans and the environment.

In this chapter, we have provided an overview of major industrial chemicals, specifically pesticides (herbicides, insecticides, fungicides and rodenticides) and metals (Cu, Cd, Pb and Hg). Finally, we provided an overview to HazMat incidents and recent developments, tools and guidance to respond to these events, protect human health, minimise the dispersion of chemicals involved in incidents and implement remedial responses to return the impacted environments to normal.

NOTES

1 CAS REGISTRY: www.cas.org/support/documentation/chemical-substances (accessed February 14, 2021).
2 REACH Registration Statistics. European Chemicals Agency: https://echa.europa.eu/registration-statistics (accessed January 28, 2021).
3 Frank R. Lautenberg Chemical Safety for the 21st Century Act (Lautenberg Chemical Safety Act): www.epa.gov/assessing-and-managing-chemicals-under-tsca/frank-r-lautenberg-chemical-safety-21st-century-act (accessed January 15, 2021).
4 Information on Chemicals, European Chemicals Agency (ECHA): https://echa.europa.eu/information-on-chemicals (accessed February 13, 2021).
5 ASPIRE decision-aided tool for PRISM. US Department of Health and Human Services: https://chemm.nlm.nih.gov/aspire.htm (accessed January 24, 2021).
6 Recovery, Remediation and Environmental Decontamination. UK Government: www.gov.uk/government/collections/recovery-remediation-and-environmental-decontamination (accessed November 12, 2020).
7 European Multiple Environmental Threats Emergency Network (EMETNET): www.rivm.nl/en/about-rivm/mission-and-strategy/international-affairs/international-projects/emetnet (accessed January 23, 2021).
8 Flash Environmental Assessment Tool (FEAT) 2.0: www.eecentre.org/resources/feat/ (accessed January 23, 2021).
9 Summary of the Federal Insecticide, Fungicide, and Rodenticide Act. EPA: www.epa.gov/laws-regulations/summary-federal-insecticide-fungicide-and-rodenticide-act (accessed February 13, 2021).
10 PAN Europe: www.pan-europe.info/press-releases/2019/12/victory-future-generations-european-governments-ban-brain-harming-pesticides (accessed January 24, 2021).
11 Basic Information about Pesticide Ingredients. EPA: www.epa.gov/ingredients-used-pesticide-products/basic-information-about-pesticide-ingredients (accessed January 28, 2021).
12 Classification and Labelling (CLP/GHS). European Commission: https://ec.europa.eu/growth/sectors/chemicals/classification-labelling_en (accessed February 13, 2021).
13 Pesticide Labels and GHS: Comparison and Samples. EPA: www.epa.gov/pesticide-labels/pesticide-labels-and-ghs-comparison-and-samples (accessed January 24, 2021).
14 EU Pesticide Residues. European Commission: https://ec.europa.eu/food/plant/pesticides/eu-pesticides-database/mrls/?event=search.pr (accessed February 13, 2021).
15 What Are Biopesticides? EPA: www.epa.gov/ingredients-used-pesticide-products/what-are-biopesticides (accessed February 13, 2021).
16 Agents Classified by the IARC Monographs, International Agency for Research on Cancer. IARC: https://monographs.iarc.who.int/agents-classified-by-the-iarc/ (accessed February 1, 2021).
17 Childhood Lead Poisoning Prevention. Blood Lead Reference Value. CDC: www.cdc.gov/nceh/lead/data/blood-lead-reference-value.htm (accessed February 2, 2021).

BIBLIOGRAPHY

Adeyinka A., E. Muco, and L. Pierre. 2020. Organophosphates. In *StatPearls*. Treasure Island, FL: StatPearls Publishing. www.statpearls.com/articlelibrary/viewarticle/26333/

Altarelli, M., N. Ben-Hamouda, A. Schneider, and M. M. Berger. 2019. Copper deficiency: Causes, manifestations, and treatment. *Nutrition Clin Practice* 34(4): 504–13.

ATSDR (Agency for Toxic Substances and Disease Registry). 1999. *Toxicological Profile for Mercury*. Atlanta, GA: U.S. Department of Health and Human Services, Public Health Service. https://wwwn.cdc.gov/TSP/ToxProfiles/ToxProfiles.aspx?id=115&tid=24 (accessed February 13, 2021).

ATSDR (Agency for Toxic Substances and Disease Registry). 2004. *Toxicological Profile for Copper.* Atlanta, GA: U.S. Department of Health and Human Services, Public Health Service. https://wwwn.cdc.gov/TSP/ToxProfiles/ToxProfiles.aspx?id=206&tid=37 (accessed February 13, 2021).

ATSDR (Agency for Toxic Substances and Disease Registry). 2012. *Toxicological Profile for Cadmium.* Atlanta, GA: U.S. Department of Health and Human Services, Public Health Service. https://wwwn.cdc.gov/TSP/ToxProfiles/ToxProfiles.aspx?id=48&tid=15 (accessed February 13, 2021).

ATSDR (Agency for Toxic Substances and Disease Registry). 2020. *Toxicological Profile for Lead.* Atlanta, GA: U.S. Department of Health and Human Services, Public Health Service. https://wwwn.cdc.gov/TSP/ToxProfiles/ToxProfiles.aspx?id=96&tid=22 (accessed February 13, 2021).

Balog, A., T. Hartel, H. D. Loxdale, and K. Wilson. 2017. Differences in the progress of the biopesticide revolution between the EU and other major crop-growing regions. *Pest Management Sci* 73(11): 2203–8.

Bello, D., C. A. Herrick, T. J. Smith, S. R. Woskie, et al. 2007. Skin exposure to isocyanates: Reasons for concern. *Environ Health Perspect* 115: 328–35.

Benbrook, C. M., and D. R. Davis. 2020. The dietary risk index system: A tool to track pesticide dietary risks. *Env Health* 19(1): 103.

Bjørling-Poulsen, M., H. R. Andersen, and P. Grandjean. 2008. Potential developmental neurotoxicity of pesticides used in Europe. *Env Health* 7: 50.

Bos, P. M. J., M. W. A. M. Ruijten, U. Gundert-Remy, S. Bull, E. Nielsen, S. M. Tissot, et al. 2013 *Human Risk Assessment of Single Exposure in Chemical Incidents: Present Situation and Emerging Chemical Incident Scenarios.* RIVM Report 320300001. www.rivm.nl/en/bibcite/reference/6441 (accessed January 20, 2021).

Boskabady, M., N. Marefati, T. Farkhondeh, F. Shakeri, A. Farshbaf, and M. H. Boskabady. 2018. The effect of environmental lead exposure on human health and the contribution of inflammatory mechanisms, a review. *Environ Int* 120: 404–20.

Boucaud-Maitre, D., M. O. Rambourg, S. Sinno-Tellier, E. Puskarczyk, X. Pineau, M. Kammerer, et al. 2019. Human exposure to banned pesticides reported to the French Poison Control Centers: 2012–2016. *Env Tox Pharmacol* 69: 51–6.

Briffa, B., E. Sinagra, and R. Blundell. 2020. Heavy metal pollution in the environment and their toxicological effects on humans. *Heliyon* 6(9): e04691.

Buah-Kwofie, A., and M. S. Humphries. 2021. Organochlorine pesticide accumulation in fish and catchment sediments of Lake St Lucia: Risks for Africa's largest estuary. *Chemosphere.* https://doi.org/10.1016/j.chemosphere.2021.129712.

Butler, D. 2019. US adopts science-bases guidance for chemical-attack response. *Nature News.* www.nature.com/articles/d41586-019-00646-4 (accessed January 24, 2021).

Calder, A., and S. Bland. 2015. Chemical, biological, radiological and nuclear considerations in a major incident. *Surgery (Oxf)* 33(9): 442–8.

CAS REGISTRY®. 2021. www.cas.org/support/documentation/chemical-substances (accessed January 15, 2021).

Charkiewicz, A. E., and J. R. Backstrand. 2020. Lead toxicity and pollution in Poland. *Int J Environ Res Public Health* 17(12): 4385.

Cheema, A. I., G. Liu, B. Yousaf, Q. Abbas, and H. Zhou. 2020. A comprehensive review of biogeochemical distribution and fractionation of lead isotopes for source tracing in distinct interactive environmental compartments. *Sci Total Environ* 719: 135658.

Chilcott, R. P., J. Larner, A. Durrant, P. Hughes, D. Mahalingam, S. Rivers, et al. 2019b. Evaluation of US Federal Guidelines (Primary Response Incident Scene Management [PRISM]) for Mass Decontamination of Casualties During the Initial Operational Response to a Chemical Incident. *Annals Emergency Med* 73(6): 671–84.

Chilcott, R. P., J. Larner, and H. Matar. 2019a. UK's initial operational response and specialist operational response to CBRN and HazMat incidents: A primer on decontamination protocols for healthcare professionals. *Emergency Med J* 36(2): 117–23.

Coalova, I., C. Ríos, and G. Chaufan. 2014. Toxicology in vitro influence of the spray adjuvant on the toxicity effects of a glyphosate formulation. *Toxicol Vitr* 28(7): 1306–11.

Daniels, R. D. 2018. Occupational asthma risk from exposures to toluene diisocyanate. A review and risk assessment. *Am J Ind Med* 61: 282–92.

Deng, F., F. Zeng, G. Chen, X. Feng, A. Riaz, X. Wu, et al. 2020. Metalloid hazards: From plant molecular evolution to mitigation strategies. *J Hazardous Mat* 124495.

Donangelo, C. M., B. T. Kerr, E. I. Queirolo, M. Vahter, F. Peregalli, N. Mañay, et al. 2021. Lead exposure and indices of height and weight in Uruguayan urban school children, considering co-exposure to cadmium and arsenic, sex, iron status and dairy intake. *Environ Res* 195: 110799.

Dórea, J. G. 2020. Neurotoxic effects of combined exposures to aluminum and mercury in early life (infancy). *Environ Res* 188: 109734.

Duarte-Davidson, R., R. Orford, S. Wyke, M. Griffiths, R. Amlôt, and R. Chilcott. 2014. Recent advances to address European Union Health Security from cross border chemical health threats. *Environ Int* 72: 3–14.

Dukare, A. S., R. K. Singh, R. K. Jangra, and B. Bhushan. 2020. Non-fungicides-based promising technologies for managing post-production penicillium induced spoilage in horticultural commodities: A comprehensive review. *Food Rev Int* 1–41.

ECHA (European Chemicals Agency). 2018. *What is an Infocard*? Reference: ECHA-18-H-21-EN. ISBN: 978-92-9020-649-1. https://echa.europa.eu/documents/10162/22177693/what_is_an_infocard_en.pdf/4960b3a4-a84f-461d-926c-b4a683b2f98f (accessed January 28, 2021).

Eckley, C. S., T. P. Luxton, B. Stanfield, A. Baldwin, J. Holloway, J. McKernan, et al. 2021. Effect of organic matter concentration and characteristics on mercury mobilization and methylmercury production at an abandoned mine site. *Environ Pollut* 271: 116369.

Engel, L. S., E. Werder, J. Satagopan, A. et al. 2017. Insecticide use and breast cancer risk among farmers' wives in the agricultural health study. *Env Health Perspect* 125(9): 097002.

Fantke, P., and R. Juraske. 2013. Variability of pesticide dissipation half-lives in plants. *Environ Sci Technol* 47(8): 3548–62.

FAO and WHO (Food and Agricultural Organisation of United Nations and World Health Organization). 2014. *International Code of Conduct on Pesticide Management, Rome*. www.fao.org/agriculture/crops/thematic-sitemap/theme/pests/code/en/ (accessed January 24, 2021).

Fedorovitch, G. 2019. Aerosol particles in lungs: Theoretical modelling of deposition and mucociliary clearance. In *Rhinosinusitis*, eds. B. S. Gendeh, and M. Turkalj. IntechOpen. www.intechopen.com/books/rhinosinusitis/aerosol-particles-in-lungs-theoretical-modeling-of-deposition-and-mucociliary-clearance.

Fletcher, S. J., P. T. Reeves, T. Hoang, and N. Mitter. 2020. A perspective on RNAi-based biopesticides. *Front Plant Sci* 11: 51.

Franchitto, N., P. Gandia-Mailly, B. Georges, A. Galinier, N. Telmon, J. Ducassé, et al. 2008. Acute copper sulphate poisoning: A case report and literature review. *Resuscitation* 78(1): 92–6.

Ganguly, B. B., S. Mandal, N. Banerjee, N. Kadam, and R. Abbi. 2018. Effect of age at exposure on chromosome abnormalities in MIC-exposed Bhopal population detected 30 years post-disaster. *Mutat Res* 809: 32–50.

Gao, Y., Q. Zhou, G. Xie, and Z. Yao. 2012. Temperature-controlled ionic liquid dispersive liquid-phase microextraction combined with HPLC with ultraviolet detector for the determination of fungicides. *J Sep Sci* 35(24): 3569–74.

García-Carmona, M., A. Romero-Freire, M. S. Aragón, and F. M. Peinado. 2019. Effectiveness of ecotoxicological tests in relation to physicochemical properties of Zn and Cu polluted Mediterranean soils. *Geoderma* 338: 259–68.

Garrido, H. 2008. *Guadiamar, Science, Technique and Restoration: The Mining Accident Ten Years After*. Consejo Superior de Investigaciones Científicas (CSIC).

Genchi, G., M. S. Sinicropi, G. Lauria, A. Carocci, and A. Catalano. 2020. The effects of cadmium toxicity. *J Environ Res Public Health* 17: 3782.

Gibb, H. J., A. Barchowsky, D. Bellinger, P. M. Bolger, C. Carrington, A. Havelaar, et al. 2019. Estimates of the 2015 global and regional disease burden from four foodborne metals—arsenic, cadmium, lead and methylmercury. *Environ Res* 174: 188–94.

Goode, E. J., S. Collins, C. Hague, R. Orford, and R. Duarte-Davidson. 2019. Rapid public health risk assessments for emerging chemical health threats. In *Chemical Health Threats: Assessing and Alerting*, eds. R. Duarte-Davidson, T. Gaulton, S. Wyke, and S. Collins, 138–51. London: Royal Society of Chemistry.

Griffiths, M. 2019. Response to chemical incident emergencies. In *Chemical Health Threats: Assessing and Alerting*, eds. R. Duarte-Davidson, T. Gaulton, S. Wyke, and S. Collins, 200–33. London: Royal Society of Chemistry.

Gupta, P. K. 2018. Epidemiology of animal poisonings in Asia. In *Veterinary Toxicology*, 3rd ed., ed. R. C. Gupta, 57–69. Academic Press: London.

Gupta, R. C. 2014. Carbamate pesticides. In *Encyclopedia of Toxicology*, 3rd ed., 661–64. London: Elsevier.

Gupta, R. C., I. R. Miller Mukherjee, J. K. Malik, R. B. Doss, W. Dettbarn, and D. Milatovic. 2019. Insecticides. In *Biomarkers in Toxicology*, 2nd ed., ed. R. C. Gupta, 455–75. London: Academic Press.

Hamdi, H., S. Abid-Essefi, and J. Eyer. 2021. Neuroprotective effects of myricetin on epoxiconazole-induced toxicity in F98 cells. *Free Radical Biol Med* 164: 154–63.

Han, Y., L. Zhu, J. Wang, J. Wang, H. Xie, and S. H. Zhang. 2014. Integrated assessment of oxidative stress and DNA damage in earthworms (Eisenia fetida) exposed to azoxystrobin. *Ecotoxicol Environ Safety* 107: 214–19.

Henriques, M. C., S. Loureiro, M. Fardilha, and M. T. Herdeiro. 2019. Exposure to mercury and human reproductive health: A systematic review. *Reprod Toxicol* 85: 93–103.

Henríquez-Hernández, L. A., D. Romero, A. González-Antuña, B. Gonzalez-Alzaga, M. Zumbado, and L. D. Boada. 2020. Biomonitoring of 45 inorganic elements measured in plasma from Spanish subjects: A cross-sectional study in Andalusian population. *Sci Total Env* 706: 135750.

Hettiarachchi, G. M., G. M. Pierzynski, F. W. Oehme, O. Sonmez, and J. A. Ryan. 2003. Treatment of contaminated soil with phosphorus and manganese oxide reduces lead absorption by Sprague-Dawley rats. *J Environ Qual* 32(4): 1335–45.

Hites, R. A. 2021. The rise and fall of chlorpyrifos in the United States. *Env Sci Technol* 55(3): 1354–8.

IARC (International Agency for Research on Cancer). 1993. *Beryllium, Cadmium, Mercury, and Exposures in the Glass Manufacturing Industry. Monographs on the Evaluation of Carcinogenic Risks to Humans*, Vol. 58. Lyon, France: IARC. https://publications.iarc.fr/76 (accessed February 2, 2021).

IARC (International Agency for Research on Cancer). 2006. *Inorganic and Organic Lead Compounds. Monographs on the Evaluation of Carcinogenic Risks to Humans*, Vol. 87. Lyon, France: IARC. https://publications.iarc.fr/105 (accessed February 2, 2021).

IARC (International Agency for Research on Cancer). 2009. *Arsenic, Metals, Fibres, and Dusts. Monographs on the Evaluation of Carcinogenic Risks to Humans*, Vol. 100C. Lyon, France: IARC (accessed February 2, 2021).

Imura, J., K. Tsuneyama, and Y. Ueda. 2019. Novel pathological study of cadmium nephropathy of Itai-itai disease. In *Cadmium Toxicity: Current Topics in Environmental Health and Preventive Medicine*, eds. S. Himeno and K. Aoshima, 39–50. Springer: Singapore.

Jardine, P. M., M. A. Stewart, M. O. Barnett, N. T. Basta, S. C. Brooks, and S. Fendorf. 2013. Influence of soil geochemical and physical properties on chromium(VI) sorption and bioaccessibility. *Environ Sci Technol* 47(19): 11241–8.

Jayaraj, R., P. Megha, and P. Sreedev. 2016. Organochlorine pesticides, their toxic effects on living organisms and their fate in the environment. *Interdiscip Toxicol* 9(3–4): 90–100.

Jayarathne, A., P. Egodawatta, G. A. Ayoko, and A. Goonetilleke. 2018. Assessment of ecological and human health risks of metals in urban road dust based on geochemical fractionation and potential bioavailability. *Sci Tot Env* 635: 609–1619.

Jung, J., Y. Bae, Y. K. Cho, X. Ren, and Y. Sun. 2020. Structural insights into conformation of amphiphilic quaternary ammonium chitosans to control fungicidal and anti-biofilm functions. *Carbohyd Polym* 228: 115391.

Kaler, S. G. 2020. Inherited disorders of human copper metabolism. In *Emery and Rimoin's Principles and Practice of Medical Genetics and Genomics*, eds. R. Pyeritz, B. Korf, and W. Grody, 413–43. London: Academic Press.

Kassouf, N., S. Syed, J. Larner, R. Amlôt, and R. P. Chilcott. 2017. Evaluation of absorbent materials for use as ad hoc dry decontaminants during mass casualty incidents as part of the UK's Initial Operational Response (IOR). *PLoS ONE* 12(2), e0170966. https://doi.org/10.1371/journal.pone.0170966.

Khademi, H., M. Gabarrón, A. Abbaspour, S. Martínez-Martínez, A. Faz, and J. A. Acosta. 2019. Environmental impact assessment of industrial activities on heavy metals distribution in street dust and soil. *Chemosphere* 217: 695–705.

Kim, J., F. A. Gobas, J. A. Arnot, D. E. Powell, R. M. Seston, and K. B. Woodburn. 2016. Evaluating the roles of biotransformation, spatial concentration differences, organism home range, and field sampling design on trophic magnification factors. *Sci Total Environ* 551–552: 438–51.

Konopelko, M., M. Kusz, J. Popiołek, A. Szafraniec, A. Alzubedi, and B. Nieradko-Iwanicka. 2018. Pyrethroids—could they be dangerous? *J Educ Health Sport* 8(9): 1191–6.

Li, Q., W. Yu, Y. Qu, J. Q. Wang, N. Mao, and H. Kang. 2021. Acute toxic encephalopathy following bromadiolone intoxication: a case report. *BMC Neurology* 21(1): 8.

Macfarlane, E., R. Carey, T Keegel, S. El Zaemay, and L. Fritschi. 2013. Dermal exposure associated with occupational end use of pesticides and the role of protective measures. *Safety Health Work* 4(3): 136–41.

Mackay, D., A. K. D. Celsie, D. E. Powell, and J. M. Parnis. 2018. Bioconcentration, bioaccumulation, biomagnification and trophic magnification: A modelling perspective. *Environ Sci Process Impacts* 20(1): 72–85.

Madejón, E., A. P. de Mora, E. Felipe, P. Burgos, and F. Cabrera. 2006. Soil amendments reduce trace element solubility in a contaminated soil and allow regrowth of natural vegetation. *Environ Pollut* 139(1): 40–52.

Mamtani, R., P. Stern, I. Dawood, and S. Cheema. 2011. Metals and disease: A global primary health care perspective. *J Toxicol* 2011: 319136. www.hindawi.com/journals/jt/2011/319136/

Martín Reina, J., J. A. Duarte, L. Cerrillos, J. D. Bautista Palomas, and I. M. Moreno Navarro. 2017. Insecticide Reproductive Toxicity Profile: Organophosphate, Carbamate and Pyrethroids. *J Toxins* 4(1): 1–7.

Martínez-Alcalá, I., and M. P. Bernal. 2020. Environmental impact of metals, metalloids, and their toxicity. In *Metalloids in Plants: Advances and Future Prospects*, eds. R. Deshmukh, D. K. Tripathi, and G. Guerriero, 451–88. New York: John Wiley & Sons.

Mesnage, R., and M. N. Antoniou. 2018. Ignoring adjuvant toxicity falsifies the safety profile of commercial pesticides. *Front Pub Health* 5: 361.

Mesnage, R., and G.-E. Séralini. 2018. Toxicity of pesticides on health and environment. *Front Pub Health* 6: 268.

Mezynska, M., and M. M. Brzóska. 2018. Environmental exposure to cadmium: A risk for health of the general population in industrialized countries and preventive strategies. *Environ Sci Pollut Res Int* 25(4): 3211–32.

Moreno-Jiménez, E., E. Esteban, R. O. Carpena-Ruiz, M. C. Lobo, and J. M. Peñalosa. 2012. Phytostabilisation with Mediterranean shrubs and liming improved soil quality in a pot experiment with a pyrite mine soil. *J Hazard Mater* 201–202: 52–9.

Murillo-Cisneros, D. A., T. Zenteno-Savín, J. Harley, A. Cyr, P. Hernández-Almaraz, P. R. Gaxiola-Robles, et al. 2021. Mercury concentrations in Baja California Sur fish: Dietary exposure assessment. *Chemosphere* 267: 129233.

Nakayama, S., A. Morita, Y. Ikenaka, H. Mizukawa, and M. Ishizuka. 2019. A review: Poisoning by anticoagulant rodenticides in non-target animals globally. *J Vet Med Sci* 81(2): 298–313.

Nisbet, A., S. Watson, and J. Brown. 2015. *The UK Recovery Handbook for Radiation Incidents.* Version 4. PHE publications gateway number: 2015067. www.gov.uk/government/publications/uk-recovery-handbooks-for-radiation-incidents-2015 (accessed December 10, 2020).

Nishijo, M., H. Nakagawa, Y. Suwazono, K. Nogawa, and T. Kido. 2017. Causes of death in patients with Itai-itai disease suffering from severe chronic cadmium poisoning: A nested case-control analysis of a follow-up study in Japan. *BMJ Open* 7(7): e015694.

OECD. 2019. *Global Chemicals Outlook II. From Legacies to Innovative Solutions: Implementing the 2030 Agenda for Sustainable Development—Synthesis Report.* United Nations Environment Programme. www.unep.org/resources/report/global-chemicals-outlook-ii-legacies-innovative-solutions (accessed February 14, 2021).

Peña-Fernández, A., S. Wyke, N. Brooke, and R. Duarte-Davidson. 2014. Factors influencing recovery and restoration following a chemical incident. *Environ Int* 72: 98–108.

Pietrzykowski, M., and W. Krzaklewski. 2018. Reclamation of mine lands in Poland. In *Bio-Geotechnologies for Mine Site Rehabilitation*, eds. M. N. V. Prasad, P. Favas, and S. K. Maiti, 496–513. Amsterdam, Oxford and Cambridge: Elsevier.

Pottage, T., E. Goode, C. Shieber, S. Wyke, S. Speight, and A. M. Bennett. 2015. *UK Recovery Handbook for Biological Incidents.* Version 1. PHE publications gateway number 2015406. www.gov.uk/government/publications/uk-recovery-handbook-for-biological-incidents (accessed December 10, 2020).

Quijano, L., V. Yusà, G. Font, and O. Pardo. 2016. Chronic cumulative risk assessment of the exposure to organophosphorus, carbamate and pyrethroid and pyrethrin pesticides through fruit and vegetables consumption in the region of Valencia (Spain). *Food Chem Toxicol* 89: 39–46.

Rafati Rahimzadeh, M., M. Rafati Rahimzadeh, S. Kazemi, and A. A. Moghadamnia. 2017. Cadmium toxicity and treatment: An update. *Caspian J Intern Med* 8(3): 135–45.

Rehder, D. 2013. The future of/for vanadium. *Dalton Trans* 42(33): 11749–61.
Rehman, K., F. Fatima, I. Waheed, and M. Akash. 2018. Prevalence of exposure of heavy metals and their impact on health consequences. *J Cell Biochem* 119(1): 157–84.
Rial-Berriel, C., A. Acosta-Dacal, M. Á. Cabrera Pérez, A. Suárez-Pérez, A. Melián Melián, M. Zumbado, et al. 2020. Intensive livestock farming as a major determinant of the exposure to anticoagulant rodenticides in raptors of the Canary Islands (Spain). *Sci Tot Env* 768: 144386. https://doi.org/10.1016/j.scitotenv.2020.144386.
Roberts J. R., and J. R. Reigart. 2013. *Recognition and Management of Pesticide Poisonings*, 6th ed. Washington, DC: U.S. Environmental Protection Agency, 2013, EPA Reports epa735k13001. www.epa.gov/pesticide-worker-safety/pesticide-poisoning-handbook-complete-document (accessed January 28, 2021).
Roda, E., A. Giampreti, S. Vecchio, P. Apostoli, and T. Coccini. 2016. Mercury vapour long-lasting exposure: lymphocyte muscarinic receptors as neurochemical markers of accidental intoxication. *Case Rep Med* 2016: 9783876. www.hindawi.com/journals/crim/2016/9783876/
Sánchez, C. E. P., S. F. M. Mora, C. E. P. Illescas, and M. P. A. Gaibor. 2019. Uso de plaguicidas y su consecuencia en la leucemia linfoide y mieloide en trabajadores agrícolas. *Mikarimin Rev Cient Multidisciplinaria* 5(1): 37–56.
Santos-Sacramento, L., G. P. Arrifano, A. Lopes-Araújo, M. Augusto-Oliveira, R. Albuquerque-Santos, P. Y. Takeda, et al. 2021. Human neurotoxicity of mercury in the Amazon: A scoping review with insights and critical considerations. *Ecotox Environ Safety* 208: 111686.
Sastre, J., G. Rauret, and M. Vidal. 2007. Sorption-desorption tests to assess the risk derived from metal contamination in mineral and organic soils. *Environ Int* 33(2): 246–56.
Satarug, S. 2019. Cadmium sources and toxicity. *Toxics* 7(2): 25.
Settimi, L., R. Orford, F. Davanzo, et al. 2016. Development of a new categorization system for pesticides exposure to support harmonized reporting between EU Member States. *Env Int* 91: 332–40.
Sharma, B. K. 2006. *Environmental Chemistry*. New Delhi: Goel Publication House.
Silberman J., and A. Taylor. 2020. Carbamate toxicity. In *StatPearls*. Treasure Island, FL: StatPearls Publishing. www.statpearls.com/articlelibrary/viewarticle/30316/
Singer, F. R. 2020. The evaluation and treatment of Paget's disease of bone. *Best Pract Res Clin Rheumatol* 34(3): 101506.
Tadeo, J. L. 2019. *Analysis of Pesticides in Food and Environmental Samples*, 2nd ed. Boca Raton, FL: CRC Press.
Taylor, A. A., J. S. Tsuji, M. R. Garry, M. E. McArdle, W. L. Goodfellow Jr., W. J. Adams, et al. 2020. Critical review of exposure and effects: Implications for setting regulatory health criteria for ingested copper. *Environ Manage* 65: 131–59.
Tchounwou, P. B., C. G. Yedjou, A. K. Patlolla, and D. J. Sutton. 2012. Heavy metal toxicity and the environment. *EXS* 101: 133–64.
Tielemans, E., E. Louwerse, J. de Cock, D. Brouwer, G. Zielhuis, and D. Heederik. 1999. Exposure to fungicides in fruit growing: Re-entry time as a predictor for dermal exposure. *Am Ind Hyg Assoc J* 60: 789–93.
Tleuova, A. B., E. Wielogorska, V. Talluri, F. Štěpánek, C. T. Elliott, and D. O. Grigoriev. 2020. Recent advances and remaining barriers to producing novel formulations of fungicides for safe and sustainable agriculture. *J Controlled Release* 326: 468–81.
Travlos, I., N. Chermona, and D. Bilalis. 2017. Glyphosate efficacy of different salt formulations and adjuvant additives on various weeds. *Agronomy* 7(3): 60.
Trumbo, P., A. A. Yates, S. Schlicker, and M. Poos. 2001. Dietary reference intakes: Vitamin A, vitamin K, arsenic, boron, chromium, copper, iodine, iron, manganese, molybdenum, nickel, silicon, vanadium, and zinc. *J Am Diet Assoc* 101(3): 294–301.
Wacewicz-Muczyńska, M., K. Socha, J. Soroczyńska, M. Niczyporuk, and M. H. Borawska. 2021. Cadmium, lead and mercury in the blood of psoriatic and vitiligo patients and their possible associations with dietary habits. *Sci Total Env* 757: 143967.
Wang, B., and Y. Du. 2013. Cadmium and its neurotoxic effects. *Oxid Med Cell Longevity* 2013: 898034. www.hindawi.com/journals/omcl/2013/898034/
Wani, J. A., A. F. Wali, S. Majid, S. Rasool, M. U. Rehman, S. M. Rashid, et al. 2020. Bio-pesticides: Application and possible mechanism of action. In *Bioremediation and Biotechnology*, Vol. 2, eds. R. A. Bhat, K. R. Hakeem, and M. A. Dervash, 97–119. New York: Springer.
Warner, M., S. Rauch, P. Brambilla, S. Signorini, P. Mocarelli, and B. Eskenazi. 2020. Prenatal dioxin exposure and glucose metabolism in the Seveso Second Generation study. *Envion Int* 134: 105286.

Wazir, S. M., and I. Ghobrial. 2017. Copper deficiency, a new triad: Anemia, leucopenia, and myeloneuropathy. *J Community Hosp Intern Med Perspect* 19;7(4): 265–8.

WHO. 2019. *Recommended Classification of Pesticides by Hazard and Guidelines to Classification, 2019 Edition*. Geneva: World Health Organization. Licence: CC BY-NC-SA 3.0 IGO. www.who.int/ipcs/publications/pesticides_hazard/en/ (accessed January 24, 2021).

Wilms, W., M. Woźniak-Karczewska, A. Syguda, M. Niemczak, Ł. Ławniczak, J. Pernak, et al. 2020. Herbicidal ionic liquids: A promising future for old herbicides? Review on synthesis, toxicity, biodegradation, and efficacy studies. *J Agric Foord Chem* 68(39): 10456–88.

Wisnewski, A. V. 2007. Developments in laboratory diagnostics for isocyanate asthma. *Curr Opin Allergy Clin Immunol* 7(2): 138–145.

Wyke, S., N. Brooke, A. Dobney, D. Baker, and V. Murray. 2020. *The UK Recovery Handbook for Chemical Incidents*. Version 1.1, 2020. PHE publications, gateway number: GW-1451. https://assets.publishing.service.gov.uk/government/uploads/system/uploads/attachment_data/file/912808/UKRHCI_2019_Update_v1.1__2_.pdf (accessed November 11, 2020).

Wyke, S., A. Peña-Fernández, N. Brooke, and R. Duarte-Davidson. 2014. The importance of evaluating the physicochemical and toxicological properties of a contaminant for remediating environments affected by chemical incidents. *Environ Int* 72: 109–18.

Yamada, D., M. Hiwatari, P. Hangoma, D. Narita, C. Mphuka, B. Chitah, et al. 2020. Assessing the population-wide exposure to lead pollution in Kabwe, Zambia: An econometric estimation based on survey data. *Sci Rep* 10: 15092.

Yaman, M., and I. Akdeniz. 2004. Sensitivity enhancement in flame atomic absorption spectrometry for determination of copper in human thyroid tissues. *Analytical Sci* 20(9): 1363–6.

Yang, J., and D. Mosby. 2006. Field assessment of treatment efficacy by three methods of phosphoric acid application in lead-contaminated urban soil. *Sci Total Environ* 366(1): 136–42.

Ye, F., Z. Xie, X. Wu, and X. Lin. 2006. Determination of pyrethroid pesticide residues in vegetables by pressurized capillary electrochromatography. *Talanta* 69: 97–102.

Yin, S., Y. Sun, J. Yu, et al. 2021. Prenatal exposure to organochlorine pesticides is associated with increased risk for neural tube defects. *Sci Total Env* 770: 145284. https://doi.org/10.1016/j.scitotenv.2021.145284.

Zhang, Y., M. A. Arugula, M. Wales, J. Wild, and A. L. Simonian. 2015. A novel layer by layer assembled multi-enzyme/CNT biosensor for discriminative detection between organophosphorus and non-organophosphorus pesticides. *Biosens Bioelectron* 67: 287–95.

14 Food Toxicology

Arturo Anadón, Irma Ares, María-Rosa Martínez-Larrañaga and María-Aránzazu Martínez

CONTENTS

14.1 INTRODUCTION

Access to sufficient, safe food is a basic requirement for human health. The European Union (EU) food safety criteria define the acceptability of food products. If the criteria are not met, the product/batch must not enter the market, or it must be withdrawn if it has been placed in the market. Food safety does not mean food uniformity. The system for ensuring food safety is common to all EU countries, but it allows for diversity. Food safety has to be the driving force in the regulation of the food supply and is an intrinsic part of food quality. The EU actively promotes diversity and quality. The EU also protects distinctive or traditional foods associated with certain regions or certain production methods from being unjustly copied by others, and it promotes "organic farming" (EU 2004). Producers and processors must also comply with a large number of rules on specific issues; the point of all these rules is to make sure that food is as safe as is technically possible, to keep consumers informed and to give them as much choice as possible (EU 2004).

Food safety efforts and monitoring are required by the World Health Organization (WHO), in collaboration with the Food and Agriculture Organization (FAO) of the United Nations, which has a central international role in developing guidelines to reinforce and harmonise food systems, in particular through the jointly managed *Codex Alimentarius* Commission. Codex standards have become the *de facto* international standards for food safety. WHO and FAO also manage the International Food Safety Authorities Network (INFOSAN), which provides opportune information during food safety emergencies and assists countries in building strong systems to prevent such incidents. In addition, WHO has established the "Global Foodborne Infections Network" to promote integrated, laboratory-based surveillance and foster multisectoral collaboration (Fukuda 2015).

In regard to international food safety, standards are set out in the application of the World Trade Organization (WTO) Agreement on the Application of Sanitary and Phytosanitary Measures (SPS Agreement), which was drawn up to ensure that countries apply measures to protect human and animal health (sanitary measures) and plant health (phytosanitary measures). The sanitary

DOI: 10.1201/9780203730584-14

measures, however, must be based on risk assessment, or in other words, based on science. The SPS Agreement therefore incorporates safety aspects of foods in trade, and risk analysis has a major component. One of the main objectives of the SPS Agreement is to protect human and animal health and the phytosanitary situation in all WTO member countries. Today the SPS Agreement is regarded as being a strong instrument which will further the goal of the *Codex Alimentarius* Commission (i.e. to harmonise food standards worldwide). For food safety, the SPS Agreement recognises, as the international reference, the standards, guidelines and recommendations established by the *Codex Alimentarius* Commission relating to food additives, veterinary drug and pesticide residues, contaminants, methods of analysis and sampling, and codes and guidelines of hygienic practices (www.fao.org/fao-who-codexalimentarius/committees/general-subject-committees/en/). For example, the Code of Practice for several animal species developed by the different Codex Committees and the basic texts on food hygiene are the major international documents useful for food safety guidance in animal products (Okocha et al. 2018).

An EU process Hygiene Criterion gives guidance on, and is an indicator of, the acceptable functioning of Hazard Analysis Critical Control Point (HACCP)-based manufacturing, handling and distribution processes. It establishes indicative contamination limits above which corrective actions are required in order to maintain the hygiene of the process in compliance with the General Food Law (EC 2002c; Fukuda 2015). The free movement of food and feed within the European Community can be achieved only if food and feed safety requirements do not differ significantly among member states. Ensuring food safety and food security in a highly globalised world is proving increasingly difficult, and problems derived should be handled using an integrated and comprehensive approach to the entire food chain. In order to ensure the safety of food, it is necessary to consider all stages of food production chain as a continuum from the level of primary production including assessing the safety of animal feeds right through to the supply of food to consumers. Food safety issues have shown the need for comprehensive traceability of food along the food production chain. Food safety and the protection of consumer interests are of increasing concern to the general public, nongovernmental organisations, professional associations, international trading partners and trade organisations. A food business operator is best placed to devise a safe system for supplying food and ensuring that the food it supplies is safe; thus, it should have the primary legal responsibility for ensuring food safety (EC 2002c).

Nowadays, food ingredients often come from multiple regions, countries or continents, with each commodity having travelled thousands of kilometres from a specific field, farm, pen of pigs, flock of broiler chickens, or factory. Contamination at one end of the whole food chain can affect populations on the other side of the world (Fukuda 2015). Microbiological risks, chemical contamination of food and public health issues arising from an unbalanced diet have been identified as some of Europe's major challenges with respect to safe food and healthy diet (Bronzwaer 2008).

The first rules on food safety date from the very early days of the EU. The food safety crises of the 1990s, as the bovine spongiform encephalopathy (BSE) crisis showed, demonstrated that it was time to replace what had become a mixed assortment of EU rules with a simpler and more comprehensive approach. These food safety issues transcend borders and require wide public health responses. In the last decades, food alerts have highlighted the risks of contamination from certain types of feed, especially those used in intensive farming. As a result, EU policymaking prioritises the protection of human and animal health and has forbidden the sale of any animal feed materials which could represent any danger to human or animal health or to the environment. Labels must show clearly what the farmer is buying (EU 2004).

Scandals have drawn worldwide attention, including rapeseed intended for industrial use (Spain, 1981); dioxin (Belgium, 1999); milk adulterated with melamine (China, 2008); methanol poisoning from the sale of illegal spirits (Czechia and Poland, 2012–2014); and horse meat in beef products (Ireland and the United Kingdom (UK), 2013) have drawn worldwide attention. Another food safety incident in the EU resulted from the illegal use of fipronil, a broad-spectrum insecticide that belongs to the phenylpyrazole chemical family, to treat animal facilities such as laying hen stables, and the subsequent discovery of fipronil in eggs in Belgium in 2017, triggered an investigation that led to the identification of fipronil in a product that was used by a Dutch service provider to control mites in laying hen facilities. Naturally, such a finding created alarm. Such food safety incidents have demonstrated the necessity to establish appropriate measures that could avoid frauds or food safety incidents, thereby ensuring that all foods, whatever their type and origin, and all feed should be subject to regulatory measures in the event of a serious risk to human health, animal health or the environment. Such a comprehensive approach to emergency food safety measures should allow effective action to be taken and avoid artificial disparities in the treatment of a

Table 14.1 Toxicity Classification Based on Lethal Dose 50 (LD_{50})

Oral LD_{50} values (single dose to rats) mg/kg b.w.	Toxicity classification
>1	extremely toxic
1–50	highly toxic
50–500	moderately toxic
500–5000	slightly toxic

Abbreviation: b.w. = body weight.

serious risk in relation to food or feed (EC 2002c). The new approach also paid closer attention to the risks from contaminated feed. The result was a new piece of umbrella legislation known as the General Food Law, which was phased in between 2002 and 2005.

Food toxicology aims to establish the safety of food rather than on specific studies of the toxic effects of food. Food toxicology is different from other areas of toxicology largely because of the nature and chemical complexity of food and deals with the ingested substances taken as food or as component of food, which begins and often ends in the gastrointestinal tract. Food toxicology demonstrates how natural or synthetic poisons/toxicants in diverse food products cause harmful, detrimental or adverse side effects in living organisms. The term "toxicity" highlights the quality of being poisonous, especially the degree of virulence of a toxic microbe or of a chemical compound can cause toxic manifestation (poisonous effects) and body injury (Table 14.1).

The degree of toxicity or the propensity of the harmful effect depends on the route and site of exposure and the duration (time) and frequency of exposure. It is important to highlight that food toxicology is a very important area because the global population depends on the food supply chain, and any residues of veterinary drugs or pesticides, contamination or toxic manifestation may cause serious widespread adverse health effects.

The term "food production chain" (or "from farm to fork") includes the responsibilities of all stakeholders (i.e. scientists, industry, farmers and consumers) to take effective measures to minimise risks to both animal and human health. A safe food chain from farm to fork, correctly regulated and effectively controlled, is the way to build consumer confidence. Any food that reaches the consumer should be free from hazards such as microbiological (biotic) or chemical (abiotic) contaminants. For each food safety issue that poses an unacceptable level of harm, risk managers identify options that could be implemented to mitigate or manage risks. The presence in food of biotic contaminants, such as bacteria, rickettsia, viruses, moulds and parasitic protozoa/metazoa that may cause human illness, is characterised as food poisoning or food intoxication if the responsible agent grew in or on food and produced a toxin, or it may cause a food infection if a parasite is ingested with the food and chooses the human organism as its host. The microorganisms most commonly associated with food contamination and illness are bacteria *Salmonella, Clostridium perfringens, Staphylococcus aureus, Bacillus cereus* and other *Bacillus* spp., *Escherichia coli, Vibrio parahaemolyticus, Yersinia enterocolitica, Campylobacter jejune* and *Listeria monocytogenes* and also viruses. Numerous processes may be involved in getting food from the intensive farming or fishery to the table; consequently, contamination can occur at multiple sites along the food production chain.

14.2 FRAUDULENT PRACTICES AND CONTAMINATION BY CHEMICALS AND RESIDUES IN FOOD AND FEED

The complex nature of our globalised food supply chains and the economic incentive to provide cheaper food products have contributed to the prevalence of food fraud. Fraudulent practices in the food sector may also lead to public health risks. Additionally, the industrialisation and globalisation of food production have introduced many additional opportunities for food to become contaminated with bacteria, viruses, moulds, biological toxins, industrial toxins and wastes, pesticides, heavy metals, parasites, and plant and marine toxins. The rapid international production and shipping of food products around the world can result in widespread health problems when a food product is contaminated with a toxic or poisonous substance, leading to delays in identifying the source and cause. In some cases the consequences of food poisoning are rapid, as may be the case for many bacterial and viral contaminations. Other cases of poisoning act more slowly and insidiously, as is the cases of drinking water contaminated with lead or that of contaminated infant formula. A particular case are the environmental contaminants that may affect the health of animals and humans when present in feed and food. Many food chemical incidents originate in contaminated animal feed that arrives finally in contaminated animal products destined for human consumption. Certain contaminants, namely persistent lipophilic compounds that bioaccumulate, will increase in concentration (biomagnify) as they move forward to higher levels of the food chain. In particular, dioxins and other persistent organic pollutants are a group of compounds that cause enormous concern (Feil et al. 2000). Depending on their properties, chemical contaminants such as dioxin/polychlorinated biphenyls (PCBs), heavy metals or natural toxins (mycotoxins and others) present in the feed may be transferred from the animal to animal food products such as meat, milk, eggs or edible tissues, where the contaminants may further accumulate (e.g. in muscles, liver, kidneys and eggs; Feil et al. 2000). Global acquisition of feed materials has become common in order to achieve economically viable production of animal feeds. At the same time, this globalisation increases the risk for potential microbiological and toxicological contaminants that compromise the quality and nutritional value of feeds, but more importantly also the safety of animal-derived foods such as milk, meat, eggs and honey. Consumers expect food to meet their nutritional needs and to be wholesome and tasty, but overall to be safe.

To avoid any risk to public health, the EU is equally strict about the amount of residues from veterinary medicines or pesticides present in food when it is put on sale to the consumer. The European food policy indicates that antimicrobial medicinal products shall not be used in animals for the purpose of promoting growth or to increase yield. For that reason, the EU in 2006 banned antibiotics growth promoters (AGPs), such as avoparcin, ardacin, zinc bacitracin, virginiamycin, tylosin phosphate, spiramycin, monensin sodium, salinomycin sodium, avilamycin, and flavophospholipol, from being added to animal feed because use of these substances at subtherapeutic levels in large groups of livestock could increase the selection pressure on antimicrobial-resistant strains of bacteria and encourage the dissemination of resistance (Anadón and Martínez-Larrañaga 1999; Anadón 2006).

Overall, consumers have expressed concern regarding the health impact of veterinary drug residues in their food, which means all pharmacologically active substances, expressed in mg/kg (ppm) or µg/kg (ppb) on a fresh weight basis, whether pharmacologically active substances, excipients (inactive substances that serve as the vehicle or medium for a drug or other active substance) or degradation products, and their metabolites which remain in food obtained from animals (EC 2009). In order to protect public health, maximum residue limits (MRLs) should be established in accordance with generally recognised principles of safety assessment, taking into account toxicological risks, environmental contamination and the microbiological and pharmacological effects of residues (Woodward 1998). Animal residue levels in animal tissues above the MRL value will clearly have an impact on human health. MRL represent the maximum level or concentration of drug residues permitted in animal tissues (including meat, milk, eggs and honey) at the time of slaughter.

A critical factor in the medication of all food-producing animals (i.e. animals bred, raised, kept, slaughtered or harvested for the purposes of producing food) is the mandatory withdrawal period, defined as the time during which a drug must not be administered prior to the slaughter of the animal for consumption (EC 2009). There may be, however, situations where no suitable authorised veterinary medicinal product is available. In those situations, by way of exception, veterinarians should be allowed to prescribe other medicinal products to the animals under their responsibility in conformity with strict rules and only in the interest of animal health or animal welfare. In that case, veterinarians should ensure that an appropriate withdrawal period is prescribed, so that harmful residues of those veterinary medicinal products do not enter the food chain, and particular care should therefore be taken when administering antimicrobials (EU 2019). The withdrawal time is the period of time between the last administration of a drug and the collection of edible tissue (including meat, fat, eggs, milk and honey) from a treated animal (Anadón et al. 2018). The withdrawal periods are based on residue depletion studies conducted using radiolabelled compounds and specific conditions of use (type of animal, age category, dosage, route of administration) as well as unlabelled substances to ensure that residues above the levels that have been shown to be safe will not be present in animal products used as human food, which are evaluated by food safety agencies. Altering the dosage regimen using a veterinary drug in a species or class of animal or category of animals for which the drug has not been approved, or otherwise deviating from the approved label directions, invalidates the labelled withdrawal times (Paige et al. 1997). The MRLs in various foodstuffs (i.e. muscles, liver, kidneys, fat, milk, eggs and honey) are the points of reference for the establishment of withdrawal periods in marketing authorisations for veterinary medicinal products to be used in food-producing animals and for the control of residues in food of animal origin in the EU member states and at border inspection posts. In EU-EMA or member states, approval of drug dosages, routes of administration, durations of treatment, withdrawal times and MRLs are designed to ensure the safety of foods derived from treated animals. For example, the time between the last doses given to broiler chickens and the time when the level of residues in the tissues are lower than, or equal to, the MRL can be observed in the published manuscripts (Anadón et al. 2008a, 2008b, 2012). Linear regression analysis of the logarithmic transformed data can be considered for the calculation of the withdrawal periods. Using this approach, the withdrawal time was determined as the time when the one-sided, 95% upper tolerance limit of the regression line with 95% confidence level was below the MRL (EMEA 2002).

14.3 RISK ASSESSMENT OF VETERINARY DRUG RESIDUES

Risk assessment has emerged as the process of assessing the possible adverse health effects in humans as a result of exposure to substances or other potential hazards. Risk analysis is a science-based process consisting of a number of important steps in the overall process: (1) risk assessment, which comprises hazard identification, hazard characterisation (dose-response assessment), exposure assessment and risk characteristics (determination); (2) risk management; and (3) risk communication (Anadón and Martínez-Larrañaga 1999). Other elements in the risk analysis associated with foodborne hazards are risk management and risk perception. In order to protect public health and facilitate international food trade, the member countries of the WTO have signed the SPS Agreement, entered into force with the establishment of the WTO on 1 January 1995.

The standard approach to assessing the safety of chemical contaminants and residues in foodstuffs intended for human consumption is the acceptable daily intake (ADI). Current recommendations with regard to "safe" levels of intake for food additives, drug and pesticide residues have been based on concepts such as that of ADI (see Table 14.2). This concept was developed by the Joint FAO/WHO Expert Committee on Food Additives (JECFA, 1958), an internationally sponsored expert committee whose original function was that of providing safety evaluations of food additives (i.e. substances intentionally added to food) and it has been modified several times since then. However, when the mandate of JECFA was extended to include, in its toxicological assessment, food contaminants including residues of veterinary drugs as well as food additives, the ADI concept was not found to be suitable as a toxicological endpoint for assessment (Vettorazzi and Kouthon 1981), and the above committee developed the concept of provisional tolerable weekly intake (PTWI), which should be applied to 10 cumulative toxic elements (WHO/FAO 1972). This concept has proven to be useful, and its application to the toxicological assessment of cadmium, mercury and lead was carried out in 1972 and still holds true today.

To determine the food safety of residues for an antimicrobial product, the drug sponsor conducts a standard battery of standard toxicity studies (after acute and chronic exposures) in laboratory animals. The ADI approach was developed to take account of effects based on classical

Table 14.2 Endpoints of Assessments

	Health-Based Guidance Value (HBGV)
Food additives, drug residues, pesticide residues:	Acceptable daily intake (ADI) (mg/kg b.w.) Acute reference dose (ARfD) (mg/kg b.w.)
Contaminants:	
(a) With cumulative properties:	Provisional tolerable weekly intake (PTWI) (mg/kg b.w.) or provisional tolerable monthly intake (PTMI) (mg/kg b.w.)
(b) With no cumulative properties:	
(i) *Essential and toxic*:	Provisional maximum tolerable daily intake (PMTDI) (mg/kg b.w.)
(ii) *Not essential but toxic*:	Provisional maximum tolerable daily intake (PMTDI) (mg/kg b.w.)
(iii) *Essential and unavoidable*:	Maximum tolerable daily intake (MTDI) (mg/kg b.w.)

Abbreviation: b.w. = body weight.

toxicology and it is applied to the results of standard toxicology studies in laboratory animals. The studies required to satisfy human food safety are extensive and include reproductive toxicity: developmental toxicity (study of the effects on reproduction, study of developmental toxicity), genotoxicity, carcinogenicity, exceptions. There are other requirements: for particular groups of active substances, or if the effects observed during repeated dose studies in animals include changes indicative of immunotoxicity, neurotoxicity, endocrine dysfunction or any other adverse effect, further testing shall be required (e.g. sensitisation studies or delayed neurotoxicity tests; Anadón et al. 2018). Furthermore, microbiological properties of residues must also be studied (potential effects on human gut flora; potential effects on the microorganisms used in industrial food processing; Anadón et al. 2018).

These studies are used to identify no observable effect levels (NOELs), no observable adverse effect levels (NOAELs), lowest adverse effect levels (LOELs), lowest observed adverse effect levels (LOAELs); the ADI was calculated by dividing one of these values by a suitable safety factor (SF). The establishment of a toxicological ADI from the determination of a NOEL/NOAEL or LOEL/LOAEL and application of an appropriate SF provide hazard identification and characterisation. Usually, the highest dose tested is the NOAEL, which serves as the point of departure (POD) for this study. When LOEL/LOAEL is used an additional SF must be applied. Hazard characterisation describes the relationship between the administered dose of, or exposure to, a chemical and the incidence of an adverse health effect. The critical effect is determined, which is "the first adverse effect observed as the dose or exposure is increased". In cases, where the toxic effect is assumed to have a threshold, hazard characterisation usually results in the establishment of the HBGV (e.g. ADI, tolerable daily intake [TDI] or acute reference dose [ARfD]; IPCS 2009).

The acute studies evaluate experimental animals for problems such as allergenicity or teratogenic effects, whereas long-term chronic studies identify whether problems such as carcinogenicity are associated with a particular drug. The establishment of the ADI from the determination of a NOEL/NOAEL and application of an appropriate SF provide the hazard identification and characterisation. The ADI provides a human HBGV for chronic or long-term exposures to residues in food, and is most often established from a POD (e.g. a NOAEL or benchmark dose identified from repeated dose exposure studies in experimental animals). The NOEL/NOAEL for the most sensitive toxicological parameter, normally in the most sensitive species of experimental animals, is used as the starting point. 118 SECTION | I General.

These studies will determine a dose level that does not create any health problems. The ADI provides a human HBGV for chronic or long-term exposures to residues in food, and is most often established from a Point of Departure (POD, e.g. NOAEL) identified from repeated dose exposure studies in experimental animals. This ADI is then compared with the chronic dietary exposure. Although most active substances have the same ADI value, in some cases the ADI may differ between authorities due to differences in assessment of toxicological data (e.g. use of different SFs, different interpretation of "marginal", in particularly the response at lower doses).

An HBGV value may be derived from the NOAEL/BMDL (benchmark-dose lower bound) for the most significant toxicological endpoint (which may include carcinogenicity by a nongenotoxic mode of action or a noncarcinogenic endpoint) from the pivotal study in the most sensitive species by application of an uncertainty factor (IPCS 2017). In circumstances in which no health-based guidance value has been proposed, it may be possible to calculate the margin of exposure (MOE) between a reference point from the dose-response relationship in animals or humans and the estimated human dietary exposure. A lower MOE would be considered health protective if the reference point was based on human data.

Therefore, if the reference point was from an animal study, but with some important gaps in the database, then the minimum MOE considered to be health protective would be typically greater than 100 and possibly up to 10,000 (Benford et al. 2010).

The information needed to establish toxicological (including pharmacological) endpoints is obtained from studies in experimental animals or (rarely) in humans. However, in those cases where a veterinary drug has also been developed as a human medicine or a candidate human medicine, pharmacological (including pharmacokinetic) studies will have been carried out in human volunteers as part of the drug development program. Where available, this information will be taken into account for the identification of a NOAEL (or similar, such as the $BMDL_{10}$ [lower 95% confidence limit on the benchmark dose for a 10% response]; WHO 2016) and the establishment of an ADI, TDI or ARfD. There has been increasing use of a benchmark dose (BMD) approach in preference to the NOAEL in setting HBGV values, because it makes more use of the dose-response relationship and provides quantification of the uncertainty and variability in the dose-response data (EFSA 2009; IPCS 2009). The BMD is a dose level, derived by statistical modelling of dose-response data, associated with a specified low but measurable change in response, the benchmark response (BMR). The BMR should be at the lower end of the observed dose-response range, because extrapolation outside the range of observation increases the dependence on the statistical models. For quantal data, the BMR is an increase in the incidence of a lesion/response compared with the background response.

For compounds that are genotoxic and carcinogenic, it is generally not considered possible to determine a threshold of effect, and the use of the NOAEL is not appropriate. For such compounds, where a BMDL cannot be determined, the T25 (the chronic daily dose in mg/kg body weight that will give 25% of the animals tumours at a specific tissue site, after correction for spontaneous incidence, in the standard life span of that species) may provide an alternative POD. However, the suitability of the T25 for a given dataset needs to be considered on a case-by-case basis in terms of carcinogenic potency (IPCS 2017).

Until recently, the exposure was estimated using a simple model based on a hypothetical diet, the standard food basket (or so-called estimated daily intake [EDI]) intended to cover high consumers of animal products in order to check that the proposed MRLs values would not result in exposures exceeding the ADI values (FAO/WHO 2009). For example, in the European Union (European Medicines Agency), this "model diet" for a 60 kg adult includes muscle (also fish, 300 g), liver (100 g [for poultry, 10 g]), kidney (50 g [for poultry, 10 g]) and fat (50 g [for poultry, 90 g]) and, if appropriate, milk (1.5 L), eggs (100 g), and honey (20 g [JECFA used 50 g). In the European Union, for poultry and pigs, the MRL for fat relates to fat and skin in natural proportions, whereas for finfish muscle it relates to muscle and skin in natural proportions. The food consumption data for estimating chronic exposure to veterinary drug residues for the general population (e.g. adult, pregnant woman) and children compared with values used for EDI includes mammalian muscle, fat and skin; mammalian offal (all, liver, kidney); fish and seafood (fish, crustaceans, molluscs); poultry (muscle, fat and skin); poultry offal (all, liver, kidney); and eggs, milk and honey. The Center for Veterinary Medicine (CVM) of the US Food and Drug Administration (FDA) uses the same consumption value (g/person/day). These consumption amounts used in the model diet are not suitable for estimating acute exposure because many individual food consumption surveys may report higher values on a single day.

For an acute risk assessment, comparing exposure with a single food consumed at a single occasion, or over 24 hours (1 day) is taken in consideration. Consumers of the injection site tissue are potentially exposed to appreciably higher residues than consumers of the muscle tissue (Sanquer et al. 2006). A further possibility is that people at a special occasion/event or a specific subpopulation may consume a large portion of food derived from an edible tissue where the veterinary drug residues may be more concentrated. It is known that consideration of acute toxicity it was not appropriate to use chronic dietary exposure estimates to compare with ARfD as part of the risk characterisation of acutely toxic veterinary drug residues (FAO/WHO 2009). The ARfD of a chemical is an estimate of the amount of the substance in food and/or drinking water, normally expressed on a body weight basis that can be ingested in a period of 24 hours or less without appreciable risk to the health of the consumer. The aim of the single-dose study is to identify the most appropriate NOAEL or LOAEL to derive an ARfD, to provide further information on the dose-response curve time to peak effects and reversibility for the acute toxic effects, and to provide a flexible approach for an adequate characterisation of relevant acute effects.

Table 14.3 indicates the HBGVs (i.e. an ADI for a chronic intake or an ARfD for an acute intake) that are the most often established using a POD from a toxicity study in experimental animals.

The ADI is calculated by dividing the NOEL, NOAEL, or BMDL by a suitable SF or uncertainty factor, usually 100, which assumes that human are 10 times more sensitive that animals, and

Table 14.3 Health-Based Guidance Values for Chronic (ADI) and Acute Intake (ARfD)

Food intake	Health-based guidance values
Chronic intake	ADI = NOEL, NOAEL, $BMDL_{10}$/SF = mg/kg body weight/day
Acute intake	ARfD = NOEL or NOAEL/SF = mg/kg body weight/day

that in the human population there is a 10-fold range of sensitivity. The ADI represents the total drug residues (parent and all metabolites) that can be safety consumed daily throughout one's life. The ADIs can also be developed from pharmacological or microbiological data based on no-pharmacological observed effect levels (Anadón and Martínez-Larrañaga 1999). The concept of the no-hormonal effect levels (pharmacological endpoint) used with the hormonal growth promoters was based on similar criteria to the toxicological ADI. However, the impact of low levels of antibiotics on the intestinal flora is not examined directly in these toxicological studies. For substances with microbiological activity, what is actually used is the major adverse microbiological effect arising from the effects of residues of antimicrobial drugs in food of animal origin and effect on the human gastrointestinal bacteria flora, particularly those acting on the colonic flora. The identification of the residue profile and its decline in the treatment animal resulting in the setting of MRLs enables the exposure assessment and risk characterisation to be determined (Anadón and Martínez-Larrañaga 1999).

The lowest NOEL/NOAEL (based on the toxicological, pharmacological or microbiological endpoints) is used to calculate an ADI according to the following formula: ADI = NOEL or NOAEL/SF(100) = mg/kg body weight per day. Higher SFs (e.g. 200, 500 and 1000) can be used depending of the quality and quantity of relevant data. When LOAEL is used, an additional SF to the factor of 100 must be used.

Based on the lowest ADI, together with metabolism and residue depletion studies, the MRL of the residues is determined for each tissue, expressed in micrograms per kilogram on a fresh weight basis. ADIs are based on total residues of drug plus all metabolites, whereas MRLs are composed of a single, quantifiable marker residue, most commonly the parent compound but in some instances a single metabolite or a mixture of compounds. The target tissue is chosen to monitor the total residue and is usually the tissue with the slowest rate of depletion. To establish the MRL for each tissue, food consumption estimates are made on the basis of an assumed standard meal.

In the EU, legislation on residues of veterinary medicines and contaminants is harmonised. The key pieces of legislation are Council Directive 96/23/EC (EC 1996b), which contains guidelines for controlling veterinary drug residues in animals and their products with detailed procedures; whereas the Commission and Regulation 37/2010/EU regulates pharmacologically active substances and their classification setting MRLs in foodstuffs of animal origin (EU 2010). The EU has set safe MRLs for these drugs and other veterinary substances for use as veterinary drugs in animal products entering the human food chain.

To protect public health, veterinary medicinal products (VMPs), on the basis of the scientific assessment of the safety of those substances, were classified in the European Union in four Annexes to Council Regulation (EEC) No. 2377/90 (EEC 1990), laying down a community procedure for the establishment of MRLs of VMPs in foodstuffs of animal origin: Annex I, full MRL (i.e. contained pharmacologically active substances for which an MRL was established); Annex II (no MRLs required on consumer safety grounds, i.e. composed of pharmacologically active substances for which it was not necessary to establish an MRL); Annex III (provisional MRLs, pending further data, i.e. contained substances for which a provisional MRL was established); and Annex IV (for drugs considered unsafe on consumer health grounds, i.e. included substances for which an MRL could not be established because residues of those substances, at whatever limit, constitute a hazard to human health). Drugs in Annex IV are prohibited for use in food-producing animals within the EU. As of December 31, 1999, marketing authorisation was withdrawn for products containing any substance included in Annexes I, II or III (Anadón 1990).

Table 14.4 Health-Based Guidance Values (Microbiological ADI) for Chronic and Acute Intake

Food intake	Health-based guidance values
Chronic intake	$ADI = \dfrac{\text{NOAEC} \times \text{mass of colon content (220 g/day)}}{\text{Fraction of oral dose available to microorganism} \times \text{60 kg person}}$ = mg/kg bw per day
Acute intake	$ARfD = \dfrac{\text{MICcalc or NOAEC} \times \text{correction factors} \times \text{colon volume (500 mL)}}{\text{Fraction of oral dose available to microorganism} \times \text{60 kg person}}$ = mg/kg bw per day

Substances included in Annex I, II, or III of the Council Regulation (EEC) No 90/2377 (EEC 1990) are listed in the Annex of Commission Regulation (EC) No. 37/2010 (EC 2010) in table 1 ("Allowed Substances"). The table lists each pharmacologically active substance; its marker residue (parent substance, metabolite or combination of substances used to monitor the depletion of total residue in a food animal tissue); the animal species for which each is permitted; the MRL value; the target tissues (i.e. the edible tissue from which the residue depletes most slowly); other provisions (according to Article 14(7) of Regulation (EC) No. 470/2009 [EC 2009]); and their therapeutic classifications. Table 2 ("Prohibited Substances") of Regulation (EU) No 37/2010 lists substances for which an MRL cannot be established; these substances are shown here in Table 14.5. The prohibited substances must not be used in veterinary medicines for food-producing animals or in biocidal products for use in animal husbandry (table 3). This classification substitutes the four annexes of the Council Regulation (EEC) No. 2377/90 (EEC 1990).

When an environmental assessment is needed, the drug sponsor conducts laboratory toxicity studies with invertebrates, plants and microorganisms representative of the environmental compartments of concern. The NOEL, or the MIC value in the case of the microbes, is divided by a SF to arrive at a predicted environmental no-effect concentration (PNEC). The predicted environmental concentration (PEC) is calculated for the drug and compared to the PNEC. When the PEC/PNEC ratio is less than 1, the agency concludes that significant environmental effects will not occur due to the use of the animal drug product (Miller and Eirkson 1997).

In 2013, the European Food Safety Authority (EFSA) introduced reference points for action (RPAs) for nonallowed pharmacologically active substances present in food of animal origin (EFSA 2013). The aim was to define an analytical concentration for a nonallowed pharmacologically active substance that can be determined by official control laboratories which is low enough to adequately protect the consumers of food commodities that contain that substance. According with Article 1 of the Regulation (EC) No. 470/2009 (EC 2009), the RPA is defined as the level of a residue of a pharmacologically active substance established for control reasons in the case of certain substances for which an MRL has not been laid down. The minimum required performance limit (MRPL) corresponds to the average limit above which the detection of a substance or its residues can be construed as methodologically meaningful. Regulation (EC) No 470/2009 (EC 2009) stipulates that for nonallowed pharmacologically active substances, a RPA may be established when it is deemed necessary to ensure official controls for food of animal origin. The RPA shall be based on the reasonably achievable lowest level that can unequivocally be determined by official control laboratories. When residues of such nonallowed substances are detected at or above the RPA, the food is considered not to comply with Community legislation. However, even where residues of such substances are detected below the RPA, the competent authority shall carry out investigations to determine whether there has been illegal administration of a nonallowed pharmacologically active substance and, where relevant, shall apply the appropriate penalties.

As an important tool to ensure a harmonised implementation of Council Directive 96/23/EC (EC 1996b), the Commission progressively established MRPLs for analytical methods for substances for which no permitted limit has been established, and in particular for those substances whose use is not allowed, or is specifically prohibited, in the Community. According to Annex I, point 1.18 of Decision 2002/657/EC, the MRPL is the minimum content of an analyte in a sample, which at least has to be detected and confirmed (EC 2002c). It is intended to harmonise the analytical performance of methods for substances for which no permitted limit has been established. The

Table 14.5 List of Prohibited Substances for which the Maximum Residue Limits (MRL) Cannot Be Established (Table 2 of Commission Regulation (EU) No 37/2010)

- *Aristolochia* spp. and preparations thereof
- Chloramphenicol
- Chloroform
- Chlorpromazine
- Colchicine
- Dapsone
- Dimetridazole
- Metronidazole
- Nitrofurans (including furazolidone)
- Ronidazole

For all of these active substances, MRLs cannot be established.

RPA shall be based on the reasonably achievable lowest residue concentration that can unequivocally be determined by official control laboratories (i.e. the reasonably achievable lowest decision limit [CCα]). The RPA of MRPLs is defined as the minimum content of an analyte in a sample, which at least has to be detected and confirmed. In contrast to MRLs for substances that may be used in food-producing animals in the EU, the MRPLs are control tools based on expert advice on feasibility of controls and other legitimate factor in risk analysis. Consignments containing residues above the MRPLs would be rejected. If guarantees are provided, they may be returned to the sender. The MRPLs laid down in Annex II to Commission Decision 2002/657/EC (EC 2002b) are used as RPAs irrespective of the matrix tested. The Commission Decision 2003/181/EC (EC 2003) amended Decision 2002/657/EC (EC 2002b), and established MRPLs for different antimicrobials (see Table 14.6). Where results of analytical tests on products are at or above the MRPLs, the consignment concerned shall be considered noncompliant with Community legislation; where these results are below the MRPLs, the products will not be prohibited from entering the food chain (Anadón et al. 2018).

Table 14.7 lists examples of nonallowed pharmacologically active substances which have been detected in food of animal origin over the past years under the National Residue Control Plans (EFSA 2018).

Table 14.6 Minimum Required Performance Limits (MRPLs) Established for Several Antimicrobials

Substance and/or metabolite	Matrices	MRPL
Chloramphenicol	Meat, eggs, milk, urine, aquaculture products and honey	0.3 μg/mL
Medroxyprogesterone acetate	Pig kidney fat	1 μg/kg
Nitrofuran metabolites: furazolidone, furaltadone, nitrofurantoin, nitrofurazone.	Poultry meat, aquaculture products	1 μg/mL for all

Table 14.7 Examples of Nonallowed Pharmacologically Active Substances which Have Been Detected in Food of Animal Origin over the Past Years under the National Residue Control Plans (European Union)

Acepromazine (antipsychotic, phenothiazine group)	Malachite green (dye solution/antimicrobial in aquaculture)
Acid-fast green B (*in vitro* diagnostic use)	Mapenterol (β-adrenergic agonist)
Azobenzene (dye solution, strongly absorbs light)	Mecarbam (organothiophosphate insecticide)
Basic blue 26 (dye solution)	Mefenamic acid (nonsteroidal anti-inflammatory)
Boldenone (anabolic-androgenic steroid)	Megestrol (progestin + promote growth)
Brilliant green (antiseptic)	Melengestrol (progestin + promote growth)
Brombuterol (β-adrenergic agonist)	Methylene blue (solution dye + methemoglobinemia treatment)
Carbadox (antimicrobial, quinoxalines group)	Methyltestosterone (androgen steroid)
Chloramphenicol (antibiotic)	Methylthiouracil (antithyroid)
Chlorbrombuterol (β-adrenergic agonist)	Methylviolet (dye solution)
Chlormadinone (progestin, antiandrogen)	Metronidazole (antibiotic)
Chlormephos (organothiophosphate insecticide)	Nandrolone (androgen steroid)
Chloroform (inhaled anaesthetic)	Naproxen (nonsteroidal anti-inflammatory)
Chlorpromazine (antipsychotic, phenotiazine group)	New methylene blue (dye solution, antimicrobial)
Cimaterol (β-adrenergic agonist)	Nile blue (dye solution)
Clembuterol (β-adrenergic agonist)	Nitenpyram (insecticide)
Clencyclohexerol (β-adrenergic agonist)	Nitrofurans (antimicrobials)
Clenpenterol (β-adrenergic agonist)	Olaquindox (antimicrobial, growth promoter)
Clenproperol (β-adrenergic agonist)	Orciprenaline (β-adrenergic agonist)

Colchicine (anti-inflammatory)
Cristal violet (antibacterial, antifungal, anthelmintic)
Dapsone (antibiotic)
Dexamethasone (anti-inflammatory)
Diclofenac (nonsteroidal anti-inflammatory)
Dienestrol (nonsteroidal estrogen)
Diethylstilbestrol (estrogen)
Dimetridazole (antiprotozoal)
Erythrosine B (food colouring)
Ethinylestradiol (estrogen)
Ethoprophos (organophosphate insecticide)
Ethylviolet (dye solution)
Fenoterol (β-adrenergic agonist)
Formothion (acaricide, insecticide)
Haloperidol (antipsychotic)
Hexaconazole (fungicide)
Hexestrol (nonsteroidal anti-inflammatory)
Hydroxymethylclenbuterol (β-adrenergic agonist)
Ibuprofen (nonsteroidal anti-inflammatory)
Isofenphos (organothiophosphate insecticide)
Mabuterol (β-adrenergic agonist)
Oxyphenbutazone (nonsteroidal anti-inflammatory)
Pararosaniline base (dye solution)
Phenylbutazone (nonsteroidal anti-inflammatory)
Propiconazole (fungicide)
Propiopromazine (antipsychotic, phenothiazine group)
Propylthiouracil (hyperthyroidism treatment)
Pyrazophos (fungicide/insecticide)
Quinalphos (organothiophosphate insecticide)
Rhodamine 6G (fluorescent dye, antineoplastic)
Ritodrin (β-adrenergic agonist)
Ronidazole (antiprotozoal)
Salbutamol (β-adrenergic agonist)
Salmeterol (β-adrenergic agonist)
Stanozolol (anabolic, androgenic steroid)
Terbutaline (β-adrenergic agonist)
Thiamazole (hyperthyroidism treatment)
Thiouracil (hyperthyroidism treatment)
Triazophos (organothiophosphate insecticide)
Tulobuterol (β-adrenergic agonist)
Ultramarine (color pigment)
Zearalanone (mycotoxin, anabolic effect)
Zeranol (mycotoxin, anabolic effect)

14.4 HUMAN HEALTH RISK FROM DRUG RESIDUES IN FOODS

Residues of some substances for which no ADI and/or MRL has been not recommended for use as veterinary drugs due to specific human health concerns include quinoxalines (carbadox and olaquindox), chloramphenicol, nitrofurans, nitroimidazoles, malachite green, chlorpromazine, and stilbenes (diethylstilbestrol, dienestrol and hexestrol). In particular, these substances are both genotoxic and carcinogenic. The ADI model does not predict quantitative risks and is not applicable to genotoxic carcinogens (i.e. those that damage DNA), for which presumably no threshold levels can be established.

14.4.1 Quinoxalines (Carbadox and Olaquindox)

The antimicrobial drugs carbadox and olaquindox, which are quinoxalines, possess genotoxic and/or carcinogenic properties. The role of oxidative stress and the metabolism of quinoxalines in their toxicity will shed new light on the use of antioxidants and scavengers of reactive oxygen species (ROS; Wang et al. 2016). Carbadox and olaquindox are antibiotics used as feed additives for pigs to promote growth. The metabolite desoxycarbadox was found to be carcinogenic, whereas the metabolites methyl carbazate and quinoxaline 2-carboxylic acid (QCA) were not, because the genotoxic and carcinogenic nature of carbadox and some of its metabolites could not be established an ADI by several jurisdictions. Appreciable amounts of residues were found in treated animals, so on the metabolism and depletion of carbadox, and the nature of the compounds were released from the bound residues. Based on these findings ADI values cannot yet be established. But their restricted use in pigs only during the first four months of life, coupled with the specified 28-day withdrawal period, may lead to negligible residues at the time of slaughter and minimise the risk of exposure for consumers. This implied safety feature requires the application of good agricultural/veterinary practice which, however, cannot completely be guaranteed. for example, when slaughtering piglets, there is a potential contradiction between the allowed usage of the drugs up to 4 months of age and a required withdrawal period of 28 days. It is questionable whether the use of genotoxic and carcinogenic feed additives is acceptable. At present, there are no indications of negative influences of antimicrobial feed additives on animal welfare, with the exception of a possible impairment of adrenal cortex function by quinoxalines in pigs. This effect needs further investigation. Proper labelling of products is required to overcome these problems. To reduce the exposure of workers handling antimicrobial feed additives, only preparations with reduced dust formation should be allowed.

The residue concentration in muscle of 4 µg/kg of the metabolite 3 methylquinoxaline-2-carboxylic acid (MQCA) is consistent with the use of olaquindox in pigs. The potential genotoxicity and carcinogenicity cannot be ruled out, and no ADI or MRL has been established for olaquindox.

Microbiological risks due to the development of bacterial resistance have been regarded as the major problem of the use of antimicrobial growth promoters feed additives, as illustrated by the glycopeptide resistance in *Enterococci*. Genetic resistance to certain antimicrobial feed additives exists, it may be carried on plasmids and may, as such, transfer from Enterobacteriacae in animals to bacteria in humans. However, evidence that the use of avoparcin in animals is the cause of the hazard of vancomycin resistance in humans is lacking (Anadón and Martínez-Larrañaga 1999). After the introduction of olaquindox in 1982, the prevalence of resistant *Escherichia coli* in the faeces of pigs increased in three years from 0.004% to 6% in farms using olaquindox as an AMGP. In farms not using olaquindox, the prevalence of resistance also increased but to a significantly lesser degree (Linton et al. 1988). An increase of resistance to carbadox in faecal *E. coli* isolates of pigs was noticed after its introduction; resistant isolates from six farms that fed carbadox continuously to pigs either as AMGP or for prevention of swine dysentery carried the same transferable plasmid conferring carbadox resistance. However, carbadox is not used in poultry, and no resistance was found in *E. coli* from poultry in the same region (Ohmae et al. 1983). Other authors found an increase in resistance to carbadox among *E. coli* isolates from pigs from 37% to 61% during a period when carbadox was used not only used as an antimicrobial growth promoter but also for prevention of swine dysentery and therapy of salmonellosis (Mills and Kelly 1986).

14.4.2 Chloramphenicol

Chloramphenicol is an antibiotic produced by some strains of *Streptomyces venezuelae*, but now it is synthetically prepared. It is a broad-spectrum antibiotic predominantly active against the main pathogenic gram-negative bacteria, and it exerts its action by inhibiting the bacterial protein biosynthesis. The target of chloramphenicol are the ribosomes. Chloramphenicol tends to accumulate in the tissues over a longer time and at a higher concentration than in serum. Research on chloramphenicol tissue residues in food-producing animals has shown that significant amounts persist in kidney, urine, and intramuscular injection sites (Anadón 1985). In broiler chickens, chloramphenicol is extensively metabolised into dehydrochloramphenicol, nitrophenylaminopropanedione, and nitroso-chloramphenicol derivatives; the results indicate that chloramphenicol and dehydrochloramphenicol residues are cleared slowly and were at or below the detection limit of 0.005 µg/ml within 12 days after dosing. However, at the time of slaughter (12 days), nitrophenylaminopropanedione and nitroso-chloramphenicol residues were detectable in the tissue (Anadón et al. 1994). In humans, chloramphenicol is well absorbed and widely distributed and is 60% bound to plasma protein, with higher placental that maternal concentrations (Nau et al. 1981). It is hepatically glucuronidated and is a potent CYP3A4 and 2C19 inhibitor. It is well tolerated by animals, but clinical use in humans can cause dose-independent, irreversible side effects such as bone marrow aplasia and aplastic anaemia. Due to neonatal toxicity, "grey baby syndrome" and agranulocytosis, use during pregnancy, especially near term, should be avoided unless absolutely necessary (WHO 2004). Chloramphenicol is now forbidden in the EC and in the United States in animal husbandry for food-producing animals. It was not appropriate to establish an ADI for chloramphenicol because it was impossible to establish a threshold for carcinogenicity given the evidence of a potential genotoxic mechanism. It was not possible to establish any dose-response relationship or threshold dose for the induction of potentially fatal aplastic anaemia, and so MRLs could not be recommended for this active substance.

14.4.3 Nitrofurans

Nitrofuran derivatives belong to a group of synthetic broad-spectrum antibiotics, all containing a 5-nitrofuran ring and different substituents in the 2-position. The 5-nitro group is a requisite for antimicrobial activity. Five pharmacologically active substances belong to this group: nitrofurazone, nitrofurantoin, nitrofural, furazolidone and furaltadone. The antibacterial spectrum of activity is very similar among these substances, as they are mainly active against gram-negative bacteria (e.g. *Salmonella* strains and *Escherichia coli*). Nitrofurans have been used in the treatment of salmonellosis and *E. coli* septicaemia in poultry and salmonella infections in pigs and calves. Furazolidone has been used in feed additives for swine and poultry. Due to the increased solubility, furaltadone, in veterinary medical practice, furaltadone is used preferably in drinking water and furazolidone in feed and in topical applications. In human medicine, furaltadone and furazolidone are not used in the EU. It has been shown (Nouws and Laurensen 1990) that furazolidone and furaltadone are very rapidly and completely degraded in tissues of animals after slaughter.

In the bodies of treated animals, furaltadone and furazolidone are reduced at the nitro-position, forming reactive intermediates capable of reacting with proteins (protein-bound metabolites at a mg/kg tissue level). The covalently bound residues cannot be separated from animal tissues by water and organic solvents, have a long half-life and, in the absence of more information on their safety, require long withdrawal periods. It was not possible to establish an ADI for nitrofural because no-effect levels have not been fixed for the tumorigenic effects. The use of nitrofuran derivatives in food-producing animals has become questionable regarding their mutagenic and carcinogenic properties and therefore have been prohibited for use in food animals in the EU. Nitrofurans, including furaltadone, are listed in table 2 of Regulation (CE) 37/2010 (EU 2010) as prohibited substances.

14.4.4 Nitroimidazoles

Dimetridazole, metronidazole, ipronidazole, and ronidazole all belong to the 5-nitroimidazole group. These substances are active against obligate anaerobic and microaerophilic bacteria, amoebas and flagellated protozoa. In animal, the 5-nitromidazoles are used for treatment of infections with anaerobic bacteria and protozoa (Prescott and Baggot 1993). In veterinary medicine, the most important indication for dimetridazole is the control of histomoniasis ("blackhead") caused by *Histomonas meleagridis* in turkeys. The antibacterial and antiprotozoal properties of 5-nitroimidazoles involves the reduction of the 5-nitro group, which results in short-lived hydroxylamine derivatives that bind covalently to proteins and DNA, as a result of which DNA strands may also break (Edwards 1977). The 5-nitro group is essential for their therapeutic action. There are strong indications of genotoxic and carcinogenic potential of 5-nitroimidazoles. Dimetridazole is absorbed from the gastrointestinal tracts in target species. About 88% of the administered dose is excreted from turkeys within 3 days, whereas about 76% is excreted from pigs within 7 days. In both animal species, the predominant metabolite is 2-hydroxymethyl-1-methyl-5-nitro-imidazole (FAO/WHO 1989). Neither ADI nor MRLs can be established. Dimetridazole, metronidazole, and ronidazole are listed in table 2 of Regulation (CE) 37/2010 (EU 2010) as prohibited substances.

14.4.5 Malachite Green

Malachite green is an *N*-methylated triphenylmethane used as an industrial dye. It has been used in the past as an antifungal and antiprotozoal agent in aquaculture. For example, malachite green has been used for the antifungal treatment of fish and crayfish eggs (Alderman 1985). Malachite green has also been used for decades as a fish fungicide in closed systems alone or in combination with other chemicals such as formaldehyde. The frequently reported concentration in fish tissue is 0.05–0.1 mg/kg. Malachite green is metabolically reduced by fish to leucomalachite green. When fish exposed to malachite green reach the consumer, the amount of leucomalachite green present in the fish is expected to be higher than that of malachite green. Leucomalachite green induces hepatocellular adenomas and carcinomas in female mice via a mutagenic mode of action. Based on these considerations, it was inappropriate to establish an ADI for malachite green and the use of malachite green was not supported for food-producing animals. It was considered hepatocarcinogenic in female mice treated with leucomalachite green to be pivotal in the risk assessment of malachite green and leucomalachite green and evaluated the sum of malachite green and leucomalachite green in food expressed as leucomalachite green. A $BMDL_{10}$ value of 20 mg leucomalachite green/kg body weight per day derived from a carcinogenicity study in female mice as the reference point was used for the MOE calculation. Assuming a daily consumption of 300 g of fish contaminated with malachite green, comparison of these dietary exposures of 0.15 μg/kg body weight per day to the sum of malachite green and leucomalachite green (mean) and 0.69 μg/kg body weight per day (97.5th percentile) with the $BMDL_{10}$ of 20 mg/kg body weight per day indicates MOEs of 130,000 and 30,000, respectively. In the case that malachite green had been used to treat fish in aquaculture and the fish had been subjected to various depletion times, comparison of these exposure levels of 9 μg/kg body weight per day (median, no depletion time) and 1 μg/kg body weight per day (median, depletion time of 500 hours) with the $BMDL_{10}$ of 20 mg/kg body weight per day indicates MOEs of 2000 and 20,000, respectively. The 97.5th percentile intakes were about 23 (no depletion time) and 2 μg/kg body weight per day (depletion time of 500 hour), respectively, providing MOEs of about 900 and 10,000, respectively. MOEs of less than 10,000 for genotoxic and carcinogenic contaminants indicate a health concern. MRLs were not recommended for malachite green and leucomalachite green, and the use of malachite green for food-producing animals was not supported due to genotoxic and carcinogenic properties of its main metabolite, leucomalachite green (JECFA 2008).

14.4.6 Chlorpromazine

Chlorpromazine is antipsychotic agent and includes phenothiazine compounds synthesised in Europe, such as propionylpromazine, acepromazine, triflupromazine and promazine. Chlorpromazine by itself does not cause loss of consciousness but diminished arousal and motility, with some tendency to promote sleep. These central actions soon became known known as ataractic or neuroleptic actions. The less potent antipsychotic drugs may weakly induce their own hepatic metabolism, because concentrations of chlorpromazine and other phenothiazines in blood are lower after several weeks of treatment with the same dosage. Chlorpromazine binds significantly to membranes and to plasma proteins. In veterinary medicine, phenothiazines exert a sedative action; they are used as tranquillising agents, preanaesthetics and muscle relaxants in cattle, swine, sheep and goats to prevent mortality and loss of meat quality during transport to the slaughterhouse. It shows synergism with other classes of central nervous system depressants. Studies indicate possible concerns for genotoxicity, adverse effects on reproduction and developmental neurotoxicity (JECFA 1991). Some reactive metabolic intermediates were found to be capable of binding to DNA. No ADI or MRLs were allocated and therefore it should not be used in food-producing animals (JECFA 1991).

14.4.7 Hormones

Growth-promoting hormones, particularly steroids and xenobiotics, have been widely used (legally or illegally) to increase weight gain and feed efficiency in various species, primarily in cattle and to a lesser extent in ovines, swine, poultry and fish. Growth-promoting hormones can be divided into four groups of molecules:

1. Natural steroids (17β-estradiol, progesterone and testosterone)
2. Xenobiotics (trenbolone acetate and zeranol)
3. Synthetic steroids (ethinylestradiol, methyltestosterone, chlormadinone acetate, etc.)
4. Stilbenes (diethylstilbestrol and the related substances dienestrol and hexestrol).

According to Directive 2008/97/EC (EC 2008), the following substances are prohibited: stilbenes, stilbene derivatives, their salts and esters; and 17β-estradiol and its ester-like derivatives.

14.4.7.1 17β-Estradiol

The hormone 17β-estradiol is a natural sex steroid hormone produced by an animal. It was often used in many of EU countries as an effective growth promoter. It acts primarily by stimulating protein synthesis, and its response is dependent upon the nutritional status of the animal. The natural steroids 17β-estradiol, testosterone, and progesterone are by nature endogenous products playing an important role in controlling reproductive functions in humans and animals. When applied exogenously, they enter the same metabolic pathways as the endogenously produced molecules. Quantitatively, 17β-estradiol is considered as the most important estrogen in blood; 17α-estradiol, estriol and estrone are present at lower concentrations, and are present under free or conjugated forms. The determination of 17β-estradiol in urine and faeces during the first weeks after treatment is useful to identify suspect animals in a farm.

Based on the Opinion of the Scientific Committee on Veterinary Measures relating to Public Health (SCVPH; EC 2002a), 17β-estradiol is considered a complete carcinogen, as it exerts both tumour-initiating and tumour-promoting effects. The mutagenic and genotoxic potential of 17β-estradiol is a consequence of its metabolic activation to reactive quinones. Experiments with heifers indicated a dose-dependent increase in residue levels of all hormones, particularly at implantation sites. Misplaced implants and repeated implants, which appear to be frequent, represent a considerable risk that highly contaminated meat can enter the food chain (EC 2002c; Anadón et al. 2018).

14.4.7.2 Diethylstilbestrol

The stilbenes (diethylstilbestrol [DES] and the related substances dienestrol and hexestrol) are synthetic sexual steroids that were used in the past as growth promoters in cattle. DES is a synthetic estrogen structurally similar to estradiol with potent estrogen activity. DES is perhaps the most widely known of the estrogenic growth-promoting agents. It is structurally different from the endogenous estrogens and exerts some of its action through the steroid hormone receptor pathway by binding to the estrogen receptor and activating gene transcription. The relative binding affinity of DES for the estrogen receptor is 141% (vs. 100% for estradiol), but DES is 10–20 times more active than estradiol by the oral route (Umberger 1975). DES is also metabolised to radical or quinone intermediates which react with cellular macromolecules, including DNA (Metzler 1984). The genotoxic and carcinogenic properties of stilbenes are well recognised and their use in food-producing animals is banned in most countries. From the 1940s to 1971, DES was used therapeutically in pregnant women in the United States to prevent spontaneous abortions (Levine and Berkowitz 1993). The beneficial clinical effects of DES are most likely a result of the drug's positive effects on placentation and trophoblast stem cell differentiation (Hoover et al. 2011). Use continued until the early 1970s, when a link between clear-cell adenocarcinoma in women and their exposure to DES *in utero* was established (Herbst et al. 1971). Additional studies showed that DES had other effects on children of mothers treated with DES, the most common being vaginal adenosis in females and gross abnormalities in the reproductive tracts of both genders (Herbst 1976). The use of these substances by producers could be explained first by the positive effect on carcass market value and second by the difficulties to detect treated animals, considering the limiting financial supports for control and the sophisticated analytical methods required. As noted by Maghuin-Rogister et al. (1991), some consequences are also a competition between EU meat producers who can not use anabolic agents and other countries where anabolic compounds are legally authorised to be used by farmers. The myotropic actions of anabolic steroids result from their ability to increase retention of dietary nitrogen through protein synthesis. Steroid hormones may act (1) via specific cell receptors, (2) by enhancing endogenous somatotropic hormones or (3) by modulation of other endocrine systems such as the hypothalamic-pituitary-adrenal axis. An ADI or recommended MRLs could not be established due to specific human health concerns.

14.4.7.3 Antithyroids

The antithyroid drugs that have clinical utility are the thioureylenes (propylthiouracil, methimazole and carbimazole), which belong to the family of thionamides. Antithyroid drugs inhibit the formation of thyroid hormones by interfering with the incorporation of iodine into tyrosyl residues of thyroglobulin; they also inhibit the coupling of these iodotyrosyl residues to form iodothyronines. This implies that they interfere with the oxidation of iodide ion and iodotyrosyl groups, and it was proposed that drugs inhibit the peroxidase enzyme, thus preventing oxidation of iodine or iodotyrosyl groups to the required active state (Taurog 2000). Antithyroid drugs interfere with the basal metabolic rate by altering the synthesis, release or peripheral action of the thyroid hormone; decreasing production of thyroid hormones; reducing basal metabolism; lessening gastrointestinal motility: and aiding extracellular water retention.

Antithyroid drugs have been used in meat-producing animals to increase weight-gain efficiency, particularly to improve the finish of the animal for market by promoting fat deposition. Furthermore, these drugs increases filling of the gastrointestinal tract and a higher water retention of the animal (Derblom et al. 1963). The highest concentration of antithyroid drugs is found out in the thyroid gland from the bovine treated with antithyroid drugs, leading moreover to the production of meat with lower quality. Screening of possible cattle treated with antithyroid drugs may be based on morphological properties of the thyroid gland. The weight of the gland may give an indication of the possible administration of antithyroid drugs.

In humans, it has been proven that propylthiouracil and methimazole cross the placenta equally and have equal chance of inducing foetal and/or neonatal hypothyroidism and goitre and foetal abnormalities. There is still a continued preference of propylthiouracil over methimazole use in pregnancy. Neonates from mothers with Graves disease (also known as toxic diffuse goitre, which is an autoimmune disease that affects the thyroid) did not demonstrate any significant teratogenic effects in those infants whose mothers took methimazole (Momotani et al. 1997).

Studies on the development mechanism of goitre (swelling of the neck resulting from enlargement of the thyroid gland) began with the observation that rabbits feed a diet composed largely of cabbage often displayed an antithyroid (goitrogenic) effect. The same is known in rats for sulphonamides which depressed thyroid activity. Those results were probably due to the presence of precursors of the thiocyanate ion in cabbage leaves. Later, two pure compounds were shown to produce goitre: sulphaguanidine and phenylthiourea. Some heifers fed on cabbage or rapeseed cakes demonstrated the presence of thiouracil (TU) as a residue in the urine. Thus there is a causal link between a diet based on cruciferous vegetables and the presence of TU in the urine of bovines (Pinel et al. 2006). Thiouracil was excreted in urine in the hours following the intake of cruciferous vegetables.

Glucosinolates are secondary plant metabolites of species belonging to the family Brassicaceae. Many common vegetables such as broccoli, Brussels sprouts, cabbage and cauliflower belong to this vegetable family. Glucosinolate-producing plants and some microorganisms contain specific β-thioglucosidases (myrosinases). Upon plant damage (including chewing during ingestion), the myrosinases are released and convert the glucosinolates into diverse breakdown products, such as isothiocyanates, oxazolidinethiones, thiocyanates, nitriles, epithionitriles and other indol-3-ylmethyl derivatives. The biological effects of plant glucosinolates in mammalian species are predominantly related to these glucosinolate-derived compounds. Biological effects can be toxic (e.g. changes in thyroid function), antinutritional or beneficial to health (anticarcinogenic), depending on the structures and concentrations of the glucosinolates.

Propylthiouracil is recommended in most species because it has fewer side effects than thiouracil or methylthiouracil. Methimazole is much longer acting than propylthiouracil. These drugs cross the placenta and will render a foetus hypothyroid.

The use of these compounds as growth promoters for animals has been prohibited in the EU since 1981 by Council Directives 81/602/EEC, 88/146/EEC and 88/299/EEC (repealed by Council Directive 96/22/EC; EC 1996a). The residues of antithyroid drugs may be harmful for human health, having described their carcinogenic and teratogenic properties (Pinel et al. 2006). Antithyroid drugs used illegally in food-producing animals include 6-methyl-2-thiouracil; 6-propyl-2-thiouracil, 6-phenyl-2-thiouracil, 1-methyl-2-mercaptoimidazole (tapazole) and 2-mercaptobenzimidazole (Pinel et al. 2006), which are very strong thyroid-inhibiting compounds.

14.4.8 Phenethanolamines or β-Adrenergic Agonists

These compounds have been categorised as β-adrenergic agonists or β-agonists because they share structural, physiological and pharmacological properties with the endogenous catecholamines (i.e. norepinephrine and epinephrine) and have been advocated as growth-promotion agents. Some of these active substances have a role in veterinary medicine. Among the most extensively studied compounds are clenbuterol, cimaterol, ractopamine, L644–969 and salbutamol (JECFA 2006), although several others have also been studied. In a variety of animal species, these β2-agonist agents lead to an increase in the proportion of lean meat to fat and improved efficiency of food conversion to meat. These β-adrenergic agonists are orally active and therefore, are administered in the feed. Most have been shown to repartition nutrient use toward enhanced skeletal muscle growth (protein deposition) and reduced lipid accretion ("repartitioning agents"). It is known that the β-agonist clenbuterol has been used illegally as a growth promoter in farm animals. Broad generalisations regarding their efficacy and mode of action cannot be uniformly applied, however, because differences exist in responsiveness among mammalian and avian species among their dose-response relationships and their fate in the body. The use of clenbuterol to enhance lean tissue growth requires a dose that is 10 times higher than the normal therapeutic dose of 0.8 μg/kg (Anadón et al. 2018). Clenbuterol has been implicated in several outbreaks of foodborne poisoning in humans following the consumption of liver, meat and lungs of bovines and lungs and liver of pigs (Anadón et al. 2018). However, some residues can give rise to an acute rather than a chronic toxicological effect. The β-agonists, for example, can induce short-lived pharmacological responses (e.g. tachycardia) but with no long-term consequences. For such drugs, some authorities have accepted an ARfD as the appropriate health standard (Sanquer et al. 2006).

14.4.9 Antimicrobial Use and Antimicrobial Resistance (Food-Producing Animals and Animal Food)

In regards to the risk to public health, food has always been regarded as an important route through which human beings may be exposed to antimicrobial resistant (AMR) bacteria. AMR is also a food safety problem; antibiotic use in food animals for growth promotion, disease prevention or treatment create ideal conditions for resistance genes to spread through the food chain from bacteria associated with food animals to bacteria in humans. For that reason, the food chain has attracted considerable attention with regards the risk of AMR transfer from animal to humans. Globally, the food chain represents different areas of activity and locations and therefore different associated risks, ranging from AMR in various animal species that are raised to produce food to AMR in food sold at retail and directly intended for human consumption. The globalisation of the food system, with increasing movement of livestock and agricultural produce combined with increasing human travel, facilitates the rapid spread and mixing of AMR genes that emerge (Robinson et al. 2016). In a One Health perspective, which is defined as the collaborative effort of multiple disciplines working locally, nationally and globally to attain optimal health for people, animals and our environment, it is recognised that the health of people is connected to the health of animals and the environment, and special attention will be paid on successful AMR plasmids disseminating genes conferring resistance to drug classes that are important for public health (i.e. extended-spectrum cephalosporins, carbapenems or colistin; AVMA 2008; Madec and Haenni 2018).

14.4.9.1 Antimicrobial Use in Food Animals

Antimicrobials have been used worldwide in animal husbandry of livestock, poultry, and fish for many decades because of their favourable economic effects in livestock production to maintain health and productivity (Anadón et al. 2018). The different applications of antibiotics in food animals have been described as prophylactic, therapeutic, metaphylactic (i.e. medicating mixed groups of healthy and infected animals in order to control outbreaks of disease) and to enhance animal growth and feed efficiency. Therapeutic use of antibacterial drugs is essential for maintaining animal health and animal welfare. For a long time, the use of antibacterial agents in animal production has facilitated confinement housing and allowed a higher density of animals to be maintained. Animal diseases and infections should primarily be prevented by ensuring biosecurity, following good production and good management practices (i.e. feeding practices, sanitation, housing and environmental control) and implementing integrated disease control programmes to minimise the occurrence of diseases and eradicate endemic disease (EC 2015). Biosecurity measures can influence animals' health for the better and reduce the potential need for antimicrobials. So, implementation of biosecurity measures, good farming practices and herd health planning that prevents infections reduces the need for antimicrobials (EC 2015).

Residues (i.e. low levels of substances having a pharmacological action of their metabolites and of other substances transmitted to animal products and likely to be harmful to human health) of antimicrobials in food have received much attention in recent years because of growing food safety and public health concerns. Their presence in food of animal origin (e.g. meat, fish, eggs and milk) constitutes socioeconomic challenges in international trade in animals and animal products (Anadón and Martínez-Larrañaga 1999). The major public health issues at a global scale of antimicrobial residues include the development of AMR, acute effects (i.e. hypersensitivity reactions, teratogenicity), chronic effects (carcinogenesis) and disruption of normal intestinal flora; their long-term ingestion, even at very low levels could have negative consequences in human health (Anadón et al. 2018). For example, allergic reactions to food-derived penicillin, aminoglycosides, sulphonamides, and tetracyclines have been described in sensitive persons. Epidemiologic and experimental data indicate that levels of penicillin as low as 5–10 IU are sufficient to produce an allergic reaction in previously sensitised individuals (Sundlof 1994).

14.4.9.2 Antibacterial Effects on Human Intestinal Microflora

Therapeutic doses of antibiotics can cause adverse effects on the intestinal microflora ecology (i.e. disruption of the intestinal microflora or effects on the metabolic activity of intestinal microflora) and in maintaining an individual's health. In contrast to the well-documented negative effects of therapeutic dose of antibiotics, the effect of low levels (in µg/kg (ppb) or mg/kg (ppm)) of antibiotics on perturbing the intestinal microflora is not well defined. It is possible that low doses of antimicrobial agents, such as those found as residues in foods, could alter intestinal enzyme activity and have an effect on certain hormones and drugs, because in most cases the lowest doses at which the perturbations in the intestinal microflora occur have not been determined. In order to ensure human food safety, The CVM considered data gathered from a large number of compounds and determined that the maximum safe concentration for antimicrobial products is 1 ppm in a total diet of 1.5 kg. This equals a maximum antibiotic dose of 1.5 mg/day from consuming residues in food. This level of an antimicrobial residue in food should produce no effects on the intestinal microflora (Paige et al. 1997).

Exposure to antimicrobials is associated with an increased prevalence of resistance among bacteria of the normal flora (i.e. resident commensal or indigenous) of both humans and animals which may be considered a good indicator for the selection pressure exerted by antimicrobial use and for the resistance problems to be expected in pathogens (Murray 1992). Resistant bacteria from the intestinal flora of food animals may spread to farm workers as a result of direct contact and poor hygiene; bacteria also contaminate carcasses of slaughtered animals and reach humans via the food chain. Investigation of the prevalence of resistance of certain indicator bacteria like *Escherichia coli* and *Enterococci* in the intestinal tracts of different populations of animals and humans makes it feasible to detect a possible transfer of resistant bacteria from animals to humans and vice versa. The normal bacterial flora in many species of animals include foodborne pathogens such as *Salmonella, E. coli,* and *Campylobacter.* All three bacteria can cause severe foodborne illness in humans. Many *E. coli* appear to behave as resident commensals of the gut of animals and humans, but may be opportunistic pathogens as well as donors of resistance genetic elements for pathogenic *E. coli* or other species of bacteria (Hammerum et al. 2014). Foodborne resident commensals are nonzoonotic bacterial species living in the intestinal content of animals that could be transmitted to humans by the food chain and that normally do not cause foodborne infections in humans. When carcass contamination is measured at the processing plant, it is estimated that 20% of US broiler chickens are contaminated with *Salmonella* and 80% are contaminated with *Campylobacter* (Tollefson and Miller 2000). This is especially important for zoonotic pathogens (i.e. *Salmonella* spp., *Campylobacter* spp., methicillin-resistant *Staphylococcus aureus* [MRSA], *E. coli*) and commensal food-borne bacteria (i.e. *E. coli, Enterococcus* spp.), as these can act as reservoirs for resistance genes in the gut microbiota and the environment (the resistome; APUA 2008).

14.4.9.3 Antimicrobial Resistance

Overuse of antimicrobials is occurring in sectors such as human, animal (farms and aquaculture) and agriculture (e.g. tetracyclines, streptomycin; O'Neill 2016). Most bacteria and their genes can move relatively easily within and between humans, animals and the environment. At every point where antimicrobials are used, there are often already large reservoirs of resistant bacteria and resistance genes. AMR is an ecological problem that is characterised by complex interactions involving diverse microbial populations affecting the health of humans, animals and the environment (Collignon and McEwen 2019). Large reservoirs of resistance and residual antimicrobials occur in water, soil, wildlife and many other ecological niches due to pollution by sewage, pharmaceutical industry waste and manure runoff from farms (Marti et al. 2014). It is needed to preserve the effectiveness of approved antimicrobials, including trying to eliminate their inappropriate use, particularly where they are used in high quantities; for example, in the mass medication of animals with critically important antimicrobials (CIAs) for humans (i.e. extended-spectrum cephalosporins and fluoroquinolones) and the long term use of antimicrobials in feed, including colistin, tetracyclines and macrolides for AGP (Collignon and McEwen 2019). The association between increased antimicrobial use and resistance has been documented for nosocomial infections (McGowan 1983). In contrast, it has been observed that in countries where the use of particular substances (e.g. fluoroquinolones such as ciprofloxacin, enrofloxacin, flumequine) is discouraged or banned in livestock, low levels of, or no resistance to, these antimicrobials in livestock populations are observed in foodborne zoonotic pathogens (Aarestrup et al. 2008). The rate of AMR emergence in ecosystems such as the human or animal gut is likely to be highly dependent on the quantity of antimicrobials used, along with the duration and frequency of exposure.

Same classes of antimicrobials are used in humans and animals, and the challenges that arise from risk and benefit in the two sectors must be covered under the paradigm of One Health antimicrobial resistance. These antibiotics are classified as "critically important" for human health. Some examples are the third-generation cephalosporins and the broad-spectrum beta-lactam antimicrobials that are widely used in humans (cefotaxime, ceftriaxone and several other members) and animals (ceftiofur, cefpodoxime, cefoperazone, cefovecin). Resistance to the third-generation cephalosporins is mainly mediated by extended-spectrum beta-lactamases (ESBLs) and AmpC beta-lactamase enzymes (WHO 2016). ESBL genes are highly mobile and transmitted on plasmids, transposons and other genetic elements. AmpC beta-lactamases were originally reported to be chromosomal but have also been identified on plasmids and to have spread horizontally among Enterobacteriaceae (AGISAR 2016). From the One Health antimicrobial resistance perspective, cephalosporins, fluoroquinolones and colistin are good examples of CIAs for both human and animal health (OIE 2018). The main concern regarding selection and spread of resistance from animals to humans is their use as mass medications in large numbers of animals, either for therapy or prophylaxis. There are parallels with fluoroquinolones, another class of CIAs, to which resistance among *Campylobacter jejuni* emerged following mass medication of poultry flocks (Endtz et al. 1991); where fluoroquinolones were never approved in food animals, fluoroquinolone resistant strains in food animals remain very rare (Cheng et al. 2012). These CIAs should not be used in animals for prevention or as a first-line treatment, and their use should ideally be based on the results of bacteriological tests.

Fluoroquinolone use in food animals is also linked to quinolone resistance in *Salmonella* (AGISAR 2016). Surveillance data compiled by WHO indicate that rates of fluoroquinolone resistance in nontyphoidal *Salmonella* vary widely by geographical region. Overall, third- and fourth-generation cephalosporins and fluoroquinolones are critical in human medicine; clear evidence exists for the importance of treatment of entire groups of animals for resistance against pathogens that spread from animals to humans (Nelson et al. 2007).

Early on in human medicine, colistin (also known as polymyxin E) was predominately restricted to topical use because it frequently caused systemic toxicity (i.e. nephrotoxicity and neurotoxicity) when administered systemically as colistin methate sulfonate sodium (colistimethate sodium; Falagas and Kasiakou 2006). Colistin, however, is now used much more frequently, as a last resort drug by injection, for treatment of multiresistant gram-negative infections, including carbapenemase-producing Enterobacteriaceae (*Escherichia coli, Klebsiella* spp.), and multidrug-resistant (MDR) *Pseudomonas* and *Acinetobacter* species. Clinicians have been forced to reintroduce systemic colistin treatment for the treatment of healthcare-associated infections, in particular in children who have MDR gram-negative infections (Falagas et al. 2009). Also, colistimethate sodium is used by inhalation for the treatment of *Pseudomonas aeruginosa* lung infections in patients with cystic fibrosis.

Colistin has been used for decades in veterinary medicine. Colistin sulphate is administered orally to groups of pigs, poultry and, in some cases, calves for treatment and prophylaxis of diarrhoea due to gram-negative infections or for growth promotion (Collignon and McEwen 2019). Until recently, limited data on colistin resistance were available; however, in 2015 the transferable plasmid-mediated colistin resistance gene *mcr*-1 was found in *E. coli* isolates obtained from animals, food (meat and vegetables) and human bloodstream infections from China (Liu et al. 2016). This recent detection of resistance to colistin in foodborne pathogens in humans, livestock and food across different countries raises the issue of the potential role of global travel and trade in the transboundary dissemination of resistance genes. Spread of the gene by conjugation has been shown in *Klebsiella pneumoniae, Enterobacter* spp. (notably *Enterobacter aerogenes*) and *P. aeruginosa* (Liu et al. 2016).

Colistin illustrates some important One Health dimensions of antimicrobial resistance that differ from those of third-generation cephalosporins. The colistin case demonstrates once again that using large quantities antimicrobials for group treatments or growth promotion in animals can lead to significant antimicrobial resistance problems for human health, even if the drug class is initially believed to be of lesser importance, because the relative importance of antimicrobials to human health can change (Collignon and McEwen 2019).

Global efforts to fight AMR include the 2016 United Nations Political Declaration on AMR and the 2015 WHO Global Action Plan on AMR; the latter was subsequently adopted by the World Organisation for Animal Health (OIE) and the FAO. AMR has also been addressed in the G7 and G20 forums. In addition, action plans on antimicrobial resistance, such as those launched by FAO, WHO and the EU (EU 2017) are also enabling and motivating countries to provide supportive actions in order to reduce infections and thus the need for antimicrobials. The industry is clearly working together to ensure our farming systems are sustainable and healthy. AMR is a good example of a One Health matter in which human health is connected to that of animals and the environment. According to the European One Health Action Plan on AMR, the European agencies' supportive actions will include infection prevention, biosecurity measures and control practices in human healthcare and in animal husbandry (including in livestock, poultry, fish farming aquaculture and bee honey) in order to reduce infections and thus the need for antimicrobials (EU 2017). Only a multidisciplinary effort can provide an adequate response. There is a major lack of knowledge about the release and spread of resistant organisms in the environment and the threats and risks this poses to human and animal health. For example, the release of antimicrobials into the environment through human, animal and manufacturing waste streams should be assessed and new technologies developed to enable efficient and rapid degradation of antimicrobials in wastewater treatment plants, organic waste streams and the environment (EU 2017).

The aquaculture sector is vulnerable to exotic, endemic and emerging disease epizootics. The responsible use of veterinary medicines, including antimicrobials, has benefits in terms of improved on-farm biosecurity and husbandry through the use of vaccines and disinfectants. In fact, prudent use of antibiotics in animal intensive farms and aquaculture under veterinary supervision is critical in ensuring safety of food products (Okocha et al. 2018). Unregulated use of antibiotics in animal intensive farms and aquaculture industry for the production of farm raised fish and shrimps could pose human health and food safety concerns (e.g. by antimicrobial residues) that remain largely unaddressed in most developing nations of the world (Okocha et al. 2018).

Extensive use of drugs may increase the risk of an adverse effect of residues on consumers, such as the development of drug resistance and disruption of normal intestinal flora. It is important to consider that nearly 50% of the fish traded in international markets comes from aquaculture; it is therefore of great significance to ensure that the aquaculture sector is producing safe food (Karunasagar 2008).

Antimicrobials are widely used in aquaculture for therapeutic, metaphylactic and prophylactic purposes. AMR in aquaculture can occur through direct exposure to antimicrobials delivered as group therapy to fish or through livestock and human effluents containing resistant bacteria, resistance genes and antimicrobial residues. These will then exert selective pressure on gut microbiota of fish and on other bacteria in the aquatic environment (FAO/OIE/WHO 2006). The extent and persistence of antimicrobial residues in these production systems is currently unknown, but they are likely to be greatly diluted in the environment.

Fish in aquaculture may be subject to a wide range of diseases, some of which are of an infectious nature mediated by viruses, fungi and bacteria, or by internal and external parasites. For infectious diseases where treatment is available, veterinary drugs may be administered by baths or by inclusion in the feed. Currently there are no effective veterinary drugs for the treatment of viral infections in aquaculture, and fungi are not considered a problem in the seawater phase, thus antifungal agents are not administered via feed. Therefore, only drugs against infection by bacteria and external or internal parasites may be added to the feed (Lunestad and Samuelsen 2008).

ACKNOWLEDGMENTS

This work was supported by Project Ref. PID 2020-115979RR-C33 from the Ministerio de Ciencia e Innovación, Spain.

BIBLIOGRAPHY

Aarestrup, F. M., H. C. Wegener, and P. Collignon. 2008. Resistance of bacteria of the food chain: Epidemiology and control strategies. *Exp Rev Anti Infect Ther* 6: 733–50.

AGISAR. 2016. WHO Advisory Group on Integrated Surveillance of Antimicrobial Resistance (AGISAR). In *Critically Important Antimicrobials for Human Medicine*, 5th revision 2016; Geneva, Switzerland: World Health Organization.

Alderman, D. J. 1985. Malachite green: A review. *J Fish Dis* 8: 289–98.

Anadón, A. 1985. Standard residue regulations for chloramphenicol in Spain. *Ann Rech Vet* 16(2): 149–53.

Anadón, A. 1990. Les residus de substances chimiques dans les aliments d'origine animale en Espagne. *Bull Acad Vet Fr* 63: 245–52.

Anadón, A. 2006. The EU ban of antibiotics as feed additives. Alternatives and consumer safety. *J Vet Pharmocol Ther* 29(Suppl 1): 41–4.

Anadón, A., P. Bringas, M. R. Martínez-Larrañaga, and M. J. Diaz. 1994. Bioavailability, pharmacokinetics and residues of chloramphenicol in the chicken. *J Vet Pharmocol Ther* 17(1): 52–8.

Anadón, A., F. Gamboa, M. A. Martínez, et al. 2012. Plasma disposition and tissue depletion of chlortetracycline in the food producing animals, chickens for fattening. *Food Chem Toxicol* 50: 2714–21.

Anadón, A., M. A. Martínez, M. Martinez, et al. 2008a. Plasma and tissue depletion of florfenicol and florfenicol-amine in chickens. *J Agr Food Chem* 56: 11049–56.

Anadón, A., M. A. Martínez, M. Martínez, C. O. de la Cruz, M. J. Díaz, and M. R. Martínez-Larrañaga. 2008b. Oral bioavailability, tissue distribution and depletion of flumequine in food producing animal, chicken for fattening. *Food Chem Toxicol* 46: 662–70.

Anadón, A., and M. R. Martínez-Larrañaga. 1999. Residues of antimicrobial drugs and feed additives in animal products: Regulatory aspects. *Livestock Prod Sci* 59: 183–98.

Anadón, A., M. R. Martínez-Larrañaga, I. Ares, and M. A. Martínez. 2018. Chapter 7. Regulatory aspects for the drugs and chemicals used in food producing animals. In *Veterinary Toxicology*, 3rd ed., ed. R. C. Gupta, 103–31. San Diego, CA: Elsevier/Academic Press.

APUA. 2008. AMROAR scientific meeting report on commensals as reservoirs of antibiotic resistance. In *ROAR*. Boston: APUA.

AVMA. 2008. One health: A new professional imperative. In *One Health Initiative Task Force Final Report*. Schaumburg, IL: American Veterinary Medical Association.

Benford, D., P. M. Bolger, P. Carthew, et al. 2010. Application of the Margin of Exposure (MOE) approach to substances in food that are genotoxic and carcinogenic. *Food Chem Toxicol* 48(Suppl 1): S2-S24.
Bronzwaer, S. 2008. EFSA scientific forum "from safe food to healthy diets". EU risk assessment—Past, present and future. *Trends Food Sci Tech* 19: S2-S8.
Cheng, A. C., J. Turnidge, P. Collignon, D. Looke, M. Barton, and T. Gottlieb. 2012. Control of fluoroquinolone resistance through successful regulation, Australia. *Emerg Infect Dis* 18: 1453–60.
Collignon, P. J., and S. A. McEwen. 2019. One health—Its importance in helping to better control antimicrobial resistance. *Trop Med Infect Dis* 4(1): 1–22.
Derblom, H., H. Johansson, and G. Nylander. 1963. Thyroid hormone activity and gastrointestinal function: An experimental study in rat. *Acta Chir Scandinavica* 126(307): 52.
EC. 1996a. Council Directive 96/22/EC of 29 April 1996 concerning the prohibition on the use in stock farming of certain substances having a hormonal or thyrostatic action and of beta-agonists, and repealing Directives 81/602/EEC, 88/146/EEC and 88/299/EEC (OJ L125, 23.5.1996).
EC. 1996b. Council Directive 96/23/EC of 29 April 1996 on measures to monitor certain substances and residues thereof in live animals and animal products and repealing Directives 85/358/EEC and 86/469/EEC and Decisions 89/187/EEC and 91/664/EEC (OJ L 125, 23.5.1996).
EC. 2002a. Opinion of the scientific committee on veterinary measures relating to public health on review of previous SCVPH opinions of 30 April 1999 and 3 may 2000 on the potential risks to human health from hormone residues in bovine meat and meat products (adopted on 10 April 2002).
EC. 2002b. COMMISSION DECISION of 12 August 2002 implementing Council Directive 96/23/EC concerning the performance of analytical methods and the interpretation of results (OJ L 221, 17.8.2002).
EC. 2002c. Regulation (EC) No 178/2002 of the European Parliament and of the Council of 28 January 2002 laying down the general principles and requirements of food law, establishing the European Food Safety Authority and laying down procedures in matters of food safety (OJ L 31, 1.2.2002).
EC. 2003. COMMISSION DECISION of 13 March 2003 amending Decision 2002/657/EC as regards the setting of minimum required performance limits (MRPLs) for certain residues in food of animal origin (OJ L 71, 15.3.2003).
EC. 2008. DIRECTIVE 2008/97/EC of the European Parliament and of the Council of 19 November 2008 amending Council Directive 96/22/EC concerning the prohibition on the use in stock farming of certain substances having a hormonal or thyrostatic action and of beta-agonists (OJ L 318, 28.11.2008).
EC. 2009. Commission Regulation (EC) No 470/2009 of the European Parliament and of the Council of 6 May 2009 concerning laying down Community procedures for the establishment of residue limits of pharmacologically active substances in foodstuffs of animal origin, repealing Council Regulation (EEC) No. 2377/90 and amending Directive 2001/82/EC of the European Parliament and of the Council and Regulation (EC) No. 726/2004 of the European Parliament and of the Council (OJ L 152, 16.6.2009).
EC. 2015. Commission Notice Guidelines for the prudent use of antimicrobials in veterinary medicine (OJ L 299, 11.9.2015).
Edwards, D. I. 1977. The action of metronidazole in DNA. *J Antimicrob Chemother* 3: 43–8.
EEC. 1990. Council Regulation 90/2377/EEC laying down a Community procedure for the establishment of maximum residue limits of Veterinary medicinal products in foodstuffs of animal origin (OJ No L 224, 12.8.90).
EFSA (European Food Safety Authority). 2009. Guidance of the Scientific Committee on a request from EFSA on the use of the benchmark dose approach in risk assessment. *EFSA J* 1150: 1–72.
EFSA (European Food Safety Authority). 2013. Guidance on methodological principles and scientific methods to be taken into account when establishing Reference Points for Action (RPAs) for non-allowed pharmacologically active substances present in food of animal origin. Scientific Opinion of the Panel on Contaminants in the Food Chain. *T EFSA J* 11: 1–24.
EFSA (European Food Safety Authority). 2018. Report for 2016 on the results from the monitoring of veterinary medicinal product residues and other substances in live animals and animal products. *EFSA Supp Publ* EN-1358: 1–75.
EMEA (The European Agency for the Evaluation of Medicinal Products). 2002. Updated Application Software Relating to Note for Guidance on Approach Towards Harmonisation of Withdrawal Periods for Meat. EMEA/CVMP/563/02, London, UK.

Endtz, H., G. Ruijs, B. van Klingeren, W. H. Jansen, T. Reijden, and R. P. Mouton. 1991. Quinolone resistance in *Campylobacter* isolated from man and poultry following the introduction of fluoroquinolones in veterinary medicine. *J Antimicrob Chemother* 27: 199–208.

EU. 2004. From farm to fork Safe food for Europe's consumers. Luxembourg: Office for Official Publications of the European Communities, 2004 ISBN 92-894-7772-5 European Communities.

EU. 2010. Commission Regulation (EU) No 37/2010 of 22 December 2009 on pharmacologically active substances and their classification regarding maximum residue limits in foodstuffs of animal origin (OJ L 15, 20.1.2010).

EU. 2017. A European One Health Action Plan against Antimicrobial Resistance (AMR). https://ec.europa.eu/health/sites/health/files/antimicrobial_resistance/docs/amr_2017_action-plan.pdf (accessed November 21, 2020).

EU. 2019. Regulation (EU) 2019/6 of the European Parliament and of the council of 11 December 2018 on veterinary medicinal products and repealing Directive 2001/82/EC (OJ of EU L4, 7.1.2019).

Falagas, M. E., and S. K. Kasiakou. 2006. Toxicity of polymyxins: A systematic review of the evidence from old and recent studies. *Crit Care* 10(1): R27.

Falagas, M. E., E. K. Vouloumanou, and P. I. Rafailidis. 2009. Systemic colistin use in children without cystic fibrosis: A systematic review of the literature. *Int J Antimicrob Agents* 33(6): 503.e1–503.e13.

FAO/OIE/WHO. 2006. *Antimicrobial Use in Aquaculture and Antimicrobial Resistance*. Report of a joint FAO/OIE/WHO expert consultation on antimicrobial use in aquaculture and antimicrobial resistance. Seoul, Republic of Korea.

FAO/OMS. 1989. *Evaluation of Certain Veterinary Drug Residues in Food*. Thirty-fourth report of the Joint FAO/WHO Expert Committee on Food Additives.

FAO/WHO. 2009. *Principles and Methods for the Risk Assessment of Chemicals in Food*. WHO Environmental Health Criteria, No. 240. Geneva.

Feil, V. J., J. K. Huwe, R. G. Zaylskie, et al. 2000. Chlorinated dibenzo-*p*-dioxin and dibenzofuran concentrations in beef animals from a feeding study. *J Agr Food Chem* 48: 6163–73.

Fukuda, K. 2015. Food safety in a globalized world. *Bull World Health Org* 93: 212.

Hammerum, A. M., J. Larsen, V. D. Andersen, et al. 2014. Characterization of Extended-Spectrum-lactamase (ESBL)-producing *Escherichia coli* obtained from danish pigs, pig farmers and their families from farms with high or no consumption of third- or fourth-generation cephalosporins. *J Antimicrob Chemother* 69: 2650–7.

Herbst, A. L. 1976. Summary of the changes in the human female genital tract as a consequence of maternal diethylstilbestrol therapy. *J Toxicol Environ Health* 1: 13–20.

Herbst, A. L., H. Ulfelder, and D. C. Poskanzer. 1971. Adenocarcinoma of the vagina: Association of maternal stilbestrol therapy with tumor appearance in young women. *N Engl J Med* 284: 878–81.

Hoover, R. N., M. Hyer, R. M. Pfeiffer, et al. 2011. Adverse health outcomes in women exposed *in utero* to diethystilbestrol. *N Engl J Med* 365: 1304–14.

IPCS. 2009. *Principles for Modelling Dose Response for the Risk Assessment of Chemicals*. Geneva: World Health Organization. International Programme on Chemical Safety. Environmental Health Criteria, No. 239.

IPCS. 2017. *Guidance Document on Evaluating and Expressing Uncertainty in Hazard Characterization*, 2nd ed. Geneva: World Health Organization. International Programme on Chemical Safety. Harmonization Project Document, No. 11.

JECFA (Joint FAO/WHO Expert Committee on Food Additives). 1991. Thirty-eighth report. *Evaluation of Certain Veterinary Drug Residues in Food*. Geneva.

JECFA (Joint FAO/WHO Expert Committee on Food Additives). 2006. Sixty-sixth report. *Evaluation of Certain Veterinary Drug Residues in Food*. Geneva.

JECFA (Joint FAO/WHO Expert Committee on Food Additives). 2008. Seventieth meeting. *Evaluation of Certain Veterinary Drug Residues in Food*. Geneva.

Karunasagar, I. 2008. Food safety and public health risks associated with products of aquaculture. In *Understanding and Applying Risk Analysis in Aquaculture. FAO Fish Aqua Tech Paper*, eds. M. G. Bondad-Reantaso, J. R. Arthur, and R. P. Subasinghe, 9–25. Rome: FAO.

Levine, R. U., and K. M. Berkowitz. 1993. Conservative management and pregnancy outcome in diethylbestrol-exposed women with and without gross genital tract abnormalities. *Am J Obstet Gynol* 169(5): 1125–9.

Linton, A. H., A. J. Hedges, and P. M. Bennett. 1988. Monitoring for the development of antimicrobial resistance during the use of olaquindox as a fed additive on commercial pig farms. *J Appl Bacteriol* 64: 311–27.

Liu, Y. Y., Y. Wang, T. R. Walsh, et al. 2016. Emergence of plasmid-mediated colistin resistance mechanism MCR-1 in animals and human beings in China: A microbiological and molecular biological study. *Lancet Infect Dis* 16(2): 161–8.

Lunestad, B. T., and O. Samuelsen. 2008. Veterinary drug use in aquaculture. In *Improving Farmed Fish Quality and Safety*, ed. Ø. Lie, 97–127. Cambridge: Woodhead.

Madec, J.-Y., and M. Haenni. 2018. Antimicrobial resistance plasmid reservoir in food and food-producing animals. *Plasmid* 99: 72–81.

Maghuin-Rogister, G., P. Gaspar, M. Vandenbroeck, et al. 1991. Anabolics: The situation in Belgium. *Ann Rech Vet* 22: 305–9.

Marti, E., E. Variatza, and J. L. Balcazar. 2014. The role of aquatic ecosystems as reservoirs of antibiotic resistance. *Trends Microbiol* 22: 36–41.

McGowan, J. E. Jr. 1983. Antimicrobial resistance in hospital organisms and its relation to antibiotic use. *Rev Infect Dis* 5(6): 1033–48.

Metzler, M. 1984. Diethylstilbestrol: Reactive metabolites derived from a hormonally active compound. In *Biochemical Basis of Chemical Carcinogenesis*, eds. H. Greim, R. Jung, M. Kramer, H. Marquardt, F. Oesch, 69–75. New York: Raven Press.

Miller, M. A., and C. E. Eirkson. 1997. Environmental assessments for animal drug products. Proceedings of the 7th European Association for Veterinary Pharmacology and Toxicology International Congress, Madrid (Spain), July 6–10, 1997. *J Vet Pharmacol Ther* 20: 323–5.

Mills, K. W., and B. L. Kelly. 1986. Antibiotic susceptibilities of swine Salmonella isolants from 1979 to 1983. *Am J Vet Res* 47: 2349–50.

Momotani, N., J. Y. Noh, N. Ishikawa, and K. Ito. 1997. Effects of propylthiouracil and methimazole on fetal thyroid status in mothers with Graves' hyperthyroidism. *J Clin Endocrinol Metabol* 82(11): 3633–6.

Murray, E. 1992. Problems and dilemmas of antimicrobial resistance. *Pharmacotherapy* 12(6): 86–93.

Nau, H., F. Welsch, B. Ulbrich, R. Bass, and J. Lange. 1981. Thiamphenicol during the first trimester of human pregnancy placental transfer *in vivo*, placental uptake *in vitro*, and inhibition of mitochondrial function. *Toxicol Appl Pharmacol* 60: 131–41.

Nelson, J. M., T. M. Chiller, J. H. Powers, and F. J. Angulo. 2007. Fluoroquinolone-resistant *Campylobacter Species* and the withdrawal of fluoroquinolones from use in poultry: A public health success story. *Clin Infect Dis* 44: 977–80.

Nouws, J. F. M., and J. Laurensen. 1990. Postmortal degradation of furazolidone and furaltadone in edible tissues of calves. *Vet Q* 12: 56–9.

O'Neill, J. 2016. Tackling drug-resistant infections globally: Final report and recommendations. *Rev Antimicrob Resist.*

Ohmae, K., S. Yonezawa, and N. Terakado. 1983. Epizootiological studies on R plasmid with carbadox resistance. *Jpn J Vet Res* 54: 165–70.

OIE (World Organisation for Animal Health). 2018. *OIE Annual Report on Antimicrobial Agents Intended for Use in Animals.* Second Report. Paris, France: OIE.

Okocha, R. C., I. O. Olatoye, and O. B. Adedeji. 2018. Food safety impacts of antimicrobial use and their residues in aquaculture. *Public Health Rev* 39(1): 21.

Paige, J. C., L. Tollefson, and M. Miller. 1997. Public health impact on drug residues in animal tissues. *Vet Hum Toxicol* 39(3): 162–69.

Pinel, G., S. Mathieu, N. Cesbron, et al. 2006. Evidence that urinary excretion of thiouracil in adult bovine submitted to a cruciferous diet can give erroneous indications of the possible illegal use of thyrostats in meat production. *Food Addit Contam* 23(10): 974–80.

Prescott, F. J., and J. D. Baggot. 1993. *Antimicrobial Therapy in Veterinary Medicine.* Ames, IA: Iowa State University Press.

Robinson, T. P., D. P. Bu, J. Carrique-Mas, et al. 2016. Antibiotic resistance is the quintessential one health issue. *Trans R Soc Trop Med Hygiene* 110: 377–80.

Sanquer, A., G. Wackowiez, and B. Havrileck. 2006. Qualitative assessment of human exposure to consumption of injection site residues. *J Vet Pharmacol Ther* 29: 345–53.

Sundlof, S. F. 1994. Human risks associated with drug residues in animal derived foods. *J Agromed* 1: 5–22.

Taurog, A. 2000. Hormone synthesis: Thyroid iodine metabolism. In *The Thyroid*, eds. L. E. Braverman, and R. D. Utiger, 8th ed., 61–84. New York: Lippincott Williams & Wilkins.

Tollefson, L., and M. A. Miller. 2000. Antibiotic use in food animals: Controlling the human health impact. *J AOAC Int* 83(2): 245–54.

Umberger, E. J. 1975. Products marketed to promote growth in food-producing animals: Steroid and hormone products. *Toxicology* 3: 3–21.

Vettorazzi, G., and F. Kouthon. 1981. *JECFA, Past, Present and Future*. Background paper No. 2 of the Twenty-fifth Session of the Joint FA0/WH0 Expert Committee on Food Additivés, Geneva (unpublished document).

Wang, X., G. Cheng, and Z. Liu. 2016. The critical role of oxidative stress in the toxicity and metabolism of quinoxaline 1,4-di-n-oxides *in vitro* and *in vivo Drug Metabol Rev* 48(2): 159–82.

WHO. 1995. *Evaluation of Certain Veterinary Drug Residues in Food*. 42nd Report, Joint FAO/WHO Expert Committee on Food Additives, WHO Technical Report Series 851.

WHO. 2004. WHO Food Additives Series 53, JECFA (WHO: Joint FAO/WHO Expert Committee on Food Additives), IPCS (International Programme on Chemical Safety) INCHEM, 7–85.

WHO. 2016. *Guidance document for WHO monographers and reviewers evaluating veterinary drug residues in food*. World Health Organization, 1–70.

WHO/FAO. 1972. *Evaluation of Certain Food Additives and the Contaminants Mercury, Lead and Cadmium*. 16th Report, WHO Technical Report Series 505.

Woodward, K. N. 1998. The use of microbiological end-point in the safety evaluation and elaboration of maximum residue limits for veterinary drugs intended for use in food producing animals. *J Vet Pharmacol Ther* 21: 47–53.

15 Nanotoxicology

Mª Victorina Aguilar, Mª del Carmen Berrocal, Domingo Ly-Pen and Mª Teresa Hernández

CONTENTS

15.1 INTRODUCTION

Nanomaterials, according to the European Commission recommendation (European Commission 2011), are natural, incidental or manufactured materials containing particles in an unbound state or as an aggregate or agglomerate and where, for 50% or more of the particles in the number size distribution, one or more external dimensions is in the size range 1–100 nm.

The number of nanomaterials designed, produced and used in a multitude of technological applications is increasing. Their applications have great potential to address social and clinical challenges ranging from artificial intelligence, robotics and autonomous transport to the development of new products for daily use such as cosmetics, drugs, food and medicine (Figure 15.1). In fact, in 2018 the economic impact of nanomaterials was $3.5 trillion (National Nanotechnology Coordination Office Nanotechnology 2018). Some of their usages are inspired by nature, such as eye implants, in which multifunctional biphotonic structures are used (e.g. based on the *Greta oto*, or glasswing butterfly; Narasimhan et al. 2018), or antimicrobial films made from polymers derived from essential oils (Al-Jumaili et al. 2018).

Although there are plenty of nanomaterials and from very diverse nature, the most commonly used are nonmetallic inorganic nanomolecules, metals and alloys, carbon-based nanomaterials, nanopolymers and dendrimers, or quantum dots and nanoclays (European Commission 2012).

At present, we are immersed in the nanotechnology revolution, integrated in the Fourth Industrial Revolution (4IR). According to the World Economic Forum in Davos, this revolution began in 2016 and shows the total integration of technologies, information and communication (TIC) in the context of areas of manufacturing and application such as personnel, domestic, automotive, entrepreneurship and infrastructure. The 4IR aims to "increase human capabilities through new technologies" and even to be able to "cure" aging, leading to a new and controversial "posthuman" species. In fact, according to Chad Mirkin, director of the Northwestern University Institute for Nanotechnology: "Nanotechnology involves the reconstruction of the Earth as we know it, atom by atom".

The great applicability of nanomaterials is due to their unique physical and chemical properties. For example, due to their high specific surface area and the greater number of atoms on their surfaces, their melting point is lower, affecting its thermodynamic behavior, and its chemical

DOI: 10.1201/9780203730584-15

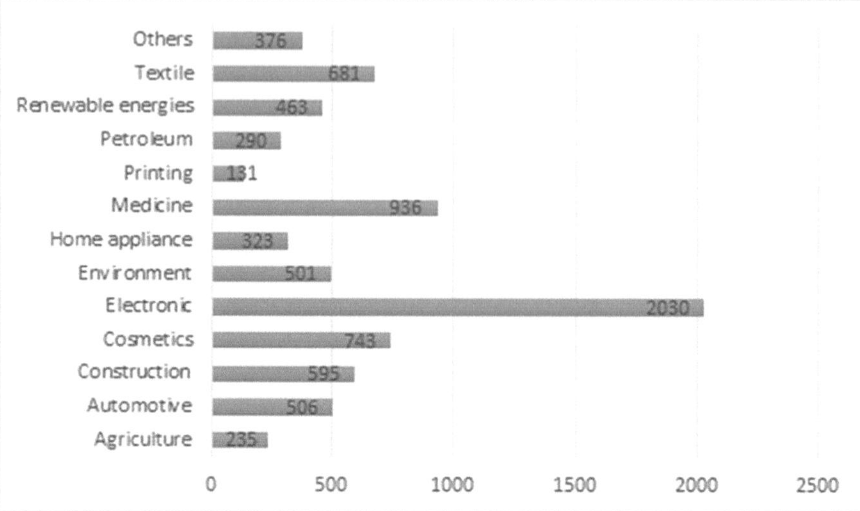

Figure 15.1 Applications of nanomaterials, with the number of products used in each application.

reactivity is increased, allowing self-assembly and exceptional properties as catalysts. Moreover, they have mechanical properties such as yield strength, hardness and strain-rate sensitivity.

It is considered that nanomaterials are a discovery of modern science, but this is not the truth, because they have a long history. Nanomaterials were used in the 4th century by artisans to dye the Lycurgus Cup, which is green in daylight but red when illuminated from the inside. In the middle ages nanomaterials were used in the elaboration of the stained glass windows of cathedrals, obtaining colors through the heating and cooling of the glass; along with the mixture of copper and silver salts with oxides and vinegar, ochre and clay, the size of their compounds was reduced. In 1857, Michael Faraday was the first author who described in scientific terms the optical properties of metals on a nanometric scale. Later, many other researchers investigated and studied the properties and applications of these nanoparticles (NPs). However, despite many published studies and more evidence, there are still many questions about their risks to human health and the environment. It has been demonstrated that both NPs of natural origin (including those expelled by volcanoes; see Table 15.1), and synthetic ones can produce deleterious effects on humans and other living organisms, such as the beneficial bacteria used in agriculture. If these risks are not properly evaluated and controlled, we are bound to block economic and social development or, even worse, to cause serious human and environmental damages.

15.2 NANOTOXICOLOGY

Because of their size and high reactivity, nanomaterials are able to interact with biological systems and the environment, producing unusual harmful effects on larger particles. The study of these deleterious effects, mechanisms of interaction, symptoms and so forth is the matter of study of nanotoxicology, which is a subspecialty of particle toxicology. Nanotoxicology has evolved over time as more knowledge has been gained. Only recently were the harmful effects of some nanoparticles studied at the pulmonary level, whereas currently nanotoxicology covers many other areas, due to nanomaterial characteristics that make it susceptible to interact with multiple fluids, cells or tissues, activating different mediators and processes.

15.2.1 Pathways of Exposure and Kinetics

Despite the high number of NPs used, there are very few studies evaluating the mechanisms of entry and their toxicokinetic processes (Figure 15.2); the few studies that do so only focus on metals and their oxides and, to a lesser extent, on polymers (Landsiedel et al. 2012; Nautiyal et al. 2017). NPs can enter the body by inhalation, dermally, by ingestion and parenterally (Docter et al. 2015).

Table 15.1 Source of Nanoparticles in the Atmosphere

	Anthropogenic	
Natural	***Unintentional***	***Intentional***
Photochemical reactions	By-products of simple combustion	Metal oxides
Volcanic eruptions	Food cooking	Nano clay
Soil erosion by wind/water	Industrial exhaust	Nanotube, spheres and wires
Forest fires	Welding	Quantum dots
Mineral dust	Vehicle exhaust	
Viruses	Coal, oil and gas boiler	

Source: Jeevanandam et al. (2018).

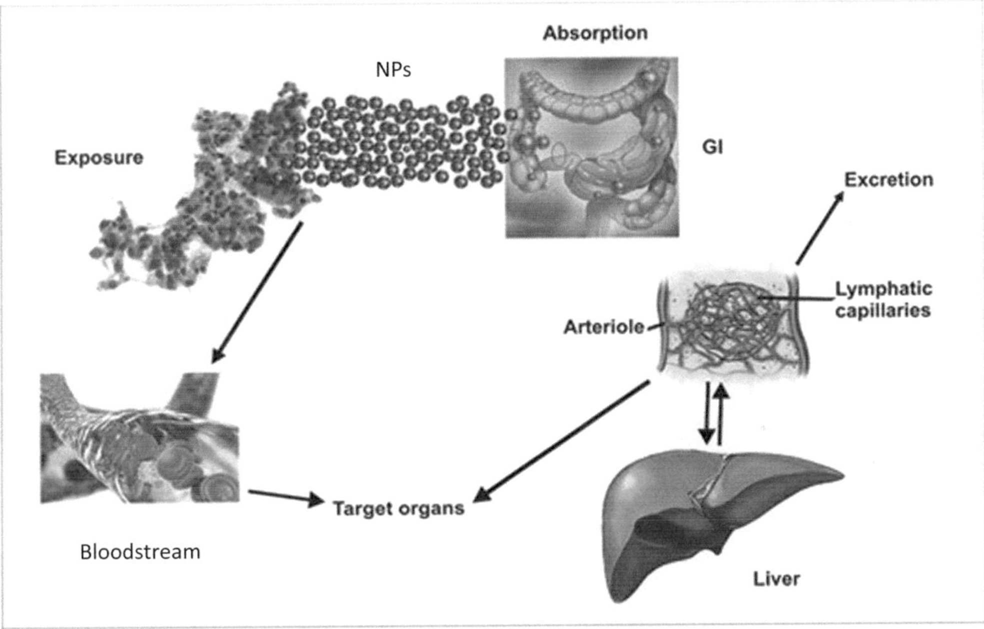

Figure 15.2 Biokinetics of nanometric particles.

NPs can enter the human body in several ways (oral ingestion, inhalation, via the bloodstream and through damaged skin). Absorption, distribution, metabolism and excretion processes subsequently take place. When translocated in the gastrointestinal tracts, NPs are transformed into their free forms and absorbed through the liver and/or the lymphatic system. They may accumulate in organs with high phagocytic activity, mainly in the liver, then the kidneys and the spleen, among others. The accumulation and retention in critical target sites, as well as potential adverse effects, largely depends upon physicochemical characteristics of the surface and core of NPs.

15.2.1.1 Dermal

One of the main entry routes of NPs is the dermal one, because many of these particles are used in the form of dermatological and cosmetic products for topical use or in textile products and dressings that come into contact with the skin. These NPs are absorbed in a size-dependent way. Nanoparticles of less than 10 nm can cross the stratum corneum through diffusion and reach the dermis by intercellular or transcellular routes, whereas particles greater than 20 nm can only be absorbed by transappendageal routes such as hair follicles or the pores of sweat gland. Larger nanoparticles do not enter the skin even through transappendageal routes (Gautam et al. 2011). This parameter also influences transport from the skin to the systemic circulation, where NPs will find other physiological barriers, such as the endothelial cells covering the inside of the blood vessels, or in their lymphatic translocation to regional lymph nodes. Other NP properties, such as their solubility, influence this transport. Thus, if they are water-soluble, they are able to cross easier into the interstitium, a process facilitated by endothelial permeability and by specific carriers (Mehta and Malik 2006). On the contrary, if NPs are liposoluble, they can transit through the intercellular space of the stratum corneum or pass through the cells of the hair follicles (as is the case for polystyrene nanoparticles; Arif and Adil 2016) or the sweat glands (Chung et al. 2019). NPs can also be stored in the nonvascularized area of the skin, from where they cannot be removed by macrophages (Gatto and Bardi 2018).

15.2.1.2 Inhalation

When NPs are inhaled (e.g. from perfumes or deodorants, or by workers in the industries where NPs are handled or produced), they can be deposited in the respiratory system, depending on their size. Ultrafine particles with sizes greater than 10 nm are deposited mostly in the alveolar region; those lesser than 10 nm are deposited mainly in the extrathoracic region and, in smaller amounts, in the trachea-bronchial region. Because of their size they can be captured by epithelial and endothelial cells, reaching the bloodstream; from there they are transported to target organs such as bone marrow, lymph nodes, the spleen or the heart (Oberdörster and Kuhlbusch 2018). Some NPs, such as carbon nanoparticles, can reach the central nervous system and ganglia by translocation along axons and dendrites of neurons. This pathway comprises sensory nerve endings of the olfactory and trigeminal nerves and an intricate network of sensory nerve endings in the tracheobronchial region. Inhaled nanoparticles can be swallowed and penetrate the gastrointestinal tract (Bunglavan et al. 2014).

15.2.1.3 Ingestion

The ingestion of NPs occurs, for example, via their use in lipsticks or in the numerous nanocomponents of commercial foods or multiple applications in food packaging (Table 15.2). Ingested NPs can be divided in the gastrointestinal tract: they can remain in free form as is the case of insulin, encapsulated in vitamin B_{12}-dextran NPs (Bunglavan et al. 2014), or they can agglomerate, solubilize, aggregate or react with some components of the gastrointestinal system, such as enzymes, intestinal microbiota or acids.

Table 15.2 Examples of the Use of Nanotechnology in Food Applications

Sector	Application	Nanomaterial
Agriculture	*Nanofertilizers*	Ag, Fe, Cu, . . .
	Nanopesticides	$Cu(OH)_2$, silicon, ZnO
	Nanoherbicides	Poly(epsilon-caprolactone)
Food processing	*Against biological deterioration:*	
	Antimicrobials	Ag/LDPE, ZnO/gelatin
	Increase bioavailability	CoQ10, vitamins, iron, calcium
	Protection chemical ingredients:	SiO_2-gallic acid, ZnO
	Antioxidants	SiO_2
	Flavors	ZnO, B_2O_6Zn
	Preservatives	TiO_2-SiO_2
	Enhance physical properties:	ZnO, iron oxide, aluminium oxide
	Color additives	SiO_2
	Additive or polymer production aid	Poly(lactic acid)
	Anticaking agent	
	Increase flame resistance	
Food contact packaging	*Edible film/coating*	Chitosan
	Active packaging	Silicate
	Flame retardation additive, gas barrier	Titanium aluminium nitrite
	Prevent abrasive wear	Montmorillonite
	Pesticide detection	Zinc oxide DQs
	Pathogen detection	Magnetic nanosensors
	Toxin detection	Plasmonic nanosensors
		Fluorescent nanosensors
		Plasmonic nanosensors

Subsequently, they are translocated through the intestinal epithelium, a process that takes place in different stages and depends on the properties of each nanomaterial, such as its surface, charge, size, presence or absence of ligands (Maisel et al. 2015). Translocation involves diffusion through intestinal mucus, transcellular or paracellular transport (Yun et al. 2013). The transcellular route is the most common route, through the enterocytes or microfold (M) cells of Peyer patches. However, macromolecules such as polymeric NPs cannot use this absorption mechanism because of their large size. The four mechanisms that can be used for the transport of polymeric NPs are phagocytosis, macropinocytosis, clathrin-mediated endocytosis and caveolin-mediated endocytosis. The paracellular route is preferred for transport of hydrophilic drugs. However, it is restricted to polymeric NPs due to the very small intercellular space and tight junctions between epithelial cells. Hence to improve paracellular transport, tight junctions can be opened reversibly using permeation/penetration enhancers such as cationic and anionic polymers (Hirlekar et al. 2017).

After being absorbed, the NPs are transported via lymphatic system to the liver (where they undergo first-pass metabolism; see below), kidneys and spleen; this has been demonstrated in polystyrene NPs of less than 100 nm. Through the portal circulation, they can also access these organs, which are the main destinations for the systemic distribution of a large number of NPs; but they are not the only ones, because they can also be transported to the kidneys, lungs, brain and so forth, as noted above. For example, gold (Au) NPs of less than 50 nm or lipid NPs, through the bloodstream, can be distributed throughout the body. Moreover, copper NPs can be retained in the gastric lumen by reacting with HCl, resulting in a metabolic alkalosis due to the fact that the resulting HCO_3 produced will return to the circulation forming large amounts of sodium bicarbonate, which increases the pH of arterial blood. The rise in blood pH can lead to a series of respiratory and renal compensatory effects, which cause anomalies such as an increase in the size of the glomerulus, a decrease in the lumen of Bowman's capsules and signs of glomerulonephritis (Anjum et al. 2015).

NPs are able to cross the human placenta, such as polystyrene NPs less than 240 nm, silica (70 nm in diameter) or gold, because they have been detected in the brain of the fetus (Yoshioka et al. 2014) and can cause embryonic death.

15.2.2 Metabolism and Kinetics

Regarding metabolism, it is essential to know the biopersistence or the degradation pathways of these materials, because this is crucial for its toxicity, especially in the long term. The ability of cells to metabolize or degrade NPs and excrete or eliminate them from the body minimizes their potential toxicity. At first, it was thought that NPs were resistant to metabolism, but studies have demonstrated that they can be degraded by enzymatic processes. Thus, for example, carbon nanotubes can undergo oxidative degradation by myeloperoxidase (MPO), so they do not induce inflammatory or oxidative stress responses (Bartneck et al. 2010; Keshavan et al. 2019). Some other NPs are nondegradable or slowly degrade and will accumulate in the organs or cells, perhaps leading to some as yet unknown long-term toxic effects.

Despite the lack of studies in this regard, it is known that the metabolism of these particles depends on factors such as surface chemical composition, animal species, medium in which these particles are found and solubility. The surface charges on nanoparticles can favor their binding of serum enzymes, leading to a so-called protein corona and effecting enzymatic regulatory mechanisms (Cagliani et al. 2019). The local pH change and the presence of degradative enzymes in the cells, such as cathepsin L, make some nanoparticles degradable and lose their coating on their surface (Uyeda et al. 2005). Protein corona formation is a process dependent on the physicochemical properties of NPs; it is dynamic, competitive and complex, and there is a continuous exchange between the proteins adsorbed in the NPs and the free proteins present in the biological environment (Simon et al. 2006). The distribution and biological effect of NPs depends on the formation of the protein crown (Mahmoudi et al. 2016). Thus the formation of the protein crown of carbon nanotubes (CNTs) with blood proteins reduces their cellular response and their cytotoxicity (Ge et al. 2011), whereas in the case of NPs of colloidal metals the formation of said crowns plays an integral role in the interactions of these NPs so that their conformational changes can induce an immune response (Nel et al. 2009; Hitler et al. 2017). When the NPs are coated with antibodies for VCAM-1 and E-selectin, in equal proportions, a more uniform vascular distribution is produced (Hossain et al. 2014). These nanoparticles can also be surrounded by lipids with an influence on their cellular uptake and, therefore, in their toxicity.

15.2.3 Excretion

Excretion is an important biological process that prevents damage and toxicity by removing unwanted materials from the body (Zhao et al. 2019). To date, information about the excretion of NPs is limited, although it is known that they use both renal and biliary routes. The renal pathway is a complex process that involves glomerular filtration, which is highly dependent on the hydrodynamic diameter. The capillary walls of the glomerulus typically have a filtration threshold (FST) of less than 5.5 nm, which means that molecules with a hydrodynamic diameter (HD) within this range (e.g. gold [AU] NPs of 1.4 nm) can be filtered quickly by the kidneys and effectively eliminated, whereas larger ones (HD > 8 nm) remain in the bloodstream. In addition, the solubility of these particles influences their elimination; insoluble NPs can remain in the tissues after prolonged exposure. They can also accumulate, mainly in the liver and spleen, where they can be retained for more than 30 days due to phagocytosis of the macrophages.

The NPs that cannot be filtered by the kidneys are excreted through the hepatobiliary pathway in the liver, appearing in the feces. There are studies showing that these NPs can be cleared in the immune system through the Kupffer cells; these are macrophages of the liver, able to phagocytose foreign materials in the hepatic blood. They play an important role in liver clearance through selective endocytosis of opsonized antigens (Lin et al. 2008). Granulocytes can also be included in this action, especially neutrophils, which also possess the phagocytic capacity of NPs.

To date, there is no evidence for the presence of NPs in human breast milk (EFSA 2009).

15.2.4 Mechanisms of Action

Although the mechanisms of toxic action after exposure to NPs have not been fully clarified, the studies published so far demonstrate its complexity and its dependence on the route of exposure (ingestion, cutaneous or respiratory), dose, susceptibility and specific physicochemical properties of the particles (i.e. the same properties that justify their applicability; Figure 15.3).

Applications include medicine, electronics and energy production as well as consumer products. However, these very properties have the potential to affect humans and the environment adversely. NPs can easily penetrate cell membranes and other biological barriers into living organisms, causing cell damage.

But we must not forget that these NPs not only involve risks as chemical agents; because of their ability to be covered with biomolecules (proteins, lipids and polysaccharides) from the environment in which they are immersed, they are transformed into biological entities with their different mechanisms of action and biological effects. Of all the factors that influence the action of NPs,

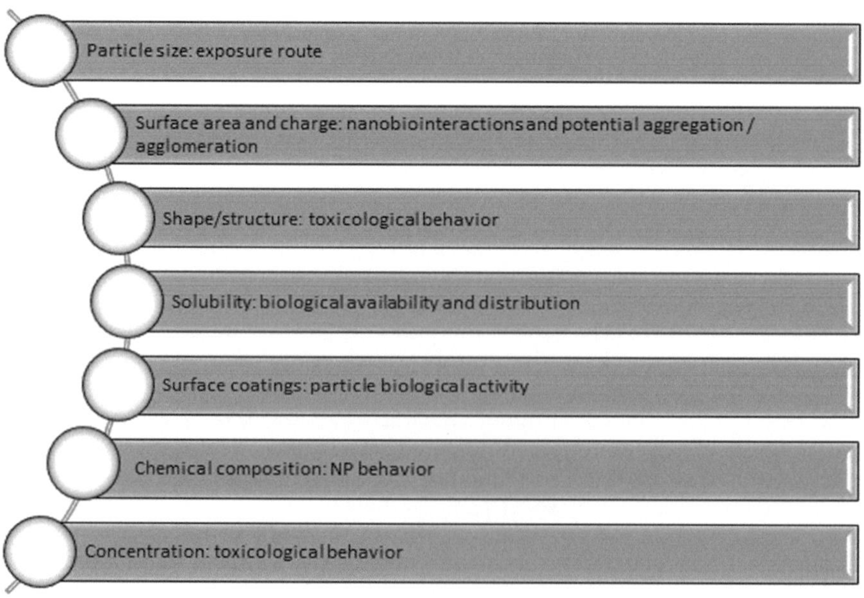

Figure 15.3 Properties which make nanometric particles (NPs) suitable for various applications.

size and chemical composition are the most influential. The existence of an inverse relationship between particle size (and therefore, a greater surface area) and toxicity has been demonstrated. In terms of chemical composition, this influences both the toxicity and the mechanism of action. For example, fullerenes generate free radicals by photosensitization, which leads to their transition to a long-lived triplet excited state and the subsequent energy or electron transfer to molecular oxygen, yielding the highly reactive singlet oxygen ($^1O_2$) or superoxide anion ($O_2^{\bullet -}$), respectively, causing lipid peroxidation at the cellular membrane level (Arif and Adil 2016); whereas graphenes cause the aggregation of particles and subsequent interaction with tissues (Qi et al. 2018). Regarding their charge, concentration, configuration (shape) and solubility, it has been shown that the addition of hydroxyl and carbonyl groups decreases the cytotoxic potential of carbon NPs.

Nevertheless, not only do the toxic effects depend on the attributes of the NPs, but other host characteristics can also influence these effects. For example, the absence of the natural pathways of excretion or hepatorenal disorders can cause carbon nanotubes or gadolinium to accumulate, thus causing deleterious effects, even when the exposure dose is small. Furthermore, genetic polymorphisms in toll-like receptors and complement activation can lead to different responses to the same particle in different individuals. This is what happens to the carbon nanotubes, which can cause immunosuppression characterized by reduced toll-dependent antibody response and proliferative capacity of T cells and increase IL-10 interleukins and levels of NAD(P)H oxidoreductase-1 mRNA (Mitchell et al. 2007). Another example is TLR1 Ile602Ser, which was linked to elevated synthesis of proinflammatory cytokines and a more effective TH1-like response in patients with Lyme disease (Rahman et al. 2016).

These factors can modify the mechanisms of action as well as their responses and cellular interactions. The main mechanisms of action are oxidative stress, inflammation, cytotoxicity, genotoxic activity and/or epigenetic alterations (basically histone methylation), with a role in the development of diseases, such as cancer (Migliore et al. 2015; Nymark et al. 2018), neurodegenerative (Ruszkiewicz et al. 2017), cardiovascular (Udali et al. 2012), autoimmune complications (Rodriguez-Cortez et al. 2011), and behavior disturbances and psychiatric disorders (Figure 15.4; Barranco 2016).

On the cell surface, or upon endocytosis, several NP formulations have been shown to trigger the production of reactive oxygen species (ROS). Elevated levels of ROS lead to the activation of the cellular stress-dependent signaling pathways mediated by mitogen-activated protein kinase

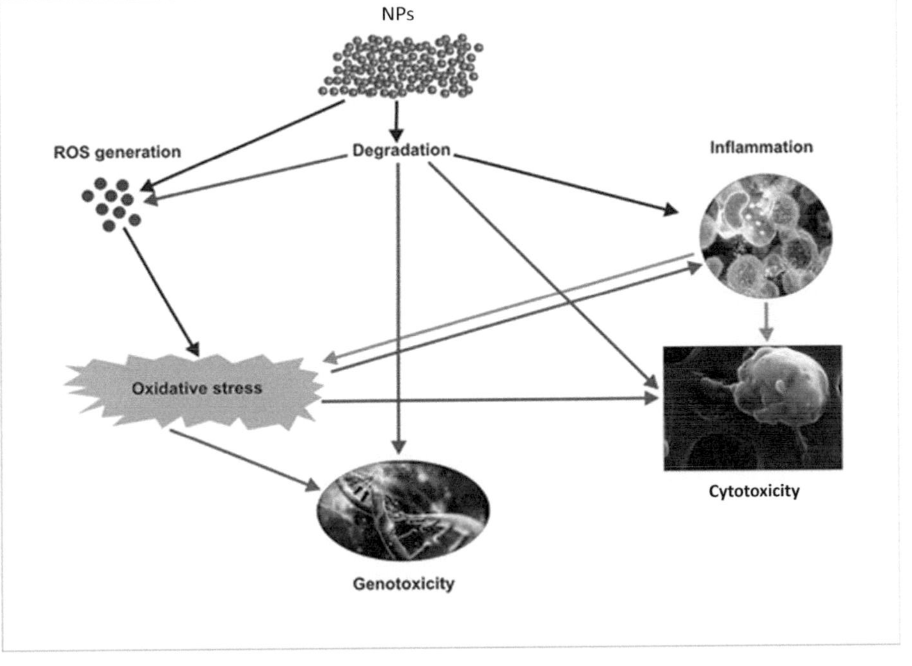

Figure 15.4 Overview of the signaling cascades mediating nanotoxicity.

(MAPK) activation and subsequent transcription factor activation. Ultimately, these transcription factors result in altered gene expression. In addition to these pathways, elevated ROS levels also lead to direct damage of organelles such as mitochondria and DNA fragmentation in the nucleus, resulting in inflammatory response and cytotoxicity.

15.2.4.1 Oxidative Stress

The main mechanism of *in vivo* nanotoxicity, both for metal- and carbon-based NPs, is the induction of oxidative stress through the generation of free radicals (Figure 15.5). Free radicals are generated from the surface of NPs when both the oxidants and free radicals are bound to the particle surface. Surface bound radicals such as $SiO^{\bullet}$ and $SiO_2^{\bullet}$ present on quartz particles, for example, are responsible for the formation of ROS such as $^{\bullet}OH\cdot$ and $O_2^{\bullet-}$. Their generation may be induced by the transcription factor Nrf2-Erk MAPK/HO-1 (e.g. at the level of human bronchial epithelial cells, BEAS-2B) by the NPs from nonmetallic oxides (Huang et al. 2010).

Moreover, some authors have detected that the TiO_2 NPs appear to be related to the levels of catalase activity (CAT), glutathione peroxidase (GPX) and glutathione S-transferase (GST). On the contrary, the activity levels of superoxide dismutase (SOD) were decreased in tissues by increasing the NP concentration of TiO_2. In addition, Tedesco et al. (2010) and Moore (2006) demonstrated signs of oxidative stress in the form of lysosomal stability, resulting from the induction of apoptosis by silver (Ag) NPs. ROS production depends on the diameter NPs, as has been observed in a study with AuNPs with sizes between 5–250 nm, in which an inverse relationship between the two parameters has been demonstrated (Misawa and Takahashi 2011). Although, in general, most cell types can tolerate small and transient increases in ROS, producing high levels or for a prolonged time, they can cause cellular damage. These ROS can interact with cellular macromolecules, including functional proteins and DNA, resulting in unregulated cell signaling, changes in cell motility, DNA damage, inflammation, increased levels of proinflammatory cytokines IL 8 and IL1.beta, microsatellite instability, inhibited methylation and accelerate telomere shortening, cytotoxicity, apoptosis by mitochondrial disruption, and production of pro-apoptotic factors (Evans et al. 2004). In fact, in some studies it has been observed that imbalances in ROS levels are correlated with the complex process of tumor progression (Sukhanova et al. 2018).

Some authors consider ROS production an adequate indicator to discriminate the adverse effects of different NPs at the cellular and molecular levels (Wang et al. 2010; Aime et al. 2009; Morimoto et al. 2013). Numerous *in vitro* studies show the ability of nanomaterials to induce oxidative stress

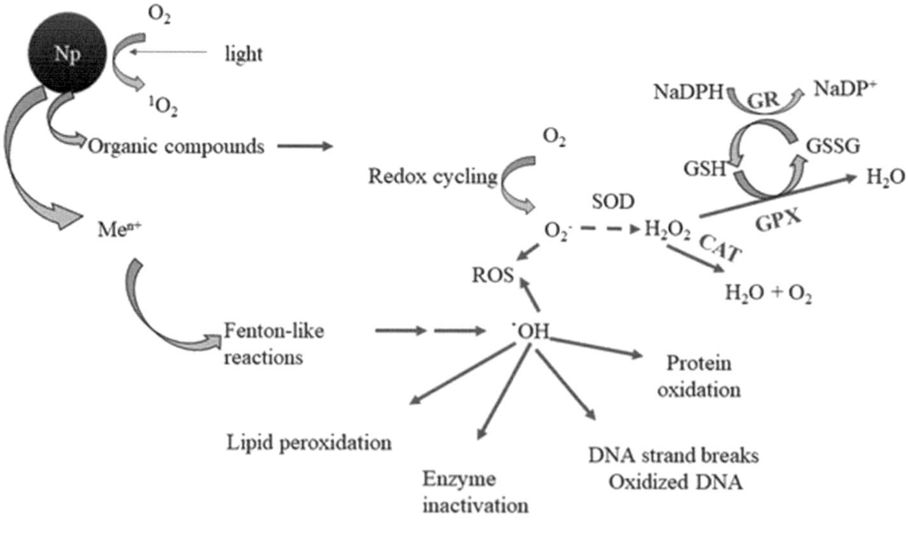

Figure 15.5 Reactive oxygen species production, effects on cellular function and defense mechanisms.

Abbreviations: CAT = catalase; GPX = glutathione peroxidase; GR = GSSG reductase; GSH = glutathione; GSSG = glutathione disulfide.

at high concentrations, and some indicate possible genotoxicity and inflammatory responses (Piperigkou et al. 2016). Therefore, the European Food Safety Authority (EFSA) has considered that for a correct characterization of the danger of nanomaterials and to establish the dose-response relationships, not only should mass be considered as a unit of measurement but also the number of particles and the surface area.

15.2.4.2 Inflammation

NPs can affect human beings by altering their immune systems and immune responses. For example, carbon nanotubes can cause immunosuppression by acting on the level of T cell–dependent antibodies. Their proliferative capacity and increased levels of interleukin-10 (IL-10) and nicotinamide adenine dinucleotide phosphate [NAD(P)H] oxidoreductase-1 levels of mRNA, or zinc oxide (ZnO) NPs that regulate by attenuation the expression of CD16 on killer NK cells, affecting immune responses mediated by immunoglobulins of the RFcg type, or NPs of titanium dioxide and fullerenes C-60 that induce proinflammatory enzymes such as IL-1, TNF-α, and IL-6 as well as chemotactic monocyte protein in the lungs of rodents.

Recent studies have observed the mediation of lysosomes in the inflammatory process because the disruption of the lysosome can trigger a cascade of inflammation in the target organ and the release of proapoptotic proteins and inflammation (Gerloff et al. 2017). If this action is combined with the generation of ROS by the mitochondria, it can trigger the activation of NLRP3 inflammasomes in phagocytic cells, thus releasing the interleukins IL-1 and IL-18 (Figure 15.6). Likewise, oxidative stress also releases proinflammatory mediators through cascades, such as nuclear factor-κB (NF-κB), MAPK and phosphoinositide 3-kinase (PI3-K) pathways (Poljak-Blaži et al. 2010; Li et al. 2010). In absence of a stimulus, NF-κB is sequestered in the cytoplasm by the inhibitor of the κB family (IκB), an inhibitor that is degraded in case of oxidative stress by releasing NF-κB that is then translocated to the nucleus to regulate the transcription of its target genes (Allen and Tresini 2000).

The MAPK pathway regulates a wide range of cellular responses such as activation (also by phosphorylation) of transcription factors, transcriptional regulations, nuclear chromatin remodeling, immediate genetic induction, cytokine production, regulation of apoptosis and progression of the cycle cell. In general, the Ras/ERK pathway mainly regulates cell growth and survival but, in some circumstances, it also determines cell differentiation, whereas the JNK and p38 pathways mainly regulate apoptosis, the inflammatory response and transmit inhibitory growth signals, as occurs with TiO_2NPs (Park et al. 2008).

15.2.4.3 Genotoxicity

NPs can cause damage to genetic material because they can enter the nucleus and interact with DNA (Li et al. 2016). This interaction can alter DNA replication and transcription in RNA (Tortiglione 2014). During mitosis, NPs can induce breaks in the chromosomes or alter the process by mechanical or chemical means, alterations of the mitotic spindle and its components and, in addition, can cause genomic instability through epigenetic mechanisms (Karlsson et al. 2015). The production of damage in the genetic material that can result in the induction or promotion of carcinogenesis or in reproductive effects, if germ cell DNA is compromised (Doak et al. 2012), is sometimes closely related to the ability of NPs to cause oxidative stress. Genotoxicity can be classified into two types: direct and indirect genotoxicity (Evans et al. 2017).

Direct genotoxicity occurs when direct action of nanomaterials with DNA occurs after cell internalization (Evans et al. 2019). Chromosomal fragmentation breaks in DNA strands, and point mutations, oxidative DNA adducts and alterations in gene expression profiles can occur. For example, fullerenes, in cell-free systems, are capable of complexing with DNA, causing breakage in strands of DNA (Cveticanin et al. 2010), mutagenicity and chromosomal damage (Folkmann et al. 2009). Likewise, some studies have observed chromosomal damage in peripheral blood lymphocytes in Syrian hamster embryo fibroblasts (SHE) and in human type B lymphocytes (WIL2-NS), or interference with chromosome segregation, centrosomal duplication, and cytokinesis (Huang et al. 2009) caused by NPTiO_2 particles used in sunscreen creams. In addition, asbestos induces micronuclei that often contain a specific chromosomal fragment, 19p, which is associated with elimination and differential gene expression in lung cancer related to asbestos. Alterations have also been observed in the expression of genes such as those produced by AuNPs in human embryonic lung fibroblasts in which there was less expression of those genes related to DNA repair such as BRCA1, Hus1, ATLD/HNGS1 and AT-V1/AT-V2 (Li et al. 2008). If the degree of DNA damage goes beyond the capacity for the cell to repair, the cells begin apoptosis, or programmed cell death.

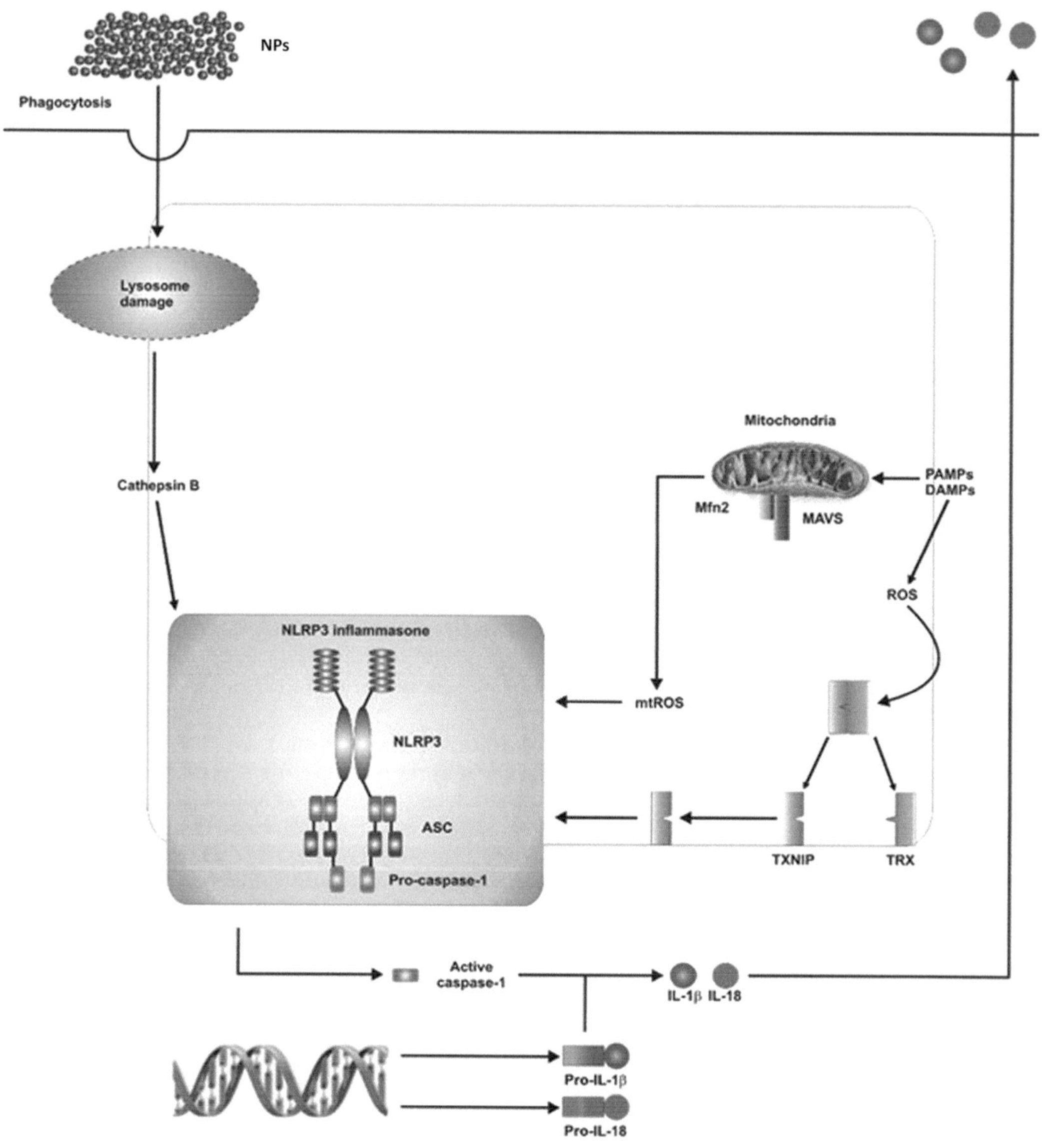

Figure 15.6 Canonical activation of the inflammasome. ROS and lysosomal rupture may collectively activate caspase 1, which mediates the release of cytokines such as IL-1β and IL-18. Increased ROS are sensed by a complex of TRX and TXNIP that can induce the dissociation of the complex. TXNIP binds to the LRR region of NLRP3, leading to NLRP3 inflammasome activation and the secretion of mature IL-1β and IL-18.

Abbreviations: ASC = apoptosis-associated speck-like protein containing a caspase-recruitment domain; DAMP = damage-associated molecular pattern; LRR = leucine-rich repeat; MAVS = mitochondrial antiviral-signaling protein; Mfn2: mitofusin 2; NLRP3 = PAMP = pathogen-associated molecular pattern; TRX = thioredoxin; TXNIP NOD (Nucleotide-binding and oligomerization like receptor pyridin domain containing 3 = TRX-interacting protein.

Indirect genotoxicity occurs because of the ability of the particles to induce an inflammatory reaction and to generate an excess of ROS (Aye et al. 2013). According to some authors, this happens with exposure to TiO_2NPs; through the activity of ROS this could produce DNA aberrations as well as polyamidoamines (PAMAM). If these processes occur in the germ line, the NP-induced genetic alterations can be transmitted to the progeny of the exposed animals (Tortiglione 2014).

These genotoxic effects can be modulated by factors such as the size of the NPs and its surface coating. For example, there is a consensus that smaller NPs produce greater reactivity and therefore, greater genotoxicity. Regarding the effect of surface coating of NPs, it has been observed that positively charged coatings of iron oxide NPs increase DNA chain breaks, whereas if they are negatively charged, their effect is insignificant (Hong et al. 2011).

In response to this DNA damage, the cells start DNA repair mechanisms or invoke cell cycle arrest and apoptosis. One of the major effector molecules activated in response to DNA damage is p53. It plays a key role in DNA repair and arrest of the cell cycle, thus avoiding the mutagenic events that favor carcinogenesis (Lane 1992).

15.2.4.4 Degradability of Nanoparticles

Dependent on the characteristics of the NPs, they can easily degrade in the microenvironment of the human body or can accumulate in organs or cells because they are not degradable or they are slowly degradable. The accumulation then can produce an unknown long-term toxic effect. This toxic effect may be due to the interaction of the accumulated NPs with the surrounding biological molecules or to their interference in certain biological processes such as enzymatic regulatory mechanisms, due to their binding to the serum enzymes due to the surface charges in the NPs.

Regarding degradable NPs, their intracellular toxic effect may be due to the change of the microenvironment in the body and to the release of toxic components. This occurs with biodegradable polymers, which produce the release of polymeric drugs (Knab et al. 2015), which in turn release pro-inflammatory cytokines activating both the response humoral as a cell (Nicolette et al. 2011). In general, NPs can be internalized by cells through endocytic mechanisms. The pH values in endosomes (pH ~6) and lysosomes (pH ~4.5) are much lower than in the extracellular environment (pH ~7). The change of local pH and the presence of degradative enzymes in cells, such as cathepsin L, make some NPs degradable and cause them to lose their surface coating.

15.2.4.5 Cytotoxicity

Another mechanism of action detected in NPs with certain physical dimensions, such as ultrafine siliceous NPs, is the alteration of cell morphology or the involvement of cellular functional components, such as the cell membrane, the protein filament network of the cytoskeleton, mitochondria and synaptic machinery. Gold NPs can alter cell membrane potential by inducing changes in calcium concentration that affect cell growth. Thus, $^{+}$AuNPs reduce the viability of normal BEC cells and ASM cells (Arvizo et al. 2010). This effect can also be due to its action in signaling routes. The cytoskeleton plays a significant role in cell shape, motility and cell division, the interaction of the extracellular matrix adhesive and the formation of neuronal architecture. Deformations of the cytoskeleton decrease the capacity of cellular functions. For example, it has been shown that silver particles induce neurotoxicity by altering the destruction of tubulin and actin cytoskeleton proteins, dissolving synaptic proteins and compromising the function of mitochondria. The effect may be independent of size, such as silica-coated iron oxide NPs that act through cytoskeletal actin (Kunzmann et al. 2011). It has been observed that AgNPs can act on mammalian germ cells, reducing mitochondrial function, increasing membrane permeability, necrosis and induction of apoptosis, which has created great concern because AgNPs are able to access human sperm through a variety of marketed products and can cause problems with fertility. Sometimes, these effects are mediated by oxidative stress, especially at the level of mitochondrial membrane potential reduction. These results are important because AgNPs are able to reach human sperm, causing fertility problems.

15.2.5 Toxic Effects on Human Health

Nanomaterials, due to their physicochemical characteristics and their interactions with biomolecules, can be potential toxic for both humans and the environment. It must be emphasized that not all nanomaterials induce toxic effects. We are facing a paradigm shift in the traditional toxicological approach, because a risk, such as that of nanomaterials, focused as a chemical, can acquire a risk of biological nature as indicated above.

Despite the increasing number of studies conducted in this regard, epidemiological data related to the toxicological effects of nanomaterials under real conditions are scarce, although different effects have been observed because NPs show different kinetic patterns. The most important adverse effects of the nanomaterials observed in animal studies and at high doses are very diverse, such as fibrosis, DNA damage, effects on the cardiovascular system and formation of tumors. For example, CNTs of length, diameter and biopersistence comparable to those of asbestos can produce inflammatory and fibrotic effects (Kavosi et al. 2018). Carbon black and titanium dioxide have been classified by the International Agency for Research on Cancer (IARC 2010) as "possibly carcinogenic to humans" (group 2B) based on animal studies. Furthermore, it has been observed that air pollution by NPs is associated with the appearance of ischemia, arrhythmias and cardiac insufficiency, peripheral thrombosis and formation of atherosclerotic lesions; this is due to the ability of NPs to directly stimulate platelet aggregation/activation or by alteration of the endothelium, to favor the formation of thrombi. Regarding DNA damage, histone methylation and other DNA alterations have been described with epigenetic effects associated with diseases such as cancer (Walters et al. 2016), neurodegenerative diseases (Migliore et al. 2015; Ruszkiewicz et al. 2017), cardiovascular complications (Udali et al. 2012) and autoimmune disorders (Rodriguez-Cortez et al. 2011), behavioral and psychiatric disorders (Barranco 2016). As the NPs can cross the placenta, they can also affect the developing tissues, producing adverse effects of great concern. However, currently there is very little knowledge in this regard, and the data that show the interaction of NPs and their characteristics with embryonic cells will contribute to assessing the risk involved in the exposure of developing tissues to NPs.

Due to the multiple applications and physicochemical properties of the NPs, they are easy to incorporate into the environment. Because of their use in the manufacture of herbicides, tires, pharmaceuticals and food and the decontamination and regeneration of soils, they can be accumulated in these soils. These nanoparticles can reach air by emission in industrial processes or in water as a result of uncontrolled discharges or soil washing. The persistence of NPs is unknown, so studies should be conducted to investigate their transformation, distribution in the different environmental compartments and effect on different animals, plants or bacterial species.

The accumulation of NPs in the roots (coming from the soil) or the aerial areas of the plants (due to atmospheric contamination), can cause transitory or permanent phytotoxicity problems in the growth and development of the plant; due to their translocation capacity, they can be distributed throughout the plant. If the plant is edible, they will integrate into the food chain with consequent health risks (Rico et al. 2011; Maharramov et al. 2019).

Several studies have also demonstrated that some nanoparticles (ZnO, TiO_2 and CuO nanoparticles) can cause ecotoxic effects (Mayne et al. 2019) because they can affect aquatic organisms (Aruoja et al. 2009; Kahru and Dubourguier 2010; Wang et al. 2019). One of the difficulties is the evaluation of the level of attainable concentration of NPs in the aquatic environment. Nevertheless, tests have been carried out with the most used aqueous phase biological models, including models of algae, micro crustaceans and fish. Thus, for example, it has been observed that NPs aggregates of TiO_2 are trapped by the algae *Pseudokirchneriella subcapitata* and thus inhibited their growth by reduction of the light available to the entrapped algal cells (Aruoja et al. 2009). These investigations showed several toxic effects, including lethal but also sublethal, biochemical, transcriptional, physiological and behavioral effects in algae, micro crustaceans and fish.

Bacteria are also subject to the toxic effects of these NPs, as is the case with AgNPs, whose antibacterial effect depends not only on size (the toxicity of AgNPs decreased with increasing primary particle size; Brown et al. 2002; Ivask et al. 2014) but also on form (Franci et al. 2015). Therefore, and given that bacteria constitute one of the main elements of the biogeochemical cycles, they could have a deleterious effect on the functioning of the entire ecosystem, of course as long as very high concentrations of nanoparticles are reached (Auffan et al. 2011).

The NPs not only act as contaminant carriers in the environment but also facilitate the entry of other contaminants into the cells, thereby enhancing the toxic effects. Baun et al. (2008) showed that the phenanthrene toxicity for *Daphnia magna* was increased by 60% in the presence of C60 aggregates.

15.3 STRATEGIES TO PREVENT TOXIC EFFECTS

Because the main mechanism of action of NPs is oxidative stress, prevention strategies have been proposed, including the introduction of antioxidant agents such as ascorbic acid inside those NPs. For example, the introduction of vitamin C in AgNPs in myeloid leukemia cells decreased their ROS production, mitochondrial damage, DNA damage and apoptosis, reducing the toxicological risk induced by these NPs (Baun et al. 2008). The same occurred in lung cells (A549) treated with L ascorbic acid and nickel ferrite NPs (26 nm in diameter; Ahamed et al. 2011). Similar effects have been observed, for example, in *Drosophila melanogaster* exposed to AgNPs or in rats exposed to ZnO NPs (21 nm) and that drank a 1% aqueous solution of ascorbic acid. In these rats, there was a decrease in oxidative stress and inflammatory processes (Fukui et al. 2015).

The preventive effect of ascorbic acid may also be due to changes in the surface of NPs, as it occurs by encapsulating ascorbic acid with poly-(L-glutamic acid)-capped silver NPs (AgNpPGA) within a polymeric matrix (lactide-co-glycolide) (PLGA) (PLGA/AgNpPGA/ascorbic acid particles). This reduced ROS production in HepG2 cells (Stevanovic et al. 2014), or with copper NPs coated with polysaccharides, such as chitosan that reduced the toxic effect and generation of ROS, although it increased the inflammatory response (Worthington et al. 2013).

Other antioxidant substances also have protective effect, such as quercetin, a flavonoid present in many plants, some of them used as food. For example, after exposure to Fe_2O_3 NPs, oxidative damage and inflammation were reduced by phosphorylation failure and Nrf2 translocation through PI3-K/Akt pathway (Sarkar and Sil 2014). The NF-κB, MAPK and PI3-K pathway facilitates inflammation induced by NPs and releases various proinflammatory cytokines and chemokines such as TNF-α, IL-6 and IL-8, producing effects. To compensate these effects, the components of the Jun/AP-1 pathway can be modulated, providing new routes for therapeutic interventions (Schonthaler et al. 2011).

15.4 CONCLUSIONS

Although nanotechnology has the potential to change our understanding of nature and life and to change society with its remarkable benefits, we must also consider the great amount of problems and risks related with its use (based upon the relatively limited number of studies to date); therefore, it is essential to take strategic preventive measures before it is too late. For this reason, the European Commission is financing projects (see https://cordis.europa.eu/project/rcn/196969/factsheet/en; https://cordis.europa.eu/project/rcn/196966/factsheet/en) to learn more about the toxicological potential of these substances individually and is supporting the development of a solid regulatory framework through informed advice based on scientific evidence. One of these projects has led to the creation of a completely free database, the Nano Health-Environment Commented Database (NHECD; https://cordis.europa.eu/project/rcn/89326/en), with dynamic and continuously developing access. In addition to including scientific documents on environmental effects and sanitary products produced by NPs, the database also allows the automatic updating of contents (https://cordis.europa.eu/project/rcn/89326/en). Also, among many outcomes, the OECD has developed guidelines for manufacturing and testing (Wilson 2018).

BIBLIOGRAPHY

Ahamed, M., M. J. Akhtar, M. A. Siddiqui, et al. 2011. Oxidative stress mediated apoptosis induced by nickel ferrite nanoparticles in cultured a549 cells. *Toxicology* 283: 101–8.

Aime, S., D. D. Castelli, S. G. Crich, E. Gianolio, and E. Terreno. 2009. Pushing the sensitivity envelope of lanthanide-based magnetic resonance imaging (MRI) contrast agents for molecular imaging applications. *Acc Chem Res* 42: 822–31.

Al-Jumaili, A., A. Kumar, K. Bazaka, and M. V. Jacob. 2018. Plant secondary metabolite-derived polymers: A potential approach to develop antimicrobial films. *Polymers* 2018, 10: 51.

Allen, R., and M. Tresini. 2000. Oxidative stress and gene regulation. *Free Radic Biol Med* 28: 463–99.

Anjum, N. A., V. Adam, R. Kizek, et al. 2015. Nanoscale copper in the soil-plant system—toxicity and underlying potential mechanisms. *Environ Res* 138: 306–25.

Arif, T., and M. Adil. 2016. Nanotechnology-dermatological perspective. *Int J Nanomed Nanosurg* 2(2).

Aruoja, V., H. C. Dubourguier, K. Kasemets, and A. Kahru. 2009. Toxicity of nanoparticles of CuO, ZnO and TiO2 to microalgae Pseudokirchneriella subcapitata. *Sci Total Environ* 407: 1461–8.

Arvizo, R. R., O. R. Miranda, M. A. Thompson, et al. 2010. Effect of nanoparticle surface charge at the plasma membrane and beyond. *Nano Lett* 10: 2543–8.

Auffan, M., E. Flahaut, A. Thill, and F. Mouchet. 2011. Ecotoxicology: Nanoparticle reactivity and living organisms. In *Nanoethics and Nanotechnology*, eds. P. Houdy, M. Lahmani, and F. Marono, 325–52. Springer.

Aye, M., C. Di Giorgio, M. Mekaouche, et al. 2013. Genotoxicity of intraperitoneal injection of lipoamphiphile CdSe/ZnS quantum dots in rats. *Mutat Res* 758: 48–55.

Barranco, C. 2016. Autoimmunity: Nanomedicine, meet autoimmune disease. *Nat Rev Rheumatol* 12: 193.

Bartneck, M., H. A. Keul, G. Zwadlo-Klarwasser, and J. Groll. 2010. Phagocytosis independent extracellular nanoparticle clearance by human immune cells. *Nano Lett* 0: 59–63.

Baun, A., S. N. Sorensen, R. F. Rasmussen, N. B. Hartmann, and C. B. Koch. 2008. Toxicity and bioaccumulation of xenobiotic organic compounds in the presence of aqueous suspensions of aggregates of nano-C(60). *Aquat Toxicol* 86: 379–87.

Brown, D. M., K. Donaldson, and V. Stone. 2002. Role of calcium in the induction of TNFa expression by macrophages on exposure to ultrafine particles. *Ann Occup Hyg* (46 Suppl 1): 219–22.
Bunglavan, S. J., A. K. Garg, R. S. Dass, and S. Shrivastava. 2014. Use of nanoparticles as feed additives to improve digestion and absorption in livestock. *Livestock Res Int* 3: 36–47.
Cagliani, R., F. Gatto, and G. Bardi. 2019. Protein adsorption: A feasible method for nanoparticle functionalization?. *Mater Basel* 12(12): pii: E1991.
Chung, M., G. Fortunato, and N. Radacsi. 2019. Wearable flexible sweat sensors for healthcare monitoring: A review. *J R Soc Interface* 16: 20190217.
Cveticanin, J., G. Joksic, A. Leskovac, S. Petrovic, A. V. Sobot, and O. Neskovic. 2010. Using carbon nanotubes to induce micronuclei and double strand breaks of the DNA in human cells. *Nanotechnology* 21: 1–7.
Doak, S. H., B. Manshian, G. J. S. Jenkin, and N. Singh. 2012. In vitro genotoxicity testing strategy for nanomaterials and the adaptation of current OECD guidelines. *Mutat Res* 745: 104–11.
Docter, D., D. Westmeier, M. Markiewicz, S. Stolte, S. K. Knauer, and R. H. Stauber. 2015. The nanoparticle biomolecule corona: Lessons learned—challenge accepted? *Chem Soc Rev* 44: 6094.
EFSA (European Food Safety Authority). 2009. Guidance of the Scientific Committee on a request from EFSA on the use of the benchmark dose approach in risk assessment. *EFSA J* 1150: 1–72.
European Commission. 2011. *2011/696/EU*. https://ec.europa.eu/research/industrial_technologies/pdf/policy/commission-recommendation-on-the-definition-of-nanomater-18102011_en.pdf.
European Commission. 2012. *Types and Uses of Nanomaterials, Including Safety Aspects*. SWD 288 final.
Evans, M. D., and M. S. Cooke. 2004. Factors contributing to the outcome of oxidative damage to nucleic acids. *Bioessays* 26(5): 533–42.
Evans, S. J., M. J. D. Clift, N. Singh, et al. 2017. Critical review of the current and future challenges associated with advanced in vitro systems towards the study of nanoparticle (secondary) genotoxicity. *Mutagenesis* 32(1): 233–41.
Evans, S. J., M. J. D. Clift, N. Singh, et al. 2019. In vitro detection of in vitro secondary mechanisms of genotoxicity induced by engineered nanomaterials. *Part Fibre Toxicol* 16, Art. N°: 8.
Faraday, M. 1857. The Bakerian Lecture. Experimental relations of gold (and other metals) to light. *Phil Trans R Soc Lond* 147: 145–81.
Folkmann, J., L. Risom, N. Jacobsen, and H. Wallin. 2009. Oxidatively damaged DNA in rats exposed by oral gavage to C60 fullerenes and single-walled carbon nanotubes. *Environ Health Perspect* 117: 703–8.
Franci, G., A. Falanga, S. Galdiero, et al. 2015. Silver nanoparticles as potential antibacterial agents. *Molecules* 20: 8856–74.
Fukui, H., H. Iwahashi, S. Endoh, et al. 2015. Ascorbic acid attenuates acute pulmonary oxidative stress and inflammation caused by zinc oxide nanoparticles. *J Occup Health* 57: 118–25.
Gatto, F., and G. Bardi. 2018. Metallic nanoparticles: General research approaches to immunological characterization. *Nanomaterials* 8: 753.
Gautam, A., D. Singh, and R. Vijayaraghavan. 2011. Dermal exposure of nanoparticles: An understanding. *J Cell Tissue Res* 11(1): 2703–8.
Ge, C., J. Du, L. Zhao, et al. 2011. Binding of blood proteins to carbon nanotubes reduces cytotoxicity. *Proc Natl Acad Sci U S A* 108(41): 16968–73.
Gerloff, K., B. Landesmann, A. Worth, S. Munn, T. Palosaari, and M. Whelan. 2017. The adverse outcome pathway approach in nanotoxicology. *Comput Toxicol* 1–3–11.
Hirlekar, R., E. J. Patil, and S. Bhairy. 2017. Oral insulin delivery: Novel strategies. *Asian J Pharm* 11: S434-S443.
Hitler, L., M. T. Nyong'a, I. Ali, et al. 2017. Mechanism, of biophysicochemical interactions and cellular uptake at the nano-bio interface: A Review. *Eur J Biophys* 5(4): 66–78.
Hong, S. C., J. H. Lee, J. Lee, et al. 2011. Subtle cytotoxicity and genotoxicity differences in the superparamagnetic iron oxide nanoparticles coated with various functional groups. *Int J Nanomed* 6: 3219–31.
Hossain, S. S., T. J. R. Hughes, and P. Decuzzi. 2014. Vascular deposition patterns for nanoparticles in an inflamed patient-specific arterial tree. *Biomech Model Mechanobiol* 13(3): 585–97.
Huang, C., R. S. Aronstam, D. Chen, and Y. Huang. 2010. Oxidative stress, calcium homeostasis, and altered gene expression in human lung epithelial cells exposed to ZnO nanoparticles. *Toxicol in Vitro* 24(1): 45–55.
Huang, Y. F., H. Liu, X. Xiong, Y. Chen, and W. Tan. 2009. Nanoparticle-mediated IgE-receptor aggregation and signaling in RBL mast cells. *J Am Chem Soc* 131: 17328–34.

IARC. 2010. *Carbon Black, Titanium Dioxide, and Talc*. IARC Monographs on the Evaluation of Carcinogenic Risks to Humans, vol. 93. Lyon, France.

Ivask, A., I. Kurvet, K. Kasemets, et al. 2014. Size-dependent toxicity of silver nanoparticles to bacteria, yeast, algae, crustaceans and mammalian cells in vitro. *PLoS ONE* 9(7): e102108.

Jeevanandam, J., A. Barhoum, Y. S. Chan, A. Dufresne, and M. K. Danquah. 2018. Review on nanoparticles and nanostructured materials: History, sources, toxicity and regulations. *J Nanotechnol* 9: 1050–74.

Kahru, A., and H. C. Dubourguier. 2010. From ecotoxicology to nanoecotoxicology. *Toxicology* 269(2–3): 105–19.

Karlsson, H., S. Di Bucchianico, A. Collins, and M. Dusinska. 2015. Can the comet assay be used reliably to detect nanoparticle induced genotoxicity? *Environ Mol Mutagen* 56: 82–96.

Kavosi, A., S. Hosseini, G. Noei, S. Madani, and K. Khalighfard. 2018. The toxicity and therapeutic effects of single-and multi-wall carbon nanotubes on mice breast cancer. *Sci Rep* 8(1): 8375.

Keshavan, S., P. Calligari, L. Stella, L. Fusco, L. G. Delogu, and B. Fadeel. 2019. Nano-bio interactions: A neutrophil-centric view. *Cell Death Dis* 10, Article number: 569.

Knab, T. D., S. R. Little, and R. S. Parker. 2015. A systems approach to modeling drug release from polymer microspheres to accelerate in vitro to in vivo translation. *J Control Release* 211: 74e84.

Kunzmann, A., B. Andersson, and C. Vogt. 2011. Efficient internalization of silica-coated iron oxide nanoparticles of different sizes by primary human macrophages and dendritic cells. *Toxicol Appl Pharmacol* 253: 81e93.

Landsiedel, R., E. Fabian, L. Ma-Hock, et al. 2012. Toxico-/biokinetics of nanomaterials. *Arch Toxicol* 86: 1021–60.

Lane, D. P. 1992. Cancer. P53, guardian of the genome. *Nature* 358: 15–6.

Li, J. J. E., S. Muralikrishnan, C. T. Ng, L. Y. L. Yung, and B. H. Bay. 2010. Nanoparticle-induced pulmonary toxicity. *Exp Biol Med* 235: 1025–33.

Li, J. J., L. Zou, D. Hartono, C. N. Ong, B. H. Bay, and L. Y. Lanry Yung. 2008. Gold nanoparticles induce oxidative damage in lung fibroblasts in vitro. *Adv Mater* 20: 138–42.

Li, J., M. Tiang, L. Cui, et al. 2016. Low-dose carbon-based nanoparticle-induced effects in A549 lung cells determined by biospectroscopy are associated with increases in genomic methylation. *Sci Rep* 6(20207).

Lin, W., I. Stayton, Y. W. Huang, X. D. Zhou, and Y. Ma. 2008. Cytotoxicity and cell membrane depolarization induced by aluminum oxide nanoparticles in human lung epithelial cells A549. *Toxicol Environ Chem* 90: 983–96.

Maharramov, A. M., U. A. Hasanova, I. A. Suleymanova, et al. 2019. The engineered nanoparticles in food chain: Potential toxicity and effects. *SN Appl Sci* 1: 1362.

Mahmoudi, M., N. Bertrand, H. Zope, and O. C. Farokhzad. 2016. Emerging understanding of the protein corona at the nano-bio interfaces. *Nano Today* 11: 817–32.

Maisel, K., L. Ensign, M. Reddy, R. Cone, and J. Hanes. 2015. Effect of surface chemistry on nanoparticle interaction with gastrointestinal mucus and distribution in the gastrointestinal tract following oral and rectal administration in the mouse. *J Control Release* 197: 48–57.

Mayne, R., J. Morgan, J. G. H. Whiting, N. Phillips, and A. Adamatzky. 2019. On measuring nanoparticle toxicity and clearance with Paramecium caudatum. *Sci Rep* 9(1): 8957.

Mehta, D., and A. B. Malik. 2006. Signaling mechanisms regulating endothelial permeability. *Physiol Rev* 86: 279–36.

Migliore, L., C. Uboldi, S. Di Bucchianico, and F. Coppede. 2015. Nanomaterials and neurodegeneration. *Environ Mol Mutagen* 56: 149–70.

Misawa, M., and J. Takahashi. 2011. Generation of reactive oxygen species induced by gold nanoparticles under X-ray and UV irradiations. *Nanomedicine* 7: 604–14.

Mitchell, L. A., J. Gao, R. V. Wal, A. Gigliotti, S. W. Burchiel, and J. D. McDonald. 2007. Pulmonary and systemic immune response to inhaled multiwalled carbon nanotubes. *Toxicol Sci* 100(1): 203–14.

Moore, M. N. 2006. Do nanoparticles present ecotoxicological risks for the health of the aquatic environment? *Environ Int* 32(8): 967–76.

Morimoto, Y., M. Horie, N. Kobayashi, N. Shinohara, and M. Shimada. 2013. Inhalation toxicity assessment of carbon-based nanoparticles. *Acc Chem Res* 46: 770–81.

Narasimhan, V., R. H. Siddique, and J. O. Lee. 2018. Multifunctional biophotonic nanostructures inspired by the longtail glasswing butterfly for medical devices. *Nat Nanotech* 13: 512–19.

National Nanotechnology Initiative. 2018. *Nanotechnology: Big Things from a Tiny World*. www.nano.gov/big-things-from-a-tiny-world (accessed November 22, 2020).

Nautiyal, P., F. Alam, K. Balani, and A. Agarwal. 2017. The role of nanomechanics in healthcare. *Adv Healthcare Mater* 7(3). doi: 10.1002/adhm.201700793.

Nel, A. E., L. Madler, and D. Velegol. 2009. Understanding biophysiochemical interactions at the nano–biointerface. *Nat Mater* 8: 543–57.

Nicolete, R., D. F. dos Santos, and L. H. Faccioli. 2011. The uptake of PLGA micro or nanoparticles by macrophages provokes distinct in vitro inflammatory response. *Int Immunopharmacol* 11(10): 1557–63.

Nohynek, G. J., and E. K. Dufour. 2012. Nano-sized cosmetic formulations or solid nanoparticles in sunscreens: A risk to human health? *Arch Toxicol* 86 1063–75.

Nymark, P., P. Kohonen, V. Hongisto, and R. C. Grafstrom. 2018. Genomic influences of inhaled nanomaterials. *Ann Am Thorac Soc* 15(Suppl 2): S91–S97.

Oberdörster, G., and T. A. J. Kuhlbusch. 2018. In vivo effects: Methodologies and biokinetics of inhaled nanomaterials. *NanoImpact* 10: 38–60.

Park, E. J., J. Yi, K. H. Chung, D. Y. Ryu, J. Choi, and K. Park. 2008. Oxidative stress and apoptosis induced by titanium dioxide nanoparticles in cultured BEAS-2B cells. *Toxicol Lett* 180: 222–9.

Piperigkou, Z., K. Karamanou, A. S. Engin, et al. 2016. Emerging aspects of nanotoxicology in health and disease: From agriculture and food sector to cancer therapeutics. *Food Chem Toxicol* 91: 42–57.

Poljak-Blaži, M., M. Jaganjac, and N. Žarković. 2010. Cell oxidative stress: Risk of metal nanoparticles. In *Handbook of Nanophysics: Nanomedicine and Nanorobotics*, ed. K. D. Sattler, 1–17. New York: CRC Press.

Qi, Y., Y. Liu, T. Xia, A. Xu, S. Liu, and W. Chen. 2018. The biotransformation of graphene oxide in lung fluids significantly alters its inherent properties and bioactivities toward immune cells. *NPG Asia Mater* 10: 385–96.

Rahman, S., M. Shering, N. H. Ogden, R. Lindsay, and A. Badawi. 2016. Toll-like receptor cascade and gene polymorphism in host-pathogen interaction in Lyme disease. *J Inflamm Res* 9: 91–102.

Rico, C. M., S. Majumdar, M. Duarte-Gardea, J. R. Peralta-Videa, and J. L. Gardea-Torresdey. 2011. Interaction of nanoparticles with edible plants and their possible implications in the food chain. *J Agric Food Chem* 59(8): 3485–98.

Rodriguez-Cortez, V. C., H. Hernando, L. de la Rica, R. Vento, and E. Ballestar. 2011. Epigenomic deregulation in the immune system. *Epigenomics* 3: 697–713.

Ruszkiewicz, J. A., A. Pinkas, B. Ferrer, T. V. Peres, A. Tsatsakis, and M. Aschner. 2017. Neurotoxic effect of active ingredients in sunscreen products, a contemporary review. *Toxicol Rep* 4: 245–59.

Sarkar, A., and P. C. Sil. 2014. Iron oxide nanoparticles mediated cytotoxicity via PI3K/Akt pathway: Role of quercetin. *Food Chem Toxicol* 71: 106–15.

Schonthaler, H. B., J. Guinea-Viniegra, and E. F. Wagner. 2011. Targeting inflammation by modulating the Jun/AP-1 pathway. *Ann Rheum Dis* 70: i109–i112.

Simon, A., C. Thiebault, C. Reynaud, B. Gouget, and M. Carriere. 2006. Toxicity of oxide nanoparticles and carbon nanotubes on cultured pneumocytes: Impact of size, structure and surface charge. *Toxicol Lett* (164): 15–41.

Stevanovic, M., I. Bracko, M. Milenkovic, et al. 2014. Multifunctional PLGA particles containing poly (L-glutamic acid)-capped silver nanoparticles and ascorbic acid with simultaneous antioxidative and prolonged antimicrobial activity. *Acta Biomater* 10: 151–62.

Sukhanova, A., S. Bozrova, P. Sokolov, M. Berestovoy, A. Karaulov, and I. Nabiev. 2018. Dependence of nanoparticle toxicity on their physical and chemical properties. *Nanoscale Res Lett* 13: 44.

Tedesco, S., H. Doyle, J. Blasco, G. Redmond, and D. Sheehan. 2010. Exposure of the blue mussel, Mytilus edulis, to gold nanoparticles and the pro-oxidant menadione. *Comp Biochem Phys Part C* 151(2): 167–74.

Tortiglione, C. 2014. The heritable effects of nanotoxicity. *Nanomedicine (Lond.)* 9(18): 2829–41.

Udali, S., P. Guarini, S. Moruzzi, S. W. Choi, and S. Friso. 2012. Cardiovascular epigenetics: From DNA methylation to microRNAs. *Mol Aspects Med* 34(4): 883–901.

Uyeda, H. T., I. L. Medintz, J. K. Jaiswal, S. M. Simon, and H. Mattoussi. 2005. Synthesis of compact multidentate ligands to prepare stable hydrophilic quantum dot fluorophores. *J Am Chem Soc* 127: 3870–8.

Walters, C., E. Pool, and V. Somerset. 2016. Nanotoxicology: A review. In *Toxicology—New Aspects to This Scientific Conundrum*, eds. S. Soloneski and M. L. Larramendy. IntechOpen.

Wang, F., W. Guan, L. Xu, et al. 2019. Effects of nanoparticles on algae: Adsorption, distribution, ecotoxicity and fate. *Appl Sci* 9(8): 1534.

Wang, X. F., P. P. Jin, Z. Tong, et al. 2010. MR molecular imaging of thrombus: Development and application of a Gd-based novel contrast agent targeting to P-selectin. *Clin Appl Thromb Hemost* 16: 177–83.

Wilson, N. 2018. Nanoparticles: Environmental Problems or Problem Solvers? *BioScience* 68(4): 241–6.

Worthington, K. L., A. Adamcakova-Dodd, A. Wongrakpanich, et al. 2013. Chitosan coating of copper nanoparticles reduces in vitro toxicity and increases inflammation in the lung. *Nanotechnology* 24(39): 395101.

Yoshioka, Y., K. Higashisaka, S. Tsunoda, and Y. Tsutsumi. 2014. The absorption, distribution, metabolism, and excretion profile of nanoparticles. In *Engineered Cell Manipulation for Biomedical Application: Nanomedicine and Nanotoxicology*, eds. M. Akashi, T. Akagi, and M. Matsusaki. Tokyo: Springer.

Yun, Y., Y. W. Cho, and K. Park. 2013. Nanoparticles for oral delivery: Targeted nanoparticles with peptidic ligands for oral protein delivery. *Adv Drug Deliv Rev* 65: 822–32.

Zhao, Y., D. Sultan, and Y. Liu. 2019. Biodistribution, excretion, and toxicity of nanoparticles. In *Theranostic Bionanomaterials*, ed. W Cui and X. Zhao, 27–53. Elsevier.

16 Human Health Risk Assessment

Sophie A. Rocks

CONTENTS

16.1 INTRODUCTION

All decisions are about balancing impact and reward, whether that impact has positive or negative results. In the risk arena, these decisions encompass uncertainty as well as a need to continually review and, potentially, revise the decision until the risk is resolved. A risk-based approach to decision-making, considering both the probability of the event as well as the impact, allows for advancement despite the possibility of harm and has been promoted by political environments and industry demands (Pollard and Rocks 2014).

In general terms, risk is formed of impact and probability (HM Treasury 2004). In the human health and environmental spheres, risk can be considered to be a function of hazard (i.e. negative impact) and exposure, where hazard is the inherent property of a substance (or a situation) that can cause harm, and exposure is the extent or likelihood that the impact can occur. How these terms are valued depends on the specific situation under investigation (Nordlander et al. 2010).

This chapter considers the specific example of human health risk assessment as applied to chemical substances, both for consumer and occupational audiences. However, the underlying principles of risk assessment, risk management, data quality and validity can be applied to other scenarios.

16.2 HUMAN HEALTH

The constitution of the World Health Organization (WHO, the UN directing and coordinating body for health) defines health as "a state of complete physical, mental and social well-being and not merely the absence of disease or infirmity" and recognises it as a fundamental human right (World Health Organisation 1948). Therefore, any assessment of human health should also include nonphysical and non-adverse effects (e.g. bad odours; see Figure 16.1). Nonphysical and nonadverse effects may not be observable (or occur) in nonhuman species, making it difficult to predict that such events will occur in humans.

It must be recognised that initial assessments of chemicals are highly unlikely to include information on human health effects unless there are historical reports of effects in the literature. When novel pharmaceuticals are developed, clinical trials are conducted only after initial chemical testing (Regulation [EC] No 1907/2006 of the European Parliament and of the Council of 18 December 2006 concerning the Registration, Evaluation, Authorisation and Restriction of Chemicals [REACH], discussed in Section 16.4). However, risks continue to be monitored, managed and assessed until they are removed. This cyclic approach to risk assessment ensures that new information is taken into consideration for an established substance, allowing for anecdotal reports to be further investigated and risk management to be revised (e.g. pharmacovigilance through the yellow card system; Medicines and Healthcare Products Regulatory Agency 2018).

DOI: 10.1201/9780203730584-16

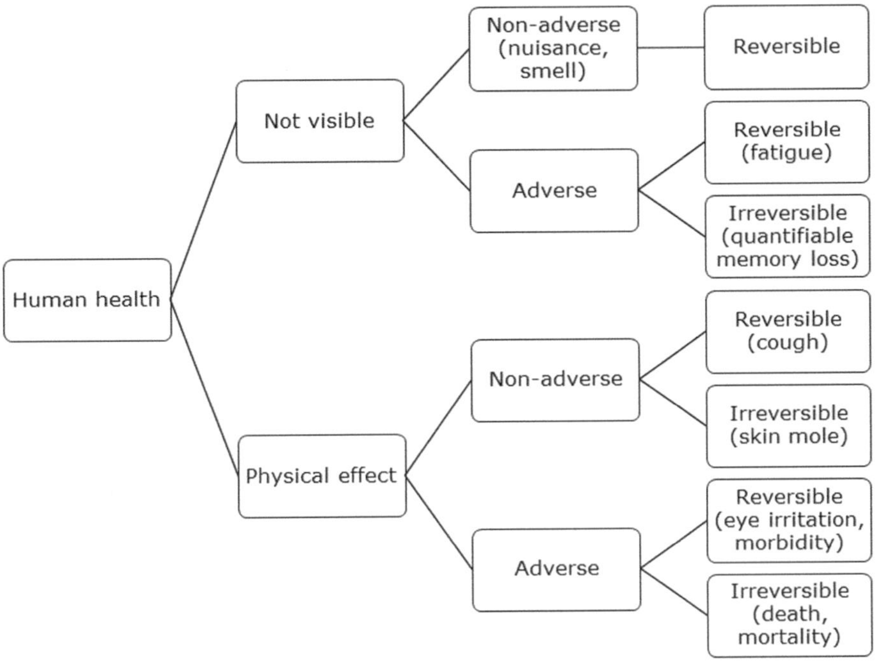

Figure 16.1

16.3 RISK

16.3.1 Overview

In order for a risk to occur, there needs to be a source (i.e. where the hazard originates from), pathway (i.e. route from source to organism of interest) and receptor (i.e. organism(s) of interest). If one of these is removed, the risk cannot occur (see Figure 16.2).

Human health risk assessments have changed over the years, from qualitative approaches to more quantitative approaches, with many now used to develop chemical exposure limits for both workers and the general population, including sensitive and vulnerable subpopulations. Information must be gathered from either the literature or from experimental data, including laboratory experiments and computer modelling, in order to determine likely effects and the doses or exposure levels at which they occur.

Once the scenario is defined (problem formulation), the process of risk assessment follows a standard series of interrogations (DEFRA 2011):

- What is the hazard?
- What are the consequences of the hazard?
- What is the exposure?
- What is the significance of the risk?

In these questions there is the consideration as to what are the uncertainties in the assessment and whether they can be addressed by obtaining additional information (additional detail on identifying and addressing different types of uncertainties in risk assessments can be found in Skinner et al. 2014).

Many risk paradigms assume that communication and data collection happen simultaneously (e.g. Guidelines for Environmental Risk Assessment and Management 2011). Such collection and dissemination can interfere with the objective assessment of collected evidence while acknowledging the inherent biases. Managers need to frame decisions with the relevant audiences, understand the magnitude and character of the risk, determine management strategies and relative costs, and then implement decisions, reducing the risk to an acceptable residual level while recognising that zero risk is not achievable.

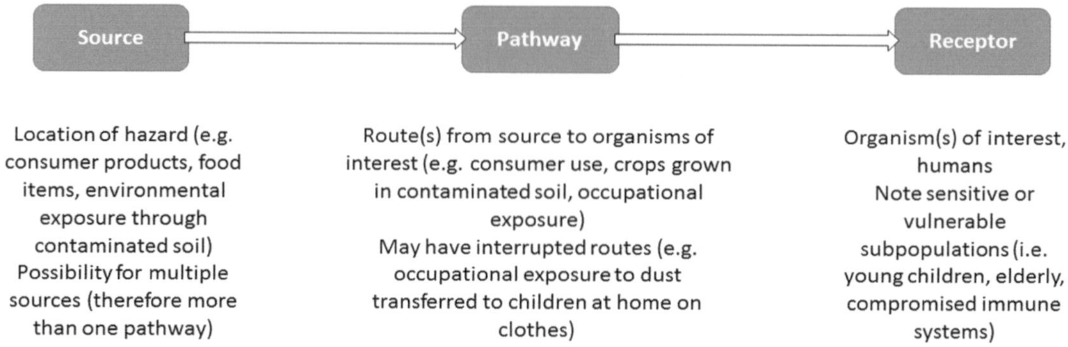

Figure 16.2

16.3.2 Quality of Data

16.3.2.1 Assessment of Data Quality

During an assessment, the risk assessor must identify, select and consider a dataset; determine the relevance, validity, quality and significance of the information collected; and then use a weight of evidence approach in order to make an assessment of the human health risk. Although all efforts are made to identify the most relevant information for the scenario under investigation, the identified evidence is unlikely to be from human exposure, so the assessor must consider effects in other relevant species using animal models to extrapolate findings to humans (see Section 16.3.3). If accidental exposure has occurred (e.g. in the workplace or environmental release), there may be reports of human health effects, but the scenario and other possible environmental exposures should be taken into account. All risk assessments should clearly state and explain assumptions made by the assessor and the uncertainties inherent in the assessment. Only when an assessment is made can management of the risk occur (see Section 16.3.4).

Risk assessors consider a wide range of information, sometimes contradictory, in order to determine the most relevant evidence on which to base their assessment (ECHA 2011). They must also consider the quality of the data including the relevance and reliability. Many toxicologists use the Klimisch scoring system (Klimisch et al. 1997; OECD 2005; see Table 16.1) to support an objective assessment of the relevance (is the data or test appropriate for the question) and the reliability (was the data collected using a standardised methodology and reported credibly and clearly). These terms are closely related and help to determine whether the data are useful for the assessment. The measurement of data quality can also be used to determine the weight given to each piece of evidence (see Linkov et al. [2009] for further discussion), influencing the final decision.

Human clinical studies that are controlled (e.g. double blind, controlled environmental exposures, using well-characterised, healthy individuals) will provide the most robust evidence linking a stressor to a resulting effect. However, such studies are frequently not available because there are significant ethical concerns associated with human testing. Clinical studies can go wrong when the stressor is poorly characterised or has unexpected effects (Butler and Callaway 2016). Epidemiological studies (i.e. statistical evaluation of existing or selected human populations to examine whether there is an association between exposure to a stressor and a human health effect) allow for a less invasive approach to gathering human data, although this tends to capture environmental exposures (e.g. developmental toxicity as a result of toxic waste exposure in Corby, UK, as reported in *Register of the Corby Group Litigation v Corby District Council*; EWHC 2009). Such studies may not have accurate exposure information and are complicated by the presence of multiple potential stressors (as the population are not under controlled conditions). Therefore, epidemiological studies can only show association and not causation (see National Research Council [1986] for further discussion).

In vivo studies using animal models are normally well characterised and controlled, with the more robust and reliable studies using a standardised test protocol (such as OECD Test Guidance documents, available at www.oecd.org/env/ehs/testing/oecdguidelinesforthetestingofchemicals.htm). However, in such models there continues to be questions as to whether the model and route of exposure are comparable to the scenario under investigation and whether the observed effects are specific to the individual animal used. *In vitro* studies (e.g. using cell lines or cultured cells) provide platforms for the study of toxicodynamics and mechanisms of toxicity but assume that the test material is representative of the chemical when it reaches the site of toxicity under investigation. Standardised testing methods (e.g. OECD Test Guidance Documents) can include provision for metabolism, however similar uncertainties exist as to whether a single cell line, or collection of cells, can be representative of organ function or the whole body.

The development of the "3 Rs" (reduce, refine and replace), embedded in the REACH regulations (EC 1907/2006; Burden et al. 2015), has meant that there is a considerable push to replace animal models with either *in vitro* (using human cell lines) or *in silico* (by computer) studies. Experimental models can be robustly designed, with controlled exposure and doses, to address specific gaps in knowledge, but assessors must account for the uncertainties associated with extrapolating results from animal subjects to humans. Assessors may rely upon analogous modes of action (i.e. sequences of key events and processes, starting with interaction of an agent with a cell, proceeding through operational and anatomical changes) to allow for extrapolation to the scenario under investigation.

Table 16.1 Summary Description of the Klimisch Scoring System Relating the Assigned Score to the Required Level of Detail

Score	Statement	Detail
1	Reliable without restrictions	Generated according to internationally (nationally) accepted testing guidelines. Generated with parameters that closely relate to a guideline method. Preferably conforms to good laboratory practice (GLP).
2	Reliable with restrictions	Generated without following accepted guidelines but are well documented and scientifically acceptable. Does not necessarily conform to GLP.
3	Not reliable	(Animal) model is not relevant to receptor, identified interference in the test protocol. Recorded information is not sufficient for assessment.
4	Not assignable	Insufficient experimental details recorded. Secondary literature (e.g. books or review articles).

Table 16.2 Example Uncertainty Factors, Standard Magnitude of the Factor and Guidance as to When the Factor May Be Included in the Calculation of Assessment Levels

Uncertainty factors	General guidance
H (intraspecies variation in human population)	10-fold factor (prolonged exposure, average healthy humans)—account for variation in sensitivity in human population
A (interspecies variation, experimental animal to man)	10-fold factor—account for uncertainty in extrapolating animal data to humans
S (subchronic to chronic exposure)	10-fold factor—account for uncertainty in extrapolating from NOAELs identified in short-term experimental models to a chronic NOAEL
L (LOAEL to NOAEL)	10-fold factor—account for uncertainty in extrapolating from LOAELs to NOAELs
D (incomplete database)	10-fold factor—account for inability of single study to adequately address all possible adverse outcomes
MF (modifying factor)	Use professional judgement to determine additional factors—magnitude depends on professional assessment of uncertainties; the default is 1

Note: This list is not exhaustive. The risk assessor must use their own expert judgement to assign a value to the factor under consideration as well as what factors to include.

Abbreviations: A = interspecies variation; D = database; H = intraspecies variation (in this case, within the human population); L = extrapolation from LOAEL (lowest observed adverse effect level) to NOAEL (no observed adverse effect level); MF = modifying factor; S = extrapolation from subchronic to chronic.

16.3.2.2 Uncertainty

In any assessment there is an acknowledged level of uncertainty, which increases as variables and unknowns are identified. Variability (observable diversity) can be measured but cannot be reduced, whereas uncertainty (imperfect knowledge) may be reduced by the acquisition of better evidence (Skinner et al. 2014). The level of uncertainty depends on many factors, including the quality of the evidence, the variability and abundance of the data, the assumptions, working hypotheses and conjectures made at each stage of the risk assessment process. Therefore, safety factors are the conventional manner to address uncertainties in a risk assessment and can be adjusted to allow for interspecies variability (Dorne and Renwick 2005).

Uncertainty factors (also known as safety factors or modifying factors) are commonly used to adjust LOAEL (lowest observed adverse effect level) or NOAEL (no observed adverse effect level) to account for data insufficiency and intraspecies and interspecies variations. Such modifications are based upon the assumption that humans (the receptors) are the most susceptible species to the hazard and that a sufficient reduction in acceptable exposure will be protective for the majority of the population under study. The standard uncertainty factor is taken as 1000 (equally split between intraspecies and interspecies variations), however additional modifying factors are included as uncertainty increases (see Table 16.2; this subject is well reviewed in Dankovic et al. [2015]).

The use of conceptual models, defined in the problem determination stage of the risk assessment, help to describe the source-pathway-receptor, target organisms and vulnerabilities, which in turn helps to clarify the context and legal requirements of the risk assessment, providing focus and boundaries (Guidelines for Environmental Risk Assessment and Management 2011).

16.3.3 Risk-Assessment Process

16.3.3.1 Problem Formulation

Problem formulation (i.e. stating the issue to be studied along with its boundaries) is a stage that is often overlooked but is essential to ensuring that relevant data are collected for the assessment and that the end assessment remains focused on the studied issue.

In this stage, the purpose for the assessment and the scope of the assessment are defined, allowing for the search terms and technical approaches to be agreed; applicable legislation and requirements are also highlighted to ensure comprehensive data collection. Stakeholder input (e.g. the public and regulators) at this stage can help to refine the study boundaries and focus the data collection and assessment. Ensuring that the problem is clearly described for all audiences will help to select the level and type of assessment and provide a foundation for risk management decisions later on in the process (see Section 16.3.4).

Assessors use a range of questions to help define or formulate the problem to be studied (see Figure 16.3), identifying who is likely to be affected by the hazard (i.e. the receptor). Identifying the receptor also allows the assessor to highlight whether vulnerable receptors (those unable to take care of themselves, e.g. babies or the elderly) or sensitive receptors (those who are likely to have an increased or decreased response to external stimuli, e.g. individuals with a compromised immune system) need to be specifically considered. Questions are then used to identify the hazard, where it occurs, how it reaches the receptor and the route of exposure (e.g. toxicokinetics). Any information on the toxicodynamics is also recorded here in order to inform the subsequent steps.

The information in the problem formulation can be used to develop a conceptual model for the scenario; this is a useful step if there is environmental exposure and there are many potential pathways to consider. Such models include the hypothesised sources of contaminants, routes of transport and exposure as well as endpoint receptors, and they are normally presented as a flow chart supported by a narrative. Additional information about conceptual models can be found in Sutter (1996) and European Commission (2010).

16.3.3.2 Hazard Identification

The questions asked in the problem identification stage are used to ensure that the hazard, source or chemical under study is identified. The hazard identification stage involves the identification of substance-intrinsic adverse effect(s) to humans. The dataset under assessment is likely to include chemical and physical properties as well as dose-response relationships measured by *in vivo* and/or *in vitro* assays and/or predicted from structure-activity relationships. The identified hazard data will serve to inform the classification and labelling of chemical substances and the risk assessment to humans, respectively. The classification and labelling schemes follow a stringent

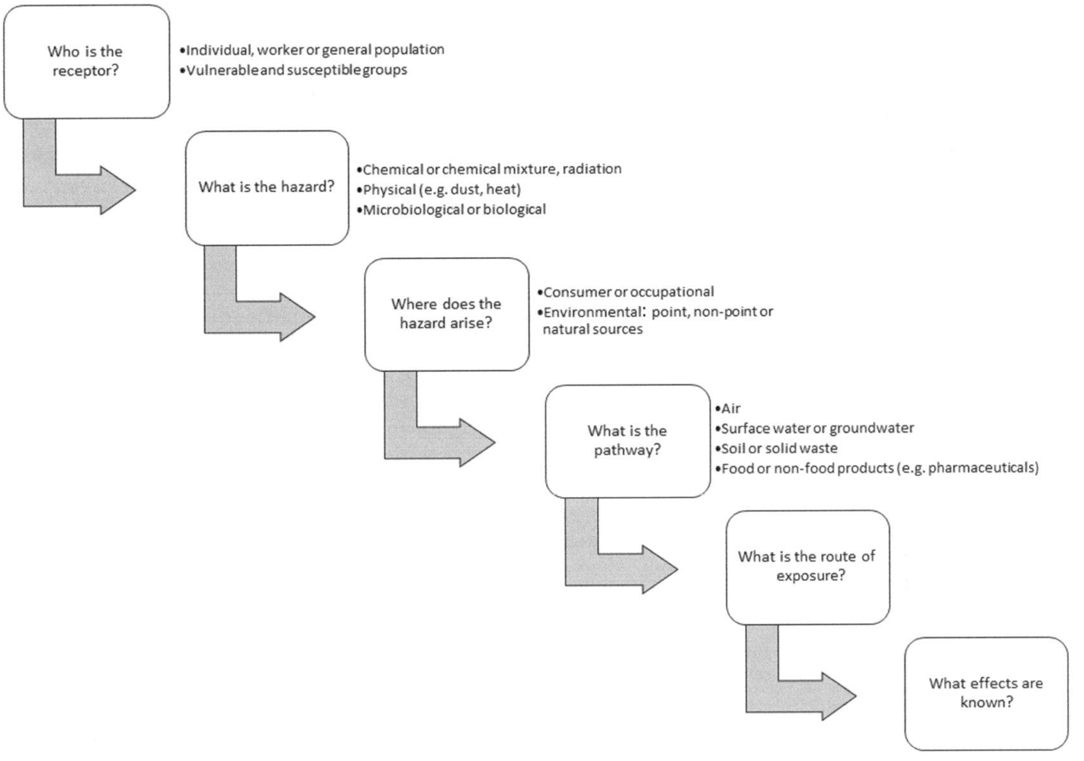

Figure 16.3

rule set as defined in legislation (e.g. European Regulation EC 1272/2008 on Classification, Labelling and Packaging of Substances and Mixtures).

The primary question in the hazard identification stage is whether the exposure to a stressor can cause an increase in the incidence of specific adverse health effects and whether those effects are likely to occur in humans. Where the stressor is an identifiable chemical, the assessor considers the available evidence either for the chemical itself or its chemical analogues (i.e. those with a similar chemical structure) and whether there are negative effects associated with the agent. This evidence should include information about the chemical agent's toxicokinetics and toxicodynamics in other, relevant animal models or computer models which will allow the assessor to draw conclusions as to the likely effects in humans. The relevant information is further investigated in the hazard characterisation step, where a decision is made whether or not additional studies are required.

The physicochemical data allow the assessor to determine what state and type of exposure is likely with the chemical, as well as the likely route into the body of main concern. This information is defined in the REACH regulations (see Section 16.4) as including the flammability and oxidation potential of the chemical, with the simple explanation that if the route of exposure is aerobic and the chemical reacts violently on exposure to oxygen, there is no point in continuing the assessment. The physicochemical information will also help to focus animal model design: if the chemical is water soluble, then oral administration is more likely than inhalation, however if the chemical is highly volatile or dusty, then inhalation administration is more representative.

16.3.3.3 Hazard Characterisation

In many instances the hazard identification and characterisation stages are entwined, making it difficult to draw a clear distinction between the two. The hazard characterisation follows closely from the information gained in the hazard identification stage, allowing the assessor to examine the relevant literature and experimental evidence that supports the scenario under investigation as defined in the problem formulation stage (as shown in Figure 16.3). Again, if human studies are unavailable, then data from representative experimental models can be used to provide a baseline from which expert judgement can be used to extrapolate likely effects and thresholds of effect into humans. *In silico* models can reduce the interspecies variability (e.g. physiologically based pharmacokinetic [PBPK] modelling) or, indeed, replace animal models completely by using quantitative structure–activity relationship models (QSAR models). As with all computer models, the robustness and reliability of the results are determined by the original data.

For many substances, threshold dose levels need to be exceeded before adverse effects are observed. Such effects can be characterised by the lowest observed adverse effect level (LOAEL) or the no observed adverse effect level (NOAEL; further discussed in other chapters). The derivation of such thresholds allows for assessors to determine a level at which adverse effects are unlikely to occur. For chemicals such as carcinogens, these thresholds do not exist, making any limits relatable only to the limit of detection of the compound, which would need to be revised as technology and detection methods evolve. Other compounds may exhibit a nonlinear or nonmonotonic dose-response relationship, as seen with endocrine disruptors (Lagarde 2015), where adverse effects may not occur in the experimental dose regime.

The final assessment should state a suggested impact and the threshold at which it is unlikely to occur, as well as recognising the inherent uncertainties in the assessment as well as the dataset that it is based on. The report should include information as to the manifestation of symptoms, duration of impact, the severity (including morbidity, mortality and sequelae), the physiopathology, epidemiological pattern, multigenerational affects (if applicable) and quality of life (Environment Agency/IEH 2000; International Life Sciences Institute 2000; World Health Organisation 2009). The potential for induction of adverse effects at the most sensitive stages of development (e.g. the embryo or foetus) and/or to the most sensitive species in the environment is particularly important (ECHA 2009; DEFRA 2011).

16.3.3.4 Exposure Assessment

The exposure assessment determines the amount of a chemical that is released in a certain scenario and available to a receptor (Environmental Protection Agency 1992). For human health risk assessment in an occupational setting, the exposure assessment usually includes an estimation of direct human exposure resulting from activities, either via air or dermal contact (however, in some cases oral exposure may also be taken into account). This information can be derived from environmental (or occupational) measurements, or it can be estimated by exposure modelling (Van Engelen et al. 2007). In many human health experiments, the exposure assessment is defined in the scenario under investigation (i.e. known dose and route). Occupational exposure can also be well described using known work patterns. Consumer exposure is more variable, and there is a continued possibility that occupational and consumer exposure will be concurrent.

A generic assessment of exposure will depend upon the inherent characteristics of the substance (including the physicochemical characteristics) and the potential sources, uses and disposal pathways that the substance may be subject to.

Exposure assessments used for the derivation of occupational exposure limits (OELs) rely on measured exposure data (Vinzents et al. 1995; Kauffer and Vincent 2007). The key exposure determinants (particularly with reference to REACH) are:

- Physical state of the substance;
- Physical state of the product handled;
- Vapour pressure for liquids, dustiness for solids;
- The level of containment;
- Presence or absence of local exhaust ventilation;
- Duration of activity; and
- What is done with the substance.

(Gabriel et al. 2010)

These variables are easily measured and characterised in the occupational setting but are more difficult to apply to a consumer scenario. Separate to these variables is the substance bioavailability. The bioavailability of the substance may be limited due to its physicochemical properties or its application and use (e.g. when incorporated in a matrix). Uncertainties in predicting exposure to man and/or the environment can be due to lack of realism of models as well as inadequate techniques or representativeness for measuring exposure in the field. Uncertainties in the exposure assessment will make it difficult to predict accurate potential risks, despite an apparent availability of large amounts of toxicity data (International Programme on Chemical Safety/ World Health Organisation 2008).

16.3.3.5 Risk Characterisation

The characterisation of risk is based on the most reliable evidence on effect, dose response and exposure available. Where there is little or no evidence, assessors rely on expert judgement (in some instances drawing upon the wider community) to interpret poor datasets in order to allow for a risk management decision. In all cases, independent of the dataset used to inform the assessment, the characterisation of the risk must include explanations as to the assumptions in the assessment, the rationale behind adjustments, and the location and severity of uncertainty in the assessment (Guidelines for Environmental Risk Assessment and Management 2011).

Where the substance under study demonstrates physical hazards (e.g. flammability or explosive properties), there is little need for a toxicological study unless it can be proven that the substance will enter the body under the conditions of use.

The aim of the risk characterisation stage is to determine whether the levels of the substance or chemical under investigation are at a magnitude at which adverse effects may occur. Therefore, the assessor compares the identified adverse effects of the substance to the NOAEL or LOAEL for human health for that particular substance, route of exposure and receptor. The NOAEL or LOAEL should exceed the estimated exposure of the substance, the extent that it needs to be exceeded by is variable, depending on:

- Intraspecies (and interspecies) variation in response to the substance
- Variability and uncertainty in the experimental data
- Nature and severity of the adverse effect
- Human population affected (e.g. vulnerable and susceptible subgroups)
- Differences in exposure (i.e. route, duration, frequency)
- Observed dose-response relationship
- Data quality and persuasiveness.

Where there is insufficient evidence of effect in humans, then a modified value needs to be calculated. This value, termed the derived no-effect level (DNEL) and known commonly in the United States as the reference dose (RfD; Environmental Protection Agency 1993), requires the modification of the defined NOAEL or LOAEL, identified in the risk characterisation stage, using uncertainty (or safety) factors (see Section 16.2), and it is itself subject to a number of uncertainties. Laboratory testing can only be a surrogate for effects in the real world, as human studies are only permitted for the development of pharmaceuticals; therefore any results are subject to extrapolation from a known species into an unknown species. The calculated DNEL, modified by conservative uncertainty factors and informed by expert judgement (see Figure 16.4), will only be a best guess scenario rather than one backed by evidence.

The DNEL, the NOAEL or LOAEL where there is sufficient human evidence, must be compared to the exposure or dose level. This comparison is used to describe the risk potential and can be termed a risk characterisation ratio (RCR), margin of exposure (MOE) or margin of safety (MOS). (It is worth noting here that many risk assessors dislike the use of the word "safety", as it suggests a certainty which is unsupported by the body of evidence used for the assessment.) These terms represent a quotient derived by comparing exposure levels to the appropriate measurement of effect and are commonly used interchangeably; however, there are differences in their calculations and measures of protection.

RCR is commonly used in chemical assessments where there are multiple routes of exposure and measures of toxicity (e.g. EC 1907/2006). For chemical exposure, particularly consumer exposure via the environment, the substance may be present in multiple compartments in the environment (i.e. multiple routes, sources, durations and populations must be considered in the assessment). Therefore the risk is only considered to be well managed (although not nonexistent) when the hazard and exposure assessments are robust and the RCRs for all exposure scenarios and adverse endpoints are below 1 (HSAC 2013; ECHA 2016).

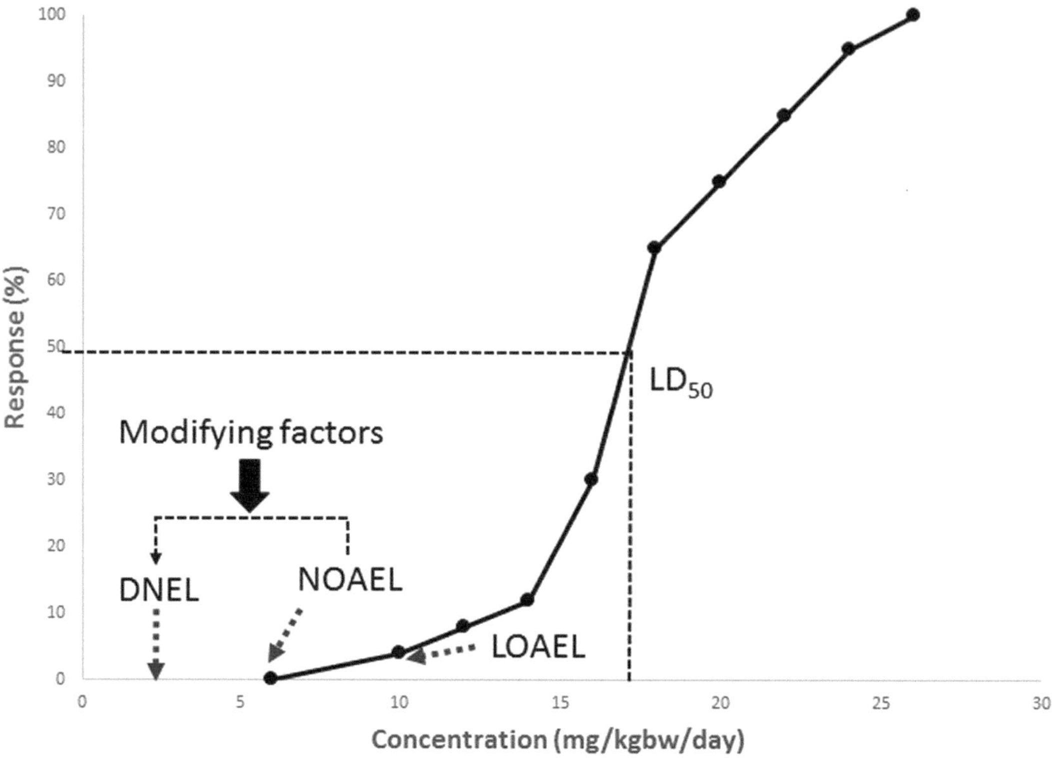

Figure 16.4

The MOE is the ratio of NOAEL (in units of weight [or body weight] per day or parts per million) obtained from animal models to the predicted human exposure level and is commonly used for cosmetic or food additives (e.g. MOE = NOAEL/exposure). If the adverse effect has a known dose-response threshold (e.g. the substance is not genotoxic or carcinogenic), the MOE is considered to be protective if it is below or equal to 100. The assessor may need to increase the MOE level if there is a mismatch between the data used to calculate the NOAEL and the potential exposure (e.g. NOAEL is derived from acute toxicity studies but the exposure is chronic); a NOAEL is not identified but a LOAEL is available; the quality of data is unclear or poor; the exposure may be underestimated; and there are subpopulations that are likely to be sensitive or vulnerable.

There are three definitions of MOS applied to different situations (Table 16.3; Environmental Protection Agency 2014). This measurement is considered to be protective and is currently used (Scientific Committee on Toxicity, Ecotoxicity and the Environment 2001). The assessment of MOS is comparable to that of MOE, with similar requirements as to the assessment of data quality. The RCR can be considered to be equivalent to MOS but is commonly used in chemical risk assessment (EC 1907/2006).

The derivation of MOE, MOS or RCR will depend on the available data. For example, for the dermal use of salicylic acid in a cosmetic product, the calculation of MOS would use a NOAEL of 75 mg/kg body weight per day derived from a developmental toxicity study in rats (Tanaka et al. 1973) identified as the most sensitive toxicological endpoint in the most sensitive species (although the route in this study was oral, salicylic acid has a high oral bioavailability in humans, therefore the NOAEL is appropriate). When the aggregate exposure is calculated to be 1.50 mg/kg body weight per day (using SCCS 2016, assuming that all the population uses products containing salicylic acid), the MOS calculation is:

$$\text{MOS} = \frac{\text{NOAEL}}{\text{predicted exposure}} = \frac{75\,\text{mg / kg body weight per day}}{1.50\,\text{mg / kg body weight per day}} = 50$$

Table 16.3 Table Showing the Definitions and Different Calculations for the Risk Characterisation Ratio (RCR), Margin of Exposure (MOE) and Margin of Safety (MOS)

Name	Description	Calculation	Situation	Reference
Risk characterisation ratio	Ratio of exposure to DNEL (or DMEL)	RCR = Exposure/ DNEL	Used in Chemical Safety Report in REACH legislation; value should be <1.	EC 1907/2006 ECHA 2016
Margin of exposure	Ratio of NOAEL to predictive human exposure	MOE = NOAEL/ exposure	Commonly used for cosmetic or food additives	WHO/IPCS, 2009 SCCS, 2016
Margin of safety	Ratio of lethal dose to 1% of population (LD_1) to effective dose of 99% of population (ED_{99})	MOS = LD_1/ED_{99}	Measurement of drug safety in pharma industry, pesticides; MOS should be as large as possible	OECD, 2010
	Ratio of NOAEL (animal model) to predicted human exposure level (equivalent to MOE)	MOS = NOAEL/ estimated exposure dose	Safety of cosmetic ingredients; MOS ≥100 is considered to be protective	CFDA, 2015 WHO/IPCS, 2009

Note: For all calculations, the different measurements should use the same units of measurement (e.g. ppm or mg/kg body weight per day).

Given that this value is less than 100, the assessor would need to make a case if they considered the values used to still be protective.

Although we have concentrated on NOAEL and LOAEL in the determination of risk, benchmark-dose calculations can be used to statistically provide similar values where a benchmark-dose lower bound (BMDL) value is used (see Davis et al. [2011] for additional information). Where it is not possible to determine a no-effect concentration (e.g. in the case of nonthreshold carcinogens), then a derived minimal effect level (DMEL) is set and is considered to be a reference risk level of very low concern, and the resulting ratio is noted as such (ECHA 2016). For legislation that requires an MOE for a substance under assessment that is either genotoxic or carcinogenic, or where an NOAEL cannot be identified, the MOE should be greater than 10,000 (European Food Standards Agency 2012).

Although there are key requirements of some legislation (e.g. EC 1907/2006) are based on risk and its associated assessments (e.g. chemical safety reports and the prioritisation of substances for evaluation using a risk-based approach [as stated in chapter 2 of Title VI in EC 1907/2006]), other requirements rely on hazard classification only. This is most notable where it is not possible to specify a no-effect level for substances, such as in the case of persistent, bioaccumulative and toxic substances or very persistent and very bioaccumulative substances (defined in Annex XIII of EC 1907/2006), both of which may be made subject to authorisation.

Following the consideration of the risk characterisation stage, there are three possible overall recommendations that the assessor can make:

1. The risk characterisation suggests that at present there is no need for additional information and no need for additional risk reduction measures, therefore the substance is of no immediate concern but should be monitored.
2. The substance is of concern and additional information is required to support either the hazard characterisation and exposure assessment, or both.
3. The substance is of high concern, and additional information must be collected or risk reduction measures (RRM) should be put into place; the risk should then be reviewed and reassessed.

These recommendations may require the assessment to be repeated once additional, newly generated experimental evidence is collected, thereby forming a naturally pragmatic, or tiered approach to the assessment (Guidelinès for Environmental Risk Assessment and Management 2000). This tiered approach helps the assessment to be proportional to level of effort and resources spent assessing it in terms of complexity and priority (in relation to other impacts and risks, respectively). Initial risk screening would consider literature evidence in order to make an initial assessment, after which the assessor would determine whether additional information (such as generic or detailed quantitative studies) would need to be considered or, indeed, generated in order to answer the questions identified in the problem formulation step.

16.3.4 Risk Management

Many risk assessments include risk management in their framework (HM Treasury 2004). This inclusion can be confusing, as the assessment of risk is an objective process (with acknowledged inherent bias) whereas risk management needs to consider the wider risk landscape (Guidelines for Environmental Risk Assessment and Management 2000). The risk-management process must consider external forces such as stakeholders, resources and reputation in the decision to ensure that the management of the risk is appropriate to the risk appetite of the organisation (or community; Guidelines for Environmental Risk Assessment and Management 2011). It is well recognised that the process of risk management is cyclic and should include openness, transparency and involvement of stakeholders in decision processes (HM Treasury and Office of Government Commerce 2004).

All risk management decisions seek to determine whether the risk is acceptable for its target audience, therefore a risk manager has a range of options (from tolerate to terminate) and must ensure that the chosen action is appropriate for the risk and for the perception of the risk in the audience (see Figure 16.5; HM Treasury 2004).

Although the technical decision process in this area concerns whether the hazard or exposure needs to be reduced so that the calculated levels are below that considered to be of concern, there is a wider and holistic decision-making process that should be acknowledged and will be influenced by objective factors (including quantitative risk, comparative risk, cost and feasibility) as well as subjective factors (including social [such as prejudice, freedom and trust], quality of life, safety, political, ethics, psychological [fear or embarrassment], and judicial; Slovic 1993). These wider subjective factors are likely to also affect the efficiency of any risk management, as these will impact the belief and confidence of the audience (e.g. general population) and will influence their compliance and behaviour with any secondary management measure.

Risk can be modulated by reducing exposure enabling the continuation of use of chemical substances. In such circumstances risk management measures (RMMs) are used to reduce the potential for exposure of the substance in question (Bruinen De Bruin et al. 2007). Risk management strategies, measures and processes should be selected using a decision process that weighs up a

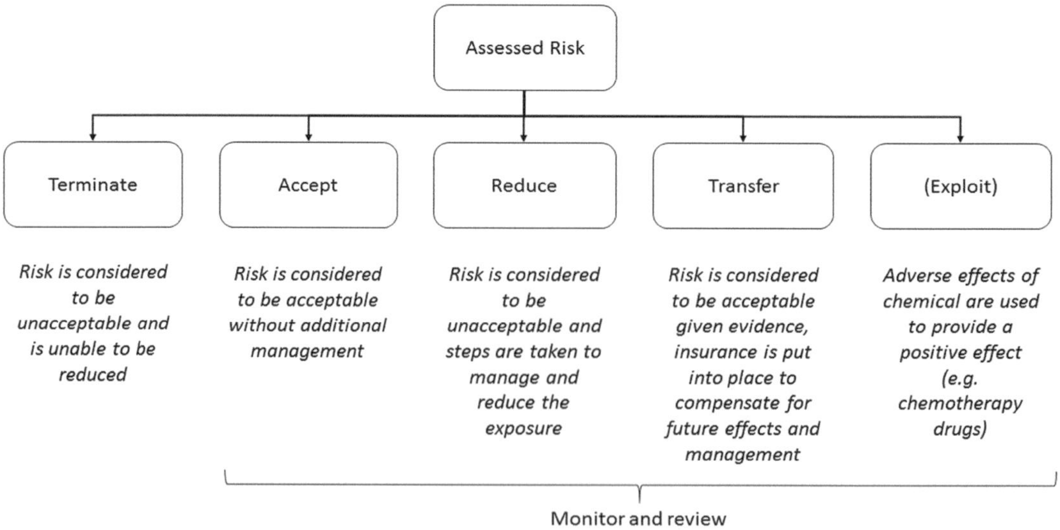

Figure 16.5

number of different options and is proportionate to the risk faced. Where the risk assessment has identified a scenario of concern (e.g. RCR is greater than, equal to, or approaching 1, dependant on the risk appetite of the organisation), RMMs can be used to reduce the ratio to a more acceptable level.

The risk itself can be reduced either by modifying the hazard (e.g. changing the substance or source) or the exposure (e.g. how much substance a receptor is exposed to). The exposure of a substance can be reduced by:

- Changing the physical form of the substance (e.g. removing the likelihood of dust from fine powders by changing into solid pellets, removing the potential for formation of aerosol droplets or volatility of substances from liquids);
- Changing the requirements from the substance (e.g. reducing the amount of the substance present in a commercial product or the amount of the product used per activity);
- Changing the use of substance in a commercial product (e.g. duration and frequency of exposure in an occupational setting; amount of applied chemical; temperature; containment of process; capacity of surroundings; ECHA 2012).

Other measures that could be used to reduce risk are aimed at controlling, limiting or avoiding exposures and helping to ensure the safe handling of the substance as part of a consumer product (Bruinen De Bruin et al. 2007). Measures used to reduce risk to a minimal level are considered in a specific hierarchy (ECHA 2012):

1. Modifying processes either through automating the process or using intrinsically safe equipment to reduce exposure;
2. Reducing the risk by limiting the concentration of a substance, changing the physical form of the substance, applying closed processes, and/or installing local exhaust ventilation that effectively reduces the exposure;
3. Increasing the availability of general area ventilation and other workplace-related measures (e.g. segregation of work areas and improved safety facilities);
4. Other collective RMMs that protect a large proportion of the workplace population (e.g. organisational measures);
5. Personal protective equipment, as a last resort.

There is a large body of literature that considers and reports the development and identification of RMMs in the chemical industry in relation to the occupational, consumer and environment sectors (e.g. Bruinen De Bruin et al. 2007; Van Engelen et al. 2007; Lee et al. 2013).

16.4 LEGISLATION

The importation and manufacture of chemical species falls under the Registration, Evaluation, Authorisation and Restriction of Chemicals (REACH) Regulation [EC] No 1907/2006 of the European Parliament and of the Council of 18 December 2006. REACH, implemented June 1, 2007, is a European Union (EU) regulation that imposes obligations on manufacturers, importers and downstream users of chemicals within the EU (including, at the time of writing, the United Kingdom of Great Britain and Northern Ireland). The regulation is designed to protect both human health and the environment from manufactured or imported chemicals and to enhance the competitiveness of the EU chemicals industry by promoting its reputation as a responsible industry and encouraging innovation. The regulation places the responsibility for the control and safe use of substances with the industry bodies, which must register all chemical substances and provide additional information for those substances that either exceed certain amounts or meet other criteria. The registration of substances may include the preparation of the technical dossier and chemical safety assessment (CSA) or chemical safety report (CSR; EC 1907/2006; Rocks et al. 2009).

REACH applies to chemical substances manufactured or imported into the EU in quantities equal to or greater than 1 tonne per year. All individual chemical substances, substances in preparations and in articles (if release can be foreseen under normal conditions of use) are covered by the legislation (EC 1907/2006). Specific exclusions from REACH include radioactive substances, substances under customs supervision (e.g. those passing through the EU and not processed in any way), substances being transported, waste, naturally occurring low-hazard substances and nonisolated intermediates. Where substances are covered by more specific legislation (e.g. human and veterinary medicines, food and foodstuff additives, plant protection products and biocides, isolated intermediates and substances used for research and development), there is provision under the legislation to exempt these substances.

16.5 CONCLUSIONS

The cyclic nature of risk assessment and management requires that data continue to be collected and risk monitored throughout the life of the risk. Although they have been addressed separately in this chapter for ease of description, in real-life scenarios they are likely to occur concurrently, and therefore it is likely that the two distinct actions will overlap.

For each risk assessment, specific legislation will define how relevant and reliable data is collected and assessed. Problem formulation is essential to ensure that the risk assessment process captures the most relevant data and that the final assessment is focused and as comprehensive as possible within resource and data availability restrictions.

BIBLIOGRAPHY

Bruinen De Bruin, Y., P. Hakkinen, M. Lahaniati, et al. 2007. Risk management measures for chemicals in consumer products: Documentation, assessment, and communication across the supply chain. *J Expo Sci Environ Epidemiol* 17(Suppl 1): S55–S66.

Burden, N., C. Mahony, B. P. Muller, C. Terry, C. Westmoreland, and I. Kimber. 2015. Aligning the 3Rs with new paradigms in the safety assessment of chemicals. *Toxicology* 330: 62.66.

Butler, D., and E. Callaway. 2016. Scientists in the dark after French clinical trial proves fatal (Nature News). *Nature* 529: 263–4.

China Food and Drug Administration. 2015. *Cosmetic Safety Evaluation Guideline*. www.sesec.eu/app/uploads/2015/12/CFDA-cosmetic-safety-evaluation-guideline_English.pdf (accessed November 11, 2020).

Dankovic, D. A., B. D. Naumann, A. Maier, M. L. Dourson, and L. S. Levy. 2015. The scientific basis for uncertainty factors used in setting occupational exposure limits. *J Occup Environ Hyg* 25(12, suppl 1): S55-S68.

Davis, J. A., J. S. Gift, and Q. J. Zhao. 2011. Introduction to benchmark dose methods and U.S. EPA's benchmark dose software (BMDS) version 2.1.1. *Toxicol Appl Pharmacol* 254: 181–91.

DEFRA (Guidelines for Environmental Risk Assessment and Management). 2011. *Guidelines for Environmental Risk Assessment and Management (Greenleaves III)*. www.gov.uk/government/uploads/system/uploads/attachment_data/file/69450/pb13670-green-leaves-iii-1111071.pdf (accessed November 11, 2020).

Dorne, J. L. C. M., and A. G. Renwick. 2005. The refinement of uncertainty/safety factors in risk assessment by the incorporation of data on toxicokinetic variability in humans. *Toxicol Sci* 86: 20–6.

Environment Agency/IEH. 2000. *Guidelines for Environmental Risk Assessment and Management*. www.iehconsulting.co.uk/IEH_Consulting/IEHCPubs/HumExpRiskAssess/guidelines-forenvironmental.pdf (accessed November 11, 2020).

Environmental Protection Agency. 1992. Guidelines for Exposure Assessment. *Federal Register* 57(104): 22888–938. US Environmental Protection Agency. http://cfpub.epa.gov/ncea/cfm/recordisplay.cfm?deid=15263#Download (accessed November 11, 2020).

Environmental Protection Agency. 1993. *Reference Dose (RfD): Description and Use in Health Risk Assessments Background Document 1A*. US Environmental Protection Agency. www.epa.gov/iris/reference-dose-rfd-description-and-use-health-risk-assessments (accessed November 11, 2020).

Environmental Protection Agency. 2014. Guidance for applying quantitative data to develop data-derived extrapolation factors for interspecies and intraspecies extrapolation. *Federal Register* R-14-002F. www.epa.gov/sites/production/files/2015-01/documents/ddef-final.pdf (accessed November 11, 2020).

European Chemicals Agency (ECHA). 2009. *Guidance in a Nutshell Chemical Safety Assessment ECHA-09-B-15-EN (2009)*. https://echa.europa.eu/support/guidance-on-reach-and-clp-implementation/guidance-in-a-nutshell (accessed November 11, 2020).

European Chemicals Agency (ECHA). 2011. *Guidance on Information Requirements and Chemical Safety Assessment*. Chapter R.4: Evaluation of available information Reference: ECHA-2011-G-13-EN. https://echa.europa.eu/documents/10162/13643/information_requirements_r4_en.pdf (accessed November 11, 2020).

European Chemicals Agency (ECHA). 2012. *Guidance on Information Requirements and Chemical Safety Assessment.* Appendix R13: Risk management measures and operational conditions ECHA-12-G-19-EN. http://echa.europa.eu/documents/10162/13632/information_requirements_r13_en.pdf (accessed November 11, 2020).
European Chemicals Agency (ECHA). 2016. *Guidance on Information Requirements and Chemical Safety Assessment Part E: Risk Characterisation.* https://echa.europa.eu/documents/10162/13632/information_requirements_part_e_en.pdf (accessed November 11, 2020).
European Commission. 2008. *European Regulation (EC) No 1272/2008 of the European Parliament and of the Council on Classification, Labelling and Packaging of Substances and Mixtures.* EC 1272/2008. http://ec.europa.eu/growth/sectors/chemicals/classification-labelling/ (accessed November 11, 2020).
European Commission. 2010. *Guidance on Risk Assessment and the Use of Conceptual Models for Groundwater Common Implementation Strategy for the Water Framework Directive (2000/60/EC).* Guidance Document No. 26. https://circabc.europa.eu/sd/a/8564a357-0e17-4619-bd76-a54a23fa7885/Guidance%20No%2026%20-%20GW%20risk%20assessment%20and%20conceptual%20models.pdf (accessed November 11, 2020).
European Food Standards Agency Scientific Committee. 2012. Scientific Opinion on the applicability of the Margin of Exposure approach for the safety assessment of impurities which are both genotoxic and carcinogenic in substances added to food/feed. *EFSA J* 10(3): 2578.
European Union (EU). 2006. *Regulation (EC) No. 1907/2006 of the European Parliament and of the Council on the Registration, Evaluation, Authorisation and Restriction of Chemicals ('REACH').*
EWHC. 2009. Register of the Corby Group Litigation vs Corby District Council; [2010] Env LR D2, [2009] EWHC 1944 (TCC), [2009] NPC 100 Case No: HT-09–63. 2009. www.bailii.org/ew/cases/EWHC/TCC/2009/1944.html (accessed November 22, 2020).
Gabriel, S., D. Koppisch, and D. Range. 2010. The MGU: A monitoring system for the collection and documentation of valid workplace exposure data. *Gefahrst Reinhalt Luft* 70(1–2): 43–49.
Hazardous Substances Advisory Committee (HSAC). 2013. *Hazard and Risk Assessment of Substances: The HSAC Approach.* https://webarchive.nationalarchives.gov.uk/20151210231753/www.gov.uk/government/uploads/system/uploads/attachment_data/file/342114/HSAC-annual-report-2013_final.pdf (accessed November 22, 2020).
HM Treasury. 2004. *The Orange Book. Management of Risk—Principles and Concepts.* The Stationary Office, London. www.gov.uk/government/uploads/system/uploads/attachment_data/file/220647/orange_book.pdf (accessed November 11, 2020).
HM Treasury and Office of Government Commerce. 2004. *Managing Risks to the Public.* The Stationary Office, London.
International Life Sciences Institute. 2000. *Revised Framework for Microbial Risk Assessment.* Washington, DC: ILSI Risk Science Institute Press.
International Programme on Chemical Safety/World Health Organisation. 2008. *Uncertainty and Data Quality in Exposure Assessment.* Geneva, Switzerland: WHO. www.who.int/ipcs/publications/methods/harmonization/exposure_assessment.pdf?ua=1 (accessed November 11, 2020).
Kauffer, E., and R. Vincent. 2007. Occupational exposure to mineral fibres: Analysis of results stored on colchic database. *Ann Occup Hyg* 51(2): 131–42.
Klimisch, H., M. Andreae, and U. Tillmann. 1997. A systematic approach for evaluating the quality of experimental toxicological and ecotoxicological data. *Regul Toxicol Pharmacol* 25: 1–5.
Lagarde, F., C. Beausoleil, S. M. Belcher, et al. 2015. Non-monotonic dose-response relationships and endocrine disruptors: A qualitative method of assessment. *Environ Health* 14: 13.
Lee, J., A. B. Pedersen, and M. Thomsen. 2013. Framework for combining REACH and national regulations to obtain equal protection levels of human health and the environment in different countries—Comparative study of Denmark and Korea. *J Environ Manag* 125: 105–16.
Linkov, I., D. Loney, S. Cormier, F. K. Satterstrom, and T. Bridges. 2009. Weight-of-evidence evaluation in environmental assessment: Review of qualitative and quantitative approaches. *Sci Total Environ* 407(19): 5199–205.
Medicines and Healthcare Products Regulatory Agency. 2018. *The Yellow Card Scheme: Guidance for Healthcare Professionals.* www.gov.uk/guidance/the-yellow-card-scheme-guidance-for-healthcare-professionals (accessed November 11, 2020).
National Research Council (US) Safe Drinking Water Committee. 1986. Chapter 7, Data on humans: Clinical and epidemiological studies. In *Drinking Water and Health: Volume 6.* Washington, DC: National Academies Press.

Nordlander, K., C. M. Simon, and H. Pearson. 2010. Hazard v. risk in EU chemicals regulation. *Eur J Risk Regul* 3: 239–50.
OECD. 2005. *Manual for Investigation of HPV Chemicals. Chapter 3—Data Evaluation*. Organization for Economic Co-operation and Development. www.oecd.org/chemicalsafety/risk-assessment/49191960.pdf (accessed November 11, 2020).
OECD. 2010. *Series on Testing and Assessment No. 124 Guidance for the Derivation of an Acute Reference Dose ENV/JM/MONO(2010)15*. www.oecd.org/officialdocuments/publicdisplaydocumentpdf/?cote=env/jm/mono(2010)15&doclanguage=en (accessed November 11, 2020).
Pollard, S. T. J., and S. A. Rocks. 2014. Chapter 3: A Recent Chronology of Public Risk Management in Government. In *Annual Report of the Government Chief Scientific Adviser 2014. Innovation: Managing Risk, Not Avoiding It. Evidence and Case Studies*. https://assets.publishing.service.gov.uk/government/uploads/system/uploads/attachment_data/file/381906/14-1190b-innovation-managing-risk-evidence.pdf (accessed November 11, 2020).
Rocks, S. A., R. Owen, S. J. T. Pollard, et al. 2009. Risk assessment of manufactured nanomaterials. In *Environmental and Human Health Effects of Nanoparticles*, eds. J. Lead and E. Smith. Blackwell Science.
Scientific Committee on Consumer Safety (SCCS). 2016. *Notes of Guidance for the Testing of Cosmetics Ingredients and their Safety Evaluation*. 9th revision. SCCS/1564/15, revised version of 25 April 2016. https://ec.europa.eu/health/scientific_committees/consumer_safety/docs/sccs_o_190.pdf (accessed November 22, 2020).
Scientific Committee on Toxicity, Ecotoxicity and the Environment (CSTEE). 2001. *Position Paper on Margins of Safety (MOS) in Human Health Risk Assessment*. 22nd CSTEE plenary meeting, Brussels, 06/07 March 2001.
Skinner, D., S. A. Rocks, S. J. T. Pollard, and G. H. Drew. 2014. Identifying uncertainty in environmental risk assessments: The development of a novel typology and its implications for risk characterisation. *Hum Ecol Risk Assess* 20: 607–40.
Slovic, P. 1993. Perceived risk, trust, and democracy. *Risk Anal* 13: 675–82.
Sutter, G. W. 1996. *Guidelines for Developing Conceptual Models for Ecological Risk Assessments*. https://rais.ornl.gov/documents/tm186.pdf (accessed November 11, 2020).
Tanaka, S., K. Kawashima, S. Nakaura, et al. 1973. Teratogenic effect of dietary salicylic acid in rats. *J Food Hyg Soc* 20(14): 549–57.
Van Engelen, J. G. M., G. Heinemeyer, and C. Rodriguez. 2007. Consumer exposure scenarios: Development, challenges and possible solutions. *J Expo Sci Environ Epidemiol* 17(Suppl 1): S26–S33.
Vinzents, P., B. Carton, P. Fjeldstad, B. Rajan, and R. Stamm. 1995. Comparison of exposure measurement stored in European databases on occupational air pollutants and definitions of core information. *Appl Occup Environ Hyg* 10(4): 351–4.
World Health Organisation. 1948. *The Constitution of the World Health Organisation including amendments (resolutions WHA26.37, WHA29.38, WHA39.6 and WHA51.23)*. http://apps.who.int/gb/bd/PDF/bd47/EN/constitution-en.pdf?ua=1 (accessed November 11, 2020).
World Health Organisation. 2009. *Library Cataloguing-in-Publication Data Principles and Methods for the Risk Assessment of Chemicals in Food*. (Environmental Health Criteria; 240) World Health Organization. II. Food and Agriculture Organization of the United Nations. http://apps.who.int/iris/bitstream/handle/10665/44065/WHO_EHC_240_eng.pdf?sequence=152&isAllowed=y (accessed November 11, 2020).
World Health Organization and Inter-Organization Programme for the Sound Management of Chemicals. 2009. *Principles for Modelling Dose-Response for the Risk Assessment of Chemicals*. Geneva, Switzerland: World Health Organization.

17 Toxicological Modelling

Gary Moss and Simon Wilkinson

CONTENTS

17.1 GENERAL INTRODUCTION

Broadly speaking, toxicological testing cascades from *in vitro* testing to *in vivo* testing. Computational (or *in silico*) methods aim to complement these methods ostensibly to increase the speed of the process and to minimise the need for animal testing, where regulatory requirements allow it to do so. This has clear beneficial impacts on animal welfare, experimental cost and, ultimately, safety assessment. Both *in vitro* and computational approaches to toxicity testing are regarded as "failing early"; that is, candidate molecular structures that are predicted to be associated with a particular toxic endpoint such as genotoxicity or sensitisation are identified at a very early stage in the discovery process, allowing them to be discarded before further time and money are expended on their testing. Computational methods are also unique in that they allow for estimates of toxicity (or, in the case of pharmaceuticals, toxicity and efficacy) to be determined before chemicals are even synthesised, further reducing costs and refining experimental designs.

Computational (or *in silico*) toxicology employs a wide range of methods to estimate toxic endpoints. These include databases of chemical information,[1] which are interrogated for information on existing chemicals and their toxicological endpoints. Models such as quantitative structure–activity relationships (QSARs) are widely used and often constructed from molecular descriptors which are generated via specialist software packages. More recently, Machine Learning methods, a form of artificial intelligence that uses datasets to facilitate automatic learning and improvement, have also been used in this field.

Generally, a range of endpoints may be used to relate the biological response to the physicochemical properties of molecules of interest (e.g. chemicals that permeate into and across the skin). These include, in the context of exposure of chemicals to skin, measures of irritation following single- or multiple-dose exposure, frequency and duration of exposure, or a quantitative measure of the amount of a chemical permeating into and/or across the skin. Toxicology of skin falls broadly into two distinct areas: dermal toxicity, associated with events observed in the skin or on the skin surface, and systemic toxicity, following topical administration and subsequent percutaneous absorption which may manifest itself elsewhere in the body. In most cases, this chapter will focus on the latter model of skin permeation as a representative endpoint. A detailed description of these methods, which is beyond the scope of this chapter, may be found elsewhere (Valerio 2009; Reisfeld and Mayeno 2013; Toropov et al. 2014; Modi et al. 2012; Milan et al. 2011; Wilson 2011; Kortagere 2013).

DOI: 10.1201/9780203730584-17

When generating a computational model, most of the approaches described above involve five main steps (Raies and Bajic 2016; Devillers 2013). These are the gathering of biological data that relate the physicochemical properties of a chemical (or its formulation) to toxicity endpoints, calculation of molecular descriptors for the chemicals of interest, generation of a model which relates the chemical properties of the chemicals of interest to the desired endpoint, evaluation of the accuracy of the predictive model including validation, and interpretation of the model in the appropriate context.

The field of *in silico* toxicology is enormous and impossible to discuss in detail in a chapter of this length. Thus, this chapter will focus on the key underlying aspects of toxicological modelling and in particular the use of algorithms and related methods used to provide estimates of toxicity. This will emphasise the general nature of how models have developed historically and the current state of the art, notably the selection of appropriate methods to predict toxicity. It will also use—as a relevant example of good and bad practice in constructing models and understanding how they should be used, particularly in understanding limitations of models—the example of models of skin permeation to illustrate the practical application of the theoretical elements of this process.

17.2 COMPUTATIONAL (*IN SILICO*) MODELLING METHODS

17.2.1 Rule-Based Models

Generally, rule-based models are classed as human-based rules (HBRs) or induction-based rules (IBRs), with the former being empirically derived from experts in the field or from literature and the latter being derived from computational methods (Valero 2009; Raies and Bajic 2016). Although HBRs might be considered more accurate, they may be subjectively derived or biased, requiring an extensive knowledge of the literature to complete a representative model (Lepailleur et al. 2013). IBRs, on the other hand, may be derived from databases and assigned a probability regarding the relationship between their chemical structure and possible toxicity. Ideally, combinations of both models would be employed and additional factors considered, such as the role of pharmacokinetics in chemical uptake and clearance based, for example, on route of administration/entry to the body.

The most prominent example of a rule-based model is a structural alert, or toxicophore, which identifies toxicity based on the presence of an atom or functional group associated with a particular toxic issue. Toxicophores have been defined by Williams and Naisbitt (2002) as "substances that are capable of causing a toxic effect contain a structural feature or moiety that bestows the toxic property. This structural feature or moiety is referred to generically as the toxicophore". For example, this often takes the form of "IF . . . THEN" statements; for example:

IF [chemical_substructure] IS [present] THEN [a specific toxic outcome is expected]

Structural alerts are simple models which are extremely easy to use and interpret, and they are widely used in drug design in order to reduce toxicity of candidate molecules. Equally, their simplicity means that they lack subtlety both in the nature of descriptors (e.g. chemical structures are either present or absent) and endpoints (e.g. normally being binary, such as indicating whether or not a structure is or is not a skin sensitiser). They provide little or no mechanistic insight into the process of toxicity and no information on the nature of the toxicity, such as the amount of a chemical required to produce a toxic event. As with all such models, the usability of outputs will depend on appropriate inputs; for example, an incomplete list of rules may result in false negative outcomes, where toxic chemicals or functional groups are predicted to be nontoxic. Such subjectivity again highlights the need for a combination of rule-based methods, such as HBRs and IBRs, rather than the use of one or the other.

Rule-based systems have also been described as "expert systems", such as DEREK, StAR and METEOR (Greene et al. 1999), Oncologic Cancer Expert Systems (OCES; Woo and Lai 2005; Woo et al. 1995, 1996) and HazardExpert (CompuDrug 2013). These provide prestructured rules and structural alerts for specific endpoints or provide structural alerts from datasets, such as Computer-Assisted Structure Elucidation (CASE; Klopman 1984) and Categorical Structure Activity Relationship (cat-SAR; Cunningham et al. 2008). Other methods which focus on particular molecular substructures are also widely used; these include pattern growth methods such as the Molecular Fragment Miner (MOFA; Borgelt and Berthold 2002) and Graph-based Substructure Pattern Mining (gSpam; Yan and Han 2002). This area has been reviewed extensively, and the reader is directed elsewhere for a more comprehensive overview of this field (Lepailleur et al. 2013). Further examples include the OECD Quantitative Structure–Activity Relationship (QSAR)

Toolbox (Benigni et al. 2013; OECD 2015). Other structural alerts have been developed for hepatotoxicity; cytotoxicity; various skin interactions such as corrosion, permeation and irritation; irritation of the eye; and sensitisation of the skin—these are reviewed extensively elsewhere (Raies and Bajic 2016).

17.2.2 Read-Across and Trend Analysis in Source Chemicals

Read-across is a "nontesting" approach that has gained much attention in recent years, notably because it can be used as a means of fulfilling data gaps under, for example, REACH (Registration, Evaluation, Authorisation and Restriction of Chemicals).[2] It aims to fill data gaps for chemicals by relating their endpoints to those of chemicals with similar structures and/or physicochemical properties. Read-across methods can be qualitative or quantitative, depending on the endpoint being examined (Reisfeld and Mayeno 2013). This is summarised in the ECHA Read-Across Assessment Framework (European Chemicals Agency 2017). Thus, REACH requirements, such as endpoints for human or environmental health, may be predicted from information conducted on reference substances by consideration of similar chemicals, hence the expression "read-across". Thus, the properties of one target chemical, usually a suitable endpoint, are derived from the properties of the reference or source chemicals. The specific endpoint will depend on the application under consideration, as described in the REACH documentation (REACH Online, Annex VI to X). Endpoints are listed either as types of study, such as carcinogenicity studies, or specific properties, such as skin irritation.

Generally, this class of methodological approaches is related to a group of chemicals whose properties, including pharmaceutical efficacy (if relevant) and toxicity follow a similar pattern. These chemicals are commonly classified, or grouped, according to the OECD Guidance on Grouping of Chemicals (OECD 2018). Read-across methods thus allow the estimation of unknown toxicity (or similar endpoint) for a chemical which is chemically from the same general class whose toxicity is already understood and quantified, being chemically analogous to the target of interest. It is common to use such approaches to characterise trends in chemical series, including mutagenicity and carcinogenicity assessment (Reisfeld and Mayeno 2013) and skin permeation based on the systematic variation of chemical structures (Abraham et al. 1990, 1995, 1997, 1999; Moss et al. 2006).

There are two ways to produce suitable read-across methods: analogue and category methods. The analogue approach is adopted when read-across is employed between a few, very structurally similar substances for which it is not possible to establish a clear trend. The structural similarity is used to predict the required endpoint, and a worst-case approach is often adopted given the scope of the available data. The categorical approach is used when a read-across method is employed between several substances which have structural similarities. The structural similarities infer that properties are proposed to be similar or the same for the target chemical (European Chemicals Agency 2017).

Identification of chemical similarities is achieved either by considering the presence or absence of certain chemical groups (i.e. a binary approach, where the presence (1) or absence of a group (0) is recorded, yielding a score for a particular chemical) or the frequency of occurrence of the groups of interest on a target molecule (e.g. the presence of a hydroxyl group, which is recorded as the number of groups of interest present on a molecule). Physicochemical measures of molecular properties, such as molecular weight, melting point or lipophilicity, can also be used in quantitative models. Descriptors may be grouped into hierarchies depending on their significance (Dimitrov and Mekenyan 2010).

Read-across methods are simple to use and easy to interpret. They can be derived from readily available descriptors and the range of descriptors that can be used allow flexibility in model development and use. They may, however, be unable to provide complex mechanistic information and may lack complexity, providing a superficial view of toxicity. As they are often based on chemical similarity, they can be constructed from small datasets, which may restrict their accuracy compared to methods which employ large datasets (e.g. QSAR approaches). The outputs of such models may therefore require expert interpretation, which may render to the model a certain degree of subjectivity.

17.2.3 Uncertainty Factor Modelling

Uncertainty factors can be used to predict the reference dose (RfD) and concentration (RfC) and provide quantitative information for use in risk assessments for health effects. As such, they are simple models that are easy to implement and understand, and provide a prediction for toxicity that accounts for variation in the data, allowing its extrapolation to more relevant endpoints. The inclusion of an uncertainty factor in the estimation of toxicity aims to provide an understanding of biological variation, notably within different populations in a particular species. For example, it aims to provide an estimate of how demographic factors (e.g. age or gender) or particular health conditions may influence the prediction of toxicity. Uncertainty factor modelling is also used in the extrapolation of exposure (e.g. from short-term exposure to long-term exposure). Uncertainty factor models are usually characterised by two descriptors: no observed adverse effects (NOAEL, which is the highest dose not resulting in toxicity) and the uncertainty factor (UF, which aims to introduce variability due to differences in populations or in the amount or duration of exposure). NOAEL is normally divided by UF to produce the estimate of risk. Values of NOAEL are limited by their associated experimental design, as they may vary based on the population size or content being tested. Further, they lack an indication of the "appreciable risk" of toxicity and do not yield a zero-effect threshold (Raies and Bajic 2016). To try and address such shortcomings, modifying factors (MFs) and safety factors (SFs) may be added to the model, the former to attempt to account for uncertainty in the data and the latter to characterise irreversible effects, such as teratogenicity. Additional factors are included to try to characterise child safety. For example, in relation to developmental ages in children, the US Environmental Protection Agency applies an additional 10-fold factor in assessing the risk in infants and children to pesticide exposure; it is also common to make similar adjustments in, for example, doses of medicinal treatments in infants and children. Normally, the MF and SF factors do not exceed 10 (Falk-Filipsson et al. 2007).

17.2.4 Pharmacokinetic and Pharmacodynamic Models

It should be recognised that pharmacokinetics (PK) and pharmacodynamics (PD) is an enormous area whose scope cannot be effectively be summarised in a chapter of this length. The reader is therefore directed elsewhere (Sung et al. 2014; Ritter et al. 2016; Golan et al. 2017) for a more comprehensive discussion of this subject, and the section below will focus on the key aspect of modelling PK/PD systems.

PK models predict the time-dependent concentration of a chemical in the body—or in more sophisticated models, in plasma and other compartments such as the liver, respiratory tract and central nervous system—from its administration to its elimination. The modelling, in essence, allows us to determine what amount of a chemical to administer (e.g. how much of a drug to give), how often to give it (including the dosage form or formulation), how different routes of administration affect the uptake and elimination of drugs and possibly how some interactions may occur. PK methods are often associated with dose-response and time-response models. PD relates a biological response to the concentration of a drug in its tissue, considering therefore the effects of drugs (or, more broadly, toxic events following uptake of chemicals) and their mechanism of action (Sung et al. 2014). PK and PD are often summarised as, respectively, the effect a chemical has on the body and the effect the body has on a chemical. As they are generally associated with the delivery, uptake and elimination of drugs via the processes of adsorption, distribution, metabolism and excretion (ADME) when considered in the context of toxicity, they are often defined as toxicokinetic and toxicodynamic models. PK/PD models are often compartmental, with each discrete compartment representing a particular environment the chemical may find itself in. For example, a drug passing through the skin would pass into the skin barrier (the stratum corneum), and then into the viable tissue; thereafter, it would pass into the vascular system and be absorbed systemically.

The simplest PK models consider the whole body as a single compartment, assume rapid equilibrium of the chemical in the body and not consider distribution time in the body. In such a case, this process may be represented by Equation 17.1:

$$C(t) = C_0 e^{-kt} \tag{17.1}$$

where $C(t)$ is the concentration at a specified time, C_0 is the initial concentration, k is the elimination constant, and a plot of log C_0 against time results in a straight line with gradient $-k$.

Further complexity is added by increasing the number of compartments, such as in Guy and Hadgraft's model (Guy and Hadgraft 1982). In the theoretical case of a two-compartment model, considering a distribution phase with initial concentration C_a and an elimination phase with initial concentration C_b, the concentration C is represented by Equation 17.2:

$$C(t) = C_a e^{-at} + C_b e^{-bt} \tag{17.2}$$

Although such models are unable to extrapolate between different chemicals or provide mechanistic insight into the ADME process, they may be modified to include physiological descriptors (physiologically based pharmacokinetic models [PBPK]) including blood volume, blood flow, chemical partitioning, protein binding, metabolism and excretion (Reisfeld and Mayeno 2013; Golan et al. 2017). PBPK models represent each organ as a separate compartment, or may often include many compartments within one tissue (e.g. the stratum corneum and viable epidermis, which are compartments 2 and 3, respectively, in Guy and Hadgraft's model). As the descriptors used to populate a PBPK model are tissue specific, they are able to yield mechanistic insight into the ADME process and to allow extrapolation of the model to other chemicals.

Diffusion-based pharmacokinetic models, for example by Cooper (1976), have been applied to the uptake and absorption of exogenous chemicals into and across the skin. Cooper used a single-compartment model and assumed diffusion through the skin according to the flux of solutes exiting the membrane per unit area, $J_s(t)$, deriving Equation 17.3.

$$J_s(t) = -D_m \frac{\partial C_m}{\partial x}|x = h_m = k_p C_v \left[1 + 2\sum_{n=1}^{\infty}(-1)^n \cdot exp\left(-\frac{t}{t_d}\pi^2 n^2\right)\right] \quad (17.3)$$

where

$J_s(t)$ is the flux of solutes exiting the membrane per unit area of the membrane
D_m is the diffusion coefficient
C_m is the solute (penetrant) concentration in the stratum corneum under equilibrium
h_m is the diffusion path length, also often defined as the membrane thickness
k_p is the permeability coefficient
C_v is the solute concentration in its vehicle under equilibrium, normally in a volume V_v, where $V_v = \infty$
t is the time
t_d is the diffusion time, and is defined as $t_d = \frac{h_m^2}{D_m}$.

One early example of this in the field of transdermal drug delivery is the Guy and Hadgraft compartmental model (Guy and Hadgraft 1982). It should be noted that in this model, passage of the chemical is possible in both directions. A more comprehensive range of compartmental models of skin absorption was presented by Roberts et al. (1999).

Subsequently, Roberts et al. critiqued the limitations of single-compartment PK models of skin permeation, suggesting that they may be limited in describing the whole process of skin absorption. They proposed Equation 17.4:

$$\hat{J}s(s) = \frac{k_p C_v}{s} \cdot \frac{\sqrt{st_d}}{\sinh\left(\sqrt{st_d}\right) + \frac{\sqrt{st_d}}{st_d\lambda_2 + \lambda_1} \cdot \cosh\left(\sqrt{st_d}\right)} \quad (17.4)$$

where

$\lambda_1 = Cl_R \cdot K_m / \left(k_p A K_R\right)$

$\lambda_2 = V_R / \left(K_R / V_m\right)$

Cl_R is the clearance (ml/min) of solution containing solute from the receptor phase
K_m is the partition coefficient between the membrane and the vehicle (C_m/C_V)
A is the area of application
K_R is the partitioning between the membrane and the receptor (C_m/C_R)
C_R is the concentration in the receptor
V_R is the volume of the receptor
V_m is the volume in the membrane
s is the Laplace variable (Roberts et al. 1999)
and other terms as defined in Equation 17.3 or discussed in (Roberts et al. 1999).[3]

Equation 17.4 is comparable to Guy and Hadgraft's model, as it takes into account the effectiveness of subcutaneous blood flow in removing the penetrant from the epidermis, as well as the accumulation of the penetrant *in vivo*, thus representing the *total* process of percutaneous absorption in the context of ADME modelling (Roberts et al. 1999).

PD models are quantitative expressions of the measurable effects of chemicals, typically at the cellular level. They relate a biological endpoint, or response, to the concentration of a chemical in a particular tissue, or compartment (Ritter et al. 2016; Golan et al. 2017). Similar to PBPK models, physiologically based pharmacodynamics (PBPD) models consider the compartmentalised environment of the body. PBPD models, like PBPK models, may be described by differential equations.

PBPK and PBPD models offer powerful methods to determine the doses within tissues, rather than the administered or applied dose. For example, it is widely understood in skin permeation that the vast majority of the applied dose does not get absorbed into and across the skin, so PBPK/PBPD models are able to consider internal doses rather than applied doses, as well as the processes of metabolism and elimination, thus presenting a more accurate and representative model of the ADME process. Such models also are route specific and allow extrapolation for chemicals via a particular route, and they have also been widely used for interspecies models (e.g. animal-to-human studies). Such biologically based dose-response models may be used to extrapolate exposure to toxic chemicals at much lower doses than those tested on animals (Ritter et al. 2016; Golan et al. 2017).

Clearly, as is common with all models, they exhibit limitations. These include the estimation of parameters which may not be measured, possibly introducing error into the estimate of toxicity (Moss et al. 2015). In the case where *in vitro* or *in vivo* descriptors are generated, the favoured approach may be to use a QSAR model rather than a PK/PD model. As with all models, extrapolation is a potential source of error, and this is particularly the case when extrapolating to higher or lower doses or between different mammalian species (e.g. from rodents to humans). A further discussion of PBPK/PBPD and biologically based dose-response models may be found elsewhere (Mager et al. 2003; Crump et al. 2010).

17.2.5 Quantitative Structure–Activity Relationships (QSARs)

QSARs have been reviewed extensively, and the reader is directed to Cronin and Schultz's work on the design of QSAR models (Cronin and Schultz 2003). Further, as the examples emphasised in this chapter focus on skin permeability, the reader is also directed to reviews of QSARs and similar methods in skin permeation (Moss et al. 2002, 2012, 2015).

QSARs relate the activity of a given endpoint (e.g. skin permeability) to the physicochemical properties of the molecules in a dataset, represented by a selection of relevant molecular descriptors. Most QSAR models present this relationship in the form of a discrete algorithm, which mathematically relates the endpoint to the relevant physicochemical descriptors, inferring that the chemicals in the dataset operate by the same mechanism. Toxicity (T) may be considered to be a function of the chemical properties of a chemical of interest (θ_p), Equation 17.5:

$$T = \theta_p \tag{17.5}$$

The beginning of this field is associated with the development of the model by Hansch et al. (1962). They related the log of chemical concentration (C) to the lipid solubility or lipophilicity (represented by the octanol-water partition coefficient, π) and the Hammett constant (σ, which is one of a number of constants which are used to characterise particular substituents in a molecule), Equation 17.6:

$$\log\left(\frac{1}{C}\right) = 4.08\pi - 2.14\pi^2 + 2.78\sigma + 3.36 \tag{17.6}$$

In such expressions, the positive or negative coefficient before a descriptor defines whether the relationship to the endpoint for a particular descriptor is positive or negative (Gramatica 2013). The initial work in the field of modelling percutaneous absorption was developed by El Tayar et al. (1991). A further example in the context of skin permeation this relationship is classically represented by the algorithm published by Potts and Guy (1992), which relates the skin permeability (as permeability coefficient, K_p) to the lipophilicity (as $\log K_{oct}$) and molecular weight (MW) of a penetrant of interest, Equation 17.7:

$$\log K_p(cm.sec^{(-1)}) = -6.3 + 0.71.\log K_{oct} - 0.0061.MW \quad [r^2 = 0.67] \qquad (17.7)^4$$

Equation 17.7 aims to represent a wide range of physicochemical properties (in the case of this expression, molecular weights and lipophilicities), or "chemical space", specifically $-3 < \log K_{oct} < +6$ and $18 < MW < 750$. As such, it is commonly defined as a global model which aims to relate the permeability of all chemicals in this chemical space to the relevant physicochemical factors. Further examples (e.g. Abraham et al. 1990, 1995, 1997, 1999; Potts and Guy 1995) and others reviewed by Moss et al. (2002) focus on specific chemicals or even congeneric series, and are commonly defined as local models relevant only to a particular chemical space or set of chemical structures; for example, an algorithm was derived (Potts and Guy 1995) ($n = 37$, based on data by Flynn 1990) from a subset of a previous study (Potts and Guy 1992) for nonelectrolytes only, Equation 17.8.

$$\log K_p\left(cm.sec^{-1}\right) = 0.0256.MV - 1.72\sum\alpha_2^H - 3.95\sum\beta_2^H - 4.85$$
$$\left[n = 37; r^2 = 0.94; s \text{ not given}; F = 165\right] \quad (17.8)$$

where
MV is molecular volume
$\sum\alpha_2^H$ is the solute hydrogen bond acidity
$\sum\beta_2^H$ is the solute hydrogen bond basicity.

Interestingly, in the context of skin permeation, this model presents a very different mechanism of permeation than Equation 17.7 in that it places a significant emphasis on hydrogen bonding rather than lipophilicity.

More broadly, in the field of percutaneous absorption a large number of QSAR models have been developed. It is not the aim of this chapter to discuss them all, but only the principles and issues in their design and their limitations.

QSARs are constructed either from relating an endpoint, such as skin permeability, to the physicochemical descriptors of a molecules which are generated by experiments or from predictive software packages. Relationships between the endpoint and the molecular properties are then sought. Although a wide range of parameters can be used to describe a molecule—in some cases, in excess of 2700 descriptors can be used (Moss et al. 2013)—the number of useful descriptors is significantly lower. This is due to redundancy in descriptors that are highly intercorrelated or biased, or with descriptors that are mechanistically difficult to interpret and use. In such cases, feature selection methods can be used to produce lists of relevant descriptors that do not introduce bias into the analysis (Lam et al. 2010). There is also the practical problem of nonexperts who might not be able to use models because they are unable to generate highly specific molecular descriptors as specialist software is often required to do this (Cronin and Schultz 2003). Thus, QSARs should be mechanistically interpretable by the end user.

QSARs are generally based on simple linear relationships between the descriptors and the endpoint. These include linear and multiple linear regression methods and partial least squares methods for continuous endpoints such as clinically measurable responses from patients. Categorical methods, such as discriminant analysis, have also been used for categorical endpoints (Pugh et al. 2005; Deeb and Goodarzi 2012; Devillers 2013). QSARs generally require large datasets to derive relationships between molecular properties and endpoints of interest, limiting their development or introducing datasets (and models) which are derived from collated data sources, introducing variance into the models and reducing their precision (Moss et al. 2015). One example of an approach to solve this issue is the application of manually derived hyperparameters in Machine Learning models of skin absorption, where in many cases models with as few as 14 chemicals are as predictive as models with 85 chemicals (Ashrafi et al. 2018).

More recently a range of nonlinear methods, in particular those associated with Machine Learning, have been employed in this field (Mitchell 2014). Commonly applied methods include artificial neural networks, support vector machines, clustering methods, Bayesian methods and K-nearest neighbour regression (Baert et al. 2007; Moss et al. 2012, 2015; Shah et al. 2016).

Linear methods are generally considered simpler and may require the use of fewer parameters than complex models (Ashrafi et al. 2018). They find more utility when applied to fields where modelling is not the core expertise and, despite limitations in predictive ability and mechanistic insight—which in the case of skin models began with the use of limited numbers of defined descriptors due to empirical findings dominating model development (Moss et al. 2015). On the other hand, these models are often criticised for being complex, lacking transparency (i.e. black box modelling), overfitting and offering little or no mechanistic insight; they have often been accused of having little or no real-world relevance, even though most processes in nature are inherently nonlinear (Mitragotri et al. 2011). Most if not all of these claims have been addressed in the wider literature, and nonlinear methods find significant use in QSAR modelling as they have been shown to provide both better estimates of skin permeability and insight to the underlying mechanism of skin permeation (Devillers 2013; Lam et al. 2010; Moss et al. 2015). Further, rudimentary visualisation methods are commonly applied to these problem domains in order to characterise the nature of the data. This, in turn, assists with method selection and dataset design (i.e. the removal of skew or other bias in the data; Lam et al. 2010; Sun et al. 2011; Moss et al. 2015).

Increasingly, many QSAR models are available as parts of various software packages, such as the OECD QSAR Toolbox (OECD 2015), HazardExpert (CompuDrug 2013), METEOR (Lhasa Ltd. 2018a) and Derek Nexus (Lhasa Ltd. 2018b). Specialist packages are also available to derive molecular descriptors, including MOE, WinMolconn, HYBOT and Dragon. Specific packages can be applied to particular endpoints to tailor model use correctly.

17.2.6 Nanoparticles and Toxicity

One of the largest emerging areas in recent years has been the application of nano-scale technologies to pharmaceutical drug delivery. Although some criticism regarding efficacy has been directed at "nanopharmaceutics" (Wilhelm et al. 2016; Park 2019), the field offers some potential. However, one significant emerging trend is the association of nanoparticles with toxicity. This is due to the change in physicochemical properties associated with nano-scale (10^{-9} m) particles, including immunological responses and the ability of nanoparticles to penetrate cell membranes because of their size, hence increasing cellular uptake. The large surface areas of nanoparticles may increase reactivity and bioavailability compared to larger particles; the shape of nanoparticles (e.g. nanotubes) may be considered to be more toxic in regions of the lung and surface modifications, including electrostatic changes, may influence toxicity. Nanoparticles have in some cases been shown to result in more significant toxicity than larger particles (Warheit 2008; Wani et al. 2011).

QSAR methods have been applied to model nanoparticle toxicity. In doing so, they apply specific descriptors which are relevant to nanoparticles, such as particle size range and zeta potential, morphology (shape), surface area, solubility, comparative bioavailability and surface properties (notably electrostatic properties). Given the very nature of nanoparticles, it is difficult to characterise them in quantitative models in the same way as small organic chemicals due to the diversity of structures and the range of mechanisms of action, as well as a significant paucity of data. In such cases, read-across methods can be used as they do not require large datasets (Gajewicz et al. 2015). Considerations of how the surface and core of the nanoparticles differ and how both can be accounted for in quantitative models are also significant (Winkler et al. 2013). QSAR models, often called quantitative structure–nanotoxicity relationships (QSNRs) have been developed for cytotoxicity (Pathakoti et al. 2014) and the EC_{50} of silver nanoparticles (Silva et al. 2014).

17.3 DEVELOPMENT OF SUCCESSFUL *IN SILICO* PREDICTIVE MODELS

Cronin and Schultz commented that the QSAR and related fields have grown over the last 20 years or so (Cronin and Schultz 2003). However, until recently there was little attempt at developing a framework for model construction, particularly for a regulatory context. The literature is replete with many models, most of which are not validated and which have little or no utility in the broader field. It may even be argued that they can be misleading and lead to incorrect inferences, both in terms of toxicity and mechanisms of toxicity. The aim of any of the models highlighted in this chapter is to develop them in such a way as to provide an accurate estimate of toxicity. Thus, model selection and model design are highly significant in this process. Models are most accurate when used within their defined limits and should be extrapolated with care; even then, it has been shown that some models have limited applicability even within their stated range of validity, so validation remains essential (Moss et al. 2006). Cronin and Schultz further commented that model transparency allows readily interpretable mechanistic insight to the biological process being modelled to be developed. Limitations of models are mostly due to their misuse or their construction from inappropriate data. Hence models should exhibit a series of essential and desirable criteria in order to ensure full utility.

The framework of Cronin and Schultz was based around three main components associated with predominately QSAR or related models: the veracity of the toxicity data, the derivation or measurement of molecular descriptors and the use of the most suitable method to develop the model. To this we have added further considerations from the more recent literature. To this list can be added the issue of model overfitting, particularly given the growth of nonlinear methods, consideration of the applicability domain (here, presented in the context of Potts and Guy's "chemical space") and model validation, summarised in Table 17.1.

17.3.1 Sources of Data: "Garbage In, Garbage Out"

The above vernacular expression sums up rather bluntly the issue of data quality in model development. Although well-defined endpoints can be used their derivation, particularly the variance associated with experimentally derived laboratory data, must be considered in the model development and in the consideration of any outputs. Hence consistency in the use of experimental protocols, as well as transparency in the source data from which models are collated and developed, is key. In the context of skin permeation, the difficulties in this matter were highlighted by Chilcott et al. (2005), who illustrated the challenges of obtaining, and by inference, modelling experimental data. A consistency of approach for model development (especially in the collection of data used to build QSAR models) may result in the adoption of good practice across a range of fields in which modelling is used. One issue that is commonly discussed is the availability and transparency of data, although those who produce models by abstracting the data of others from the literature often are unaware of particular concerns, notably any confidentiality in the data source (e.g. if it is from a regulatory submission or is of competitive value) or that the data being published are generally not being published to provide for modellers, but rather to address specific research questions in particular fields. An understanding of these issues, which were highlighted in part by Chilcott et al., is essential in the consistent and correct use of published data.

The model should be considered, with the modeller, by subject experts who are able to understand the implications of a particular dataset, particularly in regard to its source or sources,

Table 17.1 Essential and Desirable Features for a Quantitative Model for Percutaneous Absorption

Essential properties	Desirable properties
A well-defined and readily measurable endpoint (i.e. in the case of skin permeation this would be the permeability coefficient, k_p, or flux [J_{max} or J_{ss}]).	Do not extrapolate or use the model beyond its original domain (i.e. its applicability domain), both in terms of the range of physicochemical descriptors used or the class(es) of chemicals from which the model was developed.
A chemically and/or biologically diverse dataset that is equally representative of the "chemical space" associated with the model.	The model should be correctly used and interpreted within the context of the model's findings and of prior knowledge in the field. Thus, a quantitative model may require subjective interpretation by subject expert(s) in addition to presenting a quantitative model.
The dataset should be split into suitably representative training and test sets which are separate and consistent.	
Physiochemical descriptors that are, in number and type, consistent with the skin permeability endpoint (i.e. molecular features that have been shown experimentally or empirically to be significant in the permeation process).	If black box methods, which are of increasing value to the field, are to be used, then they should be transparently generated and validated.
The method of analysis should be appropriate and produce a model that is statistically valid. The statistical performance of each model should, in statistical terms, be fully described.	Appreciation of the precision of the model when interpreting its output. In the case of models featuring algorithms mean and standard deviation values for each significant descriptor should be provided, and where overfitting is a concern, r^2_{adj} should be used rather than r^2.
The dataset should be constructed to avoid issues of bias (e.g. skew) in the data.	
The model and method should have a strong mechanistic basis.	A "global" or "local" model should be used as appropriate.
Development of models by multidisciplinary groups of experts.	

Source: Modified from Cronin and Schultz (2003).

relevance to the model and overall quality, when considering both the statistical quality of the resultant model and the relevance of the model itself to represent the process being examined. Thus, pragmatism and multidisciplinary collaboration are essential in understanding the quality and value of a model in relation to its endpoint. For example, it might be sensible to use a small number of simple descriptors that have mechanistic relevance to the endpoints of interest rather than simply using all available molecular descriptors which are known to be largely irrelevant or introduce bias into the analysis due to highly correlated descriptors; thus, feature selection methods should be used to design models based on complex collections of molecular descriptors.

17.3.2 Outliers and Overfitting

Outliers are identified by their lack of statistical fit. It is often considered that this lack of fit is the result of a different mechanistic process underpinning data identified as outliers in a model. However, it is important to consider both the data and the selection of the method of analysis in this context. In the context of skin permeation, steroids were classically thought to be outliers, permeating the skin by a different mechanism of action. This postrationalisation came from the analysis of the first Potts and Guy model which showed that steroids were outliers (Potts and Guy 1992). This model was based on Flynn's dataset which was found by Johnson et al. to be erroneous by at least an order of magnitude (Flynn 1990; Johnson et al. 1995). Further, studies which demonstrate the failure of QSAR models based on linear regression even inside their applicability domains (Moss et al. 2006, 2009) clearly indicate that rather than assuming the use of a certain method of analysis, rudimentary characterisation of the data should be conducted beforehand, and this would inform the choice of method used to analyse the data and produce the model (Sun et al. 2011).

There are established and well-understood methods for the removal of outliers. For regression-based methods, these include the identification of outliers, identification based on high standard residuals. Once identified in this way, outliers are often removed, either individually and subjectively by consideration of each outlier, a process which requires subject-specific knowledge for it to be accomplished meaningfully. The benefit *and* risk with considering and removing outliers is that their removal can result in a better model, statistically speaking. Therefore, the removal of outliers has to be justified and simply not used as a tool to improve model fit. Overall, however, Johnson's consideration of the quality of the Flynn dataset should be considered when manipulating the model by selecting chemicals for removal.

It often follows that removing outliers can result in overfitting of the data by the model. Normally this is associated with an excellent fit for data in the model and a poor fit with the model for other data. The use of complex or inappropriately gathered or derived data, or even a large number of molecular descriptors, can result in overfitting. Resulting models are thus rendered meaningless, often by being unnecessarily complex, and may find random errors that become addressed as specific mechanistic points, as discussed above in the discussion of erroneous data.

Overfitting is usually associated with a high coefficient of determination (the *r*-value, normally as r^2) and standard error of the estimate (Moss et al. 2002; Gramatica 2013).[5] This can be addressed by the use of the adjusted value for fit, r^2_{adj}. This limits the impact of increases in r^2 that results from the addition of further, irrelevant molecular descriptors to the model.

17.3.3 Data from Experimental or Clinical Sources

The nature of the input data used to build a computational model clearly affects the nature of the output data which the model generates. Further, biological and clinical processes are by their very nature variable within and between populations, and many of these processes are nonlinear. Thus, standardisation of models should account for this, and models should be presented in such a manner as to build this variability into their outputs. This can be achieved, for example, by including the error associated with each term in the model, rather than offering a simple algorithm (e.g. Equation 17.7; Geinoz et al. 2004; Sun et al. 2011). Therefore, once it is accepted that biological measurements are subject to error, it should be considered that this will be reflected in the output and, if different descriptors or protocols are responsible for different degrees of error, then the consideration of particular descriptors might play a role in improving the quality of the model (Ashrafi et al. 2018).

17.3.4 Applicability Domain

The applicability domain is "a theoretical region in physicochemical space for which a QSAR model should make predictions with a given reliability" (Gramatica 2013). An example of this is the stated range of validity for the Potts and Guy (1992) model, discussed above. The applicability domain defines the type of chemicals, based on the relevant physicochemical descriptors, whose key endpoints should be well estimated by a particular model. Clearly, global (Moss et al. 2009) and local models (Shah et al. 2016) have different applicability domains, and more recent studies have indicated that despite the traditional preference for global models, local models might provide more accurate estimates of endpoints for certain chemicals (Shah et al. 2016) and that highly predictive models can be produced from small datasets (Ashrafi et al. 2018). Hence applicability domain is a key parameter in the selection of an appropriate model.

Applicability domains have been developed by Hewitt and Ellison, who defined the key domains of interest as being the descriptor choice, the nature of structural fragments present on a molecule, structural similarities within a dataset, mechanistic outcomes and methods which consider metabolic aspects of the ADME process (Hewitt and Ellison 2010). The simplified molecular-input line entry system (SMILES) has also been considered as an applicability domain (Toropov and Benfenati 2008).

17.3.5 Descriptor Selection and Interpretation and Dataset Design

As discussed above, selection of descriptors is vital to the production of a relevant and accurate model. Use of descriptors which are repetitive (i.e. chemically similar, often describing similar or even the same properties of a molecule) and/or which are highly colinear should not be used (Basak et al. 2000; Romanelli et al. 2000; Patel and Cronin 2001; Moss et al. 2013). Further, Romanelli et al. (2000) commented that regression analysis based on multivariate methods should not be developed by using colinear descriptors, as this will produce a model with an artificially high regression coefficient. There is no defined boundary at which covariance is considered to be an issue, but the general consensus would be to ensure that it is as low as is practicable. Cronin and Schultz commented that the collinearity of descriptors should be lower than the statistical fit of the model and that the r^2_{adj} value (adjusted for degrees of freedom), not the r^2 value, should be reported (Cronin and Schultz 2003). This is a somewhat theoretical concept, as the ability to apply an abstract concept to real-world data is a challenge. This is highlighted in the field of percutaneous absorption, where the majority of the available data sits within a relatively narrow "chemical space" (i.e. $-1.0 < \log P < 3.0$, or $MW < 500$), and as such it might be more realistic to consider such models of skin permeation local models rather than global models (Moss et al. 2015).

17.4 CONCLUSIONS

A wide range of *in silico* toxicological models exists for the estimation of various toxic endpoints. In this chapter, we can only introduce the reader to this field and prompt further, more specific reading in this area. To this end, a list of further reading and a comprehensive list of references are provided below which signpost specific methods in a number of fields, and the reader is encouraged to explore the relevant methods in more detail.

The key points of this chapter are that the selection of the correct method, the selection of data and descriptors with which to populate a model, the selection of the appropriate methods of analysis, followed by the expert analysis and contextualisation of output form the central platform to the successful use of any model. Most models rely on the "correct" input to yield a relevant output; not structuring a dataset correctly or adopting an inappropriate method (e.g. based on the nature of data or the size of the available dataset) may lead to the production of misleading or frankly incorrect outcomes, the implications of which might be misleading.

In his pioneering work on the development of quantitative relationships between the physicochemical properties of a molecule and its skin permeability, Professor Gordon Flynn commented that his models should be considered to be "interim tool[s]" to be continually developed, added to and modified as new data, or even new methods which might suggest a more suitable method of analysis, are understood and established. To this end, the use of explicit and unchanging algorithms might be of limited value and the newer, black box methods of Machine Learning may provide a more dynamic platform from which the field of *in silico* modelling may develop and grow to address its core aim of producing high predictive models of toxicologically relevant endpoints. However, such methods remain challenged by their perceived lack of transparency. It is possible that wider investigation and application of such models to toxicological endpoints will go some way to addressing these challenges.

NOTES

1 For example, the eChemical portal contains a comprehensive list of searchable databases: www.echemportal.org/echemportal/propertysearch/page.action?pageID=2.
2 Further information on REACH can be found at www.hse.gov.uk/reach/index.htm.
3 From Roberts et al. (1999, pp. 1–16).
4 The *r*-value, normally as *r* or r^2, which defines the strength of the relationship between variables, with the strongest relationship resulting in a value of 1.
5 The *r*-value, normally as *r* or r^2, which defines the strength of the relationship between variables, with the strongest relationship resulting in a value of 1; the standard error of the estimate (which characterises the accuracy of predictions by comparing the similarity of the regression line developed in the model with the experimental data from which it is derived).

FURTHER READING

Books

Golan, D. E., A. H. Tashjian, E. J. Armstrong, and A. W. Armstrong. 2017. *Principles of Pharmacology—The Pathophysiologic Basis of Drug Therapy*, 4th ed. Philadelphia: Wolters Kluwer Health/Lippincott Williams & Wilkins.

Moss, G. P., D. R. Gullick, and S. C. Wilkinson. 2015. *Predictive Methods in Percutaneous Absorption*. Heidelberg: Springer AG.

Reisfeld, B., and A. N. Mayeno. 2013. *Computational Toxicology*, Vol. 930. New York: Humana Press.

Ritter, J. M., R. J. Flower, G. Henderson, and H. P. Rang. 2016. *Rang & Dale's Pharmacology*. 8th ed. London: Churchill Livingstone.

Review Papers

Abraham, M. H., H. S. Chadha, and R. C. Mitchell. 1995. The factors that influence skin penetration of solutes. *J Pharm Pharmacol* 47: 8–16.

Cronin, M. T. D., and T. W. Schultz. 2003. Pitfalls in QSAR. *J Mol Struct (THEOCHEM)* 622: 39–51.

Mitchell, J. B. O. 2014. Machine learning methods in chemoinformatics. *WIREs Comput Mol Sci* 4: 468–81.

Raies, A. B., and V. B. Bajic. 2016. *In silico* toxicology: Computational methods for the prediction of chemical toxicity. *Wiley Interdiscip Rev. Computl Mol Sci* 6: 147–72.

Key Articles

Abraham, M. H., H. S. Chadha, F. Martins, R. C. Mitchell, M. W. Bradbury, and J. A. Gratton. 1999. Hydrogen bonding Part 46. A Review of the correlation and prediction of transport properties by an LFER method: physicochemical properties, brain penetration and skin permeability. *Pestic Sci* 55: 78–88.

Abraham, M. H., H. S. Chadha, and R. C. Mitchell. 1995. The factors that influence skin penetration of solutes. *J Pharm Pharmacol* 47: 8–16.

Abraham, M. H., F. Martins, and R. C. Mitchell. 1997. Algorithms for skin permeability using hydrogen bond descriptors: The problem of steroids. *J Pharm Pharmacol* 49: 858–65.

Abraham, M. H., G. S. Whiting, R. M. Doherty, and W. J. Shuely. 1990. Hydrogen bonding Part 13.A new method for characterisation of GLC stationary phases—The Laffort effect. *J Chem Soc Perkin Trans* 2(2): 1451–60.

Barratt, M. D. 1995. Quantitative structure–activity relationships for skin permeability. *Toxicol In Vitro* 9: 27–37.

Flynn, G. L. 1990. Physicochemical determinants of skin absorption. In *Principles of Route-to-Route Extrapolation for Risk Assessment*, ed. T. R. Gerrity and C. J. Henry, 93–127. New York: Elsevier.

Lepailleur, A., G. Poezvara, and R. Bureau. 2013. Automated detection of structural alerts (chemical fragments) in (eco)toxicology. *Comput Struct Biotechnol J* 5: 1–8.

Potts, R. O., and R. H. Guy. 1995. A predictive algorithm for skin permeability: the effects of molecular size and hydrogen bond activity. *Pharm Res* 12: 1628–33.

Sung, J. H., B. Srinivasan, M. B. Esch, et al. 2014. Using PBPK guided "Body-on-a-Chip" systems to predict mammalian response to drug and chemical exposure. *Exp Biol Med* 239: 1225–39.

BIBLIOGRAPHY

Abraham, M. H., H. S. Chadha, F. Martins, R. C. Mitchell, M. W. Bradbury, and J. A. Gratton. 1999. Hydrogen bonding Part 46. A Review of the correlation and prediction of transport properties by an LFER method: Physicochemical properties, brain penetration and skin permeability. *Pestic Sci* 55: 78–88.

Abraham, M. H., H. S. Chadha, and R. C. Mitchell. 1995. The factors that influence skin penetration of solutes. *J Pharm Pharmacol* 47: 8–16.

Abraham, M. H., F. Martins, and R. C. Mitchell. 1997. Algorithms for skin permeability using hydrogen bond descriptors: The problem of steroids. *J Pharm Pharmacol* 49: 858–65.

Abraham, M. H., G. S. Whiting, R. M. Doherty, and W. J. Shuely. 1990. Hydrogen bonding Part 13. A new method for characterisation of GLC stationary phases—The Laffort effect. *J Chem Soc Perkin Trans* 2(2): 1451–60.

Ashrafi, P., Y. Sun, N. Davey, R. G. Adams, S. C. Wilkinson, and G. P. Moss. 2018. Model fitting for small skin permeability data sets: Hyperparameter optimisation in Gaussian Process Regression. *J Pharm Pharmacol* 70: 361–73.

Baert, B., E. Deconinck, M. Van Gele, et al. 2007. Transdermal penetration behaviour of drugs: CART-clustering, QSPR and selection of model compounds. *Bioorgan Med Chem* 15: 6943–55.

Basak, S. C., A. T. Balaban, G. D. Grunwald, and B. D. Gute. 2000. Topological indices: Their nature and mutual relatedness. *J Chem Inf Comput Sci* 40: 891–8.

Benigni, R., C. Bossa, and O. Tcheremenskaia. 2013. Nongenotoxic carcinogenicity of chemicals: Mechanisms of action and early recognition through a new set of structural alerts. *Chem Rev* 113: 2940–57.

Borgelt, C., and M. R. Berthold. 2002. Mining molecular fragments: Finding relevant substructures of molecules. In *IEEE International Conference on Data Mining, Maebashi City, Japan*, eds. V. Kumar, S. Tsurnoto, N. Zhong, P. S. Yu, and X. Wu, 51–8. Piscataway, NJ.

Chilcott, R. P., N. Barai, A. E. Beezer, et al. 2005. Inter- and intra-laboratory variation of in vitro diffusion cell measurements: An international multicenter study using quasi-standardised methods and materials. *J Pharm Sci* 94: 632–8.

CompuDrug. 2013. *HazardExpert Pro.* www.compudrug.com/hazardexpertpro (accessed November 12, 2020).

Cooper, E. R. 1976. Pharmacokinetics of skin penetration. *J Pharm Sci* 65: 1396–7.

Cronin, M. T. D., and T. W. Schultz. 2003. Pitfalls in QSAR. *J Mol Struct: THEOCHEM* 622: 39–51.

Crump, K. S., C. Chen, W. A. Chiu, et al. 2010. What role for biologically based dose–response models in estimating low dose risk? *Environ Health Perspect* 118: 585–8.

Cunningham, A. R., S. T. Moss, S. A. Iype, G. Qian, S. Qamar, and S. L. Cunningham. 2008. Structure-activity relationship analysis of rat mammary carcinogens. *Chem Res Toxicol* 21: 1970–82.

Deeb, O., and M. Goodarzi, M. 2012. *In silico* quantitative structure toxicity relationship of chemical compounds: Some case studies. *Curr Drug Saf* 7: 289–97.

Devillers, J. 2013. Methods for building QSARs. In *Computational Toxicology*, Vol. 930, ed. B. Reisfeld and A. N. Mayeno, 3–27. New York: Humana Press.

Dimitrov, S., and O. Mekenyan. 2010. An introduction to read-across for the prediction of the effects of chemicals. In *In Silico Toxicology: Principles and Applications*, ed. M. T. D. Cronin and J. C. Madden, 372–84. Cambridge: The Royal Society of Chemistry.

El Tayar, N., R. S. Tsai, B. Testa, P. A. Carrupt, C. Hansch, and A. Leo. 1991. Percutaneous penetration of drugs—A quantitative structure-permeability relationship study. *J Pharm Sci* 80: 744–9.

European Chemicals Agency. 2017. *ECHA Read-Across Assessment Framework (RAAF).* https://echa.europa.eu/documents/10162/13628/raaf_en.pdf (accessed November 13, 2020).

Falk-Filipsson, A., A. Hanberg, K. Victorin, M. Warholm, and M. Wallén. 2007. Assessment factors—Applications in health risk assessment of chemicals. *Environ Res* 104: 108–27.

Flynn, G. L. 1990. Physicochemical determinants of skin absorption. In *Principles of Route-to-Route Extrapolation for Risk Assessment*, ed. T. R. Gerrity and C. J. Henry, 93–127. New York: Elsevier.

Gajewicz, A., M. T. D. Cronin, B. Rasulev, J. Leszczynski, and T. Puzyn. 2015. Novel approach for efficient predictions properties of large pool of nanomaterials based on limited set of species: Nano-readacross. *Nanotechnology* 26: 015701.

Geinoz, S., R. Guy, B. Testa, and P. Carrupt. 2004. Quantitative structure-permeation relationships (QSPeRs) to predict skin permeation: A critical evaluation. *Pharm Res* 21: 83–92.

Golan, D. E., A. H. Tashjian, E. J. Armstrong, and A. W. Armstrong. 2017. *Principles of Pharmacology: The Pathophysiologic Basis of Drug Therapy*, 4th ed. Philadelphia: Wolters Kluwer Health/ Lippincott Williams & Wilkins.

Gramatica, P. 2013. On the development and validation of QSAR models. In *Computational Toxicology*, Vol. 930, ed. B. Reisfeld and A. N. Mayeno, 499–526. New York: Humana Press.

Greene, N., P. N. Judson, J. J. Langowski, and C. A. Marchant. 1999. Knowledge-based expert systems for toxicity and metabolism prediction: DEREK, StAR and METEOR. *SAR QSAR Environ Res* 10: 299–314.

Guy, R. H., and J. Hadgraft. 1982. Percutaneous metabolism with saturable enzyme kinetics. *Int J Pharm* 11: 187–97.

Hansch, C., P. P. Maloney, T. Fujita, and R. M. Muir. 1962. Correlation of biological activity of phenoxyacetic acids with Hammett substituent constants and partition coefficients. *Nature* 194: 178–80.

Hewitt, M., and C. M. Ellison. 2010. Developing the applicability domain of *in silico* models: Relevance, importance and methods. In *Silico Toxicology: Principles and Applications*, eds. M. T. D. Cronin and J. C. Madden, 301–33. Cambridge: The Royal Society of Chemistry.

Johnson, M. E., D. Blankstein, and R. Langer. 1995. Permeation of steroids through human skin. *J Pharm Sci* 84: 1144–6.

Klopman, G. 1984. Artificial intelligence approach to structure-activity studies. Computer automated structure evaluation of biological activity of organic molecules. *J Am Chem Soc* 106: 7315–21.

Kortagere, S. (Ed.). 2013. *In Silico Models for Drug Discovery*, Vol. 993. New York: Humana Press.

Lam, L. T., Y. Sun, N. Davey, et al. 2010. The application of feature selection to the development of Gaussian Process models for percutaneous absorption. *J Pharm Pharmacol* 62: 738–49.

Lepailleur, A., G. Poezvara, and R. Bureau. 2013. Automated detection of structural alerts (chemical fragments) in (eco)toxicology. *Comput Struct Biotechnol J* 5: 1–8.

Lhasa. 2018a. *Meteor Nexus*. www.lhasalimited.org/products/meteor-nexus.htm (accessed November 12, 2019).

Lhasa. 2018b. *Derek Nexus*. www.lhasalimited.org/products/derek-nexus.htm (accessed November 12, 2020).

Mager, D. E., E. Wyska, and W. J. Jusko. 2003. Diversity of mechanism-based pharmacodynamic models. *Drug Metab Dispos* 31: 510–19.

Milan, C., O. Schifanella, A. Roncaglioni, and E. Benfenati. 2011. Comparison and possible use of *in silico* tools for carcinogenicity within REACH legislation. *J Environ Health Part C, Environ Carcinogen Ecotoxicol Rev* 29: 300–23.

Mitchell, J. B. O. 2014. Machine Learning methods in chemoinformatics. *WIREs Comput Mol Sci* 4: 468–81.

Mitragotri, S., Y. G. Anissimov, A. L. Bunge, et al. 2011. Mathematical models of skin permeability: An overview. *Int J Pharm*. 418: 115–29.

Modi, S., M. Hughes, A. Garrow, and A. White. 2012. The value of *in silico* chemistry in the safety assessment of chemicals in the consumer goods and pharmaceutical industries. *Drug Discov Today* 17: 134–42.

Moss, G. P., J. C. Dearden, H. Patel, and M. T. D. Cronin. 2002. Quantitative Structure-Permeability Relationships (QSPRs) for percutaneous absorption. *Toxicol in Vitro* 16: 299–317.

Moss, G. P., D. R. Gullick, P. A. Cox, et al. 2006. Design, synthesis and characterisation of captopril prodrugs for enhanced percutaneous absorption. *J Pharm Pharmacol* 58: 167–77.

Moss, G. P., D. R. Gullick, and S. C. Wilkinson. 2015. *Predictive Methods in Percutaneous Absorption*. Heidelberg: Springer AG.

Moss, G. P., A. J. Shah, R. G. Adams, et al. 2012. The application of discriminant analysis and Machine Learning methods as tools to identify and classify compounds with potential as transdermal enhancers. *Eur J Pharm Sci* 45: 116–27.

Moss, G. P., Y. Sun, N. Davey, R. Adams, W. J. Pugh, and M. B. Brown. 2009. The application of Gaussian Processes to the prediction of percutaneous absorption. *J Pharm Pharmacol* 61: 1147–53.

Moss, G. P., Y. Sun, N. Davey, R. G. Adams, S. C. Wilkinson, and D. R. Gullick. 2013. The application of Gaussian Processes in the prediction of permeability across a polydimethylsiloxane membrane. In *Advances in the Dermatological Sciences*, ed. K. R. Brain and R. Chilcott, 376–83. Cambridge: The Royal Society of Chemistry.

Moss, G. P., S. C. Wilkinson, and Y. Sun. 2012. Mathematical modelling of percutaneous absorption. *Curr Opin Colloid Interface Sci* 17: 166–72.

OECD. 2015. *The OECD QSAR Toolbox*. www.oecd.org/chemicalsafety/risk-assessment/oecd-qsar-toolbox.htm (accessed November 12, 2020).

OECD. 2018. *Grouping of Chemicals: Chemical Categories and Read-Across*. www.oecd.org/chemicalsafety/risk-assessment/groupingofchemicalschemicalcategoriesandread-across.htm (accessed November 12, 2020).

Park, K. 2019. The beginning of the end of the nanomedicine hype. *J Control Release* 305: 221–2.

Patel, H., and M. T. D. Cronin. 2001. A novel index for the description of molecular linearity. *J Chem Inf Comput Sci* 41: 1228–36.

Pathakoti, K., M. J. Huang, J. D. Watts, X. He, and H. M. Hwang. 2014. Using experimental data of *Escherichia coli* to develop a QSAR model for predicting the photo-induced cytotoxicity of metal oxide nanoparticles. *J Photochem Photobiol B: Biol* 130: 234–40.

Potts, R. O., and R. H. Guy. 1992. Predicting skin permeability. *Pharm Res* 9: 663–9.

Potts, R. O., and R. H. Guy. 1995. A predictive algorithm for skin permeability: The effects of molecular size and hydrogen bond activity. *Pharm Res* 12: 1628–33.

Pugh, W. J., R. Wong, F. Falson, B. B. Michniak, and G. P. Moss. 2005. Discriminant analysis as a tool to identify compounds with potential as transdermal enhancers. *J Pharm Pharmacol* 57: 1389–96.

Raies, A. B., and V. B. Bajic. 2016. *In silico* toxicology: Computational methods for the prediction of chemical toxicity. *Wiley Interdiscip Rev Comput Mol Sci* 6: 147–72.

REACH Online. 2009. *EcoMole Ltd*. www.reachonline.eu/ (accessed November 12, 2020).

Reisfeld, B., and A. N. Mayeno. 2013. *Computational Toxicology*, Vols. I and II. New York: Humana Press.

Ritter, J. M., R. J. Flower, G. Henderson, and H. P. Rang. 2016. *Rang & Dale's Pharmacology*, 8th ed. London: Churchill Livingstone.

Roberts, M. S., Anissimov, Y. G., and Gonsalvez, R. A. 1999. Mathematical models in percutaneous absorption. In *Percutaneous Absorption*, 3rd ed. eds. R. L. Bronaugh and H. I. Maibach, 3–56. New York: Marcel Dekker, Inc./CRC Press.

Romanelli, G. P., L. F. R. Cafferata, and E. A. Castro. 2000. An improved QSAR study of toxicity of saturated alcohols. *J Mol Struct: THEOCHEM* 504: 261–5.

Shah, A., Y. Sun, R. G. Adams, N. Davey, S. C. Wilkinson, and G. P. Moss. 2016. Support Vector Regression to estimate the permeability enhancement of potential transdermal enhancers. *J Pharm Pharmacol* 68: 170–84.

Silva, T., L. R. Pokhrel, B. Dubey, T. M. Tolaymat, K. J. Maier, and X. Liu. 2014. Particle size, surface charge and concentration dependent ecotoxicity of three organocoated silver nanoparticles: Comparison between general linear model-predicted and observed toxicity. *Sci Total Environ* 468–469: 968–76.

Sun, Y., G. P. Moss, N. Davey, R. Adams, and M. B. Brown. 2011. The application of stochastic machine learning methods in the prediction of skin penetration. *Appl Soft Comput* 11: 2367–75.

Sung, J. H., B. Srinivasan, M. B. Esch, et al. 2014. Using PBPK guided "Body-on-a-Chip" systems to predict mammalian response to drug and chemical exposure. *Exp Biol Med* 239: 1225–39.

Toropov, A. A., and E. Benfenati. 2008. Additive SMILES-based optimal descriptors in QSAR modelling bee toxicity: Using rare SMILES attributes to define the applicability domain. *Bioorg Med Chem* 16: 4801–9.

Toropov, A. A., A. P. Toropova, I. Raska Jr, D. Leszczynska, and J. Leszczynski. 2014. Comprehension of drug toxicity: Software and databases. *Comput Biol Med* 45: 20–5.

Valerio, L. G. Jr. 2009. *In silico* toxicology for the pharmaceutical sciences. *Toxicol Appl Pharmacol* 241: 356–70.

Wani, M. Y., M. A. Hashim, F. Nabi, and M. A. Malik. 2011. Nanotoxicity: Dimensional and morphological concerns. *Adv Phys Chem* 450912.

Warheit, D. B. 2008. How meaningful are the results of nanotoxicity studies in the absence of adequate material characterization? *Toxicol Sci* 101: 183–5.

Wilhelm, S., A. J. Tavares, Q. Dai, et al. 2016. Analysis of nanoparticle delivery to tumours. *Nat Rev Mater* 94: 16.

Williams, D. P., and D. J. Naisbitt. 2002. Toxicophores: Groups and metabolic routes associated with increased safety risk. *Curr Opin Drug Discov Dev* 5: 104–15.

Wilson, A. G. E. (Ed.). 2011. *New Horizons in Predictive Toxicology: Current Status and Application.* Cambridge: The Royal Society of Chemistry.

Winkler, D. A., E. Mombelli, A. Pietroiusti, et al. 2013. Applying quantitative structure—activity relationship approaches to nanotoxicology: Current status and future potential. *Toxicology* 313: 15–23.

Woo, Y. T., and D. Y. Lai. 2005. Oncologic: A mechanism based expert system for predicting the carcinogenic potential of chemicals. In *Predictive Toxicology—The Book*, ed. C. Helma, 389–419. Boca Raton, FL: CRC Press.

Woo, Y. T., D. Y. Lai, M. F. Argus, and J. C. Arcos. 1995. Development of structure activity relationship rules for predicting carcinogenic potential of chemicals. *Toxicol Lett* 79: 219–28.

Woo, Y. T., D. Y. Lai, M. F. Argus, and J. C. Arcos. 1996. Carcinogenicity of organophosphorous pesticides/compounds: an analysis of their structure activity relationships. *J Environ Sci Health, Part C Environ Carcinog Ecotoxicol Rev* 14: 1–42.

Yan, X., and J. Han. 2002. gSpan: Graph-based substructure pattern mining. In *IEEE International Conference on Data Mining, Maebashi City, Japan*, eds. V. Kumar, S. Tsurnoto, N. Zhong, P. S. Yu, and X. Wu, 721–4. Piscataway, NJ. http://dm.kaist.ac.kr/kse625/resources/Yan_2002.pdf (accessed November 12, 2020).

18 Human Biomonitoring

Antonio Peña-Fernández, María de los Ángeles Peña and Marcus S. Cooke

CONTENTS

18.1 INTRODUCTION TO HUMAN BIOMONITORING

The measure of chemical substances is called "monitoring". Thus, the measure of chemicals and/or their metabolites in human tissues, organs and fluids is known as human biological monitoring or human biomonitoring (HBM), and it is used for a better estimate of exposure from multiple sources and routes. HBM started in 1890 with the detection of lead (Pb) in exposed workers' blood and urine (Müller et al. 2014), and since then it has become critical for occupational and environmental health, as HBM can provide information on the total exposure to toxic substances (i.e. the sum of all routes of exposure to a given pollutant; Olmedo et al. 2010). The impact of human biomonitoring to protect human health and the environment has been demonstrated, for example, in the reduction and elimination of Pb in gasoline (Angerer et al. 2007), and in the prohibition of persistent organochlorine compounds that present severe consequences to human health even to very low doses, such as dichlorodiphenyltrichloroethane (DDT) or polychlorinated biphenyls (PCBs; Eskenazi et al. 2009). Therefore, human biomonitoring is an invaluable tool in the development of robust public health policies and environmental medicine when combined with available epidemiology, toxicology or pharmacokinetic modelling data (Olmedo et al. 2010). HBM also has tremendous utility in estimating exposure and risk for health prevention and therefore, for the development of preventive policies and risk management (Olmedo et al. 2010). Thus, in recent years there has been increased interest from the European Union (EU) in the development of large HBM studies across Europe, especially for the biomonitoring of metals and metalloids. Moreover, local efforts have significantly contributed to generate new HBM data, for example in the biomonitoring of bisphenols and parabens in Spanish breastfeeding mothers (Sanchis et al. 2020), plasticisers in Italian children and adolescents (Tait et al. 2020) and antibiotics in pregnant women in Shanghai, China (Zeng et al. 2020).

However, despite the relevance of this tool, HBM presents a significant challenge, which is the ability to appropriately interpret and communicate the biomonitoring data, as the detection of low concentrations of chemicals in human tissues and fluids does not necessarily mean that they imply a harmful or disease (CDC 2009). This challenge is being exacerbated by the increasingly fast development of the analytical technology, which is rapidly facilitating the study of virtually any chemical in most human fluids and tissues (Paustenbach and Galbraith 2006); however, the knowledge of their effect in human health is not evolving at the same speed. Thus, an understanding of the health risk associated with the presence of chemicals in blood or urine is needed for most of

DOI: 10.1201/9780203730584-18

the environmental pollutants or xenobiotics and/or their metabolites, particularly for emerging chemicals that have already been already detected in the environment and which are being used in novel products. Moreover, there are knowledge gaps in the current understanding of chemical risks in human health, particularly related to short-term (acute) and long-term (chronic) exposures to mixtures of chemicals (Ganzleben et al. 2017). Therefore, to be able to extract this information, toxicologists and HBM experts need to acquire a good knowledge of the physicochemical properties of the chemicals and their metabolites in conjunction with the routes of exposure and their pharmacokinetics and toxicokinetics (distribution, storing, metabolisation and excretion, concepts described in further detail in Chapter 2) and appropriate resources to analyse large amounts of data in a short time.

Another important challenge in undertaking HBM surveys is the sampling size and the robustness of the study and the statistical analysis performed. Thus, some statistical tests can reveal a significant difference of an effect just by chance (type I error), or contrarily not showing a statistical difference when there is a real effect (type II error). Therefore, selection of sampling size is of critical importance to detect real effects in an HBM study.

18.2 HBM CONCEPTS AND DEFINITIONS

Important toxicological concepts for HBM have been covered in previous chapters, including xenobiotic metabolism (Chapter 3); internal dose or amount of a chemical or their metabolites that reaches a human tissue target; pharmacokinetics and toxicokinetics (Chapter 2); and physiologically based pharmacokinetic (PBPK) models. Other relevant concepts and definitions include biomarkers.

18.2.1 Biomarkers

The term "biomarker" refers to the measurement of any chemical environmental agent or xenobiotic and their metabolites or reaction products in human fluids and tissues. Biomarkers can provide information about exposure and potential short-term health effects (WHO 2015). Biomarkers are capable of measuring the exposure to a xenobiotic (i.e. the internal dose) by incorporating information of the pharmacokinetics or toxicokinetics of the substance studied and the characteristics and susceptibilities of the individual. There are three classes of biomarkers depending on what they are measuring/evaluating (Choi et al. 2015), which will define the different approaches in HBM described below: (a) a biomarker of exposure, which can measure the interaction between a target molecule or cell with a xenobiotic; (b) a biomarker of effect, which measures an alteration in the organism (i.e. a potential health impairment or disease; and (c) a biomarker of susceptibility, which measures the ability of a human to respond to the exposure to a xenobiotic. Examples of types of biomarkers are provided in Table 18.1.

Selection of biomarkers will therefore depend on the aims of the HBM survey, selection that should be informed by availability of guidelines (reference ranges, HBM values or biomonitoring

Table 18.1 Examples of Types of Biomarkers in Different Human Matrices

	Biomarkers	Examples
Specific	Concentration of a specific chemical in blood	Lead (Pb): long half-life and will accumulate, so can be a reflect of body burden.
	Concentration of a specific metabolite of a chemical in blood	Cotinine: metabolite of nicotine that can inform of tobacco exposure.
	Protein adducts	Aflatoxin-albumin adducts to measure intake of the contamination product aflatoxin.
	DNA adducts	DNA adducts of polycyclic aromatic hydrocarbons are associated with exposure to combustion products.
Unspecific	Urinary mutagenic activity	To measure exposure to cytostatic drugs in healthcare professionals.
	DNA damage	Comet assay, to study genotoxic exposures in *in vivo* models controlling several confounding factors including diet or exercise.
	'omics	For example, microarrays to study cellular responses to chemicals.

Source: Adapted from Choi et al. (2015).

equivalents [BEs]), expected exposure ranges and analytical tools that can measure the xenobiotics or their metabolites in the population/specimen(s) subject of study.

18.3 HBM TECHNIQUES

HBMs use a myriad of techniques and processes to recruit populations and recollect all the necessary information, including ethics and participants' informed consent, and appropriate methodologies (also named standard operating procedures [SOPs]) to undertake the different processes including sampling collection, processing and analysis; and data reporting, interpretation and communication (Choi et al. 2015). Creation of robust SOPs and harmonization of protocols between the different HBM consortiums or commissions/countries is of paramount importance for HBM (Heffernan et al. 2020), as any errors in any of those steps, including in the planned experimental design, will jeopardise the results of the HBM survey. The ethical use of human specimens is regulated by the Helsinki Declaration II, which states special consideration to the information to participants, sampling consent, confidentiality and follow-up.

18.4 HBM VALUES AND INTERPRETATION OF HBM DATA

Recommendations and HBM values are usually provided for the general population, including different subgroups and/or phases (e.g. children or pregnant women), and assuming lifelong exposure (Apel et al. 2017). HBM consortiums or commissions establish HBM values or thresholds to indicate risks of adverse effects from different chemicals, which are frequently revised and updated following acquisition of knowledge. These HBM values are derived from different methods (Apel et al. 2017), which can include epidemiological studies that have recorded the relationship of adverse effects with the presence of xenobiotics or their metabolites in human matrices; following scientifically accepted tolerable intake values (e.g. PTWI; more information available in Chapter 14); and based on critical effects observed in *in vivo* animal models when there is a gap of knowledge on humans and/or when tolerable intake values are unknown.

Measuring xenobiotics and their metabolites in human matrices without characterizing their relevance or their interpretation provides little value to public health. In order to make the interpretation of the biomonitoring data as real as possible, the comparison of these results with a reference scale and exposure limit values is decisive. This scale, or reference values, can be obtained through the study of a population that can be representative of the total population or of specific groups (Seifert et al. 2000), which have been characterised under strict quality control (Schulz et al. 2009; Peña-Fernández et al. 2014). Cross-sectional studies to establish appropriate reference values require the following (WHO 2015; Saravanabhavan et al. 2017):

- A sample selection based on confounding factors and a minimum size of 120 randomly selected individuals. Moreover, reference values should be estimated for subgroups according to confounding factors such as age or sex.
- Exclusion criteria to select participants with specific factors that are not characteristic of the population studied (e.g. tobacco consumption or use of hair dyes).
- A specific time period, as exposure to xenobiotics and/or their metabolites could change. Therefore, described reference values should be rechecked over time.
- Quality analysis and data of the HBM survey should be reported, including specificity and sensitivity of the analytical method used.

Reference values (RVs) characterise the upper margin (usually the 90th or 95th percentile) of the background exposure of the population to a given environmental pollutant at a given time. The International Union of Pure and Applied Chemistry (IUPAC) recommend the nonparametric computation of the reference value within the 95% confidence interval of the 95th population percentile of the distribution of concentration of a specific compound in a human biomarker (Poulsen et al. 1997). However, the International Federation of Clinical Chemistry (IFCC) suggests the provision of reference intervals (P2.5–P97.5) for essential elements (Snoj Tratnik et al. 2019a). Although RVs cannot be used to evaluate the health status of an individual, as they have not been toxicologically derived from biological exposure limits, they can be used as thresholds to identify individuals or subpopulations that have been environmentally exposed to xenobiotics (Ewers et al. 1999; Seifert et al. 2000). Individuals/groups that have shown higher concentrations than the correspondent RV will require more studies to identify and understand exposure sources and factors that have contributed to that high exposition (Saravanabhavan et al. 2017).

The German Human Biomonitoring Commission is periodically publishing guidelines with derived reference values for different xenobiotics, which have been established following relevant toxicological data (i.e. the relationship between exposure to a chemical and its effect or derived from intake values, specifically daily intake [ADI] or tolerable daily intake [TDI]) or epidemiological data (i.e. by studying the relationship between the concentration of a biomarker with the onset of adverse health effects; WHO 2015; Ganzleben et al. 2017). Thus, the German HBM Commission publishes two levels of HBM values (Apel et al. 2017):

- HBM-I, a control value (i.e. a xenobiotic concentration threshold below which there will be no health risk assumable).
- HBM-II, a threshold above which there is need to develop and implement public health interventions to protect the health of a population.

Concentrations that fall between HBM values I and II will require more studies or continuous monitoring, as health risk cannot be ruled out (Steckling et al. 2018). HBM values are regularly updated on the German HBM Commission website, which also provides HBM reference values following the recommendations described by the IUPAC, previously discussed.

18.4.1 Biomonitoring Equivalents

A biomonitoring equivalent (BE) is the concentration of a chemical or its metabolite(s) in human biological matrices (body fluids or tissues) and relates to established reference values for intake levels (Hays and Aylward 2009; Choi et al. 2015). BEs incorporate information of the toxicokinetic properties of the chemical by including reference doses (RfD) or concentrations (RfC), minimal risk levels (MRL) and TDIs to define them (Apel et al. 2017). Apel et al. (2017) describe the different steps for establishing BEs, which should use reliable physiology-based biokinetic (PBBK) models to translate external exposure.

18.5 HBM APPROACHES AND APPLICABILITIES

According to Angerer (2002) and Angerer et al. (2007), there are three different approaches in HBM: (a) dose monitoring, or the detection and quantification of environmental chemicals (also known as xenobiotics) and their metabolites in human matrices; (b) biochemical effect monitoring, or the study of the effect of the interaction of xenobiotics with important targets in human body such as proteins or DNA; and (c) biological effect monitoring, or the determination of the effect of the interaction of cells with environmental pollutants. Recently, a fourth approach has been incorporated which consists of the determination of the susceptibility of the individual, as this disposition can modulate the previous three approaches (Müller et al. 2014).

HBM is a versatile tool with myriad applicabilities, which include the following (Müller et al. 2014; WHO 2015; Ganzleben et al. 2017; Steckling et al. 2018; Louro et al. 2019):

- Evaluation of the internal exposure from multiple pathways and sources, and the quantification of body burden (i.e. internal concentration) to environmental pollutants. The body burden is quantified based on the physicochemical properties of the xenobiotic, exposure time and individual susceptibility (i.e. the physiological characteristics). Thus, the body burden will be a reflect of the absorption, distribution, metabolism and excretion of the xenobiotic in the individual, which can be characterized by selection of appropriate biomarkers and human specimen(s) (Esteban and Castaño 2009). Moreover, HBM is particularly useful to determine exposure in vulnerable populations that are sensitive to environmental contamination, such as infants, children and pregnant women, for which there are available noninvasive methods to perform HBM surveys.
- Identification of individuals/groups exposed to chemicals at their workplace or living close to contaminated areas or hot spots. Evaluation of temporal trends and changes in exposure, as well as to identify vulnerable population subgroups.
- Efficient and cost-effective tool to identify new chemical exposures. HBM can also be used to quantify exposure, usually by requesting participants to complete validated questionnaires to recollect diary activities.
- HBM can enhance human risk assessment studies by making them more robust, contribute to regulate chemicals and inform companies and authorities about the safety of their use and risks. Thus, HBM can contribute to Regulation (EC) No 1907/2006 concerning the Registration, Evaluation, Authorisation, and Restriction of Chemicals (REACH) in Europe, which establishes

procedures for collecting evidence on the properties and hazards of substances that want to be commercialised to protect humans and the environment.

- For policymaking including consumer safety (e.g. in relation to consumption of fish to avoid exposure to mercury [Hg] in sensitive individuals) and to evaluate the effectiveness of policy actions and public health interventions. These polices can introduce measures to reduce the use of or exposure to substances and recommend bans, prohibitions and methods of use to avoid or reduce risks to humans.
- To follow up with populations affected by chemical, biological, radiation and nuclear (CBRN) incidents in the early aftermath and throughout the different recovery and restoration phases and also to identify and prevent short- and long-term effects. The release of CBRN substances or other hazardous material (HazMat) into the environment, either deliberately or involuntarily, are termed CBRN and HazMat incidents, respectively. More information about environmental recovery and restoration is described in Chapters 13, 25 and 28. HBM will also be useful to evaluate the success of the recovery plan and inform the "de-escalating" processes and lifting of bans and restrictions to protect human health after a CBRN incident.

Unfortunately, HBM does not inform about routes or sources of exposure, so environmental monitoring remains critical to tailor appropriate polices and public health interventions to protect humans and the environment (WHO 2015).

18.6 HBM COMMUNICATION

A robust HMB survey should have a communication strategy embedded withing the different milestones of the project. The communication strategy should consider ethical and political issues and should be tailored according to the group or target population the communication strategy is aimed at (WHO 2015). Table 18.2 provides some examples of communication strategies, such as promoting public awareness and safeguarding population by developing relevant polices.

Communication to HBM participants should be clear and unambiguous to avoid alarm and should include results for individuals if they have requested them when they were recruited. Communication to the public should not include means to identify individuals, and all necessary information should be presented to be able to understand the data and to inform not only policymakers but also the general public.

18.7 INTERNATIONAL APPROACHES

In the United States, the National Health and Nutrition Examination Surveys (NHANES), conducted by the National Center for Health Statistics of the Centers for Disease Control and Prevention (CDC), are cross-sectional surveys that have monitored the health and nutritional status of adults and children since 1971. The NHANES Survey II (1976–1980) began to monitor the levels of Pb in blood in the US population; a decline of Pb was noted in American children's blood as a consequence of its ban in gasoline in the mid-1970s (Choi et al. 2015). NHANES is a versatile programme that can adapt rapidly to emerging needs; since 1999 the programme is performed

Table 18.2 Examples of Communication Strategies to Participants and the Scientific Community

Target audience	Communication strategy
General population	News in different media Website
HBM participants	Letters (general or individual if requested by participant) Face-to-face Website
Policymakers	Stakeholder alerts Fact sheets and reports Conferences and/or workshops
Scientists	Communication in peer-reviewed journals Communication in scientific congresses Website reports

Source: Adapted from WHO (2015).

continuously each year (Choi et al. 2015); however, it does not provide statistical summaries according to geographical regions and does not monitor infants (WHO 2015).

In Europe, large HBM programmes were initiated in 1977 to determine the levels of Pb in blood in the general population that is not exposed occupationally. As a result of these programmes, in 1984 Germany created the national HBM programme, German Environmental Studies (GerES), which performs continual research and development (Schulz et al. 2012; WHO 2015).

Different countries in the European Union (EU) have launched large HBM programmes in recent years. For example, in 2007 Spain initiated the BIOMBIENT.ES programme (Pérez-Gómez et al. 2013), and in 2008 France launched the ENNS (Étude Nationale Nutrition Santé; Hercberg et al. 2008). The lack of a coherent approach in HBM between countries and studies in the EU necessitated the creation of an Expert team to Support Biomonitoring in Europe (ESBIO) in December 2009. The ESBIO developed a coherent approach to HBM that carried out a pilot phase known as COPHES (Consortium to Perform Human Biomonitoring). The pilot protocol has been tested in a twin project of COPHES called DEMOCOPHES (DEMOnstration of a study to Coordinate and Perform Human biomonitoring on a European Scale; Joas et al. 2012), which was implemented in 17 European countries and facilitated the implementation of a harmonised approach to HBM in Europe (WHO 2015). Following the success of the DEMOCOPHES project, a European Human Biomonitoring Initiative (named HBM4EU) was in place from 2017 to 2021; its focus was to target different substances of public health concern, including emerging chemicals and mixtures, to support policymaking in 30 European countries (Louro et al. 2019). HBM4EU will develop HBM Guidance Values (termed HBM GV) as well as guidance for operating procedures and for field work carried out.

These kinds of studies are usually limited to adult populations, use blood or urine as the most common human specimens, and can only study a limited amount of xenobiotics, which are carefully chosen during the planning of the surveys. DEMOCOPHES, for example, only used hair to monitor one chemical (Hg) in children and their mothers; similarly, the Spanish HBM survey, named BIOAMBIENT.ES, also monitored Hg in hair, but only from workers 18 years or older (WHO 2015). Large HBM programmes that target children include the GerESs. The GerES II study (1990/1992) measured metals in blood, urine and hair in children 6 to 14 years old (Seifert et al. 2000); in the latest studies, blood and morning urine samples were used from children and adolescents aged 3–14 (GerES IV; Schulz et al. 2012), and those aged 3–17 years (GerES V; Schwedler et al. 2020).

Additionally, Germany developed an archive for human specimens, the Environmental Specimen Bank (ESB), which has been used since 1985 to study exposure trends and the fate of chemicals in humans and the environment (WHO 2015; Schettgen et al. 2020). As an archive for human specimens, the ESB is of critical importance to ensure comparability of results and to allow scientists to study the effectiveness of policies and interventions created to protect the population from environmental pollutants. Owing to the importance of biobanking (i.e. the storage of human specimens under cryogenic conditions), the HBM4EU is also developing standardised procedures for all the processes related with biobanking (sampling management, shipping and reception, cryogenic storage conditions) to be able to compare exposure trends to chemicals and/or their metabolites among the 30 European participant countries (WHO 2015; Lermen et al. 2020).

18.8 BIOLOGIC MATRICES OR BIOMONITORS

Nowadays, the exposure of the general population to xenobiotics is intrinsically related to the use and production of chemicals, the presence of chemicals in different products including pharmaceuticals and personal care products, and to diet and living conditions (Kolossa-Gehring et al. 2017). Once a xenobiotic is absorbed in the human body, usually by inhalation, ingestion and/or dermal absorption, it takes different routes and can be stored in different tissues and/or excreted in different matrices such as urine, faeces or saliva, either after or without transformation. To evaluate the exposure and burden of chemicals and/or their metabolites, HBM surveys can use one or more biological matrices according to different factors, including the toxicological and toxicokinetics characteristics of the xenobiotic object of study and availability of techniques to measure small concentrations of the target chemical. Investigation of new biomarkers to measure exposure and effect and the development of biomarkers to study groups of chemicals are of critical importance (Esteban and Castaño 2009). HBM surveys usually use blood and urine as human specimens, although hair and other tissues are also recommended to study some chemicals and long-term exposures.

Human specimens should be easily accessible in necessary amounts under normal conditions without producing harm or discomfort for the participant (Domingo et al. 2017). Thus, blood and urine are among the more common materials in HBM surveys, although collection of blood uses an invasive method which can impact the sampling size and adherence of participants to the programme. Materials such as hair, nails and saliva have not been extensively used until recently due to confounding factors and limitations that can affect the presence and detection of chemicals and/or their metabolites. Moreover, access to appropriate peer-reviewed SOPs developed under strong quality methods limits the use of these noninvasive human specimens (Angerer et al. 2007).

18.8.1 Blood, Cord Blood and Serum

Blood is commonly used in HBM studies of different factors: (a) most chemicals and their metabolites use blood to travel within the human body, and blood is in contact with the entire body; (b) concentrations of chemicals in blood and target tissues/organs are in equilibrium; and (c) blood collection, management and analytical methods are standardised and have appropriate guidelines (SOPs). Moreover, blood is crucial to validate the use of other human specimens, which is carried out by studying correlations between matrices, as blood is generally considered to be a good indicator of internal chemical concentration (Esteban and Castaño 2009). Blood presents some disadvantages: it is an invasive matrix, and religious and cultural beliefs can prevent individuals from having their blood drawn (Paustenbach and Galbraith 2006) and can affect compliance rates (WHO 2015).

Blood, serum or plasma, and umbilical or cord blood, are recommended to study the following:

- Certain metals and metalloids, such as Pb. However, blood would not be appropriate when monitoring hexavalent chromium (Cr) or chemicals that can interact with erythrocytes (Hassan et al. 2020). Moreover, blood does not provide information about historical exposure to Hg or total body burden of cadmium (Cd) (WHO 2015). Careful attention should be given to the specimen, as levels of metals could be higher in whole blood than in serum (Hassan et al. 2020), or in lyophilised versus fresh blood samples (Snoj Tratnik et al. 2019b).
- Persistent organic pollutants (POPs) like dioxins (polychlorinated dibenzo-p-dioxins and polychlorinated dibenzofurans [PCDD/Fs]), polychlorinated biphenyls (PCBs), and perfluorinated compounds and brominated flame retardants (Fromme et al. 2016). POPs can be either hydrophilic or lipophilic; therefore HBM values for lipophilic POPs should be expressed per amount of lipids in serum (Drage et al. 2019). Organochlorine pesticides (e.g. aldrin, chlordane, dichlorodiphenyltrichloroethane [DDT] and toxaphene), which are highly environmentally persistent and can bioaccumulate and biomagnify in the food web, can also be monitored in plasma (Haines et al. 2017).
- Some food contaminants such as mycotoxins (secondary metabolites produced by pathogenic fungi such as *Aspergillus* or *Penicillium*; Arce-López et al. 2020).
- Cord blood can be used to study prenatal exposures, which are quite important due to the special vulnerability of the foetus during development. For example, the levels of Pb in cord blood have shown to be correlated with the maternal blood lead levels, and the blood lead levels in new-borns is a reflection of that found in their mothers (Sanders et al. 2009). Thus, cord blood has been used to determine prenatal exposure to Pb (Saoudi et al. 2018) and other chemicals, such as perfluorinated substances (Colles et al. 2020).

18.8.2 Urine

Urine is a noninvasive sample that is accessible in large volumes, which makes urine a very popular human specimen in large HMB surveys. Urine can be collected in two manners: a spot sample or a 24-hour sample. Spot samples are easier to collect but should be adjusted, usually with creatinine, to correct variabilities in the volume and the concentration of the xenobiotic (Esteban and Castaño 2009). However, urine is not a reliable matrix to determine exposure to most chemicals due to different factors, including toxicokinetics of the chemicals of interest and lack of discrimination to rule out unexposed individuals, as chemicals may also appear in those individuals. Thus, Pb in urine is only a reliable biomarker to study a recent intoxication, although it appears to be appropriate to determine the total body burden of Cd (WHO 2015).

Urine is recommended to study the following:

- Rapidly metabolised xenobiotics and nonpersistent organic compounds (e.g. phthalates, bisphenol A, pyrethroid insecticides, herbicides) and organophosphorus or organophosphate pesticides (e.g. malathion, parathion, dimethoate, chlorpyrifos and diazinon), which are highly hazardous for human health (WHO 2015). Moreover, urine is applicable to monitoring exposure to parabens, important preservatives in cosmetics and personal care products, which are described in more detail in Chapters 11 and 12.
- Volatile organic compounds (VOCs) such as benzene, toluene, ethylbenzene, xylenes, styrene and terpenes, which are used in myriad house products, such as detergents and paints. VOCs present different toxicological effects and can be carcinogenic (e.g. benzene).
- Although not recommended to measure environmental persistent organic compounds, urine has been used to monitor polycyclic aromatic hydrocarbons (PAHs), which carcinogenic chemical substances that are formed during the incomplete combustion of food.
- Exposure to some metals and metalloids, such as arsenic (As), Cd and organic compounds of Hg.

18.8.3 Saliva

Current analytical advantages can enhance the use of saliva as a reliable noninvasive method in HBM, particularly for the study of hormones, tobacco consumption (cotinine), pharmaceutical drugs and pesticides. Although saliva was first used to determine occupational exposure to certain chemicals (e.g. asbestos, metals), it showed inconsistencies due to the quality of individual specimens which impacted the reliability of the data (Paustenbach and Galbraith 2006).

18.8.4 Breast Milk

Breast milk is used to study exposures in both the mother and her child. This matrix is recommended to study lipophilic xenobiotics, concentrations that should be adjusted with the content of fat in breast milk as it is not constant. Metals such as Cd and Pb and persistent organic pollutants have been monitored in breast milk. The WHO/United Nations Environment Programme (UNEP) Human Milk survey studies dioxins, PCBs and DDT globally (van den Berg et al. 2017). Additionally, different pesticides and phthalates have been studied in breast milk (Esteban and Castaño 2009).

18.8.5 Tissues

The new analytical technologies, which allow to determine xenobiotics in virtually any sample (to levels of ppm or ppb), are impacting the use of human tissues in HBM surveys. For example, the use of adipose tissue has become irrelevant because lipid soluble xenobiotics such as dioxins can be detected in serum in ppb. However, to study chronic exposures, autopsy samples of fat tissue, bone, liver, brain or kidneys would be still relevant (Domingo et al. 2017).

18.8.6 Human Hair

Human hair has a number of indisputable advantages over other biological specimens: the sample collection is noninvasive (except for new-borns), trauma-free for the individual, easy to collect and store, and low cost. It is also an inert material which can be stored for long periods of time and used for subsequent reanalysis. It is able to reflect prolonged exposures over time, and it is a matrix that normally presents higher levels of chemicals than urine or blood, which makes it more sensitive from the point of view of elemental analysis (Unkiewicz-Winiarczyk et al. 2009). Thus, human hair has increased significantly in anthropological research particularly related with metals and metalloids, as a possible indicator of the nutritional and health status of the population (Rodrigues et al. 2008) and to assess environmental (Lemos and de Carvalho 2010) and occupational exposure (Wang et al. 2009).

Moreover, scalp hair and hair from other locations in the body (e.g. from the pubis or armpit) have been used to monitor occupational exposure to metals (Olmedo et al. 2010), as hair constitutes a minor excretory organ for metals and metalloids (Sukumar 2002). However, there is controversy in the scientific community regarding the use of hair in HBM, especially at the level of individual monitoring (Esteban and Castaño 2009). The discrepancies between the different authors arise due to (1) the lack of reference values (Barbosa et al. 2005); (2) the lack of correlation with other biomarkers (Kordas et al. 2010); (3) the lack of scientific knowledge of the kinetics of the incorporation of metals and other chemicals in hair as well as the lack of epidemiological data that explain the possible effects derived from these on human health (Rodrigues et al. 2008); and (4) the fact that unlike other human specimens, hair presents a higher risk of exogenous environmental contamination, mainly due to deposition of sebum, sweat and contaminants in suspension, as well as residues of cosmetic and/or pharmaceutical products, which make it difficult to differentiate between endogenous and exogenous deposition (Peña-Fernández et al. 2014).

Despite these limitations, and the fact that the minimum amount of hair necessary for chemical analysis (0.05–0.1 g) might be a limiting factor for some individuals (e.g. babies and individuals with little or no scalp hair), scalp hair has been considered a good specimen to estimate environmental exposure and as a first step in more extensive and expensive studies (Lemos and de Carvalho 2010).

Scalp hair is an appropriate material for study of the following:

- Metals and metalloids, such as As, Cd, Cr, Hg and Pb, and bioindicators of exposure to trace elements belonging to the group of rare earth elements (Tong et al. 2004). Human hair is considered an excellent human specimen to study dietary exposure to methylmercury, and there is consensus within the scientific community that total Hg content in hair reflects dietary exposure to methylmercury (Castaño et al. 2019). Thus, the US Environmental Protection Agency has proposed a biological limit of Hg in scalp hair in children and child-bearing women of 1 µg/g (NCR 2000), although Bellanger et al. (2013) have suggested to lower this threshold to 0.58 µg/g following recalculations and acquisition of evidence related with the neurotoxicological effects of this metal.
- Organic pollutants such as dioxins, PCBs and DDT, as different studies have reported moderate and strong correlations between the concentrations of these pollutants in hair and those found in blood (Esteban and Castaño 2009).
- Medical treatments (Nicolis et al. 2000) and estimating the use or abuse of a broad spectrum of drugs and their metabolites in a single hair of the individual (e.g. cocaine, opiates, amphetamines, benzodiazepines, ketamine and cannabinoids; Wada et al. 2010). Pragst et al. (2010) have described that hair could also be used to determine alcohol consumption through the analysis of the metabolite ethyl glucuronide.
- Carrying out archaeological and forensic investigations (Harrison et al. 2010) to study chronic environmental exposure and poisoning in individuals. For example, it has been deduced that Napoleon suffered acute As poisoning (Kintz et al. 2006).
- Continuous monitoring of specific communities or subpopulations, especially preschools and schools, which contain population groups that are more sensitive and vulnerable to environmental pollution (WHO 2015; Choi et al. 2015).

Owing to the undoubted advantages of human hair in HBM, efforts are being undertaken to standardise methods and develop SOPs to acquire a better knowledge of the physiology and characteristics of hair and to provide tentative RVs (Peña-Fernández et al. 2014; Llorente Ballesteros et al. 2017) to avoid the possible disadvantages of hair analysis previously explained.

18.8.7 Nails

Fingernails and toenails have been traditionally used in forensic science to identify As poisoning (Esteban and Castaño 2009) and other metals. Accumulation of chemicals in nails remains largely unchanged due to low metabolic activity of this keratinised tissue, making nails an ideal matrix to study long-term exposure to metals and metalloids, although they present similar disadvantages as those described for hair, such as the risk of external contamination (e.g. nail polish and clippers; WHO 2015). Moreover, the validation of nails to monitor metals and metalloids is difficult due to the lack of correlation with the levels in blood found in some studies (Gutiérrez-González et al. 2019). Authors are exploring the applicability of nails to measure exposure to different substances (e.g. flame retardants; Chen et al. 2019) and different endocrine disruptors such as bisphenol A and its analogues, parabens, perfluoroalkyl compounds and triclocarban (an antimicrobial agent; Martín-Pozo et al. 2020). However, the use of nails in large HBM surveys, though promising, will require the development of appropriate and standardised SOPs that describe their collection, laboratory methods and processes, and reporting procedures (Gutiérrez-González et al. 2019).

18.9 CONCLUSIONS AND CHALLENGES

The exposure to chemicals and other environmental pollutants is related to their production and human use, including dietary and lifestyle habits and conditions. Available data from large international HBM programmes have pointed out a decline in human exposure to important hazardous metals in the environment (e.g. Pb), although exposure to other metals and metalloids (e.g. As, Cd and Hg), particularly through food, is still a public health concern. Moreover, large HBM surveys have also detected a decline in the exposure to targeted POPs. On the contrary, exposure to emerging chemicals including plasticisers and pesticides is increasing.

Reviews also have described important future challenges, including the following (WHO 2015):

- Developing new methods that combine computational modelling with *in vitro, in silico* and high-throughput screening (HTP; i.e. the use of data-processing software, robotics, or other techniques to conduct millions of *in vitro* tests to predict bioactivity, exposure and pharmacokinetics and toxicokinetics; Wegner et al. 2020) to extract HBM guidance values due to limitations in the use of animals in experimentation (the 3Rs: replacement, reduction and refinement).
- Enhancing capabilities and research on xenobiotics that target male and female reproductive systems.
- Identifying methods and biomarkers to characterise risks to chemical mixtures during early life stages, particularly during the development of the foetus (prenatal exposures) and in infancy.
- Investigating novel biomarkers to study the links between exposure to food contaminants and internal exposure levels for food safety (Louro et al. 2019).
- Facilitating harmonisation for a consistent approach to HBM, which includes the development of appropriate guidelines (SOPs) for sampling; laboratory methods of analysis, detection and validation; and HBM data evaluation and extraction of values for identification of risks and interpretation, which in turn will enhance the development of more robust interventions and policies to protect human health.

BIBLIOGRAPHY

Angerer, J. 2002. Internal exposure and haemoglobin adducts. In *Biological Monitoring, Prospects in Occupational and Environmental Medicine*, 5–15.Weinheim: Wiley-VCH Verlag GmbH & Co. KGaA.

Angerer, J., U. Ewers, and M. Wilhelm. 2007. Human biomonitoring: State of the art. *Int J Hyg Environ Health* 210: 201–28.

Apel, P., J. Angerer, M. Wilhelm, and M. Kolossa-Gehring. 2017. New HBM values for emerging substances, inventory of reference and HBM values in force, and working principles of the German Human Biomonitoring Commission. *Int J Hyg Environ Health* 220(2 Pt A): 152–66.

Arce-López, B., E. Lizarraga, A. Vettorazzi, and E. González-Peñas. 2020. Human biomonitoring of mycotoxins in blood, plasma and serum in recent years: A review. *Toxins (Basel)* 12(3). pii: E147.

Barbosa, F. Jr., J. E. Tanus-Santos, R. F. Geriach, and P. J. Parsons. 2005. A critical review of biomarkers used for monitoring human exposure to lead: Advantages, limitations, and future needs. *Environ Health Perspect* 113: 1669–74.

Bellanger, M., C. Pichery, D. Aerts, et al. 2013. Economic benefits of methylmercury exposure control in Europe: Monetary value of neurotoxicity prevention. *Environ Health* 12: 3.

Castaño, A., S. Pedraza-Díaz, A. I. Cañas, et al. 2019. Mercury levels in blood, urine and hair in a nation-wide sample of Spanish adults. *Sci Total Environ* 670: 262–70.

CDC (Centers for Disease Control and Prevention). 2009. *Fourth National Report on Human Exposure to Environmental Chemicals*. Atlanta, GA: U.S. Department of Health and Human Services, Centers for Disease Control and Prevention. www.cdc.gov/exposurereport/pdf/fourthreport.pdf (accessed December 11, 2020).

Chen, Y., Z. Cao, A. Covaci, C. Li, and X. Cui. 2019. Novel and legacy flame retardants in paired human fingernails and indoor dust samples. *Environ Int* 133(Pt B): 105227.

Choi, J., T. Aarøe Mørck, A. Polcher, L. E. Knudsen, and A. Joas. 2015. *Review of the State of the Art of Human Biomonitoring for Chemical Substances and Its Application to Human Exposure Assessment for Food Safety*. EFSA Supporting Publications EN-724. https://efsa.onlinelibrary.wiley.com/doi/pdf/10.2903/sp.efsa.2015.EN-724 (accessed December 11, 2020).

Colles, A., L. Bruckers, E. Den Hond, et al. 2020. Perfluorinated substances in the Flemish population (Belgium): Levels and determinants of variability in exposure. *Chemosphere* 242: 125250.

Domingo, J. L., F. García, M. Nadal, and M. Schuhmacher. 2017. Autopsy tissues as biological monitors of human exposure to environmental pollutants. A case study: Concentrations of metals and PCDD/Fs in subjects living near a hazardous waste incinerator. *Environ Res* 154: 269–74.

Drage, D. S., F. A. Harden, T. Jeffery, J. F. Mueller, P. Hobson, and L. L. Toms. 2019. Human biomonitoring in Australian children: Brominated flame retardants decrease from 2006 to 2015. *Environ Int* 122: 363–8.

Eskenazi, B., J. Chevrier, L. G. Rosas, et al. 2009. The Pine River statement: Human health consequences of DDT use. *Environ Health Perspect* 117(9): 1359–67.

Esteban, M., and A. Castaño. 2009. Non-invasive matrices in human biomonitoring: A review. *Environ Int* 35(2): 438–49.

Ewers, U., C. Krause, C. Schulz, and M. Wilhelm. 1999. Reference values and human biological monitoring values for environmental toxins. *Int Arch Occup Environ Health* 72: 255–60.

Fromme, H., B. Hilger, M. Albrecht, W. Gries, G. Leng, and W. Völkel. 2016. Occurrence of chlorinated and brominated dioxins/furans, PCBs, and brominated flame retardants in blood of German adults. *Int J Hyg Environ Health* 219(4–5): 380–8.

Ganzleben, C., J. P. Antignac, R. Barouki, et al. 2017. Human biomonitoring as a tool to support chemicals regulation in the European Union. *Int J Hyg Environ Health* 220(2 Pt A): 94–7.

Gutiérrez-González, E., E. García-Esquinas, N. F. de Larrea-Baz, et al. 2019. Toenails as biomarker of exposure to essential trace metals: A review. *Environ Res* 179(Pt A): 108787.

Haines, D. A., C. Khoury, G. Saravanabhavan, K. Werry, M. Walker, and M. Malowany. 2017. Human biomonitoring reference values derived for persistent organic pollutants in blood plasma from the Canadian Health Measures Survey 2007–2011. *Int J Hyg Environ Health* 220(4): 744–56.

Harrison, A. P., I. Cattani, and J. M. Turfa. 2010. Metallurgy, environmental pollution and the decline of Etruscan civilisation. *Environ Sci Pollut Res Int* 17(1): 165–80.

Hassan, L., A. Moser, E. Rorman, et al. 2020. Human biologic monitoring based on blood donations to the National Blood Services. *BMC Public Health* 20(1): 469.

Hays, S. M., and L. L. Aylward. 2009. Using biomonitoring equivalents to interpret human biomonitoring data in a public health risk context. *J Appl Toxicol* 29(4): 275–88.

Heffernan, A. L., M. J. Gomez-Ramos, C. Symeonides, et al.; BIS Investigator Group. 2020. Harmonizing analytical chemistry and clinical epidemiology for human biomonitoring studies. A case-study of plastic product chemicals in urine. *Chemosphere* 238: 124631.

Hercberg, S., S. Chat-Yung, and M. Chaulia. 2008. The French National Nutrition and Health Program: 2001–2006–2010. *Int J Public Health* 53: 68.

Joas, R., L. A. Casteleyn, P. Biot, et al. 2012. Harmonised human biomonitoring in Europe: Activities towards and EU HBM framework. *Int J Hyg Environ Health* 215: 172–5.

Kintz, P., M. Ginet, and V. Cirimele. 2006. Multi-element screening by ICP-MS of two specimens of Napoleon's hair. *J Anal Toxicol* 30(8): 621–3.

Kolossa-Gehring, M., U. Fiddicke, G. Leng, J. Angerer, and B. Wolz. 2017. New human biomonitoring methods for chemicals of concern-the German approach to enhance relevance. *Int J Hyg Environ Health* 220(2 Pt A): 103–12.

Kordas, K., E. I. Queirolo, A. Ettinger, R. O. Wright, and R. J. Stoltzfus. 2010. Prevalence and predictors of exposure to multiple metals in preschool children from Montevideo, Uruguay. *Sci Total Environ* 408(20): 4488–94.

Lemos, V. A., and A. L. de Carvalho. 2010. Determination of cadmium and lead in human biological samples by spectrometric techniques: A review. *Environ Monit Assess* 171(1–4): 255–65.

Lermen, D., F. Gwinner, M. Bartel-Steinbach, et al. 2020. Towards harmonized biobanking for biomonitoring: A comparison of human biomonitoring-related and clinical biorepositories. *Biopreserv Biobank* 18(2): 122–35.

Llorente Ballesteros, M. T., I. Navarro Serrano, and S. Izquierdo Álvarez. 2017. Reference levels of trace elements in hair samples from children and adolescents in Madrid, Spain. *J Trace Elem Med Biol* 43: 113–20.

Louro, H., M. Heinälä, J. Bessems, et al. 2019. Human biomonitoring in health risk assessment in Europe: Current practices and recommendations for the future. *Int J Hyg Environ Health* 222(5): 727–37.

Martín-Pozo, L., S. Cantarero-Malagón, F. Hidalgo, A. Navalón, and A. Zafra-Gómez. 2020. Determination of endocrine disrupting chemicals in human nails using an alkaline digestion prior to ultra-high performance liquid chromatography-tandem mass spectrometry. *Talanta* 208: 120429.

Müller, M., K. Schmiechen, D. Heselmann, L. Schmidt, and T. Göen. 2014. Human biological monitoring—A versatile tool in the aftermath of a CBRN incident. *Toxicol Lett* 231(3): 306–14.

NRC (National Research Council). 2020. *Toxicological Effects of Methylmercury*. Washington, DC: National Academy Press.

Nicolis, I., P. Deschamps, and E. Curtis. 2000. Using hair as a diagnostic tool: X-ray fluorescence monitoring of the Zn content of hairs of patients under parenteral nutrition. *J Trace Microprobe Technol* 18: 511–16.

Olmedo, P., A. Pla, A. F. Hernández, O. López-Guarnido, L. Rodrigo, and F. Gil. 2010. Validation of a method to quantify chromium, cadmium, manganese, nickel and lead in human whole blood, urine, saliva and hair samples by electrothermal atomic absorption spectrometry. *Anal Chim Acta* 659: 60–7.

Paustenbach, D., and D. Galbraith. 2006. Biomonitoring and biomarkers: Exposure assessment will never be the same. *Environ Health Perspect* 114(8): 1143–9.

Peña-Fernández, A., M. J. González-Muñoz, and M. C. Lobo-Bedmar. 2014. "Reference values" of trace elements in the hair of a sample group of Spanish children (aged 6–9 years)—Are urban topsoils a source of contamination? *Environ Toxicol Pharmacol* 38(1): 141–52.

Pérez-Gómez, B., R. Pastor-Barriuso, M. Cervantes-Amat, et al. 2013. BIOAMBIENT.ES study protocol: Rationale and design of a cross-sectional human biomonitoring survey in Spain. *Environ Sci Pollut Res Int* 20(2): 1193–202.

Poulsen, O. M., E. Holst, and J. M. Christensen. 1997. Calculation and application of coverage intervals for biological reference values—A supplement to the approved IFCC recommendation (1987) on the theory of reference values. *Pure Appl Chem* 69(7): 1601–11.

Pragst, F., M. Rothe, B. Moench, M. Hastedt, S. Herre, and D. Simmert. 2010. Combined use of fatty acid ethyl esters and ethyl glucuronide in hair for diagnosis of alcohol abuse: Interpretation and advantages. *Forensic Sci Int* 196(1–3): 101–10.

Rodrigues, J. L., B. L. Batista, J. A. Nunes, C. J. S. Passos, and F. Barbosa Jr. 2008. Evaluation of the use of human hair for biomonitoring the deficiency of essential and exposure to toxic elements. *Sci Total Environ* 405: 370–6.

Sanchis, Y., C. Coscollà, F. Corpas-Burgos, M. Vento, M. Gormaz, and V. Yusà; Bettermilk project. 2020. Biomonitoring of bisphenols A, F, S and parabens in urine of breastfeeding mothers: Exposure and risk assessment. *Environ Res* 185: 109481.

Sanders, T., Y. Liu, V. Buchner, and P. B. Tchounwou. 2009. Neurotoxic effects and biomarkers of lead exposure: A review. *Rev Environ Health* 24(1): 15–45.

Saoudi, A., C. Dereumeaux, S. Goria, et al. 2018. Prenatal exposure to lead in France: Cord-blood levels and associated factors: Results from the perinatal component of the French Longitudinal Study since Childhood (Elfe). *Int J Hyg Environ Health* 221(3): 441–50.

Saravanabhavan, G., K. Werry, M. Walker, D. Haines, M. Malowany, and C. Khoury. 2017. Human biomonitoring reference values for metals and trace elements in blood and urine derived from the Canadian Health Measures Survey 2007–2013. *Int J Hyg Environ Health* 220(2 Pt A): 189–200.

Schettgen, T., M. Rüther, T. Weber, T. Kraus, and M. Kolossa-Gehring. 2020. N-methylmalonamic acid (NMMA) as metabolite of methylisothiazolinone and methylchloroisothiazolinone in 24-h urine samples of the German Environmental Specimen Bank from 2000 to 2017—Exposure and time trends. *Chemosphere* 246: 125743.

Schulz, C., J. Angerer, U. Ewers, U. Heudorf, and M. Wilhelm; Human Biomonitoring Commission of the German Federal Environment Agency. 2009. Revised and new reference values for environmental pollutants in urine or blood of children in Germany derived from the German environmental survey on children 2003–2006 (GerES IV). *Int J Hyg Environ Health* 212(6): 637–47.

Schulz, C., A. Conrad, K. Becker, M. Kolossa-Gehring, M. Seiwert, and B. Seifert. 2007. Twenty years of the German Environmental Survey (GerES): Human biomonitoring—Temporal and spatial (West Germany/East Germany) differences in population exposure. *Int J Hyg Environ Health* 210(3–4): 271–97.

Schulz, C., M. Seiwert, W. Babisch, et al. 2012. Overview of the study design, participation and field work of the German Environmental Survey on children 2003–2006 (GerES IV). *Int J Hyg Environ Health* 215(4): 435–48.

Schwedler, G., E. Rucic, R. Lange, et al. 2020. Phthalate metabolites in urine of children and adolescents in Germany. Human biomonitoring results of the German Environmental Survey GerES V, 2014–2017. *Int J Hyg Environ Health* 225: 113444.

Seifert, B., K. Becker, D. Helm, C. Krause, C. Schulz, and M. Seiwert. 2000. The German Environmental Survey 1990/1992 (GerEs II): Reference concentrations of selected environmental pollutants in blood, urine, hair, house dust, drinking water and indoor air. *J Exp Anal Environ Epidemiol* 10: 552–65.

Snoj Tratnik, J., I. Falnoga, and D. Mazej. 2019a. Results of the first national human biomonitoring in Slovenia: Trace elements in men and lactating women, predictors of exposure and reference values. *Int J Hyg Environ Health* 222(3): 563–82.

Snoj Tratnik, J., D. Mazej, and M. Horvat. 2019b. Analytical Quality Requirements in Human Biomonitoring Programs: Trace Elements in Human Blood. *Int J Environ Res Public Health* 16(13): pii: E2287.

Steckling, N., A. Gotti, S. Bose-O'Reilly, et al. 2018. Biomarkers of exposure in environment-wide association studies—Opportunities to decode the exposome using human biomonitoring data. *Environ Res* 164: 597–624.

Sukumar, A. 2002. Factors influencing levels of trace elements in human hair. *Rev Environ Contam Toxicol* 175: 47–78.

Tait, S., F. Carli, L. Busani, et al. 2020. Biomonitoring of Bis(2-ethylhexyl)phthalate (DEHP) in Italian children and adolescents: Data from LIFE PERSUADED project. *Environ Res* 185: 109428.

Tong, S. L., W. Z. Zhu, Z. H. Gao, Y. X. Meng, R. L. Peng, and G. C. Lu. 2004. Distribution characteristics of rare earth elements in children's scalp hair from a rare earths mining area in southern China. *J Environ Sci Health A Tox Hazard Subst Environ Eng* 39(9): 2517–32.

Unkiewicz-Winiarczyk, A., K. Gromysz-Kałkowska, and E. Szubartowska. 2009. Aluminium, cadmium and lead concentration in the hair of tobacco smokers. *Biol Trace Elem Res* 132(1–3): 41–50.

van den Berg, M., K. Kypke, A. Kotz, et al. 2017. WHO/UNEP global surveys of PCDDs, PCDFs, PCBs and DDTs in human milk and benefit-risk evaluation of breastfeeding. *Arch Toxicol* 91(1): 83–96.

Wada, M., R. Ikeda, N. Kuroda, and K. Nakashima. 2010. Analytical methods for abused drugs in hair and their applications. *Anal Bioanal Chem* 397(3): 1039–67.

Wang, T., J. Fu, Y. Wang, C. Liao, Y. Tao, and G. Jiang. 2009. Use of scalp hair as indicator of human exposure to heavy metals in an electronic waste recycling area. *Environ Pollut* 157(8–9): 2445–51.

Wegner, S. H., C. L. Pinto, C. L. Ring, and J. F. Wambaugh. 2020. High-throughput screening tools facilitate calculation of a combined exposure-bioactivity index for chemicals with endocrine activity. *Environ Int* 137: 105470.

WHO (World Health Organisation). 2015. *Human Biomonitoring: Facts and Figures*. Copenhagen: WHO Regional Office for Europe. www.euro.who.int/__data/assets/pdf_file/0020/276311/Human-biomonitoring-facts-figures-en.pdf (accessed December 11, 2020).

Yu, D., X. Liu, X. Liu, et al. 2019. Polychlorinated Dibenzo-p-Dioxins, polychlorinated dibenzofurans, and dioxin-like polychlorinated biphenyls in umbilical cord serum from pregnant women living near a chemical plant in Tianjin, China. *Int J Environ Res Public Health* 16(12): pii: E2178.

Zeng, X., L. Zhang, Q. Chen, et al. 2020. Maternal antibiotic concentrations in pregnant women in Shanghai and their determinants: A biomonitoring-based prospective study. *Environ Int* 138: 105638.

19 Introduction to Environmental Toxicology

Christian Khalil

CONTENTS

19.1 INTRODUCTION

The environment is an integral component of our lives, providing us with natural resources that are essential for our survival and wellbeing. The environment supports us with food, clothing, and shelter in addition to being a repository for the waste we generate. The natural environment has been crucial for our success and development as a species through its magical ability to convert our wastes into new resources using the natural power of the sun. Humans have managed to exploit this important characteristic, which if used sustainably manner could help us in indefinitely exploiting the natural environment. Unfortunately, there are significant concerns that arise when the natural environment is treated as a sink to all our wastes without any consideration for nature limitations. Throughout history, we have treated the environment as a dumping arena for all kind of unwanted products and wastes. In the old days, wastes from production processes were buried in the ground. During that period, the phrase "out of sight, out of mind" was common. We have learned over time that buried wastes will escape containment and contribute to polluting soils and waters, ecosystem and species and eventually find their way to the human food supply.

Promoting human health by using pharmaceuticals has contributed to significant contamination of air groundwater and soils worldwide. Pharmaceutical products as well as chemical leakages, spills, disposal of industrial and household chemicals, petroleum fuels, chlorinated solvents, pesticides, and excess nutrient input are all contributing to environmental degradation.

Environmental toxicology is an emerging discipline and one of the important subbranches of toxicology. The term "environmental toxicology" is self-explanatory, as it is attempting to clarify the toxicological impacts of chemical compounds in the natural environment. Unlike traditional toxicology, which addresses the impacts of chemicals in humans, environmental toxicology focuses mainly on the environmental impacts on ecosystems and their inhabitants. Therefore, assessing the toxic potential of chemicals in the natural environment is quite complex due to the large number of species affected and various toxicological sensitivities between test species and wildlife counterpart.

The early roots of environmental toxicology can be traced back to Egyptian civilization. The Egyptians used plant-derived chemicals in assassination. The native Indians in the Amazon forest also harnessed their knowledge of local plants to extract a poisonous chemical from a plant known as curare they successfully used for hunting through prey immobilization. Perhaps the discipline of environmental toxicology gained momentum in the early 1950s with the use of pesticides in the environment. Chlorinated and methylmercury derivatives in pesticides created havoc in the

DOI: 10.1201/9780203730584-19

natural environment due to their persistence and long half-lives. They bioaccumulated and biomagnified in aquatic ecosystems and affected pest species in addition to nontarget species such as birds. This phenomenon has been extensively described in Rachel Carson's book, *Silent Spring*, in which she cited pristine areas in the United States with no birds due to extensive use of long-lived pesticides such as DDT. Environmental toxicology knowledge gained momentum in the 1960s and '70s, where this knowledge was further exploited in warfare through US development of the biological weapon known as Agent Orange, used in the Vietnam War.

Finally, our modern society has managed to eradicate and control diseases by using pharmaceutical products. Pharmaceutical products are a part of everyday life. They have a wide range of applications, such as in agriculture, livestock production and controlling human diseases. They have been widely accepted as being major contributors to health and wellbeing. The main issues arising from pharmaceutical drugs use stems from their improper administration and the potential impacts of their metabolites in the natural environment. These metabolites, once released into the natural environment, pose significant risks to natural ecosystems.

19.2 ENVIRONMENTAL CONTAMINANTS, THEIR METABOLITES, HEALTH AND RISKS

Environmental contaminants refer to a wide range of unwanted chemicals that enter the natural environment. These chemicals and their metabolites can trigger significant environmental problems and future legacies. A classic example of improper chemical disposal that has been well documented in the literature is the mercury contamination of Minamata Bay in Japan (Harada 1995). This event highlighted the severe environmental and human toxicological problems associated with mercury disposal into the sea. Along the Japanese coast, next to the Minamata Bay, a chemical plant known as the Chisso factory was located. One of the unwanted products of the manufacturing process was mercury. The plant was disposing of the mercury as methylmercury in the sewage effluent to the sea which contaminated the local aquatic environment. This led to severe contamination in local communities that consumed local seafood contaminated with mercury, creating numerous adverse symptoms and eventually causing death.

The metabolites of disposed chemicals in the natural environment can also pose significant toxicological challenges. Surfactants are among the most common ingredients widely used in domestic and commercial cleaning processes, paints, pesticides and other applications (Thiele et al. 1997). These substances undergo biological treatment in sewage treatment plant, resulting in toxic by-products that have the potential to bioaccumulate and biomagnify in the aquatic environment. Toxic by-products of surfactants include 4-alkylphenols 4-alkylphenoxy acetic acid, 4-alkylphenoxy ethoxy acetic acid, 4-alkylphenol monoethoxylate and 4-alkylphenol diethoxylate (Acir and Guenther 2018). The exposure to surfactant metabolites, whether in wildlife or humans, can trigger significant toxicological challenges at the immune and nervous system levels.

The pharmaceutical industry is a multibillion-dollar industry (Moorkens et al. 2016) playing an important role in developing drugs for human and animal applications. Pharmaceutical drugs are critical in preserving the health of consumers, saving their lives, curing illnesses and reducing fatalities. Pharmaceutical drug usage is an acceptable practice and one of the pillars of healthcare programs in most countries around the globe.

Pharmaceutical drugs (pills, tablets, formulations and powders) are usually composed of chemical compounds extracted from plant or animal sources in addition to synthetic molecules developed in the laboratory. These compounds can interact with different body systems to provide the therapeutic outcomes required. Pharmaceuticals undergo extensive testing, including animal and human trials, prior to getting the seal of approval by regulators (e.g. the US Food and Drug Administration [FDA]) for commercialization of these products. Pharmaceutical drugs have been the miracle cure for many diseases and are acceptable today in healthcare, animal and plant management.

The most striking aspects of pharmaceutical products are their biological activity at very low doses and their ability to target specific enzymatic and signaling pathways in the human body. This characteristic makes pharmaceuticals and their derivatives very potent to ecological species when released in the environment. Pharmaceuticals have contributed to promoting human health and wellbeing and to reducing pain and suffering throughout history (Taylor 2016). Physicians usually prescribe pharmaceuticals upon diagnosis of symptoms in patients.

Brand and generic pharmaceutical drugs capture a large portion of pharmaceutical drug market (Grabowski et al. 2016). Generics are considered a safe and inexpensive alternative to brand-name pharmaceuticals as they use the same active ingredients, even though they have not been through the rigorous approval process (Zhong et al. 2018). Many studies in the literature have investigated the differences and effectiveness between using brand and generic pharmaceuticals for controlling diseases (Sicras-Mainar et al. 2018; Kim et al. 2019; Loch et al. 2017).

Another important issue associated with pharmaceuticals is misuse, which is a common practice extensively reported extensively in the literature (Mojtabai et al. 2019; Murnion and Conigrave 2019; Chen et al. 2019). Misuse has resulted in numerous ambulance calls (Crossin et al. 2019), hospital admissions (Schuring et al. 2018) and overdose deaths (Hedegaard et al. 2018; Colon-Berezin et al. 2019).

19.2.1 Pharmaceuticals: Environmental Risks

Pharmaceuticals drugs consumption and eventual disposal over the last few years has triggered serious environmental problem leading to regulators worldwide to establish environmental risk assessment (ERA) for drugs. Awareness of pharmaceutical impacts in the environment started around the 1980 where they were detected in surface waters (Nakamura et al. 2010). Today pharmaceuticals and their derivatives can be found in rivers (Pereira et al. 2017), wetlands (Andreu et al. 2016), coastal waters (Gaw et al. 2014) and even in the marine environment (Trombini et al. 2019). They pose significant challenges and risks to species and ecosystems.

The sources of pharmaceuticals in the environment are quite diverse and include wastes from manufacturing processes, disposal of expired medicines, and excretion of pharmaceutical metabolites from patients (Figure 19.1; Taylor 2016; Lundborg and Tamhankar 2017).

Examples from Germany indicate that 10% of assessed pharmaceutical drugs pose significant risks to the environment (Küster and Adler 2014). The release of pharmaceutical drugs in the natural environment can trigger significant undesirable impacts by affecting species functions such as metabolism and reproduction (Fabbri and Franzellitti 2016). There are some studies referring to no significant impacts of pharmaceutical presence on aquatic life (Taylor 2016), whereas other studies raise significant concerns from the presence of trace pharmaceuticals in the environment and their biomagnification potential (Zenker et al. 2014; Puckowski et al. 2016).

The key terms in environmental toxicology to help understand the fate of chemicals in the natural environment are bioaccumulation and biomagnification. Bioaccumulation is the process of chemical accumulation in the tissue of exposed species, whereas biomagnification refers to increased concentration of the chemical in the various trophic levels of the food chain. More details about these key processes are referred to in Section 19.4.1. These processes were illustrated in recent findings of the presence of pharmaceuticals and personal care products (more than 110

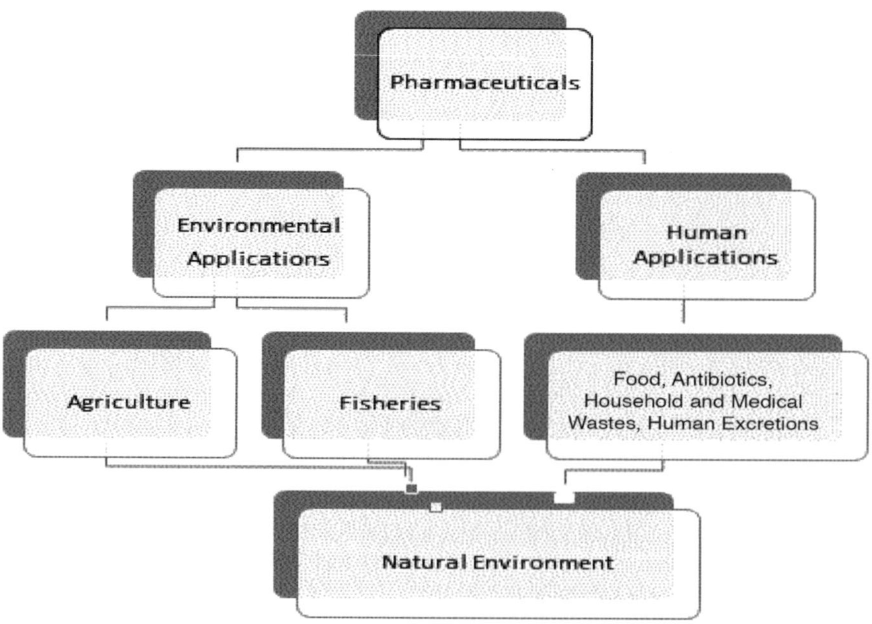

Figure 19.1 Pharmaceuticals and their residues routes of entry into the natural environment.

identified) in pristine environments such as the Arctic (Kallenborn et al. 2018). A number of factors contributes to the presence of these pharmaceuticals, such as improper treatments of sewage and its release into the cold Arctic environment, which extends the stability and half-life of these residues, making them more persistent in such colder environments (Gunnarsdóttir et al. 2013).

The natural environment is the ultimate sink for these pharmaceuticals and their by-products, which can lead to accumulation of pharmaceutical residues in different species and enhancing various resistances to them (Figure 19.1; Berkner et al. 2014). The presence of pharmaceuticals in the natural environment is promoting species developing resistance to these chemicals. We have seen the rise of many bacterial strains that developed resistance to human antibiotics (Smart et al. 2019). Multidrug bacterial resistant strains (*Mycobacterium tuberculosis, Staphylococcus aureus, Escherichia coli* [Harris et al. 2014] and various *Enterococci* species [Varela et al. 2013]) are on the rise, and our arsenal of antibiotics to deal with them is rapidly shrinking (Berkner et al. 2014). Reports in the literature of decimation of the vulture population from consumption of livestock carcasses containing diclofenac (Oaks et al. 2004), fish feminization as a result of exposure to estrogen (Corcoran et al. 2010), intersexual amphibians from estrogen exposure (Safholm et al. 2014) and in some cases mortality through behavioral changes (Brodin et al. 2014).

19.3 ENVIRONMENTAL TOXICOLOGY

Toxicology (or the science of poisons, as described by Paracelsus) is mainly focused on determining the toxic potential of a chemical by comparison to other known chemicals using laboratory tests and field observations. Environmental toxicology attempts to explore the relationships between toxic substances, humans and the health of the natural environment (Figure 19.2). Environmental toxicology has commonality with traditional toxicology and adds a further dimension to the environment and its inhabitant species.

19.3.1 Pharmaceuticals in the Natural Environment

19.3.1.1 Surface and Wastewater

The presence of pharmaceuticals in surface waters worldwide is well documented in the literature (Quesada et al. 2019). This is mainly the result of human consumption of these compounds and the inability of sewage treatment plants to properly degrade them (Clara et al. 2004). The global magnitude of this problem (aus der Beek et al. 2016) is illustrated in Figure 19.3.

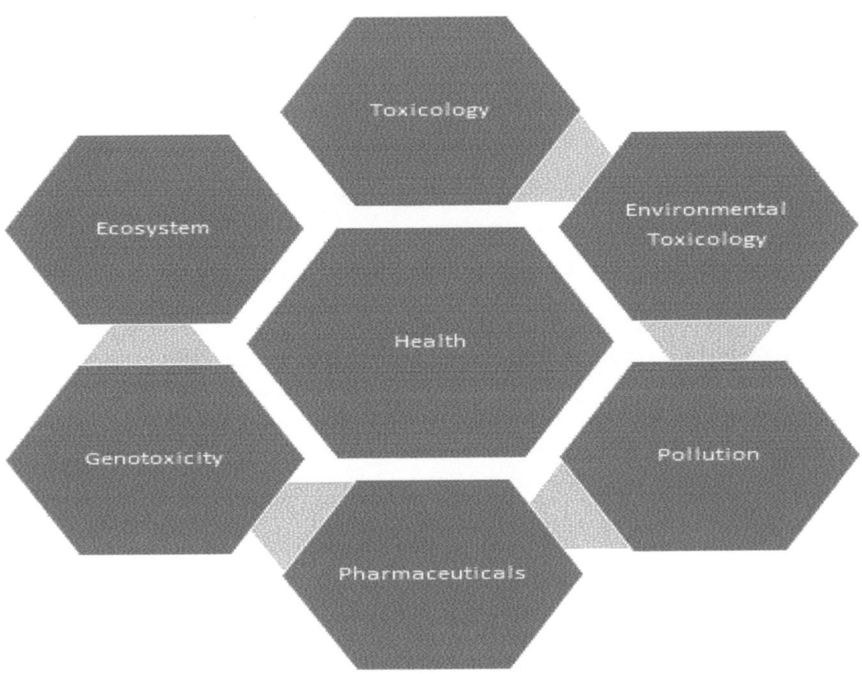

Figure 19.2 Health linkages to pollution environment, pharmaceuticals and toxicology.

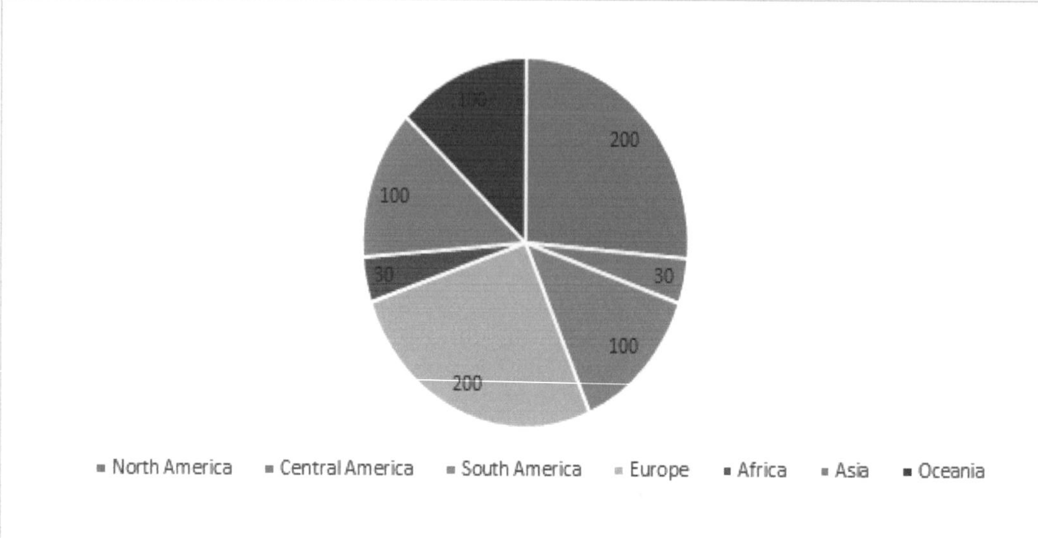

Figure 19.3 Number of pharmaceuticals detected in surface waters, groundwater, or tap/ drinking water worldwide.

Source: Adapted from aus der Beek et al. (2016).

The presence of pharmaceuticals and their derivatives in wastewater has also been extensively described in the contemporary scientific literature. Studies on European wastewater treatment effluents have identified numerous pharmaceutical compounds such as carbamazepine, tramadol, telmisartan, venlafaxine, irbesartan, fluconazole, oxazepam, fexofenadine, diclofenac, citalopram, codeine, bisoprolol, eprosartan and the antibiotics trimethoprim, ciprofloxacin, sulfamethoxazole and clindamycin (Loos et al. 2013). US studies also identified various pharmaceuticals in wastewater treatment plants, including hydrochlorothiazide, atenolol, carbamazepine, valsartan, lisinopril, enalaprilat and metoprolol (Kostich et al. 2014). These drugs have many applications in the medical field to treat diseases and relieve symptoms:

- Carbamazepine is an anticonvulsant used mainly to treat bipolar disorder.
- Tramadol is a pain reliever.
- Telmisartan and bisoprolol are prescribed for hypertension treatment.
- Venlafaxine and citalopramare are antidepressants.
- Irbesartan prevents narrowing blood vessels, lowers blood pressure and improves blood flow.
- Fluconazole is an antifungal drug.
- Oxazepam is prescribed for anxiety or alcohol withdrawal symptoms.
- Fexofenadine is an antihistamine for allergy treatment.
- Diclofenac is an anti-inflammatory drug.
- Codeine is used in pain medication.
- Eprosartan is prescribed to treat high blood pressure.
- Trimethoprim, ciprofloxacin, sulfamethoxazole and clindamycin are antibiotics used for treating infections.

Some studies on various pharmaceuticals have indicated that aquatic life exposure to pharmaceuticals and mixtures posed low risks (Kostich et al. 2014). The low risks of pharmaceuticals were mainly attributed to their presence in very low concentrations rather than their established therapeutic recommended intake.

19.3.1.2 Landfills

Another source of pharmaceuticals and their derivatives is the disposal of expired medicines, pharmaceutical derivatives and drugs into the municipal waste stream and the potential toxicological challenges this can pose to the natural environment. A number of studies of landfills (Lu et al. 2016; Sui et al. 2017; Masoner et al. 2016) identified pharmaceuticals in landfill leachates with significant problems in their removal by biological treatment (Lu et al. 2016). Previous published work on landfill leachates (Khalil et al. 2018) clearly indicated the seemingly low toxicity risks of landfill leachates in chronically exposed mice indicator species. Further exploration of other parameters in exposed mice species showed enzymatic changes in the blood and livers, indicating upregulation of liver enzymes from exposure together with severe genotoxic damage (Khalil et al. 2018).

19.3.1.3 Soils

The municipal and livestock wastewater treatment process generates a concentrated solid in liquid called sludge, which is mainly used as soil fertilizer after further treatment (Bai et al. 2017; Willen et al. 2017). This sludge can contain numerous residual hazardous compounds including pharmaceuticals (Balbo and Stepanov 2018). The application of sludge to soil can lead to soil contamination with pharmaceuticals. A number of studies highlighted the accumulation of pharmaceuticals in soils irrigated with wastewater (Dalkmann et al. 2012) and sludge (Jing et al. 2017).

19.3.1.4 Plants and Animals

The presence of pharmaceuticals in the natural environment can also lead to its uptake by the plant and animals living in affected ecosystems. Many studies have highlighted the ability of plants to take up a range of pharmaceuticals and metalloids under various growing conditions (Herklotz et al. 2010). These compounds accumulate in the plants and the crops they produce (Sullivan et al. 2010; Shen et al. 2017). Studies have confirmed, for example, the uptake and accumulation of arsenic in tomato plants at concentrations above acceptable levels for human consumption (Burlo et al. 1999). The accumulation phenomenon (Li et al. 2019) witnessed in the case of pharmaceuticals can also impact the proteins of various plant tissues at the cellular level (Hofbauer and Stoger 2013). The consequences of accumulating certain chemicals is mainly at the level of toxicity; for example, arsenic uptake and accumulation by plants can interfere with metabolic processes, and arsenic stunts plant growth by inducing phytotoxicity (Shrivastava et al. 2015).

Animal species are also affected by the presence of pharmaceuticals in their natural environment (Fabbri and Franzellitti 2016). The literature confirms the fate and uptake of these chemicals in affected species. A US study on antidepressant pharmaceuticals confirms the occurrence of these compounds in the water and sediment and highlights their uptake into fish neural tissue (Schultz et al. 2010). These compounds have also the ability to bioaccumulate in exposed fish species (Valdes et al. 2014), posing further toxicological challenges. Numerous recent studies in the literature described the impacts of antidepressant on fish cholinergic system and behavioral responses such as predation and aggressiveness (Theodoridi et al. 2017; Duarte et al. 2019; Wolkers et al. 2017).

19.3.2 Systems Used in Environmental Toxicity Determination

19.3.2.1 Mammalian In Vitro *Systems*

There are numerous examples of pollutants toxicity establishment using *in vitro* systems using human- and animal-derived cell cultures exposed to one or mixture of chemical pollutants. The toxicity determination of individual chemicals and their mixtures using *in vitro* cell systems in the environment has also been reported (Khalil et al. 2019; Khalil et al. 2018; Al Hageh et al. 2018).

There are numerous ethical issues with animal use, and therefore there is a push to use alternatives in toxicity testing (such as *in vitro* toxicology determination) in human-derived cell cultures or even cells derived from ecological species such as bats and chicks (Tani et al. 2019; Katayama et al. 2018) (Figure 19.4).

Some important parameters can be derived from the dose-response curves: the lethal concentration 50% (LD_{50}), the no-observed-effect concentration (NOEC) and the lowest-observed-effect concentration (LOEC) are the usual calculations reported from toxicity tests. It is important to note that tests performed in animals do pose challenges with respect to extrapolation to humans. Animals have different metabolism than humans, hence the need to multiply the obtained results

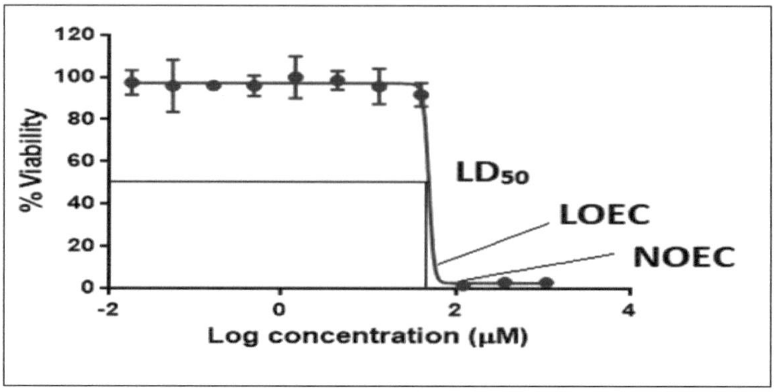

Figure 19.4 Representative cytotoxicity measurement of chemical using human derived cells plotted as percent viability versus log of the compound concentration in μM.

Table 19.1 Calculated Toxicity Parameters

Toxicity tests	*In vitro* toxicity data (IC_{50} in ppm)	EC_{50} (ppm) from daphnids (EC_{50} = 3030 ppm)
Fibroblasts	1809.88 ± 466	–
HepG2	1100.23 ± 95	–
Lung (A549)	3465.7 ± 0.017	–
Oyster development	–	1515 ± 350
Sea urchin development	–	1614 ± 50

by a factor. The FDA proposed the use of a 100-fold factor to account for several sources of uncertainties (Dankovic et al. 2015).

The correlation between ecological and human toxicity using unknown chemicals was presented in research conducted in a natural river in Tasmania (Khalil and Winder 2008). The authors compared contaminated water toxicity using daphnia, development tests among sea urchins and primary human-derived fibroblasts (Khalil and Winder 2008). The data presented in Table 19.1 indicates a good correlation in the toxic response to an environmental insult between environmental indicator species and human derived cells *in vitro*. The paper reported a good correlation (R^2 = 0.75) between the IC_{50} *in vitro* data using human immortalized cells (1100.23 ± 95 ppm) and the EC_{50} from sea urchin (1614 ± 50 ppm) and oyster development tests (1515 ± 350 ppm).

19.3.2.2 Ecological Systems

An alternative to using animal- or human-derived cells to establish toxicity of pharmaceutical products involves using indicator species in the natural environment. Examples of such tests include:

- Sediment toxicity.
- Effluent toxicity.
- Bioconcentration.
- Amphibian toxicity.
- Earthworm toxicity.
- Plant tests.

19.3.2.2.1 Sediment Toxicity Testing

The traditional approach in measuring toxicity depends on collecting samples and subjecting them to analytical chemistry to determine their constituents (Khalil et al. 2018) in order to establish overall toxicity resulting from chemicals present in the mixture.

This approach provides a good snapshot of the pollutants present in the water column, but unfortunately it does not account for toxic components that can accumulate over time in the sediment. Sediment toxicity testing is important because the biological processes used for cleaning wastewater and sewage are not very effective in removing pharmaceuticals and their derivatives.

Sediment toxicity testing is conducted in a controlled laboratory environment using a single species (freshwater or marine) of amphipods, freshwater invertebrates or planktonic species with numerous published methods (Simpson et al. 2016) for collecting data and analysis.

Pharmaceuticals and their derivatives pose significant environmental challenges in the sediment in aquatic ecosystems, where they have the potential to bioaccumulate (Kurth et al. 2017). Researchers have investigated the toxicity, accumulation potential, stress induction and impaired gonad development of various environmental contaminants, using a range of animals such as zebrafish (Babic et al. 2018; Kurth et al. 2017) and midges (Nieto et al. 2017).

19.3.2.2.2 Effluent Toxicity Testing

The EPA has developed effluent toxicity testing methods that have been extensively used to determine industrial effluents and stormwater impacts on aquatic life. These methods test for acute and chronic toxicity of effluents using approved methodologies. The test species recommended for these methods include different species of algae, invertebrates and fish.

Results of testing of leachate effluents was presented in Figure 19.2. More specific literature exists on effluents containing pharmaceuticals (Maselli Bde et al. 2015; Tamura et al. 2017), antibiotics and veterinary products (Maselli Bde et al. 2015).

19.3.2.2.3 Bioconcentration Toxicity Testing

Bioconcentration testing is mainly undertaken in fish and other aquatic species (algae and mollusks) to determine the ability of the species to bioconcentrate contaminants present in material they use as their food source. Researchers reported the bioconcentration potential of pharmaceuticals in trout (Grabicova et al. 2017), fish (McCallum et al. 2019) and invertebrates (Miller et al. 2019).

19.3.2.2.4 Amphibian Testing

Amphibian testing involves frogs and their larvae. The selection of amphibian testing is the result of their sensitivity to chemicals including pharmaceuticals in addition to their complex lives at the surface and within aquatic ecosystems. The EPA has developed various detailed testing protocols in this respect.

The frog embryo teratogenesis assay-Xenopus (FETAX) is an example of standardized bioassays that can be used for assessing developmental toxicity (Mouche et al. 2011). The presence of pharmaceuticals in aquatic systems does pose some challenges affecting development toxicity in species. The impact of the hormonal (Safholm et al. 2014) and teratogenic (Melvin et al. 2017; Islas-Flores et al. 2018) effects of selected pharmaceuticals has been described in the literature.

19.3.2.2.5 Earthworm Testing

Soil toxicity is important for the ecological populations inhabiting it. Direct measures of soil toxicity can be undertaken by leaching soil samples and subjecting them to analytical assessment. This very useful approach, however, does not provide information on toxicity of soil to terrestrial organisms. A range of tests have therefore been developed to test using various organisms, such as the earthworm test. Plant germination and growth tests also are available for determination of soil toxicity. These tests have been used to assess the uptake of pharmaceuticals by earthworms (Carter et al. 2016b, 2016a).

19.3.2.2.6 Plant Testing

Several tests using plants have been developed for assessing chemical presence in the natural environment, including the seed germination test. Pharmaceuticals in the environment pose significant stress on plants and can be absorbed by plants (Madikizela et al. 2018; Ristroph et al. 2017) and affect their development and hormone homeostasis (Carter et al. 2015). Plant testing is part of ecotoxicological assessment, whereby the impact of chemicals in the natural environment is assessed. It is important to note that such toxicity assessment could also be correlated to mammalian toxicity if humans consume the plants or animals exposed to chemicals.

19.4 ECOSYSTEM POLLUTION

Environmental toxicology requires a basic understanding of how ecosystems interact within the context of a food web. This is important, as there is a need to understand the movement of contaminants throughout the food web and how they can affect plants and animal species. The release of pollutants into the natural environment, even in minute quantities, can impair ecological function. The pollutants, whether single or a mixture of chemicals, have to be bioavailable for uptake by animals for the process to work.

One of the main areas of impacts of pharmaceutical by products is at the water level. Therefore, the focus of the next section will be on water pollution to introduce the pharmaceuticals and the impact of their by-products on natural systems.

19.4.1 Water Pollution Ecosystem Toxicological Impacts

The availability of a chemical will depend on its lipophilic properties and it ability to cross cellular membranes. This is the result of the phenomena of bioaccumulation and biomagnification in the food chain. Pollutants have different target organs in ecological species, with some pollutants like pesticides having a strong affinity to fat deposits. Animals exposed to pesticides will store them in their fat deposits (bioaccumulation). As predators in the upper trophic level, they will consume these animals and biomagnify the levels in their bodies, as shown in DDT high levels in birds feeding on fish (see Figure 19.5).

There are numerous case studies illustrating the significant risks posed by exposure to contaminants biomagnified in a system. Examples from Japan include mercury poisoning in Minamata Bay (Sakamoto et al. 2018). Other studies also highlighted the fate and toxicological impacts of pharmaceutical products presence in surface waters (Maldonado-Torres et al. 2018).

19.4.2 Chemical Mixtures in the Environment

The assessment of chemical mixtures and their cytotoxicity is very challenging. This is mainly the result of chemical interactions in mixtures that can lead to an overall toxic profile independent from an individual toxicity profile. The EPA approach to mixture is to assume an additive effect in calculating the toxicity that does not really reflect the overall toxicity. Chemical interactions between chemicals can lead to additive, synergistic, potentiative or antagonistic effects. An additive interaction occurs when two chemicals can trigger a toxic effect that is equal to the sum of the LD_{50} of each chemical. A synergistic interaction occurs when the two chemicals trigger a toxic impact that is much greater than the sum of their individual LD_{50}. A potentiative interaction involves an overall toxic response greater than the sum of their individual LD_{50}, with one of the chemicals enhancing the toxicity of the other chemical. An antagonistic interaction occurs when one of the chemicals in solution counteracts the effect of the other chemical, so no toxicity is

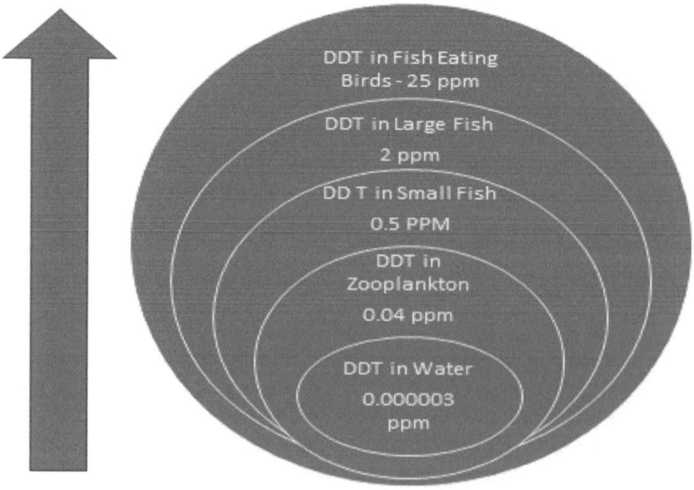

Figure 19.5 DDT Bioaccumulation and bio magnification in food web.

Source: Adapted from Miller and Spoolman (2012).

reported. Humans, animals and plants species are rarely exposed to single chemicals but rather to a cocktail of chemicals. This renders the establishment of toxicity more problematic. These exposures are the result of accepting the consensus on use of chemicals for pest management and disease control.

In the natural environment, it is very challenging to identify single chemicals in pollutants, let al.ne mixtures of chemicals and their potential interactions. Researchers have designed a number of models to assess overall mixture toxicity before trying to fingerprint the toxicity to individual chemicals present in the chemical cocktail (Bakand et al. 2005).

Studies on multiple chemical exposures in natural systems have confirmed the complex interactions between pharmaceutical entities. For example, a recent study (Simu et al. 2018) on pharmaceutical mixtures highlighted the concerns resulting from combining more than one drug. The study highlighted the potentiative interaction between fluoxetine (an antidepressant that had no effect on water fleas) and clofibric acid (a pesticide), leading to significant mortalities in water fleas. Other similar studies on mixtures investigated the interaction between paracetamol (acetaminophen) and hypromellose, in mixture of these chemicals affected paracetamol crystallinity (Leyk and Wesolowski 2019). Multiple chemical exposures in the natural environment can also lead to oxidative stress in exposed fish (McRae et al. 2019) and various toxicity profiles. A good example of toxicity profile alteration are the synergistic interactions between various bisphenol analogue mixtures that were reported as potential endocrine disruptors (Owczarek et al. 2018).

19.4.3 Chemical Genotoxic Potential in the Natural Environment

Chemicals can be assessed for toxicity using the dose-response concept presented previously, with the exception of carcinogenic chemicals. Carcinogenic chemicals can trigger cancer, and there is a long induction period between exposure and symptoms in exposed individuals. Scientists have developed a range of tests such as the Ames test and Comet assays to establish the mutagenic potential of chemicals and their mixtures. The Ames test assesses the potential of a chemical to trigger mutations using various strains of *Salmonella typhimurium*. These bacteria carry a special genetic mutation in histidine synthesis genes and therefore cannot grow in histidine-free media. The tested substance ability to mutate the genetics of the bacteria, allowing it to grow in histidine-free media, is assessed. The Comet assay, in contrast, is based on cell cultures and evaluates single- and double-strand DNA damage triggered by tested substances. Affected cells undergo a series of manipulations to unwind the DNA before electrophoresis. The DNA damaged cells will migrate from the cell in the form of a comet. The comet parameters are measured to establish levels of genotoxic damages. In cells with no DNA damage, the DNA will not migrate in the form of a comet but rather form a halo around the cell (Soni et al. 2010).

A good example of genotoxic damage is the measure of single- and double-strand DNA breaks in animal organs and cell cultures resulting from exposure to environmental pollutants using the comet assay system (Khalil et al. 2018). Individual pharmaceuticals and their mixtures have also been assessed for their toxicity in the environment (Bakare et al. 2009; Wieczerzak et al. 2019), and some researchers suggested that they also contributed to oxidative stress in affected species (Adeoye et al. 2015).

19.5 CONCLUSION

This chapter introduced environmental toxicology and various testing methodologies that elucidated some of the important ecological risks posed by releasing pharmaceuticals and their by-products on species and their natural habitats. A wealth of scientific publications was presented highlighting the significant toxicological impacts triggered by chemical products, including pharmaceuticals in our natural environment. Chemical extensive use in health promotion worldwide has created important challenges for regulators globally. The need to promote health and at the same time preserve the natural system requires innovative thinking and immediate actions from regulators and pharmaceutical manufacturers worldwide. Strategic global policies should be developed and adopted to attenuate and reduce the risks of human and environmental harm triggered by these products. The presence of various chemicals and their derivatives in drinking water, plants and meat products in areas around the globe should raise alarm bells. Perhaps the most challenging task is to better research and understand the interactive nature of chemicals and the potentiative and synergistic toxicological impacts triggered when they are present in complex mixtures, which is the case in nature. We need to develop better technologies to treat our effluents to remove pharmaceuticals prior to discharge into natural systems. We need to draft better regulation for properly administering these pharmaceuticals in the human population with minimal environmental impacts. More funded research is needed to better understand the significant environmental and ecological impacts posed by pharmaceuticals and their by-products. There are pressing needs to innovate and develop environmentally friendly pharmaceuticals with shorter half-lives that can provide benefits for the users and degrade quickly enough before triggering hormonal and reproductive changes and mortalities in affected species. We should strive to meet our obligations to the environment and future generations through adopting the precautionary principle regarding pharmaceuticals. We should aim for using our existing knowledge to preserve the natural systems that are essential for our existence, health and wellbeing.

BIBLIOGRAPHY

Acir, I.-H., and K. Guenther. 2018. Endocrine-disrupting metabolites of alkylphenol ethoxylates—A critical review of analytical methods, environmental occurrences, toxicity, and regulation. *Sci Total Environ* 635: 1530–46.

Adeoye, G. O., C. G. Alimba, and O. B. Oyeleke. 2015. The genotoxicity and systemic toxicity of a pharmaceutical effluent in Wistar rats may involve oxidative stress induction. *Toxicol Rep* 2: 1265–72.

Al Hageh, C., M. Al Assaad, and Z. El Masri, et al. 2018. A long-lived cuprous bis-phenanthroline complex for the photodynamic therapy of cancer. *Dalton Trans* 47(14): 4959–67.

Andreu, V., E. Gimeno-Garcia, J. A. Pascual, P. Vazquez-Roig, and Y. Pico. 2016. Presence of pharmaceuticals and heavy metals in the waters of a Mediterranean coastal wetland: Potential interactions and the influence of the environment. *Sci Total Environ* 540: 278–86.

aus der Beek, T., F.-A. Weber, A. Bergmann, et al. 2016. Pharmaceuticals in the environment—Global occurrences and perspectives. *Environ Toxicol Chem* 35(4): 823–35.

Babic, S., J. Barisic, D. Stipanicev, et al. 2018. Assessment of river sediment toxicity: Combining empirical zebrafish embryotoxicity testing with in silico toxicity characterization. *Sci Total Environ* 643: 435–50.

Bai, Y., C. Zang, M. Gu, et al. 2017. Sewage sludge as an initial fertility driver for rapid improvement of mudflat salt-soils. *Sci Total Environ* 578: 47–55.

Bakand, S., C. Winder, C. Khalil, and A. Hayes. 2005. Toxicity assessment of industrial chemicals and airborne contaminants: Transition from in vivo to in vitro test methods: A review. *Inhal Toxicol* 17(13): 775–87.

Bakare, A. A., A. A. Okunola, O. A. Adetunji, and H. B. Jenmi. 2009. Genotoxicity assessment of a pharmaceutical effluent using four bioassays. *Genet Mol Biol* 32(2): 373–81.

Balbo, S., and I. Stepanov. 2018. The wild west of e-cigarettes. *Chem Res Toxicol* 31(9): 823–4.

Berkner, S., S. Konradi, and J. Schönfeld. 2014. Antibiotic resistance and the environment—there and back again: Science & Society series on Science and Drugs. *EMBO Rep* 15(7): 740–4.

Brodin, T., S. Piovano, J. Fick, J. Klaminder, M. Heynen, and M. Jonsson. 2014. Ecological effects of pharmaceuticals in aquatic systems—Impacts through behavioural alterations. *Philos Trans R Soc Lond B Biol Sci* 369(1656).

Burlo, F., I. Guijarro, A. A. Carbonell-Barrachina, D. Valero, and F. Martinez-Sanchez. 1999. Arsenic species: Effects on and accumulation by tomato plants. *J Agric Food Chem* 47(3): 1247–53.

Carter, L. J., J. J. Ryan, and A. B. Boxall. 2016a. Does uptake of pharmaceuticals vary across earthworm species? *Bull Environ Contam Toxicol* 97(3): 316–22.

Carter, L. J., J. J. Ryan, and A. B. Boxall. 2016b. Effects of soil properties on the uptake of pharmaceuticals into earthworms. *Environ Pollut* 213: 922–31.

Carter, L. J., M. Williams, C. Bottcher, and R. S. Kookana. 2015. Uptake of pharmaceuticals influences plant development and affects nutrient and hormone homeostases. *Environ Sci Technol* 49(20): 12509–18.

Chen, Q., M. R. Larochelle, D. T. Weaver, et al. 2019. Prevention of prescription opioid misuse and projected overdose deaths in the United States. *JAMA Netw Open* 2(2): e187621.

Clara, M., B. Strenn, and N. Kreuzinger. 2004. Carbamazepine as a possible anthropogenic marker in the aquatic environment: Investigations on the behaviour of Carbamazepine in wastewater treatment and during groundwater infiltration. *Water Res* 38(4): 947–54.

Colon-Berezin, C., M. L. Nolan, J. Blachman-Forshay, and D. Paone. 2019. Overdose deaths involving fentanyl and fentanyl analogs—New York City, 2000–2017. *MMWR Morb Mortal Wkly Rep* 68(2): 37–40.

Corcoran, J., M. J. Winter, and C. R. Tyler. 2010. Pharmaceuticals in the aquatic environment: A critical review of the evidence for health effects in fish. *Crit Rev Toxicol* 40(4): 287–304.

Crossin, R., D. Scott, S. Arunogiri, K. Smith, P. M. Dietze, and D. I. Lubman. 2019. Pregabalin misuse-related ambulance attendances in Victoria, 2012–2017: Characteristics of patients and attendances. *Med J Aust* 210(2): 75–9.

Dalkmann, P., M. Broszat, C. Siebe, et al. 2012. Accumulation of pharmaceuticals, Enterococcus, and resistance genes in soils irrigated with wastewater for zero to 100 years in central Mexico. *PLoS One* 7(9): e45397.

Dankovic, D. A., B. D. Naumann, A. Maier, M. L. Dourson, and L. S. Levy. 2015. The scientific basis of uncertainty factors used in setting occupational exposure limits. *J Occup Environ Hyg* 12(Suppl 1): S55–68.

Duarte, I. A., M. P. Pais, P. Reis-Santos, H. N. Cabral, and V. F. Fonseca. 2019. Biomarker and behavioural responses of an estuarine fish following acute exposure to fluoxetine. *Marine Env Res* 147: 24–31.

Fabbri, E., and S. Franzellitti. 2016. Human pharmaceuticals in the marine environment: Focus on exposure and biological effects in animal species. *Environ Toxicol Chem* 35(4): 799–812.

Gaw, S., K. V. Thomas, and T. H. Hutchinson. 2014. Sources, impacts and trends of pharmaceuticals in the marine and coastal environment. *Phil Trans R Soc Lond B Biol Sci* 369(1656).

Grabicova, K., R. Grabic, G. Fedorova, et al. 2017. Bioaccumulation of psychoactive pharmaceuticals in fish in an effluent dominated stream. *Water Res* 124: 654–62.

Grabowski, H., G. Long, R. Mortimer, and A. Boyo. 2016. Updated trends in US brand-name and generic drug competition. *J Med Econ* 19(9): 836–44.

Gunnarsdóttir, R., P. D. Jenssen, P. E. Jensen, A. Villumsen, and R. Kallenborn. 2013. A review of wastewater handling in the Arctic with special reference to pharmaceuticals and personal care products (PPCPs) and microbial pollution. *Ecol Eng* 50: 76–85.

Harada, M. 1995. Minamata disease: Methylmercury poisoning in Japan caused by environmental pollution. *Crit Rev Toxicol* 25(1): 1–24.

Harris, S., C. Morris, D. Morris, M. Cormican, and E. Cummins. 2014. Antimicrobial resistant Escherichia coli in the municipal wastewater system: Effect of hospital effluent and environmental fate. *Sci Total Environ* 468–469: 1078–85.

Hedegaard, H., B. A. Bastian, J. P. Trinidad, M. Spencer, and M. Warner. 2018. Drugs most frequently involved in drug overdose deaths: United States, 2011–2016. *Natl Vital Stat Rep* 67(9): 1–14.

Herklotz, P. A., P. Gurung, B. Vanden Heuvel, and C. A. Kinney. 2010. Uptake of human pharmaceuticals by plants grown under hydroponic conditions. *Chemosphere* 78(11): 1416–21.

Hofbauer, A., and E. Stoger. 2013. Subcellular accumulation and modification of pharmaceutical proteins in different plant tissues. *Curr Pharm Des* 19(31): 5495–502.

Islas-Flores, H., I. Perez-Alvarez, and L. M. Gomez-Olivan. 2018. Evaluation of Teratogenicity of Pharmaceuticals Using FETAX. *Methods Mol Biol* 1797: 299–307.

Jing, X., G. Yao, D. Liu, et al. 2017. Effects of wastewater irrigation and sewage sludge application on soil residues of chiral fungicide benalaxyl. *Environ Pollut* 224: 1–6.

Kallenborn, R., E. Brorstrom-Lunden, L. O. Reiersen, and S. Wilson. 2018. Pharmaceuticals and personal care products (PPCPs) in Arctic environments: Indicator contaminants for assessing local and remote anthropogenic sources in a pristine ecosystem in change. *Environ Sci Pollut Res Int* 25(33): 33001–13.

Katayama, M., T. Hirayama, T. Tani, K. Nishimori, M. Onuma, and T. Fukuda. 2018. Chick derived induced pluripotent stem cells by the poly-cistronic transposon with enhanced transcriptional activity. *J Cell Physiol* 233(2): 990–1004.

Khalil, C., C. Al Hageh, S. Korfali, and R. S. Khnayzer. 2018. Municipal leachates health risks: Chemical and cytotoxicity assessment from regulated and unregulated municipal dumpsites in Lebanon. *Chemosphere* 208: 1–13.

Khalil, C., J. B. Chahine, B. Chahla, T. Hobeika, and R. S. Khnayzer. 2019. Characterization and cytotoxicity assessment of nargile smoke using dynamic exposure. *Inhal Toxicol*: 1–14.

Khalil, C., and C. Winder. 2008. Surface water toxicity assessment by ecotoxicological and in vitro toxicological assays. *WIT Trans Ecol Environ* 110: 253–62.

Kim, D. H., V. M. Addis, W. Pan, and B. L. VanderBeek. 2019. Comparative effectiveness of generic latanoprost versus branded prostaglandin analogs for primary open angle glaucoma. *Ophthalmic Epidemiol* 26(1): 63–71.

Kostich, M. S., A. L. Batt, and J. M. Lazorchak. 2014. Concentrations of prioritized pharmaceuticals in effluents from 50 large wastewater treatment plants in the US and implications for risk estimation. *Environ Pollut* 184: 354–9.

Kurth, D., S. Lips, R. Massei, et al. 2017. The impact of chemosensitisation on bioaccumulation and sediment toxicity. *Chemosphere* 186: 652–9.

Küster, A., and N. Adler. 2014. Pharmaceuticals in the environment: Scientific evidence of risks and its regulation. *Phil Trans R Soc London. Ser B, Biol Sci* 369(1656): 20130587.

Leyk, E., and M. Wesolowski. 2019. Interactions between paracetamol and hypromellose in the solid state. *Frontiers Pharmacol* 10: 14.

Li, Y., J. B. Sallach, W. Zhang, S. A. Boyd, and H. Li. 2019. Insight into the distribution of pharmaceuticals in soil-water-plant systems. *Water Res* 152: 38–46.

Loch, A., J. P. Bewersdorf, D. Kofink, D. Ismail, I. Z. Abidin, and R. S. Veriah. 2017. Generic atorvastatin is as effective as the brand-name drug (LIPITOR((R))) in lowering cholesterol levels: A cross-sectional retrospective cohort study. *BMC Res Notes* 10(1): 291.

Loos, R., R. Carvalho, D. C. Antonio, et al. 2013. EU-wide monitoring survey on emerging polar organic contaminants in wastewater treatment plant effluents. *Water Res* 47(17): 6475–87.

Lu, M. C., Y. Y. Chen, M. R. Chiou, M. Y. Chen, and H. J. Fan. 2016. Occurrence and treatment efficiency of pharmaceuticals in landfill leachates. *Waste Manag* 55: 257–64.

Lundborg, C. S., and A. J. Tamhankar. 2017. Antibiotic residues in the environment of South East Asia. *BMJ (Clinical research ed.)* 358: j2440–j2440.

Madikizela, L. M., S. Ncube, and L. Chimuka. 2018. Uptake of pharmaceuticals by plants grown under hydroponic conditions and natural occurring plant species: A review. *Sci Total Environ* 636: 477–86.

Maldonado-Torres, S., R. Gurung, H. Rijal, et al. 2018. Fate, transformation, and toxicological impacts of pharmaceutical and personal care products in surface waters. *Environ Health Insights* 12: 1178630218795836.

Maselli Bde, S., L. A. Luna, O. Palmeira Jde, et al. 2015. Ecotoxicity of raw and treated effluents generated by a veterinary pharmaceutical company: A comparison of the sensitivities of different standardized tests. *Ecotoxicology* 24(4): 795–804.

Masoner, J. R., D. W. Kolpin, E. T. Furlong, I. M. Cozzarelli, and J. L. Gray. 2016. Landfill leachate as a mirror of today's disposable society: Pharmaceuticals and other contaminants of emerging concern in final leachate from landfills in the conterminous United States. *Environ Toxicol Chem* 35(4): 906–18.

McCallum, E. S., A. Sundelin, J. Fick, et al. 2019. Investigating tissue bioconcentration and the behavioural effects of two pharmaceutical pollutants on sea trout (Salmo trutta) in the laboratory and field. *Aquat Toxicol* 207: 170–8.

McRae, N. K., S. Gaw, B. W. Brooks, and C. N. Glover. 2019. Oxidative stress in the galaxiid fish, Galaxias maculatus, exposed to binary waterborne mixtures of the pro-oxidant cadmium and the anti-oxidant diclofenac. *Environ Pollut* 247: 638–46.

Melvin, S. D., L. J. Habener, F. D. L. Leusch, and A. R. Carroll. 2017. (1)H NMR-based metabolomics reveals sub-lethal toxicity of a mixture of diabetic and lipid-regulating pharmaceuticals on amphibian larvae. *Aquat Toxicol* 184: 123–32.

Miller, G. T., and S. Spoolman. 2012. *Living in the Environment: Principles, Connections, and Solutions*. Belmont, CA: Brooks/Cole.

Miller, T. H., M. D. Gallidabino, J. I. MacRae, S. F. Owen, N. R. Bury, and L. P. Barron. 2019. Prediction of bioconcentration factors in fish and invertebrates using machine learning. *Sci Total Environ* 648: 80–9.

Mojtabai, R., M. Amin-Esmaeili, E. Nejat, and M. Olfson. 2019. Misuse of prescribed opioids in the United States. *Pharmacoepidemiol Drug Saf.*

Moorkens, E., C. Jonker-Exler, I. Huys, P. Declerck, S. Simoens, and A. G. Vulto. 2016. Overcoming barriers to the market access of biosimilars in the European Union: The case of biosimilar monoclonal antibodies. *Front Pharmacol* 7: 193.

Mouche, I., L. Malesic, and O. Gillardeaux. 2011. FETAX assay for evaluation of developmental toxicity. *Methods Mol Biol* 691: 257–69.

Murnion, B., and K. M. Conigrave. 2019. Pregabalin misuse: The next wave of prescription medication problems. *Med J Aust* 210(2): 72–3.

Nakamura, A., T. Osonoi, and Y. Terauchi. 2010. Relationship between urinary sodium excretion and pioglitazone-induced edema. *J Diabetes Investig* 1(5): 208–11.

Nieto, E., C. Corada-Fernandez, M. Hampel, P. A. Lara-Martin, P. Sanchez-Arguello, and J. Blasco. 2017. Effects of exposure to pharmaceuticals (diclofenac and carbamazepine) spiked sediments in the midge, Chironomus riparius (Diptera, Chironomidae). *Sci Total Environ* 609: 715–23.

Oaks, J. L., M. Gilbert, M. Z. Virani, et al. 2004. Diclofenac residues as the cause of vulture population decline in Pakistan. *Nature* 427(6975): 630–3.

Owczarek, K., B. Kudlak, V. Simeonov, Z. Mazerska, and J. Namiesnik. 2018. Binary mixtures of selected bisphenols in the environment: Their toxicity in relationship to individual constituents. *Molecules* 23(12).

Pereira, A., L. J. G. Silva, C. S. M. Laranjeiro, L. M. Meisel, C. M. Lino, and A. Pena. 2017. Human pharmaceuticals in Portuguese rivers: The impact of water scarcity in the environmental risk. *Sci Total Environ* 609: 1182–91.

Puckowski, A., K. Mioduszewska, P. Lukaszewicz, et al. 2016. Bioaccumulation and analytics of pharmaceutical residues in the environment: A review. *J Pharm Biomed Anal* 127: 232–55.

Quesada, H. B., A. T. A. Baptista, L. F. Cusioli, D. Seibert, C. de Oliveira Bezerra, and R. Bergamasco. 2019. Surface water pollution by pharmaceuticals and an alternative of removal by low-cost adsorbents: A review. *Chemosphere* 222: 766–80.

Ristroph, K. D., C. E. Astete, E. Bodoki, and C. M. Sabliov. 2017. Zein nanoparticles uptake by hydroponically grown soybean plants. *Environ Sci Technol* 51(24): 14065–71.

Safholm, M., A. Ribbenstedt, J. Fick, and C. Berg. 2014. Risks of hormonally active pharmaceuticals to amphibians: A growing concern regarding progestogens. *Philos Trans R Soc Lond B Biol Sci* 369(1656).

Sakamoto, M., N. Tatsuta, K. Izumo, et al. 2018. Health impacts and biomarkers of prenatal exposure to methylmercury: Lessons from Minamata, Japan. *Toxics* 6(3).

Schultz, M. M., E. T. Furlong, D. W. Kolpin, et al. 2010. Antidepressant pharmaceuticals in two U.S. effluent-impacted streams: Occurrence and fate in water and sediment, and selective uptake in fish neural tissue. *Environ Sci Technol* 44(6): 1918–25.

Schuring, C. A., D. L. Dodge, T. J. Whitcomb, G. C. Wall, H. L. Smith, and G. A. Hicklin. 2018. Overdoses and substance toxicity in patients admitted to intensive care units in a Midwestern U.S. City. *J Trauma Nurs* 25(2): 87–91.

Shen, Z. J., C. Xu, Y. S. Chen, and Z. Zhang. 2017. Heavy metals translocation and accumulation from the rhizosphere soils to the edible parts of the medicinal plant Fengdan (Paeonia ostii) grown on a metal mining area, China. *Ecotoxicol Environ Saf* 143: 19–27.

Shrivastava, A., D. Ghosh, A. Dash, and S. Bose. 2015. Arsenic contamination in soil and sediment in India: Sources, effects, and remediation. *Curr Pollut Rep* 1(1): 35–46.

Sicras-Mainar, A., L. Sanchez-Alvarez, R. Navarro-Artieda, and J. Darba. 2018. Treatment persistence and adherence and their consequences on patient outcomes of generic versus brand-name statins routinely used to treat high cholesterol levels in Spain: A retrospective cost-consequences analysis. *Lipids Health Dis* 17(1): 277.

Simpson, S. L., O. Campana, and K. T. Ho. 2016. Chapter 7—Sediment toxicity testing. In *Marine Ecotoxicology*, eds. J. Blasco, P. M. Chapman, O. Campana, and M. Hampel, 199–237. Academic Press.

Simu, G. M., J. Atchana, C. M. Soica, D. E. Coricovac, S. C. Simu, and C. A. Dehelean. 2018. Pharmaceutical mixtures: Still a concern for human and environmental health. *Curr Med Chem* 27(1): 121–53.

Smart, A., B. de Lacy Costello, P. White, et al. 2019. Sniffing out resistance—Rapid identification of urinary tract infection-causing bacteria and their antibiotic susceptibility using volatile metabolite profiles. *J Pharm Biomed Anal* 167: 59–65.

Soni, B., N. P. Visavadiya, N. Dalwadi, D. Madamwar, C. Winder, and C. Khalil. 2010. Purified c-phycoerythrin: Safety studies in rats and protective role against permanganate-mediated fibroblast-DNA damage. *J Appl Toxicol* 30(6): 542–50.

Sui, Q., W. Zhao, X. Cao, et al. 2017. Pharmaceuticals and personal care products in the leachates from a typical landfill reservoir of municipal solid waste in Shanghai, China: Occurrence and removal by a full-scale membrane bioreactor. *J Hazard Mater* 323(Pt A): 99–108.

Sullivan, J., J. Greenfield, G. Cumberford, J. Grant, and J. Stewart. 2010. Extraction efficiencies of heavy metals in hydroethanolic solvent from herbs of commerce. *J AOAC Int* 93(2): 496–8.

Tamura, I., Y. Yasuda, K. I. Kagota, et al. 2017. Contribution of pharmaceuticals and personal care products (PPCPs) to whole toxicity of water samples collected in effluent-dominated urban streams. *Ecotoxicol Environ Saf* 144: 338–50.

Tani, T., T. Eitsuka, M. Katayama, et al. 2019. Establishment of immortalized primary cell from the critically endangered Bonin flying fox (Pteropus pselaphon). *PLoS One* 14(8): e0221364.

Taylor, D. 2016. The pharmaceutical industry and the future of drug development. In *Pharmaceuticals in the Environment*, 1–33. The Royal Society of Chemistry.

Theodoridi, A., A. Tsalafouta, and M. Pavlidis. 2017. Acute exposure to fluoxetine alters aggressive behavior of zebrafish and expression of genes involved in serotonergic system regulation. *Front Neurosci* 11: 223.

Thiele, B., K. Gunther, and M. J. Schwuger. 1997. Alkylphenol Ethoxylates: Trace analysis and environmental behavior. *Chem Rev* 97(8): 3247–72.

Trombini, C., M. Hampel, and J. Blasco. 2019. Assessing the effect of human pharmaceuticals (carbamazepine, diclofenac and ibuprofen) on the marine clam Ruditapes philippinarum: An integrative and multibiomarker approach. *Aquat Toxicol* 208: 146–56.

Valdes, M. E., M. V. Ame, L. Bistoni Mde, and D. A. Wunderlin. 2014. Occurrence and bioaccumulation of pharmaceuticals in a fish species inhabiting the Suquia River basin (Cordoba, Argentina). *Sci Total Environ* 472: 389–96.

Varela, A. R., G. Ferro, J. Vredenburg, et al. 2013. Vancomycin resistant enterococci: From the hospital effluent to the urban wastewater treatment plant. *Sci Total Environ* 450–451: 155–61.

Wieczerzak, M., J. Namiesnik, and B. Kudlak. 2019. Genotoxicity of selected pharmaceuticals, their binary mixtures, and varying environmental conditions—Study with human adenocarcinoma cancer HT29 cell line. *Drug Chem Toxicol*: 1–11.

Willen, A., C. Junestedt, L. Rodhe, M. Pell, and H. Jonsson. 2017. Sewage sludge as fertiliser—Environmental assessment of storage and land application options. *Water Sci Technol* 75(5–6): 1034–50.

Wolkers, C. P. B., M. Serra, A. Barbosa Junior, and E. C. Urbinati. 2017. Acute fluoxetine treatment increases aggressiveness in juvenile matrinxa (Brycon amazonicus). *Fish Physiol Biochem* 43(3): 755–9.

Zenker, A., M. R. Cicero, F. Prestinaci, P. Bottoni, and M. Carere. 2014. Bioaccumulation and biomagnification potential of pharmaceuticals with a focus to the aquatic environment. *J Environ Manag* 133: 378–87.

Zhong, H., G. Chan, Y. Hu, H. Hu, and D. Ouyang. 2018. A comprehensive map of FDA-approved pharmaceutical products. *Pharmaceutics* 10(4).

20 Exposomics and Environmental Monitoring

Sonia Dagnino and Jessica Laine

CONTENTS

20.1 HISTORY AND DEFINITION OF THE EXPOSOME

In the past decades, it has been well recognized that environmental exposures play a key role in disease etiology, yet there was a critical need to take a more comprehensive approach to better understand these exposures by collecting more complete exposure assessments and to better understand mechanisms underlying exposure-disease responses in human populations. To address this, Christopher Wild introduced the concept of the "exposome" in 2005, which was defined as the totality of all "life-course environmental exposures (including lifestyle factors), from the prenatal period onwards" that can influence and trigger consequences on our health (Wild 2005). This concept was created to complement the genome. Advances in the study of the genome had promised to reveal key causes for understanding noncommunicable diseases; however, from the first successful genome-wide association study (GWAS) performed in 2002 to today, over 3000 studies have examined over 1800 diseases and traits and have revealed that only a small portion of diseases can be explained by genetics. Other factors, such as diet, physical activity, lifestyle, environmental exposures, and socioeconomic status, have a strong influence on health outcomes (Lichtenstein et al. 2000; Saracci and Vineis 2007; Willett 2002). For example, in 2010 it was estimated that for over 50 million global deaths, approximately 50% could be attributed to a small set of exposures, where tobacco smoke, alcohol use, diet, air pollution (internal and ambient) and low physical activity were ranked among the top 10 risk factors attributable for disease burdens (Lim et al. 2012). These findings emphasize the importance of considering multiple exposures in determining causality and prevention of diseases and early mortality.

20.2 WHAT IS THE EXPOSOME?

Since 2005, the definition of the exposome has evolved. The most recent description was defined by Miller and Jones as "the cumulative measure of environmental influences and associated biological responses throughout the lifespan, including exposures from environment, diet,

DOI: 10.1201/9780203730584-20

behavior, and endogenous processes" (Miller and Jones 2014). With this definition, the exposome takes into account (1) general external exposures (climate, urban environment, etc.), which are common to a great number of individuals; (2) the specific external exposome (e.g. diet, lifestyle, behavior), which is more individual-specific; and the (3) internal exposome (e.g. DNA mutations, metabolism), which encompasses processes internal to the body (Figure 20.1; Rappaport and Smith 2010). Collectively, this view promotes a more comprehensive understanding of the exposome by considering a broader view of exposures and the cumulative biological responses and endogenous processes that occur from various exposures.

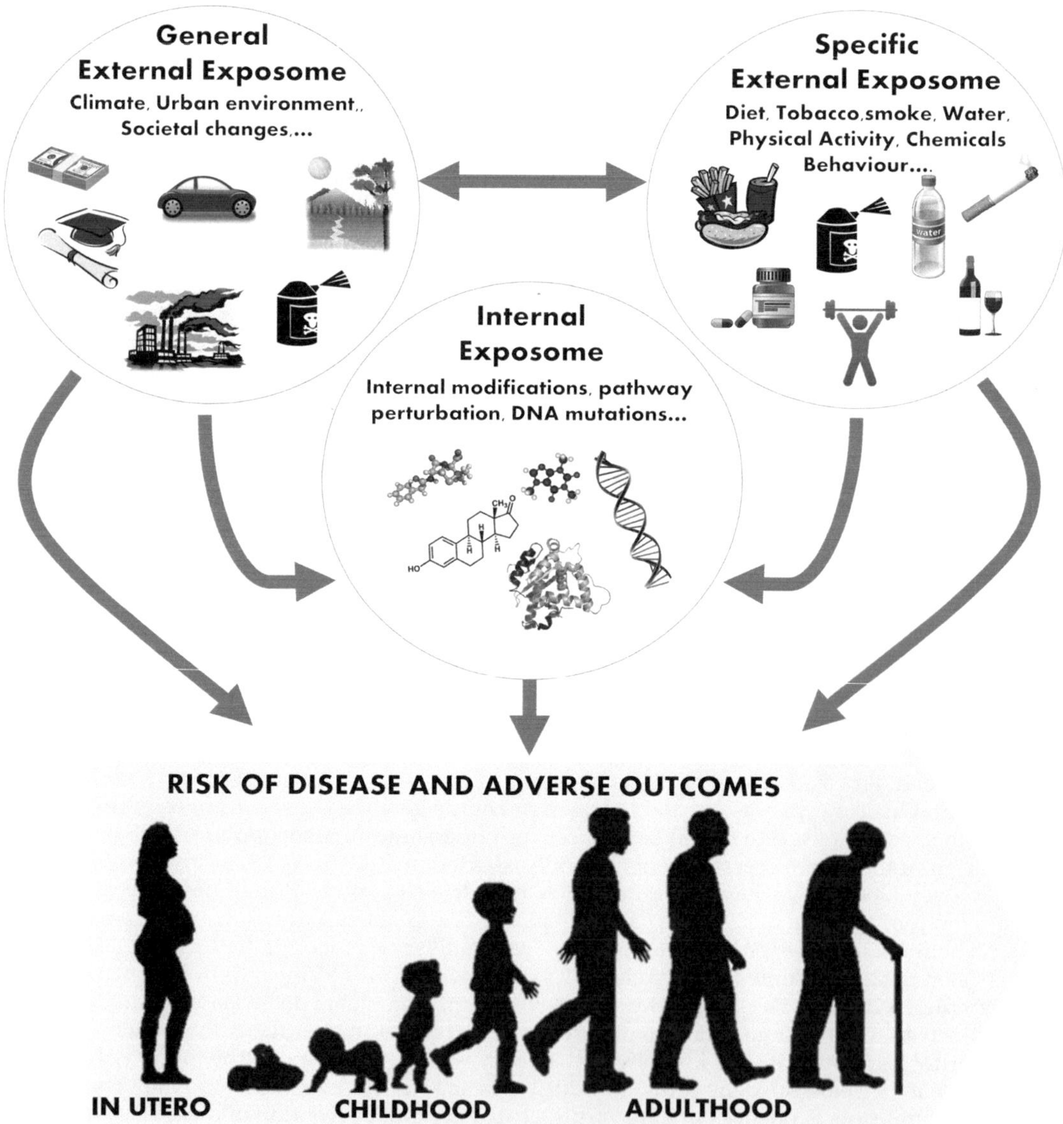

Figure 20.1 The three layers of factors that can influence the exposome from conception onwards throughout the life course.

Source: Adapted from Dagnino (2019).

20.3 HOW TO MEASURE THE EXPOSOME

The peculiarity of the exposome is that, unlike the static genome, it is ever changing. In 2011, Rappaport proposed two methods to measure the exposome (Rappaport 2011). The first approach ("bottom up") consists of measuring all exposures of the external exposome by analyzing potential exposures from environmental samples (air, water, food, etc.) and linking them to health outcomes. The advantage of this approach is that it identifies the exposure source, but it does not consider the internal processes, such as metabolism, oxidative stress, and inflammation. The second approach ("top-down") consists of measuring the internal exposome by measuring either biomarkers of exposure(s) or molecular responses represented by 'omics approaches in an individual's blood, urine, or other body fluids and assessing their relationship to health outcomes (Figure 20.2). The success of this second approach has been led by the development of 'omics techniques which allow for the measure of a great number of internal features (e.g. metabolomics). The disadvantage of the top-down method is that sources of exposure may not be easily identified.

In this chapter, we will explore current and new developments for the evaluation of the internal and external exposome. The advancement of the field of exposome science has been propelled by improvements in exposure science, molecular epidemiology, and big-data statistical approaches.

20.3.1 Measuring the External Exposome: Technologies for Measuring the External Exposome

To fully understand the exposome, internal and external measurements must be combined. 'Omics technologies provide information on exposure to environmental chemicals by measuring the chemical itself in the body (e.g. metabolomics), its metabolites, or its molecular signatures subsequent to exposure(s). However, assessing the internal exposome alone cannot always provide information on the nature of exposures, including the duration (i.e. frequency and timing), magnitude (i.e. exposure concentration and dose), interactions of mixtures, and complex dose-response relationships of exposure(s). The measurement of the external exposome, on the other hand, requires more detailed exposure assessments to permit the understanding and the nature of exposures. Specifically, in measuring the external exposome we want to consider the timing and location for when and where the exposure occurred, exposure durations, and the complexity of exposure mixtures. Additionally, exposures must be evaluated at a high time and space resolution and at the individual level. Such details may lead to less exposure misclassification and improvements in estimations of associations between exposures and health outcomes. The measure of the external exposome has greatly evolved in the past decade. Traditional exposure assessments relied on questionnaires and monitors for single or a few exposures, whereas new exposure assessments are aimed to capture a large quantity of exposures in the range of thousands to hundreds

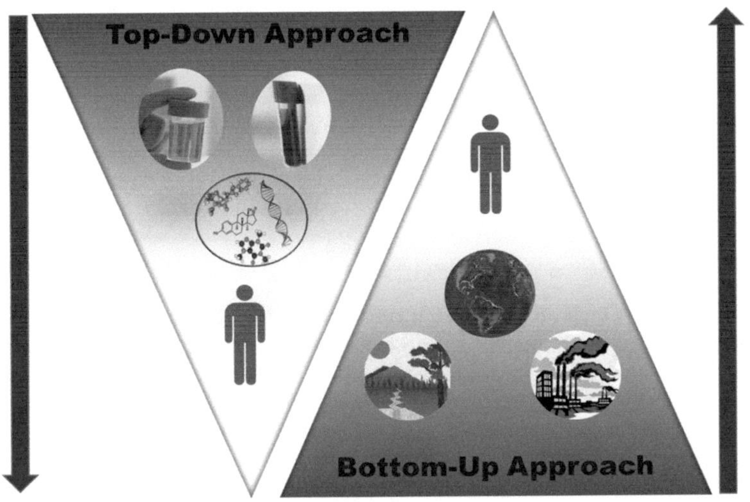

Figure 20.2 The two main approaches for the analysis of the exposome. The bottom-up approach identifies exogenous exposures from external measurements (air, water, food, etc.); the top-down approach identifies exposures from internal measurements (blood, urine, etc.).

of thousands. The development of new sensor-based technologies and geocoding (e.g. with geographic information systems [GIS]) has allowed for the estimation of exposures over broad geographical areas and promoted the establishment of large epidemiological investigations that link many external exposures to health outcomes.

20.3.2 Personal Exposure Assessment

20.3.2.1 Traditional Analysis of the External Exposome

Personal exposure can also be assessed by measuring specific xenobiotics in environmental matrices through from a variety of samples. An exposome approach can be particularly useful when environmental matrices might contain hundreds or thousands of chemicals, such as those found in household dust and food, among other samples. This form of exposure assessment is more traditional and has been applied in numerous epidemiological studies. Assessing environmental samples may be targeted or untargeted.

Analysis of household dust has been used in multiple studies to evaluate indoor exposure to specific chemicals. Dust is an important environmental media, as compounds originating from the indoors and those brought home from the outdoors can concentrate in dust samples. For example, many studies have analyzed household dust to evaluate exposure to flame retardants (Castorina et al. 2017; Hoffman et al. 2015). Recently, multiple samples of household dust were analyzed with an "untargeted" approach. Analysis revealed the presence of 3,228 possible chemicals in dust samples, most of them never described before in the literature as having been present in the household (Rager et al. 2016).

The dietary exposome, which quantifies either toxic exposures from food or food compounds such as nutrients, can provide important information for continual and ubiquitous exposure in humans via their food intake. Recently, Scalbert et al. reviewed the importance of the food exposome and how it contributes to the understanding of dietary exposures. Dietary exposures can be evaluated by extracting information from detailed questionnaires, but also by analyzing food components from metabolomics data (Scalbert et al. 2019). Multiple biomarkers of food intake have been identified as part of the metabolome, which can be compounds present in food or metabolites of these compounds. Metabolomic dietary biomarkers are less prone to errors and misclassification compared to questionnaires, as the assessment eliminates report biases that arise from self-reported diet, such as recall bias. Using this novel approach, Khakimov et al. found that in a Danish cohort, 33 dietary metabolites measured in plasma discriminated individuals with a healthier diet from those who had an unhealthy diet. This indicates the importance of such information on lifestyle behaviors for the evaluation of the exposome (Khakimov et al. 2016).

20.3.2.2 Sensors

A range of new personal sensors are being created in order to evaluate personal exposure. Recent advances in personal sensing technologies was reviewed by Nieuwenhuijsen et al. The review highlighted that available sensors could monitor an individual's exposure to noise, air pollution (black carbon), temperature, green space and physical activity but also biophysical measurements, such as blood pressure and heart rate. Importantly, multiple studies indicated variability of an individual's exposure within a 24-hour period, with large differences observed from day to night and from assessments carried out outdoors versus indoors, stressing the need to account for such variation in external exposome assessments (Nieuwenhuijsen et al. 2014).

In the past couple of years, a new range of passive samplers have been developed for assessing individual's exposure to chemicals. These sensors are based on silicone wristbands or pins, which are given to an individual to wear for a determined (usually short) amount of time. Following the exposure period, the band is analyzed with high-resolution mass spectrometry–based techniques (targeted or untargeted) to evaluate the amount of selected chemical compounds one has been exposed to. With this technique, O'Connell et al. were able to detect exposure to 49 compounds in 30 individuals, including polycyclic aromatic hydrocarbons (PAHs), pesticides, phthalates, and other industrial compounds (O'Connell et al. 2014). Using this technique, Dixon et al. evaluated different exposure patterns for a range of 1530 chemicals in populations from Peru, Senegal, and North America. The study revealed that chemical exposures varied greatly between individuals but also that many individuals were exposed to similar chemical mixtures, highlighting the ubiquity of exposures, where notably 13 potentially endocrine-disrupting chemicals were detected in over 50% of all studied populations across continents (Dixon et al. 2019). Stapleton et al. identified reasonable correlation between chemicals measured in wristbands and concentrations of the same chemicals in matched urine samples, indicative that these passive sensors can provide semiquantitative accurate information on exposure (Hammel et al. 2016).

20.3.2.3 Smartphone-Based Sensors

With the explosion of success of smartphones and other mobile communication devices (tablets, smartwatches, smart bands), it is possible to collect personal data on a wide variety of external measures from a large number of individuals, as over 65% of the worldwide population owns a smartphone. These devices can integrate sensors for GPS localization, climate (e.g. barometer, and thermometer), light sensors, accelerometers, and more, providing a large amount of data on external exposures. Smartphone technologies can also be used in combination with personal sensors and specific apps for the detection of exposure to specific pollutants or monitoring of health status. For example, Chen et al. developed a sensor to evaluate volatile organic compounds exposure based on a smartphone interface (Chen et al. 2012). Other applications are being developed by using existing available technology on smartphones, such as using smartphone camera sensors to estimate heart rate based on a video of bare skin (Majumder et al. 2017).

20.3.3 Remote Sensing

Remote sensing broadly refers to the use of satellite and aircraft to obtain information about areas or objects. The technique is based on a sensor that is carried by a satellite or aircraft that is capable to detect electromagnetic radiation from Earth. The signal is processed with complex algorithms in order to derive the requested parameters (NASA 1999).

Sensing instruments can provide information on the earth's atmospheric chemistry, considering spatial distribution and temporal trends. It is possible to collect information on the urban environment such as air pollution measurements (e.g. nitrous oxides, particulate matter, ozone), climate, green space, land use, and so on. The TERRA satellite from the National Aeronautics and Space Administration (NASA) was one of the first launched that was equipped with sensor instruments to measure air pollution of particulate matter (PM) and carbon monoxide (CO; Pan et al. 2014). Through the years, additional sensors have been launched to evaluate spatial distribution and temporal trend of various air pollutants, such as nitrogen dioxide (NO_2), sulfur dioxide (SO_2), CO, PM, and ozone (Boys et al. 2014). These methods were used by van Donkelaar et al. in a study where information on three satellites was combined with a chemical transport model and ground-based sun photometer observation (van Donkelaar et al. 2015). The remote sensing data was correlated to available data measured on the ground ($R^2 = 0.81$). Remote sensing was also used by Robinson et al. to evaluate parts of the "urban exposome" and how it affects pregnancy. Values of air pollution, green space, and noise were modelled while also considering socioeconomic determinants to determine differences in urban exposure to environmental hazards for pregnant women of lower and higher socioeconomic status (Robinson et al. 2018). New developments in remote sensing are also being developed for a more comprehensive assessment of urban environments. For example, satellite data are being combined with data from Google Street View images (Larkin and Hystad 2019). One limitation of remote sensing applications is that they do not offer individual-level exposure information; to overcome this, remote sensing data often need to be integrated with individual exposure measurements.

20.4 MEASURING THE INTERNAL EXPOSOME: Biomonitoring

Biomonitoring has been a feature of environmental epidemiology and toxicology for some time and is considered as a way to obtain information on the exposure itself and also on the internalized dose and potential metabolism of exposures. Samples such as saliva, urine, blood, and other tissues are collected, and xenobiotics are detected using a targeted (i.e. a specific chemical, metabolite, or reaction product) or untargeted (thousands of chemicals, metabolites, or reaction products) approach. Targeted analysis is the more traditional approach, and although limited in the breadth of exposures that can be measured, it can provide accuracy and depth for specific chemicals of interest. However, biomonitoring within the field of the exposome requires going a step beyond just targeted and other traditional approaches, as the aim of exposome exposure assessment is to capture all exposures experienced across the life course. A full review of biomonitoring for exposome studies has been previously carried out (Dennis et al. 2017). In order to measure a large number of exposures, many exposome studies take an untargeted approach, using high-resolution metabolomics, for example, which is discussed later in this chapter. Often, taking an untargeted approach may lead to an agnostic view for detection of emerging exposures of concern, which can generate hypotheses for exposure-disease responses. An advantage of taking an untargeted approach are that these approaches can overcome limitations of traditional biomonitoring, including detection of short-lived exposures versus the ability to detect an exposure based on its half-life; small quantities of biospecimens are needed; and costs are relatively similar to detect a large number of features (>10,000) compared to traditional biomonitoring analysis. However, untargeted approaches often result in many unknown analytes and the need for advanced bioinformatics for data analyses and time-consuming annotation processes. It may also be difficult to link measures to the exposure source. Other approaches include suspect screening (analyses of suspected chemicals of concern) or semitargeted analyses (e.g. initial discovery of exposures using metabolomics followed a quantitative targeted approach; Dennis et al. 2017). For many studies of the exposome, it is suggested to use a combination of targeted, untargeted, and hybrid methods.

20.4.1 Measuring the Internal Exposome: 'Omics Technologies for Exposome Analysis

A defining characteristic of exposome studies is that the effects from external exposures (the external exposome) should be detectable inside the body (the internal exposome; Rappaport and Smith 2010). By assessing the internal exposome, we can determine whether and to what extent exposures enter the body. This aids in quantification of exposures and how they are metabolized and allows us to determine the influences of exposures on molecular and biological functions. An important feature of assessments of the internal exposome is a focus on an aggregate of extensive biological markers that represent internal doses of exposures and the effects/biological responses of exposures, using a variety of biological tissues.

It is well known that exposure to a variety of substances, including environmental toxicants, diet, and pharmaceuticals, can initiate local and global changes in gene transcription, enzyme activity, metabolite pathway alterations, and protein synthesis/folding (Niedzwiecki et al. 2019). Therefore, in measuring the internal exposome there is a strong emphasis on assessing 'omics, represented by metabolites (metabolomics), proteins (proteomics), miRNA and mRNA (transcriptomics), DNA (genomics and epigenomics), and adducts (including adductomics) in various biological tissues. This may be conducted through targeted or nontargeted analysis of 'omics via high-throughput platforms. Often, the study of 'omics is approached at a systems level, where they are measured and quantitatively analyzed as large networks of molecular and functional changes that occur in multiple levels of biological organization. Borrowing from systems biology, this approach offers strategies to obtain mechanistic knowledge, combining molecular data acquisition on a large scale, through multi-omic platforms with data integration.

Assessments of the internal exposome can reveal current effects of exposures but also provide historical pictures and potential predictive road maps of exposures and health trajectories. Molecular fingerprints of past exposures can be indicative of the effects of environmental stressors occurring years or decades before assessments. Additionally, as the advances of precision medicine continue, individual-level data may be used to predict an individual's risk of future disease, based on current and/or potential future exposures.

The application of 'omics within an exposome paradigm, where multiple exposures have been assessed in relation to 'omic profiles, is limited; however, the Human Exposome Project in the United States; the Human Early-Life Exposome (HELIX) in Europe (Vrijheid et al. 2014); and the Enhanced Exposure Assessment and 'Omic Profiling for High Priority Environmental Exposures in Europe (EXPOsOMICS; Vineis et al. 2016) projects have investigated multi-omics in relation to the external exposome. The following sections highlight 'omics profiling common to exposome studies (Figure 20.3).

20.4.2 Genomics

The role of genetics in disease and health has been investigated in several large-scale clinical and epidemiological studies, mostly via GWAS, where many genetic polymorphisms have been associated with health outcomes. However, variations in the genome at the population level can also occur via copy number variations and structural variations. A similar approach to GWAS is to assess the effects from multiple exposures via environment-wide association studies (EWAS) designed to identify multiple environmental factors associated with disease on a broad scale (Patel et al. 2010). Yet, these two approaches do not consider the interactions of genes and the environment, and it is clear that for nearly all complex human diseases, neither genes nor environmental factors are the exclusive cause. With this in mind, gene by environment studies have been implemented. Most studies of gene-environment interaction have examined how variations in DNA sequences interact with a single exposure to influence disease risk. Certain genotoxic substances can cause alterations in DNA sequences, and one's genomic profile can be used to

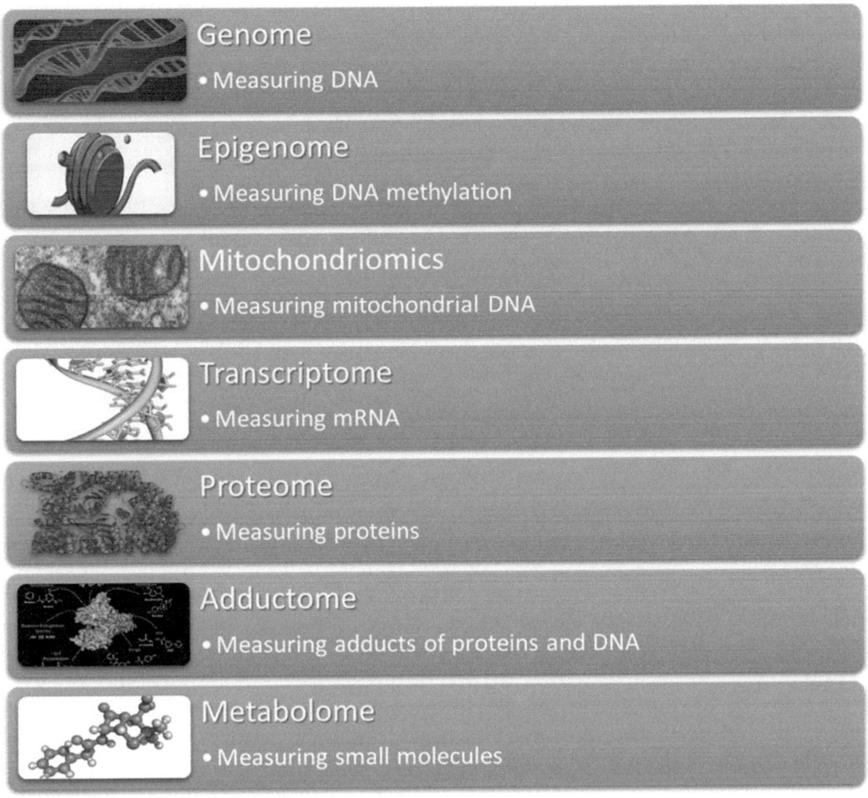

Figure 20.3 'Omics technologies used for the measurement of the exposome.

identify those whose genetic background makes them more or less susceptible to adverse effects exposures.

Often, in these studies, genetic variants have been assessed with a single exposure, such as dietary components, pharmaceuticals, or a chemical toxicant. Rarely have multiple exposures been assessed, which presents a challenge for exposome studies to implement integration of multiple exposures with the sizeable genome. Recently, epidemiological studies have attempted to evaluate the interaction between multiple genes and multiple exposures via genome-environment-wide interaction studies (GEWIS; Thomas 2010). Implementation of GEWIS has been limited by the complexity of statistical analyses; however, a recent study put forth methods with application to investigate 21 lifestyle environmental risk factors along with genes in relation to body mass index (Moore et al. 2019). Another carried out a GEWIS for type 2 diabetes, obesity, hypertension, and coronary heart disease with multiple exposures of smoking, physical activity, diets, and total energy intake (Kim et al. 2019).

20.4.3 Mitochondriomics

Another promising area for exposome studies is mitochondriomics, an investigation of the properties of mitochondrial DNA (mtDNA). Certain characteristics of mtDNA differ from nuclear DNA, such as the lack of histone-wrapping protection and limited repair mechanisms, which makes mtDNA vulnerable to accumulating damage from exposures. Mitochondrial DNA is particularly sensitive to environmental toxicants, but there is still a poor understanding of how these environmental changes in mitochondrial DNA, adducts, and metabolomics influence human diseases (Walker et al. 2018).

20.4.4 Epigenomics

The effects of the environment can impact gene regulation by leaving marks on the epigenome. Epigenetics is the study of processes of heritable regulators of gene expression that do not alter

DNA sequence. Most commonly assessed in human populations are DNA methylation and histone modifications, although emerging studies are investigating noncoding RNAs (e.g. microRNAs) and acetylation. Environmental exposures can affect epigenetic transgenerational inheritance via changes in the germ line. Numerous environmental factors have been associated with alterations in the epigenome, including toxicants (Baccarelli and Bollati 2009) such as air pollution, metals, endocrine-disrupting chemicals, nutritional factors, pharmacological compounds, and social exposures, among others.

Importantly, epigenetic alterations may result in gene silencing. For example, an increase in methylated CpGs can downregulate expression of a gene. However, alterations in DNA methylation are not always associated with changes in gene expression, which is an important factor to consider when evaluating the functionality of epigenetic shifts from environmental exposures. Some of these molecular changes can persist over time even after the environmental stressor is removed. However, the longevity of and downstream functional changes from methylation modifications have a wide array of temporal variability, ranging from minutes to years (Messerlian et al. 2017). For studies of the exposome, we are just starting to unravel what these changes mean in terms of their consequences on health.

20.4.5 Transcriptomics

The transcriptome is the complete set of RNA transcripts that are produced by the genome comprising messenger RNA (mRNA) molecules expressed by an organism. The transcriptome actively changes and can be influenced by developmental stages and/or environmental factors. By measuring the transcriptome, we can determine the amount of gene activity through gene expression.

Exposures to environmental toxicants have been linked to gene expression profiles in epidemiological studies, animal studies, and cell culture. The field of toxicogenomics has advanced our understanding of how multiple genes are involved in responses of biological models to chemical exposure (Hamadeh et al. 2002). Additionally, the application of transcriptomics in nutritional epidemiology and dietary intervention studies has provided an understanding into mechanistic pathways involved in disease prevention by dietary factors (van Breda et al. 2015).

20.4.6 Metabolomics

The study of metabolomics may be a promising approach in exposome research, as it can potentially evaluate both exogenous and endogenous chemicals, allowing both external and internal exposome profile evaluations in a single assessment. Metabolomics is the systematic study of small-molecule metabolites resulting from cellular processes, where the metabolome is a dynamic and sensitive measure of the phenotype at the molecular level. The field of metabolomics enables the identification of potential disease biomarkers in the form of endogenous metabolites (gene-derived metabolites) and exogenous metabolites (environmentally derived metabolites) and may provide insights into the causes of diseases (Wishart 2016). Current metabolomics platforms can measure more than 20,000 small molecules in each biological sample, representing endogenous metabolites, dietary chemicals, microbiome-derived metabolites, environmental chemicals, commercial products, and drugs (Niedzwiecki et al. 2019).

In metabolomic studies, metabolomic features often are considered downstream of transcription and translation; however, these features also influence and interact with protein activity, RNA metabolism, and epigenetic regulation and gene expression. Metabolites can regulate DNA, RNA, and protein functions mainly through chemical modification and metabolite-macromolecule interaction. Therefore, in exposome metabolomic studies it is important to consider that metabolites are viewed and analyzed as active modulators of gene and protein activity, as bioactive reagents.

The use of metabolomics in relation to exposures has been established in separate fields, with disciplines of pharmaceutical metabolomics (Kaddurah-Daouk et al. 2014), nutritional metabolomics (Ulaszewska et al. 2019), and toxicants (toxicometabolomics; Ukolov et al. 2017). Studies from these fields have identified metabolomic features associated with dietary intakes such as meat and fish, and toxicants such as air pollution and dioxins (as reviewed by Bloszies and Fiehn [2018]). As the exposome field continues to grow, it will be an impressive feat to integrate multiple exposures and the thousands of measurable metabolomic features.

20.4.7 Proteomics

Proteomics is the characterization of the proteome, including expression, structure, functions, interactions, and modifications of proteins at any stage (Domon and Aebersold 2006). A proteome-wide dataset may be considered as the most relevant to characterize a biological system, based on proteins being effectors of biological function where their levels are dependent on mRNA levels and translational control and regulation (Aslam et al. 2017). Proteomics assessments are particularly important because mRNA and protein levels for a given gene are imperfectly correlated.

There have been few larger untargeted proteomic studies. However, the proteome has been assessed in relation to prenatal exposure to tobacco smoke (Colquhoun et al. 2009), prenatal exposure to arsenic (Bailey et al. 2014), and exposure to DDT (Tadi Uppala 2012). These limited studies signify the need for more proteomic-wide studies of the exposome.

20.4.8 Adductomics

Electrophiles present in blood are highly reactive species that have been long suspected of causing cancer and other diseases because of their ability to bind DNA and proteins. These electrophiles enter the organism by direct exposure (inhalation, ingestion) or via internal mechanisms such as metabolism of exogenous compounds, oxidation of lipids, ionizing radiation, and inflammation associated with prior disease. The challenge for their measurement is due to their very short half-life in the body and their high reactivity, and although it is difficult to measure electrophiles directly in the blood, it is possible to measure the adducts resulting from their binding to other stable and bigger molecules such as DNA, glutathione (a tripeptide involved in antioxidation processes and xenobiotic metabolism) or proteins (hemoglobin and albumin).

Protein adducts have a longer life span in the blood than glutathione (mean residence time of 28 days vs. hours for glutathione), and they are present in much higher concentrations (30 mg vs. 0.003–0.008 mg of DNA per mL of blood); therefore, they provide suitable markers to measure exposure to toxic electrophiles. In the past 50 years, studies have focused on the targeted analysis of certain specific adducts to hemoglobin (Hb) or human serum albumin (HSA). During the last decade, analytical advances have allowed the development of new untargeted approaches called "adductomics". Adductomics is based on the untargeted analysis of adducts bound to a specific nucleophilic site. For protein, most of the development has been made on the measuring of adducts of the HAS-Cysteine 34 (Cys34 HSA) site. This free cysteine residue is regarded as the target of oxidation and modification. The first study reporting Cys34 HAS adductomics was published by Li et al.; based on a semi-untargeted method, they were able to measure up to 100 adduct features in serum samples of smokers and nonsmokers (Li et al. 2011). The method has been further developed by Grigoryan et al. (2016) using high-resolution mass spectrometry that allows the potential identification of the measured adducts (Grigoryan et al. 2016). Currently, over 70 distinct Cys34 HSA adducts have been identified in human serum, some being adducts of specific compounds (e.g. ethylene oxide, benzene, mercaptoacetic acid) providing information on specific exposures, and others being adducts of reactive oxygen species which can inform on the internal redox homeostasis and oxidative stress related to exposure or disease status. Recent work with this technology has shown promising results. Different Cys34 adduct profiles were detected between smokers and nonsmokers (Grigoryan et al. 2016) and between offspring of smoking and nonsmoking mothers (Yano et al. 2019). In a study comparing normal and occupationally exposed workers to benzene, five adducts generated directly and indirectly by benzene exposure were identified and found to be significantly higher in exposed workers (Grigoryan et al. 2018). In another study, adducts of glutathione discriminated individuals exposed to indoor combustion of either coal or wood (Lu et al. 2017). As part of the EXPOsOMICs project, this method was applied to a small sample size of the Oxford Street II study ($n = 50$). Adducts profiles revealed a perturbation in the glutathione pathway in chronic obstructive pulmonary disease and ischemic heart disease cases exposed to air pollution (Liu et al. 2018). Interesting findings have been reported by applying this technique to a prospective cohort of lung cancer cases and controls, where the authors identified lower levels of N-acetylcysteine in future cases, years before diagnosis, indicating the potential implication of the oxidative stress in lung cancer onset (Dagnino et al. 2019). Other developments have been made to identify adducts of Hb. With a similar untargeted approach, authors identified up to 24 Hb adducts in human serum, some of which were able to discriminate between smokers and nonsmokers (Carlsson et al. 2014). Additionally, Hb adducts in children's serum has been correlated to potential exposure to genotoxic chemicals (Carlsson et al. 2017).

Other techniques involving the measure of adducts of DNA have shown promising results in the context of the exposome in very recent studies. DNA damage by addition of adducts can lead to mutations and ultimately adverse health outcomes, such as the development of cancer. The measurement of these DNA adducts can be key to understand mechanisms involved in the development of cancer and to evaluate the carcinogenic effects of environmental exposures. The formation of DNA adducts to evaluate genotoxicity has been widely used to determine potential carcinogenic effects of a specific compound or mixture of compounds (Food and Drug Administration HHS 2012). Thus, methods to agnostically detect DNA adducts in human samples have been poorly investigated. New developments based on high-resolution mass-spectrometry analysis have allowed great advancement in this field (Balbo et al. 2014). The use of these methods has shown promising results for the detection of adducts related to exposure to a chemotherapeutic drug and a bacterial toxin (Stornetta et al. 2015; Wilson et al. 2019). This method has not yet been applied in an exposome context. An interesting approach has been recently developed by Cooke et al. to detect DNA adducts in urine. This method consists of the analysis of targeted and untargeted DNA adduct in a single urine sample with liquid chromatography-triple quadrupole tandem mass spectrometry (Chang et al. 2018; Cooke et al. 2018). Preliminary applications of these methods to *in vitro* and *in vivo* studies have allowed the detection of DNA adducts indicative of exposure to methanesulfonate (MMS) and five N-nitrosamines, as well as the discovery of five new DNA adducts induced by the exposure to MMS. The use of a noninvasive matrix and the possibilities of measuring DNA adducts of exogenous and endogenous processes indicate this method as a potential promising tool for exposome analysis (Cooke et al. 2019).

20.4.9 Tissue- and Cell-Specific 'Omics

Heterogeneity in expression of different 'omics that is not due to the influence of the external exposome can arise and result from a variety of reasons, such as other environmental or technical confounding factors and/or tissue and cell type. It is important to determine and correct for these differences because without accounting for them, biased estimates and erroneous associations may occur.

Many biospecimens used in human population studies, such as whole blood, contain a mixture of different cell types with potentially different epigenome and transcriptome signatures. It is clear that there are tissue-specific epigenetic and gene expression processes occurring at different organs and cells that influence each other. Multitissue samples and cell-specific assessments are needed to fully understand the complexity of exposure/disease responses. Clearly, this is not always feasible in human studies because of potential invasive procedures; however, there are certain tissue types and cell-specific approaches that may be possible.

One approach for future exposome studies is to focus on cell-specific 'omics. Recent methods for single-cell analysis (scRNASeq) have been developed that are mostly unsupervised and compare whole transcriptional profiles to separate hundreds or thousands of single cells into several distinct populations. As these technologies become more accessible, they will be vital for better understanding the biological processes that underlie environmentally driven diseases.

20.4.10 Analyzing 'Omics Data

Most studies of the exposome, where multi-omics data are collected, address the genome, transcriptome, proteome, and metabolome separately. Additionally, often one exposure and one 'omic platform is considered at a time, even when multiple environmental exposures are available. However, this fails to take into account the interconnections between different biological functions of 'omics and the potential for different environmental risk factors to influence 'omics collectively, either synergistically or antagonistically. Traditionally, this limitation has been due to the lack of availability of multi-omics data and until recently limited methodologies for high-dimensional analysis.

Although there are several challenges for such large-scale exposome analyses, as outlined by Chadeau-Hyam et al. (2013), some of the main challenges in the analysis of large, complex datasets produced by untargeted 'omics analyses in studies of the exposome are simultaneous testing of multiple hypotheses, consideration of multiple correlated exposures, exposure interactions, nonlinear exposure-response relations, and temporal factors of exposures. Often univariate approaches are carried out, where each predictor of an exposure is separately assessed with each 'omic and corrected for multiple testing. These limitations may be overcome by using emerging approaches of dimensionality reduction techniques and variable selection approaches that take into account the correlation structure between multiple 'omics and/or multiple exposures and multi-omics. Systems and network analyses are also being used to better assess 'omics in terms of one another and taking into account their biological functionality. Such approaches are promising to address cross-omics relationships and their complex dynamics.

Future exposome studies may also excel by combining human population 'omics with biologically derived *in silico* models of cell function and physiological processes to enhance our understanding of the complex biological mechanisms underlying exposure-disease responses.

20.5 INTEGRATING EXTERNAL AND INTERNAL EXPOSOMES AND HEALTH

In most exposome studies, multiple exposures are assessed in relation to a single 'omic or a health outcome, or separately a single 'omic is assessed in association with a health outcome. This approach is informative of how exposures influence biological processes and potential disease progressions. For example, an impressive recent study by Warembourg et al. (2019) assessed 89 prenatal maternal exposures and 128 postnatal child exposures and found that early-life exposure to several chemicals, as well as built environment and meteorological factors, influenced blood pressure in children. Yet, these approaches do not fully embody the exposome paradigm nor identify the full causal relationship between exposures and disease. To fully understand the exposome, we must account for the external exposome of multiple exposures assessed simultaneously and multiple components of the internal exposome as they relate to health and disease states (i.e. as they lie on the causal path).

One approach put forth to combine the external and internal exposome along with health outcomes is the meet-in-the-middle (MITM) approach (Vineis et al. 2013). The MITM approach consists of an integration of the bottom-up and the top-down methodologies of the exposome to uncover causal linkages along the exposure-outcome connection. Briefly, this approach demonstrated in a nested case-control study (Figure 20.4) the mechanistic chain linking air pollution to coronary heart disease risk (CCVD), where results (among nonsmokers) identified inflammatory and oxidative pathways from DNA methylation markers that were associated with both air pollution measures and coronary heart disease.

However, this approach does not consider mediating mechanisms in one causal model. Ignoring mediating processes in statistical and causal models may lead to misspecified models, potential false interpretations of causality, and missing (i.e. not seeing associations) important exposure/disease responses (VanderWeele 2016).

It is important that we start to view and analyze 'omics as they reflect biological processes that lie on the causal path between an exposure and a disease. To date, there are a few studies that have approached high-dimensional mediation models in relation to one or a few exposures and health outcomes. One example is a study conducted by Huang et al. (2018), where DNA methylation was assessed as a mediator of the association of mid-childhood body mass index and cardiometabolic risk in early adolescence. Another study, by Laine et al. (2020), assessed the effects of prenatal exposure to multiple air pollutants and the mediating role of cross-omics on birthweight. Future studies of the exposome will benefit from integrating exposures and molecular and biological processes with phenotypic data.

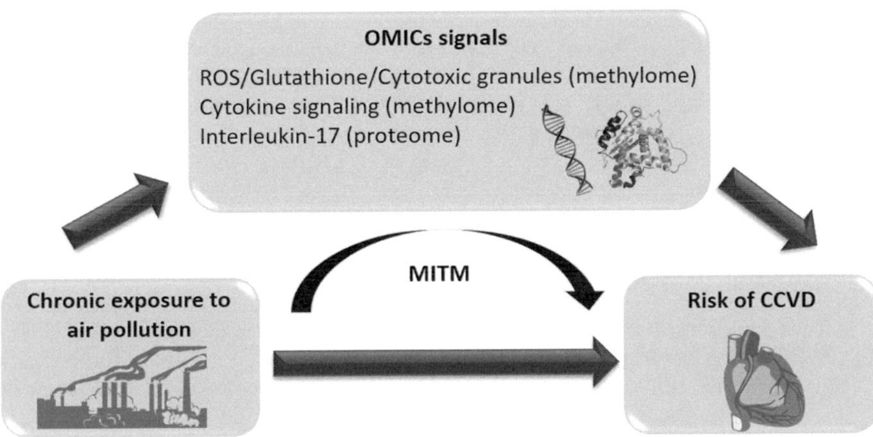

Figure 20.4 Identification of biological pathways associated with both air pollution and incident CCVD with a meet-in-the-middle design.

Source: Adapted from Fiorito et al. (2018).

20.6 CONCLUSION

The field of exposome science is relatively young and is constantly evolving. The concept has taken our view of environmental health from that of focusing on single to a few exposures to mirror the real world of multiple exposures. However, the exposome field is still limited by costs of both external and internal exposome assessment, and thus most studies focus on key windows of susceptibility and targeted hypotheses. As we continue to make improvements in exposure assessment, molecular biology, and big-data statistical approaches in the field, we will be more able to accurately capture the dynamic features of the exposome.

BIBLIOGRAPHY

Aslam, B., M. Basit, M. A Nisar, M. Khurshid, and M. H. Rasool. 2017. Proteomics: Technologies and their applications. *J Chromatogr Sci* 55(2): 182–96.

Baccarelli, A., and V. Bollati. 2009. Epigenetics and environmental chemicals. *Curr Opin Pediatr* 21(2): 243–51.

Bailey, K. A., J. Laine, J. E. Rager, et al. 2014. Prenatal arsenic exposure and shifts in the newborn proteome: Interindividual differences in tumor necrosis factor (TNF)-responsive signaling. *Toxicol Sci* 139(2): 328–37.

Balbo, S., S. S. Hecht, P. Upadhyaya, and P. W. Villalta. 2014. Application of a high-resolution mass-spectrometry-based DNA adductomics approach for identification of DNA adducts in complex mixtures. *Anal Chem* 86(3): 1744–52.

Bloszies, C. S., and O. Fiehn. 2018. Using untargeted metabolomics for detecting exposome compounds. *Curr Opin Toxicol* 8: 87–92.

Boys, B. L., R. V. Martin, A. van Donkelaar, et al. 2014. Fifteen-year global time series of satellite-derived fine particulate matter. *Environ Sci Technol* 48(19): 11109–18. https://doi.org/10.1021/es502113p.

Carlsson, H., J. Aasa, N. Kotova, et al. 2017. Adductomic screening of hemoglobin adducts and monitoring of micronuclei in school-age children. *Chem Res Toxicol* 30(5): 1157–67.

Carlsson, H., H. von Stedingk, U. Nilsson, and M. Tornqvist. 2014. LC-MS/MS screening strategy for unknown adducts to N-terminal valine in hemoglobin applied to smokers and nonsmokers. *Chem Res Toxicol* 27(12): 2062–70.

Castorina, R., C. Butt, H. M. Stapleton, et al. 2017. Flame retardants and their metabolites in the homes and urine of pregnant women residing in California (the CHAMACOS cohort). *Chemosphere* 179: 159–66.

Chadeau-Hyam, M., G. Campanella, T. Jombart, et al. 2013. Deciphering the complex: Methodological overview of statistical models to derive OMICS-based biomarkers. *Environ Mol Mutagen* 54(7): 542–57.

Chang, Y. J., M. S. Cooke, C. W. Hu, and M. R. Chao. 2018. Novel approach to integrated DNA adductomics for the assessment of in vitro and in vivo environmental exposures. *Arch Toxicol* 92(8): 2665–80.

Chen, C., K. D. Campbell, I. Negi, et al. 2012. A new sensor for the assessment of personal exposure to volatile organic compounds. *Atmos Environ* (1994) 54: 679–87.

Colquhoun, D. R., L. R. Goldman, R. N. Cole, et al. 2009. Global screening of human cord blood proteomes for biomarkers of toxic exposure and effect. *Environ Health Perspect* 117(5): 832–8.

Cooke, M. S., C.-W. Hu, Y.-J. Chang, and M.-R. Chao. 2018. Urinary DNA adductomics—A novel approach for exposomics. *Environ Int* 121: 1033–8.

Cooke, M. S., C.-W. Hu, and M.-R. Chao. 2019. Editorial: Mass spectrometry for adductomic analysis. *Front Chem* 7(794).

Dagnino, S. 2019. Unravelling the exposome: Conclusions and thoughts for the future. In *Unraveling the Exposome: A Practical View*, eds. S. Dagnino and A. Macherone, 425–37. Cham: Springer International Publishing.

Dagnino, S., B. Bodinier, H. Grigoryan, et al. 2019. Agnostic Cys34-albumin adductomics and DNA methylation: Implication of N-acetylcysteine in lung carcinogenesis years before diagnosis. *Int J Cancer* 146(12): 3294–303.

Dennis, K. K., E. Marder, D. M. Balshaw, et al. 2017. Biomonitoring in the era of the exposome. *Environ Health Perspect* 125(4): 502–10.

Dixon, H. M., G. Armstrong, M. Barton, et al. 2019. Discovery of common chemical exposures across three continents using silicone wristbands. *R Soc Open Sci* 6(2): 181836.

Domon, B., and R. Aebersold. 2006. Mass spectrometry and protein analysis. *Science* (New York, NY) 312(5771): 212–17.

Food and Drug Administration HHS. 2012. International Conference on Harmonisation; guidance on S2(R1) Genotoxicity Testing and Data Interpretation for Pharmaceuticals intended for Human Use; availability. Notice. *Fed Regist* 77(110): 33748–9.

Grigoryan, H., W. M. B. Edmands, Q. Lan, et al. 2018. Adductomic signatures of benzene exposure provide insights into cancer induction. *Carcinogenesis* 39(5): 661–8.

Grigoryan, H., W. Edmands, S. S. Lu, et al. 2016. Adductomics pipeline for untargeted analysis of modifications to Cys34 of human serum albumin. *Anal Chem* 88(21): 10504–12.

Hamadeh, H. K., P. R. Bushel, S. Jayadev, et al. 2002. Gene expression analysis reveals chemical-specific profiles. *Toxicol Sci* 67(2): 219–31.

Hammel, S. C., K. Hoffman, T. F. Webster, K. A. Anderson, and H. M. Stapleton. 2016. Measuring personal exposure to organophosphate flame retardants using silicone wristbands and hand wipes. *Environ Science Technol* 50(8): 4483–91.

Hoffman, K., S. Garantziotis, L. S. Birnbaum, and H. M. Stapleton. 2015. Monitoring indoor exposure to organophosphate flame retardants: Hand wipes and house dust. *Environ Health Perspect* 123(2): 160–5.

Huang, J. V., A. Cardenas, E. Colicino, et al. 2018. DNA methylation in blood as a mediator of the association of mid-childhood body mass index with cardio-metabolic risk score in early adolescence. *Epigenetics* 13(10–11): 1072–87.

Kaddurah-Daouk, R., R. M. Weinshilboum, and P. R. Network. 2014. Pharmacometabolomics: Implications for clinical pharmacology and systems pharmacology. *Clin Pharmacol Ther* 95(2): 154–67.

Khakimov, B., S. K. Poulsen, F. Savorani, et al. 2016. New Nordic Diet versus Average Danish Diet: A randomized controlled trial revealed healthy long-term effects of the new Nordic Diet by GC-MS blood plasma metabolomics. *J Proteom Res* 15(6): 1939–54.

Kim, J., A. Ziyatdinov, V. Laville, et al. 2019. Joint analysis of multiple interaction parameters in genetic association studies. *Genetics* 211(2): 483–94.

Laine, J., B. Bodinier, O. Robinson, et al. 2020. Prenatal exposure to multiple air pollutants, mediating molecular mechanisms, and shifts in birthweight. *Environ Sci Technol* 54(22): 14502–13.

Larkin, A., and P. Hystad. 2019. Evaluating street view exposure measures of visible green space for health research. *J Expos Sci Environ Epidemiol* 29(4): 447–56.

Li, H., H. Grigoryan, W. E. Funk, et al. 2011. Profiling Cys34 adducts of human serum albumin by fixed-step selected reaction monitoring. *Mol Cell Proteomics* 10(3): M110.004606.

Lichtenstein, P., N. V. Holm, P. K. Verkasalo, et al. 2000. Environmental and heritable factors in the causation of cancer—analyses of cohorts of twins from Sweden, Denmark, and Finland. *N Eng J Med* 343(2): 78–85.

Lim, S. S., T. Vos, A. D. Flaxman, et al. 2012. A comparative risk assessment of burden of disease and injury attributable to 67 risk factors and risk factor clusters in 21 regions, 1990–2010: A systematic analysis for the Global Burden of Disease Study 2010. *Lancet* (London, England) 380(9859): 2224–60.

Liu, S., H. Grigoryan, W. M. B. Edmands, et al. 2018. Cys34 adductomes differ between patients with chronic lung or heart disease and healthy controls in central London. *Environ Sci Technol* 52(4): 2307–13.

Lu, S. S., H. Grigoryan, W. M. Edmands, et al. 2017. Profiling the Serum Albumin Cys34 adductome of solid fuel users in Xuanwei and Fuyuan, China. *Environ Sci Technol* 51(1): 46–57.

Majumder, S., T. Mondal, and M. J. Deen. 2017. Wearable sensors for remote health monitoring. *Sensors (Basel)* 17(1): 130.

Messerlian, C., R. M. Martinez, R. Hauser, and A. A. Baccarelli. 2017. 'Omics' and endocrine-disrupting chemicals—new paths forward. *Nat Rev Endocrinol* 13(12): 740–8.

Miller, G. W., and D. P. Jones. 2014. The nature of nurture: Refining the definition of the exposome. *Toxicol Sci* 137(1): 1–2.

Moore, R., F. P. Casale, M. Jan Bonder, et al. 2019. A linear mixed-model approach to study multivariate gene-environment interactions. *Nat Genet* 51(1): 180–6.

NASA. 1999. *Remote Sensing Methods*. https://earthobservatory.nasa.gov/features/RemoteSensing/remote_08.php (accessed November 15, 2020).

Niedzwiecki, M. M., D. I. Walker, R. Vermeulen, M. Chadeau-Hyam, D. P. Jones, and G. W. Miller. 2019. The exposome: Molecules to populations. *Annu Rev Pharmacol Toxicol* 59: 107–27.

Nieuwenhuijsen, M. J., D. Donaire-Gonzalez, M. Foraster, D. Martinez, and A. Cisneros. 2014. Using personal sensors to assess the exposome and acute health effects. *Int J Environ Res Public Health* 11(8): 7805–19.

O'Connell, S. G., L. D. Kincl, and K. A. Anderson. 2014. Silicone wristbands as personal passive samplers. *Environ Sci Technol* 48(6): 3327–35.

Pan, L., D. Tong, P. Lee, H. C. Kim, and T. Chai. 2014. Assessment of NOx and O3 forecasting performances in the U.S. National Air Quality Forecasting Capability before and after the 2012 major emissions updates. *Atmos Environ* 95: 610–19.

Patel, C. J., J. Bhattacharya, and A. J. Butte. 2010. An Environment-Wide Association Study (EWAS) on type 2 diabetes mellitus. *PLoS One* 5(5): e10746.

Rager, J. E., M. J. Strynar, S. Liang, et al. 2016. Linking high resolution mass spectrometry data with exposure and toxicity forecasts to advance high-throughput environmental monitoring. *Environ Int* 88: 269–80.

Rappaport, S. M. 2011. Implications of the exposome for exposure science. *J Expos Sci Environ Epidemiol* 21(1): 5–9.

Rappaport, S. M., and M. T. Smith. 2010. Epidemiology. Environment and disease risks. *Science* 330(6003): 460–1.

Robinson, O., I. Tamayo, M. de Castro, et al. 2018. The urban exposome during pregnancy and its socioeconomic determinants. *Environ Health Perspect* 126(7): 077005–077005.

Saracci, R., and P. Vineis. 2007. Disease proportions attributable to environment. *Environ Health* 6: 38.

Scalbert, A., I. Huybrechts, and M. J. Gunter. 2019. The food exposome. In *Unraveling the Exposome: A Practical View*, eds. S. Dagnino and A. Macherone, 217–45. Cham: Springer International Publishing.

Stornetta, A., P. W. Villalta, S. S. Hecht, S. J. Sturla, and S. Balbo. 2015. Screening for DNA alkylation mono and cross-linked adducts with a Comprehensive LC-MS(3) adductomic approach. *Anal Chem* 87(23): 11706–13.

Tadi Uppala, P. 2012. Health effects of environmental DDT exposure: An epidemiological and serum proteomic study. *J Community Med Health Educ* 2.

Thomas, D. 2010. Gene—Environment-wide association studies: Emerging approaches. *Nat Rev Genet* 11(4): 259–72.

Ukolov, A. I., E. D. Kessenikh, A. S. Radilov, and N. V. Goncharov. 2017. Toxicometabolomics: Identification of markers of chronic exposure to low doses of aliphatic hydrocarbons. *J Evolut Biochem Physiol* 53(1): 25–36.

Ulaszewska, M. M., C. H. Weinert, A. Trimigno, et al. 2019. Nutrimetabolomics: An Integrative Action for Metabolomic Analyses in Human Nutritional Studies. *Mol Nutr Food Res* 63(1): 1800384.

van Breda, S. G., L. C. Wilms, S. Gaj, et al. 2015. The exposome concept in a human nutrigenomics study: Evaluating the impact of exposure to a complex mixture of phytochemicals using transcriptomics signatures. *Mutagenesis* 30(6): 723–31.

van Donkelaar, A., R. V. Martin, M. Brauer, and B. L. Boys. 2015. Use of satellite observations for long-term exposure assessment of global concentrations of fine particulate matter. *Environ Health Perspect* 123(2): 135–43.

VanderWeele, T. J. 2016. Mediation analysis: A practitioner's guide. *Annu Rev Public Health* 37: 17–32.

Vineis, P., K. van Veldhoven, M. Chadeau-Hyam, and T. J. Athersuch. 2013. Advancing the application of omics-based biomarkers in environmental epidemiology. *Environ Mol Mutagen* 54(7): 461–7.

Vineis, P., M. Chadeau-Hyam, H. Gmuender, et al. 2016. The exposome in practice: Design of the EXPOsOMICS project. *Int J Hyg Environ Health.*

Vrijheid, M., R. Slama, O. Robinson, et al. 2014. The human early-life exposome (HELIX): Project rationale and design. *Environ Health Perspect* 122(6): 535–44.

Walker, D. I., K. D. Pennel, and D. P. Jones. 2018. The mitochondrial exposome. In *Mitochondrial Dysfunction Caused by Drugs and Environmental Toxicants*, ed. J. A. Dykens, and Y. Will, 613–37. Wiley.

Warembourg, C., L. Maitre, I. Tamayo-Uria, et al. 2019. Early-life environmental exposures and blood pressure in children. *J Am Coll Cardiol* 74(10): 1317–28.

Wild, C. P. 2005. Complementing the genome with an "exposome": The outstanding challenge of environmental exposure measurement in molecular epidemiology. *Cancer Epidemiol Biomark Prev* 14(8): 1847–50.

Willett, W. C. 2002. Balancing life-style and genomics research for disease prevention. *Science* 296(5568): 695–8.
Wilson, M. R., Y. Jiang, P. W. Villalta, et al. 2019. The human gut bacterial genotoxin colibactin alkylates DNA. *Science* 363(6428).
Wishart, D. S. 2016. Emerging applications of metabolomics in drug discovery and precision medicine. *Nat Rev Drug Discov* 15(7): 473–84.
Yano, Y., H. Grigoryan, C. Schiffman, et al. 2019. Untargeted adductomics of Cys34 modifications to human serum albumin in newborn dried blood spots. *Anal Bioanal Chem* 411(11): 2351–62.

21 Pesticide Toxicology

Elliot Gordon

CONTENTS

DOI: 10.1201/9780203730584-21

21.1 INTRODUCTION

Welcome to the world of pesticide toxicology. This chapter is written with the newly minted toxicologist in mind. Its primary focus is the Office of Pesticides in the US Environmental Protection Agency (EPA), but topics generally apply equally to other international agencies. The successful pesticide toxicologist melds the science of toxicology with an understanding and ability to work with his or her regulatory colleagues. One view is that you simply provide necessary data to support registrations and their subsequent defense at EPA or other regulatory bodies; in effect, a technician for the respective registration director. In reality, the pesticide toxicologist is a key member of the director's team. The regulatory team may set up a Gantt chart for obtaining a registration, based on "typical" requirements necessary for completion of the required studies. When it comes to study assignment, conduct and reporting, typical often does not apply. If you understand the characteristics of the proposed active ingredient and its unique properties, you will be in a position to counsel the team on how the study list needs to be modified. We provide some of the tools that provide this counsel.

21.2 THE ART AND SCIENCE OF PESTICIDE TOXICOLOGY

"Art and Science" is a well-worn term for a number of disciplines. It applies also to the pesticide toxicologist. It rests on two components: (1) technical proficiency and (2) a command of regulatory landscape. Melded together, these constitute your professional foundation.

Technical proficiency integrates board certification, experience in conducting studies, and an understanding of the active ingredients. Certification as a diplomate of the American Board of Toxicology (DABT) is a "union card" necessary for advancement. Other certifying bodies include the American Board of Applied Toxicology (ABAT), the Forensic Toxicologist Certification Board, the American College of Medical Toxicology (ACMT), and European Registered Toxicologist (ERT). Technicians may be members of American Association of Laboratory Animal Science (AALAS). Table 21.1 notes such organizations, Table 21.2 lists a number of international resources, and Table 21.3 notes toxicity agencies within the United States.

The vision of the ABT is to establish a globally recognized credential in toxicology representative of competency and commitment to human health and environmental sciences. The mission of the ABT is to identify, maintain, and evolve a standard for professional competency in the field of toxicology.

Board certification is necessary but not sufficient. Technical competency develops over time as one studies details of guideline protocols, is physically present and observes technicians perform procedures, and authors documents such as final reports, waiver application, position papers, and papers submitted to peer-reviewed journals. Study protocols range from acute, "six-pack" studies through subchronic and chronic studies. Studies targeting specific endpoints include mutagenicity, developmental, reproductive, neurotoxicity, and carcinogenicity. Routes of administration include oral, dermal, and inhalation. Other routes may be used, such as intradermal, intraperitoneal, and intravenous. Table 21.4 directs you to documents related to pesticide registrations. These include FIFRA 6(a)(2) reporting and the electronic submission portal.

One cannot easily be an expert in all things toxicological; in reality, one often emerges as a generalist in most phases of toxicology and a recognized expert in a particular phase. Some of these areas follow.

Table 21.1 Toxicology-Related Associations and Societies

Organization	Link
American Academy of Clinical Toxicology (AACT)	www.clintox.org/
American Association of Laboratory Animal Science (AALAS)	www.aalas.org
American Board of Applied Toxicology (ABAT)	www.clintox.org/resources/abat
American Board of Toxicology (ABT)	www.abtox.org
American Chemical Society (ACS)	www.acs.org/content/acs/en.html
American Chemistry Council (ACC)	www.americanchemistry.com/
American College of Medical Toxicology (ACMT)	www.acmt.net/
American College of Toxicology (ACT)	www.actox.org/index.asp
American Conference of Governmental Industrial Hygienists (ACGIH)	www.acgih.org/
American Industrial Hygiene Association (AIHA)	www.aiha.org/Pages/default.aspx
European Registered Toxicologist (ERT)	www.eurotox.com/ert/
Federation of American Societies for Experimental Biology (FASEB)	www.faseb.org
Forensic Toxicologist Certification Board, the American College of Medical Toxicology (ACMT)	www.acmt.net
Society for Risk Analysis (SRA)	www.sra.org/
Society of Forensic Toxicologists (SOFT)	www.soft-tox.org/
Society of Toxicologic Pathology (STP)	www.toxpath.org/index.asp
Society of Toxicology (SOT)	www.toxicology.org/index.asp
Teratology Society	www.teratology.org/

Table 21.2 International Resources

Agency	Link
Canadian Centre for Occupational Health and Safety (CCOHS)	www.ccohs.ca/
Chemical Safety and Biosafety (OECD)	www.oecd.org/chemicalsafety/
Ecotoxicology and Models (EFSA)	www.ecotoxmodels.org/
Federation of European Toxicologists and European Societies of Toxicology (Eurotox)	www.eurotox.com/
Health and Safety Executive (UK)	www.hse.gov.uk/
International Agency for Research on Cancer (IARC)	www.iarc.fr/
International Forum on Chemical Safety (IFCS)	www.who.int/ifcs/en/
International Occupational Safety and Health Information Center (ILO-CIS)	www.ilo.org/safework/lang—en/index.htm
International Programme on Chemical Safety (IPCS)	www.who.int/ipcs/en/
Inter-Organization Programme for the Sound Management of Chemicals	www.who.int/iomc/en/
United Nations Environment Programme (UNEP) Chemicals and Waste	www.unenvironment.org/explore-topics/chemicals-waste

Table 21.3 US Toxicology-Related Agencies

Agency	Link
Agency for Toxic Substances and Disease Registry (ATSDR)	www.atsdr.cdc.gov
Centers for Disease Control and Preventions (CDC)	www.cdc.gov
US Chemical Safety Hazard and Investigation Board	www.csb.gov
US Consumer Product Safety Commission (CPSC)	www.cpsc.gov

Table 21.4 EPA Registration References

Reference	Link
Data Requirements for Pesticide Registration	www.epa.gov/pesticide-registration/data-requirements-pesticide-registration
Electronic Submission Portal	https://cdx.epa.gov
EPA Pesticide Registration	www.epa.gov/pesticide-registration
FIFRA 6(a)(2) Incident Reporting	www.epa.gov/pesticide-incidents/incident-reporting-pesticide-manufacturers-registrants
Guiding Principles for Data Development	www.epa.gov/pesticide-registration/guiding-principles-data-requirements
Harmonized Test Guidelines	www.epa.gov/test-guidelines-pesticides-and-toxic-substances/final-test-guidelines-pesticides-and-toxic
Registration Manual	www.epa.gov/pesticide-registration/pesticide-registration-manual
Series 870 Health Effects Guidelines	www.epa.gov/test-guidelines-pesticides-and-toxic-substances/series-870-health-effects-test-guidelines
Study Profile Templates	www.epa.gov/pesticide-registration/study-profile-templates

21.3 PESTICIDE PATH TO THE MARKET

There are four stages in the path to establishing a successful pesticide product: (1) efficacy is first established, (2) safety studies are conducted, (3) formulation and manufacturing challenges are solved, and (4) the product is commercialized. The pesticide toxicologist will focus on the second stage. Absent success in any of these stages will result in delayed commercialization.

21.4 AREAS OF SPECIALIZATION

Areas of expertise include mutagenicity, immunotoxicity, reproductive and developmental toxicity, neurological toxicity, inhalation toxicity, and endocrine toxicity, to name a few. For each, there is a wealth of references available that can help the generalist become expert. Rather than include an extensive reference list, the reader may use standard internet searches to gather these as needed.

21.4.1 Mutagenicity

Mutagenic insults can have health consequences. There are physical, chemical, or biological mutagens that all lead to changes in an organism's DNA. These changes can initiate the development of cancer. Although the public's concern with cancer is well founded, there are other mutagen-related diseases such as cystic fibrosis, sickle cell anemia, and Tay-Sachs disease. Known mutagenic agents include x-rays, ultraviolet radiation, and reactive chemicals.

There are many tools used to detect mutagenic events. These include the Ames assay, DNA repair, the mouse lymphoma TK assay, the mammalian *in vivo* bone marrow chromosomal aberration test, and the comet assay (single-cell gel electrophoresis assay). The Ames assay was one of the first widely used detection methods and is described below.

21.4.1.1 Ames Assay

The Ames assay assesses the mutagenic potential of a test substance. *Salmonella typhimurium*, which cannot grow in histidine-free medium due to mutations involved in histidine synthesis, is exposed to the test material. Mutagenic changes induced by the test material which result in the bacteria to once again grow in a histidine-free media are noted by counting the number of colonies compared to the control plate.

Mutagenicity in the Ames assay may be relevant to human risk or it may simply indicate the *potential* for mutagenic changes. By example the fungicide captan is mutagenic in Ames but is not mutagenic *in vivo* due to its rapid degradation and inability to reach cell nuclei intact.

21.4.1.2 Other Assays

Other assays used in evaluating mutagenicity include sister chromatid exchange in CHO cells, unscheduled DNA synthesis in WI-38 cells, *Drosophila* sex-linked recessive, mouse-dominant lethal assays, *Escherichia coli* EP-2 uvrA reverse mutation, and *Saccharomyces cerevisiae* D3 mitotic recombination. These are briefly outlined below.

Sister chromatid exchange: This assay for genotoxicity detects alterations to the biochemistry underlying cellular homologous recombination. Cells are grown in the presence of 5′-bromodeoxyuridine for two rounds of DNA replication followed by collecting metaphase spreads on glass slides. Analysis records the exchanges that are present.

Unscheduled DNA synthesis: This assay measures a cell's ability to perform global genomic nucleotide excision repair (NER).

Drosophila sex-lined recessive (SLRL): This assay tests for transmissible mutations in the germ cells of a eukaryote. It detects both point mutations and small deletions in the germ line of the insect.

Mouse-dominant lethal assays: This assay notes chromosomal aberrations (structural and numerical anomalies) that result fetal deaths.

E. coli EP-2 uvr reverse mutation: This assay detects back mutations (reversion of a point mutation). The test bacteria (e.g. *Salmonella* or *E. coli*) have various mutations that inactivate a gene involved in the synthesis of an essential amino acid (e.g. histidine or tryptophan). When mutations take place that reverse such mutations, the bacteria are able to grow in media that do not have these amino acids.

S. cerevisiae D3 mitotic recombination: This assay measures whether test substance increase mitotic recombination in *S. cerevisiae* D3.

21.4.2 Immunotoxicity

Immune dysfunction in humans can be manifested by immunosuppression which, in turn, plays a critical role in combating disease. Other endpoints include allergy, autoimmunity or inflammatory reactions.

21.4.2.1 SRBC Assay

Rats and/or mice are exposed to a pesticide for at least 30 days (OPPTS 879.7800). They are then immunized by IV injection of sheep red blood cells (SRBC) 4 days prior to the end of exposure. At the end of exposure, the anti-SRBC plaque-forming cell assay or enzyme-linked immunosorbent assay (ELISA) is performed to determine the effects of the pesticide on either splenic immunoglobin M (IgM) plaque-forming cell (PFC) assay or serum IgM levels. This test is considered the gold standard for the T-cell dependent antibody response (TDAR).

Flow cytometry can be used to evaluate lymphocyte populations: total T cluster of differentiation (CD)-3; total B CD-45R, and natural killer (NK) and T subpopulations CD4 and CD8. These relate to the splenic or peripheral-blood lymphocyte populations. EPA has reviewed over 170 immunology studies with their relevance to regulatory risk assessment and has concluded: "Consistent with the Agency's commitment of the 3Rs (Replacement, Refinement and Reduction) of animal testing," waivers will be considered for these studies (Rowland 2013).

21.4.3 Reproductive and Developmental Toxicity

The ability to successfully reproduce without teratogenic outcomes is important. We know humans are susceptible to teratogens. Thalidomide is a potent teratogen that causes severe limb defects. Other teratogens include alcohol, phenytoin (Dilantin), and infection with varicella virus. Outcomes include fetal alcohol syndrome, fetal hydantoin syndrome, and varicella syndrome. Developmental studies are designed to identify potential teratogens before they become apparent in humans. OPPTS 870.3700 describes the procedure to determine developmental defects in rats. In essence, the pregnant dam is exposed to graded doses of the test material and the fetuses are harvested approximately 1 day prior to expected delivery. Gross observations of limbs and organs are made.

21.4.4 Neurological Toxicity

Adverse neurotoxic effects may be neurochemical, anatomical, physiological, or behavioral. Endpoints may include inhibition of neurotransmission; alterations in nerve cell bodies, axon, or myelin sheath; or changes in sensations of sight, hearing, or touch. Important aspects are cognitive functions such as learning, memory, and mood. Examples of neurotoxins include lead, ethanol, manganese, glutamate, and tetanus toxin.

The blood-brain barrier is composed of protective cells astrocytes that surround the capillaries; they absorb nutrients from the blood and transport them to the neurons.

21.4.4.1 DNT Study

The developmental neurotoxicity (DNT) study starts with pregnant rats from day 6 of gestation through day 21 postpartum. The test material is incorporated in the feed and subjects the pups to exposure *in utero*, through the dam's milk, and finally in the feed itself as the pups mature. Subsequent observations include motor activity, learning, and memory.

21.4.5 Inhalation Toxicity

Successful inhalation studies are in part dependent on the physical nature of the test material. A key parameter that must be obtained is the establishment of respirable particles, either as dust or aerosol. Acceptable dust particles are generally 1–4 µm MMAD (mass median aerodynamic diameter; Whalan and Redden 1994). Nasopharyngeal deposition will increase as the particles get larger. Pulmonary deposition in Fischer 344 rats decreased from 13.3% for 0.29 µm particles to 4.8% for 4 µm particles. If it is technically not possible to generate respirable particles (dust or aerosol), consider basing a waiver on this. If respirable particles are not generated in the application of the product, occupational workers will not be exposed.

21.4.6 Endocrine Toxicity

An intact endocrine system is critical for development and maintaining health. The pituitary gland, dubbed the master gland, integrates effects of the ovaries, testes, pancreas, and the adrenal, thyroid, and pineal glands. The messengers in this system are the hormones.

The EPA launched its Endocrine Disruptor Screening Program (EDSP) in the 1990s. The Endocrine Disruptor Screening and Testing Advisory Committee (EDSTAC), the expert committee formed to develop the endocrine testing program, issued its report in 1998. Subsequently, EPA issued call-in notices to conduct screens: Table 21.5 notes these screens.

Tier 1 screens included 52 chemicals. Based on the weight of evidence analysis of the data from these screens, in 2015 EPA recommended 18 of these pesticides for Tier 2 testing.

Table 21.5 Endocrine Screens

Study	Reference
ER binding assay (rat uterine cytosol); OCSPRP 890.1250	www.epa.gov/sites/production/files/2015-07/documents/final_890.1250_er_binding_assay_sep_10.4.11.pdf
Erα transcriptional activation assay (human cell line HeLa 9903); OCSPP 890.1300	https://ntp.niehs.nih.gov/iccvam/suppdocs/feddocs/epa/epa-890-1300.pdf
AR binding assay (rat prostate cytosol); OCSPP 890.1150	www.epa.gov/sites/production/files/2015-07/documents/final_890.1150_ar_bindng_assay_sep_10.5.11.pdf
Steroidogenesis assay (human cell line H295R); OCSPP 890.1550	www.epa.gov/sites/production/files/2015-07/documents/final_890.1550_steroidogenesis_assay_sep_8.1.11.pdf
Aromatase assay (human recombinant microsomes); CSPP 890.1200	www.epa.gov/sites/production/files/2015-07/documents/final_890.1200_aromatase_assay_sep_8.1.11.pdf
Uterotrophic assay (rat); OCSPP 890.1600	www.epa.gov/sites/production/files/2015-07/documents/final_890.1600_uterotrophic_assay_sep_9.22.11.pdf
Hershberger assay (rat); OCSPP 890.1400	www.epa.gov/sites/production/files/2015-07/documents/final_890.1600_hershberger_assay_sep_10.6.11.pdf
Pubertal female assay (rat); OCSPP 890.1450	www.epa.gov/sites/production/files/2015-07/documents/final_890.1450_female_pubertal_assay_sep_8.24.11.pdf
Pubertal male assay (rat); OCSPP 890.1500	www.epa.gov/sites/production/files/2015-07/documents/final_890.1500_male_purbertal_assay_sep_8.24.11.pdf
Fish short-term reproduction assay; OCSPP 890.1350	www.epa.gov/sites/production/files/2015-07/documents/final_890.1350_fish_short_term_reproduction_assay_sep_10.6.11.pdf
Amphibian metamorphosis assay (frog); OCSPP 890.1100	www.epa.gov/sites/production/files/2015-07/documents/final_890.1100_ama_sep_9.30.11.pdf

21.5 RISK ASSESSMENT

21.5.1 General

All regulatory actions depend on a risk-assessment foundation. The shorthand procedure involves establishing an adverse endpoint, a NOAEL for this endpoint, an evaluation of the exposure (both systemic and dermal), and determining the relationship to exposure. Typically, if systemic exposure is 100 times lower that the NOAEL, the exposure of product is considered acceptable. The 100x is the uncertainty factor (UF) that incorporates translation from animals to humans (10x UF). Where additional uncertainty is present due to the absence of a NOAEL or particular sensitivity to children (or other stated factors), an additional UF of 10x may be applied. This brings the total UF to 1000x.

21.5.2 Occupational Risk

The key factor in establishing occupational risk is a clear understanding of the exposure. Here EPA publishes unit exposure tables based on field data collected in the various exposure task forces (US EPA 2016). The EPA uses imperial units (lbs) versus metric units (kg). For each pound of active ingredient handled and the procedure employed, a µg/lb a.i. is provided. The dermal absorption factor (DAF) is derived from *in vivo* rat studies or, more accurately, from triple-pack dermal studies in which the rat absorption factor is adjusted by the ratio of rat to human absorption determined in an *in vitro* study. It is accepted that rats generally have higher dermal absorption than humans.

21.5.3 Dietary Risk

Dietary risk is determined by two factors: the residues contained on produce and the typical amount of these products ingested. Dermal exposure is not generally a factor.

21.6 TECHNICAL AIDS

Authoring white papers (e.g. waiver requests, summary documents) is an integral part of the pesticide toxicologist's job. Having aids that facilitate the art of writing is helpful. Two such aids are a reference manager and file coordinator between your computers (e.g. desktop and laptop). The following programs below are examples and are not intended as endorsements.

21.6.1 EndNote

EndNote is a reference manager that provides ready access to your saved documents and automatically generates bibliographies in a variety of formats (Clarivate Analytics 2018). It saves time both in finding key documents and writing your paper. Other available programs include Mendeley, Zotero, Desalegn Belay, Citavi and JabRef.

21.6.2 Dropbox

Dropbox is one of a number of available file coordination programs that synchronize files between computers (Dropbox 2018). It is particularly helpful when traveling with a laptop and ensuring that the changes made will automatically convey to your desktop computer. Other available programs include FolderShare, SyncToy, rsync, and Box Sync.

21.6.3 ProWorkFlow

Organizing your projects, tracking tasks, and knowing key personnel is helpful in meeting your target goals. ProWorkFlow is just one example of a helpful management program (ProWorkFlow 2018). Other available programs include Wrike, Smartsheet, Asana, Zoho Projects, FunctionFox, Slack, LiquidPlanner, Accelo, and Workfront.

21.7 EPA TOXICITY PLATFORMS

21.7.1 Tox21

Toxicology Testing in the 21st Century (Tox21) is the US Environmental Protection Agency's approach to future toxicology testing (NIH 2018b; US EPA 2008). It is a collaborative effort by EPA, the National Toxicology Program at the National Institute of Environmental Health Sciences (NIEHS) and the Food and Drug Administration (FDA). The Toxicity Forecaster (ToxCast) is one of the programs that provides input data (NIH 2018a). In essence, preliminary screening approaches are designed to identify which compounds need further testing and which should not unnecessarily burden current testing programs. Since 2008 approximately 10,000 chemicals have been subject to around 70 quantitative high-throughput screening assays.

Table 21.6 Other Helpful References

Identity	Link
Acute Exposure Guideline Levels (AEGLs)	www.epa.gov/aegl
Chemical Right-to-Know Initiative	http://apps.americanbar.org/dch/committee.cfm?com=nr351500
Chemicals under the Toxic Substances Control Act (TSCA)	www.epa.gov/chemicals-under-tsca
ChemView	https://chemview.epa.gov/chemview
Envirofacts	https://www3.epa.gov/enviro/
EPA-Expo-Box	www.epa.gov/expobox

21.7.2 EPA CompTox Chemicals Dashboard

This is a web-based data integration hub for environmental and toxicology data (US EPA 2018). The national center for computational toxicology was established in 2005 and was designed to integrate high-throughput and high-content technologies, modern molecular biology, data mining and statistical modeling, and computational biology and chemistry. Collaboration exists between NERL, OPPT, and Tox21. Over 750,000 chemicals with property data are included, over 1000 chemicals with biological assay data, and over 4500 chemicals have data that have been hand-curated and validated. This is an excellent portal for a "first look" at your chemical in question. Other helpful references are noted in Table 21.6.

21.8 INFLUENTIAL BODIES

National and international organizations shape the pesticide toxicology landscape and continually move the goalposts of what is expected as the science of risk assessment evolves.

21.8.1 IARC

The International Agency for Research on Cancer (IARC) is the specialized agency of the World Health Organization (WHO; IARC 2018). The objective of the IARC is to promote international collaboration in cancer research. The agency is interdisciplinary, bringing together skills in epidemiology, laboratory sciences, and biostatistics to identify the causes of cancer so that preventive measures may be adopted and the burden of disease and associated suffering reduced.

21.8.2 NTP

The National Toxicology Program was founded in 1978 (NTP 2018). Its goal is to generate, interpret, and share toxicological information about potentially hazardous substances in the environment. It is a division of NIEHS. There are three core agencies in NTP: NIEHS, the National Institute for Occupational Safety and Health (NIOSH), and the National Center for Toxicological Research (NCTR).

21.8.3 TERA

Toxicology Excellence for Risk Assessment (TERA) is based in Cincinnati, Ohio (TERA 2018). It is a nonprofit independent organization that conducts scientific research and development on risk issues "in a transparent and collaborative fashion."

21.8.4 OECD

The mission of the Organization for Economic Co-operation and Development (OECD) is to promote policies that will improve the economic and social wellbeing of people around the world (OECD 2019). They set standards on a wide range of things, from agriculture and tax to the safety of chemicals. Pesticides are among the chemicals addressed.

21.8.5 Codex Alimentarius

Codex Alimentarius develops and adopts food standards that serve as a reference for interaction food trade (Codex 2018). All European countries are members. The Codex Committee on Pesticide Residues (CCPR) deals with dietary risk assessments that incorporate on residues with anticipated exposures.

21.8.6 EFSA

The European Food Safety Authority, set up in 2002, and legally established under the General Food Law (Regulation 178/2002), operates independently of European legislative and executive institutions (EFSA 2018). EFSA produces scientific opinions and advice that form the basis for European policies and legislation. Plant protection is one of the areas covered. Generally, EFSA's work in undertaken in response to requests for scientific advice from the European Commission, the European Parliament, and EU Member States. There are a number of scientific panels and scientific committees. The Panel on Plant Protection Products and their Residues (PPR) has members who are experts in chemistry, toxicology, ecotoxicology, exposure assessment, and environmental sciences. They provide scientific advice and risk assessment of pesticides for operators, workers, consumers, and the environment. Evaluation of active substances used in pesticides is carried out by Rapporteur Member States and reviewed by EFSA staff (PPR 2018).

21.9 TASK FORCES

Task forces provide an efficient approach to collecting data and developing assessments on issues that affect pesticide companies generically. They can be formed after data call-ins or in response to actions or proposed actions by regulatory agencies. Task forces are used by both the pesticide industry as well as opponents of pesticide companies. Table 21.7 lists exposure task forces.

21.9.1 Industry Task Forces

The task forces noted in Table 21.7 are summarized below.

21.9.1.1 Spray Drift Task Force (SDTF)

The SDTF was formed to measure exposure to occupational workers associated with the application of pesticides by aircraft. With the development of global positioning systems (GPS), one group of workers, the flaggers, are no longer needed. Dissipation downwind from the application site was measured. It has completed its work and is disbanded.

21.9.1.2 Agricultural Re-Entry Task Force (ARTF)

ARTF established exposure to farm workers who enter fields previously treated with pesticides. All studies have been completed, and the group holds annual update conference calls.

21.9.1.3 Outdoor Residential Exposure Task Force (ORETF)

ORETF targeted homeowners who apply products to lawns. All studies have been submitted to EPA but the task force is still functional.

21.9.1.4 Agricultural Handler Exposure Task Force (AHETF)

AHETF is the fourth of the exposure task forces that developed data for occupational exposure to mixer/loaders and applicators. It has concluded its studies and has entered a holding phase.

21.9.2 Anti-Industry Task Forces

Some view pesticides with disdain. They refer to pesticides as poisons that cause harm to adults, children, and the environment. Like-minded individuals have formed their own task forces to promote reduced use of pesticides.

Table 21.7 Occupational Exposure Task Forces

Task Force	Link
Spray Drift Task Force (SDTF)	www.ncbi.nlm.nih.gov/pubmed/11878479
Agricultural Re-Entry Task Force (ARTF)	www.exposuretf.com/Home/ARTF/tabid/57/Default.aspx
Outdoor Residential Exposure Task Force (ORETF)	www.exposuretf.com/Home/ORETF/tabid/58/Default.aspx
Agricultural Handler Exposure Task Force (AHETF)	www.exposuretf.com/Home/AHETF/tabid/59/Default.aspx

21.9.2.1 Earth Justice

Earth Justice is a nonprofit public interest environmental law organization (EarthJustice 2020a). It notes that "true and lasting change happens when the power of the law is on your side." Their mission is to hold those who break our nation's laws accountable for their actions. They consider themselves "the legal backbone for thousands of organizations, large and small." In addition, they have an international program (EarthJustice 2020b).

21.9.2.2 Beyond Pesticides

Beyond Pesticides (formerly the National Coalition against the Misuse of Pesticides) is a 501(c)3 non-profit organization headquartered in Washington, DC. It works with allies in protecting public health and the environment to lead the transition to a world free of toxic pesticides. The founders established Beyond Pesticides as a nonprofit membership organization in 1981 and felt that without the existence of such an organized, national network, local, state, and national pesticide policy would become, under chemical industry pressure, increasingly unresponsive to public health and environmental concerns.

21.9.2.3 NRDC

The Natural Resources Defense Council is a US-based nonprofit international environmental advocacy group, with its headquarters in New York City. They combine online activists with the expertise of some 600 scientists, lawyers, and policy advocates across the globe to ensure the rights of all people to the air, the water, and the wild.

21.9.2.4 Sierra Club

Conservationist John Muir founded the Sierra Club in 1892. Its mission is to "explore, enjoy, and protect the wild places of the earth; practice and promote the responsible use of the earth's ecosystems and resources; and, to educate and enlist humanity to protect and restore the quality of the natural and human environment; and to use all lawful means to carry out these objectives." Among their missions is to solve the climate crisis in a way that protects the environment and also is enduring, fair, and equitable. To the extent pesticides contribute to the climate crisis, they are fair game. The Sierra Club helped pass the Clean Air Act, Clean Water Act, and Endangered Species Act.

21.9.2.5 Task Force on Systemic Pesticides

This task force focuses on the neonicotinoids such as imidacloprid. The neonics are implicated in colony collapse disorder among honeybees. These systemic pesticides are taken up by plants and expose insects that forage on flowers. Although direct toxicity studies are not conclusive with regard to adverse effects to bees, the apparent linkage (epidemiology) of the neonics to colony collapse begs for further study.

21.10 REGULATORY LANDSCAPE

The regulatory landscape includes those for and against agrichemicals. Depending on which side of the fence your organization resides, the goal will be either to promote the use of plant protection products or to eliminate their use. Both sides must provide reliable data to meet their stated goals.

There are a number of guidelines published that form the basis of data requirements. Some guidelines and platforms supported by EPA follow.

21.10.1 Acute Exposure Guideline Levels (AEGLs)

Acute exposure guideline levels (AEGLs) describe the human health effects from once-in-a-lifetime or rare exposure to airborne chemicals. Used by emergency responders when dealing with chemical spills or other catastrophic exposures, AEGLs are set through a collaborative effort of the public and private sectors worldwide.

21.10.2 Chemical Right-to-Know Initiative

On June 22, 2016, the president signed into law the Frank R. Lautenberg Chemical Safety for the 21st Century Act, which amends TSCA, the nation's primary chemicals management law. The new law addressed several issues, including the following:

- Mandatory requirement for EPA to evaluate existing chemicals with clear and enforceable deadlines
- A new risk-based safety standard

- Increased public transparency for chemical information
- A consistent source of funding for EPA to carry out the responsibilities under any new law.

For historical reference regarding TSCA, in 2014 the American Bar Association Section of Environment, Energy, and Resources Pesticide, Chemical Regulation and Right-to-Know Committee, with the support of the section's Special Committee on Congressional Relations, developed a variety of resources for TSCA stakeholders interested in participating in the TSCA reform debate. These resources promoted greater awareness, understanding, and dialogue on TSCA as the reform efforts were underway, and included the following:

- Management of trade secrets and confidential business information
- Development of federal preemption policies governing state tort law
- Development of federal preemption policies governing state statutes and regulations
- Development of appropriate legal standards to guide regulatory action.

21.10.3 Chemicals under the Toxic Substances Control Act (TSCA)

Chemicals can be found in many of the products we use in our everyday lives. Some chemicals pose risks to humans and the environment. Under TSCA and the Pollution Prevention Act, the EPA evaluates potential risks from new and existing chemicals and finds ways to prevent or reduce pollution before they enter the environment. TSCA provides information on chemicals and their regulation; ways to prevent and reduce pollution; and safer chemicals, products and practices.

21.10.4 ChemView

The ChemView database provides information on chemical health and safety data received by the EPA and its assessments and regulatory actions for specific chemicals under TSCA. ChemView contains no confidential business information (CBI).

21.10.5 Envirofacts

The Envirofacts Multisystem Search integrates information from a variety of databases and includes latitude and longitude information. Each of these databases contains information about facilities that are required to report activity to a state or federal system. Using this search form, you can retrieve information about hazardous waste (including the biennial report), toxic and air releases, Superfund sites, and water discharge permits. Facility information and a map of their locations is provided.

21.10.6 EPA ExpoBox (A Toolbox for Exposure Assessors)

EPA ExpoBox is a toolbox created to assist individuals from within government, industry, and academia as well as the general public with assessing exposure. It is a compendium of exposure assessment tools that links to guidance documents, databases, models, reference materials, and other related resources. EPA ExpoBox is organized according to the following tool sets: approaches, exposure, media, routes, tiers and types, lifestages and populations, and chemical classes.

21.11 LAWS, REGULATIONS, GUIDANCE, AND DOCKET

Become familiar with the laws and associated guidance that determines what data are required, what issues are important to governments, and current issues that drive regulatory actions.

21.11.1 Laws

21.11.1.1 Clean Air Act

The Clean Air Act (CAA) is the comprehensive federal law that regulates air emissions from stationary and mobile sources. Among other things, this law authorizes the EPA to establish National Ambient Air Quality Standards (NAAQS) to protect public health and public welfare and to regulate emissions of hazardous air pollutants (EPA 2017).

21.11.1.2 Clean Water Act

The Clean Water Act (CWA) establishes the basic structure for regulating discharges of pollutants into US waters and regulating quality standards for surface waters. The basis of the CWA was enacted in 1948 and was called the Federal Water Pollution Control Act, but the Act was renamed (with amendments) and significantly reorganized and expanded in 1972 (EPA 2018).

21.11.1.3 Toxic Substances Control Act

TSCA of 1976 provides the EPA with authority to require reporting, record-keeping and testing requirements, and restrictions relating to chemical substances and/or mixtures. Certain substances are generally excluded from TSCA, including food, drugs, cosmetics, and pesticides, among others.

21.11.1.4 Executive Orders

In addition to environmental laws (just a few noted above), the EPA is authorized to issue executive orders. Examples are the Regulatory Planning and Review and Reducing Regulation and Controlling Regulatory Costs.

21.11.2 Policy and Guidance

EPA maintains a list of significant guidance documents as called for by the Office of Management and Budget (OMB). Documents can be accessed at www.regulations.gov/.

21.12 CONCLUSION

The successful pesticide toxicologist combines the technical aspects of risk assessment with the requirements of the regulatory world. Each new challenge presents an opportunity to master a key component that increases your value to your company or client.

BIBLIOGRAPHY

Clarivate Analytics. 2018. *EndNote*. https://endnote.com (accessed November 21, 2020).

Codex. 2018. *Codex Alimentarius*. https://ec.europa.eu/food/safety/international_affairs/standard_setting_bodies/codex_en (accessed November 21, 2020).

Dropbox. 2018. *Dropbox*. www.dropbox.com (accessed November 21, 2020).

EarthJustice. 2020a. *About Us*. https://earthjustice.org/about (accessed November 21, 2020).

EarthJustice. 2020b. *International Program*. https://earthjustice.org/about/offices/international (accessed November 21, 2020).

EFSA. 2018. *European Food Safety Authority*. www.efsa.europa.eu (accessed November 21, 2020).

EPA. 2017. *Summary of the Clean Air Act*. www.epa.gov/laws-regulations/summary-clean-air-act (accessed November 21, 2020).

EPA. 2018. *Summary of the Clean Water Act*. www.epa.gov/laws-regulations/summary-clean-water-act (accessed November 21, 2020).

IARC. 2018. *ARC's Mission: Cancer Research for Cancer Prevention*. www.iarc.fr/about-iarc-mission/ (accessed November 21, 2020).

NIH. 2018a. *Assay Development & Screening*. National Center for Advancing Translational Sciences. https://ncats.nih.gov/preclinical/drugdev/assay (accessed November 21, 2020).

NIH. 2018b. *Tox21 Program Goals*. National Center for Advancing Translational Sciences. https://ncats.nih.gov/tox21/about/goals (accessed November 21, 2020).

NTP. 2018. *National Toxicology Program*. https://ntp.niehs.nih.gov/ (accessed November 21, 2020).

OECD. 2019. *Organisation for Economic Co-operation and Development: About*. www.oecd.org/about/ (accessed November 21, 2020).

PPR. 2018. *In*. www.efsa.europa.eu/en/panels/ppr (accessed November 21, 2020).

ProWorkFlow. 2018. *ProWorkFlow*. www.proworkflow.com (accessed November 21, 2020).

Rowland, J. 2013. *A Retrospective Analysis of immunotoxicity Studies (879.7800)*. Health Effect Division, Office of Pesticide Programs, U.S. Environmental Protection Agency, 25 pages.

TERA. 2018. *Toxicology Excellence for Risk Assessment*. Cincinnati, OH: TERA.

US EPA. 2008. *Toxicology Testing in the 21st Century (Tox21)*. www.epa.gov/chemical-research/toxicology-testing-21st-century-tox21 (accessed November 21, 2020).

US EPA. 2016. *Occupational Pesticide Handler Unit Exposure Surrogate Reference Table*. US Environmental Protection Agency, Office of Pesticide Programs, 17 pages.

US EPA. 2018. *CompTox Chemicals Dashboard*. www.epa.gov/chemical-research/comptox-chemicals-dashboard (accessed November 21, 2020).

Whalan, J. E., and J. C. Redden. 1994. *Interim Policy for Particle Size and Limit Concentration Issues in Inhalation Toxicity Studies*. Health Effects Division, Office of Pesticide Products, United States Environmental Protection Agency, 21 pages.

22 Pharmaceuticals and Cosmetics as Emerging Environmental Contaminants

Carmen José Mateos Vega and María José González Muñoz

CONTENTS

22.1 INTRODUCTION

The terms "emerging contaminants" or "contaminants of emerging concern" (CECs) are used to cover "any synthetic or naturally occurring chemical or any microorganism that is not commonly monitored in the environment but has the potential to enter the environment and cause known or suspected adverse ecological and/or human health effects" (US Geological Survey [USGS]). In other words, CECs are any substances that we are beginning to suspect could cause harm. Of all the emerging contaminants, those that have been the subject of possibly more studies in recent years are pharmaceutical and personal care products (PPCPs), which include pharmaceuticals from diverse families and substances used as additives in cosmetics, personal hygiene products and food (Daughton and Ternes 1999; Jones et al. 2001).

Pharmaceuticals are defined as prescription, over-the-counter and veterinary therapeutic drugs used to prevent or treat human and animal diseases, whereas personal care products are used mainly to improve the quality of daily life (Boxall et al. 2012) "cosmetic product means any substance or mixture intended to be placed in contact with the external parts of the human body (epidermis, hair system, nails, lips and external genital organs) or with the teeth and the mucous membranes of the oral cavity with a view exclusively or mainly to cleaning them, perfuming them, changing their appearance, protecting them, keeping them in good condition or correcting body odours" (European Union 2009).

The presence in the environment of these types of compounds went unnoticed until the publication, in the late 1990s, of a pioneering work by Daughton and Ternes, in which it became clear that the problem of drug contamination had reached a dimension comparable to that of organochlorine pesticides (Daughton and Ternes 1999). Moreover, interest in the safety issue of PPCPs has steadily increased over the past 30 years (Schumock et al. 2014) due to great worldwide production (greater than 2×10^7 metric tons), and it continues to increase due to the high demand of PPCPs in preventing or curing disease and sustaining economic development via routes such as aquaculture and livestock farming, resulting in their continuous release into the environment (Nikolaou et al. 2007). Therefore, PPCPs are becoming ubiquitous due to their extensive application and poor removal by conventional biological wastewater treatment (WWT) plants (Wang and Wang. 2016).

PPCPs are known to be released into aquatic environments through multiple pathways, including domestic wastewater, hospital discharges, improper manufacturer disposal, sewage treatment plants (STPs) and water treatment plants (WTPs; Leung et al. 2012; Liu and Wong 2013). The released PPCPs may either retain their original concentrations and structures or be mobilized and converted into other active (or inactive) compounds during their life spans in aquatic matrices (Yang et al. 2017). Although most PPCPs are designed to perform specific actions on specific therapeutic targets, and many of them turn out to be substances of low toxicity according to traditional

 DOI: 10.1201/9780203730584-22

analyses, in most cases their mechanisms of toxic action are unknown. Furthermore, their potential toxicity for nontarget organs and organisms and, especially, the problems derived from the chronic exposure to which we are subject are all unknown (Brausch et al. 2012). Regulation of PPCPs has been strictly enforced and implemented to minimize their consumption (by the US Environmental Protection Agency [EPA], the US Food and Drug Administration [FDA], the National Institute of Environmental Health Sciences [NIEHS], and the Organisation for Economic Co-operation and Development [OECD]; Daughton 2002). Regarding the routes of exposure to these types of compounds, it is important to note that as far as voluntary administration is concerned, the general population is increasingly sensitive to the responsible consumption of pharmaceuticals, some of which is restricted by prescription. This situation is not applicable to personal care products, in whose formulations we find a large amount of chemical additives to which we expose ourselves daily through multiple channels (and, in most cases, completely unnoticed). It is also important to note that, as a result of this continued use, these substances are detected in the environment more frequently and at higher concentrations than pharmaceuticals, representing the group of environmental pollutants most commonly detected in surface waters around the world (Ternes et al. 2004; Peck 2006).

22.2 CLASSIFICATION OF PPCPS

PPCPs can be classified into several classes based on their various purposes (Daughton and Ternes 1999). Pharmaceuticals generally include antibiotics, hormones (e.g. contraceptives), analgesics, anti-inflammatory drugs, blood lipid regulators, beta blockers and cytostatic drugs. Personal care products include preservatives, bactericides/disinfectants, insect repellents, fragrances, and sunscreen ultraviolet (UV) filters, which sometimes can account for up to 20% of sunscreen formulations (Chisvert and Salvador 2007; Liu and Wong 2013). The specific classes and corresponding purposes of PPCPs are summarized in Table 22.1 (Wang and Wang 2016; Juliano and Magrini 2017; Yang et al. 2017).

Table 22.1 The Classification and Functions of Pharmaceutical and Personal Care Products (PPCPs)

Classification of PPCPs	Functions	Frequently detected representative compounds
Pharmaceuticals		
Antibiotics	Fight bacterial infections; inhibit or stop bacterial growth Treat and prevent bacterial infections	Amoxicillin Azithromycin Cefaclor Cephalexin Difloxacin Doxycycline Enoxacin Levofloxacin Mecillinam Norfloxacin Ofloxacin Penicillin Sulfamethazine Sulfamethoxazole Sulfapyridine Tetracycline Trimethoprim Tylosin
Anticonvulsants	Control (prevent) seizures (convulsions) or stop a series of seizures	Carbamazepine Gabapentin Phenobarbital Phenytoin Primidone
Antineoplastics	Prevent, inhibit or halt the development of a neoplasm	Cyclophosphamide Epirubicin Ifosfamide Methotrexate Tamoxifen

(*Continued*)

Table 22.1 (Continued)

Classification of PPCPs	Functions	Frequently detected representative compounds
Beta blockers	Bind to beta-adrenoceptors and block the binding of norepinephrine and epinephrine to these receptors	Acebutolol Atenolol Metoprolol Nadolol Pindolol Propranolol Sotalol
Diagnostic contrast media	Enhance the visibility of internal structures in x-ray-based imaging techniques, such as computed tomography (contrast CT), projectional radiography and fluoroscopy	Diatrizoate acid Iomeprol Iopamidol
Hormones	A group of chemical substances, usually steroids, whose purpose is to stimulate certain reproductive functions Regulate metabolism; control reproductive function and sexual development; maintain homeostasis	17-β Estradiol Androstenedione Estriol Estrone Mestranol Testosterone
Lipid regulators	Regulate triglycerides and cholesterol in blood Treat dyslipidaemias, primarily elevated cholesterol	Bendroflumethiazide Bezafibrate Clofibrate Clofibric acid Furosemide Gemfibrozil Simvastatin
Nonsteroidal anti-inflammatory drugs (NSAIDs)	Relieve or reduce pain and inflammation	Acetaminophen (paracetamol) Aspirin Diclofenac Ibuprofen Indomethacin Ketoprofen Naproxen Nimesulide Phenazone Salicylic acid
Psychiatric drugs	Affect the mind, emotions, and behaviour Improve psychiatric disorders	Amitriptyline Diazepam Doxepin Fluoxetine Imipramine Meprobamate Oxazepam Primidone Thioridazine
Personal care products		
Bactericides/disinfectants	A chemical agent used on nonliving surfaces to destroy, neutralize or inhibit the growth of disease-causing microorganisms	2-Phenylphenol 4-Chlorocresol 4-Chloroxyenol Bromoprene Chloroprene Triclosan
Fragrances/insect repellents	Chemical substances that impart a sweet or pleasant odour; deter insects from approaching or settling	Galaxolide Musk ketone Musk xylene N,N-diethyl-m-toluamide (DEET) Toxalide fragrance
Plastic microbeads	Remove dead skin and enable deep cleansing Provide a decorative role in personal care products	Nylon Polyethylene Polyethylene terephthalate Polymethyl methacrylate Polypropylene

Table 22.1 (Continued)

Classification of PPCPs	Functions	Frequently detected representative compounds
Preservatives	Prevent decomposition by microbial growth or by undesirable chemical changes	2-Phenoxyethanol Butyl 4-hydroxybenzoate Ethyl 4-hydroxybenzoate Isobutyl 4-hidroxybenzoate Isopropyl 4-hydroxybenzoate Methyl paraben Parabens Propyl 4-hydroxybenzoate
Sunscreen agents	Protect skin from the sun's UV radiation; reduce sunburn and other skin damage	*Organic Filters* 2-Ethyl-hexyl-4-trimethoxycinnamate 4-Methyl-benzylidene-camphor Ethylhexyl methoxycinnamate Octocrylene Octyl-methoxycinnamate Octyl-triazone Oxybenzone *Inorganic filters* TiO_2 ZnO

22.3 SOURCES

PPCPs may enter the aqueous environment directly or indirectly (Figure 22.1) through anthropogenic activities, resulting in their presence in surface water and groundwater at concentrations of nanograms to micrograms per litre (Sui et al. 2015).

Many PPCPs rapidly dissipate in the environment, but their large utilization results in their pseudo-persistence in water environments and has serious ecological impacts on aquatic organisms (Kostich et al. 2014). The high-frequency detection of PPCPs in STPs, including sewage effluent and reclaimed water (Chen et al. 2013), is caused by their universal consumption (Liu and Wong 2013), low human metabolic capability (Borova et al. 2014), improper disposal (Ternes et al. 2004) and biologically active structures (McClellan and Halden 2010). These contaminants can enter the surface water via directly discharging into the surface water by industries, hospitals, households and WTPs and through land runoff (in animal farming and manure applications). In the case of biosolids spread on agricultural land, the contaminants can reach the groundwater by leaching or bank filtration (Price et al. 2010; Boxall et al. 2012; Lambropoulou and Nollet 2014). PPCP residues can also enter the environment through the natural hydrologic cycle (Petrović et al. 2003; Mompelat et al. 2009). Another source of PPCPs is via their manufacture, when wastewater from production facilities goes directly into STPs (Fick et al. 2009). Mitigating the availability of PPCPs is their adsorption by various environmental sinks. For example, within the surface water compartment, sediment can adsorb the PPCPs because it has a variety of binding sites (Kästner et al. 2014). Soil can also be a sink for PPCPs (a) by passing into the soil following the irrigation with contaminated treated or untreated wastewater and (b) because the sludge arising from the treatment process in STPs is deposited onto soil as fertilizer (Yang and Toor 2015). In addition, some PPCPs can transfer to soil by atmospheric wet deposition (e.g. rain; Kallenborn et al. 2004).

The major sources of PPCPs in the environment are via STPs (Daughton and Ternes 1999), WTPs and landfill leaching. PPCPs are often not completely or consistently removed during conventional WWT processes due to their physicochemical properties (Chen and Zhou 2014). Nonmetabolized or dissolved pharmaceutical ingredients in some drugs used by humans or animals (e.g. methotrexate) are excreted from the body via feces and urine (Montforts 1999; Kim et al. 2011) and are finally discharged into sewerage systems (Kimura et al. 2007). As a consequence, pharmaceuticals can also reach freshwater through runoff from land treated with digested sludge for agricultural purposes (Nikolaou et al. 2007). Likewise, veterinary drugs are released into the environment when animal waste, either in solid or liquid states, is sprayed onto agricultural field as fertilizer. These veterinary drugs, together with their metabolites, pollute the soil and can enter the food chain (Kaczala and Blum 2016). Consequently, agricultural runoff can enter freshwater systems and leach to groundwater (Farré et al. 2008). Other PPCP exposure pathways include the disposal

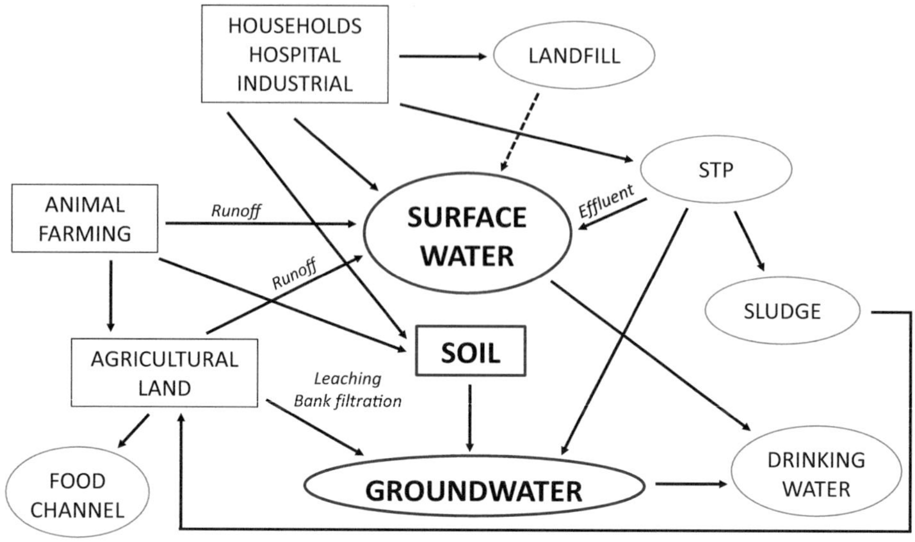

Figure 22.1

of unused medicines into landfill sites, the disposal of the carcasses of treated animals, and irrigation using reclaimed water (Fick et al. 2009; Awad et al. 2014). Similarly, personal care products, including shampoos, body washes, toothpastes, sunscreens, cosmetics and hand lotions, are discharged into sewerage systems and surface water through daily washing activities. Thus, they can pass through WTPs and reach the environment (Figure 22.2; Peck 2006). Additionally, the sloughing off of skin cells during swimming and other recreational activities can contribute to discharge of personal care products, as in the case of UV filters (e.g. TiO_2 and ZnO; Brausch and Rand 2011). In fact, UV filters have been detected in surface waters such as seas, oceans and coastal waters (Jiang et al. 2014; Sánchez Rodríguez et al. 2015), rivers (Ekpeghere et al. 2016), lakes (Balmer et al. 2005), groundwater (Jurado et al. 2014) and sediments (Amine et al. 2012). The release of TiO_2 and ZnO from cosmetics and personal care products applied on skin into the environment can occur due to immersion or abrasion with sand (Botta et al. 2011). TiO_2 is also released into aquatic environments from sources other than sunscreens, such as paint and food coloring. WTPs remove the majority of TiO_2 present in the influent sewage, but a small amount is released into natural water bodies (Shi et al. 2016).

Parabens are a class of widely used preservatives present in various PPCPs, such as shampoos, commercial moisturizers, shaving gels and personal lubricants. They are also used in topical/parenteral pharmaceuticals, suntan products, makeup and toothpaste and as food preservatives. They have been linked to the development of allergic reactions and breast cancer (Golden et al. 2005). EU law allows the use of parabens in cosmetics, and one or more of them can be present in a given product. The maximum total concentration allowed in such consumer products is 8 g of parabens per kg of cosmetic product, with no single paraben having a higher concentration than 4 g/kg. For the longer paraben molecules (propyl- and butyl-parabens), the Scientific Committee on Consumer Safety recommends to lower the limit to a maximum total concentration of 1.9 g/kg parabens. Despite restrictions from European Union in 2009 and 2014 (European Union 2009), parabens are still extensively employed and can be detected in air, dust, soil and water (Błedzka et al. 2014). They are continuously released at relatively high levels into urban and hospital wastewaters as a result of the widespread use of personal care products, pharmaceuticals and household chemicals containing them. Fortunately, WWT plants effectively remove most parabens present in the aqueous phase, with an average removal rate of over 90% (González-Mariño et al. 2011; Haman et al. 2015).

Triclosan (TCS) is a disinfectant with a potential carcinogenic effect and toxicity to ecosystems. It is a controversial chemical antibacterial component present in toothpastes and other products such as hand or body soaps, shower gels and deodorants. Although during the water purification process it is eliminated between 80% and 90% (Ricart et al. 2010), the volume of TCS discharged into the environment is so important and constant because this compound has been detected in countless environmental compartments (Dann and Hontela 2011). The principal pollution source of TCS is domestic sewage, where this compound is released with the rinsing of products containing it (Yueh and Tukey 2016).

Another detrimental contaminant is plastic debris, both macro- and microplastics. Specifically, the latter in the marine environment may come from a variety of different sources including the fragmentation of larger items, cosmetics, clothing and industrial processes. Plastic microbeads are present as abrasive scrubbers in a number of personal care and cosmetic products such as hand cleansers, toothpastes, face scrubs, bubble bath, shampoos and soaps (Cole et al. 2011), where they replace natural exfoliating materials previously utilized (e.g. pumice, oatmeal, apricot husks). In addition to the function of dead skin removal and deep cleansing, plastic microbeads can also play a decorative role in personal care products. Around 93% of the microbeads used in personal care products are made of polyethylene (Gouin et al. 2015), but sometimes they are made from is polypropylene, nylon, polyethylene terephthalate, or polymethyl methacrylate (Napper et al. 2015).

It is important to note that the management and use practices of PPCPs vary in different regions of the world, hence the significance of different exposure pathways also varies geographically.

22.4 OCCURRENCE OF PPCPS IN WATER ENVIRONMENTS

22.4.1 PPCPs in Surface Water

As previously noted, effluent from STPs is the predominant pathway through which PPCPs enter surface water and accumulate in aquatic environments (Luo et al. 2014). With respect to pharmaceutical agents, analgesic and anti-inflammatory drugs are one of the most common residues in sewage because of their high consumption. Likewise, antibiotics constitute approximately half of PPCP contamination, confirming that these stable drugs are difficult to remove in conventional STPs (Ashton et al. 2004; Ziylan and Ince 2011). Indeed, the levels of some antidepressant drugs like sulpiride and carbamazepine are useful for tracing contamination sources in rivers (Ashton et al. 2004; Hernando et al. 2006) due to their apparent omnipresence.

The UV filters most frequently identified in riverine and marine sediment are ethylhexyl methoxycinnamate, octocrylene, butylmethoxydibenzoylmethane, octyldimethyl-p-aminobenzoic acid and benzophenone derivatives (Amine et al. 2012; Barón et al. 2013; Kim and Choi 2014). The presence of these PPCPs has been detected even in the Arctic region, albeit at relatively low concentrations (Tsui et al. 2014). Logically, the concentrations of UV filters suffer seasonal fluctuations, being higher in summer (Tsui et al. 2014). Swimming pools are the places where the highest concentrations of these filters are found, due to the washing off from the skin of bathers during swimming, especially in baby pools, where levels 10 times higher than normal have been detected (Zwiener et al. 2007). On the other hand, TiO_2 residues released from sunscreen produce submicronic aggregates that remain in suspension in freshwater, whereas in seawater conditions (with high salt concentrations) they aggregate and gradually settle, being confined to the sediment (Botta el al. 2011). The most common parabens present in cosmetics, methylparaben and propylparaben, are also the parabens most frequently detected in surface waters (Błedzka et al. 2014), especially rivers (Peng et al. 2006; Kimura et al. 2014).

With respect to TCS, it is found ubiquitously in the aquatic environment, where it partitions between sediment and water and can be thus detected in both phases (Yueh and Tukey 2016; Roberts et al. 2016), with a preferential accumulation in sediments due to its lipophilicity. Finally, the microplastics present in cosmetics are too small to be trapped in sewage treatment plants, so they enter waterways via domestic drainage systems and are transported to seas and oceans (Browne et al. 2008; Derraik 2002).

22.4.2 PPCPs in Groundwater

Vulliet et al. (2011) found that the frequencies and concentrations of PPCPs are lower in groundwater than in surface water and do not show significant trends or seasonal variations. This is good news, because it implies that the contamination has not reached the subsoil with such intensity. Among the various pharmaceuticals, antibiotics, anti-inflammatories and analgesics, lipid regulators and N,N-diethyl-m-toluamide (DEET) are frequently detected in groundwater (Sui et al. 2015). Sulfonamides are the most extensively studied pharmaceutical drugs because they are one of the antibiotic families most frequently found in all kinds of environmental waters, and have been found at high concentrations (between 11 ng L^{-1} and 112 ng L^{-1}) in several studies (López-Serna et al. 2013; Peng et al. 2014; Schaider et al. 2014). Veterinary antibiotics are frequently reported in the groundwater around breeding facilities and farmland (Bartelt-Hunt 2011). The antibiotics detected in groundwater from different countries are variable in concentrations and species, suggesting different consumption patterns. The most commonly detected anti-inflammatories and analgesics in groundwater include ibuprofen, diclofenac, salicylic acid, carbamazepine and acetaminophen (paracetamol) because they are widely and frequently consumed. Exfiltration of wastewater to groundwater might be responsible for the high concentration levels of anti-inflammatories and analgesics (Ellis et al. 2002). Carbamazepine (an antiepileptic drug and tricyclic antidepressant) is frequently detected in groundwater and at relatively high concentrations (Osenbrück et al. 2007; Loos et al. 2010). It is not subjected to degradation or adsorption (Clara et al. 2004), which might be one of the reasons for their ubiquity in the groundwater. In general, the reported detection frequencies of lipid regulators and metabolites such as bezafibrate, gemfibrozil and clofibric acid in groundwater is lower than those of some antibiotics, anti-inflammatories and carbamazepine (Peng et al. 2014; Tran et al. 2014). Several studies have detected in groundwater continuous occurrence, but relatively low concentrations, of other PPCPs such as x-ray contrast media (derived from radiological examinations from hospitals), beta blockers, musks and sunscreen agents (Zemann et al. 2015).

22.5 TRANSPORT

Once released into the environment, there exists the possibility of long-range transport of some PPCPs depending on the physicochemical properties of the compound and the characteristics of the receiving environment. PPCPs generally have low volatility and are highly polar and hydrophilic in nature; therefore, their distribution through the environment will primarily occur through aqueous transport and food chain dispersal (Calamari et al. 2003). Several groups of PPCPs can be found in sludge samples of STPs through adsorption in the target organisms (e.g. humans). This creates a potential pathway for PPCPs into the environment by direct release, or application of sludge to agricultural land as fertilizer (Van Wieren et al. 2012). Sorption in sediment is another mechanism through which PPCPs are transported to the aquatic environment. The sediment acts as a sink and accumulates these environmental contaminants, which may be released back to the aquatic environment (Zhao et al. 2013). For instance, once in the environment, microbeads made of high-density plastics (polyester, polyvinyl chloride) settle out of the water column and accumulate in the sediment, whereas low-density microplastics (polyethylene, polystyrene) float on the sea surface (Van Cauwenberghe et al. 2015).

22.6 ENVIRONMENTAL DEGRADATION AND TRANSFORMATION

PPCPs may be partially metabolized or incompletely biodegraded in artificial water matrices; model constructed wetlands were used to assess wastewater contaminant removal (Cardinal et al. 2014). Biodegradation, photodegradation and other abiotic transformation processes such as hydrolysis (Blair et al. 2013) may reduce concentrations of PPCPs in the environment and result in the formation of other compounds (Alexy et al. 2004). Thus, carbamazepine is transformed by photodegradation to acridine, a toxic metabolite which is also mutagenic and carcinogenic. However, tetracycline cannot be photodegraded because of its adsorption onto sediment (Tolls 2001). On the other hand, diclofenac, ibuprofen, metronidazole, acetaminophen (paracetamol) and several other PPCPs could be easily and rapidly degraded through direct photolysis (Robinson et al. 2007), suggesting photolysis as one of the major degradation pathways of PPCPs in surface waters (Carlson et al. 2015), although degradation does not necessarily lessen the potential toxicity.

Organic UV filters can undergo photodegradation and can react with the chlorine used to disinfect the water, with the formation of by-products that are potentially harmful (Kockler et al. 2012; Santos et al. 2012). In aqueous environments, organic UV filters can degrade by direct or indirect photolysis; the first process depends on the absorption of UV radiation by the organic compound, whereas the second implies the participation of different photosensitizer agents, such as dissolved organic matter or reactive oxygen species (ROS). Under UV radiation, some organic filters are also able to produce ROS (Sánchez-Quiles and Tovar-Sánchez 2015). Nano TiO_2 and ZnO can undergo photocatalytic reactions when exposed to UV radiation, with the generation of ROS such as $O_2^{\bullet-}$, $^{\bullet}OH$, and H_2O_2 (Lewicka et al. 2013).

Biodegradation arises from the reaction of contaminants with natural microbial flora in the environment. Certain microorganisms (during WWT and in the environment) utilize PPCP substrates either as a carbon or energy source, resulting in the formation of transformation products (Helbling et al. 2010). However, the increase in PPCP concentrations could inhibit biodegradation (enzyme saturation), therefore becoming toxic to the naturally occurring microorganisms (Onesios-Barry et al. 2014).

During WWT, PPCPs may be completely destroyed, partially transformed to metabolites or in some instances left unchanged (Xia et al. 2005). Some of these transformation products are toxic but with unknown mechanisms of toxicity and persistence.

In WWT plants, parabens can adhere to the sewage sludge or can be degraded; in particular, they can undergo fast biodegradation, with the production of p-hydroxybenzoic acid as a main degradation product, or can react easily with free chlorine, forming several chlorinated by-products (Canosa et al. 2006). These halogenated derivatives can also be generated when paraben-containing cosmetics are mixed with chlorinated tap water at home (Canosa et al. 2006). Unfortunately, little is known about the potential risk of these halogenated derivatives for human health (Haman et al. 2015). During the WWT process, TCS can also be converted into chlorinated derivatives, which are more toxic and more persistent than their parent compounds (Yueh and Tukey 2016). Microbeads are highly resistant to degradation.

22.7 ENVIRONMENTAL RISK OF PPCPS

The adverse effects of many PPCPs and their metabolites are still unknown, although some authors have studied their harmful properties in detail (Hernando et al. 2006; Verlicchi et al. 2014). It has been shown that continuous exposure to low, subtoxic concentrations of certain PPCPs can cause unexpected consequences and unintended effects on nontarget species (Heberer 2002). Thus, their presence in the environment may pose a threat to human and ecological health.

Pharmaceuticals designed to be biologically active may also affect nontarget organisms (Burkina et al. 2015; Steinbach et al. 2016) and induce undesirable effects on humans and ecosystems (Heberer 2002). Consequently, the anti-inflammatory drug diclofenac and certain antibiotics (azithromycin, erythromycin and clarithromycin) have recently been included on the "watch list" of priority substances under the Water Framework Directive (European Commission 2015).

Between 2000 and 2003, there was an unusual increase in the mortality rate of adults and subadults of a species of vultures in Pakistan, drastically decreasing the population and placing it on the list of animals in danger of extinction. The vultures would have been exposed to diclofenac by feeding on corpses of cattle previously treated with this NSAID, causing acute renal failure and death. It is the first documented case of ecological disaster due to pharmacontamination.

Today more examples of drugs with toxic effects in various species are known, such as alteration in the reproduction of fish exposed to ethinyl estradiol, selection of resistant strains in environmental bacteria exposed to antibiotics discharged from manufacturing plants, stimulation of the reproduction of zebra mussels exposed to different selective serotonin reuptake inhibitors (SSRIs) and reduction of the predation capacity of a variant of bass exposed to fluoxetine.

Due to insufficient knowledge in terms of toxicity, impacts and behaviours of PPCPs, few are routinely monitored in the environment and many are unregulated (Lapworth et al. 2012). But as their potential for long-term risk to environment is increasingly recognized, relevant regulations and standards can be expected over the next decades. Cosmetics pose the most pressing ecological problems compared to pharmaceuticals because they are used in much larger quantities throughout the course of life and, as they intended for external application, they are not subjected to metabolic transformation. Therefore they are introduced unaltered into the environment in large amounts during washing, showering or bathing (Ternes et al. 2004). Their ecological and environmental impacts are associated with the large amounts used and with the fact that sometimes they are environmentally persistent, bioactive and potentially able to bioaccumulate (Brausch et al. 2011).

In order to classify the toxicity of PPCPs, a combination of traditional risk assessment procedures, persistence, bioaccumulation and toxicity (PBT) criteria can be used.

22.8 PERSISTENCE

The inability to effect complete removal of PPCPs from WWT plants, due to their physicochemical properties, render some of them persistent (Snyder 2008). Pharmaceuticals, and the compounds from their degradation processes, and particulate sorption have greater potential for environmental persistence because of their constant release into the environment (Houtman et al. 2004). Due to their physicochemical characteristics, the drugs are retained in the solids coming from the purification processes and then to the environment. So, they are considered "pseudo-persistent". Likewise, the ecological and environmental impact of cosmetics is associated with the large amounts used, constant release and with the fact that sometimes they are environmentally persistent (i.e. they cannot be eliminated or transformed from the environment and remain in it forever; Brausch et al. 2011).

22.9 BIOACCUMULATION

PPCPs may also exhibit adverse cumulative effects on terrestrial and aquatic ecosystems (Hernando et al. 2006). Certain PPCPs can lead to bioaccumulation in fish and other aquatic creatures. Many PPCPs and their metabolites are biologically active and can impact nontarget aquatic organisms. For example, carbamazepine and the antimicrobial agents triclocarban and TCS have been detected in algae, the high lipid content of which provides an entry point for the trophic transfer of lipophilic organic contaminants (Coogan et al. 2007). Likewise, 43 pharmaceuticals from different classes have been detected in mussel tissues. Other organisms in which PPCPs have been detected are crustaceans, snails and fish, such as goldfish (*Carassius auratus*) and the Eurasian perch (*Perca fluviatilis*; Brodin et al. 2014; Du et al. 2015). Due to their stability and lipophilicity, TCS and its metabolite methyl triclosan have the potential to bioaccumulate in algae (Coogan et al. 2007), plants (Prosser et al. 2014) and animals such as earthworms (Higgins et al. 2011), marine mussels (Kookana et al. 2013), snails (Coogan and La Point 2008), amphibian larvae (Palenske et al. 2010), fish (Houtman et al. 2004) and other marine mammals (Fair et al. 2009). Organic UV filters have been found in several aquatic organisms (invertebrates, fish, mammals and birds; Bachelot et al. 2012; Gago-Ferrero et al. 2013). Some of these compounds are able to bioaccumulate in the muscle and lipids of these organisms and are likely to enter marine food chains (Gago-Ferrero et al. 2015). If so, they will be ingested by man and may cause harmful effects which are as yet unknown.

Parabens have been reported in fish tissues (Xue and Kannan 2016). The presence of parabens and their metabolite p-hydroxybenzoic acid has been established in marine mammals (Xue et al. 2015), the tissues of marine birds and their eggs (Xue and Kannan 2016) and even in animals from remote marine locations, such as polar bears from Alaska (Xue et al. 2015) and albatrosses from Midway Atoll in the northwestern Pacific Ocean (Xue and Kannan 2016). Parabens also have been detected in human urine, serum, adipose tissue, milk, amniotic fluid and placental tissue (Valle-Sistac et al. 2016; Guo et al. 2017), and in human breast cancer tissues (Barr et al. 2012). Finally, ingestion and accumulation of microplastics has been well documented in different marine organisms, including copepods (Cole et al. 2011), bivalves (Vandermeersch et al. 2015), fish (Davison and Asch 2011) and seabirds (Van Franeker et al. 2011).

22.10 TOXICITY

The toxicity of PPCPs depends on several factors: the exposed organism, duration of exposure, contaminant concentration and the developmental window during which exposure occurs (Wilkinson et al. 2016). Several studies have examined the effect of PPCPs on different organisms. Thus, the toxicity of organic UV filters has been evaluated in several aquatic organisms, concluding that these filters are lethal for phytoplankton species, microalgae, protozoa and crustaceans (Sánchez-Quiles and Tovar-Sánchez 2015).

Studies on the lethal effects of nano TiO_2 on algae, freshwater and marine invertebrates, and fish were reported in recent years (Menard et al. 2011). As far as ZnO is concerned, its toxicity in zebrafish (Chen et al. 2014), marine algae (Zhang et al. 2016), sea urchins (Manzo et al. 2013) and other marine organisms is well documented. Specifically, ZnO aggregates were found to exert dose-dependent toxicity on zebrafish embryos and larvae, reducing the hatching rate and causing pericardial edema. As a consequence, nano ZnO is classified as "extremely toxic" (Ma et al. 2011). Experimental data demonstrated that the toxic effect of ZnO is partly ascribable to the release of Zn^{2+} ions and partly due to oxidative stress following ROS generation (Xiong et al. 2011). TCS exhibits toxicity to algae species (Yang et al. 2008), alters benthic bacterial communities' composition (fostering cyanobacteria over algae; Drury et al. 2013), exhibits teratogenic responses and mortality in the embryos and larvae of zebrafish (Oliveira et al. 2009) and exhibits endocrine disruption in fish (Crofton et al. 2007). Exposure to TCS can produce negative effects on human health, including thyroid function impairment, endocrine disruption, oxidative stress and liver carcinogenesis (Roberts et al. 2016).

A recent study (Sussarellu et al. 2016) provides evidence that microplastics induce feeding modification and reproductive disruption in oysters. Indeed, marine organisms can be at risk of starvation or intestinal blockage, impairment of reproduction and feeding ability, and reduced growth rates consequent to microplastic ingestion (Wright et al. 2013). Another concern is the potential of microplastics to aid the delivery (after ingestion) of persistent organic pollutants (POPs, including polychlorinated biphenyls, polycyclic aromatic hydrocarbons, organochlorine pesticides and alkylphenols), which can be transferred along the food chain and can harm fish and other aquatic organisms (Wardrop et al. 2016).

Pharmaceuticals are designed specifically to maximize their biological activity at low doses and to target certain metabolic, enzymatic, or cell-signalling mechanisms. Thus, these compounds could be pharmacologically active in nontarget organisms and, therefore, raise the risk of ecotoxicological effects (Fabbri and Franzellitti 2016). Especially sensitive are aquatic biota, which are unintentionally exposed to pharmaceuticals in their natural environment.

An important factor to consider is that metabolite concentrations from pharmaceutical may be more significant than that of parent compounds in the environment because they are introduced after human and veterinary use. Furthermore, the toxicity may be different, as in some acetylated metabolites of antibiotics (such as N4-acetylsulfapyridine; see Chapter 3), which were found to be more toxic than the parent compound (sulfapyridine) in algae (García-Galán et al. 2012). As well, the toxicological properties of drugs can be changed by the presence of active pharmaceutical agents under undesirable conditions in the aquatic environment. For example, the photodegradation products of naproxen have been shown to have more toxic effects than the parent compound in algae, rotifers and microcrustaceans (Isidori et al. 2005).

The main problems derived from the presence of these PPCPs in the environment are their ability to interfere with the endocrine system to produce undesired effects/disruption of homeostasis and the creation of antibiotic resistance. In relation to the first aspect, a number of studies also suggest that some organic UV filters possess endocrine-disruption capacities, similarly to many other xenobiotics, in both *in vitro* and *in vivo* models. Benzophenone-3, for example, displays a weak estrogenicity effect *in vitro* and strong anti-androgenic and weak estrogenic (but also anti-estrogenic) activity in murine models. Moreover, adverse effects of this compound on the reproduction capability of aquatic organisms has been reported (Kim and Choi 2014). Parabens also display endocrine-disrupting activity both *in vitro* and *in vivo*, and they have been associated with alterations of the reproductive system of male experimental animals (Darbre and Harvey 2008). These compounds exhibit a binding affinity for the estrogen receptors and are able to interfere with the plasma vitellogenin concentration in some organisms, such as rainbow trout and Japanese medaka (rice fish), although at concentrations largely exceeding the maximum concentrations detected in surface waters (Błedzka et al. 2014). As far as humans are concerned, several studies have been published claiming a link between the endocrine-disrupting activity of parabens and breast cancer (Darbre 2006), although these studies are not conclusive.

It is important to consider that although PPCPs may be present at low concentrations that do not elicit significant toxic effects, complex mixtures of them that could lead to synergistic interactions can potentially be hazardous (Cleuvers 2003). For example, carbamazepine (an antiepileptic) and clofibric acid (a lipid-lowering drug) exhibited much stronger effects on *Daphnia magna* than as single compounds at the same concentration. Likewise, the mixture effect of estradiol (E2) and 4-tert-nonylphenol can result in an additive/synergistic reaction, the consequence of which is the production of vitellogenin in juvenile rainbow trout (Thorpe et al. 2001).

Extensive use of antibiotics in human medicine and animal husbandry is the major cause of the long-term persistence of antibiotics at low levels and the promotion and proliferation of antibiotic-resistant bacteria (WHO 2015). This fact is another important concern related to the presence of these drugs in the environmental because the presence of antibiotics could have a detrimental effect on naturally occurring bacteria present in the environment (Novo et al. 2013). Some studies have reported adverse effects on aquatic organisms, including toxicity of ciprofloxacin (a broad-spectrum antimicrobial carboxyfluoroquinoline) in green algae (Halling-Sörensen et al. 2000), toxicity of oxolinic acid (a feed additive used in fish farms) in *D. magna* and toxicity of fluoroquinolone antibiotics (ciprofloxacin, lomefloxacin, ofloxacin, levofloxacin, enrofloxacin and flumequine) on five aquatic organisms (the cyanobacterium *Microcystis aeruginosa*, duckweed, green alga, *Pseudokirchneriella*, *Daphnia magna* and fathead minnow; Robinson et al. 2005). Interestingly, the bactericidal action of ciprofloxacin results from inhibition of the enzymes topoisomerase II (DNA gyrase) and topoisomerase IV, which are required for bacterial DNA replication, transcription, repair, strand supercoiling repair and recombination—of course, analogous processes occur in other organisms.

In conclusion, the number of emerging contaminants released into the environment as a consequence of human activities is increasing day by day and reflects the growing consumption and use of a wide range of products, including PPCPs (Juliano and Magrini 2017). These compounds are a unique group of emerging environmental contaminants due to their inherent ability to induce physiological effects in humans and animals at low doses (Ebele et al. 2017). Therefore, it is imperative to further the current understanding of the toxicological implications of acute and chronic exposure to PPCPs and their complex mixtures to allow a more precise assessment of their actual ecological and health risks.

BIBLIOGRAPHY

Alexy, R., A. Schöll, T. Kümpel, and K. Kümmerer. 2004. What do we know about antibiotics in the environment? In *Pharmaceuticals in the Environment*, ed. K. Kümmerer. Berlin Heidelberg: Springer.

Amine, H., E. Gómez, J. Halwani, C. Casella, and H. Fenet. 2012. UV filters, ethylhexyl methoxycinnamate, octocrylene and ethylhexyl dimethyl PABA from untreated wastewater in sediment from Eastern Mediterranean river transition and coastal zones. *Mar Pollut Bull* 64: 2435–42.

Ashton, D., M. Hilton, and K. V. Thomas. 2004. Investigating the environmental transport of human pharmaceuticals to streams in the United Kingdom. *Sci Total Environ* 333(1): 167–84.

Awad, Y. M., S. C. Kim, S. A. A. El-Azeem, et al. 2014. Veterinary antibiotics contamination in water, sediment, and soil near a swine manure composting facility. *Environ Earth Sci* 71(3): 1433–40.

Bachelot, M., D. Munaron, P. Le Gall, C. Casellas, H. Fenet, and E. Gomez. 2012. Organic UV filter concentrations in marine mussels from French coastal regions. *Sci Total Environ* 420: 273–79.

Balmer, M. E., H. R. Buser, M. D. Muller, and T. Poiger. 2005. Occurrence of some organic UV filters in wastewater, in surface waters, and in fish from Swiss lakes. *Environ Sci Technol* 39: 953–62.

Barón, E., P. Gago-Ferrero, M. Gorga, et al. 2013. Occurrence of hydrophobic organic pollutants (BFRs and UV-filters) in sediments from South America. *Chemosphere* 92: 309–16.

Barr, L., G. Metaxas, C. A. J. Harbach, L. A. Savoy, and P. D. Darbre. 2012. Measurement of paraben concentrations in human breast tissue at serial locations across the breast from axilla to sternum. *J Appl Toxicol* 32: 219–32.

Bartelt-Hunt, S., D. D. Snow, T. Damon-Powell, and D. Miesbach. 2011. Occurrence of steroid hormones and antibiotics in shallow groundwater impacted by livestock waste control facilities, *J Contam Hydrol* 123: 94–103.

Blair, B. D., J. P. Crago, C. J. Hedman, and R. D. Klaper. 2013. Pharmaceuticals and personal care products found in the Great Lakes above concentrations of environmental concern. *Chemosphere* 93: 2116–23.

Błedzka, D., J. Gromadzinska, and W. Wasowicz. 2014. Parabens. From environmental studies to human health. *Environ Int* 67: 27–42.

Borova, V. L., N. C. Maragou, P. Gago-Ferrero, C. Pistos, and N. S. Thomaidis. 2014. Highly sensitive determination of 68 psychoactive pharmaceuticals, illicit drugs, and related human metabolites in wastewater by liquid chromatography—tandem mass spectrometry. *Anal Bioanal Chem* 406(17): 4273–85.

Botta, C., J. Labille, M. Auffan, et al. 2011. TiO_2-based nanoparticles released in water from commercialized sunscreens in a life-cycle perspective: Structures and quantities. *Environ Pollut* 159: 1543–50.

Boxall, A. B., M. A. Rudd, B. W. Brooks, et al. 2012. Pharmaceuticals and personal care products in the environment: What are the big questions? *Environ Health Perspect* 120: 1221–9.

Brausch, J. M., K. A. Connors, B. W. Brooks, and G. M. Rand. 2012. Human pharmaceuticals in the aquatic environment: A review of recent toxicological studies and considerations for toxicity testing. *Rev Environ Contamin Toxicol* 218: 1–99.

Brausch, J. M., and G. M. Rand. 2011. A review of personal care products in the aquatic environment: Environmental concentrations and toxicity. *Chemosphere* 82(11): 1518–32.

Brodin, T., S. Piovano, J. Fick, J. Klaminder, M. Heynen, and M. Jonsson. 2014. Ecological effects of pharmaceuticals in aquatic systems impacts through behavioural alterations, Philosophical Trans. *Proc R Soc Lond. Ser B, Biol Sci* 369.

Browne, M. A., A. Dissanayake, T. S. Galloway, and D. M. Lowe. 2008. Ingested microscopic plastic translocates to the circulatory system of the mussel, Mytilus edulis. *Environ Sci Technol* 42: 5026–31.

Burkina, V., S. Sakalli, M. K. Rasmussen, et al. 2015. Does dexamethasone affect hepatic CYP450system of fish? Semi-static in-vivo experiment on juvenile rainbow trout. *Chemosphere* 139: 155–62.

Calamari, D., E. Zuccato, S. Castiglioni, R. Bagnati, and R. Fanelli. 2003. Strategic survey of therapeutic drugs in the rivers Po and Lambro in northern Italy. *Environ Sci Technol* 37: 1241–8.

Canosa, P., I. Rodríguez, E. Rubí, N. Negreira, and R. Cela. 2006. Formation of halogenated by-products of parabens in chlorinated water. *Anal Chim Acta* 575: 106–13.
Cardinal, P., J. C. Anderson, J. C. Carlson, et al. 2014. Macrophytes may not contribute significantly to removal of nutrients, pharmaceuticals, and antibiotic resistance in model surface constructed wetlands. *Sci Total Environ* 482–483: 294–304.
Carlson, J. C., M. I. Stefan, J. M. Parnis, and C. D. Metcalfe. 2015. Direct UV photolysis of selected pharmaceuticals, personal care products and endocrine disruptors in aqueous solution. *Water Res* 84: 350–61.
Chen, K., and J. L. Zhou. 2014. Occurrence and behavior of antibiotics in water and sediments from the Huangpu River, Shanghai, China. *Chemosphere* 95: 604–12.
Chen, T. H., C. C. Lin, and P. J. Meng. 2014. Zinc oxide nanoparticles alter hatching and larval locomotor activity in zebrafish (Danio rerio). *J Hazard Mater* 277: 134–40.
Chen, W., S. Lu, W. Jiao, M. Wang, and A. C. Chang. 2013. Reclaimed water: A safe irrigation water source? *Environ Dev* 8: 74–83.
Chisvert, A., and A. Salvador. 2007. UV filters in sunscreen and other cosmetics. Regulatory aspects and analytical methods. In *Analysis of Cosmetic Products*, ed. A. Chisvert and A. Salvador, 83–120. Amsterdam, The Netherlands: Elsevier.
Clara, M., B. Strenn, and N. Kreuzinger. 2004. Carbamazepine as a possible anthropogenic marker in the aquatic environment: Investigations on the behavior of carbamazepine in wastewater treatment and during groundwater infiltration. *Water Res* 38: 947–54.
Cleuvers, M. 2003. Aquatic ecotoxicity of pharmaceuticals including the assessment of combination effects. *Toxicol Lett* 142: 185–94.
Cole, M., P. Lindeque, C. Halsband, and T. Galloway. 2011. Microplastics as contaminants in the marine environment: A review. *Mar Pollut Bull* 62: 2588–97.
Coogan, M. A., R. E. Edziyie, T. W. La Point, and B. J. Venables. 2007. Algal bioaccumulation of triclocarban, triclosan, and methyl-triclosan in a North Texas wastewater treatment plant receiving stream. *Chemosphere* 67: 1911–18.
Coogan, M. A., and T. W. La Point. 2008. Snail bioaccumulation of triclocarban, triclosan, and methyltriclosan in a North Texas, USA, stream affected by wastewater treatment plant runoff. *Environ Toxicol Chem* 27: 1788–93.
Crofton, K. M., K. B. Paul, M. J. DeVito, and J. M. Hedge. 2007. Short-term in vivo exposure to the water contaminant triclosan: Evidence for disruption of thyroxine. *Environ Toxicol Pharmacol* 24: 194–7.
Dann, A. B., and A. Hontela. 2011. Triclosan: Environmental exposure, toxicity and mechanisms of action. *J Appl Toxicol* 31(4): 285–311.
Darbre, P. D. 2006. Environmental oestrogens, cosmetics and breast cancer. *Best Pract Res Clin Endocrinol Metabol* 20: 121–43.
Darbre, P. D., and P. W. Harvey. 2008. Paraben esters: Review of recent studies of endocrine toxicity, absorption, esterase and human exposure, and discussion of potential human health risks. *J Appl Toxicol* 28: 561–8.
Daughton, C. G. 2002. Environmental stewardship and drugs as pollutants. *Lancet* 360(9339): 1035–6.
Daughton, C. G., and T. A. Ternes. 1999. Pharmaceuticals and personal care products in the environment: Agents of subtle change? *Environ Health Perspect* 107: 907–38.
Davison, P., and R. G. Asch. 2011. Plastic ingestion by mesopelagic fishes in the North Pacific Subtropical Gyre. *Mar Ecol Prog Ser* 432: 173–80.
Derraik, J. G. B. 2002. The pollution of the marine environment by plastic debris: A review. *Mar Pollut Bull* 44: 842–52.
Drury, B., J. Scott, E. J. Rosi-Marshall, and J. J. Kelly. 2013. Triclosan exposure increases triclosan resistance and influences taxonomic composition of benthic bacterial communities. *Environ Sci Technol* 47: 8923–30.
Du, B., S. P. Haddad, W. C. Scott, C. K. Chambliss, and B. W. Brooks. 2015. Pharmaceutical bioaccumulation by periphyton and snails in an effluent-dependent stream during an extreme drought. *Chemosphere* 119: 927–34.
Ebele, A. J., M. A. E. Abdallah, and S. Harrad. 2017. Pharmaceuticals and personal care products (PPCPs) in the freshwater aquatic environment. *Emerg Contam* 3: 1–16.
Ekpeghere, K. I., U. J. Kim, S. H. O. H. Y. Kim, and J. E. Oh. 2016. Distribution and seasonal occurrence of UV filters in rivers and wastewater treatment plants in Korea. *Sci Total Environ* 542: 121–8.

Ellis, J. B., D. M. Revitt, P. Lister, C. Willgress, and A. Buckleyet. 2002. Experimental studies of sewer exfiltration. *Water Sci Technol* 47: 61–7.

European Commission. Implementing Decision 2015/495 of 20 March 2015 Establishing a Watch List of Substances for Union-wide Monitoring in the Field of Water Policy Pursuant to Directive 2008/105/EC and Amending. Directive 2000/60/EC. *Official Journal of the European Union*, L 78/40 2015.

European Union. 2009. Regulation (EC) No 1223/2009 of the European Parliament and of the Council of 30 November 2009 on Cosmetic Products.

Fabbri, E., and S. Franzellitti. 2016. Human pharmaceuticals in the marine environment: Focus on exposure and biological effects in animal species. *Environ Toxicol Chem* 35: 799–812.

Fair, P. A., H. B. Lee, J. Adams, et al. 2009. Occurrence of triclosan in plasma of wild Atlantic bottlenose dolphins (Tursiops truncatus) and in their environment. *Environ Pollut* 157: 2248–54.

Farré, M. L., S. Pérez, L. Kantiani, and D. Barceló. 2008. Fate and toxicity of emerging pollutants, their metabolites and transformation products in the aquatic environment. *TrAC Trends Anal Chem* 27: 991–1007.

Fick, J., H. Söderström, R. H. Lindberg, C. Phan, M. Tysklind, and D. G. Larsson. 2009. Contamination of surface, ground, and drinking water from pharmaceutical production. *Environ Toxicol Chem* 28(12): 2522–7.

Gago-Ferrero, P., M. B. Alonso, C. P. Bertozzi, et al. 2013. First determination of UV filters in marine mammals. Octocrylene levels in Franciscana dolphins. *Environ Sci Technol* 47: 5619–25.

Gago-Ferrero, P., M. S. Díaz-Cruz, and D. Barceló. 2015. UV filters bioaccumulation in fish from Iberian river basins. *Sci Total Environ* 518–519: 518–25.

García-Galán, M. J., S. González Blanco, R. López Roldán, S. Díaz-Cruz, and D. Barceló. 2012. Ecotoxicity evaluation and removal of sulfonamides and their acetylated metabolites during conventional wastewater treatment. *Sci Total Environ* 437: 403–12.

Golden, R., J. Gandy, and G. Vollmer. 2005. A review of the endocrine activity of parabens and implications for potential risks to human health. *Crit Rev Toxicol* 35(5): 435–58.

González-Mariño, I., J. B. Quintana, I. Rodríguez, and R. Cela. 2011. Evaluation of the occurrence and biodegradation of parabens and halogenated by-products in wastewater by accurate mass liquid chromatography-quadrupole-time-of-flight-mass spectrometry (LC-QTOF-MS). *Water Res* 45: 6770–80.

Gouin, T., J. Avalos, I. Brunning, et al. 2015. Use of micro-plastic beads in cosmetic products in Europe and their estimated emissions to the North Sea environment. *SOFW J* 3: 1–33.

Guo, J., C. Wu, D. Lu, et al. 2017. Urinary paraben concentrations and their associations with anthropometric measures of children aged 3 years. *Environ Pollut* 222: 307–14.

Halling-Sörensen, B., H. C. H. Lützhöft, H. R. Andersen, and F. Ingerslev. 2000. Environmental risk assessment of antibiotics: Comparison of mecillinam, trimethoprim and ciprofloxacin *J Antimicrob Chemother* 46: 53–8.

Haman, C., X. Dauchy, C. Rosin, and J. F. Munoz. 2015. Occurrence, fate and behavior of parabens in aquatic environments: A review. *Water Res* 68: 1–11.

Heberer, T. 2002. Occurrence, fate, and assessment of polycyclic musk residues in the aquatic environment of urban areas e a review. *Acta Hydrochim Hydrobiol* 30: 227–43.

Helbling, D. E., J. Hollender, H. P. E. Kohler, H. Singer, and K. Fenner. 2010. High through put identification of microbial transformation products of organic micropollutants, *Environ Sci Technol* 44: 6621–7.

Hernando, M. D., M. Mezcua, A. R. Fernández-Alba, and D. Barceló. 2006. Environmental risk assessment of pharmaceutical residues in wastewater effluents, surface waters and sediments. *Talanta* 69(2): 334–42.

Higgins, C. P., Z. H. Paesani, T. E. Chalew, R. U. Halden, and L. S. Hundal. 2011. Persistence of triclocarban and triclosan in soils after land application of biosolids and bioaccumulation in Eisenia foetida. *Environ Toxicol Chem* 30: 556–63.

Houtman, C. J., A. M. Van Oostveen, A. Brouwer, M. H. Lamoree, and J. Legler. 2004. Identification of estrogenic compounds in fish bile using bioassay-directed fractionation. *Environ Sci Technol* 38: 6415–23.

Isidori, M., M. Lavorgna, A. Nardelli, A. Parrella, L. Previtera, and M. Rubino. 2005. Ecotoxicity of naproxen and its phototransformation products. *Sci Total Environ* 348: 93–101.

Jiang, J. J., C. L. Lee, and M. D. Fang. 2014. Emerging organic contaminants in coastal waters: Anthropogenic impact, environmental release and ecological risk. *Mar Pollut Bull* 85: 391–9.

Jones, P. D., N. Voulvoulis, and J. N. Lester. 2001. Human pharmaceuticals in the aquatic environment a review. *Environ Technol* 22: 1383–94.
Juliano, C., and G. A. Magrini. 2017. Cosmetic ingredients as emerging pollutants of environmental and health concern. A mini-review. *Cosmetics* 4: 11.
Jurado, A., P. Gago-Ferrero, E. Vázquez-Suñé, et al. 2014. Urban groundwater contamination by residues of UV filters. *J Hazard Mater* 271: 141–9.
Kaczala, F., and E. Blum. 2016. The occurrence of veterinary pharmaceuticals in the environment: A review. *Curr Anal Chem* 12(3): 169–82.
Kallenborn, R. K., U. Berger, and U. Järnberg. 2004. Perfluorinated Alkylated Substances (PFAS) in the Nordic environment. *Nord Counc Minist.*
Kästner, M., K. M. Nowak, A. Miltner, S. Trapp, and A. Schaeffer. 2014. Classification and modelling of nonextractable residue (NER) formation of xenobiotics in soil— a synthesis. *Crit Rev Environ Sci Technol* 44: 2107–71.
Kim, K. R., G. Owens, S. I. Kwon, K. H. So, D. B. Lee, and Y. S. Ok. 2011. Occurrence and environmental fate of veterinary antibiotics in the terrestrial environment. *Water Air Soil Pollut* 214(1–4): 163–74.
Kim, S., and K. Choi. 2014. Occurrences, toxicities, and ecological risks of benzophenone-3, a common component of organic sunscreen products: A mini-review. *Environ Int* 70: 143–57.
Kimura, K., H. Hara, and Y. Watanabe. 2007. Elimination of selected acidic pharmaceuticals from municipal wastewater by an activated sludge system and membrane bioreactors. *Environ Sci Technol* 41(10): 3708–14.
Kimura, K., Y. Kameda, H. Yamamoto, et al. 2014. Occurrence of preservatives and antimicrobials in Japanese rivers. *Chemosphere* 107: 393–9.
Kockler, J., M. Oelgemöller, S. Robertson, and B. D. Glass. 2012. Photostability of sunscreens. *J Photochem Photobiol C* 13: 91–110.
Kookana, R. S., A. Shareef, M. B. Fernandes, S. Hoare, S. Gaylard, and A. Kumar. 2013. Bioconcentration of triclosan and methyl-triclosan in marine mussels (Mytilus galloprovincialis) under laboratory conditions and in metropolitan waters of Gulf St Vincent, South Australia. *Mar Pollut Bull* 15: 66–72.
Kostich, M. S., A. L. Batt, and J. M. Lazorchak. 2014. Concentrations of prioritized pharmaceuticals in effluents from 50 large wastewater treatment plants in the US and implications for risk estimation. *Environ Pollut* 184: 354–9.
Lambropoulou, D. A., and L. M. Nollet. 2014. *Transformation Products of Emerging, in the Environment: Analysis, Processes, Occurrence, Effects and Risks.* John Wiley and Sons.
Lapworth, D. J., N. Baran, M. E. Stuart, and R. S. Ward. Emerging organic contaminants in groundwater: A review of sources, fate and occurrence. *Environ Pollut* 163: 287–03.
Leung, H. W., T. B. Minh, M. B. Murphy, et al. 2012. Distribution, fate and risk assessment of antibiotics in sewage treatment plants in Hong Kong. South China. *Environ Int* 42: 1–9.
Lewicka, Z. A., W. W. Yu, B. L. Oliva, E. Q. Contreras, and V. L. Colvin. 2013. Photochemical behavior of nanoscale TiO_2 and ZnO sunscreen ingredients. *J Photochem Photobiol A* 263: 24–33.
Liu, J. L., and M. H. Wong. 2013. Pharmaceuticals and personal care products (PPCPs): A review on environmental contamination in China. *Environ Int* 59: 208–24.
Loos, R., G. Locoro, S. Comero, et al. 2010. Pan-European survey on the occurrence of selected polar organic persistent pollutants in ground water. *Water Res* 44: 4115–26.
López-Serna, R., A. Jurado, E. Vázquez-Suñé, J. Carrera, M. Petrović, and D. Barceló. 2013. Occurrence of 95 pharmaceuticals and transformation products in urban groundwaters underlying the metropolis of Barcelona, Spain. *Environ Pollut* 174: 305–15.
Luo, Y. L., W. S. Guo, H. H. Ngo, et al. 2014. A review on the occurrence of micropollutants in the aquatic environment and their fate and removal during wastewater treatment. *Sci Total Environ* 473: 619–41.
Ma, H., N. J. Kabengi, P. M. Bertsch, J. M. Unrine, T. C. Glenn, and P. L. Williams. 2011. Comparative phototoxicity of nanoparticulate and bulk ZnO to a free-living nematode Caenorhabditis elegans: The importance of illumination mode and primary particle size. *Environ Pollut* 159: 1473–80.
Manzo, S., M. L. Miglietta, G. Rametta, S. Buono, and G. Di Francia. 2013. Embryotoxicity and spermiotoxicity of nanosized ZnO for Mediterranean Sea urchin Paracentrotus lividus. *J Hazard Mater* 254–255: 1–9.

McClellan, K., and R. U. Halden. 2010. Pharmaceuticals and personal care products in archived U.S. biosolids from the 2001 EPA national sewage sludge survey. *Water Res* 44(2): 658–68.

Menard, A., D. Drobne, and A. Jemec. 2011. Ecotoxicity of nanosized TiO2. Review of in vivo data. *Environ Pollut* 159: 677–84.

Mompelat, S., B. Le Bot, and O. Thomas. 2009. Occurrence and fate of pharmaceutical products and by-products, from resource to drinking water. *Environ Int* 35(5): 803–14.

Montforts, M. H. M. M. 1999. *Environmental Risk Assessment for Veterinary Medicinal Products. Part 1. Other Than GMO-containing and Immunological Products*. Bilthoven, The Netherlands: National Institute for Public Health and the Environment (RIVM).

Napper, I. E., A. Bakir, S. J. Rowland, and R. C. Thompson. 2015. Characterisation, quantity and sorptive properties of microplastics extracted from cosmetics. *Mar Pollut Bull* 99: 178–85.

Nikolaou, A., S. Meric, and D. Fatta. 2007. Occurrence patterns of pharmaceuticals in water and wastewater environments. *Anal Bioanalyt Chem* 387: 1225–34.

Novo, A., S. Andre, P. Viana, O. C. Nunes, and C. M. Manaia. 2013. Antibiotic resistance, antimicrobial residues and bacterial community composition in urban wastewater. *Water Res* 47: 1875–87.

Oliveira, R., I. Domingues, G. C. Koppe, and A. M. Soares. 2009. Effects of triclosan on zebrafish early-life stages and adults. *Environ Sci Pollut Res Int* 16: 679–88.

Onesios-Barry, K. M., D. Berry, J. B. Proescher, I. K. Sivakumar, and E. J. Bouwer. 2014. Removal of pharmaceuticals and personal care products during water recycling: Microbial community structure and effects of substrate concentration. *Appl Environ Microbiol* 80: 2440–50.

Osenbrück, K., H. R. Gläser, K. Knöller, et al. 2007. Sources and transport of selected organic micropollutants in urban groundwater underlying the city of Halle (Saale), Germany. *Water Res* 41: 3259–70.

Palenske, N. M., G. C. Nallani, and E. M. Dzialowski. 2010. Physiological effects and bioconcentration of triclosan on amphibian larvae. *Comparat Biochem Physiol C Toxicol Pharmacol* 152: 232–40.

Peck, A. M. 2006. Analytical methods for the determination of persistent ingredients of personal care products in environmental matrices. *Anal Bioanal Chem* 386: 907–39.

Peng, X., K. Adachi, C. Chen, et al. 2006. Discovery of a marine bacterium producing 4-hydroxybenzoate and its alkyl esters, parabens. *Appl Environ Microbiol* 72: 5556–61.

Peng, X., W. Ou, C. Wang, Q. Huang, J. Jin, and J. Tan. 2014. Occurrence and ecological potential of pharmaceuticals and personal care products in groundwater and reservoirs in the vicinity of municipal landfills in China. *Sci Total Environ* 490: 889–98.

Petrović, M., S. González, and D. Barceló. 2003. Analysis and removal of emerging contaminants in wastewater and drinking water. *TrAC Trends Anal Chem* 22(10): 685–96.

Price, O. R., G. O. Hughes, N. L. Roche, and P. J. Mason. 2010. Improving emissions estimates of home and personal care products ingredients for use in EU risk assessments. *Integr Environ Assess Manag* 6(4): 677–84.

Prosser, R. S., L. Lissemore, E. Topp, and P. K. Sibley. 2014. Bioaccumulation of triclosan and triclocarban in plants grown in soils amended with municipal dewatered biosolids. *Environ Toxicol Chem* 33: 975–84.

Ricart, M., H. Guasch, M. Alberch, et al. 2010. Triclosan persistence through wastewater treatment plants and its potential: Toxic effects on river biofilms. *Aquat Toxicol* 100(4): 346–53.

Roberts, J., A. Kumar, J. Du, et al. 2016. Pharmaceuticals and personal care products (PPCPs) in Australia's largest inland sewage treatment plant, and its contribution to a major Australian river during high and low flow. *Sci Total Environ* 541: 1625–37.

Robinson, A. A., J. B. Belden, and M. J. Lydy. 2005. Toxicity of fluoroquinolone antibiotics to aquatic organisms. *Environ Toxicol Chem* 24: 423–30.

Robinson, P. F., Q. T. Liu, A. M. Riddle, and R. Murray-Smith. 2007. Modeling the impact of direct phototransformation on predicted environmental concentrations (PECs) of propranolol hydrochloride in UK and US rivers. *Chemosphere* 66: 757–66.

Sánchez Rodríguez, A., M. Rodrigo Sanz, and J. R. Betancort Rodríguez. 2015. Occurrence of eight UV filters in beaches of Gran Canaria (Canary Islands). An approach to environmental risk assessment. *Chemosphere* 131: 85–90.

Sánchez-Quiles, D., and A. Tovar-Sánchez. 2015. Are sunscreens a new environmental risk associated with coastal tourism? *Environ Int* 83: 158–70.

Santos, A. J. M., M. S. Miranda, and J. C. G. Esteves da Silva. 2012. The degradation products of UV filters in aqueous and chlorinated aqueous solutions. *Water Res* 46: 3167–76.

Schaider, L. A., R. A. Rudel, J. M. Ackerman, S. C. Dunagan, and J. G. Brody. 2014. Pharmaceuticals, perfluorosurfactants, and other organic wastewater compounds in public drinking water wells in a shallow sand and gravel aquifer. *Sci Total Environ* 468–469: 384–93.

Schumock, G. T., E. C. Li, K. J. Suda, et al. 2014. National trends in prescription drug expenditures and projections for 2014. *Am J Health Syst Pharm* 71(6): 482–99.

Shi, X., Z. Li, W. Chen, et al. 2016. Fate of TiO_2 nanoparticles entering sewage treatment plants and bioaccumulation in fish in the receiving streams. *NanoImpact* 3–4: 96–103.

Snyder, S. A. 2008. Occurrence, treatment, and toxicological relevance of EDCs and pharmaceuticals in water. *Ozone Sci Eng* 30: 65–9.

Steinbach, C., R. Grabic, G. Fedorova, et al. 2016. Bioconcentration, metabolism and half-life time of the human therapeutic drug diltiazem in rainbow trout Oncorhynchus mykiss. *Chemosphere* 144: 154–9.

Sui, Q., X. Cao, S. Lu, W. Zhao, Z. Qiu, and G. Yu. 2015. Occurrence, sources and fate of pharmaceuticals and personal care products in the groundwater: A review. *Emerg Contam* 1(1): 14–24.

Sussarellu, R., M. Suquet, Y. Thomas, et al. 2016. Oyster reproduction is affected by exposure to polystyrene microplastics. *Proc Natl Acad Sci USA* 113: 2430–5.

Ternes, T. A., A. Joss, and H. Siegrist. 2004. Peer reviewed: Scrutinizing pharmaceuticals and personal care products in wastewater treatment. *Environ Sci Technol* 38(20): 392A-399A.

Thorpe, K. L., T. H. Hutchinson, M. J. Hetheridge, M. Scholze, J. P. Sumpter, and C. R. Tyler. 2001. Assessing the biological potency of binary mixtures of environmental estrogens using vitellogenin induction in juvenile rainbow trout (Oncorhynchus mykiss). *Environ Sci Technol* 35: 2476–81.

Tolls, J. 2001. Sorption of veterinary pharmaceuticals in Soils: A review. *Environ Sci Technol* 35: 3397–406.

Tran, N. H., J. Li, J. Hu, and S. L. Ong. 2014. Occurrence and suitability of pharmaceuticals and personal care products as molecular markers for raw wastewater contamination in surface water and groundwater. *Environ Sci Pollut Res* 21: 4727–40.

Tsui, M. M. P., H. W. Leung, T. C. Wai, et al. 2014. Occurrence, distribution and ecological risk assessment of multiple classes of UV filters in surface waters from different countries. *Water Res* 67: 55–65.

Valle-Sistac, J., D. Molins-Delgado, M. Díaz, L. Ibáñez, D. Barceló, and M. S. Díaz-Cruz. 2016. Determination of parabens and benzophenone-type UV filters in human placenta. First description of the existence of benzyl paraben and benzophenone-4. *Environ Int* 88: 243–9.

Van Cauwenberghe, L., L. Devriese, F. Galgani, J. Robbens, and C. R. Janssen. 2015. Microplastics in sediments: A review of techniques, occurrence and effects. *Mar Environ Res* 111: 5–17.

Vandermeersch, G., L. Van Cauwenberghe, C. R. Janssen, et al. 2015. A critical view on microplastic quantification in aquatic organisms. *Environ Res* 143: 46–55.

Van Franeker, J. A., C. Blaize, J. Danielsen, et al. 2011. Monitoring plastic ingestion by the northern fulmar Fulmarus glacialis in the North Sea. *Environ Pollut* 159: 2609–15.

Van Wieren, E. M., M. D. Seymour, and J. W. Peterson. 2012. Interaction of the fluoroquinolone antibiotic, ofloxacin, with titanium oxide nanoparticles in water: Adsorption and breakdown. *Sci Total Environ* 441: 1–9.

Verlicchi, P., M. Al Aukidy, A. Jelic, M. Petrovic, and D. Barceló. 2014. Comparison of measured and predicted concentrations of selected pharmaceuticals in wastewater and surface water: A case study of a catchment area in the Po Valley (Italy). *Sci Total Environ* 470–471: 844–54.

Vulliet, E., C. Cren-Olivé, and M. F. Grenier-Loustalot. 2011. Occurrence of pharmaceuticals and hormones in drinking water treated from surface waters. *Environ Chem Lett* 9(1): 103–14.

Wang, J., and S. Wang. 2016. Removal of pharmaceuticals and personal care products (PPCPs) from wastewater: A review. *J Environ Manag* 182: 620–40.

Wardrop, P., J. Shimeta, D. Nugegoda, et al. 2016. Chemical pollutants sorbed to ingested microbeads from personal care products accumulate in fish. *Environ Sci Technol* 50: 4037–44.

WHO. 2015. *Antibiotic Resistance: Multi-Country Public Awareness Survey*. World Health Organisation.

Wilkinson, J. L., P. S. Hooda, J. Barker, S. Barton, and J. Swinden. 2016. Ecotoxic pharmaceuticals, personal care products, and other emerging contaminants: A review of environmental, receptor-mediated, developmental, and epigenetic toxicity with discussion of proposed toxicity to humans. *Crit Rev Environ Sci Technol* 46: 336–38.

Wright, S. L., R. C. Thompson, and T. S. Galloway. 2013. The physical impacts of microplastics on marine organisms: A review. *Environ Pollut* 178: 483–92.

Xia, B. A., K. K. Das, and G. Pillar. 2005. Occurrence and fate of pharmaceuticals and personal care products (PPCPs) in biosolids. *J Environ Qual* 34: 91–104.
Xiong, D., T. Fang, L. Yu, X. Sima, and W. Zhu. 2011. Effects of nano-scale TiO2, ZnO and their bulk counterparts on zebrafish: Acute toxicity, oxidative stress and oxidative damage. *Sci Total Environ* 409: 1444–52.
Xue, J., and K. Kannan. 2016. Accumulation profiles of parabens and their metabolites in fish, black bear, and birds, including bald eagles and albatrosses. *Environ Int* 94: 546–53.
Xue, J., N. Sasaki, M. Elangovan, G. Diamond, and K. Kannan. 2015. Elevated accumulation of parabens and their metabolites in marine mammals from the United States coastal waters. *Environ Sci Technol* 49: 12071–79.
Yang, L. H., G. G. Ying, H. C. Su, J. Stauber, M. S. Adams, and M. T. Binet. 2008. Growth-inhibiting effects of 12 antibacterial agents and their mixtures on the freshwater microalga Pseudokirchneriella subcapitata. *Environ Toxicol Chem* 27: 1201–8.
Yang, Y., Y. S. Ok, K. H. Kim, E. E. Kwon, and Y. F. Tsang. 2017. Occurrences and removal of pharmaceuticals and personal care products (PPCPs) in drinking water and water/sewage treatment plants: A review. *Sci Total Environ* 596–597: 303–20.
Yang, Y., and G. S. Toor. 2015. *Contaminants in the Urban Environment. Pharmaceuticals and Personal Care Products (PPCPs) Part 2*. University of Florida Extension.
Yueh, M. F., and R. H. Tukey. 2016. Triclosan: A widespread environmental toxicant with many biological effects. *Annu Rev Pharmacol Toxicol* 56: 251–72.
Zemann, M., L. Wolf, F. Grimmeisen, et al. 2015. Tracking changing X-ray contrast media application to an urban-influenced karst aquifer in the Wadi Shueib, Jordan. *Environ Pollut* 198: 133–43.
Zhang, C., J. Wang, L. Tan, and X. Chen. 2016. Toxic effects of nano-ZnO on marine microalgae Skeletonema costatum: Attention to the accumulation of intracellular Zn. *Aquat Toxicol* 178: 158–64.
Zhao, J. L., Q. Q. Zhang, F. Chen, et al. 2013. Evaluation of triclosan and triclocarban at river basin scale using monitoring and modeling tools: Implications for controlling of urban domestic sewage discharge. *Water Res* 47: 395–405.
Ziylan, A., and N. H. Ince. 2011. The occurrence and fate of anti-inflammatory and analgesic pharmaceuticals in sewage and fresh water: Treatability by conventional and nonconventional processes. *J Hazard Mater* 187(1): 24–36.
Zwiener, C., S. D. Richardson, D. DeMarini, and F. H. Frimmel. 2007. Drowning in disinfection byproducts? Assessing swimming pool water. *Environ Sci Technol* 41: 363–72.

23 Environmental Recovery and Remediation

The role of soil pollution on human health

M. Carmen Lobo-Bedmar and M. Mar Gil-Díaz

CONTENTS

23.1 INTRODUCTION

Soil is considered the superficial layer of the Earth's crust. Its main constituents are mineral particles, organic matter, water, air and living organisms. Thus, soil is a live system and a nonrenewable resource due to the slow speed of the formation process (Lobo et al. 2009). In the last decades, environmental policies have recognised the importance of maintaining soil health for its important relationship with air and water quality and human health. Because of this, soil pollution has been identified as one of the 10 major soil threats (FAO and ITPS 2015; FAO 2016). Local or diffuse sources from anthropogenic activities are responsible for the loss of functionality of the soil. Landrigan et al. (2018) concluded that pollution is the largest environmental cause of disease and premature death in the world today, especially in low- and middle-income countries, where it is the largest cause of disease and death with nearly 92% of pollution-related deaths (Landrigan et al. 2018; Prüss-Ustün et al. 2016). The World Health Organization (WHO) estimated that 12.6 million people died in 2012 as a result of living or working in an unhealthy environment (23% of all deaths). According to the data of 2015, pollution was responsible for three times as many deaths as AIDS, tuberculosis and malaria combined. In fact, in the most affected countries, pollution is responsible for more than 1 in 4 deaths (Landrigan et al. 2018). Pollution negatively affects ecosystems and is responsible for global climate change.

Pollution can be due to natural sources or anthropogenic activities, which have increased in the last 200 years with industrialisation. Pollution is caused by electricity generation, industrial processes, manufacturing, mining, road transport, military activities, use of agrochemicals and land application of domestic sludge (van Liedekerke et al. 2014). In the EU, there are about 2.5 million potentially contaminated sites, of which about 14% (350,000 sites) require urgent remediation (van Liedekerke et al. 2014). According to the European Environment Agency, the management of contaminated land in Europe costs an estimated €6.5 billion per year. In the United States, the Environmental Protection Agency (EPA) calculates that tens of thousands of sites in the country require remediation, and its National Priorities List for urgent remediation currently contains more than 1300 sites (US EPA 2018a). The EPA estimated that it will cost up to $170–250 billion to clean up that nation's hazardous waste sites (US EPA 2004). With regards to the developing world, the studies show that to date, more than 2000 contaminated sites have been identified in 47 countries exposing an estimated population of 71.5 million to risk (Ericson et al. 2013). Chemical pollution implies the presence of certain chemical elements or potentially harmful substances in

DOI: 10.1201/9780203730584-23

amounts that exceed recommended levels for the health of humans, animals and plants or that cause a detriment to soil normal functioning (Orgiazzi et al. 2016; Pierzynski et al. 2005). Pollutants can be organic or inorganic. Organic pollutants include mineral oils (total petroleum hydrocarbons [TPHs]), chlorinated hydrocarbons (dioxins, furans, polychlorinated biphenyls [PCBs], other chlorinated solvents), polycyclic aromatic hydrocarbons (PAHs), phenols, aromatic hydrocarbons, pesticides and the so-called emerging pollutants such as pharmaceutical and veterinary products, illicit drugs, personal care products, food additives and engineered nanomaterials. More than 140,000 new chemicals have been developed in the last century, and only a small number of these compounds (fewer than 2500) have undergone systematic tests to evaluate their potential toxicity on humans and ecosystems (Landrigan et al. 2018). Inorganic pollutants include metals such as silver (Ag), cadmium (Cd), cobalt (Co), chromium (Cr), copper (Cu), mercury (Hg), nickel (Ni), lead (Pb), zinc (Zn); the metalloids arsenic (As), antimony (Sb), selenium (Se); and radionuclides, phosphates, ammonia, nitrates, sulphates and cyanide. The presence of pollutants in soil and water conduces their accumulation on food and feed, introduction into the trophic chain and the subsequent accumulation in the human body over time, causing health problems such as physical and mental disorders, organ dysfunction, cancer, reduced life expectancy, weakening of the body's immune system and even the death (Kuppusamy et al. 2016 and references therein). In addition, pathogenic organisms are also considered soil and water pollutants (Rodrigues and Römkens 2018). Some of these chemicals, such as certain pesticides, metals, additives in food, and personal care products, can affect the endocrine (hormonal) system. Endocrine disruptors have been associated with altered reproductive function in males and females, increased incidence of breast cancer, abnormal growth patterns and neurodevelopmental delays in children, and changes in immune function (Bergman et al. 2012). Figure 23.1 shows the different types of main pollutants of global concern.

23.2 ORGANIC POLLUTANTS

Many of the organic pollutants are persistent organic pollutants (POPs), which are chemicals of global concern due to their potential for long-range transport, persistence in the environment, ability to biomagnify and bioaccumulate in ecosystems, and significant negative effects on human health and the environment (WHO 2018a). The Stockholm Convention on POPs, in force since 2004, has as its objectives to protect human health and the environment from those chemicals. The Stockholm Convention collects a list of POPs establishing their elimination, restriction or unintentional production (Stockholm Convention 2008). Figure 23.1 collects the POPs listed in the Stockholm Convention: Annex A includes those whose production and use must be eliminated, and Annex B includes those whose production and use must be restricted. For those POPs in Annex C, measures must be taken to reduce their unintentional release with the goal of continuing minimisation and, where feasible, ultimate elimination.

In addition, the Rotterdam Convention, in force since 2004, is a multilateral treaty which promotes shared responsibility and cooperative efforts among parties in the international trade of certain hazardous chemicals and pesticides. The Convention also contributes to the environmentally sound use of those hazardous chemicals by facilitating information exchange about their characteristics, by providing for a national decision-making process on their import and export, and by disseminating these decisions to Parties to the Convention (Rotterdam Convention 2010; FAO 2018a).

The main types of organic pollutants are described below.

- Mineral oils are the main organic pollutants found in soils and groundwater in the EU (van Liedekerke et al. 2014). Mineral oils include TPHs, which may enter the environment through accidents, from industrial releases, or as by-products from commercial or private uses. Some TPH fractions will float on the water and form surface films, whereas other ones will sink to the bottom sediments. They present high ecotoxicity and low biodegradability, thus they are a considerable threat to the environment.

- Aromatic hydrocarbons such as benzene, toluene, ethylbenzene and xylene (BTEX) are petroleum hydrocarbons used in solvents and in many fuels (Figure 23.2). Among BTEX, benzene is reported as the most hazardous human carcinogen and it is known for its leukaemogenic potential (Mazzeo et al. 2011). Toluene and xylene are not carcinogenic; however, it has been reported that toluene can enhance carcinogenesis when it is in combination with other compounds (Varjani et al. 2017). BTEX have a characteristic strong odour, low solubility in water, and are highly lipophilic and flammable. They can be introduced into the environment by gasoline leakage from storage tanks, transportation, pipelines and petrochemical wastewaters.

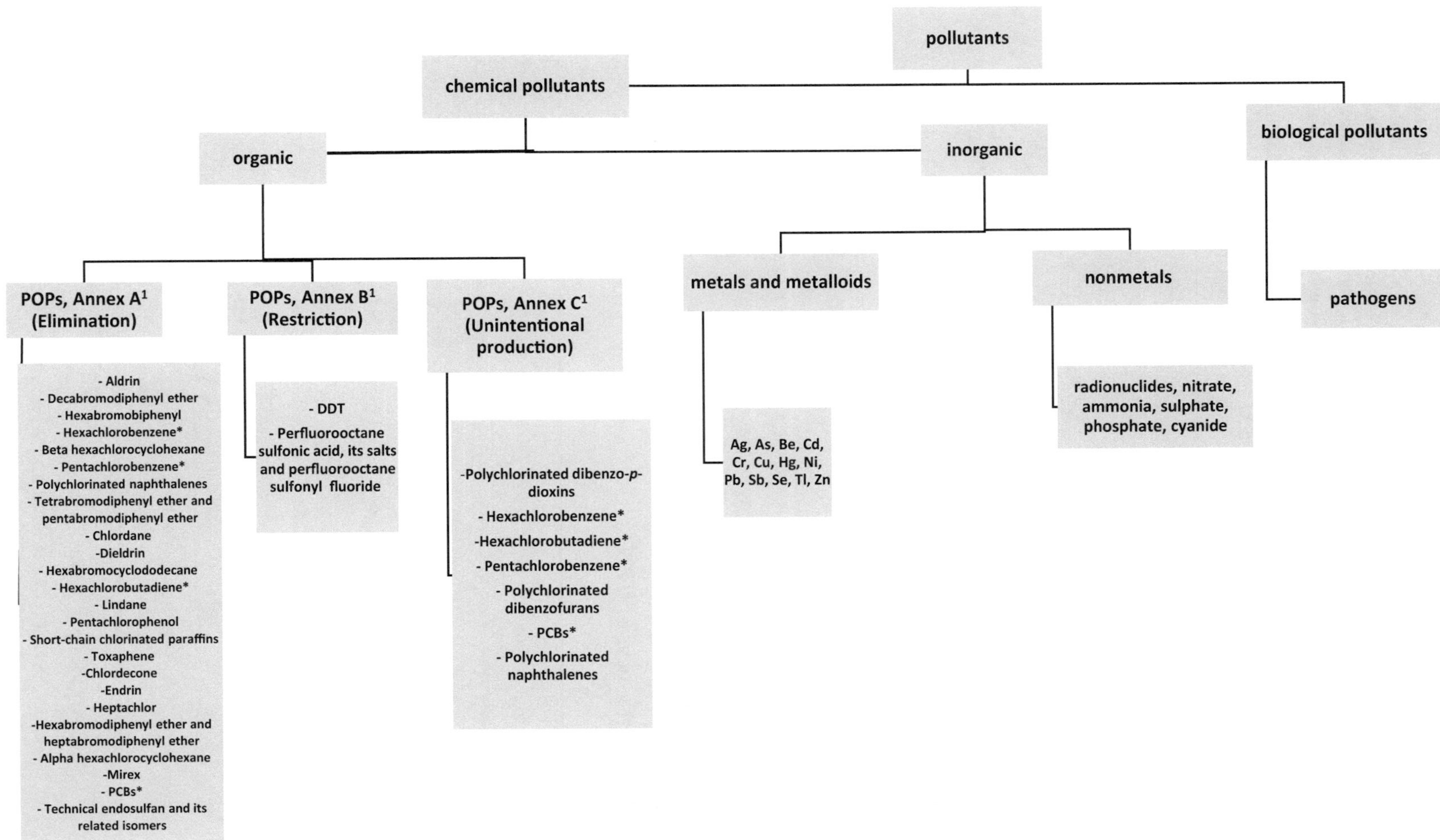

Figure 23.1 Chemical and biological pollutants of global interest.

Notes:

[1] POPs (persistent organic pollutants) listed by the Stockholm Convention (2010) in Annex A (to eliminate the production and use); Annex B (to restrict the production and use); and Annex C (to reduce unintentional release).

* POP included in two annexes.

- Pesticides. The main types of pesticides are herbicides, insecticides and fungicides, grouped according to function. They can also be disinfectants, attractants, plant defoliants and plant growth regulators. They are spread on soils to control agricultural pests. Based on their chemical structure, they are classified in organo-phosphorus, organo-chlorine, nitrogen-benzene, phenol, metallo-organic and other compounds (Sun et al. 2018). The pesticide market grew very rapidly in the 1960s, and their use led to significant increases in agricultural productivity (Tarradellas and Bitton 1997). Yield averages of wheat, rice and maize (the major sources of human nutrition) reached increases of more than 100% (Popp et al. 2013). The total amount of pesticides used between 1988 and 1991 in the European Union corresponded to a total arable crop surface of 78 × 106 ha (Tarradellas and Bitton 1997). Currently, China is the world's largest pesticide producer and consumer. Figure 23.3 shows the amount of pesticides used per continent in 2011 (FAO 2018b).

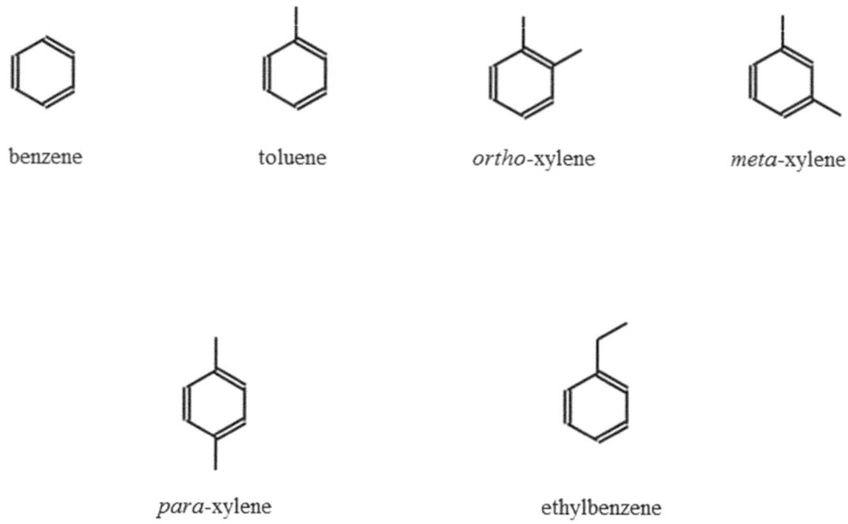

Figure 23.2 Chemical structures of BTEX.

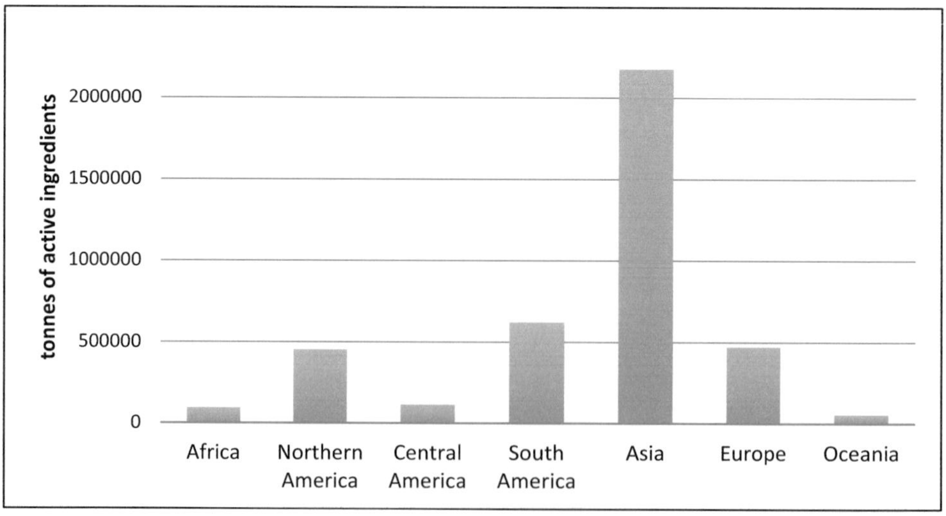

Figure 23.3 Use of pesticides per continent in 2011.

Source: FAOSTAT (Statistics Division of the Food and Agriculture Organization of the United Nations; www.fao.org/faostat/en/?#data/RP).

However, as little as 1% of an applied pesticide reaches the target pest; the remainder ends up in soil, water and air, affecting flora, fauna and soil biological activity, and thus entering the food chain (Sun et al. 2018). The use of persistent organic pollutants as pesticides led to their long-term accumulation in soils and deleterious effects on animal and human health. The organochlorine pesticide 1,1,1-trichloro-2,2-bis(4-chlorophenyl)ethane (DDT) was widely used during World War II to protect humans from malaria, typhus and other diseases spread by insects (Figure 23.4). It was also used as a pesticide for a wide variety of crops. However, its effects on human health and environment had not been previously evaluated. DDT and its metabolites have a high stability and persistence (as much as 50% can remain in the soil 10–15 years after application), which has allowed their bioaccumulation in trophic chains. DDT is considered an endocrine disrupter, produces cancer, accumulates in adipose tissue of humans and animals, and has been detected in breast milk. Therefore, it was banned in many countries during the 1970s. Notwithstanding this, DDT residues can be found anywhere in the world, including Antarctica, and in food throughout the world (Stockholm Convention 2008).

Lindane is also an organochloride pesticide which corresponds to the gamma isomer of 1,2,3,4,5,6-hexachlorocyclohexane (HCH; Figure 23.5). Lindane is synthetised from photochlorination of benzene, which yields a mixture of five main isomers: alpha-HCH (53%–70%) in two enantiomeric forms, (+)-alpha-HCH and (–)-alpha-HCH; beta-HCH (3%–14%); gamma-HCH (11%–18%);

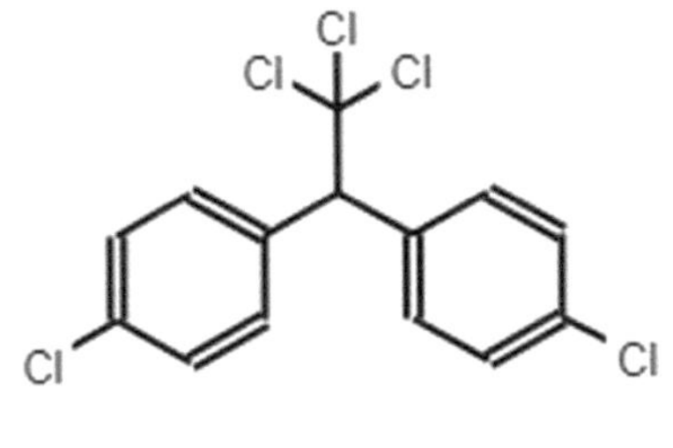

1,1,1-trichloro-2,2-bis(4-chlorophenyl)ethane (DDT)

Figure 23.4 Chemical structure of 1,1,1-trichloro-2,2-bis(4-chlorophenyl)ethane (DDT).

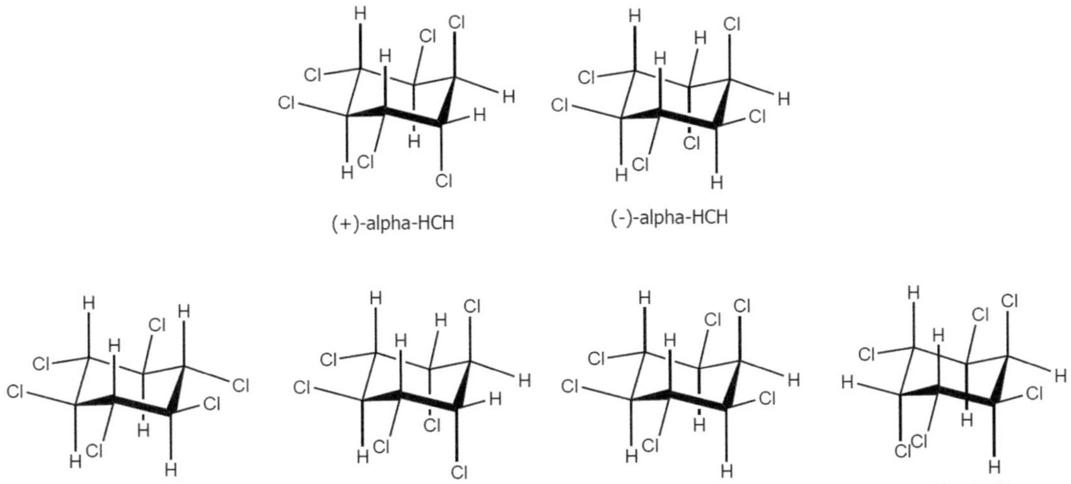

Figure 23.5 Chemical structure of isomers of hexachlorocyclohexane (HCH) obtained by chlorination of benzene.

Source: Buser and Mueller (1995).

delta-HCH (6%–10%); and epsilon-HCH (3%–5%). The mixture of HCH isomers is called technical HCH. Thus, the synthesis of 1 ton of lindane produces between 6 and 10 tons of inactive waste isomers (Vijgen 2006). The gamma isomer is the only isomer showing strong insecticidal properties. Lindane has been used as a broad-spectrum insecticide for seed and soil treatment, foliar applications, tree and wood treatment, and against ectoparasites in both veterinary and human applications. Lindane and its isomers are persistent and bioaccumulate easily in the food chain; there is also evidence for long-range transport and toxic effects in laboratory animals and aquatic organisms. According to that, lindane is considered a persistent pollutant, and its alpha, beta and gamma HCH isomers are included in the Stockholm Convection. Lindane was industrially produced mainly from the 1950s to the 1970s or 1990s all over the world, and it is estimated that global lindane use was around 600,000 tons during the period 1950–2000 (Vijgen 2006). Compared to other organochlorine pesticides such as DDT, lindane is more soluble in water and more volatile. Furthermore, lindane is stable to light. For these reasons, lindane is ubiquitous. Lindane is the most toxic HCH isomer, is classified as carcinogenic to humans and affects the central nervous and endocrine systems. Acute exposure can induce effects ranging from mild skin irritation to dizziness, headaches, diarrhoea, nausea, vomiting, and even convulsions and death (Vega et al. 2016). It is also highly toxic to aquatic organisms.

Glyphosate (N-(phosphonomethyl) glycine) is the most frequently used herbicide worldwide both in industrialised and developing countries, and it has been used for several decades. However, there is some controversy about its use due to the different considerations in relation to its carcinogenic effects to humans (Agostini et al. 2020). In this regard, the International Agency for Research on Cancer (IARC) concluded in 2015 and confirmed in 2018 that glyphosate is probably carcinogenic to humans. In contrast, the European Food Safety Authority claimed it is unlikely that glyphosate poses a human cancer risk, and WHO and the FAO classify it as unlikely to pose a carcinogenic risk to humans from exposure through the diet (Agostini et al. 2020). Consequently, some citizens' initiatives call for a ban on the use of glyphosate-based herbicides. It is allowed in the EU until December 15, 2022 (EU 2020). In the United States, however, after considering public comments on glyphosate use, the EPA (2020) finds that there are no risks of concern for human health when glyphosate is used in accordance with its current label. EPA also found that glyphosate is unlikely to be a human carcinogen (US EPA 2020). However, taking into account its intensive use and its slow decomposition, the accumulation of glyphosate in the environment, including plants, animals and humans, has become quite worrisome (Van Bruggen et al. 2018). In this regard, some chronic effects and indirect effects as shifts in microbial communities in soil, plants, water and intestinal tracts can affect resistance mechanisms inducing severe impacts on plant, animal and human health (Hoffman et al. 2015; Van Bruggen et al. 2018).

Other pesticides currently allowed, such as organophosphates and neonicotinoid insecticides, have also shown negative effects on human and animal health. In fact, neonicotinoids affect the behaviour and health of bees and other pollinators (Landrigan et al. 2018). These results reveal the lack of adequate tests to evaluate new pesticides.

- PAHs are produced by the incomplete combustion and pyrolysis of organic materials, mainly fossil fuels. PAHs present at least two fused aromatic rings in different structural arrangements (Figure 23.6). They are relatively stable, hydrophobic (higher with increasing benzene rings) and volatile (Gao et al. 2018). The annual emission of benzo(a)pyreno in the United States is near 1300 tons (Tarradellas and Bitton 1997). PAHs are ubiquitous in the environment, including air, soil, water, sediment, dust and food (Gao et al. 2018). They reach the human body through ingestion, inhalation, and dermal contact, although the diet is responsible for 88%–98% exposure (Singh and Agarwal 2018). PAHs are carcinogenic and can induce other health problems such as cardiovascular disease, diabetes, infertility, oxidative stress and poor foetal development (Gao et al. 2018). There are several hundred PAHs, but only 16 are listed as priority contaminants by the EPA (Banger et al. 2010).

- Dioxins and furans include two groups of chlorinated hydrocarbons, namely, polychlorinated dibenzo-para dioxins (PCDD) and polychlorinated dibenzofurans (PCDF; Berntssen and Lundebye 2008). Certain dioxin-like polychlorinated biphenyls (PCBs) with similar toxic properties are also included under the dioxin umbrella. Some 419 types of dioxin and furan compounds have been identified, but only about 30 of these are considered to have significant toxicity, with TCDD (2,3,7,8-tetrachlorodibenzo-p-dioxin) being the most toxic. The toxicity of the other congeners is assigned using toxic equivalency factors (TEF) relative to TCDD toxicity (Berntssen and Lundebye 2008). They are produced when mixtures of hydrocarbons and

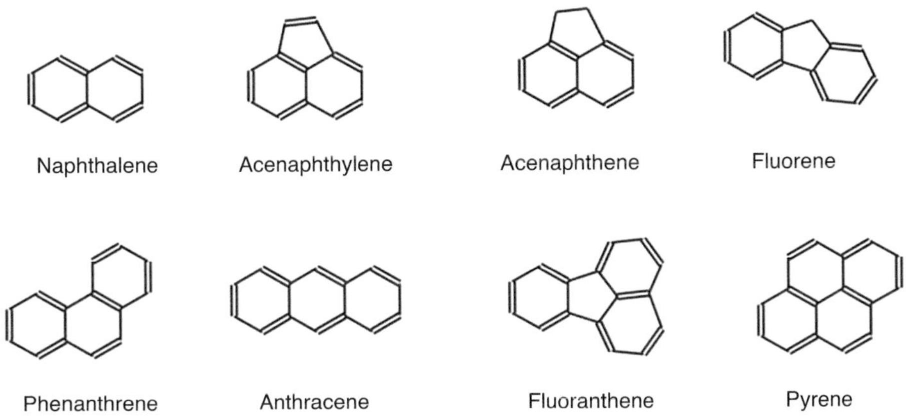

Figure 23.6 Chemical structure of polycyclic aromatic hydrocarbons (PAHs).

chlorine are exposed to high temperatures, including through different industrial processes (smelting, oil refining, chlorine bleaching of paper, chemical manufacturing) and in incineration due to incomplete combustion. Currently, there are available technologies to control waste incineration with low dioxin and furan emissions. These toxins also originate from natural processes, such as volcanic eruptions and forest fires (Berntssen and Lundebye 2008; WHO 2018b). They are very persistent compounds and highly lipid soluble which increases with the molecular weight; therefore, they are considered POPs (Figure 23.1). Dioxins and furans can bioaccumulate in the food chain, mainly in the fatty tissues of animals and humans. The higher an animal is in the food chain, the higher the concentration of dioxins, and studies have shown that their half-life in an animal body is 7–11 years. More than 90% of human exposure is through food, mainly meat and dairy products, fish and shellfish (WHO 2018b). Although they can be found all over the world in the environment, the highest levels are found in some soils, sediments and food. Very low levels are found in plants, water and air (WHO 2018b). Dioxins are very toxic for human and ecosystems and are classified as human carcinogens. They can also induce negative effects on the immune and nervous systems, reproductive disorders and foetal deformities, liver damage and neurobehavioral alterations (Berntssen and Lundebye 2008).

- PCBs have been used industrially since 1929 in insulating fluids in transformers and capacitors, as well as in hydraulic systems, surface coatings, flame retardants, inks and fluorescent light ballasts due to their nonflammability, chemical stability, high boiling point and electrical insulating properties. The estimated production of PCBs between 1930 and 1975 was 700,000 tons in the United States and 1.2 million tons worldwide (Bursian et al. 2011). They have a range of toxicity and vary in consistency from thin, light-coloured liquids to yellow or black waxy solids (US EPA 2018b). There are 209 congeners depending on the total number of chlorine substituents and their position, but commercial formulations contain only 130 of them. Aroclor is the most common commercial mixture of PCBs; there are many types of Aroclors, and each has a distinguishing four-digit suffix number that indicates the degree of chlorination. The first two digits indicate the number of carbon atoms in the phenyl rings, and the second two digits indicate the percentage of chlorine by mass in the mixture. For example, the name Aroclor 1254 means that the PCB has two phenyl rings and the mixture contains approximately 54% chlorine by weight (US EPA 2018b). PCBs are persistent; can remain for long periods cycling between air, water and soil; and can travel long distances. Research performed between 1960 and 1970 detected adverse effects associated with various organisms that were exposed to and incorporated PCBs into their tissues, including seabirds, minks, chickens, and humans fed with rice polluted with PCBs. For these reasons, PCBs were progressively banned by developed countries during the 1970s and are included in the Stockholm Convention (Figure 23.1). Studies about their effects on human health have concluded that PCBs can induce cancer and affect the immune, reproductive, nervous and endocrine systems.

- Chlorinated solvents such as tetrachloroethylene, trichloroethylene, carbon tetrachloride, and vinyl chloride are also known as chlorinated paraffins (Tarradellas and Bitton 1997). They are used in the plastic industry, aerosol spray containers, highway paint, machinery degreasing and dry cleaning. They are harmful to humans and ecosystems and can induce cancer. Spills and leaks of chlorinated solvents have caused widespread subsurface pollution in the environment. Commonly these pollutants are present in the subsurface in the form of nonaqueous phase liquids (NAPL), and because they have a density greater than water, they tend to sink to the bottom of aquifers or pool on top of low-permeability units. They are also present in aquifer sediments and as vapours in the unsaturated zone.
- Emerging pollutants include those which have been in the environment for a while but for which concerns have been raised only more recently, including pharmaceutical and veterinary products, illicit drugs, personal care products, food additives, surfactants and gasoline additives. Emerging pollutants correspond in most cases to unregulated pollutants, which may be candidates for future regulation depending on research on their potential health effects and monitoring data regarding their occurrence (Barceló 2003). There is a wide variety of pharmaceutical compounds consumed by humans which are excreted either not metabolised or partially metabolised into wastewater, especially in high-income and middle-income countries. In Germany alone, it is estimated that between 6000 and 7000 tons of active pharmaceutical compounds are potentially released per year (0.45 kg per capita per year). Typical sewage treatment systems cannot eliminate them, so they can accumulate in the environment. In fact, antibiotics and other pharmaceutical compounds have been detected in soil, groundwater, rivers, seas and other environmental samples (Knapp et al. 2010; Stumpe and Marschner 2007; Huerta et al. 2013, 2016; Rodriguez-Mozaz et al. 2015; Kummerer 2001). A consequence is the increasing antibiotic resistance in soils (Knapp et al. 2010). In addition, there is increasing evidence that pharmaceutical compounds can be bioaccumulated by plants, soil and aquatic organisms, and therefore they can transfer to the trophic chain (Huerta et al. 2018; Madikizela et al. 2018; Marti et al. 2018; Ben Mordechay et al. 2018; Wu et al. 2015; Xie et al. 2017).

Veterinary medicines are widely used to prevent disease in livestock and treat illness. As in humans, they are excreted as the parent compound or metabolites and can enter the environment both directly (e.g. via the use of medicines in fish farms; quantitatively the most important) and indirectly (via the application of animal manure to soil; Boxall et al. 2003). Nevertheless, the continued use of these kind of wastes as organic amendments can produce an accumulation of veterinary compounds in the soil. Monitoring studies have found veterinary medicines in soils, groundwater, surface waters, sediment or biota, their persistence being variable across and within types (Boxall et al. 2004). In this sense, antibiotics are often more toxic than other veterinary compounds; avermectins are toxic to many organisms at low concentrations (ng L^{-1} to µg L^{-1}). For other types of veterinary medicines, the toxicity values are generally in the mg L^{-1} range (Boxall et al. 2004). However, due to lack of knowledge regarding the toxicity data, it is difficult to evaluate the long-term effects after continued exposure even at low concentrations (Boxall et al. 2004).

A contaminant may be "emerging" because a new source or a new pathway to humans has been discovered or a new detection method or treatment technology has been developed (US EPA 2010). Engineered nanomaterials are considered emerging pollutants. Nanomaterials refer to material with one dimension smaller than 100 nm, whereas nanoparticles are materials with at least two dimensions between 1 and 100 nm. The nanoscale size is responsible for unique chemical, biological and physical properties as compared to larger particles of the same material. In the last 20 years, the global market of nanotechnology has grown considerably, and a wide range of applications in a variety disciplines has been developed, including biotechnology, electronics, medicine, chemical synthesis, manufacturing, agriculture, food, personal care and environmental remediation (Klaine et al. 2008; PEN 2020). Currently, more than 1800 nanotechnology-based consumer products have been inventoried (PEN 2020; Vance et al. 2015). The most common nanomaterials found in products are Ag, carbon (C), titanium (TiO_2), silicon/silica (SiO_2), zinc oxide (ZnO), copper and copper oxide (Cu, CuO), iron (Fe^0, Fe_3O_4), and cerium (CeO_2). Engineered nanomaterials may be released into the environment primarily through industrial and environmental applications, improper handling or consumer waste. According to the consumption of Ag nanoparticle and ZnO nanoparticles in Europe per person, their release has been assumed to be significant and broadly distributed in Europe (Dumont et al. 2015). The presence of nanoparticles can induce phytotoxicity depending on their physical (surface area, shape, electrical charge, porosity, magnetic properties) and chemical properties (composition, hydrophobicity, concentration). Some studies show positive effects of some nanoparticles on plant development at determined experimental conditions, including higher seed germination and crop biomass, improved photosynthesis, inhibited plant pathogens and increased resistance against stress, which explains their use in agriculture (Pacheco and Buzea 2018). In contrast, many studies have found detrimental effects of nanoparticles in different plant species depending on the experimental conditions. In general, hormesis responses can occur, which are positive effects for a low dose of the nanoparticles, whereas a higher dose is toxic and inhibits the growth. Pacheco and Buzea (2018) have reviewed the beneficial and detrimental effects of different types of nanoparticles on plants. Nevertheless, the continued application of nanoparticles to soil can increase their concentration and availability for plants and other soil organisms, affecting soil biota and entering the food chain. Their effects in the long term are unknown. Regarding the impact on mammals, Pacheco and Buzea (2018) concluded that nanoparticles are toxic for humans and animals mainly due to their small size, which allows entry into the organism, reaching the circulatory system and organs such as liver, kidneys, brain and heart, and even individual cells and organelles (Buzea et al. 2007). Nanoparticles have been related to respiratory, cardiovascular and neurodegenerative diseases and various cancers in humans and animals (Pacheco and Buzea 2018).

23.3 INORGANIC POLLUTANTS

- "Metals and metalloids are the main pollutants affecting soil and groundwater" (van Lidekerke et al. 2014) by "Metals and metalloids are the main pollutants affecting soil and groundwater" (van Lidekerke et al. 2014) (Figure 23.1). Metals and metalloids are introduced into the environment due to natural and human activities including mining, industrial and military activities, road transport, urban development, the use of agrochemicals and the land application of industrial or domestic sludge (Adriano 2001; Kabata-Pendias 2011). Unlike organic contaminants, metals and metalloids in the environment cannot be chemically degraded and consequently persist in soil for extended periods after their introduction. Due to this, the persistence of pollutants in soil is much longer than in other compartments of the biosphere and, especially by metals and metalloids, appears to be virtually permanent. Metals and metalloids accumulated in soils can decrease slowly by leaching, plant uptake or erosion. Plants can absorb the metals and metalloids accumulated in the soil and be transferred into the food chain, which poses serious hazards for animal and human health. The capacity of plants for absorption of metals and metalloids depends on the plant species and availability of metals and metalloids as well as environmental factors. The total concentration of metals and metalloids in soil provides little information on their potential impact on soil organism; the bioavailability, referred to as their accessible fraction which can be assimilated by an organism, is a better indicator of the risk for biota (Kumpiene et al. 2017). In this sense, there is a certain tendency in using the concept of bioavailability in risk assessment and management of contaminated sites (Kumpiene et al. 2017). There are a wide variety of analytical methodologies to determine the available fraction of metals and metalloids in soils, such as leachability methods, the addition of neutral salt solutions, strong complexing agents (diethylenetriaminepentaacetic [DTPA], ethylendiamintetraacetic acid [EDTA]), or a mixture of low-molecular-weight organic acids (lactic, acetic, formic, citric and malic acids) which simulates the composition of soil solution in the rhizosphere. The bioavailability of metals and metalloids depends on the degree of pollution, metal characteristics and soil properties including pH, electrical conductivity, cation exchange capacity, nature of organic and inorganic ligand ions, texture and redox condition and may vary depending on the environmental conditions. Therefore, in acidic conditions, the availability of Zn, Pb, Ni, Cd and Cu could increase, whereas for As and Cr it could decrease (Kabata-Pendias 2011). Metals and metalloids can be found in soils under different forms, including (in decreasing order of availability) solubilised in the soil solution; as exchangeable ions in the exchange complex; adsorbed onto oxides and hydroxides of aluminium, iron, and manganese; complexed with organic matter; and in the mineral fraction of soil. The chronic exposure of humans and animals to metals and metalloids causes negative effects due to the potential bioaccumulation in different parts of the body. The main sources of exposure of these elements for humans are occupational setting and diet.
- Radionuclides are radioactive forms of elements. They can occur naturally in the environment, as minerals present in Earth's crust and by cosmic rays hitting atoms in Earth's atmosphere, or come from human activity. Some radionuclides have been present in rocks since

the formation of Earth (potassium-40, rubidium-87, uranium and thorium) and others are products of their decay (radium and radon; US EPA 2017a). More half of the average annual radiation exposure of people in the United States comes from natural sources, and radon is the single largest natural source of exposure. Radon comes from the breakdown of radium. Breathing indoor air containing radon is the most common exposure route. Regarding human activities, the use of nuclear weapons, the nuclear power industry, improper disposal of radioactive waste, and nuclear accidents such as Chernobyl in 1986 and Fukushima in 2011 have caused radionuclide contamination worldwide. Cesium-137 and strontium-90 are typical fission radionuclides. Both of them were released into the environment during the Chernobyl and Fukushima nuclear accidents. Cesium-137 moves easily through the air, dissolves easily in water and binds strongly to soil, concentrating in the soil surface, whereas strontium-90 can travel through the soil profile (US EPA 2017b). Radionuclides may be taken up by plants and introduced into the food chain. Consumption of agricultural products contaminated with radionuclides represents the principal route of intake of radionuclides in humans (Zhu and Shaw 2000). Strontium-90 acts like calcium and is readily incorporated into bones and teeth, where it can cause cancers of the bone, bone marrow and soft tissues around the bone. Because of this, special attention in milk monitoring must be taken into account after a large release. Exposure to cesium-137 can increase the risk for cancer because of the presence of high-energy gamma radiation. In addition, the ingestion or inhalation of cesium-137 causes its accumulation in the soft tissues, especially muscle tissue, which increases the risk of cancer. Large amounts of cesium-137 can cause burns, acute radiation sickness and even death (US EPA 2017b).

- Cyanide (–C≡N) compounds are released in the mining of Ag and gold (Au), galvanic and chemical industries (plastic production), exhaust gas from vehicles, as a result of fires at industrial workshops and houses, through wastewater and from tobacco smoke, which is the most significant source of cyanide emissions to the air (Jaszczak et al. 2017). They are also found naturally in plants and processed foods as cyanogenic glycosides. Their environmental fate and ecotoxicity depend on their chemical form, and they can be found in water, soil, exhaled air, food and biological samples (Jaszczak et al. 2017). Most cyanides reached the environment as complexes of $[Fe(CN)_6]^{3-}$ and $[Fe(CN)_6]^{4-}$, which show low toxicity, but due to the light they convert into highly toxic and volatile free cyanides. In soil, without the light, this process is very slow (Jaszczak et al. 2017). In air, cyanide ions are present mainly as hydrogen cyanide, which can be transported over long distances from the emission source. Its duration in the atmosphere is approximately 5 months (Jaszczak et al. 2017). Liquid or gaseous hydrogen cyanide and alkali salts of cyanide are extremely toxic to humans. In fact, hydrogen cyanide is among the most rapidly acting of all known poisons. The mechanism of action is inhibition of cellular respiration, by inhibition of cellular oxygenation with resultant tissue anoxia. Long-term inhalation exposure affects the central nervous, cardiovascular and respiratory systems. Skin contact and oral ingestion are other sources of cyanides in humans and animals. Other effects in humans include an enlarged thyroid gland and irritation to the eyes and skin. The lethal oral dose of cyanide compounds is estimated to be 1.5 mg CN– per kg of body weight (Jaszczak et al. 2017).
- Fertilisers and amendments (raw or composted materials derived from animals [manure], plants [pruning waste, straw], municipal biosolids and waste from manufacturing processes). Some agriculture practices can contribute to nutrient pollution when not properly managed. Fertilisers and animal manure, rich in nitrogen and phosphorus, are the primary sources of nutrient pollution from agricultural sources. Excess nutrients can impact water quality when it rains or when water and soil containing nitrogen and phosphorus wash into nearby waters or leach into groundwaters (US EPA 2017c). Fertilised soils and livestock can be significant sources of gaseous, nitrogen-based compounds like ammonia and nitrogen oxides. Ammonia can be harmful to aquatic life if large amounts are deposited in surface waters. Nitrous oxide is a potent greenhouse gas, with 300 times more impact on global warming than a molecule of carbon dioxide (Jain et al. 2000). Phosphorus (P) losses from farms are generally small in relation to the amounts present in soils. However, small quantities of P have a great impact on water quality because P is the main nutrient for aquatic weeds and algae, which produce eutrophication problems. The use of sustainable fertiliser doses and their application at the optimum time of year and with the right method can significantly reduce their potential pollution.

23.4 REMEDIATION STRATEGIES

Pollutants can have a different impact on the soil depending on the site characteristics. Factors such as soil pH, organic matter content or soil granulometry will be key factors to minimise the potential risk of toxicity due to contaminants. In this sense, the soil properties will condition different processes; pollutant leaching will suppose their movement through the soil profile to groundwater or their absorption by the plant and as a consequence, their introduction in the trophic chain. Processes of degradation may lead to the elimination of the pollutant or, on the contrary, to the formation of toxic intermediary metabolites which suppose a new contaminant source. In any case, soil parameters could condition pollutant persistence in the soil and therefore the risk to soil health.

The treatment of soil contamination involves facing different challenges that will depend on the type of pollutant, its dimension, the properties of the soil, and its future use. Over the past decades, different technologies for soil remediation have been evaluated. Most of them have been performed at lab scale, but some studies are focused on the application at field scale. Biological and physicochemical technologies can be used for the treatment of polluted soils. In order to maintain soil functionality, biological technologies present advantages, are sustainable and can be applied *in situ*, although their application is limited by the pollution size and the process time. Physicochemical alternatives cannot always be applied *in situ*, which supposes the loss of the soil resources, however the required time is lower and their efficiency for high pollutant content is demonstrated.

In general, the limits of a remediation process are conditioned by the soil's future use and the time required for the clean-up (Lobo et al. 2009). Traditionally, soil clean-up criteria has been set using concentration-based standards that often require the application of a remediation technology to reach background levels or to other specified levels that are considered acceptable according to specific regulation at the regional or national level. However, soil clean-up criteria now includes risk-based criteria. In this situation, the approach used recognises that each pollutant poses a risk on human health due to its capacity to be mobilised and transported to the receptor. The risk-based approach considers the need to establish the concentration of each pollutant that is acceptable in each soil. In this situation, it is necessary to define the degree of availability of the chemical substances for transport and the adverse impact (Loehr 1996).

23.4.1 Biological Remediation Strategies

Among biological technologies, microorganisms are widely used in soil polluted with organic compounds. Different authors have tested the capability of bacteria and fungus to degrade organic molecules (EPA 2005; Varjani 2017; Chen et al. 2015; Masciandaro et al. 2013; Rodríguez-Rodríguez et al. 2012; Mansour et al. 2012; Abioye et al. 2010).

Biological strategies, known as bioremediation, use living systems, plants or microorganisms to enable the subsequent recovery of the site, although their efficiency is limited by the size of the contamination and the longest process time. Bioremediation, in the literal sense of the word, means "to give remedy" through biological processes. Therefore, bioremediation processes are based on the elimination of organic or inorganic contaminants from the environment through the action of living beings, such as microorganisms (bacteria, unicellular algae, yeasts), fungi, and plants and enzymes derived from them. When plants are responsible for transforming or eliminating contaminants, the term "phytoremediation" is used. The metabolic activities of certain organisms allow the degradation or elimination of contaminants, resulting in products that are less toxic to the environment and human health. In this sense, microorganisms are able to use compounds present in their environment and transform them into precursors of their cellular constituents, obtaining from them the energy they need to perform biosynthetic processes. During the process of biodegradation of organic compounds, the metabolic intermediates which are produced must be identified in order to evaluate their potential toxicity and avoid a new contaminant in the soil system (Figure 23.7).

A microorganism's ability to metabolise contaminants using them as a carbon source is known as soil natural attenuation capacity. Therefore, in addition to the decontamination process, the incorporation of carbon, nitrogen and other elements contained in the pollutant is achieved in the soil's biological cycle (Wilson et al. 1983). The microorganisms are able to mineralise/degrade practically all of the natural organic compounds. However, certain organic compounds of anthropogenic origin that contain nonbiotic functional groups may be persistent because, due to the novelty of these compounds, the microorganisms have not yet developed mechanisms to carry out their degradation.

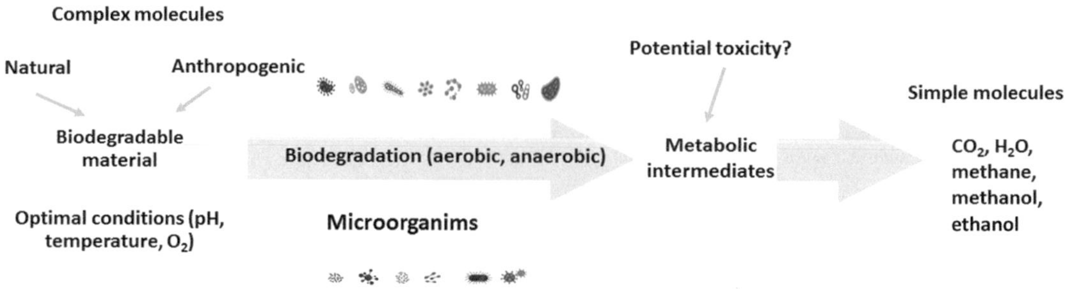

Figure 23.7 General scheme of bioremediation of organic pollutants.

Natural attenuation is often used as part of the clean-up of a polluted site where the source of contamination has been removed (EPA 542-F-96–026). It can be viable in case of contamination with volatile or semivolatile organic compounds, for certain pesticides and with certain metals, whose availability may decrease due to adsorption phenomena or changes in the oxidation state. However, natural attenuation is not an appropriate option for any site because natural processes are generally slow. Long-term monitoring and surveillance vigilance are needed to ensure that the concentration of contaminants decreases continuously and sufficiently so that it does not become a threat to health. If not, consideration should be given to the application of more aggressive corrective measures (EPA 542-F-96–026). In this sense, to enhance or accelerate natural attenuation processes, the autochthonous (native, indigenous) microbial activity of the soil can be stimulated by optimizing environmental conditions, such as nutrient content, oxygen content, pH, temperature, redox conditions and contaminant availability. This strategy is called biostimulation or assisted bioremediation. In general, aerobic degradation processes are faster than anaerobic processes, so one of the biostimulation techniques consists of soil aeration, or injection of air, pure oxygen or hydrogen peroxide (H_2O_2).

The development of the native population can also be stimulated by the addition of fertilisers or organic amendment, based on C, nitrogen (N) and P. In situations of recalcitrant pollution, often the bioavailability of the contaminant is a more limiting factor than the degradation capacity of the microorganisms. Then, to increase their availability to microorganisms, different surfactants or biosurfactants of microbial origin could be applied (Pinilla et al. 2006; Gonzini et al. 2010). Surfactants are molecules of amphipathic nature, that is to say they are constituted by two parts: a hydrophobic one, formed by amino acids, peptides (anions or cations) or mono- or polysaccharides; and another that is hydrophilic, containing saturated and unsaturated fatty acids. This characteristic allows them to reduce the superficial and interfacial tensions between two immiscible liquids, increasing the solubility of hydrophobic compounds, such as pesticides and hydrocarbons, in an aqueous medium (Banat et al. 2000). It is recommended to use biosurfactants against those of synthesis due mainly to their lower toxicity and greater degradability by microorganisms, although in general they tend to have a higher cost. The biosurfactants can be produced by bacteria, fungi and yeasts; those of bacterial origin (*Pseudomonas, Acinetobacter, Bacillus and Arthrobacter*) are the most abundant in the literature.

The processes of natural attenuation can also be favoured by bioaugmentation of microbial populations with degradative capacity. Usually, microbial cultures isolated from the own polluted site or other with similar characteristics are applied. One of the most important drawbacks of this strategy may be the lack of adaptation of exogenous microorganisms to the polluted soil. In this case, the problem can be solved by carrying out several inoculations throughout the bioremediation process, or alternatively, an inoculum with activated soil can be made; that is, the population grows in soil of the same site but with a lower degree of contamination with the purpose to improve the adaptation of microorganisms to contaminated soil. Another strategy to increase the viability of the bacteria is to add them immobilised on an inert mineral support, such as sepiolite, bentonite or zeolite (Lobo et al. 2002). In this case, the bacteria are progressively released to the soil, which is equivalent to continuous applications of the inoculum. Previously it was necessary to evaluate the kinetics of release of microorganisms to the soil.

In recent years, the use of transgenic organisms in bioremediation assays is being developed. It involves manipulating the genes involved in the degradation of a particular contaminant to increase the production of specific enzymes. A modification in the sequence can also be introduced, for example, degradation capacity of a compound, resistance to high concentrations of the contaminant. However, currently the use of genetically modified organisms is limited to those processes in which the conditions are controlled so that the release of these organisms does not occur in the environment, so it is not possible to apply this strategy in soils.

The main advantages of bioremediation strategies are that, being natural processes, they are more respectful to the environment compared to physicochemical techniques; can be carried out in the same location (*in situ*), being in general simple procedures, so they are cheaper than physicochemical techniques. All these advantages have created a good image in public opinion, which is also an advantage. Among the drawbacks are the slowness of the procedures, and even the difficulty of determining *a priori* the duration of a process, as well as the incomplete degradation of pollutants in some occasions. Prior to the application of a bioremediation procedure, it will be necessary to evaluate its effectiveness in a pilot trial using soil from the polluted site, because the result will depend on to a large extent on the soil characteristics.

The efficiency of a bioremediation strategy will depend on different factors affecting biological activity of degrading microorganisms as:

- Microorganism-contaminant interactions.
- Nature and concentration of pollutants (bioavailability, aging of contamination), and soil-contaminant interactions.
- Soil characteristics and environmental conditions. Both the physicochemical characteristics of the soil (pH, organic matter content, nutrients, texture, etc), as well as the environmental conditions such as temperature, humidity and degree of insolation, have a decisive influence on the availability of contaminants, as well as on the indigenous microbial population of the soil (Bento et al. 2005).

Bioremediation techniques can be classified according to the form of application: *in situ* (on site) or *ex situ* (off site). The on-site processes are carried out without the excavation of the land in the same location, which implies a considerably lower cost and less impact on the ground with respect to *ex situ* techniques, which involve excavation and transport to the treatment site (off site). When the decontamination process is carried out after the excavation in the same site, it is termed "on site". It is worth highlighting the processes of landfarming or treatment in beds, biopiles or composting, biofilters and bioreactors.

In relation to soil polluted with metal(loid)s, the use of microorganisms is focused on their migration or the change of their properties as oxidation state. In this sense, the mechanisms include processes of extracellular complexation, precipitation, oxidation-reduction reaction and intracellular accumulation (Yao et al. 2012). Some microorganisms are able to favour metal lixiviation. This leaching process is commonly used in mining, for the treatment of mineral industrial waste products and for sewage sludge detoxification (Bosecker 2001).

Phytoremediation focuses on the use of plants capable of surviving in polluted sites that can extract, accumulate, immobilise or transform the contaminants present in the environment, decreasing their toxicity and consequently the risk for human health. Its efficiency can sometimes be improved by agronomic techniques or by the application of amendments (Chaney et al. 1997; Ortiz Bernad et al. 2007). Phytotechnologies also have a wide spectrum of action because they are applicable in cases of both organic and inorganic contamination in soil, water or air.

This type of technology has many advantages, including its possible application *in situ*; its minimal impact on the environment, which can even improve soil quality; low visual impact; and reduced cost (Cunningham et al. 1995; Salt et al. 1995, 1998; Kabata-Pendias and Pendias 2000; Alkorta et al. 2004), because it can be on average about 10 times cheaper compared to some traditional decontamination technologies such as excavation, soil washing or "pump and treat" (Pilon-Smits 2005). But the application of phytotechnologies also has certain limitations, such as a low rate of decontamination, seasonal growth, the sensitivity of the plant to certain concentrations of contaminants (phytotoxicity) and the greater susceptibility to environmental factors. Moreover, not all plant species are suitable for phytoremediation. It should meet certain characteristics: the ability to accumulate metals, to tolerate the accumulated metal concentrations, to present rapid growth, to have a large effective biomass, and to be easily collected (Kärenlampi et al. 2000). It is desirable that they also be plants adapted to the area to be rehabilitated. Different studies have evaluated the behaviour of plants in soils contaminated by metals, and resistant ecotypes have been located (Baker and Brooks 1989; Sanitá di Toppi and Gabbrielli 1999; Ernst et al. 2002). They have even seen characteristics in these plants that allow them to adapt to soils with low levels of macronutrients (Ernst et al. 2004). Metal-resistant plants offer an excellent opportunity to analyse the impact of pollutants during their life cycle, so that the knowledge acquired can be applied in techniques for recovering contaminated soils (Ernst 1999).

Depending on the process or mechanism used by the plant, there are different types of phytotechnologies: phytoextraction, phytostabilisation, phytodegradation, phytostimulation, phytofiltration and phytovolatilisation (Figure 23.8).

Phytoextraction is based on the uptake of contaminants and their translocation and concentration in the aerial part and roots (if the root is easily harvested) of the plant for harvesting and further removal. In this type of phytotechnology, accumulative (or preferably hyperaccumulative) plant species will be selected (Pilipović et al. 2019). The efficiency of phytoextraction will depend on different factors, such as plant biomass, the type of soil and the availability of contaminants. The crop material harvested will have a determined volume that will be later reduced by drying, calcination or composting with their respective safety measures to avoid the transfer of the contaminant to another environmental compartment. The vegetable waste enriched in the metal can be stored as contaminated material or, if it is feasible, it can be used for the recovery of the metal or be valued as biomass (in the case of energy crops) for the production of electrical energy. The

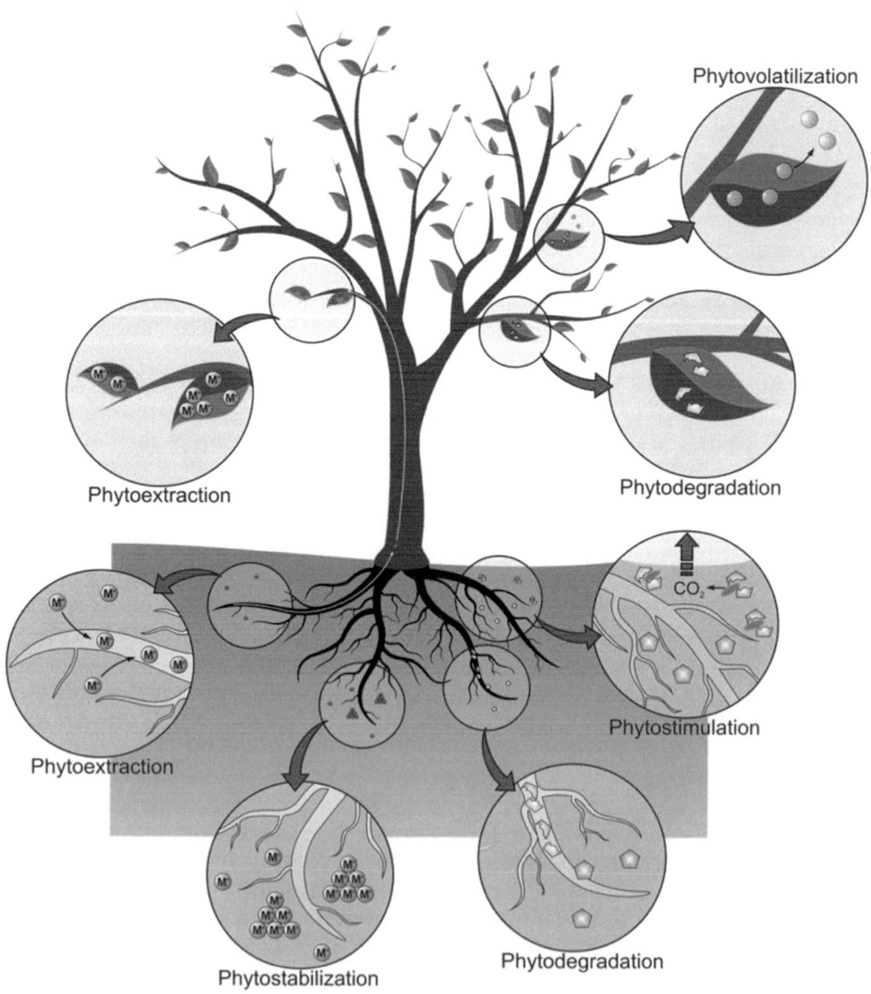

Figure 23.8 General scheme of different types of phytoremediation.

Source: Favas et al. (2014).

phytoextraction process could be improved by applying soil conditioning agents (chelates) that favour the solubility of the metal and facilitate its absorption by the plant.

Phytostabilisation favours the immobilisation of the contaminant in the soil, thus reducing its mobility and its bioavailability, controlling the transfer risk to the food chain. There are several mechanisms: the pollutant can be absorbed and accumulated inside the root without translocation to the aerial part, it can be adsorbed on the surface of the roots, or it can be precipitated by the union to some root exudate. Phytostabilising plants can be used together with amendments to modify the physicochemical state of soil contaminants, reducing their mobility and bioavailability (Pérez-Sanz et al. 2012). Among the amendments used are cyclonic ash, slag, phosphate, zeolite or compost. In addition, the vegetation cover effectively prevents the expansion of the contaminant in the soil, improving the stability of the aggregates and minimizing erosion processes in the soil due to water or air, avoiding surface runoff as well as reducing percolation to the groundwater (Moreno-Jimenez et al. 2012). The plants would also act as a sink for carbon dioxide (CO_2) and other greenhouse gases, thus improving the quality of the air.

Phytodegradation or phytotransformation involves the degradation of pollutants absorbed by plants through metabolic processes in the plant itself or from contaminants external to the plant through the effect of compounds produced by plants (Tangahu et al. 2011). On one hand, the rate of degradation of an organic pollutant in the soil depends on its bioactivity. Thus, a symbiosis mechanism is generated between plants and microorganisms in the rhizosphere. In general, the roots (rhizodegradation) excrete exudates like sugars, carbohydrates, organic acids that are primary sources of C that increase the microbial activity and, therefore, the potential degradation of this pollutant, and the roots can also excrete enzymes which have a direct effect on degradation. On the other hand, the plant can absorb a contaminant and then degrade it by the action of its metabolism. On certain occasions, the pollutants are decomposed in the flow of plant transpiration by the action of enzymes present inside the plant and by photosynthetic oxidation/reduction in simpler molecules that can be useful products for their metabolism, although, as a drawback, on some occasions metabolites even more toxic can be produced by the plant during the process. Currently, phytostimulation is being promoted by using endophytic bacteria from contaminated plants or soils to increase tolerance to metals, which increases the biomass of the plants, their tolerance and even the accumulation of the metal (Asad et al. 2019; Montalbán et al. 2017).

Phytofiltration consists on the use of roots to absorb, precipitate and/or degrade contaminants from water and other aqueous effluents (Favas et al. 2014; Garg et al. 2018). This technique can be applied using terrestrial or aquatic plants, the first ones being more efficient removing contaminants, although it is necessary to implement mechanical support. In addition, the phytofiltration can be applied *in situ* or *ex situ*, taking into account different factors such as the concentration of the effluent, the plant species to be used or the options for the elimination of the periodic harvests of the contaminated plants. The most suitable plants for this application are those with higher root production. Prior to the implantation of these species in the contaminated site, the acclimatisation of the plants to the pollutant and the development of the root system in hydroponic culture are required.

Phytovolatilisation uses plants that absorb contaminants and transport them to the aerial part transformed into volatile compounds, which can then be released to the atmosphere through the leaves. The use of tree species is widespread due to their high water-pumping capacity and evapotranspiration of the leaves. This technique has some limitations due to the risk of emission of pollutants to the atmosphere, the possible deposition of emissions in noncontaminated areas, and the fact that transformation reactions can be produced towards more toxic forms (Limmer et al. 2016).

In general, phytotechnologies have a great potential, and in the last decades, studies on this type of techniques have been increasing significantly in order to evaluate the different strategies and mechanisms used by the plants in the presence of pollutants in order to solve their different limitations and increase the areas in which these processes can be applied (Saxena et al. 2019), taking into account full consideration of plant-soil interactions when plant species are chosen for developing phytoremediation processes (Favas et al. 2014).

23.4.2 Physicochemical Remediation Strategies

23.4.2.1 Soil Excavation

The excavation of the soil supposes the total or partial elimination of contaminated soil and then must be replaced by uncontaminated soil to reduce or eliminate contamination. The elimination of the soil supposes the loss of the resource, and it is only viable for small volumes due to the high cost of transport to the landfill or treatment plant.

When a polluted soil is eliminated, replacing it with clean soil will be necessary, which also implies a high cost. The excavation and manipulation of contaminated soil requires taking a series of precautions in order to avoid dispersion of the contamination in the form of dust or vapor, so monitors should be placed on the perimeter of the excavation in order to control it. During transportation and storage, security measures must be maintained in the same way.

23.4.2.2 Isolation and Containment

Pollutants can be isolated and contained in order to prevent their movement. The use of physical barriers made of steel, cement or bentonite favour vertical and horizontal capping to reduce water infiltration (Mulligan et al. 2001). Solidification/stabilisation technologies are commonly used: in these cases pollutants are contained not in the contaminated area like using barriers (Conner 1990). The process of solidification involves a solid matrix, whereas stabilisation implies chemical reactions to reduce the mobility of the pollutants.

23.4.2.3 Thermal Treatments

Different technologies can be applied to *ex situ* soil decontamination. These technologies could be used in permanent installations or in mobile ones. In the latter case, problems associated to the soil transport would be avoided. The use of these types of technologies is adequate for organic pollutants because heavy metals would remain in the solid waste.

Thermal treatments suppose a high cost because energy is needed for the heat supply, particularly to get higher temperatures. This can be solved using waste heat generated for the generation of electrical energy (thermal recovery). This technology represents an interesting alternative for the rehabilitation of petroleum hydrocarbon–polluted soils (Vidonish et al. 2016).

23.4.2.4 Soil Incineration

Polluted soil is treated by a controlled combustion at high temperatures, between 900 and 1300°C using a rotary kiln (Figure 23.9). The gases generated in the process should be treated to ensure that they are free of dust and contaminants before releasing them into the atmosphere. When the soil contains metals, they can react to form other species, and the less volatile will remain in the soil. Metals in ash, in the scrubber sludge or produced from stack emissions, if improperly managed, can result in potential exposures and adverse health effects (US EPA 1992).

23.4.2.5 Thermal Desorption

This process uses a range of temperatures between 250 and 550°C to transfer the pollutants to the gas phase. The systems are similar to those used for incineration.

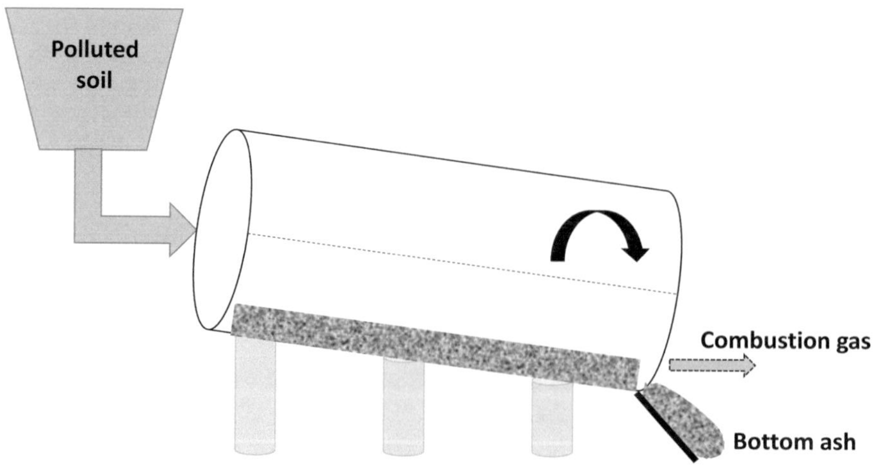

Figure 23.9 General scheme of a rotary kiln.

This technology can be used for soil polluted with relatively volatile organic pollutants (gas oil, light fuels, mineral oils and BTEX). It could be used to remove elements that may volatilise under these conditions, such as Hg, As and phosphorus.

23.4.2.6 Vitrification

Through this process, the fusion of the soil takes place using an electric current at high temperatures, between 1600–2000°C. Under these conditions, a vitreous mass forms on the soil. This process implies solidification and stabilisation, because a very stable residue is produced avoiding the migration of contaminants in the soil. The pollutants' destruction mechanism is either pyrolysis (in an oxygen-poor environment) or oxidation (when oxygen is present; US EPA 1995). Vitrification can either be performed *in situ* or *ex situ*. This technology could be defined as a solidification or stabilisation process, which requires thermal energy (Mulligan et al. 2001).

23.4.3 Extraction Treatments

23.4.3.1 Washing Treatments

This technology is based on the separation of contaminants present in the soil using a suitable extractant fluid. It is adequate for organic and inorganic substances depending on the contaminant characteristics and those from the extractant. A limiting factor in the efficiency of these technologies is the presence of a complex mixture of pollutants or the presence of humic acids in the soil or a large amount of small-sized particles.

The extraction techniques can be applied to contaminated soils with a wide range of substances, both organic and inorganic. Therefore, the choice of the washing fluid and its composition will depend on the contaminant to be eliminated.

The washing fluid must have a good coefficient of extraction, low volatility and low toxicity and should be easy and safe for handling. It is also important to ensure its possibility of treatment after removal. The primary stages in soil washing are soil preparation, washing, soil and water separation, wastewater treatment, and vapour treatment when required (US Army Corps of Engineers 2003).

23.4.3.2 Electrokinetic

Electrokinetic remediation is a developing technique that has a significant potential for *in situ* remediation of low-permeability soils contaminated with heavy metals and/or organic compounds (Lobo et al. 2009). The technique involves applying a low direct current or a low potential gradient to electrodes that are inserted into the soil (Acar and Alshawabkeh 1993; Ugaz et al. 1994; Pazos et al. 2012). The pollutants are transported to either the cathode or the anode due to electromigration and electro-osmotic processes (Figure 23.10).

The advantages of electrokinetic remediation are the low cost, the nonintrusive character, the applicability to a wide range of contaminants and the insensitivity to the pore size of the soil, which makes it suitable for fine-grained soils.

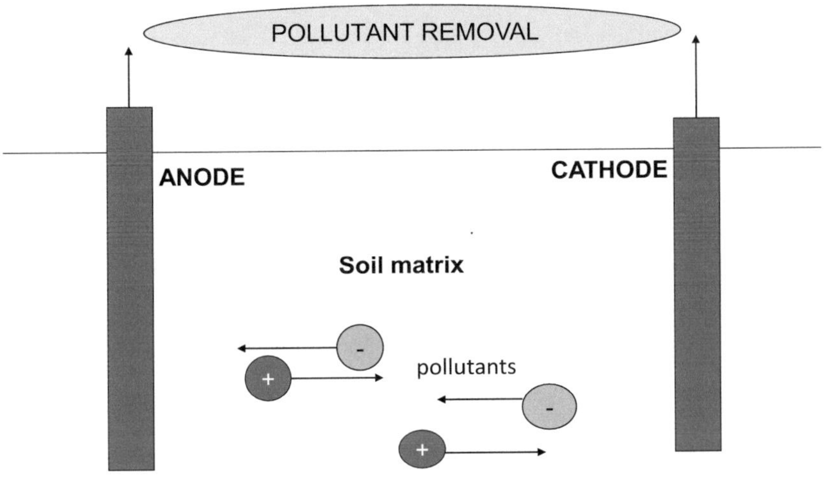

Figure 23.10 General scheme of electrokinetic process.

23.4.3.3 Stripping Treatments

Stripping treatments involve extraction in the vapor phase. They can be applied to cases of soil contamination by volatile substances. In general, they can be applied *in situ*, so that the soil is not altered (Figure 23.11). The equipment is simple, and when applying air the biodegradation of the pollutants is favoured at the same time. The air flow carries the vaporised pollutants found in the soil to the extraction wells and from there to the gas treatment units. The application of this technique should consider the location of the pollutant, whether above or below the water table.

The use of these treatments is effective when the contaminants have low molecular weight, low water solubility and high volatility, as do trichlorethane, methylene chloride, light hydrocarbons and aromatic components of gasolines (BTEX; US EPA 1992).

It is possible to favour vapour extraction, introducing vapour into the soil which increases the treatment temperature and favours the pollutant volatilisation.

23.5 COMBINING TECHNOLOGIES

Currently, the most advanced techniques use combined technologies, which add benefits and mitigate the limitations of each principle/strategy remediation (Lobo et al. 2012).

Recently, and for this purpose, physicochemical and biological technologies have been combined. Thus, after the application of an on-site soil washing strategy, the last fraction of smaller size could be treated by landfarming on the site itself. Another example is chemical oxidation, in which the application of H_2O_2 releases oxygen that facilitates the process of bioremediation by the indigenous micro-organisms of the soil (Sutton et al. 2011). Other physicochemical strategies that are currently being developed may also be able to increase their effectiveness with a second phase in which a biological technique is applied. In the case of nanoremediation, in which Fe^0 is used at a nano-scale for the immobilisation of metals and metalloids (Gil-Diaz et al. 2014, 2016, 2017, 2018, 2019), its combined use with plants also allows the control of erosion and the recovery of vegetation cover in the treated soil. Technologies in development, such as the application of electric fields (electrokinetic), could be associated with the use of hyperaccumulator plants that contribute to the elimination of metals from the soil. Some studies have combined the use of an electric field

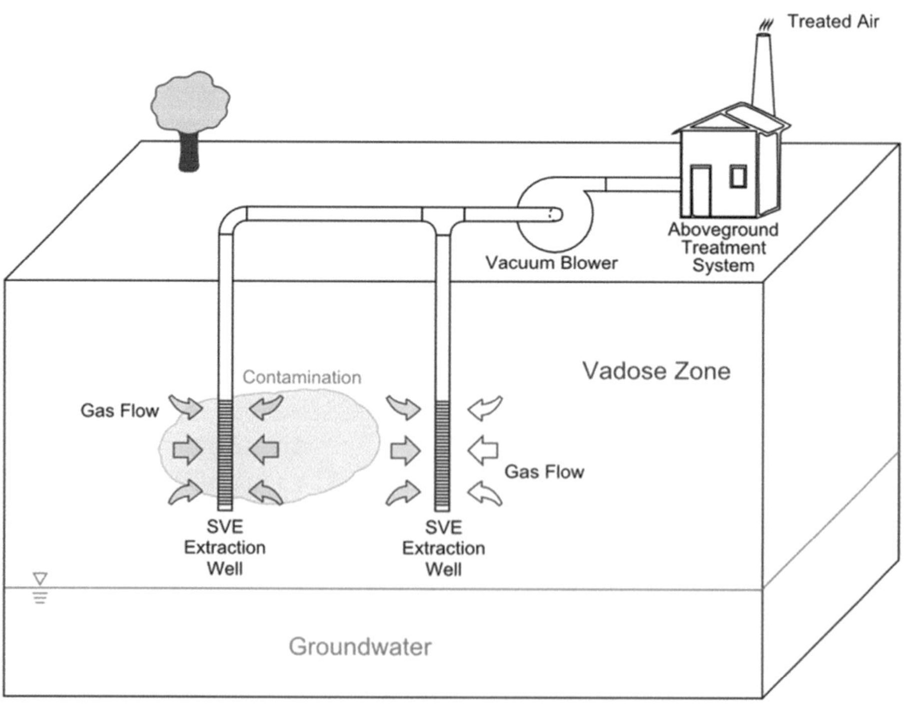

Figure 23.11 Scheme of soil vapor extraction.

Source: www.enviro.wiki/index.php?title=Soil_Vapor_Extraction_(SVE).

and a biosurfactant that enhances bioremediation in order to remove gas oil from a polluted soil (Gonzini et al. 2010). In all cases, this combination could only be carried out if it was available for the recovery of the site for the time required for the application of a biological strategy.

On the other hand, the combination of nanoremediation with electrokinetics has given good results for decontamination of soils polluted with metals and metalloids at lab scale (Gomes et al. 2014, 2015).

23.6 CONCLUDING REMARKS

In general, contaminated soil can be considered a vehicle for the introduction of pollutants into the food chain, so it is necessary to take action to prevent and remediate soil contamination in order to avoid harmful effects on human health. Over the past decades, different technologies have been applied to soil remediation in an effort to remove pollutants according to economic criteria and time. With regard to the variability of soil types and the degree of disturbance of soil functions, there are considerable differences among the remediation methods. Methods without soil excavation, such as soil-air stripping, soil flushing or hydraulic treatments, may influence natural soil functions adversely by changing soil structure, porosity and water balance. Other methods based on soil excavation, such as thermal treatment and soil washing, influence soil functions to a greater extent. At the contaminated site, the excavation of topsoil causes a reduction of soil organic matter, nutrients and cationic exchange capacity with the consequences that some properties as filtering, buffering and biodegradation are reduced. Thus, the soil will lose its function as an archive of natural and cultural history. Remediation processes lead to changes in the soil properties to different degrees depending on the type of treatment. Whenever possible, the use of biological technologies that favour the recovery of soil properties and functions would be recommended. In this case, the limiting factor will be the pollution degree and the process time.

BIBLIOGRAPHY

Abioye, O. P., A. Abdul Aziz, and P. Agamuthu. 2010. Enhanced biodegradation of used engine oil in soil amended with organic wastes. *Water Air Soil Pollut* 209: 173–9.

Acar, Y. B., and A. N. Alshawabkeh. 1993. Principles of electrokinetic remediation. *Environ Sci Technol* 27(13): 2638–47.

Adriano, D. C. 2001. *Trace Elements in Terrestrial Environments: Biogeomistry, Bioavailability, Risk of Metals*, 2nd ed. New York: Springer-Verlag.

Agostini, L. P., R. S. Dettogni, R. S. dos Reis, et al. 2020. Effects of glyphosate exposure on human health: Insights from epidemiological and in vitro studies. *Sci Total Environ* 705: 135808.

Alkorta, I., J. Hernandez-Allica, J. M. Becerril, I. Amezaga, I. Albizu, and C. Garbisu. 2004. Recent findings on the phytoremediation of soils contaminated with environmentally toxic heavy metals and metalloids such as zinc, cadmium, lead and arsenic. *Rev Environ Sci Biotechnol* 3: 71–90.

Asad, S. A., M. Farooq, A. Afzal, and H. West. 2019. Integrated phytobial heavy metal remediation strategies for a sustainable clean environment—A review. *Chemosphere* 217: 925–41.

Baker, A. J. M., and R. R. Brooks. 1989. Terrestrial higher plants which hyperaccumulate metallic elements. A review of their distribution, ecology and phytochemistry. *Biorecovery* 1(2): 81–126.

Banat, I. M., R. S. Makkar, and S. S. Cameotra. 2000. Potential commercial application of microbial surfactants. *Appl Microbiol Biot* 53: 495–508.

Banger, K., G. S. Toor, T. Chirenje, and L. Ma. 2010. Polycyclic aromatic hydrocarbons in urban soils of different land uses in Miami, Florida. *Soil Sediment Contam* 19(2): 231–43.

Barceló, D. 2003. Emerging pollutants in water analysis. *Trac Trends Anal Chem* 22(10): XIV-XVI.

Ben Mordechay, E., J. Tarchitzky, Y. N. Chen, M. Shenker, and B. Chefetz. 2018. Composted biosolids and treated wastewater as sources of pharmaceuticals and personal care products for plant uptake: A case study with carbamazepine. *Environ Pollut* 232: 164–72.

Bento, F. M., F. A. O. Camargo, B. C. Okekeb, and W. T. Frankenberger. 2005. Comparative bioremediation of soils contaminated with diesel oil by natural attenuation, biostimulation and bioaugmentation. *Bioresour Technol* 96: 1049–55.

Bergman, A., J. J. Heindel, S. Jobling, K. A. Kidd, and R. T. Zoeller. 2012. *State of the Science of Endocrine Disrupting Chemicals*. Inter-Organization Programme for the Sound Management of Chemicals. www.who.int/ceh/publications/endocrine/en/ (accessed November 26, 2020).

Berntssen, M. H. G., and A. K. Lundebye. 2008. Environmental contaminants in farmed fish and potential consequences for seafood safety. In *Improving Farmed Fish Quality and Safety*, ed. O. Lie. Cambridge: Woodhead Publ Ltd.

Bosecker, K. 2001. Microbial leaching in environmental clean-up programmes. *Hydrometallurgy* 59(2–3): 245–8.

Boxall, A. B. A., L. A. Fogg, P. A. Blackwell, P. Kay, E. J. Pemberton, and A. Croxford. 2004. Veterinary medicines in the environment. *Rev Environ Contamin Toxicol*, 180: 1–91.

Boxall, A. B. A., L. A. Fogg, P. Kay, P. A. Blackwell, E. J. Pemberton, and A. Croxford. 2003. Prioritisation of veterinary medicines in the UK environment. *Toxicol Lett* 142(3): 207–18.

Bursian, S. J., J. L. Newsted, and M. J. Zwiernik. 2011. Chapter 41—Polychlorinated biphenyls, polychlorinated dibenzo-p-dioxins and polychlorinated dibenzofurans. In *Reproductive and Developmental Toxicology*, ed. R. C. Gupta. Academic Press.

Buser, H. R., and M. D. Mueller. 1995. Isomer and enantioselective degradation of hexachlorocyclohexane isomers in sewage sludge under anaerobic conditions. *Environ Sci Technol* 29: 664–72.

Buzea, C., I. I. Pacheco, and K. Robbie. 2007. Nanomaterials and nanoparticles: Sources and toxicity. *Biointerphases* 2(4): MR17-MR71.

Chaney, R. L., M. Malik, Y. M. Li, et al. 1997. Phytoremediation of metals. *Curr Opin Biotechnol* 8: 279–84.

Chen, M., P. Xu, G. Zeng, C. Yang, D. Huang, and J. Zhang.2015. Bioremediation of soils contaminated with polycyclic aromatic hydrocarbons, petroleum, pesticides, chlorophenols and heavy metals by composting: Applications, microbes and future research needs. *Biotechnol Adv* 1(33): 745–55.

Conner J.R, 1990. *Chemical Fixation and solidification of hazardous wastes.* Van Nostrand Reinhold, New York. ISBN:0442205112.

Cunningham, S. D., W. R. Berti, and J. W. Huang. 1995. Phytoremediation of contaminated soils. *Trends Biotechnol* 13(9): 393–7.

Dumont, E., A. C. Johnson, V. D. J. Keller, and R. J. Williams. 2015. Nano silver and nano zinc-oxide in surface waters—Exposure estimation for Europe at high spatial and temporal resolution. *Environ Pollut* 196: 341–9.

EPA. 2005. *EPA Guideline (EPA 589/05) Soil bioremediation*. Adelaide: EPA.

Ericson, B., J. Caravanos, K. Chatham-Stephens, P. Landrigan, and R. Fuller. 2013. Approaches to systematic assessment of environmental exposures posed at hazardous waste sites in the developing world: The Toxic Sites Identification Program. *Environ Monitor Assess* 185(2): 1755–66.

Ernst, W. H. O. 1999. Biomarkers in plants. In *Biomarkers: A Pragmatic Basis for Remediation of Severe Pollution in Eastern Europe*, eds. D. B. Peakall, C. H. Walker, and P. Migula, 135–51. Dordrecht: Kluwer Academic Publishers.

Ernst, W. H. O., A. G. L. Assuncao, J. A. C. Verkleij, and H. Schat. 2002. How important is apoplastic zinc xylem loading in Thlaspi caerulescens. *N Phytol* 155: 4–5.

Ernst, W. H. O., F. Knolle, S. Kratz, and E. Schnug. 2004. Aspects of ecotoxicology of Heavy metals in the Harz region—A guided excursion. *Landbauforsch Völkenrode* 2(54): 53–71.

EU. 2020. *Glyphosate.* https://ec.europa.eu/food/plant/pesticides/glyphosate_en (accessed November 26, 2020).

FAO. 2018a. *Global Symposium on Soil Pollution.* www.fao.org/about/meetings/global-symposium-on-soil-pollution (accessed November 26, 2020).

FAO. 2018b. *Pesticides Use.* www.fao.org/faostat/en/?#data/RP (accessed November 26, 2020).

FAO, and ITPS. 2015. *Status of the World's Soil Resources (SWSR)—Technical Summary.* Food and Agriculture Organization of the United Nations and Intergovernmental Technical Panel on Soils, Rome, Italy. www.fao.org/global-soil-partnership/intergovernmental-technical-panel-soils/en/ (accessed November 26, 2020).

Favas, P., J. Pratas, M. Varun, R. D'Souza, and S. P. Manoj. 2014. Phytoremediation of soils contaminated with metals and metalloids at mining areas: Potential of native flora. In *Environmental Risk Assessment of Soil Contamination*, ed. M. C. Hernandez-Soriano. IntechOpen.

Gao, P., E. da Silva, L. Hou, N. D. Denslow, P. Xiang, and L. Q. Ma. 2018. Human exposure to polycyclic aromatic hydrocarbons: Metabolomics perspective. *Environ Int* 119: 466–77.

Garg, M., C. Valeo, R. Gupta, S. Prasher, N.R. Sharma, and P. Cosntabel. 2018. Integrating natural and engineered remediation strategies for water quality management within a low-impact development (LID) approach. *Environ Sci Pollut Res* 25: 29304–13.

Gil-Díaz, M., A. Gonzalez, J. Alonso, and M. C. Lobo. 2016. Evaluation of the stability of a nanoremediation strategy using barley plants. *J Environ Manag* 165: 150–8.

Gil-Diaz, M., L. López, J. Alonso, and M. C. Lobo. 2018. Comparison of nanoscale zero-valent iron, compost, and phosphate for Pb immobilization in an acidic soil. *Water Air Soil Pollut* 229(10).

Gil-Diaz, M., A. Pérez-Sanz, M. A. Vicente, and M. C. Lobo.2014. Immobilisation of Pb and Zn in soils using stabilised zero-valent iron nanoparticles. Effects on soil properties. *CLEAN—Soil, Air, Water* 42(12): 1776–84.

Gil-Diaz, M., P. Pinilla, J. Alonso, and M. C. Lobo. 2017. Viability of a nanoremediation process in single or multi-metal(loid) contaminated soils. *J Hazard Mater* 321: 812–19.

Gil-Diaz, M., E. Rodriguez-Valdez, J. Alonso, D. Baragaño, J. L. Rodriguez Gallego, and M. C. Lobo. 2019. Nanoremediation and long-term monitoring of brownfield soil highly polluted with As and Hg. *Sci Total Environ* 675.

Gomes, H. I., C. Dias-Ferreira, L. M. Ottosen, and A. B. Ribeiro. 2014. Electrodialytic remediation of polychlorinated biphenyls contaminated soil with iron nanoparticles and two different surfactants. *J Colloid Interface Sci* 433: 189–95.

Gomes, H. I., L. M. Ottosen, A. B. Ribeiro, and C. Dias-Ferreira. 2015. Treatment of a suspension of PCB contaminated soil using iron nanoparticles and electric current. *J Environ Manag* 151: 550–5.

Gonzini, O., A. Plaza, L. Di Palma, and M. C. Lobo. 2010. Electrokinetic remediation of gasoil contaminated soil enhanced by rhamnolipid. *J Appl Electrochem* 40: 1239–48.

Hoffman, A. R., L. M. Proctor, M. G. Surette, and J. S. Suchodolski. 2015. The microbiome: The trillions of microorganisms that maintain health and cause disease in humans and companion animals. *Vet Pathol* 53: 10–12.

Huerta, B., E. Marti, M. Gros, et al. 2013. Exploring the links between antibiotic occurrence, antibiotic resistance, and bacterial communities in water supply reservoirs. *Sci Total Environ* 456: 161–70.

Huerta, B., S. Rodriguez-Mozaz, J. Lazorchak, et al. 2018. Presence of pharmaceuticals in fish collected from urban rivers in the US EPA 2008–2009 National Rivers and Streams Assessment. *Sci Total Environment* 634: 542–9.

Huerta, B., S. Rodriguez-Mozaz, C. Nannou, et al. 2016. Determination of a broad spectrum of pharmaceuticals and endocrine disruptors in biofilm from a waste water treatment plant-impacted river. *Sci Total Environ* 540: 241–9.

Jain, A. K., B. P. Briegleb, K. Minschwaner, and D. J. Wuebbles. 2000. Radiative forcings and global warming potentials of 39 greenhouse gases. *J Geophys Res* 105: 20773–90.

Jaszczak, E., Ż. Polkowska, S. Narkowicz, and J. Namieśnik. 2017. Cyanides in the environment—analysis—problems and challenges. *Environ Sci Pollut Res* 24(19): 15929–48.

Kabata-Pendias, A. 2011. *Trace Elements in Soils and Plants*, 4th ed. Boca Raton, FL: CRC Taylor & Francis Group.

Kabata-Pendias, A., and H. Pendias. 2000. *Trace Elements in Soils and Plants*, 416. Boca Raton, FL: CRC Press.

Kärenlampi, S., H. Shat, J. Vangronsveld, et al. 2000. Genetic engineering in the improvement of plants for phytoremediation of metal polluted soils. *Environ Pollut* 107: 225–31.

Klaine, S. J., P. J. J. Alvarez, G. E. Batley, et al. 2008. Nanomaterials in the environment: Behavior, fate, bioavailability, and effects. *Environ Toxicol Chem* 27(9): 1825–51.

Knapp, C. W., J. Dolfing, P. A. I. Ehlert, and D. W. Graham. 2010. Evidence of Increasing Antibiotic Resistance Gene Abundances in Archived Soils since 1940. *Environ Sci Technol* 44(2): 580–7.

Kummerer, K. 2001. Drugs in the environment: Emission of drugs, diagnostic aids and disinfectants into wastewater by hospitals in relation to other sources—a review. *Chemosphere* 45(6–7): 957–69.

Kumpiene, J., L. Giagnoni, B. Marschner, et al. 2017. Assessment of methods for determining bioavailability of trace elements in soils: A review. *Pedosphere* 27(3): 389–406.

Kuppusamy, S., T. Palanisami, M. Megharaj, K. Venkateswarlu, and R. Naidu. 2016. Ex-situ remediation technologies for environmental pollutants: A critical perspective. *Rev Environ Contam Toxicol* 236: 117–92.

Landrigan, P. J., R. Fuller, N. J. R. Acosta, et al. 2018. The Lancet Commission on pollution and health. *Lancet* 391(10119): 462–512.

Laszlova, K., H. Dudasova, P. Olejnikova, et al. 2018. The application of biosurfactants in bioremediation of the aged sediment contaminated with polychlorinated biphenyls. *Water Air Soil Pollut.*

Limmer, M. A., and J. G. Burken. 2016. Phytovolatilization of organic contaminants. *Environ Sci Technol* 50(13): 6632–43.

Lobo, M. C., A. Pérez-Sanz, M. Gil-Diaz, and A. Plaza. 2012. Soil remediation: Use of combined (coupled) technologies. In *Encyclopedia of Applied Electrochemistry*, ed. R. F. Savinel, K. I. Ota, and G. Kreysa. Springer.

Lobo, M. C., A. Pérez-Sanz, M. J. Martínez-Iñigo, and A. Plaza. 2009. Influence of coupled electrokinetic–phytoremediation on soil remediation: Chapter 20. In *Electrochemical Remediation Technologies for Polluted Soils, Sediments and Groundwaters*, ed. K. R. Reddy and C. Cameselle. Wiley and Sons Publications

Lobo, M. C., M. Sanchez, C. Garbi, et al. 2002. Immobilized native bacteria as a tool for bioremediation of soils and waters: Implementation and modeling. *Sci World J* 2: 1361–8.

Loehr, R. C. 1996. *The Environmental Impact of Soil Contamination: Bioavailability, Risk Assessment, and Policy Implications*. Policy Study No. 211. Los Angeles, CA: The Reason Foundation.

Madikizela, L. M., S. Ncube, and L. Chimuka. 2018. Uptake of pharmaceuticals by plants grown under hydroponic conditions and natural occurring plant species: A review. *Sci Total Environ* 636: 477–86.

Mansour, H. B., R. Mosrati, D. Barillier, K. Ghedira, and L. Chekir-Ghedira. 2012. Bioremediation of industrial pharmaceutical drugs. *Drug Chem Toxicol* 35(3): 235–40.

Marti, E., B. Huerta, S. Rodriguez-Mozaz, D. Barcelo, R. Marce, and J. L. Balcazar. 2018. Abundance of antibiotic resistance genes and bacterial community composition in wild freshwater fish species. *Chemosphere* 196: 115–19.

Masciandaro, G., C. Macci, E. Peruzzi, B. Ceccanti, and S. Doni. 2013. Organic matter-microorganism-plant in soil bioremediation: A synergic approach. *Rev Environ Sci Bio/Technol* 12(4).

Mazzeo, D. E. C., T. C. C. Fernandes, and M. A. Marin-Morales. 2011. Cellular damages in the Allium cepatest system, caused by BTEX mixture prior and after biodegradation process. *Chemosphere* 85: 13–18.

Montalbán, B., S. Thijs, M. C. Lobo, et al. 2017. Cultivar and Metal-Specific Effects of Endophytic Bacteria in Helianthus tuberosus Exposed to Cd and Zn. *Int J Mol Sci* 18(2026).

Moreno-Jiménez, E., E. Esteban, R. O. Carpena-Ruiz, M. C. Lobo, and J. Peñalosa. 2012. Phytostabilisation with Mediterranean shrubs and liming improved soil quality in a pot experiment with a pyrite mine soil. *J Hazard Mater* 201–202: 52–9.

Mulligan, C. N., R. N. Young, and B. F. Gibbs. 2001. Remediation technologies for metal-contaminated soils and groundwater: An evaluation. *Eng Geol* 60: 193–207.

Orgiazzi, A., R. Bardgett, E. Barrios, et al. 2016. *Global Soil Biodiversity Atlas*. European Commission, Publications Office of the European Union, Luxembourg, 176 pp. https://esdac.jrc.ec.europa.eu/content/global-soil-biodiversity-atlas (accessed November 26, 2020).

Ortiz Bernad, I., J. Sanz García, M. Dorado Valiño, and S. Villar Fernández. 2007. *Técnicas de recuperación de suelos contaminados. Informe de vigilancia tecnológica*. Depósito Legal: M-5.839–2007.

Pacheco, I., and C. Buzea. 2018. Nanoparticle uptake by plants: Beneficial or detrimental? In *Phytotoxicity of Nanoparticles*, ed. M. Faisal, Q. Saquib, A. A. Alatar, and A. A. Al-Khedhairy. Cham: Springer International Publishing.

Pazos, M., A. Plaza, M. Martín, and M. C. Lobo. 2012. The impact of electrokinetic treatment on a loamy-sand soil properties. *Chem Eng J* 183: 231–7.

PEN (Project on Emerging Nanotechnologies). 2020. *Inventories*. www.nanotechproject.tech/cpi/ (accessed November 26, 2020).

Pérez-Sanz, A., R. Millán, and M. J. Sierra, et al. 2012. Mercury uptake by Silene vulgaris grown on contaminated spiked soils. *J Environ Manag* 95: 233–7.

Pierzynski, G. M., J. T. Sims, and G. F. Vance. 2005. *Soils and Environmental Quality*, 3rd ed., 569. Boca Raton, FL: Taylor & Francis Group.

Pilipović, A., R. S. Zalesny, S. Rončevićc, et al. 2019. Growth, physiology, and phytoextraction potential of poplar and willow established in soils amended with heavy-metal contaminated, dredged river sediments. *J Environ Manag* 239: 352–65.

Pilon-Smits, E. 2005. Phytoremediation. *Annu Rev Plant Biol* 56: 15–39.

Pinilla, P., M. C. Lobo, and M. J. Martinez-Iñigo. 2006. Effect of the surfactant sorbitan trioleate on the bioremediation of pendimethalin-contaminated soil microcosms. In *Environmental Biotechnology and Engineering*, ed. H. M. Poggi-Varaldo, E. Rios-Leal, J. Garcia-Mena et al. Mexico: 2IMEBE.

Pinilla, P., J. Ruiz, M. C. Lobo, and M. J. Martínez-Iñigo. 2008. Degradation of oxadiazon in a bioreactor integrated in the water closed circuit of a plant nursery. *Bioresour Technol* 99(7): 2177–83.

Popp, J., K. Peto, and J. Nagy. 2013. Pesticide productivity and food security. A review. *Agron Sustain Dev* 33(1): 243–55.

Prüss-Ustün A., J. Wolf, C. Corvalán, R. Bos, and M. Neira. 2016. *Preventing Disease Through Healthy Environments: A Global Assessment of the Burden of Disease from Environmental Risks*. World Health Organization, France. http://apps.who.int/iris/bitstream/handle/10665/204585/9789241565196_eng.pdf;jsessionid=E03F1D1CB4A8B2975BFF3342871AAF1D?sequence=1 (accessed November 26, 2020).

Rodrigues, S. M., and P. F. A. M. Römkens. 2018. Chapter 9. Human health risks and soil pollution. In *Soil Pollution*, ed. A. C. Duarte, A. Cachada, and T. Rocha-Santos, 217–50. Cambridge, MA: Academic Press.

Rodriguez-Mozaz, S., S. Chamorro, E. Marti, et al. 2015. Occurrence of antibiotics and antibiotic resistance genes in hospital and urban wastewaters and their impact on the receiving river. *Water Res* 69: 234–42.

Rodríguez-Rodríguez, C. E., A. Jelic, M. A. Pereira, et al. 2012. Bioaugmentation of Sewage Sludge with Trametes versicolor in Solid Phase Biopiles Produces Degradation of Pharmaceuticals and Affects Microbial Communities. *Environ Sci Technol* 46: 12012–12020.

Rotterdam Convention. 2010. www.pic.int/Home/tabid/855/language/en-US/Default.aspx (accessed November 26, 2020).

Salt, D. E., M. Blaylock, N. P. B. A. Kumar, et al. 1995. Phytoremediation—A novel strategy for the removal of toxic metals from the environment using plants. *Bio/technology* 13(5) pps: 468–74.

Salt, D. E., R. D. Smith, and I. Raskin. 1998. Phytoremediation. *Annu Rev Plant Physiol Plant Mol Biol* (49): 643–68.

Sanitá di Toppi, L. L., and R. Gabbrielli. 1999. Response to Cd in higher plants. *Environ Exp Bot* 41: 105–30.

Saxena, G., D. Purchase, S. I. Mulla, G. D. Saratale, and R. N. Bharagava. 2019. Phytoremediation of heavy metal-contaminated sites: eco-environmental concerns, field studies, sustainability issues, and future prospects. *Rev Environ Contam Toxicol* 249: 71–131.

Singh, L., and T. Agarwal. 2018. Polycyclic aromatic hydrocarbons in diet: Concern for public health. *Trends Food Sci Technol* 79: 160–70.

Stockholm Convention. 2008 http://chm.pops.int/TheConvention/ThePOPs/AllPOPs/tabid/2509/Default.aspx (accessed November 26, 2020).

Stumpe, B., and B. Marschner. 2007. Long-term sewage sludge application and wastewater irrigation on the mineralization and sorption of 17 beta-estradiol and testosterone in soils. *Sci Total Environ* 374(2–3): 282–91.

Sun, S. X., V. Sidhu, Y. H. Rong, and Y. Zheng. 2018. Pesticide Pollution in Agricultural Soils and Sustainable Remediation Methods: A Review. *Curr Pollut Rep* 4(3): 240–50.

Sutton, N. B., J. T. C. Grotenhuis, A. A. M. Langenhoff, and H. M. M. Rijnaarts. 2011. Efforts to improve coupled *in situ* chemical oxidation with bioremediation: A review of optimization strategies. *J Soils Sediments* (2011) 11: 129–40.

Tangahu, B. V., S. R. S. Abdullah, H. Hassan Basri, M. Idris, N. Anuar, and M. Mukhlisin. 2011. A review on heavy metals (As, Pb, and Hg) uptake by plants through phytoremediation. *Int J Chemi Eng* 2011(1687–806X).

Tarradellas, J., and G. Bitton, 1997. Chemical pollutants in soils. In *Soil Ecotoxicology*, ed. J. Tarradellas, G. Bitton, and D. Rossel, 3–32. Boca Raton, FL: CRC Press.

Ugaz, A., S. Puppah, R. J. Gale, and Y. B. Acar. 1994. Electrokinetic soil processing. Complicating features of electrokinetic remediation of soils and slurries: Saturation effects and the role of the cathode electrolysis. *Chem Eng Commun* 129: 183–200.

US Army Corps of Engineers. 2003. *Safety and Health Aspects of HTRW Remediation Technologies*. EM1110-1-4007.

US EPA 2020. *Glyphosate*. www.epa.gov/ingredients-used-pesticide-products/glyphosate (accessed November 26, 2020).

US EPA. 1991. *Survey of Materials-Handling Technologies Used at Hazardous Waste Sites*. https://nepis.epa.gov/ (accessed November 26, 2020).

US EPA. 1992. *Considerations for Evaluating the Impact of Metals Partitioning during the Incineration of Contaminated Soils from Superfund Sites*. EPA/540/5-92/014, U.S.

US EPA. 1995. *Geosafe Corporation in situ Vitrification-Innovative Technology Evaluation Report*. EPA/540/R-94/520. Cincinnati, OH: U.S. Environmental Protection Agency.

US EPA. 1996. *542-F-96-026, Citizen's Guide to Natural Attenuation*. https://nepis.epa.gov/ (accessed November 26, 2020).

US EPA. 2004. *Cleaning Up the Nation's Waste Sites: Markets and Technology Trends*. https://nepis.epa.gov/Exe/ZyPDF.cgi/30006II3.PDF?Dockey=30006II3.PDF (accessed November 26, 2020).

US EPA. 2010. *Emerging Contaminant—Nanomaterials*. www.epa.gov/fedfac/emerging-contaminant-nanomaterials (accessed November 26, 2020).

US EPA. 2017a. *RadTown*. https://www3.epa.gov/radtown/ecosystems.html (accessed November 26, 2020).

US EPA. 2017b. *Radionuclides*. www.epa.gov/radiation/radionuclides (accessed November 26, 2020).

US EPA. 2017c. *Nutrient Pollution—The Sources and Solutions: Agriculture*. www.epa.gov/nutrientpollution/sources-and-solutions-agriculture (accessed November 26, 2020).

US EPA. 2018a. *Superfund: National Priorities List*. www.epa.gov/superfund/superfund-national-priorities-list-npl (accessed November 26, 2020).

US EPA. 2018b. *Learn about Polychlorinated Biphenyls*. www.epa.gov/pcbs/learn-about-polychlorinated-biphenyls-pcbs (accessed November 26, 2020).

Van Bruggen, A. H. C., M. M. He, K. Shin, et al. 2018. Environmental and health effects of the herbicide glyphosate. *Sci Total Environ* 616–617: 255–68.

van Liedekerke, M., G. Prokop, S. Rabl-Berger, M. Kibblewhite, and G. Louwagie. 2014. *Progress in the Management of Contaminated Sites in Europe*. Joint Research Centre, Report EUR 26376 EN. http://bookshop.europa.eu/en/progress-in-the-management-of-contaminated-sites-in-europe-pbLBNA26376/ (accessed November 26, 2020).

Vance, M. E., T. Kuiken, E. P. Vejerano, et al. 2015. Nanotechnology in the real world: Redeveloping the nanomaterial consumer products inventory. *Beilstein J Nanotechnol* 21: 1769–1780.

Varjani, S. J., E. Gnansounou, and A. Pandey. 2017. Comprehensive review on toxicity of persistent organic pollutants from petroleum refinery waste and their degradation by microorganisms. *Chemosphere* 188: 280–91.

Vega M., D. Romano, and E. Uotila. 2016. *Lindane (Persistent Organic Pollutant) in the EU*. www.europarl.europa.eu/RegData/etudes/STUD/2016/571398/IPOL_STU(2016)571398_EN.pdf (accessed November 26, 2020).

Vidonish, J. E., K. Zygourakis, C. A. Masiello, G. Sabadell, and P. J. Alvarez. 2016. Thermal treatment of hydrocarbon-impacted soils: A review of technology innovation for sustainable remediation. *Engineering* 2: 426–37.

Vijgen, J. 2006. *The Legacy of Lindane HCH Isomer Production*. Main Report. A Global Overview of Residue Management, Formulation and Disposal. International HCH & Pesticides Association.

WHO. 2018a. www.who.int/foodsafety/areas_work/chemical-risks/pops/en/ (accessed November 26, 2020).

WHO. 2018b. www.who.int/en/news-room/fact-sheets/detail/dioxins-and-their-effects-on-human-health (accessed November 26, 2020).

Wilson, J. T., J. McNabb, D. L. Balkwill, and W. C. Ghiorse. 1983. Enumeration and characterization of bacteria indigenous to a shallow water-table aquifer. *Ground Water* 21: 134–42.

Wu, X. Q., L. K. Dodgen, J. L. Conkle, and J. Gan. 2015. Plant uptake of pharmaceutical and personal care products from recycled water and biosolids: A review. *Sci Total Environ* 536: 655–66.

Xie, Z. X., G. H. Lu, Z. H. Yan, J. C. Liu, P. F. Wang, and Y. H. Wang. 2017. Bioaccumulation and trophic transfer of pharmaceuticals in food webs from a large freshwater lake. *Environ Pollut* 222: 356–66.

Yao, Z., J. Li, H. Xie, and C. Yu. 2012. Review on remediation technologies of soil contaminated by heavy metals. *Proc Environ Sci* 16: 722–9.

Zhu, Y. G., and G. Shaw. 2000. Soil contamination with radionuclides and potential remediation. *Chemosphere* 41(1): 121–8.

24 Toxicology and Climate Change

John P. Berry

CONTENTS

24.1 INTRODUCTION

In a landmark 1896 paper, the Swedish chemist and Nobel Laureate Svante Arrhenius provided the first quantitative framework to explain how carbon dioxide in the atmosphere is capable of effecting—by way of the absorbance of infrared radiation—change in the temperature of the Earth's surface (Arrhenius 1896). Although this proposed "greenhouse effect" (as it was later coined) was first met with skepticism, a subsequent preponderance of evidence specifically including well over a century of highly accurate and precise measurements of carbon dioxide (CO_2) and temperature (Figure 24.1) has converged to paint a clear picture of both unprecedented decadal changes in global climate and contribution of anthropogenic input of CO_2 and other so-called greenhouse gases to these changes. Global increases in temperatures have, in turn, been attributed to a wide range of physicochemical impacts on the environment ranging from alterations of meteorological patterns, including increased frequency of extreme weather events and drought, to dramatic melting of polar ice sheets resulting in a rapid rise in sea levels (Figure 24.2).

When considering *environmental* toxicology and the contribution of environmental pollutants to human health, it is perhaps not surprising, therefore, that changes in climate—and consequences of these changes—at global and "local" scales have been proposed to have far-reaching implications for the occurrence of, and risk of exposure to, a wide range of toxicants. And, indeed, a growing number of studies, and considerations of the existing data, bear this out in terms of both increased prevalence of environmental toxicants and elevated potential for exposure to toxicants. As such, the assessment of the contribution of environmental toxicants to human health is effectively a moving target, whereby the frame of reference, including considerations of toxicological baselines, is shifting on this same (i.e. decadal) time scale with documented changes in the environment.

In this chapter, the implications of climate change for human exposure to environmental toxicants will be considered. The goal is to provide a conceptual framework for understanding how climate change may (and will likely continue to) affect exposure risk in relation to various environmental toxicants and to serve as a reference for toxicologists with respect to *emerging* toxicants and routes of exposure. In order to understand how climate change may directly or indirectly influence occurrence of, and potential for exposure to, toxic contaminants in the environment, we will first endeavor to briefly define what climate change entails with respect to possible toxicologically relevant consequences. Subsequently, as the role of climate change includes both increased levels of toxic substances in the environment and increased risk for exposure to toxicants, these two aspects will be considered sequentially. In the former case, naturally occurring biogenic toxins whose presence and prevalence has been linked to climate change will be explored to provide a brief overview of these biotoxins as emerging concerns within the changing toxicological context. And finally, the relatively limited information available regarding increased potential for

DOI: 10.1201/9780203730584-24

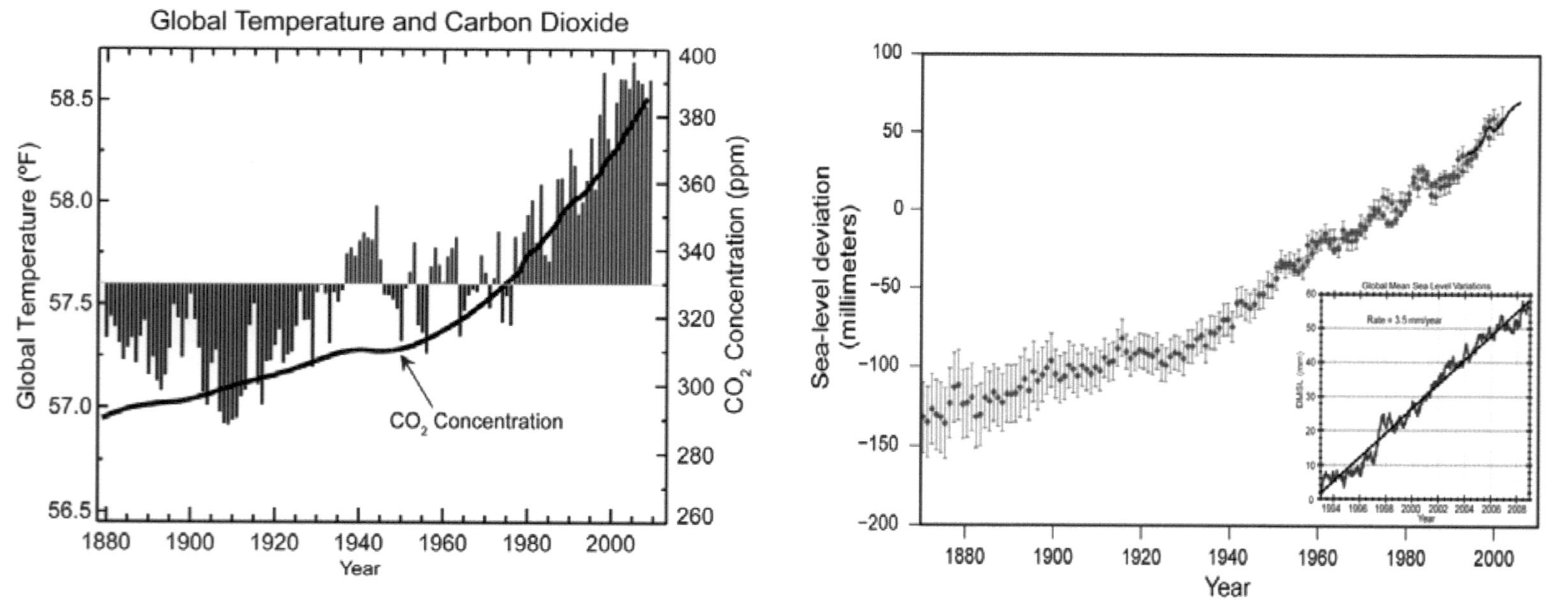

Figure 24.1 Global increase in temperature and atmospheric CO_2 levels, and concomitant sea level rise, over the past 140 years.

Source: From Karl et al. (2009).

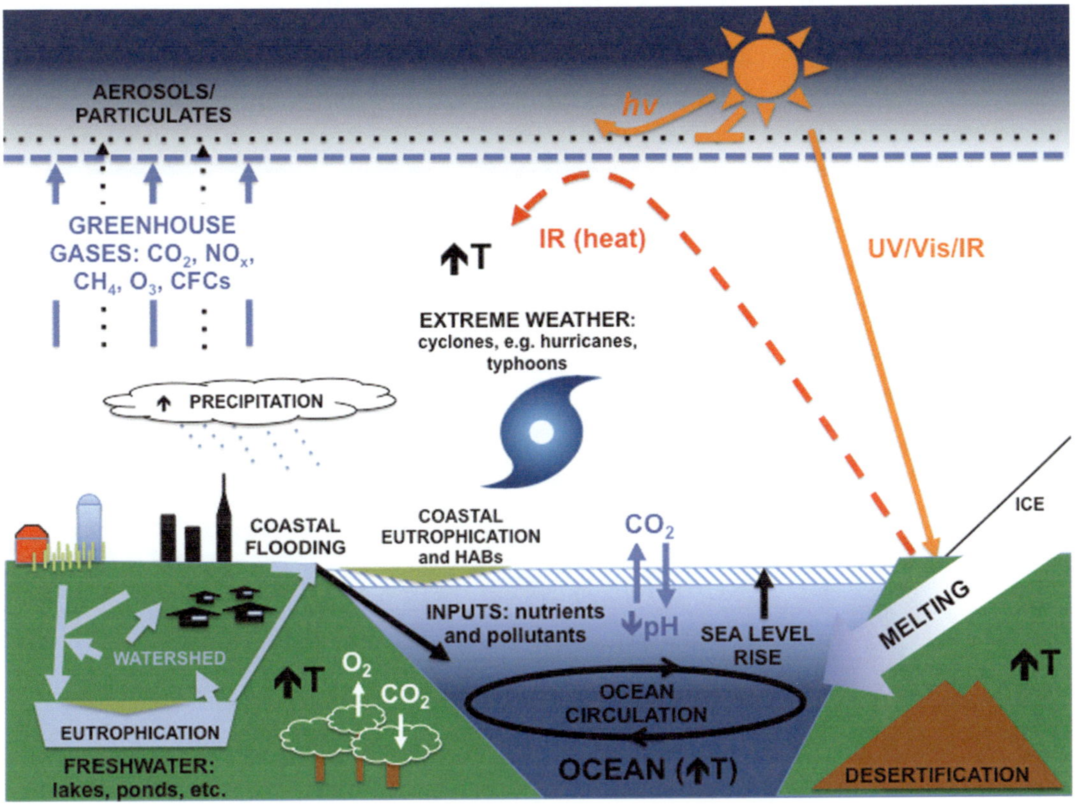

Figure 24.2 Interrelated causes and consequences of global climate change.

exposure to toxicants—including both naturally occurring and anthropogenic representatives—with respect to various consequences of climate change will be discussed. The goal is to provide a current perspective and primer for practitioners in relation to the rapidly changing global landscape that underlies potential human health in relation to environmental toxicants.

24.2 GLOBAL CLIMATE CHANGE AND RELEVANT CONSEQUENCES

In order to understand how climate change may directly or indirectly influence the presence of toxic contaminants in the environment, or the potential for exposure to these, it is worth first briefly defining what "climate change" entails before examining possible toxicological considerations in this regard. To begin with, climate change in this context will be specifically considered with respect to those changes over decadal or century time scales relevant to human health within contemporary time periods, specifically in order to discern from changes known to occur over geological time scales. As such, climate change is ostensibly synonymous with "global warming," which refers to increased global temperatures—as well as various climatological and meteorological consequences of this change—that have, indeed, been concluded to be extremely likely due to anthropogenic factors over the past century and a half (IPCC 2013).

Increased global temperature, and associated climatological and meteorological consequence, have been specifically attributed to several *greenhouse gases* (GHG). Defined as GHG based on their capability to absorb and subsequently reemit infrared radiation, they include both those gases that occur naturally in the atmosphere (i.e. CO_2, methane, nitrous oxide) and those strictly manufactured by industrial processes (i.e. halogenated hydrocarbons). Although several of the recognized GHG are indeed produced by a variety of natural biological and geochemical processes, the current levels (and the decadal increases) of these are unprecedented and, in turn, closely correlate with corresponding global increases in industrial and agricultural activities (and more generally, to imbalance between relevant sources and "sinks"). As the most abundant GHG in the atmosphere (aside from water; see below), anthropogenic sources for CO_2 primarily include combustion

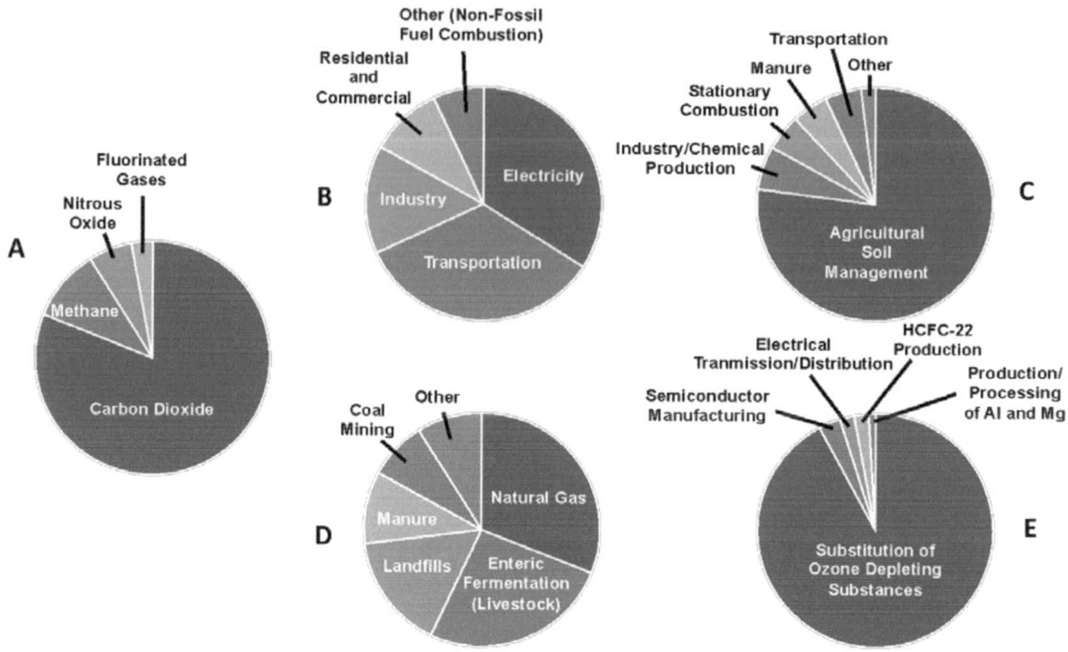

Figure 24.3 Primary greenhouses gases (not including water vapor) and their respective sources in the United States. Given is the contribution of four main greenhouse gases to total emissions (6.5×10^6 metric tons of CO_2 equivalents) in the United States in 2016 (A) and main sources of each, including CO_2 (B), nitrous oxide (C), methane (D) and fluorinated gases (E).

Source: Data from US Environmental Protection Agency (EPA), *Inventory of U.S. Greenhouse Gas Emissions and Sinks 1990–2016* (EPA 430-R-18–003).

of fossil fuels (Figure 24.3), but also lesser, yet still significant, contributions of other sources such as cement and steel production (~15%; van Ruijven et al. 2016) and petrochemical manufacturing. Owing to these sources, levels of CO_2 in the atmosphere have increased more than 40% relative to preindustrial levels and continue to increase annually (Figure 24.1). For example, measured CO_2 levels increased from approximately 280 parts per million (ppm) prior to 1750 to 379 ppm in 2005 (i.e. 35%; IPCC 2007), and to over 410 ppm in the most recent monthly measurements (NOAA 2019). Increases in CO_2 from these sources are, moreover, matched by equally rapid disappearance of key CO_2 sinks, including deforestation, and altered ocean chemistry (e.g. pH, carbon cycling).

Despite the particular attention historically paid to atmospheric CO_2 resulting from combustion of fossil fuels (which does, for example, represent approximately 76% of GHG emissions in the United States; US EPA 2018) other relevant gases have seen parallel dramatic increases over the same time scales. As a striking example, levels of nitrous oxide—which compared to CO_2 has 300 times the potential as a GHG, and for which as much as 50% of input can be attributed to human activity (Hirsch et al. 2006)—have increased approximately 20% over the last century due to both combustion of fossil fuels and increased agricultural inputs (i.e. nitrogen-based fertilizer; IPCC 2007). Even larger (as much as 170%) increases over the same time frame have been observed for methane (25 times more active than CO_2 as a GHG; IPCC 2007) in association with human activities owing to a wide range of sources, including agricultural (i.e. livestock) production, anaerobic processes in landfills, burning of biomass, wastewater, and utilization (as fuel) of methane-rich natural gas. Notably, water vapor is in fact the most abundant GHG in the atmosphere, and has been likewise increasing alongside others; however, increase in *condensable* water vapor concentrations is unique, as it is in effect a "feedback" of increased temperature resulting in other noncondensable GHGs (e.g. CO_2, nitrous oxide, methane). That said, it has been recently estimated that concomitant increases in water vapor may more than double warming effect of these other GHGs (Dessler et al. 2013). Finally, alongside these biogeochemically produced gases, a chemical diversity of "unnatural" greenhouse gases, including various halogenated gases in particular (e.g. chlorofluorocarbons, hydrofluorocarbons), are exclusively generated through industrial production for a range of applications such a cooling and refrigeration, aerosols/propellants, electronics and other manufacturing purposes (e.g. foam blowing). And, their anthropogenic emission is thereby quite clearly linked to their increasing concentrations in the atmosphere—and contribution to recent global warming—since the mid-20th century (IPCC 2007).

Decadal and century time-scale increases in global temperature (due to GHG) have in turn been attributed to a wide range of interrelated impacts on the environment (Figure 24.2). Perhaps the most commonly cited consequence is the anomalously rapid melting rates of the Earth's polar ice sheets and consequently unprecedented rise in sea level. Sea level has in fact risen over the past century (paralleling increase in global temperature) at an increasing rate, with current estimates of approximately 2.8–3.2 mm (one-eighth of an inch) per year and more than 200 mm over the past century (Figure 24.1; Church and White 2011). The potential contribution of sea level rise to human health and toxicology is underscored by the fact that approximately 40% of the human population lives in densely populated coastal areas. Notably, alongside effects on global temperature, atmospheric CO_2 additionally alters the pH of ocean waters (so-called ocean acidification), as it is also a major sink for this gas.

Accompanying increases in sea level is considerable evidence which suggests that climate change may alter patterns of global ocean circulation that have been otherwise imprinted over geological time scales. Numerous meteorological phenomena, and various other impacts on the physicochemical environment in turn, have been consequently linked to changes in global climate. Notable examples relevant to environmental toxicants, and potential for exposure to these, include increased frequency of extreme weather events, particularly tropical cyclones (hurricanes, typhoons) associated with abnormally high rainfall (e.g. monsoons) that can directly impact hydrology patterns between land and sea (e.g. input of nutrients, flooding). Simultaneously, unprecedented rates of desertification and drought with direct impacts, ranging from modified agricultural practices to inadequate access to clean drinking water, are linked to these global changes in climate.

Although not direct products of climate change, it is worth noting that these impacts are in many cases exacerbated by numerous additional (and sometimes related) anthropogenic contributors (e.g. deforestation, increased nutrient inputs, population growth, industrial pollutants) which can be difficult to separate from global warming and its many consequences. The far-reaching intricacies of climate change and its numerous associated impacts on the environment are beyond the scope of this chapter; however, as we shall see, they underlie numerous examples of the shifting environmental toxicological frame of reference including elevated levels of naturally occurring toxicants, and they increase potential for human exposure to these.

24.3 CLIMATE CHANGE AND BIOTOXINS

Although the risk of exposure to a wide range of toxicants in the environment is likely influenced by climate change (as discussed in the final section of this chapter), a myriad of naturally occurring *biotoxins*—and perhaps more accurately, the organisms producing these—have been suggested to specifically increase, with respect to their *quantifiable levels* (i.e. prevalence) in the environment, as a direct consequence of climate change. One such area of environmental toxicology to receive significant attention in this regard has been toxic *secondary metabolites* (metabolites not associated with primary cellular functions) from microbes. Levels of these secondary metabolites are influenced both directly (via altered biosynthesis) and indirectly (via increased abundance of the producing organism) by climate change including, in particular, elevated temperature and altered precipitation patterns as well as consequent effects on hydrology and nutrient inputs. Arguably the most significant of these include naturally occurring *aquatic toxins* produced by marine and freshwater microorganisms; and within terrestrial environments, fungal and bacterial toxins associated with foodstuffs through either agricultural contamination and/or spoilage. That said, a range of organisms (including animals and plants) associated with toxins relevant to human health have likewise been proposed to be affected either directly or indirectly by changes in climate, and these warrant discussion.

24.3.1 Algae and Cyanobacteria as a Source of Aquatic Biotoxins

As 70% of the Earth's surface is covered by water, the role of the marine and freshwater systems in climate change cannot be overstated. Conversely, however, impacts of climate change on aquatic biota, including potentially *toxigenic* organisms in the ocean and freshwaters, have in turn been widely proposed and consequently investigated. Of the far-reaching contributions of climate change to toxicology, in this regard, perhaps the most notable is the documented increase in numerous toxigenic members of phytoplankton communities (i.e. algae and other photosynthetic microbes) due to global warming and related environmental drivers (see Figure 24.2).

Although not, strictly speaking, a true (*monophyletic*) taxonomic classification, "algae" are operationally defined as photosynthetic, aquatic *nonplant* organisms, specifically including members of the eukaryotic protists. This classification often includes both plant-like *macroalgae* (or "seaweeds") and single-celled or colonial *microalgae*. However, it is largely the latter, and particularly various marine phytoplanktonic species, which have been most widely associated with toxins relevant to human health. Several microalgal taxa including, in particular, members of the dinoflagellate (phylum Dinoflagellata) and diatom (class Bacillariophyta) classifications, as well as a few phylogenetically affiliated taxa (e.g. Raphidophyta, Prymnesiophyta), have been thus linked to production of toxins relevant to human health (Van Dolah 2000).

Typically distinguished taxonomically from algae (a distinction which is utilized here), relevant photosynthetic aquatic microbes also include a wide diversity of prokaryotic *cyanobacteria* ("blue-green algae"), which are producers of an equally diverse repertoire of bioactive, and potentially toxic, secondary metabolites. Whereas the eukaryotic algae are generally associated with marine habitats, the cyanobacteria—despite being ubiquitous in the environment—are most frequently linked to toxigenic blooms within freshwater and, to a lesser extent, coastal phytoplanktonic systems. Cyanobacteria represent, in fact, one of the oldest groups of organisms on Earth, with fossil records dating back more than 3 billion years, and indeed they are recognized as the evolutionary origin of oxygenic photosynthesis. And, in the modern oceans, cyanobacteria are the primary global contributors to both photosynthetic production of oxygen and nitrogen cycling, underscoring their profound linkage to the environment.

Indeed, numerous examples of bioactive secondary metabolites from both algae and cyanobacteria—the majority of which, despite their structural diversity, share biosynthetic origins—have been characterized as environmental toxicants in relation to human health. The chemical and toxicological diversity of these is very briefly summarized below. However, the topic could easily constitute (and would in fact require) an entire chapter or chapters unto itself. That said, further details on the chemical toxicology are presented in a number of good reviews (see Tables 24.1 and 24.2).

Furthermore, much remains to be investigated with respect to the chemical diversity and toxicology of algal and cyanobacterial toxins. The list of toxic metabolites continues to grow. And as such, what is presented here represents only a *current* overview of relevant and recognized toxic compounds that will surely change as more of these compounds are identified and/or linked to human health effects. On the same note, recognized "toxins" from algae and cyanobacteria have been mostly characterized as part of *acute* toxicoses, but it is likely a range of subacute effects (e.g. cancer, teratogenicity, immunomodulation, neurodegeneration), resulting from chronic and/or indirect exposure to typically low levels, that can be attributed to these same compounds. At present, however, the latter largely remains to be fully elucidated.

24.3.1.1 Marine Algal Biotoxins: Chemistry and Toxicology

Within marine and coastal habitats, toxic secondary metabolites, notably including a structurally diverse array of polyketides (Figure 24.4), produced by a growing number of dinoflagellate and diatom species have been linked to human intoxications. Toxicological aspects of recognized marine algal biotoxins are briefly summarized in Table 24.1, and further details of both chemistry and toxicology, and implications to human health, can be found in several current reviews referenced therein.

Exposure to marine algal toxins is most conspicuously linked to human health through proliferative toxigenic blooms—or so-called harmful algal blooms (HABs)—although exposure sufficient for even acute health effects can, in fact, occur in the absence of obvious bloom conditions, and particularly through biomagnification of toxins (as discussed below). Such blooms are frequently evidenced by discoloration of waters (due to algal pigments), leading to the commonly used term "red tides." Accordingly, many such blooms are commonly categorized using this terminology.

Arguably, however, the most prominent example is the so-called Florida Red Tide (Figure 24.5) that specifically refers to blooms of the toxigenic dinoflagellate species, *Karenia brevis*, which primarily occur along coastal areas of the Gulf of Mexico. In this case, the brevetoxins (PbTx) produced by *K. brevis* are a suite of congeners, further subdivided into two subfamilies (PbTx A and B; Figure 24.4), which are potent neurotoxins specifically acting by way of the effective activation of voltage-gated sodium channels. Red tide events in the Gulf of Mexico are frequently

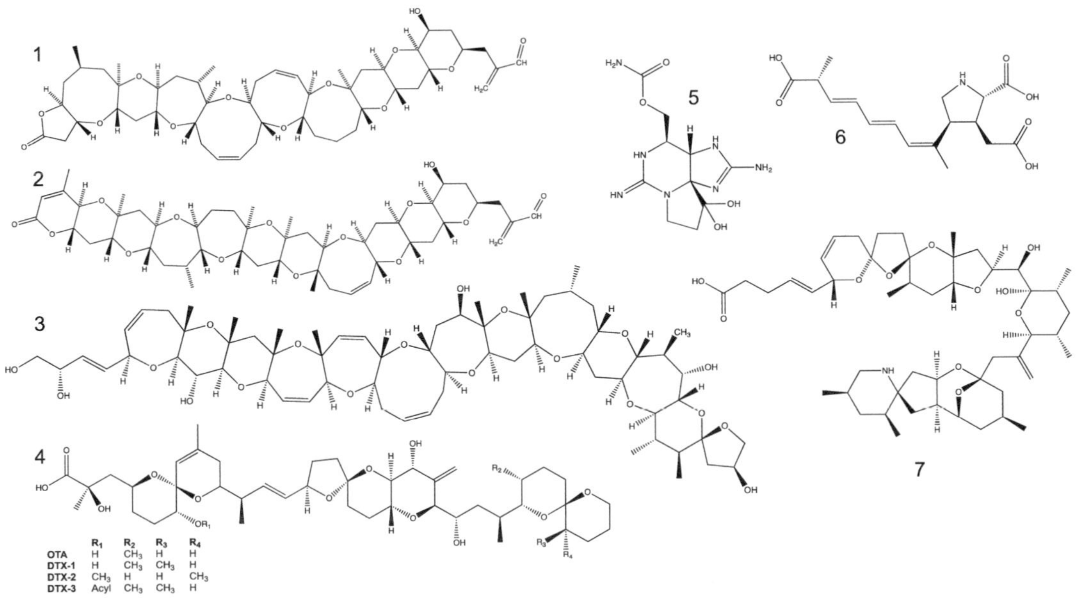

Figure 24.4 Structural diversity of biotoxins from marine algae. Shown are representatives of type A and B brevetoxins, and respectively, PbTx 1 (1) and PbTx 2 (2), ciguatoxins (e.g. CTX1B) (3), okadaic acid and other toxins (i.e. dinophysistoxins [DTX]) associated with DSP (4) and paralytic shellfish toxins (PSTs, e.g. saxitoxin) (5), as well as domoic acid (6) and azaspiracid (7).

Figure 24.5 Florida Red Tide. As shown in the image on the left, *Karenia brevis* forms large blooms in coastal systems of the Gulf of Mexico, and occasionally along the Atlantic Coast of the United States. Reports of Florida Red Tide generally peak in October; shown on the right are reported measurements of *K. brevis* cells along coastal Florida.

Source: NOAA (left); Florida Fish and Wildlife Conservation Commission (right).

linked to major fish kills and spikes in marine mammal (e.g. dolphin, manatee) deaths and strandings (Flewelling et al. 2005). In the context of human health, however, direct exposure to PbTx in coastal Gulf waters occurs primarily via inhalation of aerosolized toxin in fine water droplets ("sea spray"), and interestingly manifests primarily as pulmonary and respiratory distress. The toxicological link between this symptomology and well-known neurotoxicological mode of action remains to be clarified (Fleming et al. 2011). However, numerous studies have reported extensive mammalian metabolism of PbTx via *phase 1* (i.e. cytochrome P450 [CYP]) and *phase 2* (i.e. glutathione [GSH] conjugation) pathways and have suggested possible metabolic activation (particularly by the former pathway) to reactive epoxides capable of forming DNA adducts, and consequent genotoxicity, which may contribute to respiratory and pulmonary effects associated with inhalation (Radwan and Ramsdell 2006; Radwan and Ramsdell 2008).

With respect to human health, intoxications associated with marine algal biotoxins are perhaps most commonly recognized in relation to foodborne illness, specifically consumption of toxin-contaminated fish, shellfish and other seafood. Exposure risk is in some cases associated with discernible algal blooms, but in other cases it occurs via *biomagnification* of toxins by relevant seafood vectors in the absence of detectable blooms. Foodborne illnesses associated with marine algal toxins are generally classified (see Table 24.1) based on symptomology and/or mechanism of toxicity and relevant transmission vector (fish, shellfish). With respect to toxicology, symptoms are typically categorized based on either acute effects on the gastrointestinal system or a range of neurological effects, but in reality, most intoxications present as an overlapping range of symptoms.

Foodborne illness associated with algal biotoxins includes a number of neurotoxicological presentations linked to consumption of both fish and shellfish (Table 24.1). The potential for seafood to act as vectors for these toxins is generally mediated by the capacity of a given vector species to physiologically tolerate toxicity. For example, it is known that finfish and marine mammals are susceptible to the potent toxicity of PbTx (as frequently seen during red tide events), whereas many species of shellfish, including regionally important mollusks (e.g. oysters, clams, mussels), are largely tolerant and therefore can bioaccumulate the toxin, and thus serve as a potential vector for human dietary exposure (Watkins et al. 2008), which is specifically distinguished as "neurotoxic shellfish poisoning" (NSP; see Table 24.1). Moreover, extensive metabolism of PbTx by shellfish has been reported, and several consequent end products including amino acid and lipid conjugates

Table 24.1 Foodborne Toxicoses Associated with Marine Algal Biotoxins; Current and/or Useful Review Articles for Each Are Given

Toxicosis	Biotoxin	Source organism	Mechanism of toxicity	Symptoms	Reference
Neurotoxic shellfish poisoning (NSP)	Brevetoxins	*Karenia brevis*	Neurotoxic: Activates voltage-gated sodium channels	Nausea/vomiting, diarrhea, oral paresthesia, dysarthria, dizziness, ataxia	Watkins et al. (2008)
Ciguatera fish poisoning (CFP)	Ciguatoxins, maitotoxin Others: gambieric acid, scaritoxin	*Gamberdiscus toxicus*	Neurotoxic: Activates voltage-gated sodium channels	Nausea/vomiting, diarrhea, abdominal pain, paresthesia, dysesthesia, weakness, pruritis, myalgia	Friedman et al. (2017); Dickey and Plakas (2010)
Paralytic shellfish poisoning (PSP)	Saxitoxin and other paralytic shellfish toxins (PSTs)	*Alexandrium* spp., *Gymnodinium* spp., *Pyrodinium* spp.	Neurotoxic: Inhibits voltage-gated sodium channels	Nausea/vomiting, numbness/tingling, dizziness, respiratory difficulty	Etheridge (2010)
Amnesic shellfish poisoning (ASP)	Domoic acid	*Pseudonitzschia* spp.	Neurotoxic: Glutamate receptor agonist	Nausea/vomiting, memory impairment, seizures, dizziness	Todd (1993); Bates et al. (2018)
Diarrhetic shellfish poisoning (DSP)	Okadaic acid Dinophysistoxins, pectenotoxins	*Dinophysis* spp., *Prorocentrum* spp.	Inhibits protein phosphatases	Diarrhea, nausea/vomiting, abdominal pain	Dominguez et al. (2010); Reguera et al. (2014)
Azaspiracid shellfish poisoning (AZP)	Azaspiracid	*Azadinium spinosum*	Unknown	Diarrhea, nausea/vomiting, abdominal pain	Furey et al. (2010)

(likely associated with elimination strategies by the vector) have been suggested to contribute to uptake and distribution of the toxin and possibly toxicity (Leighfield et al. 2014). On the other hand, whereas finfish are susceptible to PbTx, many species are seemingly tolerant of a chemically complex suite of so-called ciguatoxins (CTX; see Figure 24.4; Dickey and Plakas 2010) responsible for ciguatera fish poisoning (CFP; see Table 24.1). This is notable, as PbTx and CTX share the same toxicological mechanism (activation of voltage-gated sodium channel inhibitor) and even the same binding site. As such, the bioaccumulation of CTX by a number of top-predator fish species serves as a vector for the neurotoxin. Moreover, biomagnification of acutely toxic levels of CTX by several important game fish species typically occurs both in the absence of detectable blooms and without discernibly moribund fish; CFP is in fact considered the most common nonbacterial seafood related illness, with more than 50,000 cases reported annually (Dickey and Plakas 2010).

In contrast to regionally important syndromes such as NSP and CFP, other marine algal biotoxins are associated with cosmopolitan foodborne illness. Arguably, the most prominent of these is paralytic shellfish poisoning (PSP; see Table 24.1). Caused by potently neurotoxic *saxitoxin* and several congeners (collectively referred to as so-called paralytic shellfish toxins) produced by multiple phylogenetically related marine dinoflagellates genera, and particularly species of *Alexandrium*, PSP has worldwide distribution (Figure 24.6). Similarly, although initially identified in association with regionally specific intoxications, and particularly an outbreak of what was coined amnesic shellfish poisoning (ASP; see Table 24.1; Todd 1993), occurrence—and bioaccumulation by a variety of fish and shellfish—of the neurotoxic glutamate receptor agonist, *domoic acid*, produced by the diatom genus *Pseudnitzschia*, is now likewise reported more or less globally (Figure 24.6).

Alongside neurotoxicity, exposure to algal biotoxins via consumption of seafood is widely linked to gastrointestinal disorders. The most prominent of these is diarrhetic shellfish poisoning (DSP; see Table 24.1). Toxicologically, DSP is linked to the action of several toxins—including okadaic acid and its so-called dinophysistoxin congeners—produced primarily by species of the marine dinoflagellate genus *Dinophysis* and generally attributed to compromised cellular integrity, and membrane permeability, of intestinal cells via well-described inhibition of protein phosphatases by these compounds (Dominguez et al. 2010). Notably, however, the role of this mechanism has been recently challenged (Munday 2013; Abal et al. 2018). Interestingly, bivalve mollusks not only bioaccumulate (via filter-feeding) but additionally metabolize (i.e. acylate) toxins consequently affecting both their uptake and toxicity (Reguera et al. 2014). First identified in Europe, DSP is now reported essentially worldwide (Figure 24.6).

Aside from recognized biotoxins, an ever-growing number of biologically active and potentially toxic metabolites have been (and continue to be) identified and characterized from marine algae. In some cases, these have been now linked to human intoxications and health concerns. A particularly notable example in this regard is what is now known as azaspiracid shellfish poisoning (AZP; see Table 24.1), and its causative agent azaspiracid (AZA; Figure 24.4), which was only identified within the last two decades (and the biosynthetic origin of which, i.e., *Azadinium spinosum*, has only been positively identified within the last 10 years). Symptomology of AZP is quite similar to DSP (Table 24.1), and although its toxicological targets in this regard remain to be elucidated, recent studies have identified a possible role of potassium channel antagonism (Twiner et al. 2012), and it has been suggested that metabolic biotransformation of AZA by shellfish (as for other shellfish-associated toxins) may, in addition, play an important role in its toxicity (O'Driscoll et al. 2011). However, in many other cases the possible contribution of a chemically diverse range of bioactive metabolites from marine algae to human health remains to be clarified.

24.3.1.2 Cyanobacterial Toxins: Chemistry and Toxicology

Cyanobacteria are prodigious producers of a wealth of bioactive secondary metabolites (Figure 24.7). Blue-green algae are both ubiquitous components of aquatic systems and capable of forming sporadic, seasonal or even persistent blooms in freshwater (lakes, ponds) and coastal systems. Accordingly several bioactive cyanobacterial metabolites—the so-called *cyanotoxins*—have been linked to human health, including both acute toxicoses (Carmichael et al. 1977; Griffiths and Saker 2003; Svircev et al. 2017) and long-term outcomes, such as cancer and neurodegeneration (Fujiki and Suganuma 2011; Caller et al. 2018). Potential exposure routes have been identified, including drinking water, recreational exposure and dietary routes (e.g. bioaccumulation in seafood, contamination of agricultural irrigation waters). Biologically active secondary metabolites (including recognized toxins) from cyanobacteria are structurally diverse and include alkaloids, peptides, polyketides, fatty acids, terpenoids, nonproteinogenic amino acids and many others (Figure 24.7). Toxicologically, however, the known cyanotoxins are often operationally classified based on target

Marine Algal Biotoxins

Region	Ocean/Sea	NSP	CFP	PSP	ASP	DSP	AZP
Africa							
	Indian Ocean		✔				
	Atlantic Ocean			✔		✔	
	Red Sea						
Asia							
	Indian Ocean		✔	✔		✔	
	South Pacific Ocean		✔	✔		✔	
	North Pacific Ocean			✔		✔	
Australia							
	South Pacific Ocean		✔	✔	✔	✔	
	Indian Ocean			✔	✔	✔	
Europe							
	North Atlantic Ocean			✔	✔	✔	✔
	Baltic Sea			✔		✔	
	Mediterranean Sea			✔	✔	✔	
North America							
	North Atlantic Ocean			✔	✔	✔	
	Mid-Atlantic Ocean	✔	✔			✔	
	Caribbean Sea		✔				
	Gulf of Mexico	✔	✔	✔			
	Pacific Ocean			✔	✔	✔	
	North Pacific Ocean			✔	✔		
South America							
	South Pacific Ocean			✔	✔	✔	
	Atlantic Ocean			✔		✔	

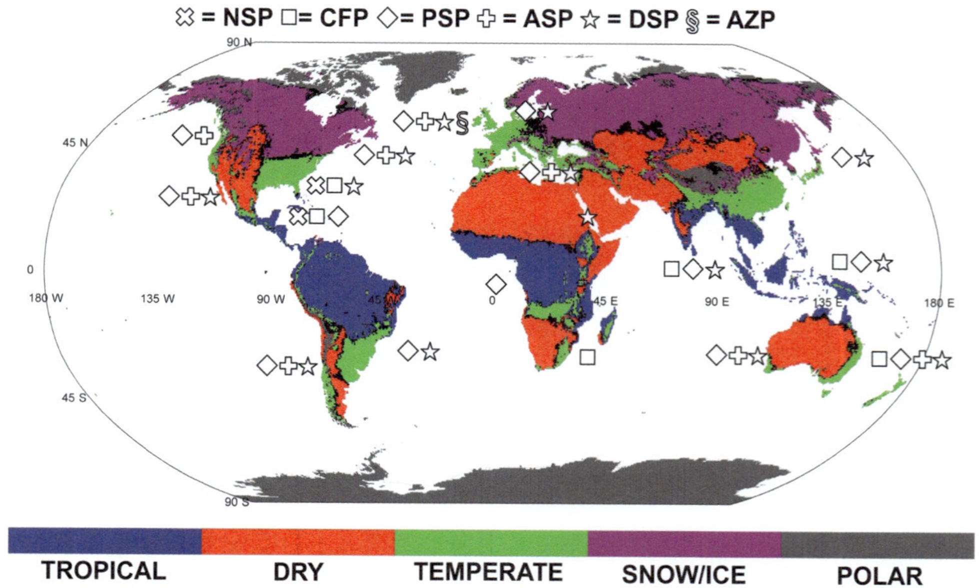

Figure 24.6 Global distribution of marine algal biotoxins.

Abbreviations: ASP = amnesic shellfish poisoning; AZP = azaspiracid shellfish poisoning; CFP = ciguatera fish poisoning; DSP = diarrhetic shellfish poisoning; NSP = neurotoxic shellfish poisoning; PSP = paralytic shellfish poisoning.

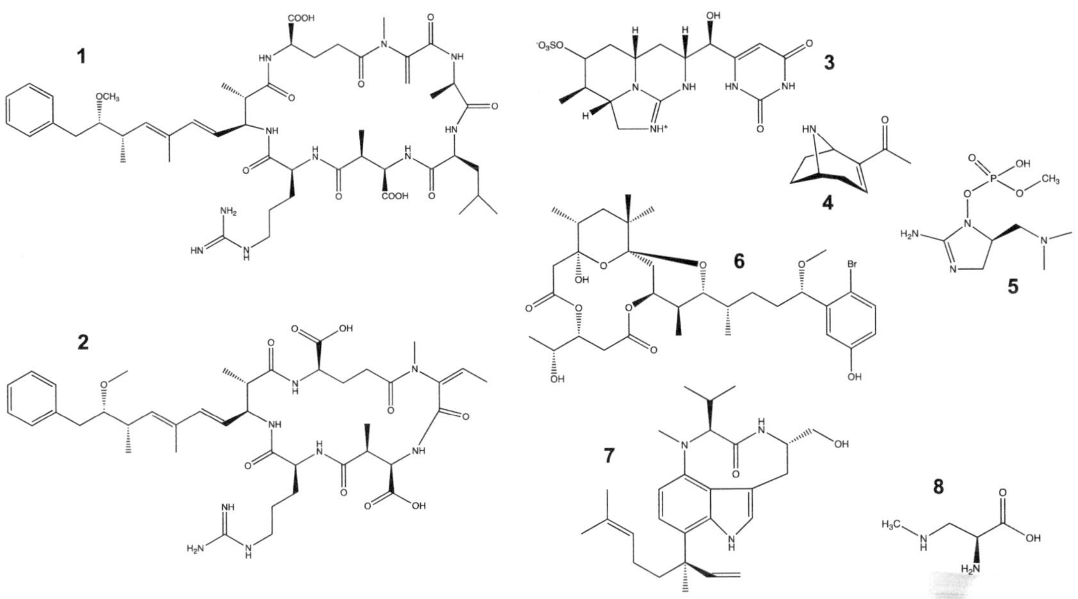

Figure 24.7 Structural diversity of recognized cyanotoxins. Shown are representatives of the microcystins—and specifically MC-LR (1)—which include >100 variants, and nodularins (e.g. NOD-R, 2), as well as cylindrospermopsin (3), anatoxin-a (4), anatoxin-a(s) (5), dermatoxic aplysiatoxin (6) and lyngbyatoxin (7), and the nonproteinogenic amino acid β-methylamino-L-alanine (BMAA, 8).

organ systems and thereby most frequently categorized as *hepatotoxins, neurotoxins* and *dermatotoxins*. Current reviews of bioactive cyanobacterial metabolites, including recognized toxins, are available (see Table 24.2).

With respect to hepatotoxins, two structurally and toxicologically related (see Figure 24.7 and Table 24.2) classes of peptides known as the microcystins (MC)—which include nearly 100 reported variants—and nodularins (NOD) are associated with acute and possible chronic toxicity worldwide. These nonribosomal peptides are produced by several cyanobacterial species, and in particular, the namesake genera *Microcystis* and *Nodularia*, respectively, which are globally distributed. And, likewise, the toxins are reported in freshwater and estuarine systems worldwide. Both MC and NOD are potent inhibitors of Type 1/2a serine/threonine protein phosphatases and are readily concentrated within hepatocytes by organic anion transporter proteins (OATP), such that exposure primarily manifests as hepatoxicity, with acute gastroenteritis and organ failure (Pearson et al. 2010; Steiner et al. 2016). Although intoxication is most frequently associated with domestic animals (e.g. livestock, pets), cases of human intoxications including fatalities have been historically (albeit rarely) reported. In one particularly noteworthy case, improper treatment of dialysis water at a clinic in Caruaru, Brazil, resulted in acute liver failure among 100 (of the 131) patients—with 76 of these cases resulting in death—due to exposure to MC; exposure in this case was retroactively estimated to be approximately 19.5 µg MC/L, or approximately 20 times the guideline for safe drinking water for this toxin (Carmichael et al. 2001). Alongside acute toxicity, both MC and NOD are tumor promoters (via protein phosphatase inhibition), and numerous studies have suggested a potentially contributing role in cancers (including possible reported cases in humans, and in particular, liver cancers. This is due to their primary distribution in hepatocytes at subacute and indeed quite low (<0.5 ng MC/L) concentrations (Nishiwaki-Matsushima et al. 1992; Ueno et al. 1996; Yu et al. 2001; Fujiki and Suganuma 2011).

The hepatotoxic alkaloid cylindrospermopsin (CYN) is most prominently recognized as a metabolite of the filamentous freshwater cyanobacterial species *Cylindrospermopsis raciborskii*, but it is in fact produced by a taxonomically diverse range of cyanobacteria. CYN generally targets hepatocytes (although it does accumulate in other organs to a lesser extent), but unlike MC and NOD, the mechanism for transport remains unclear. Regardless, the toxin has been linked to acute hepatotoxicity; in fact, it was first identified in association with the "Palm Island mystery disease"

Table 24.2 Recognized Cyanobacterial Toxins (Cyanotoxins) and Their Mechanisms of Toxicity; Current and/or Useful Review Articles for Each Are Given

Toxicological classification	Toxin(s)	Mechanism of toxicity	Reference
Hepatotoxins	Microcystins	Inhibition of protein phosphatase	Dawson (1998); Svircev et al. (2017); Massey et al. (2018)
	Cylindrospermopsin	Inhibition of protein synthesis	Kinnear et al. (2010); de la Cruz et al. (2013); Moreira et al. (2013)
Neurotoxins	Anatoxin-a	Acetylcholine receptor agonist	Osswald et al. (2007)
	Anatoxin-a (s)	Inhibits acetylcholinesterase	Aráoz et al. (2010)
	Saxitoxin/PSTs	Inhibits voltage-gated sodium channel	Wiese et al. (2010)
	BMAA	Glutamate receptor agonist, protein misincorporation and others	Caller et al. (2018); Cox et al. (2018)
Dermatotoxins	Lyngbyatoxin	Protein kinase C	Osborne et al. (2001)
	Aplysiotoxin	Protein kinase C	Osborne et al. (2001)

which affected 148 members of indigenous groups in Queensland, Australia, specifically presenting as severe hepatoenteritis, dehydration and bloody diarrhea (Griffiths and Saker 2003). The toxic mechanism of CYN remains to be completely clarified, although inhibition of protein synthesis, followed by cytotoxicity, generally is considered a primary mode of action. That said, it has been suggested by multiple comparative toxicological studies that genotoxicity (and possibly cytotoxicity) of CYN may be associated with phase 1 activation, although no reactive end product has been yet identified in this regard (see Pichardo et al. 2017).

Some of the most acute effects of exposure to toxigenic cyanobacteria involve potent neurotoxins. Of these, two chemically distinct (but toxicologically related) compounds, *anatoxin-a* and *anatoxin-a(S)*, are perhaps the best recognized. First identified as the so-called very fast death factor in association with livestock poisonings (Gorham 1960; Carmichael et al. 1977), the bicyclic alkaloid anatoxin-a (Figure 24.7) was the first cyanobacterial toxin to be isolated and chemically and toxicologically characterized (Devlin et al. 1977). Originally isolated from *Anabaena flos-aquae*, it has now been identified in a wide range of globally distributed cyanobacterial genera, and it has been linked to human and animal intoxications including fatalities (Osswald et al. 2007). Toxicologically, anatoxin-a is a highly potent agonist of nicotinic acetylcholine receptors (nAChR) in the central nervous system (CNS) and neuromuscular junctions, whereby the toxin binds irreversibly to the receptor (with 20 times the affinity of the neurotransmitter acetylcholine); unlike acetylcholine, it is not degraded by synaptic acetylcholinesterase enzymes, leading to overstimulation and eventual desensitization (Thomas et al. 1993; Osswald et al. 2007). Interestingly, a second neurotoxin, *anatoxin-a(S)*, occasionally co-occurs and converges toxicologically with anatoxin-a: specifically, the chemically unique (as an organophosphate; Figure 24.7) anatoxin-a(S) potently inhibits acetylcholinesterase, which would otherwise regulate (via degradation) synaptic levels of the neurotransmitter acetylcholine. And accordingly, anatoxin-a(S) can potentially work in concert with anatoxin-a (with the latter acting as an agonist of the neurotransmitter), leading to a synergistic toxicity when the toxins co-occur (as it does in certain species of *Anabaena*).

Interestingly, although STX is most commonly associated with PSP (as discussed above), and specifically marine eukaryotic algae (i.e. dinoflagellates) as a source organism, the same toxin and its many congeners have been curiously reported from numerous species of prokaryotic cyanobacteria (Jackim and Gentile 1968; Deeds et al. 2008). Notably, the chemical profiles of PSTs in cyanobacteria seem to be highly species and strain specific (Testai et al. 2016). Although documented human intoxications from exposure to cyanobacterial STX are relatively limited (compared to PSP and marine algal sources of the toxin), a few possible instances of intoxication have been reported (e.g. Rapala et al. 2005). This is perhaps due to the relatively limited consumption (compared to ocean-derived seafood) of freshwater fish and shellfish as potential vectors for the toxin.

In addition to potently neurotoxic compounds, associated with acute intoxications, a growing body of knowledge suggests that cyanobacterial metabolites may be linked to subacute and chronic effects on the CNS. Perhaps the most well-investigated example is the nonproteinogenic amino acid, β-N-methylamino-L-alanine (BMAA; Figure 24.7), which has been linked to neurodegenerative disease (Caller et al. 2018). A link between BMAA and neurodegenerative disease was first identified in association with remarkably high incidence of amyotrophic lateral sclerosis (ALS) among the indigenous Chamorra of Guam and the apparent role of this "unusual" amino acid in this regard (Ince and Codd 2005). Specifically, BMAA is produced by a cyanobacterial endosymbiont of a cycad species, and in turn it is accumulated to high levels by a species of fruit bat consumed regionally in Guam as a delicacy. The epidemiological link between diet-derived BMAA and ALS among the Chamorro was subsequently reinforced by the reported detection of the amino acid in ALS patients (Bradley and Mash 2009) and, indeed, a plethora of studies that followed. Moreover, since then BMAA has been reported worldwide and found to be produced by a taxonomically and ecologically diverse array of cyanobacteria (Cox et al. 2005), and has been purportedly linked to the occurrence of a broad range of neurodegenerative diseases including Alzheimer's disease and dementia similar to Parkinson's disease (Bradley and Mash 2009). As far as mechanism of toxicity, BMAA is known to act as an agonist—albeit a relatively weak one—of glutamate receptors (similar to other known neurotoxins, e.g., domoic acid). A more recent body of evidence suggests that BMAA, as an amino acid, may be erroneously incorporated (in place of canonical proteinogenic amino acids) via ribosomal translation during protein synthesis, and in turn result in aberrant proteins similar to protein aggregates (e.g. neurofibrillary tangles, amyloid plaques), which are in fact hallmarks of many neurodegenerative diseases (although the role of aggregates in the etiology of these diseases remains unclear). Very recent studies, however, suggest that BMAA neurotoxicity is perhaps a consequence of several interrelated pathways of toxicity including (alongside GluR activation and possible protein misincorporation) a role of mitochondrial disruption and oxidative stress, and impairment of several energetic and metabolic pathways, including lipids, in particular, which are crucial to neural function (Roy et al. 2017).

Finally, as one of the most prodigious cyanobacterial producers of bioactive secondary metabolites, the marine species *Moorea producens* (formerly *Lyngbya majuscula*) is known to secrete potent dermal irritants (dermatotoxins) which are linked to so-called seaweed dermatitis, characterized by severe skin blistering, following contact with the filamentous blue-green alga (which is typically found in shallow coastal waters). Often found as mixtures of bioactive metabolites from the species, two of these—namely *lyngbyatoxin A* and *aplysiatoxins*—are generally associated with contact toxicity, although other potentially co-occurring congeners (e.g. antillatoxin) have been suggested to contribute (Osborne et al. 2001). Furthermore, these metabolites have been linked to carcinogenicity. And both acute dermal toxicity and carcinogenicity have been generally attributed to the activation of protein kinase C (Basu et al. 1992).

The chemical and toxicological diversity of toxic cyanobacterial metabolites is clearly understated by the limited number of cyanotoxins which have been clearly characterized relative to human intoxications. Cyanobacteria, in fact, produce a myriad of bioactive secondary metabolites, and a considerable amount of research on these highly diverse metabolites (including, in particular, a rich repertoire of nonribosomal peptides) for which several good reviews are available (e.g. Welker and von Döhren 2006; Liu and Rein 2010). Although these metabolites have been characterized in terms of a range of relevant biological activity, particularly in relation to possible biomedical (drug discovery) potential, their possible role as toxins remains by and large unexplored. At the same time, the contribution of both recognized, and currently uncharacterized, cyanobacterial toxins to long-term or otherwise subacute health effects including cancer, neurodegeneration and others (e.g. immunosuppression) remains understudied, as is the case for many of the naturally occurring biotoxins, and it represents a critical endeavor for future toxicological investigation.

24.3.2 Climate Change and Toxic Algae and Cyanobacteria

Regardless of taxa or habitat (marine vs. freshwater), climate change has been widely proposed to both directly and indirectly influence population size and the structure of marine and freshwater planktonic communities including, in particular, blooms of toxigenic species in terms of frequency, severity and duration of occurrence (for reviews, see Paerl and Huisman 2009; Paerl and Paul 2012; Wells et al. 2015). Moreover, many of the recognized toxins from algae and cyanobacteria, despite a high degree of structural variability, are known to share common biosynthetic origins, including, in particular, *nonribosomal peptide synthetase* and *polyketide synthase* pathways (see Rein et al. 2006; Jones et al. 2009; Dittmann et al. 2013), which have in turn been found to be influenced directly by a range of environmental parameters including those relevant to climate change and its impacts (e.g. temperature, salinity, nutrient levels, light intensity). And, both prevalence of toxigenic algae and cyanobacteria, and production of their toxins, are thereby tied intrinsically to climate (and thus climate change).

As fundamental parameters for phytoplankton communities, changes in temperature and precipitation patterns (via altered hydrology and nutrient inputs) are clearly anticipated to directly impact algal communities. Higher temperatures favor growth of many algal and cyanobacterial species in terms of both overall population growth (i.e. size of blooms) and seasonal timing, including earlier appearance and/or more persistent blooms. Simultaneously, changes in hydrology, including increased local flow of nutrients into aquatic systems due to higher precipitation, as well as climate change–induced alterations of ocean currents, conspire to drive freshwater and coastal eutrophication and shifts in geographical distribution of marine HABs. However, it is rather likely that a suite of intricately intertwined physicochemical consequences of climate change including temperature-driven stratification within water column, altered light regimes and input of nutrients from watersheds which, in fact, drive ecological shifts within these communities. Most notably, increased input of nutrients, including nitrates and phosphates from agricultural fertilizers, household products (e.g. detergents) and animal (e.g. livestock) waste, leading to freshwater and coastal *eutrophication* (i.e. overgrowth of algae and cyanobacteria), are considered major contributors. But frankly, we are only beginning to understand how these interrelated factors may come together to trigger blooms or otherwise increase the occurrence, persistence and toxigenicity of algae in the environment (Wells et al. 2015).

That said, these ecological shifts have been widely proposed to lead to the increased presence of numerous toxigenic species of algae and cyanobacteria, and in turn, to present a growing health concern in relation to the many ways in which human populations interface with aquatic environments (e.g. drinking water, fish and shellfish consumption, recreational exposure). Increased incidence of the Florida Red Tide, for example, has notably been linked to extreme weather events, and particularly drastic increases in precipitation levels associated with hurricanes (Hu et al. 2006). Although blooms of *Karenia brevis* occur nearly every year in the eastern Gulf of Mexico, the most severe red tide events in recent decades have followed active hurricane seasons which, it has been proposed, facilitate elevated outflow of nutrients into the Gulf to fuel blooms (Brand and Compton 2007; Hu et al. 2006). Moreover, these recent blooms have been frequently longer in duration (extending well beyond the typical seasonal windows). At the same time, it has been proposed that deposition of otherwise limiting micronutrients (e.g. iron) from Saharan dust—which may be worsened by increased desertification—may similarly fuel red tide blooms either directly or indirectly (via increased nitrogen-producing *Trichodesmium* blooms; Walsh and Steidinger 2001). Likewise, the prevalence of several HAB species, including those associated with PSP, ASP and DSP, have been linked to global weather patterns, particularly El Niño–Southern Oscillation cycles (Hallegraeff 1993, 2010). Similarly, although ciguatera has been long reported in marine systems, its reported frequency, particularly in the South Pacific, has reached epidemic proportions in recent decades (Hallegraeff 2010).

In freshwater systems, climate change and its many consequences have been proposed to catalyze expansion of cyanobacterial blooms both spatially and in terms of the seasonal appearance and duration of blooms. Increased water temperature can in fact selectively promote cyanobacteria, enabling them to outcompete microalgae, and along with related environmental consequences of climate change (e.g. vertical stratification of water) to contribute to both increased blooms and shifts in cyanobacterial populations, potentially toward toxigenic species (Paerl and Paul 2012). With respect to global increase of blooms, a particularly striking example is the measured warming and concomitant appearance of cyanobacterial blooms in the Great Lakes within the last decade, including a record-setting toxigenic *Microcystis*-containing bloom in Lake Erie (Carmichael and Boyer 2016), and more recently, the appearance of blooms in the historically oligotrophic Lake Superior (Hauser 2018). As an example of both expansion of range, and differential toxin production, the toxigenic species *Cylindrospermopsis raciborskii* was originally thought to be limited to tropical waterways; however, it is now found worldwide, including in temperate systems, and a preponderance of evidence suggests this migration may be related to the effects of climate change, including elevated temperature in particular (Sinha et al. 2012). Moreover, *C. raciborskii* produces (as discussed above) multiple recognized toxins, specifically CYN and STX, and the toxin profile of the species has been shown closely tied to geographical distribution of strains of the species; in particular, whereas strains from Australia and New Zealand produce CYN but not STX, isolates from South American produce STX but not CYN, and neither toxin has been confirmed so far from *C. raciborskii* in North America or Europe (Sinha et al. 2012).

In light of the geographic expansion of toxic algal and cyanobacterial species, and the increased magnitude, duration and toxigenicity of toxigenic blooms anticipated with climate change, the challenges to practitioners in the field of toxicology will be multifold. Generally speaking, increased awareness of the health concerns associated with algal and cyanobacterial toxins will be most critical for extensive human populations that reside near and/or have regular contact with waterbodies, including marine, coastal and freshwater systems. Toxicological training related to diagnosis and detection in relation to seafood and waterborne toxic contaminants will be essential. As *emerging* concerns, more frequent monitoring for toxins in water and fish, shellfish and other seafood will be generally key to preventative measures. Increased drought (as anticipated with current climate change models) would simultaneously lead to increased reliance on increasingly limited water supplies and pose concerns with respect to cyanobacterial biotoxins in particular. More generally, continued investigation of the highly diverse bioactive metabolites from algae and cyanobacteria in relation to adverse human health effects (including the still largely uncharacterized long-term, chronic and subacute health effects) of both known and yet unknown toxins will be fundamental to understanding environmental health impacts.

24.3.3 Climate Change and Mycotoxins as Agricultural Contaminants

Within the terrestrial environment, perhaps the most notable examples of increased biotoxin prevalence with respect to climate change are the numerous *mycotoxins* produced by fungal contaminants of agricultural crops, including food and feed plants. These toxins, in turn, pose concerns for human exposure through direct or indirect (i.e. meat, milk, egg) consumption of contaminated agricultural products. Although hundreds of bioactive and thus potentially toxic fungal metabolites have been identified, a relatively small subset (Figure 24.8) of these have been linked so far to possible human health impacts (see, e.g., review by Marin et al. 2013).

Foodborne illness associated with contamination of foodstuffs by mycotoxigenic fungi has been recognized for hundreds if not thousands of years, and introduction of these toxins to human health likely coincided with humankind's transition to agriculture (cultivation, harvest and storage of crops) approximately 10,000 years ago. A good review on the history of mycotoxin research can be found in Pitt and Miller (2017). It is assumed, in fact, that increased awareness of these mycotoxins may have led to mitigation of these health effects through improved harvesting and postharvesting (e.g. storage) practices. That said, however, it has been most recently estimated that approximately 25% of crops worldwide are still contaminated with one or more mycotoxins (Alshannaq and Yu 2017). Furthermore, co-occurrence of mycotoxins (due to the production of multiple toxins by the same species, and wide distribution of this limited number of recognized toxigenic species) has been reported and is suggested to lead to additive and interactive toxic effects (Smith et al. 2016). Moreover, a growing body of evidence suggests that both temperature and moisture (i.e. water levels) can significantly influence the species composition and abundance of toxigenic fungal microflora by way of both preharvest and postharvest (storage, transport and processing) contamination. And, indeed, emerging evidence (Paterson and Lima 2010; Magan et al. 2011) suggests that changes with regards to these parameters, specifically in the context of climate change over decadal scales, may be driving recent increases in the prevalence of toxigenic species and their corresponding toxic metabolites in a host of agricultural products.

Mycotoxins are a structurally diverse array of fungal secondary metabolites including terpenoids, alkaloids, polyketides and peptides (Figure 24.8) that likewise have been linked to a wide range of toxicities, but primarily associated with subacute health effects (e.g. carcinogenicity), although cases of acute toxicosis have been documented (Marin et al. 2013). Taxa of molds that produce mycotoxins primarily include species of *Aspergillus*, *Penicillium*, *Fusarium* and *Claviceps*, as well as a few other phylogenetically affiliated genera. And, given the chemical diversity within the mycotoxins, they are typically categorized based upon these recognized toxin-producing genera.

The genera *Aspergillus* and *Penicillium* are, for example, producers of two of the most common and potently toxic mycotoxins, *aflatoxins* and *ochratoxin*. Perhaps the best studied of the mycotoxins (and first to be chemically and toxicologically characterized) are the cytotoxic and mutagenic aflatoxins (AFs). With more than 14 variants (Figure 24.8), AFs were first identified in the 1960s and have since been linked to both acute intoxications (i.e. aflatoxicosis) including numerous fatal outbreaks (>100 deaths) in economically developing nations, and chronic subacute health effects, including well-documented teratogenicity and carcinogenicity (Williams et al. 2004). Generally

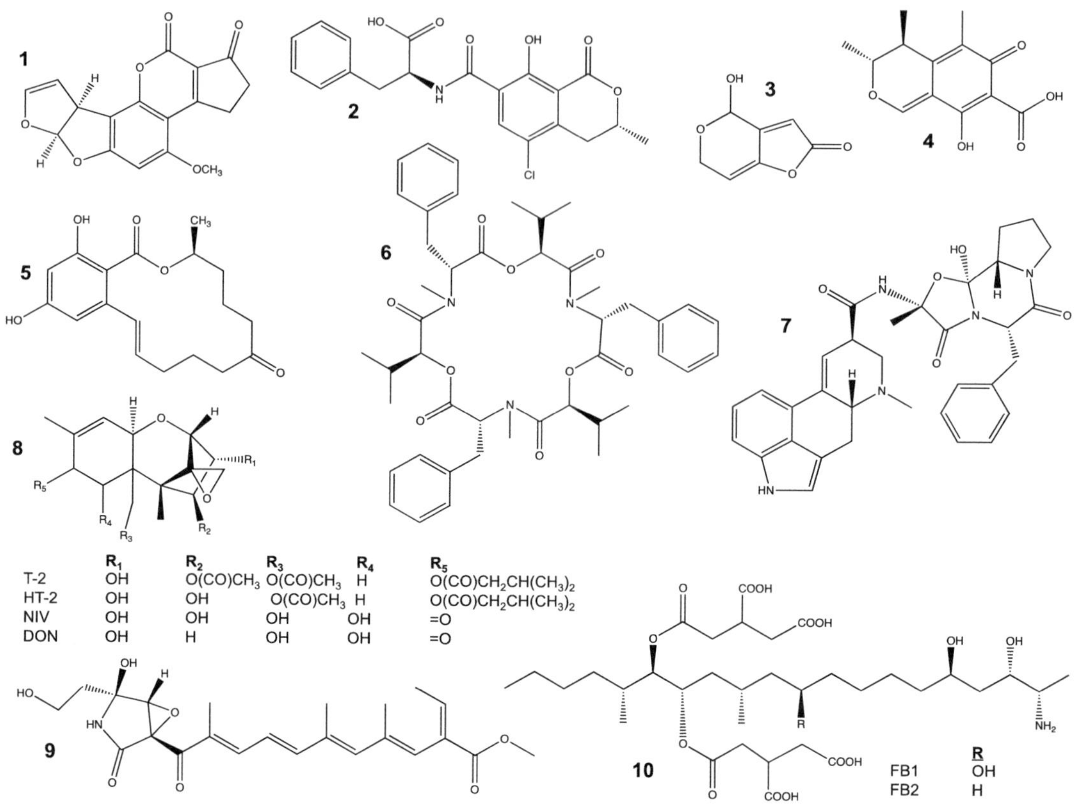

Figure 24.8 Chemical structures of common and/or representative mycotoxins, including aflatoxin B1 (1); ochratoxin A (2); patulin (3); citrinin (4); zearalenone (5); beauvericin (6); ergotamine (7); trichothecenes (8), specifically including T-2, HT-2, nivalenol (NIV) and deoxynivalenol (DON); fusarin C (9); and fumonisins (10), including fumonisin B1 and B2 (FB1 and FB2, respectively).

considered the most toxic congener, AFB_1 is classified, for example, by the International Agency for Research on Cancer (IARC) as a Group 1 carcinogen. Exposure to AFs primarily occurs via contamination of grains, nuts and legumes, although their presence in other agricultural products (e.g. fruits) has been reported. Given the relatively long history of AF as toxic contaminants of agricultural products, they are the only mycotoxins currently regulated by US Food and Drug Administration (FDA) action levels, specifically in the parts per billion (ppb) range; other mycotoxins are only subject currently to advisory levels. Notably, it has been well established that metabolic activation, specifically via several species-specific CYP isoforms to a highly reactive epoxide (i.e. AFB_1-8,9-epoxide), is essential to the toxicity of AFB_1 (Deng et al. 2018).

Also among the most commonly detected and thus best studied mycotoxins are the ochratoxins, of which *ochratoxin A* (OTA) is by far the most common and most toxic. Produced by species of *Penicillium* and *Aspergillus* (including *Aspergillus niger* and *A. ochraceus* in particular), OTA contaminates a wide range of crop plants from grains and nuts to fruits; consequently it is one of the most commonly found mycotoxin contaminants of a wide range of commodities (e.g. cereal, coffee, wine, dried fruits, meats). The mechanism of OTA toxicity remains unclear but includes inhibition of protein synthesis, induction of oxidative stress and formation of DNA adducts, leading to both cytotoxicity (i.e. apoptosis and necrosis) and cell-cycle arrest, and carcinogenicity (Koszegi and Poór 2016). Toxicity, however, has been primarily linked to kidney and associated organ systems, and it is widely considered *nephrotoxic*. Although the metabolism of OTA is well known, the most common metabolic by-product—specifically following hydrolytic cleavage of the amide bond to the phenylalanine moiety (Figure 24.8) of the molecule—is the isocoumarin ochratoxin α, which is generally considered to have low toxicity (Haq et al. 2016; Kupski et al. 2016). Given its considerable

Table 24.3 Major Toxigenic Fungal Contaminants of Agricultural Products; Current and/or Useful Review Articles for Each Are Given

Primary fungal source(s)	Biotoxin	Mechanisms of toxicity	Human toxicoses
Aspergillus spp. (*Penicillium spp.*)	Aflatoxin	Genotoxicity	Carcinogenicity/teratogenicity, hepatoxicoses
	Ochratoxins	Genotoxicity, oxidative damage	Nephrotoxicity, carcinogenicity, neurodevelopmental toxicity
	Patulin	Genotoxicity, oxidative stress (ROS, decreased GSH), interaction with enzyme thiols, inhibition of immunological pathways	Gastrointestinal toxicity, hepatotoxicity; neurotoxicity; carcinogenicity
	Citrinin	Inhibition of cytokine production, RNA/DNA synthesis and nitric oxide gene expression; induction of oxidative stress (ROS) and caspase pathway; mutagenicity	Nephrotoxicity, carcinogenicity
Fusarium spp.	Trichothecenes	DNA, RNA and protein synthesis inhibition, cytotoxicity	Gastroenteritis, growth retardation, neuroendocrine, immunosuppression
	Fumonisins	Ceramide synthase inhibition	Teratogenicity, carcinogenicity
	Zearalenone	Endocrine disruption: estrogenic	Reproductive toxicity
Claviceps spp.	Ergot alkaloids	Agonist of 5-HT1 (i.e. serotonin) and other neurotransmitter (e.g. dopamine) receptors	Gangrene (vasoconstriction), convulsion and hallucination

toxicity, regulatory guidelines have been established for OTA in various foods and feed, at low ppb levels, in the European Union and a few other countries; however, many others (e.g. United States, Canada) have yet to set such guidelines at the time of writing.

As *Aspergillus* and *Penicillium* are producers of multiple mycotoxins, it is known that their contamination of agricultural products can lead to co-occurrence of mycotoxins. Indeed, OTA and AFs (both potently toxic mycotoxins) have been shown to co-occur, although interestingly their combined toxicity ranges from additive or synergistic to antagonistic, depending on the toxicological system (Klaric et al. 2013). On the other hand, both OTA and AFs are also known to commonly co-occur with two relatively less toxic metabolites, namely, *patulin* and *citrinin* (Figure 24.8), which are produced by the same genera, *Aspergillus* and *Penicillium*, among others. In the case of citrinin in particular, not only is co-occurrence with OTA relatively common, but a number of toxicological studies have suggested synergistic interactions between the two toxins (Klaric et al. 2013). That said, as the mechanism of toxicity for each of the toxins remains to be fully elucidated, understanding interactions between them remains far from clear and is part of ongoing toxicological studies.

As one of the most frequent toxigenic fungal contaminants of agricultural products, the genus *Fusarium* is known to produce a very diverse range of mycotoxins which are commonly classified together as the so-called Fusarium toxins (Munkvold 2017). By far, the largest family of *Fusarium* toxins—and, in fact, the most chemically diverse of all mycotoxins with more than 170 structural variants—are the trichothecenes (Figure 24.8), which are primarily found as contaminants of grain crops and their products (e.g. cereals, beer), although they have been occasionally detected in other food crops (e.g. legumes, potatoes, fruits). Despite the chemical diversity, trichothecenes generally share inhibition of protein synthesis as a common mechanism of action (Arunchalam and Doohan 2013), although other mechanisms (e.g. DNA damage, oxidative stress) have been demonstrated and investigated. The two most common subtypes (Figure 24.8) are deoxynivalenol (DON)—which is sometimes reported "vomitoxin," in association with acute toxicity—and T-2, with the latter generally considered to be more toxic (and specifically shown to inhibit DNA synthesis and protein synthesis). Alongside T-2, its most abundant metabolite HT-2 (i.e. acetylated at the C-4 position; see Figure 24.8) is often considered to contribute significantly to toxicity (Yang et al. 2018). Both subtypes are associated with acute effects, including fatal cases of intoxication.

One historically notable example was a large outbreak of "alimentary toxic aleukia" in Russia in the 1940s, characterized by gastrointestinal distress (e.g. nausea/vomiting, diarrhea) and effects on blood, including aleukia (a decrease in white blood cells) and hemorrhaging in particular. This outbreak resulted from contamination of overwintered grains with *Fusarium* spp. (and T-2), and, in fact, it led to the first characterization of this ailment (see Pitt and Miller 2017). Subacute and chronic effects include teratogenicity and "anorexic effects." However, evidence generally suggests minimal contribution to cancer: DON is, for example, classified by IARC as Group 3 ("not classifiable as to its carcinogenicity to humans"). Trichothecenes, however, are considered to be quite chemically stable, and as such they are not typically removed by food preparation (e.g. cooking, fermentation).

Alongside the trichothecenes, *fumonisins* and *zearalenone* are produced by various species of *Fusarium*; likewise, they are found most commonly as heat-stable contaminants of grains (corn in particular), and thereby linked to health concerns. Fumonisins, including three major congeners (fumonisin B1, B2 and B3), have been associated with neurotoxicity, hepatotoxicity and nephrotoxicity by way of documented oxidative stress, cytotoxicity and alteration of cytokine expression (Stockmann-Juvala and Savolainen 2008); FB1 is generally considered to be the most toxic variant. Zearalenone (ZEN), on the other hand, has well-documented estrogenic activity, and accordingly it has been widely linked to reproductive disorders among livestock including infertility, hormone dysfunction, reduced sperm production and increased gestation length; it may also affect humans (e.g. hyperestrogenism; Zinedine et al. 2007). Metabolism is known to play a key role in the toxicity of ZEN; specifically, phase 1 metabolic reduction of ZEN leads to two 7-hydroxylated epimers, α- and β-zearalenol; the former has been shown to be nearly two orders of magnitude more estrogenic than the former (Shier et al. 2001; Tatay et al. 2018). Along with these, numerous other minor bioactive metabolites (e.g. beauvericin, moniliformin, enniatins, equisetin, fusarins) have also been identified from *Fusarium*; however, little is currently known regarding their contributions to human health. Moreover, these toxins have been reported to co-occur with other *Fusarium* toxins (as well as toxins from other fungal species, e.g., OTA, aflatoxins), and the toxicity of these mixtures has been investigated in a growing number of recent studies (Speijers and Speijers 2004; Smith et al. 2016; Allasane-Kpembi et al. 2017; Yang et al. 2017).

Toxigenic species of the genus *Claviceps*, as contaminants of grains, represent an historically notable source of mycotoxins in agriculture. The genus—and particularly the species *Claviceps purpurea*—is specifically associated with a structurally diverse group of toxic metabolites collectively referred to as the "ergot alkaloids" (Figure 24.8 shows a few examples), which have in turn been linked (via consumption of contaminated rye and other grains) to *ergotism*. Ergotism is primarily characterized by gangrenous effects, owing to vasoconstrictive properties of the alkaloids, as well as neurotoxicoses. It is, in fact, one of the oldest known examples of *mycotoxicosis*, including both epidemics of St. Anthony's Fire during the middle ages, and historically documented accusations of witchcraft likely resulting from the neurotoxic effects (e.g. convulsions, delirium) of the ergot alkaloids (Liegl and McGrath 2016; Pitt and Miller 2017). Although reports of ergotism are far less common, particularly in economically developed parts of the world, it has been proposed that unpredictability due to altered climate may lead to a resurgence (Coufal-Majewski et al. 2016).

In reality, hundreds of bioactive metabolites from fungal species have been identified, specifically associated with contamination of agricultural products. A particularly notable example is a remarkably diverse repertoire of bioactive, and possibly toxic, metabolites from species of the genus *Alternaria*, including alternariol, tenuazoic acid, altertoxin, tentotoxin and several others (Fraeyman et al. 2017). Found as contaminants of grains, fruits and vegetables, the *in vitro* biological activity of the *Alternaria* toxins is as diverse as their chemistry (e.g. estrogenic activity, genotoxicity, cytotoxicity, inhibition of protein synthesis). Their link to human health effects, however, largely remains to be clarified.

In addition to agricultural products as source of mycotoxins, it is worth mentioning that other routes of exposures to these toxins may exist. It has been shown, in particular, that airborne molds, and particularly species of the genus *Stachybotrys*, associated with mold and mildew contamination of damp buildings (or so-called sick building syndrome), have been found to produce trichothecenes which have been linked to health effects including respiratory disorders and headaches. This example is particularly germane given its association with water intrusion and thus possible links to climate change by way of increased moisture (e.g. flooding) and temperature and humidity.

A growing body of evidence does point to the impact of climate change on the prevalence of mycotoxins in agricultural products and other possible routes of exposure (e.g. airborne molds).

The growth of molds as contaminants of crop plants (and elsewhere) is generally favored by elevated precipitation and temperature, and would be expected to directly increase the prevalence of most species. Indirectly, it has been shown that drought stress, on the other hand, can increase susceptibility to infection (Wu et al. 2011). Moreover, these factors have been shown to not only contribute to the growth of toxigenic fungi but also their production of toxins. As an example, experimental studies demonstrate interactive effects of temperature, moisture and CO_2 on growth of *A. flavus* and increased biosynthesis (including gene expression and measured levels) of aflatoxins (Medina et al. 2015, 2017).

Supporting the predicted role of altered climate on mycotoxins, assessments of mycotoxins and their producing species have indeed suggested a close link between climate and mycotoxigenic species (and their toxins), including possibly recent (i.e. decadal) increases in their prevalence. The role of temperature and humidity is illustrated by a number of studies that have investigated occurrence of OTA in European wines, for example (via fungal contamination of grapes), and have found a general trend toward higher levels in wines sourced from warmer and more humid climates (e.g. Mediterranean climates) compared to more temperate climates including those with significant periods of freezing temperatures (Paterson et al. 2018). Likewise supporting a shift in range and increased prevalence of mycotoxin contaminants with changes in climate, a recent study—the first in more than a decade—of OTA in US wines showed not only geographic trends (toward higher OTA levels in warmer climates) but also greatly elevated levels overall (compared to previous studies), suggesting increases on decadal scales (De Jesus et al. 2017). On a larger global scale, species of *Aspergillus* previously associated with tropical and subtropical regions have emerged only within the last two decades, as common fungal contaminants, in temperate regions of Europe (Battilani et al. 2016; Van der Fels-Klerx et al. 2016). Toward predicting future scenarios, one recent modeling study (Van der Fels-Klerx et al. 2012) projected geographically specific increases in DON, aflatoxins and *Alternaria* toxins based on various climatic scenarios relevant to anticipated changes (e.g. temperature increases of 2–5°C).

24.3.4 Climate Change and Foodborne Bacterial Toxins

Pathogenic heterotrophic bacteria are a widespread concern as both infectious and toxigenic contaminants of food and water. Clearly, the perishability of food due to bacterial growth is not a strictly contemporary health issue: spoilage due to improper storage of food can occur in nearly any environmental context, and it could be argued, historically represents one of the most long-standing toxicological threats to human health. Similar to the awareness of mycotoxins as contaminants of agricultural crops (as discussed above), recognition of the role of bacterial agents in foodborne illness has consequently led to improvements in food preparation, processing and storage practices that have greatly reduced incidences of food poisoning. That said, foodborne disease associated with microbial and particularly bacterial pathogens remains one of the most important threats to human health. Underscoring the magnitude of the impact on human health, the World Health Organization (WHO) estimated more than 600 million cases of foodborne illness, including approximately 420,000 fatalities, in 2010 alone (Kirk et al. 2015). Human illness associated with bacterial toxins and/or infection, however, not only remains a ubiquitous health concern in general, but a growing body of evidence suggests a possible contribution of climate change to global prevalence (Lake and Barker 2018).

Although other microorganisms, including viruses, protozoans and helminths in particular, impact human health through contamination of food, the pathological mechanisms in these cases are largely due to the infectious nature of these organisms and not toxicity, strictly speaking. Several pathogenic bacterial species, however, are linked to human health effects by way of their intricate and often still not fully elucidated toxicity (e.g. endotoxins, enterotoxins). Yet even in these cases, teasing apart the infectious versus toxigenic natures of these organisms complicates the toxicology.

A multitude of studies have suggested that climate change is contributing to the increased prevalence of foodborne bacterial disease, although the mechanisms remain to be clarified. Potential culprits that have been proposed include not only shifts in geographical distribution, levels, diversity and seasonality of bacterial pathogens in association with changes in environmental conditions, but also a host of agricultural practices in relation to elevated temperature, increased drought and other consequences of climate change (Lake 2017; Lake and Barker 2018). For example, the transition to indoor husbandry of livestock, due to elevated temperatures, is predicted to enable spread of pathogenic bacteria. Alongside proposed increases in algal biotoxins, elevated water temperatures likewise have been linked to the increased contamination of fish and shellfish by a number of pathogenic marine bacterial species (e.g. *Vibrio* spp.) in both fisheries and aquaculture contexts. Changes in precipitation patterns, and specifically increased flooding and runoff as observed with extreme weather events (e.g. hurricanes), have been linked to increased bacterial contamination of crops. Similarly, climate change induced drought has been linked—specifically through use of poor-quality irrigation water—to bacterial (e.g. *Salmonella*) contamination of produce and consequent human illness (see Lake and Barker 2018).

24.3.5 Venomous Animals and Climate Change

Elevated global temperatures associated with climate change are anticipated, and in a growing number of cases have been shown to either increase populations sizes or alter (i.e. expand) the geographic range of venomous animal species (Needleman et al. 2018a, 2018b). Indeed, most venomous species are *ectothermic*, and thus dependent on ambient environmental temperature to regulate many aspects of their physiology; as such, they are directly and significantly affected by even small changes in temperature. Alongside direct effects of temperature on physiology, it is likely that increases in populations and/or expansions of range are compounded by climate change–related alterations in habitat. For example, increased rates of plant biomass production (linked to both climate and CO_2 levels; see below) may effectively provide additional habitat structure to harbor venomous species (e.g. spiders, snakes). Simultaneously, elimination of potentially competing and predatory species (due to ecophysiological consequences of elevated temperature, etc.) may enable the emergence of venomous species. Relevant species, in this regard, span invertebrate and vertebrate taxa and include both aquatic and terrestrial representatives.

Within terrestrial habitats, a particularly relevant example, with respect to toxicology and human health, are venomous snakes. Approximately 2.5 million snakebites by venomous species are reported each year; 85,000 deaths are estimated annually, with the vast majority of these (>80,000) occurring in parts of Asia and Africa (Gutierrez et al. 2010). Toxicologically, the venom of snakes primarily represents highly complex mixtures (~90%–95% dry weight) comprising enzymatic proteins and other polypeptides—including thousands of known representatives—with a wide range of activities including cytotoxicity, neurotoxicity, cardiotoxicity and "hemotoxicity." Changes in temperature, at levels consistent with climate change, have been proposed to increase both the activity and distribution of venomous snakes. In one example (Tasoulis and Isbister 2017), elevated temperature (i.e. +3°C) was suggested, through modeling studies, to impact the physiology, in terms of metabolic rate, digestive function and activity levels (e.g. amount of time above ground), of the highly venomous pit viper (*Trimeresurus gracilis*). With respect to relevant changes in distribution, multiple studies investigating a host of venomous snake species have documented and/or forecast significant shifts and expansions of geographic range on decadal scales, and particularly migration into regions where venomous snakes do not currently occur, including more highly populated areas (Nori et al. 2014; Wu 2016; Yanez-Arenas et al. 2016). Estimates, therefore, project increases on the order of several million more snakebites per year (Yanez-Arenas et al. 2016). The projected increase in venomous snakebites poses a unique problem to medical practitioners, as the supply of antivenom is, in general, highly limited—and particularly so in regions where the majority of envenomations occur (e.g. rural Africa; Theakston and Warrell 2000).

Similar studies have supported increases in activity and range of venomous invertebrate taxa, including insects and arachnids in particular (e.g. spiders). Beginning with the latter, spiders and other arachnids (e.g. scorpions) are frequently associated with envenomation. However, a rather small fraction (<0.1%) of the more than 40,000 known species of spiders are effectively poisonous to humans, due to the size of the dose relative to its toxic potency. That said, several spiders are known to produce sufficiently potent venom comprising neurotoxic and "necrotic" (i.e. cytotoxic) polypeptides as well as several small molecules, which are medically relevant, and manifest as both localized reactions and systemic responses including neurotoxicity and hemotoxicity (Table 24.4). As ectotherms, the survival of spiders and therefore their geographic range of distribution are directly linked to temperature. And, indeed, a number of studies have demonstrated a link between climate change and the distribution of venomous spider species, indicating an overall shift northward, but otherwise a limited expansion of overall range (Cramer et al. 2008; Saupe et al. 2011; Leroy et al. 2014). Beyond spiders, it is expected that other venomous arachnids may be similarly affected with respect to distribution and consequent risk of human exposure; a notable example would be scorpions, whose distribution is largely limited to warm and particularly arid habitats, which are expanding due to climate change (i.e. desertification). However, little is currently known regarding the impacts of climate change on their distribution and population sizes.

Table 24.4 Venomous Species of Spiders Associated with Reported Human Envenomations, and Their Geographical Distribution and Reported Toxicological Effects

Species/taxa	Common name	Geographic range	Toxicity
Loxosceles reclusa	Brown recluse	United States: South/Southeast and Midwest	Necrotic lesions
Phoneutrial spp.	Brazilian wandering spiders	South and Central America	Neurotoxic
Cheiracanthium inclusum	Yellow sac spider	United States, Mexico and South America	Necrotic lesions
Family Lycosidae	Wolf spiders	Worldwide	Allergenic
Latrodectus spp.	Black/brown/red widow and redback spiders	Worldwide	Pain, nausea, mild paralysis
Family Dipluridae	Funnel-web spiders	North/South America and Australia	Neurotoxic

As for arachnids, temperature (specifically, cold weather) acts as an effective barrier for distribution of a number of stinging insects, particularly species of the order Hymenoptera; these include bees, wasps and ants, which are relevant to human health through potentially serious adverse reactions. Although the number of stinging ant species are rare, several do produce venoms, chemically comprising proteins and small molecules (e.g. alkaloids, formic acid), which act as painful cutaneous irritants. In the United States alone, approximately 17,000 people annually seek medical attention for ant stings. Stinging bees and wasps are among some of the most commonly reported animal envenomation cases. Compared to the venom of ants, those of bees and wasps are largely proteins (e.g. phospholipase A, melittin), along with a few small biological active molecules including neurotransmitters (e.g. acetylcholine) and allergens (e.g. histamine). In the case of both ants and bees/wasps, lethal doses have been determined and usually require multiple (estimated range from ~10 to 1000) stings; most acute lethal presentations generally relate to the hypersensitive *anaphylaxis* response that affects approximately 1% of the population.

With respect to climate change, expansion of the range of ants and bees has been investigated and found to be impacted particularly by elevated temperature. In the case of ants, both negative (decrease) and positive (increase) effects of higher temperatures, dependent on species, have been reported (Bertelsmeier et al. 2013). However, one of the more notable stinging ant species, *Solenpsis invicta* (the stinging red ant), has been predicted to see a considerable increase in geographic range—specifically estimated to be approximately 5% within the next three decades and as much as 21% within the next century—in the United States alone, where it was historically introduced as an invasive species (Needleman et al. 2018a). Similarly, the highly aggressive hybridized species commonly known as Africanized honeybees has not only been documented to have spread (over the past several decades) from the neotropics into the United States, but it is projected to continue northern expansion of its range due to warming trends within North America.

Moving the focus from terrestrial to aquatic (specifically ocean) habitats, venomous marine species of both invertebrate and vertebrate taxa have been proposed to be shifting in either range or overall population sizes within oceans due to climate change. Indeed, given the fundamental role of the world's oceans in climate, and thus climate change, it is perhaps not surprising that several marine animals, including toxicologically and medically relevant venomous species, have been found to be affected in this regard.

Arguably, the most compelling example in marine habitats is a number of species of venomous jellyfish, or simply "jellies." These are members of the phylum Cnidaria, which also includes both corals and sea anemones, and Anthozoans their relatives. Jellyfish refer *sensu stricto* to the exclusively marine species of the class Scyphozoa, which are characterized as existing primarily as the motile *medusae* stage of the cnidarian life cycle. This designation, however, is often extended to include other medusoid classes (e.g. Hydrozoa, Cubozoa) and even related phyla (e.g. Ctenophora, or "comb jellies"). Approximately 100 of the 10,000 known species of jellyfish and their allies are associated with human envenomation, including highly venomous (and indeed potentially deadly) representatives, such as several species of box jellyfish, and Irukandji jellyfish (*Carukia barnesi* and *Malo* spp.). Envenomation by jellies occurs via discharge of harpoon-like nematocysts containing venom chemically and toxicologically based on *porins* (which create polypeptide "pores" in cell membranes) and a few similarly cytolytic proteinaceous toxins, which rapidly lead to the painful response associated with stings. In extreme cases, venom exposure can be fatal due to cellular release of potassium that can, in turn, affect both neurophysiology and heart function.

Given their natural history, both the population size and distribution of jellyfish and their phylogenetic allies are well known to be affected by a complex interaction of environmental parameters including temperature, salinity, pH, water currents and weather anomalies, all of which are directly or indirectly linked to climate change. Indeed, a considerable body of evidence suggests that jellyfish are becoming more abundant within their current ranges, and moreover, that geographic ranges of a number of species are expanding globally (Needleman et al. 2018b). Warmer temperatures have led to larger "blooms," or populations, and prolonged reproductive seasons, with jellyfish appearing earlier and persisting longer. Aside from temperature, other consequences of climate change additionally favor larger populations, including reduced pH (resulting from temperature and CO_2 input), specifically owing to tolerance and reduced competition (from pH-sensitive species), and a positive response to extreme weather events. Meanwhile, changes in ocean currents and circulation patterns are expected to lead to expansion and/or shift of the geographic range. Accompanying the increased occurrence of jellyfish, there have been consequent increases in human exposure and medically relevant encounters reported (Oiso et al. 2005; Brotz et al. 2012). A notable study, for example, found that the occurrence of stings by the highly venomous box jelly, *Chironex fleckeri*, and the Irukandji jellyfish species was closely correlated with sea surface temperature (Jacups 2010). Increased awareness by and training of medical practitioners with respect to toxicology and treatment practices associated with jellyfish envenomation will likely be increasingly important given the emergence of this growing health concern.

Furthermore, it is likely—and has been similarly proposed—that other venomous marine invertebrates may be increasing with climate change via related mechanisms (Needleman et al. 2018a, 2018b). One notable example cited in this regard is the observed increase in populations of the venomous crown-of-thorns starfish, which has been shown to respond positively to changes in both temperature and pH owing to many of the similar effects documented for jellyfish (e.g. increased growth rates, reduced competition). It has likewise been proposed that other venomous invertebrates (e.g. the highly toxic blue-ringed octopus) may increase with changes in climate, but at present, this remains to be confirmed by solid quantitative evidence.

A number of venomous marine vertebrates may also be regionally increasing, due to expansion or shift of geographic range, as threats to human health. Numerous (>250) species of venomous species of fish are known. Examples include stingrays, whose stinging barbs not only cause severe physical damage but comprise both enzymatic proteins (e.g. 5-nucleotidase, phosphodiesterase) and serotonin, which facilitate both painful responses and potentially lethal effects. Stingray encounters are the most commonly reported of fish envenomations; although it has been suggested that their frequency in coastal waters will increase with climate change, there is currently a lack of solid quantitative evidence to confirm this. Other venomous fish include a number of pufferfish species, lionfish (*Pterois* spp.), scorpionfish (*Scorpaena* spp.) and stonefish (*Synanceja* spp.). Lionfish, characterized by long venomous spines, are endemic to the South Pacific but have become globally invasive. Moreover, several aspects of their growth, life cycle and distribution are directly linked to temperature, which currently limits their range. At least one study has suggested that climate change will consequently enable expansion of this range—with an estimated increase of about 45% in the Southeastern United States, where they have been introduced as invasive—within the next century. Increased numbers of lionfish in this region, and the Caribbean, have already been accompanied by increased envenomation reports (Needleman et al. 2018b).

24.3.6 Plant Toxins and Climate Change

Rising temperatures and altered precipitation patterns, accompanied by elevated levels of CO_2 vis-á-vis climate change have been well documented to impact plant communities specifically including increased *primary production* (i.e. biomass) and shifts in both distribution (i.e. geographic range) and growth habit and life-cycle (e.g. phenology and flowering). Alongside these ecophysiological changes, the evidence points to possible alterations in biochemistry including toxic secondary metabolites. Compared to microbial contaminants and venomous animals, the contribution of climate change to toxic plants in relation to human health impacts is, however, relatively difficult to discern. This is due to both general avoidance of potentially poisonous plants by humans and rather limited examples of unintentional exposure (e.g. ingestion, contact) to their toxins. In the current absence of sufficient human health data, effects of climate change on toxic plants and possible exposure to their toxins is perhaps best revealed by documented poisoning of nonhuman animals (e.g. domestic pets, livestock, wildlife) that may, in effect, serve as portents of increased risk regarding plant toxins. And, indeed, a growing body of knowledge does suggest that climate change may, for instance, be increasing the toxic potential of plants toward a number of herbivores due to not only increased prevalence but also altered physiology under stress of increased temperature (Dearing 2013; Connors et al. 2017).

With respect to human health, one particularly compelling example is that of the genus *Toxicondendron* including poison ivy (*Toxicondendron radicans*) and related species (Figure 24.10). In a 6-year experimental study, it was found that elevated CO_2 levels both selectively increased biomass production by *T. radicans* (nearly twofold compared to ambient CO_2) and stimulated production of more toxic *unsaturated* variants (Figure 24.9) of the associated irritant, urushiol (Mohan et al. 2006). Approximately 80% of the world's human population is sensitive to urushiol, and particularly these unsaturated (e.g. diene and triene) variants. Accordingly, it is anticipated that increased potential for human exposure will drive increased reports of this already common dermatological affliction. Furthermore, it is possible that effects on growth and toxicity, observed for *T. radicans*, can be extended to other species of the globally distributed genus, *Toxicondendron*, and perhaps more broadly to other members of the family Anacardiaceae (e.g. mango, pistachio, cashew) which similarly produce allergenic oleoresins.

In another example, investigators have suggested that climate change may drive spread of brackenfern (*Pteridium aquilinum*) and consequently exposure to the recognized carcinogen, ptaquiloside, produced by the species (Kollmann et al. 2009). It has been shown that leaching of ptaquiloside can lead to both contamination of groundwater and introduction into the diet particularly via contamination of milk. Even quite low levels of the toxin (≤ 50 ng/L) are carcinogenic.

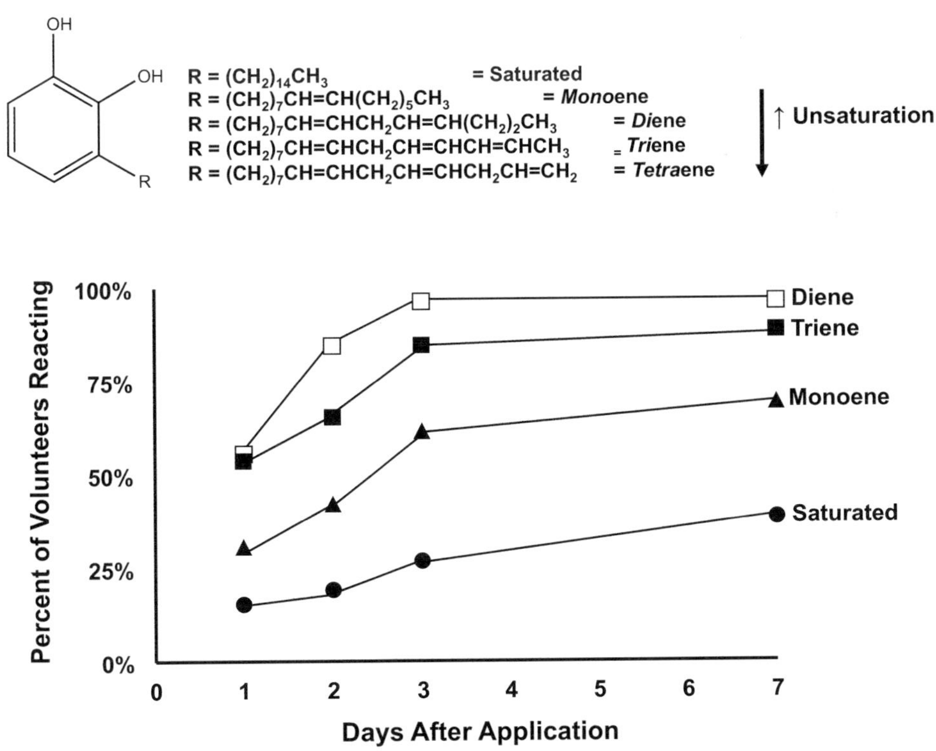

Figure 24.9 Effect of unsaturation (i.e. number of double bond) on urushiol dermatotoxicity. Shown are saturated and increasingly unsaturated (i.e. monoene, diene, triene and tetraene) congeners of C_{15} urushiol, and relative dermatotoxicity of each (measured by number of volunteers reacting to each variant over a 7-day period).

Source: Data adapted from Johnson et al. (1972).

Figure 24.10 Examples of toxic plants. Shown on the left is the brackenfern (*Pteridium aquilinum*), which produces the carcinogen ptaquiloside (inset). On the right is poison ivy (*Toxicondendron* spp.), which produces the irritant urushiol (inset).

Although a link to human health remains to be established, a rather convincing body of evidence (Rasmussen et al. 2008; Micheloud et al. 2017) supports a link between occurrence of brackenfern-derived ptaquiloside to cancer and related disorders among livestock (e.g. bovine enzootic hematuria). Moreover, the well-described potential for ptaquiloside to contaminate groundwater is influenced directly by multiple factors related to climate change including increased precipitation, pH and temperature. Even the production of the toxin has been suggested to be under climate control (Kollmann et al. 2009). When combined with potential increases in primary production (i.e. more brackenfern biomass), this emerging body of knowledge suggests a possible impact on human health via increased contribution of the toxin to cancers.

24.4 CLIMATE CHANGE AND EXPOSURE RISK

Alongside the contribution of human activity to the greenhouse gases that fuel climate change, innumerable chemical products from many of the same anthropogenic sources—including the energy, agriculture and industrial sectors—can themselves act as environmental toxicants. Climate change, in turn, has been proposed to increase *risk of exposure* to a wide range of both these anthropogenic toxicants and naturally occurring biotoxins (discussed in the previous sections). The role of climate change and its many consequences, with respect to increasing future risk of human exposure, largely remains unquantified at present, but a growing number of assessments indicate a potentially significant contribution (Noyes et al. 2009; LaKind et al. 2016; Doherty et al. 2017).

One area that has received considerable attention in this regard has been increased exposure to airborne environmental pollutants including gases, aerosols and particulates that, in some cases, contribute (as greenhouse gases) to climate change, or are emitted alongside greenhouse gases but can also themselves act directly as toxicants. Of these, ozone (O_3) and particulate matter (PM) have been perhaps best studied, and thus most clearly linked to adverse human health effects (Doherty et al. 2017). Within the *troposphere* (the lowest level of the Earth's atmosphere), O_3 is produced from other greenhouse gas "precursors" (e.g. methane, NO_x), and itself is approximately 1000 times more active, as a greenhouse gas, compared to CO_2. Ozone, however, is directly linked to adverse human health effects, including respiratory and cardiovascular illness, through its considerable oxidative potential in relation to numerous cellular and biochemical pathways (Pryor et al. 1995).

Particulate matter, on the other hand, includes fine aerosolized particles that are typically classified based on size (e.g. <2.5 μm [$PM_{2.5}$] and ≤10 μm [PM_{10}]), and are products of both human activities (e.g. combustion of fossil fuels, incinerators, agricultural burning, industrial emissions) and natural processes ranging from fires and volcanic emissions to mineral dusts. The contribution of the latter, however, is also likely impacted, to some extent, by human activities and climate change (e.g. increased dusts associated with desertification, forest fires). Anthropogenic PM globally represents only about 10% of all aerosols; however, these have been clearly linked to profound impacts on human health including short- and long-term cardiovascular mortality and morbidity, adverse birth outcomes and childhood respiratory disease. The molecular mechanism of PM toxicity remains unclear, but demonstrated cellular toxicity (evidenced by cell death) is generally thought to involve oxidative stress, DNA damage and inflammatory pathways (Peixoto et al. 2017). Particulate matter is, moreover, associated with adsorption, and thus transport, of known toxicants that may contribute to adverse health impacts: a globally important example is *polycyclic aromatic hydrocarbons* (PAHs) produced by combustion (e.g. fossil fuels, burning) and, indeed, a wide range of human activities. Ubiquitous in the environment, PAHs are transported by multiple mechanisms including aerosolization via PM, and moreover they have been clearly linked to carcinogenicity, cardiovascular disease and developmental toxicity (Cho et al. 2018). Illustrating the severity of the possible human health risks of these airborne toxicants, one study (Fann et al. 2012) estimated PM and O_3 were responsible for 130,000 and 4700 deaths, respectively, in a single year. And the most recent reports by WHO (2018) estimate that 93% of children in the world are exposed to polluted air which threatens their health, and that an estimated 600,000 children died from respiratory infection (related to air pollution) in 2016 alone.

Risk of exposure to toxicants such as O_3 and PM is not only controlled by levels of emissions but also transformation, transport and deposition, which are in turn directly or indirectly affected by climate and weather. In the absence of required long-term datasets, the deductive argument has been made that emission levels of both O_3 and PM may increase in the environment owing to increased energy consumption, including combustion of fossil fuels, and adaptive practices within industrial, agricultural and household sectors; for example, higher demand for cooling and refrigeration (emission of hydrofluorocarbons) and increased use of nitrogen-based fertilizers (and consequently higher levels of NO_x as precursors of O_3) to meet the agricultural challenges of changing climate. On the other hand, a rather solid and growing body of evidence supports a role of climate change in affecting transformation of airborne pollutants. With respect to ozone, for example, elevated temperature is well documented to increase levels through photochemical processes that generate O_3 from NO_x and volatile organic carbon (VOC) precursors. Annually, this phenomenon is evidenced by seasonal peaks during the summer months, but recent studies have suggested this "high-ozone season" may in fact be extending into later months due to climate change (Zhang and Wang 2016).

That said, in addition to increased levels (via emission and/or transformation of atmospheric gases), it is known that meteorological patterns associated with climate change may influence transport and deposition, and consequently exposure risk. For O_3, for example, transport and deposition are determined by temperature and air circulation patterns. Hot, high-pressure systems can increase deposition (commonly referred to as "stagnation") of so-called ground-level ozone, which is the most immediate exposure route for human health. Interestingly, although enrichment of ground-level O_3 due to stagnation tends to be higher in cities, subsequent wind dispersion tends to transport this ozone to outlying suburban and rural areas and may explain health impacts in these outlying areas (Figure 24.11; Tibbets 2015). Similarly, PM deposition (and, as a consequence, exposure risk) is increased by both stagnation and *air inversion* events (vertically "inverted" increases of temperature with elevation) that are responsible for exposure to a wide range of air pollutants—and which have been suggested by recent data (Fei et al. 2018) to be intensifying with changing climate.

Despite relatively little quantitative evidence (compared to air pollution), it has been increasingly, and perhaps unsurprisingly, proposed that climate change may potentiate risk of exposure to a wide range of pollutants by way of increased mobilization and/or transport through water (Scheringer 2009; Bogdal et al. 2010; Grannas et al. 2013; Milner et al. 2017). A comprehensive discussion of the list of environmental toxicants, in this regard, is well beyond the scope of this chapter, but they include numerous examples from agriculture (pesticides, pharmaceuticals), mining (toxic metals), industry (*numerous* persistent organic pollutants), energy and transportation (PAHs, toxic metals) and household (pharmaceuticals, personal care products) sectors. Of particular note is a long list of *persistent organic pollutants* (POPs), including numerous pesticides and industrially discharged chemical wastes such as polychlorinated biphenyls (PCBs), polybrominated and polychlorinated diphenyl ethers (PBDE/PCDE), dioxins and perfluorinated compounds (PFCs). Despite being, in many cases, effectively banned, restricted or otherwise no longer used, these POPs evade degradation (owing to chemical stability) and thus persist as so-called legacy pollutants in the environment. Although they have been often associated with regional environmental contaminations, their mobilization and global transport is now documented (Scheringer 2009). Their future fate and transport in the environment, particularly in relation to climate change, largely remains to be seen.

With respect to mobilization of toxicants into the environment, the Earth's *cryosphere* represents a globally important reservoir, on decadal and century scales, for a wide range of toxic environmental pollutants (Figure 24.11). Operationally defined as *all* places on Earth where water exists in its frozen state (as ice and snow), the global cryosphere includes, in particular, expansive high-latitude (polar) continental ice sheets/caps, glaciers and permafrost, as well as frozen regions of the polar oceans including Arctic ice and "drift ice," and high-altitude (mountain) ice and snow, which collectively encompass over 10% of the planet's surface (and notably represents more than 75% of the world's freshwater). The cryosphere can effectively serve as a "sink" for various pollutants that can be globally transported through either ocean waters or, more commonly, atmospheric deposition (via volatilization or aerosolization and subsequent "cold trapping" in polar climates). And, indeed, toxicants including POPs and heavy metals (e.g. lead, mercury) have been measured at considerable levels in polar ice. With continued melting, however, this reservoir has been increasingly recognized as a major contributor, through consequent *remobilization*, to transport and fate of these pollutants (Bogdal et al. 2010; Grannas et al. 2013; Milner et al. 2017). Illustrating the magnitude of this potential, a recent study found that Arctic permafrost contains mercury (Hg), which is readily transported by atmospheric deposition, at levels 10 times higher than *all* of the Hg emitted into the environment over the past 30 years (Schuster et al. 2018).

One of the first indicators (and currently the best) of the release of environmental pollutants from the cryosphere into the environment is trophic bioaccumulation and biomagnification. A growing number of studies have measured levels of POPs and toxic metals (e.g. Hg) within Arctic, Antarctic and other high-latitude food webs and have suggested possible decadal increases in bioaccumulation levels due to ecological and/or species-specific physiological mechanisms (McKinney et al. 2015; Alava et al. 2018). Such studies have, moreover, identified biomagnification by a number of important commercial fish and seafood species; thus they not only point to remobilization from cryosphere, in general, but also imply a direct (i.e. dietary) link to human health, and a globally heightened risk of exposure to these toxicants by way of the international commercial fishing industry.

Transport of pollutants beyond the local environment of the cryosphere remains to be seen, and is primarily limited by the current lack of requisite long-term datasets, but is expected to occur both through re-release into ocean waters (and subsequent transport via global ocean circulation patterns), and documented (Ma et al. 2011) revolatilization into the atmosphere (Figure 24.11). The potential contribution of polar ice melting to the reintroduction of environmental toxicants is underscored by documented "Arctic amplification," whereby warming of the Arctic has been shown to be nearly twice the global average (Serreze and Barry 2011). Furthermore, although melting of continental Antarctic ice is generally thought to be occurring at a relative slower pace compared to the Arctic, a recent study (IMBIE 2018) found that melting rates in Antarctica have actually tripled over the period 1992–2017. And not only does Antarctica represent a larger fraction of the cryosphere, but it is also a fundamental component of global ocean circulation: specifically, the Southern Ocean provides key mixing of waters for all of the world's major oceans (Figure 24.11; discussed further below). As such, the contribution of remobilization of pollutants from Antarctic melting may be increasingly important in future climatological (and toxicological) scenarios.

This increased remobilization of pollutants will, in turn, be coupled with expected changes in global ocean circulation patterns that are predicted by most future climate models. Perhaps the most prominent, and so far best documented, of these is the apparent slowdown (and possible future shutdown) of the *Atlantic Meridional Overturning Circulation* (AMOC), which is a foundational component of the ocean's larger "conveyor belt" (Figure 24.11). It is driven by *thermohaline circulation* that effectively begins in the North Atlantic, with colder and more saline (due to evaporative cooling and evaporative removal of water at these latitudes), and thus denser, water sinks. Consequent deep circulation moves water southward to the Southern Ocean, where circumpolar mixing and "recharging" of currents (by sinking cold water of the Antarctic) occur, and subsequently continues northward toward the Indian Ocean and North Pacific where it eventually

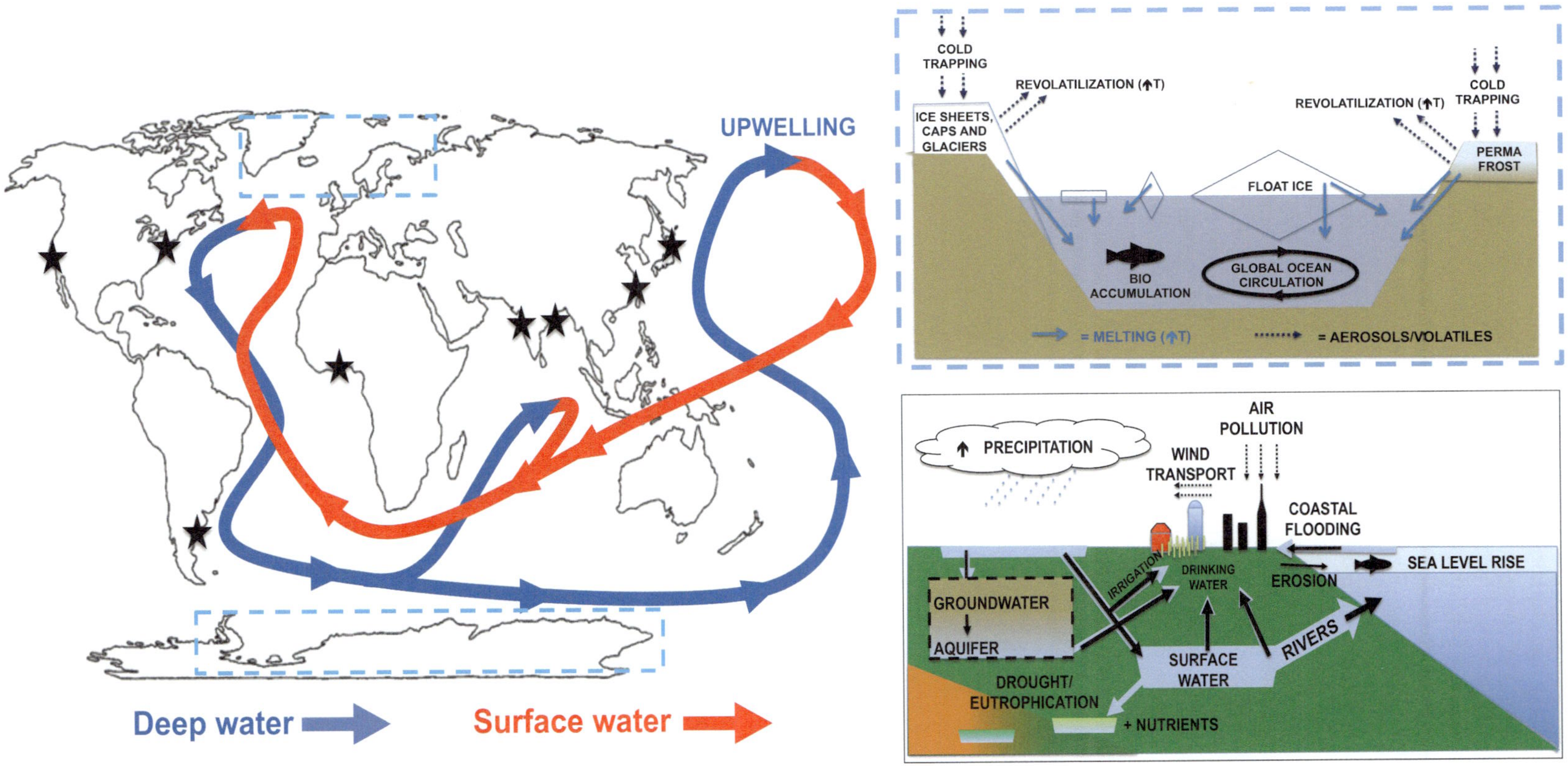

Figure 24.11 Global and local routes of environmental pollutants relevant to human health. Global circulation by the so-called conveyor belt (shown on left) is driven by sinking water in the North Atlantic, followed by subsequent transport as "deep water" (blue line) carrying chemical constituents to the Indian and Pacific Oceans (via Antarctic "common water" of the Southern Ocean), where it upwells and circulates back as surface water (red line) to the Atlantic Ocean. In high latitudes (shown by blue dashed boxes and upper right inset), the cryosphere including ice caps/sheets, floating ice and permafrost can accumulate pollutants by "cold trapping," and following warming and melting can remobilize into water (blue lines/arrows) where they can be bioaccumulated locally, and join global ocean circulation, as well as through revolatilization (dashed black lines) and subsequent atmospheric recirculation. Also shown (by stars) are 8 of the 10 most populated cities in the world that are situated in coastal areas. In coastal areas and inland, various routes can deposit and mobilize pollutants (shown in the lower right inset). High precipitation (via surface water and transport to coastal areas, e.g., river input) and sea level rise can conspire to reintroduce pollutants into coastal waters. Both surface and groundwater can introduce pollutants into drinking and irrigation waters. Drought (e.g. reduced drinking and irrigation water sources), coupled with eutrophication (i.e. increased nutrients in waters), can exacerbate human exposure to pollutants. Meanwhile, meteorological patterns (e.g. inversions) can deposit air pollution into urban environments, and transport these to outlying (i.e. rural and suburban) areas.

upwells. After upwelling to the surface, water returns (southward and westward) from the North Pacific to the Atlantic and rejoins (via the Gulf Stream) the AMOC. Water masses following this so-called conveyor belt are imprinted by chemical composition of both their source waters and by chemical constituents "collected" en route by downwelling/convection, particle sinking and mixing. As such, the AMOC is not only fundamental to the hydrology of the ocean in relation to regulation of temperature (including about 25% of the Earth's heat flux), transport of nutrients and carbon cycling, but it also potentially represents a means of global circulation for environmental pollutants.

Although data remains scarce, due in part to difficulty in sampling, several recent measurements of POPs have suggested that deep ocean water may accommodate considerable levels of these chemically persistent pollutants, and it has been proposed, in fact, that it may represent an effective "tracer" of deep water circulation (Lohmann et al. 2006; Ma et al. 2015; Sun et al. 2016). Interestingly, the return path of the conveyor belt (via surface water from the North Pacific to the North Atlantic) has been recently suggested to transport microplastics to the extreme polar Arctic (Cózar et al. 2017). Although toxicity of microplastics remains largely uncharacterized, and primarily associated with ecotoxicity at present (rather than human health effects), a growing number of studies suggest that microplastics may adsorb various toxicants relevant to human health, including a wide range of organic and metal pollutants (Bakir et al. 2012, 2014; Brennecke et al. 2016; Munier and Bendell 2018). With respect to a contribution to human exposure and health impacts, the chemical stability (and thus the environmental persistence) of these pollutants (e.g. POPs) suggests that despite being mostly phased out in terms of production and industrial use, when coupled to deposition and global circulation, they may represent a continued concern in the future.

On this note, a growing body of evidence suggests that climate change is slowing down the AMOC; this is attributed to decreased salinity in the North Atlantic (due to melting of the Greenland ice sheet and "freshening" of Arctic and North Atlantic waters). These studies have measured unprecedented weakening in the AMOC over decadal to century time scales in the past 1000 years (e.g. including a 15% reduction since the mid-20th century; Caesar et al. 2018; Thornalley et al. 2018). The effect of this observed, and future, slow down (or possible shutdown) of this fundamental component of ocean circulation with respect to environmental pollutants—in addition, of course, to its possible role in global climate change—remains to be seen.

Aside from the global picture, altered hydrology due to climate change will likely increase exposure risk at "local" levels where human society and environment interface. Altered precipitation patterns and sea level rise, in particular, are expected to increase human exposure to environmental toxicants through extensive interactions between humans and aquatic systems, both direct (i.e. drinking water, recreational exposure) and indirect (e.g. irrigation of agricultural crops, aquaculture and fisheries).

Among anticipated scenarios, sea-level rise specifically coupled to more frequent and extreme weather events (particularly precipitation) is expected to increase coastal flooding (Figures 24.2 and 24.11) and consequently risk of exposure to toxic pollutants. The elevated potential for human exposure in this regard is highlighted by the fact that approximately 40% of the Earth's population lives in densely populated coastal regions, including 8 of the 10 most populated cities in the world (Figure 24.11). Alongside many projected economic impacts of coastal flooding, a growing number of studies have identified elevated outflow of toxic pollutants including toxic heavy metals (Santschi et al. 2001; Saniewska et al. 2014; Coates-Marnane et al. 2016; Kwasigroch et al. 2018); POPs and pesticides (Santschi et al. 2001; Moreno-Gonzalez et al. 2013; Leon et al. 2017; Moreno-Gonzalez and Leon 2017; Reoyo-Prats et al. 2017); PAHs (Kuo et al. 2014; Leon et al. 2017); and others (e.g. disinfection by-products; Ruecker et al. 2017) into coastal waters following major precipitation events and the consequent flooding. Introduction of terrigenous pollutants into coastal waters is largely thought to occur via sediments that are transported primarily through riverine input. One such alternative route is coastal erosion, whereby physical abrasion of coastal landforms (e.g. cliffs, beaches) remobilizes pollutants into coastal waters. Indeed, shoreline erosion is widely regarded as one of the key consequences associated with climate change due to the convergence of sea level rise and extreme weather events (e.g. cyclones) in coastal environments. Supporting this, a recent study showed that coastal erosion was responsible for a small (~5%, compared to 80% riverine input), yet likely significant source of weather-based introduction of Hg into coastal waters (Beldowska et al. 2016). Regardless of the route of introduction into coastal waters, increased tidal flooding due to sea level rise redirects sediments (and pollutants) toward coastal habitats, including estuarine (e.g. embayments, lagoons) and nearshore waters. Introduction of pollutants into coastal and nearshore marine waters is thereby expected to potentiate risk through both ambient and recreational exposure and, perhaps more importantly, consequent contamination of coastal fisheries. Indeed, given the both local and global importance of these fisheries and, moreover, the extensive global commercial distribution of fish and seafood, coastal flooding as an emerging source of pollutants represents a worldwide health concern.

Upstream from coastal environments, precipitation associated with increasingly frequent extreme weather events is anticipated to amplify input of a wide range of both pollutants and nutrients (as fuel for eutrophication and potentially toxic blooms; see previous section) into watersheds including rivers, lakes and wetlands, and aquifers (Figure 24.11). Surface waters are likely to be most directly affected by release or leaching of toxic pollutants into the watershed. On the other hand, transport of chemically diverse toxicants into groundwater is complicated by numerous interactions between water and soils in relation to mobility and chemical transformations (i.e. degradations) and is highly dependent upon on a range of factors including chemistry of the pollutants—and of the groundwater, surface water and soils (and their interactions)—hydrology, land use and climate (e.g. temperature; Ciszewski and Grygar 2016). Increased mobilization and transport of toxicants within watersheds is likely to have many consequent implications for human exposure risk, including contamination of drinking and recreation water, as well as crop plants (by way of irrigation waters) and inland fisheries.

The most obvious ramification of watershed contamination is the potential for contamination of drinking water by way of both aquifers (groundwater) and surface waters. Although sources of drinking water vary regionally, approximately half of the world's population relies on groundwater, whereas the other half utilizes surface water including rivers, lakes and reservoirs. One recent study (Xiao et al. 2015) of environmentally ubiquitous perfluorinated compounds found that although subsurface soils effectively slowed their mobility, these chemically stable and highly persistent compounds could reach, and subsequently leach into, aquifers where they rapidly mobilize and potentially contaminate drinking water for many years after being discharged (disposed) into the environment. Deterioration of drinking water quality in this regard is likely to be exacerbated by increased reliance on the otherwise globally decreasing sources of clean water as a consequence of climate change, including expanded desertification (via droughts) and salt water intrusion (due to sea level rise) in coastal environments and other anthropogenic inputs including accelerated freshwater eutrophication (and presence of toxic algae and cyanobacteria) due to nutrient inputs.

At the same time, a number of studies have suggested potential contamination of irrigation waters—or, alternatively, deposition of pollutants into agricultural soils—by flooding that in turn represents a route of dietary exposure through bioaccumulation of toxicants by food crops (Ciesielczuk et al. 2014). Although extreme weather events (which are predicted, in most future climate scenarios, to be more frequent and/or intense) are associated with elevated precipitation levels, models of future climate change suggest widespread uncertainty in precipitation patterns including increased drought. And, as such, adaptive agricultural practices will increasingly rely on irrigation for crops, and thus, possibly increased potential for exposure, uptake and bioaccumulation of toxicants by crops (and consequent agricultural products) through use of contaminated groundwater. On the same note, adaptive agricultural practices will likely include increased use of pesticides and fertilizers (nutrients) that will directly or indirectly (e.g. by fueling toxic algal blooms) increase levels of toxicants in the local environment. Similar impacts via direct bioaccumulation/biomagnification of pollutants, or increased nutrient inputs, in the contexts of inland freshwater fisheries and aquaculture have been suggested (Jacobs et al. 2002).

Finally, alongside increased vulnerability due to risk of exposure, emerging (albeit currently very limited) evidence suggests that climate change may increase physiological *susceptibility* to environmental toxins (Noyes et al. 2009; Holmstrup et al. 2010; Hooper et al. 2013). Most insight in this regard is derived from ecotoxicology data, which specifically measured interactive effects (between climate and chemical stressors) among invertebrate and poikilothermic vertebrate taxa (e.g. fish, amphibians). Consideration of these data suggests *synergistic* interactions between climatic parameters (e.g. temperature, humidity, oxygen levels) and various toxic pollutants (ranging from toxic metals to pesticides, PAHs and other POPs) in more than 50% of cases (Holmstrup et al. 2010). Although such studies have almost exclusively focused on animal models of toxicity, particularly phylogenetically "lower" species, many of the same relevant pathways are evolutionarily conserved. Likewise, one might assume that impacts of environmental parameters such as temperature, humidity and water quality (specifically in relation to aquatic species) would be considerably less for humans as compared to wildlife, as they are both capable of physiologically regulating temperature (as homeotherms) and considerably modifying their immediate surroundings with respect to these factors (e.g. air conditioning, constructed shelter). However, this assumption may hold less true for economically developing populations of the world and thereby highlights a potentially important economic disparity with respect to human health in a globally changing climate.

24.5 CONCLUSIONS

Over the past 150 years, following the Industrial Revolution, the Earth's atmosphere has seen an unprecedented increase (>30%) in carbon dioxide, as well as many other "greenhouse gases," from human activity (e.g. fossil fuel combustion, industrial discharge, agriculture), accompanied by well-documented attendant elevation of global temperature (+1°C) and consequent sea level rise (+200 mm), with far-reaching repercussions including altered atmospheric, hydrological and meteorological patterns. Continued climate change, coupled with a wide range of other—and indeed often interrelated—human impacts (e.g. increased nutrient inputs, deforestation, emission of industrial pollutants, population growth), therefore will represent an extraordinary global shift in the future physical, chemical and biological landscape. Paralleling our understanding of the complex interrelatedness of the diverse physicochemical factors which drive climate change, an emerging body of scientific knowledge points to equally intricate impacts of this change on the planet and its inhabitants, including human health in particular. A central tenet of this latter aspect is a shifting "toxicological frame of reference," including relevant baselines, and emerging health concerns with respect to environmental toxicants themselves and their associated routes (i.e. risk) of exposure.

Deductively, changes in relevant environmental parameters such as increased temperature and uncertain hydrology (due to increased precipitation, and conversely, drought), would be expected to alter, and most likely increase, the prevalence and spatial and temporal distribution of a diverse range of "toxic" (i.e. biotoxigenic, venomous, poisonous) species—and their respective toxins. And indeed, growing evidence and model-based predictions bears this out. Concurrently, however, global and local physicochemical alterations due to climate change are likely to increase potential for mobilization, transport and deposition (and, thus, the risk of exposure to) a wide range of both anthropogenic and naturally occurring toxicants in the environment. Adequate data in this regard, however, remains limited.

With this shifting toxicological framework, the challenge for practitioners will be effective measurement, including both monitoring and long-term datasets regarding prevalence and exposure risk, and investigation of the possible adverse human health impacts, including chronic and subacute effects of toxicants in particular. With respect to monitoring, emphasis on relevant human habitats and *sectors* including coastal populations where both prevalence (e.g. algal/cyanobacterial toxins) and risk of exposure to environmental pollutants are elevated, and agriculture and fishery/aquaculture industries (i.e. food safety), will be particularly critical. On the most immediate (i.e. decadal) scale, increases in various biotoxins have been reported; not only does this demonstrate the effectiveness of monitoring efforts, but it also provides the first evidence of "short-term" effects of climate change on these toxicants. On longer time scales (i.e. century or more), however, understanding the potential of changes in global atmospheric and ocean circulation (mobilization, transport and deposition) with respect to environmental toxicants represents a significantly more challenging endeavor, and to date it largely remains to be documented by sufficiently long-term datasets. However, despite this lack in the body of toxicological evidence, a growing body of research does suggest that a myriad of toxicants do persist in the environment (many despite being phased out), and with changes in global (i.e. atmospheric, ocean) and local (i.e. hydrological) circulation patterns, they will potentially (if not likely) pose human health concerns in the (possibly) near future. Alongside, assessment of risk (due to either increased prevalence, or otherwise elevated potential for exposure), studies aimed at understanding the adverse health impacts will be critical, including increased recognition and documentation of toxicity (e.g. biomarkers, epidemiology) and elucidation of less apparent long-term (e.g. cancer, neurodegeneration, immunotoxicity, reproductive and developmental) toxicity.

Given the complex nature of climate change and its projected impacts on human health, the road forward will accordingly require a highly interdisciplinary expertise and approach which will span—and, moreover, integrate—toxicology including chemical (e.g. analytical chemistry), molecular (i.e. DNA/RNA/protein), cellular and physiological aspects, public health sciences including epidemiologists and clinical practitioners, and of course, environmental sciences including oceanography/limnology, climatology and other relevant earth sciences. Given the clear contribution of human activity and the consequent role of economics and socioeconomics with respect to both climate change and its many implications (including toxicological ones), incorporation of policymakers will be uniquely relevant to addressing what is expected to be a continued interaction between climate change and human health.

BIBLIOGRAPHY

Abal, P., M. C. Louzao, T. Suzuki, et al. 2018. Toxic action reevaluation of okadaic acid, dinophysistoxin-1 and dinophysistoxin-2: Toxicity equivalency factors based on the oral toxicity study. *Cell Physiol Biochem* 49: 743–57.

Alava, J. J., A. M. Cisneros-Montemayor, U. R. Sumaila, and W. W. L. Cheung. 2018. Projected amplification of food web bioaccumulation of MeHg and PCBs under climate change in the Northeastern Pacific. *Sci Rep* 8: 13460.

Allasane-Kpembi, I., O. Puel, P. Pinton, A. M. Cossalter, T. C. Chou, and I. P. Oswald. 2017. Co-exposure to low doses of the food contaminants deoxynivalenol and nivalenol has a synergistic inflammatory effect on intestinal explants. *Arch Toxicol* 91: 2677–87.

Alshannaq, A., and J.-H. Yu. 2017. Occurrence, toxicity, and analysis of major mycotoxins in food. *Int J Environ Res Public Health* 14: 632.

Aráoz, R., J. Molgo, and N. Tandeu de Marsac. 2010. Neurotoxic cyanobacterial toxins. *Toxicon* 56: 813–28.

Arrhenius, S. 1896. On the influence of carbonic acid in the air upon the temperature of the ground. *Phil Mag J Sci* 41: 237–76.

Arunchalam, C., and F. M. Doohan. 2013. Trichothecene toxicity in eukaryotes: Cellular and molecular mechanisms in plants and animals. *Toxicol Lett* 217: 149–58.

Bakir, A., S. J. Rowland, and R. C. Thompson. 2012. Competitive sorption of persistent organic pollutants onto microplastics in the marine environment. *Mar Pollut Bull* 64: 2782–9.

Bakir, A., S. J. Rowland, and R. C. Thompson. 2014. Enhanced desorption of persistent organic pollutants from microplastics under simulated physiological conditions. *Environ Pollut* 185: 16–23.

Basu, A., A. Kozikowski, and J. S. Lazo. 1992. Structural requirements of lyngbyatoxin A for activation and downregulation of protein kinase C. *Biochemistry* 31: 3824–30.

Bates, S. S., K. A. Hubbard, N. Lundholm, M. Montresor, and C. P. Leaw. 2018. *Pseudo-nitzschia, Nitzschia* and domoic acid: New research since 2011. *Harmful Algae* 79: 3–43.

Battilani, P., P. Toscano, H. J. Van der Fels-Klerx, et al. 2016. Aflatoxin B1 contamination in maize in Europe increases due to climate change. *Sci Rep* 6: 24328.

Beldowska, M., A. Jedruch, L. Leczynski, D. Saniewska, and U. Kwasigroch. 2016. Coastal erosion as a source of mercury into the marine environment along the Polish Baltic shore. *Environ Sci Pollut Res Int* 23: 16372–82.

Bertelsmeier, C., G. M. Luque, and F. Courchamp. 2013. Increase in quantity and quality of suitable areas for invasive species as climate changes. *Conserv Biol* 27: 1458–67.

Bogdal, C., D. Nikolic, M. P. Lüthi, U. Schenker, M. Scheringer, and K. Hungerbühler. 2010. Release of legacy pollutants from melting glaciers: Model evidence and conceptual understanding. *Environ Sci Technol* 44: 4063–9.

Bradley, W. G., and D. C. Mash. 2009. Beyond Guam: The cyanobacterial/BMAA hypothesis of the cause of ALS and other neurodegenerative disease. *Amyotroph Later Scler* 10(Suppl 2): 7–20.

Brand, L. E., and A. Compton. 2007. Long-term increase in *Karenia brevis* abundance along the Southwest Florida Coast. *Harmful Algae* 6: 232–52.

Brennecke, D., B. Duarte, F. Paiva, I. Cacador, and J. Canning-Clode. 2016. Microplastics as vector for heavy metal contamination from the marine environment. *Estuar Coast Shelf Sci* 178: 189–95.

Brotz, L., W. W. L. Cheung, K. Kleisner, E. Pakhomov, and D. Pauly. 2012. Increasing jellyfish populations: Trends in large marine ecosystems. *Hydrobiologia* 690: 3–20.

Bui-Klimke, T. R., and F. Wu. 2015. Ochratoxin A and human health risk: A review of the evidence. *Crit Rev Food Sci Nutr* 55: 1860–9.

Caesar, L., S. Rahmstorf, A. Robinson, G. Feulner, and V. Saba. 2018. Observed fingerprint of a weakening Atlantic Ocean overturning circulation. *Nature* 556: 191–6.

Caller, T., P. Henegan, and E. Stommel. 2018. The potential role of BMAA in neurodegeneration. *Neurotoxicol Res* 33: 222–6.

Carmichael, W. W., S. M. Azevedo, J. S. An, et al. 2001. Human fatalities from cyanobacteria: Chemical and biological evidence for cyanotoxins. *Environ Health Perspect* 109: 663–8.

Carmichael, W. W., and G. L. Boyer. 2016. Health impacts from cyanobacteria harmful algal blooms: Implications for the North American Great Lakes. *Harmful Algae* 54: 194–212.
Carmichael, W. W., P. R. Gorham, and D. F. Biggs. 1977. Two laboratory case studies on the oral toxicity to calves of the freshwater cyanophyte (blue-green algae) *Anabaena flos-aquae* NRC-44–1. *Can Vet J* 18: 71–5.
Cho, C. C., W. Y. Hsieh, C. H. Tsai, C. Y. Chen, H. F. Chang, and C. S. Lin. 2018. In vitro and in vivo experimental studies of $PM_{2.5}$ on disease progression. *Int J Environ Res Public Health* 15: pii: E1380.
Church, J. A., and N. J. White. 2011. Sea-level rise from the late 19th to the early 21st century. *Surv Geophys* 32: 585–602.
Ciesielczuk, T., G. Kusza, J. Poluszynska, and K. Kochanowska. 2014. Pollution of flooded arable soils with heavy metals and polycyclic aromatic hydrocarbons (PAHs). *Water Air Soil Pollut* 225: 2145.
Ciszewski, D., and T. M. Grygar. 2016. A review of flood-related storage and remobilization of heavy metal pollutants in river systems. *Water Air Soil Pollut* 227: 239.
Coates-Marnane, J., J. Olley, J. Burton, and A. Grinham. 2016. The impact of a high magnitude flood on metal pollution in a shallow subtropical estuarine embayment. *Sci Total Environ* 569–570: 716–30.
Codd, P. A., S. A. Banack, S. J. Murch, et al. 2005. Diverse taxa of cyanobacteria produce beta-N-methylamino-L-alanine, a neurotoxic amino acid. *Proc Nat Acad Sci USA* 102: 5074–8.
Connors, P. K., J. R. Malenke, and M. D. Dearing. 2017. Ambient temperature-mediated changes in hepatic gene expression of a mammalian herbivore (*Neotoma lepida*). *Mol Ecol* 26: 4322–38.
Coufal-Majewski, S., K. Stanford, T. McAllister, et al. 2016. Impacts of cereal ergot in food animal production. *Front Vet Sci* 3: 15.
Cózar, A., E. Martí, C. M. Duarte, et al. 2017. The Arctic Ocean as a dead end for floating plastics in the North Atlantic branch of the thermohaline circulation. *Sci Adv* 3: e1600582.
Cramer, K. L., and A. V. Maywright. 2008. Cold temperature tolerance and distribution of the brown recluse spider *Loxosceles reclusa* in Illinois. *J Arachnol* 36: 136–9.
Crews, C. 2015. Analysis of ergot alkaloids. *Toxins* 7: 2024–50.
Dawson, R. M. 1998. The toxicology of microcystins. *Toxicon* 36: 953–62.
Dearing, M. D. 2013. Temperature-dependent toxicity in mammals with implications for herbivores: A review. *J Compar Physiol B* 183: 43–50.
Deeds, J. R., J. H. Landsberg, S. M. Etheridge, G. C. Pitcher, and S. W. Longan. 2008. Non-traditional vectors for paralytic shellfish poisoning. *Mar Drugs* 6: 308–48.
De Jesus, C. L., A. Bartley, A. Z. Welch, and J. P. Berry. 2017. High incidence and levels of ochratoxin A in wines sourced from the United States. *Toxins* 10: pii: E1.
De la Cruz, A. A., A. Hiskia, T. Kaloudis, et al. 2013. A review on cylindrospermopsin: The global occurrence, detection, toxicity and degradation of a potent cyanotoxin. *Environ Sci Process Impact* 15: 1979–2003.
Deng, J., L. Zhao, N. Y. Zhang, et al. 2018. Aflatoxin B_1 metabolism: Regulation by phase I and II metabolizing enzymes and chemoprotective agents. *Mutat Res* 778: 79–89.
De Oliveira Filho, J. W. G., M. T. Islam, E. S. Ali, et al. 2017. A comprehensive review on biological properties of citrinin. *Food Chem Toxicol* 110: 130–41.
Dessler, A. E., M. R. Schoeberl, T. Wang, S. M. Davis, and K. H. Rosenlof. 2013. Stratospheric water vapor feedback. *Proc Nat Acad Sci USA* 110: 18087–91.
Devlin, J. P., O. E. Edwards, P. R. Gorham, N. R. Hunter, R. K. Pike, and B. Stavric. 1977. Anatoxin-a, a toxic alkaloid from *Anabaena flos-aquae* NRC-44h. *Can J Chem* 55: 1367–71.
Dickey, R. W., and S. M. Plakas. 2010. Ciguatera: A public health perspective. *Toxicon* 56: 123–36.
Dittmann, E., D. P. Fewer, and B. Neilan. 2013. Cyanobacterial toxins: Biosynthetic routes and evolutionary routes. *FEMS Microbiol Rev* 37: 23–43.
Doherty, R. M., M. R. Heal, and F. M. O'Connor. 2017. Climate change impacts on human health over Europe through its effect on air quality. *Environ Health* 16(Suppl 1): 118.
Dominguez, H. J., B. Paz, A. H. Daranas, M. Norte, J. M. Franco, and J. J. Fernandez. 2010. Dinoflagellate polyether within the yessotoxin, pectenotoxin and okadaic acid toxins groups: Characterization, analysis and human health implications. *Toxicon* 56: 191–217.
Eaton, D. L., and E. P. Gallagher. 1994. Mechanisms of aflatoxin carcinogenesis. *Annu Rev Pharmacol Toxicol* 34: 135–72.
Escrivá, L., G. Font, and L. Manyes. 2015. In vivo toxicity studies of fusarium mycotoxins in the last decade: A review. *Food Chem Toxicol* 78: 185–206.

Etheridge, S. M. 2010. Paralytic shellfish poisoning: Seafood safety and human health perspectives. *Toxicon* 56: 108–22.
Fann, N., A. D. Lamson, S. C. Anenberg, K. Wesson, D. Risley, and B. J. Hubbell. 2012. Estimating the national public health burden associated with exposure to ambient PM2.5 and ozone. *Risk Anal* 32: 81–95.
Fei, J., J. P. Evans, A. Di Luca, et al. 2018. Projected change in characteristics of near surface temperature inversion for southeast Australia. *Clim Dyn* 52: 1487–503.
Fleming, L. E., B. Kirkpatrick, L. C. Backer, et al. 2011. Review of Florida Red Tide and human health effects. *Harmful Algae* 10: 224–33.
Flewelling, L. J., J. P. Naar, J. P. Abbot, et al. 2005. Red tides and marine mammal mortalities. *Nature* 435: 755–6.
Fraeyman, S., S. Croubels, M. Devreese, and G. Antonissen. 2017. Emerging *Fusarium* and *Alternaria* mycotoxins: Occurrence, toxicity and toxicokinetics. *Toxins* 9: pii: E228.
Friedman, M. A., M. Fernandez, L. C. Backer, et al. 2017. An updated review of ciguatera fish poisoning: Clinical, epidemiological, environmental, and public health management. *Mar Drugs* 15: 72.
Fujiki, H., and M. Suganuma. 2011. Tumor promoters—Microcystin-LR, nodularin and TNF-alpha and human cancer development. *Anticancer Agents Med Chem* 11: 4–18.
Furey, A., S. O'Doherty, K. O'Callaghan, M. Lehane, and K. J. James. 2010. Azaspiracid poisoning (AZP) toxins in shellfish: Toxicological and health considerations. *Toxicon* 56: 173–90.
Gorham, P. R. 1960. Toxic waterblooms of blue-green algae. *Can Vet J* 1: 235–45.
Grannas, A. M., C. Bogdal, K. J. Hageman, et al. 2013. The role of the global cryosphere in the fate of organic contaminants. *Atmos Chem Phys* 13: 3271–301.
Griffiths, D. J., and M. L. Saker. 2003. The Palm Island mystery disease 20 years on: A review of research on the cyanotoxin cylindrospermopsin. *Environ Toxicol* 18: 78–93.
Gutierrez, J. M., D. Williams, H. W. Fan, and D. A. Warrell. 2010. Snakebite envenoming from a global perspective: Towards an integrated approach. *Toxicon* 56: 1223–35.
Hallegraeff, G. M. 1993. A review of harmful algal blooms and their apparent global increase. *Phycol Rev* 32: 79–99.
Hallegraeff, G. M. 2010. Ocean climate change, phytoplankton community responses, and harmful algal blooms: A formidable predictive challenge. *J Phycol* 46: 220–35.
Haq, M., N. Gonzalez, K. Mintz, et al. 2016. Teratogenicity of ochratoxin A and the degradation product, ochratoxin a, in the zebrafish (*Danio rerio*) embryo model of vertebrate development. *Toxins* 8: 40.
Hauser, C. 2018. Algae bloom in Lake Superior raises worries on climate change and tourism. *New York Times*, August 29.
Hirsch, A. I., A. M. Michalak, L. M. Bruhwiler, W. Peters, E. J. Dlugokencky, and P. P. Tans. 2006. Inverse modeling estimates of the global nitrous oxide surface flux from 1998–2001. *Glob Biogeochem Cycles* 20: GB1008.
Holmstrup, M., A.-M. Bindesbol, G. J. Oostingh, et al. 2010. Interactions between effects of environmental chemicals and natural stressors: A review. *Sci Total Environ* 408: 3746–62.
Hooper, M. J., G. T. Ankley, D. A. Cristol, L. A. Maryoung, P. D. Noyes, and K. E. Pinkerton. 2013. Interactions between chemical and climate stressors: A role for mechanistic toxicology in assessing climate change risks. *Environ Toxicol Chem* 32: 32–48.
Hu, C., F. E. Muller-Karger, and P. W. Swarzenski. 2006. Hurricanes, submarine groundwater discharge, and Florida's red tides. *Geophys Res Lett* 33: L11601.
IMBIE. 2018. Mass balance of the Antarctic ice sheet from 1992 to 2017. *Nature* 558: 219–22.
Ince, P. G., and G. A. Codd. 2005. Return of the cycad hypothesis—Does the amyotrophic lateral sclerosis/parkinsonism dementia complex (ALS/PDC) of Guam have new implications for global health? *Neuropathol Appl Neurobiol* 31: 345–53.
Intergovernmental Panel on Climate Change. 2007. *Climate Change 2007: The Physical Science Basis. Contribution of Working Group I to the Fourth Assessment Report of the Intergovernmental Panel on Climate Change*, ed. S. Solomon, D. Qin, M. Manning, et al. Cambridge, UK: Cambridge University Press.
Intergovernmental Panel on Climate Change. 2013. Summary for policymakers. In *Climate Change 2013: The Physical Science Basis. Contribution of Working Group I to the Fifth Assessment Report of the Intergovernmental Panel on Climate Change*, eds. T. F. Stocker, D. Qin, G.-K. Plattner, et al. Cambridge, UK: Cambridge University Press.

Jackim, E., and J. Gentile. 1968. Toxins of a blue-green alga: Similarity to saxitoxin. *Science* 162: 915–16.

Jacobs, M. N., A. Covaci, and P. Schepens. 2002. Investigation of selected persistent organic pollutants in farmed Atlantic salmon (*Salmo salar*), salmon aquaculture feed, and fish oil components of the feed. *Environ Sci Technol* 36: 2797–807.

Jacups, S. P. 2010. Warmer waters in the Northern Territory herald an earlier onset to the annual *Chironex fleckeri* stinger season. *Ecohealth* 7: 14–17.

Johnson, R. A., H. Baer, C. H. Kirkpatrick, C. R. Dawson, and R. G. Khurana. 1972. Comparison of the contact allergenicity of the four pentadecylcatechols derived from poison ivy urushiol in human subjects. *J Allergy Clin Immunol* 49: 27–35.

Jones, A. C., L. Gu, C. M. Sorrels, D. H. Sherman, and W. H. Gerwick. 2009. New tricks from ancient algae: Natural products biosynthesis in marine cyanobacteria. *Curr Opin Chem Biol* 13: 16–23.

Karl, T. R., J. M. Melillo, and T. C. Peterson. 2009. *Global Climate Change Impacts in the United States*, 196. Cambridge, UK and New York: Cambridge University Press.

Kensler, T. W., B. D. Roebuck, G. N. Wogan, and J. D. Groopman. 2011. Aflatoxin: A 50-year odyssey of mechanistic and translational toxicology. *Toxicol Sci* 120(Suppl 1): S28–S48.

Kinnear, S. 2010. Cylindrospermopsin: A decade of progress on bioaccumulation research. *Mar Drugs* 8: 542–64.

Kirk, M. D., S. M. Pires, R. E. Black, et al. 2015. World Health Organization estimates of the global and regional disease burden of 22 foodborne bacterial, protozoal, and viral diseases, 2010: A data synthesis. *PLoS Med* 12: e1001921.

Klaric, M. S., D. Rasic, and M. Peraica. 2013. Deleterious effects of mycotoxin combinations involving ochratoxin A. *Toxins* 5: 1965–87.

Kollmann, J., B. W. Strobel, and H. C. B Hansen. 2009. Climate change, invasive species and toxic plant substances in soil and water. *IOP Conf Ser Earth Environ Sci* 6: 302022.

Koszegi, T., and M. Poór. 2016. Ochratoxin A: Molecular interactions, mechanisms of toxicity and prevention at the molecular level. *Toxins* 8: 111.

Krska, R., and C. Crews. 2008. Significance, chemistry and determination of ergot alkaloids: A review. *Food Addit Contam Part A Chem Anal Control Expo Risk Assess* 25: 722–31.

Kupski, L., M. Freitas, D. Ribeiro, E. B. Furlong, and E. Fernandes. 2016. Ochratoxin A activates neutrophils and kills these cells through necrosis, an effect eliminated through its conversion into ochratoxin α. *Toxicology* 368–9: 91–102.

Kwasigroch, U., M. Beldowska, A. Jedruch, and D. Saniewska. 2018. Coast erosion—A "new" land-based source of labile mercury to the marine environment. *Environ Sci Pollut Res Int* 25: 28682–94.

Lake, I. R. 2017. Food-borne disease and climate change in the United Kingdom. *Environ Health* 16(Suppl 1): 117.

Lake, I. R., and G. C. Barker. 2018. Climate change, foodborne pathogens and illness in higher-income countries. *Curr Environ Health Rep* 5: 187–96.

LaKind, J. S., J. Overpeck, P. N. Breysse, et al. 2016. Exposure science in an age of rapidly changing climate: Challenges and opportunities. *J Expo Sci Environ Epidemiol* 26: 529–38.

Leighfield, T. A., N. Muha, and J. S. Ramsdell. 2014. Tissue distribution of amino acid- and lipid-brevetoxins after intravenous administration to C57BL/6 mice. *Chem Res Toxicol* 27: 1166–75.

Leon, V. M., R. Moreno-Gonzalez, V. Garcia, and J. A. Campillo. 2017. Impact of flash flood events on the distribution of organic pollutants in surface sediments from a Mediterranean coastal lagoon (Mar Menor, SE Spain). *Environ Sci Pollut Res Int* 24: 4284–300.

Leroy, B., C. Bellard, N. Dubos, et al. 2014. Forecasted climate and land use changes, and protected areas: The contrasting case of spiders. *Divers Distrib* 20: 686–97.

Li, Y., Z. Wang, R. C. Beier, et al. 2011. T-2 toxin, a trichothecene mycotoxin: Review of toxicity, metabolism, and analytical methods. *J Agric Food Chem* 59: 3441–53.

Liegl, C. A., and M. A. McGrath. 2016. Ergotism: Case report and review of the literature. *Int J Angiol* 25: e8–e11.

Liu, L., and K. S. Rein. 2010. New peptides isolated from *Lyngbya* species: A review. *Mar Drugs* 8: 1817–37.

Lohmann, R., E. Jurado, M. E. Pilson, and J. Dachs. 2006. Oceanic deep water formation as a sink of persistent organic pollutants. *Geophys Res Lett* 33: L12607.

Ma, J., H. Hung, C. Tian, and R. Kallenborn. 2011. Revolatilization of persistent organic pollutants in the Arctic induced by climate change. *Nat Clim Change* 1: 255.

Ma, Y., C. J. Halsall, J. D. Crosse, et al. 2015. Persistent organic pollutants in ocean sediments from the North Pacific to the Arctic Ocean. *J Geophys Res* 120: 2723–35.

Magan, N., A. Medina, and D. Aldred. 2011. Possible climate-change effects on mycotoxin contamination of food crops pre- and postharvest. *Plant Pathol* 60: 150–63.

Malir, F., V. Ostry, A. Pfohl-Leszkowicz, J. Malir, and J. Toman. 2016. Ochratoxin A: 50 years of research. *Toxins* 8: pii: E191.

Marin, S., A. J. Ramos, G. Cano-Sancho, and V. Sanchis. 2013. Mycotoxins: Occurrence, toxicology, and exposure assessment. *Food Chem Toxicol* 60: 218–37.

Massey, I. Y., F. Yang, Z. Ding, et al. 2018. Exposure routes and health effects of microcystins on animals and humans: A mini-review. *Toxicon* 151: 156–62.

McKinney, M. A., S. Pedro, R. Dietz, et al. 2015. A review of ecological impacts of global climate change on persistent organic pollutants and mercury pathways and exposures in arctic marine ecosystems. *Curr Zool* 61: 617–28.

Medina, A., M. K. Gilbert, B. M. Mack, et al. 2017. Interactions between water activity and temperature on the *Aspergillus flavus* transcriptome and aflatoxin B1 production. *Int J Food Microbiol* 256: 36–44.

Medina, A., M. Schmidt-Heydt, A. Rodriguez, R. Parra, R. Geisen, and N. Magan. 2015. Impacts of environmental stress on growth, secondary metabolite biosynthetic gene clusters and metabolite production of xerotolerant/xerophilic fungi. *Curr Genet* 61: 325–34.

Micheloud, J. F., L. A. Colque-Caro, O. G. Martinez, E. J. Gimeno, D. da Silva Freitas Ribeiro, and B. S. Blanco. 2017. Bovine enzootic heamaturia from consumption of *Pteris deflexa* and *Pteris plumula* in northwestern Argentina. *Toxicon* 134: 26–9.

Milner, A. M., K. Khamis, T. J. Battin, et al. 2017. Glacier shrinkage driving global changes in downstream systems. *Proc Natl Acad Sci USA* 114: 9770–8.

Mohan, J. E., L. H. Ziska, W. H. Schlesinger, et al. 2006. Biomass and toxicity response of poison ivy (*Toxicondendron radicans*) to elevated atmospheric CO_2. *Proc Natl Acad Sci USA* 103: 9086–9.

Moreira, C., J. Azevedo, A. Antunes, and V. Vasconcelos. 2013. Cylindrospermopsin: Occurrence, methods of detection and toxicology. *J Appl Microbiol* 114: 605–20.

Moreno-Gonzalez, R., J. A. Campillo, V. Garcia, and V. M. Leon. 2013. Seasonal input of regulated and emerging oceanic pollutants through surface watercourses to a Mediterranean coastal lagoon. *Chemosphere* 92: 247–57.

Moreno-Gonzalez, R., and V. M. Leon. 2017. Presence and distribution of current use pesticides in surface marine sediments from a Mediterranean coastal lagoon (SE Spain). *Environ Sci Pollut Res Int* 24: 8033–48.

Munday, R. 2013. Is protein phosphatase inhibition responsible for toxic effects of okadaic acid in animals? *Toxins* 5: 267–85.

Munier, B., and L. I. Bendell. 2018. Macro and micro plastics sorb and desorb metals and act as a point source of trace metals to coastal ecosystems. *PLoS One* 13: e0191759.

Munkvold, G. P. 2017. Fusarium species and their associated mycotoxins. *Methods Mol Biol* 1542: 51–106.

National Oceanic and Atmospheric Administration (NOAA). 2019. *Global Greenhouse Gas Reference Network: Mean Monthly CO_2 at Mauna Loa, January 2019.* Earth System Research Laboratory, Global Monitoring Division. www.esrl.noaa.gov/gmd/ccgg/trends/index.html (accessed November 26, 2020).

Needleman, R. K., I. P. Neylan, and T. Erickson. 2018a. Environmental and ecological effects of climate change on venomous marine and amphibious species in the wilderness. *Wilderness Environ Med* 29: 343–56.

Needleman, R. K., I. P. Neylan, and T. Erickson. 2018b. Potential environmental and ecological effects of global climate change on venomous terrestrial species in the wilderness. *Wilderness Environ Med* 29: 226–38.

Nishiwaki-Matsushima, R., T. Ohta, S. Nishiwaki, et al. 1992. Liver tumor promotion by the cyanobacterial cyclic peptide toxin microcystin-LR. *J Cancer Res Clin Oncol* 118: 420–4.

Nori, J., P. A. Carrasco, and G. C. Leynaud. 2014. Venomous snakes and climate change: Ophidism as a dynamic problem. *Clim Change* 122: 67–80.

Noyes, P. D., M. K. McElwee, H. D. Miller, et al. 2009. The toxicology of climate change: Environmental contaminants in a warming world. *Environ Int* 35: 971–86.

O'Driscoll, D., Z. Skrabakova, J. O'Halloran, F. N. A. M. van Pelt, and K. J. James. 2011. Mussels increase xenobiotic (azaspiracid) toxicity using a unique bioconversion mechanism. *Environ Sci Technol* 45: 3102–8.

Oiso, N., K. Fukai, M. Ishii, T. Ohgushi, and S. Kubota. 2005. Jellyfish dermatitis caused by *Porpita pacifica*, a sign of global warming? *Dermatitis* 52: 232–3.

Osborne, N. J., P. M. Webb, and G. R. Shaw. 2001. The toxins of *Lyngbya majuscula* and their human and ecological health effects. *Environ Int* 27: 381–92.

Osswald, J., S. Rellán, A. Gago, and V. Vasconcelos. 2007. Toxicology and detection methods of the alkaloid neurotoxin produced by cyanobacteria, anatoxin-a. *Environ Int* 33: 1070–89.

Paerl, H. W., and J. Huisman. 2009. Climate change: A catalyst for global expansion of harmful cyanobacterial blooms. *Environ Microbiol Rep* 1: 27–37.

Paerl, H. W., and V. J. Paul. 2012. Climate change: Links to global expansion of harmful cyanobacteria. *Water Res* 46: 1349–63.

Pal, S., N. Singh, and K. M. Ansari. 2017. Toxicological effects of patulin mycotoxin on the mammalian system: An overview. *Toxicol Res* 6: 765–71.

Paterson, R. R. M., and N. Lima. 2010. How will climate change affect mycotoxins in food? *Food Res Int* 43: 1902–14.

Paterson, R. R. M., A. Venancio, N. Lima, M. Guilloux-Benatier, and S. Rousseaux. 2018. Predominant mycotoxins, mycotoxigenic fungi and climate change related to wine. *Food Res Int* 103: 487–91.

Payros, D., I. Alassane-Kpembi, A. Pierron, N. Loiseau, P. Pinton, and I. P. Oswald. 2016. Toxicology of deoxynivalenol and its acetylated and modified forms. *Arch Toxicol* 90: 2931–57.

Pearson, L., T. Mihali, M. Moffitt, R. Kellmann, and B. Neilan. 2010. On the chemistry, toxicology and genetics of the cyanobacterial toxins, microcystin, nodularin, saxitoxin and cylindrospermopsin. *Mar Drugs* 8: 1650–80.

Peixoto, M. S., M. F. de Oliveira Galvão, and S. R. Batistuzzo de Medeiros. 2017. Cell death pathways of particulate matter toxicity. *Chemosphere* 188: 32–48.

Pitt, J. I., and J. D. Miller. 2017. A concise history of mycotoxin research. *J Agric Food Chem* 65: 7021–33.

Plakas, S. M., and R. W. Dickey. 2010. Advances in monitoring and toxicity assessment of brevetoxins in molluscan shellfish. *Toxicon* 56: 137–49.

Pryor, W. A., G. L. Squadrito, and M. Friedman. 1995. A new mechanism for the toxicity of ozone. *Toxicol Lett* 82–83: 287–93.

Radwan, F. F. Y., and J. S. Ramsdell. 2006. Characterization of *in vitro* oxidative and conjugative metabolic pathways for brevetoxin (PbTx-2). *Toxicol Sci* 89: 57–65.

Radwan, F. F. Y., and J. S. Ramsdell. 2008. Brevetoxin forms covalent DNA adducts in rat lung following intratracheal exposure. *Environ Health Perspect* 116: 930–6.

Rapala, J., A. Robertson, A. P. Negri, et al. 2005. First report of saxitoxin in Finnish lakes and possible associated effects on human health. *Environ Toxicol* 20: 331–40.

Rasmussen, L., D. Lauren, B. Smith, and H. Hansen. 2008. Variation in ptaquiloside content in bracken (*Pteridium esculentum*) in New Zealand. *N Z Vet J* 56: 304–9.

Reguera, B., P. Riobó, F. Rodriguez, et al. 2014. Dinophysis toxins: Causative organisms, distribution and fate in shellfish. *Mar Drugs* 12: 394–461.

Rein, K. S., and R. V. Snyder. 2006. The biosynthesis of polyketide metabolites by dinoflagellates. *Adv Appl Microbiol* 59: 93–125.

Reoyo-Prats, B., D. Aubert, C. Menniti, et al. 2017. Multicontamination phenomena occur more often than expected in Mediterranean coastal watercourses: Study case of the Tet River (France). *Sci Total Environ* 579: 10–21.

Roy, U., L. Conklin, J. Schiller, J. Matysik, J. P. Berry, and A. Alia. 2017. Metabolic profiling of zebrafish (*Danio rerio*) embryos by NMR spectroscopy reveals multifaceted toxicity of beta-methylamino-L-alanine (BMAA). *Sci Rep* 7: 17305.

Ruecker, A., H. Uzun, T. Karanfil, M. T. K. Tsui, and A. T. Chow. 2017. Disinfection byproduct precursor dynamics and water treatability during an extreme flooding event in a coastal blackwater river in southeastern United States. *Chemosphere* 188: 90–8.

Saniewska, D., M. Beldowska, J. Beldowski, and L. Falkowska. 2014. Mercury in precipitation at an urbanized coastal zone of the Baltic Sea (Poland). *Ambio* 43: 871–7.

Santschi, P. H., B. J. Presley, T. L. Wade, B. Garcia-Romero, and M. Baskaran. 2001. Historical contamination of PAHs, PCBs, DDTs, and heavy metals in Mississippi River Delta, Galveston Bay and Tampa Bay sediment cores. *Mar Environ Res* 52: 51–79.

Saupe, E. E., M. Papes, P. A. Selden, and R. S. Vetter. 2011. Tracking a medically important spider: Climate change, ecological niche modeling, and the brown recluse (*Loxosceles reclusa*). *PLoS One* 6: e17731.

Scheringer, M. 2009. Long-range transport of organic chemicals in the environment. *Environ Toxicol Chem* 28: 677–90.
Schuster, P. F., K. M. Schaefer, G. R. Aiken, et al. 2018. Permafrost stores globally significant amount of mercury. *Geophys Res Lett.*
Serreze, M. C., and R. G. Barry. 2011. Processes and impacts of Arctic amplification: A research synthesis. *Glob Planet Change* 77: 85–96.
Shier, W. T., A. C. Shier, W. Xie, and C. J. Mirocha. 2001. Structure-activity relationships for human estrogenic activity in zearalenone mycotoxins. *Toxicon* 39: 1435–8.
Sinha, R., L. A. Pearson, T. W. Davis, M. A. Burford, P. T. Orr, and B. A. Neilan. 2012. Increased incidence of *Cylindrospermopsin raciborskii* in temperate zones—Is climate change responsible? *Water Res* 46: 1408–19.
Smith, M.-C., S. Madec, E. Coton, and N. Hymery. 2016. Natural co-occurrence of mycotoxins in foods and feeds and their *in vitro* combined toxicological effects. *Toxins* 8: 94.
Speijers, G. J., and M. H. Speijers. 2004. Combined toxic effects of mycotoxins. *Toxicol Lett* 153: 91–8.
Steiner, K., L. Zimmermann, B. Hagenbuch, and D. Dietrich. 2016. Zebrafish OATP-mediated transport of microcystin congeners. *Arch Toxicol* 90: 1129–39.
Stockmann-Juvala, H., and K. Savolainen. 2008. A review of the toxic effects and mechanisms of action of fumonisin B1. *Human Exp Toxicol* 27: 799–809.
Sun, C., T. Soltwedel, E. Bauerfeind, D. A. Adelman, and R. Lohmann. 2016. Depth profiles of persistent organic pollutants in the North and Tropical Atlantic Ocean. *Environ Sci Technol* 50: 6172–9.
Svircev, Z., D. Drobac, N. Tokodi, B. Mijovic, G. A. Codd, and J. Meriluoto. 2017. Toxicology of microcystins with reference to cases of human intoxications and epidemiological investigations of exposures to cyanobacteria and cyanotoxins. *Arch Toxicol* 91: 621–50.
Tasoulis, T., and G. K. Isbister. 2017. A review and database of snake venom proteomes. *Toxins* 9: pii: E290.
Tatay, E., S. Espin, A.-J. Garcia-Fernandez, and M.-J. Ruiz. 2018. Estrogenic activity of zearalenone, α-zearalenol and β-zearalenol assessed using the E-screen assay in MCF-7 cells. *Toxicol Mech Methods* 28: 239–42.
Testai, E., S. Scardal, S. Vichi, F. M. Buratti, and E. Funari. 2016. Risk to human health associated with the environmental occurrence of cyanobacterial neurotoxic alkaloids anatoxins and saxitoxins. *Crit Rev Toxicol* 46: 385–419.
Theakston, R. D., and D. A. Warrell. 2000. Crisis in snake antivenom supply for Africa. *Lancet* 16: 2014.
Thomas, P., M. Stephens, G. Wilkie, et al. 1993. (+)-Anatoxin-a is a potent agonist at neuronal nicotinic acetylcholine receptors. *Neurochemistry* 60: 2308–11.
Thornalley, D. J. R., D. W. Oppo, P. Ortega, et al. 2018. Anomalously weak Labrador Sea convection and Atlantic overturning during the past 150 years. *Nature* 556: 227–230.
Tibbets, J. H. 2015. Air quality and climate change: A delicate balance. *Environ Health Perspec* 123: A148–A153.
Todd, E. C. D. 1993. Domoic acid and amnesic shellfish poisoning—A review. *J Food Protection* 56: 69–83.
Twiner, M. J., G. J. Doucette, A. Rasky, X.-P. Huang, B. L. Roth, and M. C. Sanguinetti. 2012. Marine algal toxin azaspiracid is an open-state blocker of hERG potassium channels. *Chem Res Toxicol* 25: 1975–84.
U.S. Environmental Protection Agency. 2018. *Inventory of U.S. Greenhouse Gas Emissions and Sinks 1990–2016 (EPA 430 R-18-003).* www.epa.gov/sites/production/files/2018-01/documents/2018_complete_report.pdf (accessed November 26, 2020).
Ueno, Y., S. Nagata, T. Tsutusumi, et al. 1996. Detection of microcystins, a blue-green algal hepatotoxin, in drinking water sampled in Haimen and Fusui, endemic areas of primary liver cancer in China, by highly sensitive immunoassay. *Carcinogenesis* 17: 1317–21.
Van der Fels-Klerx, H. J., C. Liu, and P. Battilani. 2016. Modelling climate change impacts on mycotoxin contamination. *World Mycotoxin J* 9: 717–26.
Van der Fels-Klerx, H. J., P. W. Goedhart, O. Elen, T. Börjesson, V. Hietaniemi, and C. J. Booij. 2012. Modeling deoxynivalenol contamination of wheat in northwestern Europe for climate change assessments. *J Food Prot* 75: 1099–106.
Van Dolah, F. 2000. Marine algal toxins: Origins, health effects, and their increased occurrence. *Environ Health Perspect* 108(Suppl 1): 103–41.

Van Ruijven, B., D. P. van Vuuren, W. Boskaljonk, M. L. Neelis, D. Saygin, and M. K. Patel. 2016. Long-term model-based projections of energy use and CO_2 emissions from the global steel and cement industries. *Resour Conserv Recycl* 112: 15–36.

Walsh, J. J., and K. A. Steidinger. 2001. Saharan dust and Florida red tides: The cyanophyte connection. *J Geophys Res* 106: 11597–612.

Watkins, S. M., A. Reich, L. E. Fleming, and R. Hammond. 2008. Neurotoxic shellfish poisoning. *Mar Drugs* 6: 431–55.

Welker, M., and H. von Döhren. 2006. Cyanobacterial peptides—Nature's own combinatorial biosynthesis. *FEMS Microbiol Rev* 30: 530–63.

Wells, M. L., V. L. Trainer, T. J. Smayda, et al. 2015. Harmful algal blooms and climate change: Learning from the past and present to forecast the future. *Harmful Algae* 49: 68–93.

WHO. 2018. *Air Pollution and Child Health: Prescribing Clean Air.* Summary (WHO/CED/PHE/18.01). Geneva: World Health Organization.

Wiese, M., P. M. D'Agostino, T. K. Mihali, M. C. Moffitt, and B. A. Neilan. 2010. Neurotoxic alkaloids: Saxitoxins and its analogs. *Mar Drugs* 8: 2185–211.

Williams, J. H., T. D. Phillips, P. E. Jolly, J. K. Stiles, C. M. Jolly, and D. Aggarwal. 2004. Human aflatoxicosis in developing countries: A review of toxicology, exposure, potential health consequences, and interventions. *Am J Clin Nutr* 80: 1106–22.

Wu, F., D. Bhatnagar, T. Bui-Klimke, et al. 2011. Climate change impacts on mycotoxin risks in US maize. *World Mycotoxin J* 4: 79–93.

Wu, J. 2016. Detecting and attributing the effects of climate change on the distribution of snake species over the past 50 years. *Environ Manag* 57: 207–19.

Xiao, F., M. F. Simcik, T. R. Halbach, and J. S. Gulliver. 2015. Perfluorooctane sulfonate (PFOS) and perfluorooctanoate (PFOA) in soils and groundwater of a U.S. metropolitan area: Migration and implications for human exposure. *Water Res* 72: 64–74.

Yanez-Arenas, C., A. T. Peterson, K. Rodriguez-Medina, and N. Barve. 2016. Mapping current and future potential snakebite risk in the new world. *Clim Change* 134: 697–711.

Yang, S., H. Zhang, M. De Boevre, et al. 2018. Toxicokinetics of HT-2 toxin in rats and its metabolic profile in livestock and human liver microsomes. *J Agric Food Chem* 66: 8160–8.

Yang, Y., S. Yu, Y. Tan, N. Liu, and A. Wu. 2017. Individual and combined toxic effects of co-occurring deoxynivalenol family of mycotoxins on human gastric epithelial cells. *Toxins* 9: pii: E96.

Yu, S., N. Zhao, and Z. Zi. 2001. The relationship between cyanotoxin (microcystin, MC) in pond-ditch water and primary liver cancer in China. *Chi J Oncol* 23: 96–9.

Zhang, Y., and Y. Wang. 2016. Climate-driven ground level ozone extreme in the fall over the Southeast United States. *Proc Natl Acad Sci USA* 113: 10025–30.

Zinedine, A., J. M. Soriano, J. C. Moltó, and J. Mañes. 2007. Review on the toxicity, occurrence, metabolism, detoxification, regulations and intake of zearalenone: An oestrogenic mycotoxin. *Food Chem Toxicol* 45: 1–18.

25 Chemical Warfare and Terrorism

Hazem Matar and Robert P. Chilcott

CONTENTS

25.1 HISTORICAL USE AND DEVELOPMENT OF CHEMICAL WARFARE AGENTS

Chemicals have been used to debilitate armies and gain tactical advantage on the battlefield since at least AD 256 (Simon 2011). Chemical weapons gained prominence during the First World War when chlorine, phosgene and sulphur mustard (HD), along with other blister agents, were used to incapacitate troops, most notably at the battle of Flanders near Ypres, Belgium, in July 1917 (Chauhan et al. 2008). Fritz Haber, who is considered by some to be the father of chemical warfare, played a major role in the First World War by developing nonballistic dissemination of chemical agents. Haber also formulated the initial mathematical relationship between exposure time and gas concentration to result in death (LCt_{50}), known as Haber's rule. By the end of that war, the consensus was that deployment of toxic chemicals was an undesirable form of warfare; this led to the 1925 signing of the Geneva Protocol for the Prohibition of the Use in War of Asphyxiating, Poisonous or Other Gases, and of Bacteriological Methods of Warfare. However, the German chemist Gerhard Schrader (of the I.G. Farben Corporation), while carrying out experiments to develop improved insecticides during the 1930s, accidently discovered the first class of nerve agents, which later became to be known as the "G-agents": tabun (GA) and sarin (GB). These agents were quickly appropriated by the German military in 1937, and in 1944 sarin became the foundation for soman (GD), which was manufactured but never deployed during the Second World War (Pitschmann 2014). Thereafter, the "V-agents" emerged from similar insecticide studies by British chemist Ranajit Ghosh (of Imperial Chemical Industries; ICI). These agents were found to be too toxic for agricultural use and were sent to the British chemical warfare establishment at Porton Down. One agent from this "Venomous" or "V" series of agents, VX, was selected for its potent toxicity and later became the foundation for many isomers developed by the Russians and Chinese (VR and CVX, respectively). Subsequent developments have led to the synthesis and use of the "Novichok" series of agents (Nepovimova and Kuca 2018; Vale et al. 2018).

Despite legal treaties prohibiting their use, chemical warfare agents have been deployed on several occasions, such as the Iran-Iraq War (1980), the attack on the Tokyo Metro (1995), the Syrian civil war (2017), the assassination of Kim Jong-nam (2017) and the attempted execution of Russian ex-spy Sergei Skripal in 2018, to name but a few. In addition, other chemical compounds have been used, such as ricin[1] in the assassination of the Bulgarian dissident Georgi Markov (Musshoff and Madea 2009), Zyklon B during the holocaust in Nazi Germany (Rotenberg 2003) and the dioxin poisoning of Viktor Yushchenko (Schecter et al. 2006).

DOI: 10.1201/9780203730584-25

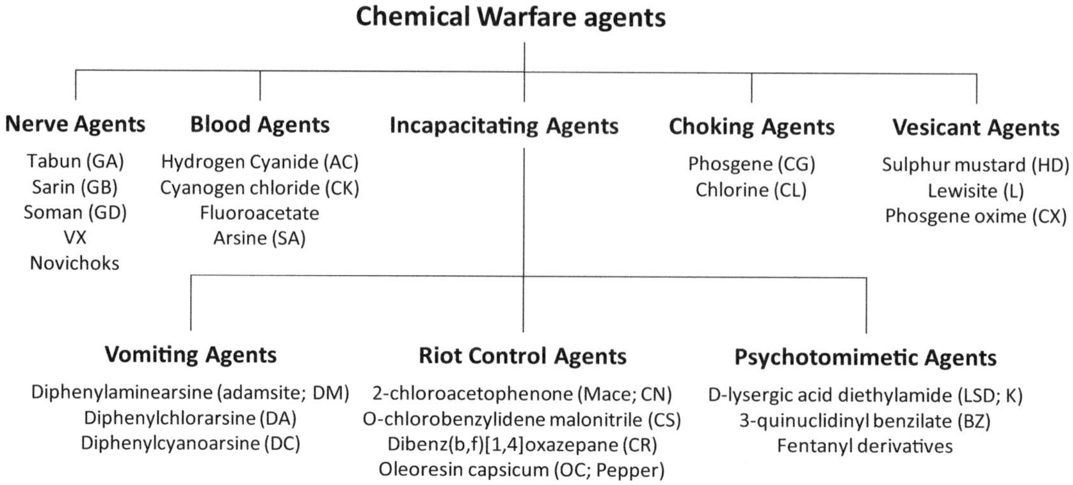

Figure 25.1 *Summary of chemical warfare agents.* Some compounds such as fluoroacetate have been added but are technically not CWC scheduled chemicals.

25.2 TYPES OF CHEMICAL WARFARE AGENTS

As briefly mentioned above, there are several types of chemical warfare and incapacitating agents. These agents are generally classified based upon their mode of action (Figure 25.1). Blister agents (vesicants) primarily affect the skin, lungs, eyes and mucosal membranes. Blood agents interfere with the body's metabolic processes, impeding the function of vital biochemical pathways. Choking agents target and damage the lungs, causing pulmonary oedema and, in sufficient doses, death due to asphyxiation. Nerve agents achieve toxicity by overstimulating the body's nervous system, ultimately resulting in respiratory paralysis and death. Incapacitating agents can vary in their mode of action but are designed to temporarily render individuals incapable of performing their tasks by exerting either physiological or mental effects. Toxic industrial chemicals (TICs) may fall under several different categories, including nerve agents (organophosphorus pesticides). The final class of compounds, which often blurs the lines between chemical and biological weapons, includes substances known as toxins. These are typically produced by living organisms, such as bacteria, fungi, plants and animals (Madsen 2001). These toxins can be categorised according to their mode of action (cardiotoxins, hepatotoxins, neurotoxins, etc.; Ganesan et al. 2010). The most potent are neurotoxins, such as botulinum toxin, which at present is the most potent poison known to exist (Arnon et al. 2001; Gill 1982).

25.3 ROUTES OF EXPOSURE

In general, the routes of exposure depend upon the physicochemical properties of chemical warfare agents; an obvious example is volatility. An inherent ability to evaporate and disperse will result in an inhalational hazard. However, that does not mean that chemical vapours do not also pose a dermal hazard, as vesicants can still elicit a response (Chilcott et al. 2000). Typically, nonvolatile chemicals of a molecular weight less than 500 Da are generally regarded as dermal hazards, given their ability to persist and be absorbed through the body via direct skin contact. The oral route of exposure is generally considered to be not as significant as inhalation or cutaneous exposure, in view of the properties of chemical agents and the likely methods used to disseminate them. However, it is not inconceivable for nonvolatile agents and toxins to be ingested from contaminated food sources. Typically, the main exposure routes for toxins is via injection, ingestion, inhalation and to some degree dermal contact, depending upon the toxin. Generally, the onset of symptoms is dependent upon the route of exposure, not taking into account injection; the rate of absorption is fastest via inhalation and generally slowest via percutaneous exposure.

25.4 NERVE AGENTS

Nerve agents are representative of a class of compounds known as organophosphates, which also encompasses pesticides such as malathion, parathion and dichlorvos. As briefly mentioned above, nerve agents such as the G and V series achieve greater acute mammalian toxicity via

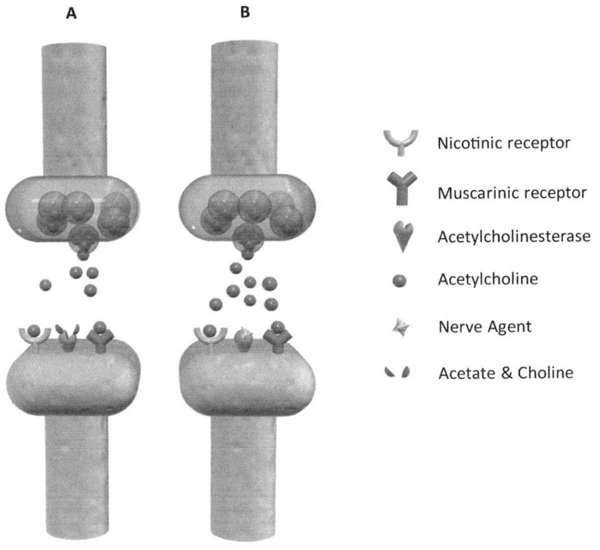

Figure 25.2 *Representation of functional a nerve (A) and nerve agent inhibited (B).* An electrical nerve impulse (action potential) travels down to the end of the pre-synaptic neurone (left) releasing vesicles containing acetylcholine (ACh) into the synaptic cleft. The acetylcholine binds and stimulates nicotinic and muscarinic receptors. Acetylcholinesterase (AChE) hydrolyses ACh to choline and acetate. Nerve agents bind to AChE via phosphorylation rendering it unable to degrade ACh, which results in overstimulation

inhalation and cutaneous routes of absorption (Evison et al. 2002). The mechanism of action of nerve agents is the inhibition of the enzyme acetylcholinesterase (AChE), which results in an accumulation of the neurotransmitter acetylcholine (ACh) at neural synapses and neuromuscular junctions (Figure 25.2). Nerve agents typically bind AChE via phosphorylation, rendering it unable to degrade ACh. This excessive accumulation leads to overstimulation and potential paralysis of muscarinic and nicotinic receptors, which may result in death due to respiratory failure.

Because of their mechanism of toxicity, nerve agents, irrespective of the specific chemical, will exhibit the same clinical symptoms as those resulting from overstimulation of cholinergic pathways (Koelle 1981). Exposure to toxic concentrations will result in a variety of effects, such as hypersalivation, bronchorrhoea, incontinence, miosis (or mydriasis), bronchospasm, bradycardia, muscle fasciculations, paralysis, convulsions and depression of the respiratory system (Grob 1956; Munro 1994). The lowest recorded percutaneous LD_{50} (lethal dose required to kill 50% of the test population) in humans are 18 mg/kg and 0.086 mg/kg for GD and VX, respectively (Maynard and Beswick 1992). The lowest LD_{50} for inhalation have been estimated as 0.07 mg/kg and 0.015 mg/kg, respectively. It must be noted that there is great variation in the estimated toxicity due to the lack of human exposure data (Maynard and Beswick 1992).

Treatment following exposure to nerve agents will vary depending upon the chemical agent (Worek et al. 2004). The standard treatment for nerve agent exposure is in the form of an antidote that is administered intramuscularly via an autoinjector, which typically contains atropine (anticholinergic), an oxime such as pralidoxime (to reactivate inhibited AChE) and a benzodiazepine (anticonvulsant; Eyer and Worek 2007; Hobbiger 1963). There is some scepticism concerning the clinical benefits of oxime treatment following acute exposure to organophosphates, primarily because of its poor affinity for the particular organophosphate-AChE complex and the potential "ageing" of the AChE (Eddleston et al. 2002; Sirin et al. 2012). Ageing of AChE is compound-specific: GD ages in minutes, whereas VX ages over several days post exposure. This is attributed to monodealkylation of the dialkyl-phosphonyl enzyme, which is then irreversibly inhibited by the increased stability resulting from resistance to common oxime reactivators (Millard et al. 1999). Therefore, for some nerve agents the efficacy of antidotes is limited. Depending upon the route of exposure and chemical properties, victims may exhibit symptoms of nerve agent poisoning if an appropriate amount is inhaled. Conversely, the onset of symptoms via dermal exposure may be delayed by hours post exposure when applied to normal skin. This highlights the importance of decontamination to remove the source of contamination and prevent further absorption, as once the agent is absorbed, treatment with autoinjectors is necessary. Once large amounts have been absorbed via the skin, systemic absorption may be sustained for several hours (Chilcott 2007).

25.5 VESICANTS

A vesicant is an animal-, plant- or synthetically derived material that can cause vesicles or blisters. There are two main types of vesicant agents used in chemical warfare: mustards and arsenicals. The most noteworthy is sulphur mustard, which can cause severe dermal, respiratory and systemic damage following exposure (Kehe and Szinicz 2005). It is often referred to incorrectly as mustard gas, whereas in actual fact it is a liquid. It was first noted for its toxic properties by dye chemists in the 1880s (Pechura and Rall 1993). Sulphur mustard was initially deployed in the First World War at Ypres, Belgium, in 1917 (Bismuth et al. 2004). Exposure to 40–100 $\mu g\ cm^{-2}$ has been quoted as the dose required to elicit vesication (Papirmeister et al. 1991). However, historic human volunteer trials have quoted a dose of approximately 6 $\mu g\ cm^{-2}$ at 22°C ($n = 60$) following vapour contamination (Renshaw 1946), and this correlates well with other values reported in the literature (Dalton et al. 2006; Nagy et al. 1946).

The exact mechanism of toxicity of sulphur mustard is still not fully understood (Jowsey et al. 2012). However, there are several theories regarding its mechanism of action. It has been suggested that glutathione depletion may be responsible for increasing intracellular calcium, resulting in cell death (Orrenius et al. 1989). An alternative theory is the cellular depletion of NAD^+ (nicotinamide adenine dinucleotide) by poly[ADP-ribose] polymerase (PADPRP; Papirmeister et al. 1985). There is evidence that specific skin cell types, such as melanocytes, might be sensitive to sulphur mustard (Brown and Rice 1997). Sulphur mustard may also undergo metabolic activation, which ultimately results in the unregulated phosphorylation of the keratinocyte cytoskeleton, causing detachment from the basement membrane (Brimfield et al. 1998). However, none of these theories coherently links the biochemical effects of mustard to its potent vesicating ability (Maynard and Chilcott 2009).

Sulphur mustard is a potent alkylating agent that is capable of binding a range of molecules, in particular thiol-containing proteins (Byrne et al. 1996) and DNA (Fox and Scott 1980). It is therefore cytotoxic to all dividing cells, with rapidly dividing epithelial tissue being particularly susceptible (Dacre and Goldman 1996). Sulphur mustard–induced skin lesions initially appear as erythema within 2 to 24 hours of exposure. This latency period is related to the length of exposure and the dose absorbed (Vogt et al. 1984). Later signs of exposure include the development of small vesicles filled with yellow serous fluid, which coalesce to form large, pendulous blisters (Momeni et al. 1992). However, histological changes in human skin are apparent within 30 minutes following exposure to little as 2 μL of liquid contamination (Warthin et al. 1918). There is currently no standard treatment for sulphur mustard exposure, as its mechanism of action is not fully understood (Kehe et al. 2008). Therefore, current treatments for sulphur mustard are based on generic therapies for alleviating skin burns and include the application of antibacterial ointments and debridement of necrotic tissue (Rice 2003).

Derivatives of sulphur mustard, known as nitrogen mustards, were also developed as vesicating agents but were never deployed in war. There are three main analogues of nitrogen mustards: HN1, HN2 and HN3. As a result of their inherent potent alkylating properties, nitrogen mustards led the way as the first chemotherapeutic agents (Singh et al. 2018).

Lewisite is another well-known vesicant which elicits clinical effects similar to those of sulphur mustard; however, the dermal response is more rapid and painful. Furthermore, the lesions caused by lewisite exposure more resemble thermal burns rather than those of sulphur mustard (Chilcott et al. 2000). Lewisite can be mixed with sulphur mustard, for which it can act as an antifreeze; this concoction is often referred to as Agent Yellow (Nelson et al. 2006; Stone et al. 2016). Unlike sulphur mustard, a specific treatment is available for lewisite and is still used for accidental arsenical accidents. The treatment, 2,3-dimercaptopropanol, also known as British anti-Lewisite (BAL), is a very effective chelating agent (Vilensky and Redman 2003).

Phosgene oxime (CX), often referred to nettle gas (nettle agent), which on contact with skin results in erythema and urticaria (hives), is also classified as a vesicant even though it does not produce blisters. It has been produced and stockpiled alone or as a mixture with lewisite and sulphur mustard to enhance their penetration (Goswami et al. 2018). The mechanism of action is still not fully understood, but the onset of physiological effects is rapid, extremely painful and may also cause visual impairment or result in blindness after ocular exposure. There is currently no antidote for phosgene oxime; treatment is primarily supportive and symptomatic. Rapid decontamination may lessen the severity of deleterious effects.

25.6 CHOKING AGENTS

Choking agents, as the name suggests, target and disrupt the respiratory system, ultimately resulting in death due to increased pulmonary secretions and inefficient gaseous exchange in the lungs. Two notable choking agents are phosgene and chlorine, both of which were deployed during the First World War. Phosgene induces acute lung damage due to the reactivity of its carbonyl group, ultimately disrupting the epithetical lining of the lungs and resulting in pulmonary oedema (Diller 1985). Phosgene has a long latency period even after exposure to a lethal dose. No treatment is available following phosgene exposure. The main form of treatment is supportive and is usually longer due to the latent onset of pulmonary oedema. Chlorine is an order of magnitude less toxic than phosgene and results in an immediate onset of signs and symptoms such as blurred vision and breathing difficulties, and death may occur within 4–6 hours following sufficient exposure.

25.7 BLOOD AGENTS

The most infamous blood agent is hydrogen cyanide (HCN). There are other cyanide compounds, such as cyanogen chloride (NCCl) and potassium cyanide (KCN), that can liberate the cyanide ion (CN^-), which is the toxic moiety (Simeonova and Fishbein 2004). CN^- exerts its acute toxicity by inhibiting cytochrome oxidase (via binding of haem iron), which is the terminal enzyme of the mitochondrial electron transport chain. This results in cellular hypoxia, which causes a shift from aerobic to anaerobic cellular respiration culminating in lactic acidosis. Therefore, tissues that are highly dependent on oxidative metabolism and have limited anaerobic capacity, such as the central nervous system, are particularly vulnerable to cyanide intoxication (Way 1984).

Another blood agent is sodium fluoroacetate (marketed as "compound 1080"), which was developed as a rodenticide. Unlike HCN and NCCl, which are listed in the chemical weapons convention (CWC Schedule 3), compound 1080 is not a scheduled chemical but is banned in many countries for its toxicity. Sodium fluoroacetate exhibits its toxicity by ultimately inhibiting the citric acid (Krebs) cycle, as fluoroacetate combines with coenzyme A to form fluoroacetyl CoA. This is substituted for acetyl CoA (Gal et al. 1956) and reacts with citrate synthase to produce fluorocitrate, which binds to aconitase with great affinity, thus inhibiting the cycle (Proudfoot et al. 2006). Even though the mechanism of toxicity is well understood, there is no formal antidote for intoxication with fluoroacetate. The most acceptable treatment, which has been used for decades, is ethanol, if administered immediately after poisoning by increasing blood acetate levels via oxidation of ethanol and therefore inhibiting fluorocitrate production (Goncharov et al. 2006).

The toxic industrial chemical arsine is the hydride of arsenic (AsH_3, called "SA") and is a colourless, nonirritating gas with a mild, garlic-like odour. Arsine is typically used in the semiconductor and metal-refining industries. It exhibits acute toxicity by targeting the haematopoietic system, particularly erythrocytes (red blood cells). Arsine ultimately induces haemolysis, causing haemoglobinuria and subsequent kidney damage (Czerczak and Fishbein 2002). There is no antidote following arsine exposure. Treatment consists of supportive medical care; some cases require blood transfusions, and dialysis may be required if kidneys are damaged.

25.8 INCAPACITATING AGENTS

Incapacitating agents are nonlethal and are deployed to debilitate individuals by transiently inducing physiological or mental effects. As such, theoretically, any chemical may serve as an incapacitating agent (depending on dose and the ratio between effect and toxicity). However, there are some notable examples of incapacitating agents, which are classified either as psychotomimetics or riot control agents (RCAs).

Psychotomimetic compounds such as *D*-lysergic acid diethylamide (LSD) and 3-quinuclidinyl benzilate (BZ) are compounds that disorientate, confuse and cause hallucinations. LSD acts on the brain as an agonist at serotonin receptors and at higher doses stimulates dopamine receptors. It was trialled by the US military as an incapacitating agent, although its effects were too unreliable for warfare purposes. In contrast, BZ is a potent anticholinergic psychotomimetic that can readily cross the blood-brain barrier (BBB). It is a competitive inhibitor of ACh at postsynaptic muscarinic receptors in the peripheral and central nervous system. As such, the mechanism of treatment is to increase the levels of ACh by inhibiting AChE. The effects of nerve agents as treatments were evaluated (Sidell et al. 1973), but the effects of BZ can be treated with physostigmine (Sidell 1976).

The four most common RCAs are 2-chloroacetophenone (CN; Mace), *o*-chlorobenzylidene malonitrile (CS), dibenz(*b,f*) [1,4]oxazepine (CR) and oleoresin capsicum (OC; pepper spray). These chemicals are commonly deployed for managing times of civil unrest, hostage situations and in some instances by civilians as personal defence devices (pepper spray). Furthermore, militaries often deploy these agents for training purposes to provide service personnel with an appreciation of their personal protective equipment (Thomas et al. 2002). These agents are commonly referred to as tear gases due to fact they result in immediate ocular pain and lacrimation. They are typically dispersed as fine powder particles using fogging devices as aerosols or spray solutions; as such, they are not gases. Effects following exposure to RCAs are intense irritation of the eyes and marked irritation to the nose, throat and lungs and to some extent the skin (Olajos and Salem 2001). The margin of safety between effect and serious adverse effects is large (Olajos and Salem 2001), and RCAs are generally regarded as safe when used appropriately, despite the fact their mechanism of action is largely not fully understood (Schep et al. 2015). There is no formal antidote for RCAs apart from treating symptoms and decontamination with emphasis on irrigation of the eyes.

The use of chemicals to serve as incapacitating agents is not limited to the ones briefly discussed. They could also be pharmaceutical-type agents such as the mixture used in the Moscow theatre siege in 2002, in which an alleged undisclosed mixture of an opioid and sedative was deployed to subdue the aggressors and calm the hostages. The mixture was believed to be either fentanyl/carfentanyl or halothane/isoflurane (Riches et al. 2012; Schiermeier 2002). Based upon their potency and narrow therapeutic window, this led to 125 deaths, likely attributed to respiratory depression.

25.9 COUNTERMEASURES

25.9.1 Pharmaceutical

In view of the acute toxicity of chemical warfare agents, several countermeasures are in place for reacting to and dealing with nefarious releases in both military and civilian scenarios. There are distinct differences between military and civilian countermeasures. The obvious advantage for military personnel over civilians is that they are generally prepared for reacting to a chemical, biological, radiological or nuclear (CBRN) incident through training and have access to personal protective equipment (PPE) such as respirators and suits. Furthermore, antidotes are more readily available; for instance, soldiers may carry oxime autoinjectors. Additionally, a chemical warfare agent may be detected and identified much more rapidly by military personnel using readily deployed detection equipment. Further countermeasures that are more suitable for military applications involve the use of prophylactics. There are two main forms of prophylactic treatments. The first is treatment with a nerve agent pretreatment set (NAPS), usually delivering pyridostigmine which was designed to reversibly occupy a proportion of available AChE, thus limiting the amount of AChE available for nerve agent inhibition. However, a disadvantage of pyridostigmine is that it is unable to cross the BBB, and as a result it only protects the peripheral nervous system. Therefore, physostigmine was developed as a replacement with the ability to cross the BBB. However, its half-life is significantly shorter than that of pyridostigmine (~30 minutes compared to 8 hours). The combination of physostigmine with hyoscine (scopolamine, an anticonvulsant) has been suggested, but because of their relatively short half-lives they are administered via transdermal patches for sustained release (Scott 2007). There is controversy as to whether the use of NAPS may have led to the onset of Gulf War syndrome following Operation Desert Storm in Kuwait in 1991 (Myhrer and Aas 2016). Gulf War syndrome is a multisymptom condition that manifested as fatigue and cognitive disorders. Its cause is not fully understood, but suspicion of a causative effect has been placed on the use of NAPS, vaccinations administered to soldiers and potential exposure to chemical warfare agents and organophosphate pesticides. For military application, promising therapies were often fast-tracked to the battlefield if the benefit of specific treatments was deemed to outweigh both the acute and chronic adverse effects. This approach has now changed as a result of Gulf War syndrome: treatments now have to be compliant with extant regulations and authorities such as the MHRA, EMA and FDA (Scott 2007).

25.9.2 Topical Skin Protectants

An alternative prophylactic to drugs is the use of topical skin protectants. As the name suggests, these are formulations designed to be put on the skin of military personal to act as a barrier to chemical warfare agents. In some cases, this practice has also been adopted for industrial workers (Wang et al. 2006) and domestic settings (e.g. to mitigate incontinence-associated dermatitis; Matar et al. 2017; Pather et al. 2017). In the context of chemical warfare countermeasures they are designed to primarily form a protective barrier, with the option of incorporating excipients that provide specific neutralising capacity (Braue et al. 2005). Typically, these creams have been found to be very effective in reducing dermal penetration and the formation of blisters following chemical warfare agent exposure (Chilcott et al. 2007; Millerioux et al. 2009). However, even though some of them have received FDA approval (e.g. SERPCWA), they have never been proven against chemical warfare agents under operational conditions.

25.10 DECONTAMINATION

One common strategy (process) that applies to both military and civilian situations is the implementation of decontamination. Decontamination can be defined as "the process of removing hazardous material(s) both on or available to the external surfaces of the body in order to reduce local or systemic exposure to a contaminant and thus minimise the risk of subsequent adverse health effects" (Chilcott 2013). Typically, decontamination is conducted by physical removal or by chemical neutralisation. Physical removal methodologies aim to absorb/sequester contaminants, whereas chemical neutralisation aims to modify the structure of the contaminant in order to reduce or eliminate toxicity by hydrolysis, oxidation or other chemical means (Chan et al. 2013). However, there are distinct differences in the decontamination methods available to military personal compared to civilians.

25.10.1 Military Considerations

The lack of clean water on the battlefield has prompted the development of personal products to decontaminate skin following exposure to chemical warfare agents (Yang et al. 1992). There are certain desirable qualities that a decontaminant should possess: effectiveness, rapid mode of action, minimal skin irritation or disruption to the barrier function of the skin, ease of disposal, minimal residual contamination and readily available and affordable (Chan et al. 2013; Hurst 1997; Roberts and Maynard 2007; Yang et al. 1992). Products that decontaminate by physical removal include (but are not limited to) Fuller's earth, Dutch powder, Ambergard and M291; their mode of action is fundamentally based upon physical absorption, although they may exhibit some elements of chemical neutralisation (Roberts and Maynard 2007). Fuller's earth is a natural form of aluminium silicate (clay) sourced from the ground. It contains roughly twice as many silicon as aluminium atoms and an abundance of exchangeable (loosely bonded) minerals such as calcium. Fuller's earth is processed to a fine powder, which increases the efficacy of its absorptive properties by producing an increased surface area, and it is effective at absorbing a wide range of liquid contaminants. *In vitro* studies corroborate these absorbing properties (Chilcott et al. 2001; Matar et al. 2016; Roul et al. 2017), and *in vivo* studies have demonstrated that Fuller's earth is at least as effective as a commercially available reactive (chemically neutralising) skin decontaminants against sulphur mustard and VX (Taysse et al. 2007; Taysse et al. 2011). Fuller's earth has long been used by military organisations and is regarded as the gold standard against which other decontaminants are compared.

There is debate regarding the relative importance of physical removal and chemical neutralisation. Some researchers state that the primary objective should be physical removal, as the neutralising properties of some decontaminants are not as rapid as physical removal (Hurst 1997). In contrast, the rationale of rapid neutralisation is that it should prevent the absorption of intact toxic agent, which is regarded as the primary objective (Gordon and Clarkson 2009). At present, the main disadvantage surrounding chemical neutralising agents is the lack of a single substance capable of neutralising compounds with a wide range of physical and chemical properties (i.e. there is no "magic bullet", as Paul Ehrlich stated in 1900).

Historically, 0.5% hypochlorite solution has been used as the universal reactive decontaminant of choice because of its alleged ability to denature chemical and biological warfare agents (Yang et al. 1992). However, it has been demonstrated that hypochlorite solutions should not be used to decontaminate wounds affected by sulphur mustard, as they may exacerbate sulphur mustard–induced lesions (Gold et al. 1994). Other neutralising products include a polyurethane sponge impregnated with a specific enzyme (recombinant acetylcholinesterase; rAChE) that can catalyse the hydrolysis of organophosphate nerve agents (Gordon et al. 1999). However, this particular product has been shown to undergo stoichiometric reactions with nerve agents and so may be prone to rapid saturation (Gordon and Clarkson 2009). There is a plethora of "neutralising" decontamination products (Thors et al. 2017), although none as commercially successful as Reactive Skin Decontamination Lotion (RSDL), which is currently in service in a number of countries. RSDL was designed to absorb and neutralise chemical warfare agents and can be administered as a lotion or via a sponge soaked with the lotion. The first generation of RSDL comprised two main components: a solvent and an oxime. The solvent was designed to solubilise chemical warfare agents away from the skin while the oxime reacted with chemical warfare agents to produce less toxic breakdown products (Gerecke et al. 2009). RSDL has been shown to be more efficacious than Fuller's earth against sulphur mustard and VX in domestic swine if decontamination is commenced within 5 minutes of exposure (Taysse et al. 2007). However, a number of studies have indicated equivalence with Fuller's earth (Matar et al. 2016).

25.10.2 Civilian Considerations

The potential threat of a deliberate release of chemical warfare agents, targeting civilian populations, as exemplified by the 1995 Tokyo sarin incident (Okumura et al. 1996), has led to the development of more effective decontamination strategies for first responders and contaminated individuals. Responding to civilian incidents following a terrorist attack is a complex task, and many issues need to be considered in order to develop an effective strategy for managing contaminated casualties. Terrorist incidents can be further divided into those with prior warning of a release and those where unexpected release has already occurred; in the former case, there may be an indication of where a release may take place and what substance may be used in the attack. This may allow sufficient time to coordinate a response. The latter type of incident is more complicated with regard to chemical identification, as patients may not necessarily present signs or symptoms for several hours post exposure (e.g. sulphur mustard, phosgene, VX and Novichoks). In addition, there may be a considerable delay in the emergency response with respect to logistics, chemical hazard reduction and decision-making processes. Furthermore, the release of a chemical warfare agent in civilian environments may not be associated with an overt event, as the agent may be secretly dispersed.

25.11 RESPONDING TO CBRN INCIDENTS: UNITED KINGDOM

The manner in which emergency services respond to CBRN or hazardous material (HAZMAT) events may vary between countries. In the United Kingdom, the Home Office has published guidelines for first responders (Home Office 2015) based on safety triggers known as STEP 1–2–3 Plus.

Step 1: Assess the situation. Where one casualty is incapacitated with no obvious reason, then emergency services may administer care using standard protocols.
Step 2: Where two casualties incapacitated with no obvious reason, then approach with caution using standard protocols.
Step 3: Where three or more casualties incapacitated in close proximity with no obvious reason, then use caution and follow PLUS.

The main function of PLUS is to initiate disrobing and improvised decontamination. PLUS is a CBRN first-responder flow chart, according to which emergency services must not enter the scene but should withdraw to a safe distance; cordons must be set up and a hazardous area response team (HART), HAZMAT, and fire and rescue teams are deployed. However, while awaiting these specialist responses, first responders should begin to evacuate casualties away from the scene of contamination, offering them reassurance and advice in addition to instructions for self-care involving removal of clothing (disrobe) and dry decontamination.

It is worth drawing attention to the fact that, even though this is a robust operating procedure, it is not perfect. This deficiency is related to the onset of symptoms (which, as previously discussed, is dictated by the route of exposure), as some chemical warfare agents have a delayed physiological reaction. This was evidenced recently by the Novichok incident in Salisbury in 2018, in which two victims were identified as unwell and behaving in a strange manner on a park bench. A police officer was first on scene; the officer became exposed but did not exhibit symptoms of nerve agent poisoning until sometime after exposure and also required hospitalisation. The likely route of exposure was deemed to be via dermal contact.

If a CBRN incident has been declared (Step 3), the main objectives are to remove casualties from the scene and to administer decontamination and treatment as soon as possible. Incident response facilities, such as decontamination showers and on-scene medical facilities, are placed upwind from the incident. This is to prevent the wind from carrying any volatile chemicals or substances towards treatment areas and causing additional chemical exposure.

The incident scene is generally divided into three areas; hot, warm and cold zones (Figure 25.3; Han et al. 1999). These zones should be monitored to prevent the uncontrolled movement of contaminated individuals.

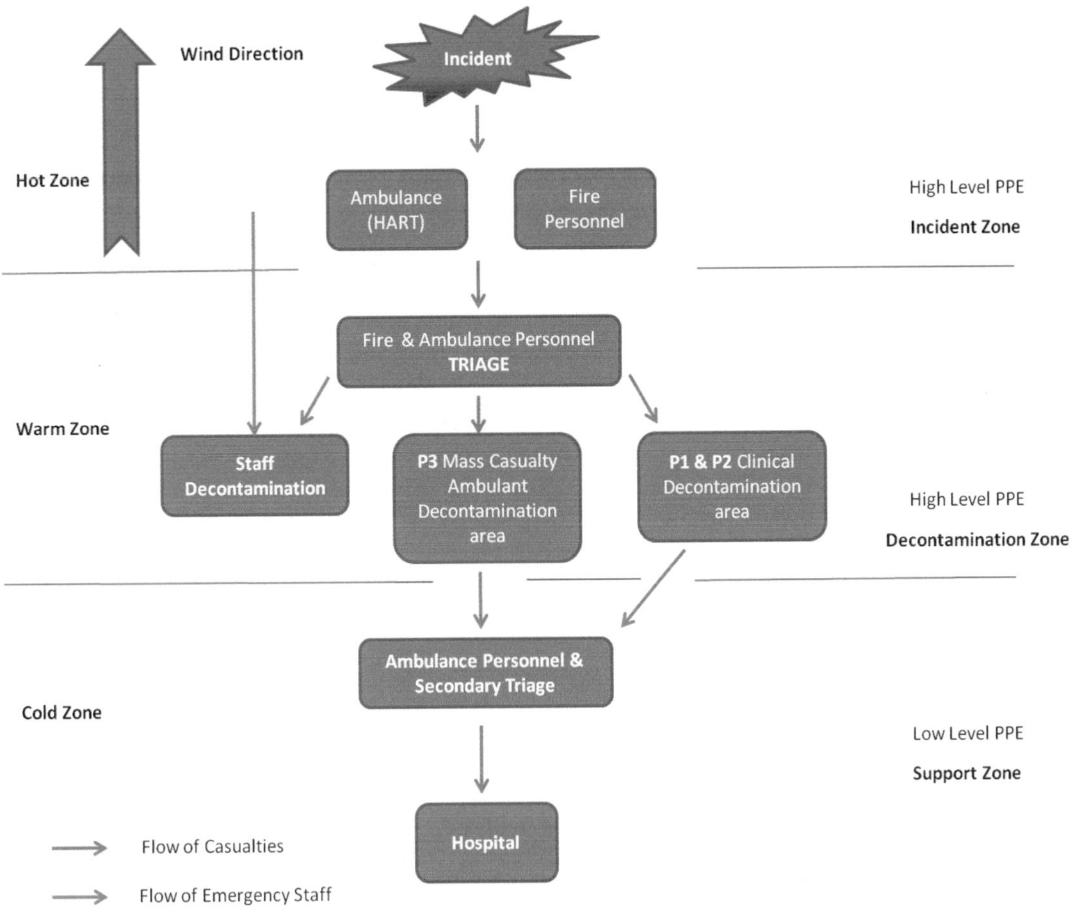

Figure 25.3

The hot zone is only accessible to fire and rescue service and ambulance HART responders in full PPE comprising chemically impervious suits with respirators or self-contained breathing apparatus. The role of hot-zone responders is to evacuate and rescue casualties and to provide life-saving medical interventions, such as the administration of antidotes and resuscitation. HART is unusual in that it allows for advanced medical treatment in the hot zone, as they can intubate, manage spinal injuries and perform intravenous and intraosseous injections. An initial triage is conducted in the hot zone to identify and prioritise the clinical management of casualties prior to their transfer to the warm zone. At this point, casualties undergo a second triage into three categories: P1, P2 and P3. The first two include heavily contaminated or severely injured casualties who require urgent medical care, such as resuscitation, or victims who are stretcher-bound. Casualties categorised as P3 are termed the "walking wounded", who are able to walk unassisted and perform self-decontamination. A fourth category (P4) is assigned to casualties who are either deceased or "expectant" (beyond medical intervention).

The warm zone is subdivided into two decontamination areas: mass casualty and clinical. Mass casualty decontamination is operated by the fire and rescue service using a specialised showering tent, a "mass decontamination unit" (MDU). This unit is primarily used for P3 casualties as it comprises a high-throughput system, designed to decontaminate casualties as quickly and as efficiently as possible. These units are also equipped with disrobe and rerobe packs at either side of the MDU. The clinical decontamination tents are reserved for incapacitated casualties (P1 and P2), who are physically decontaminated by ambulance personnel. The labour-intensive decontamination process means that this system has a much lower throughput of casualties. Typically, casualties undergoing clinical decontamination are scrubbed with water and detergent solution.

Once casualties have been decontaminated, they are taken to a holding bay where further triage and prioritisation of casualties is performed prior to admission to hospital or transfer to a community holding centre.

25.11.1 Disrobing

Disrobing is the procedure of carefully removing clothing following exposure to contaminants. It is consistent with the physical process of decontamination and is considered to be the most important component of decontamination for all casualties (Chilcott et al. 2018; Levitin et al. 2003; Okumura et al. 1996). The removal of clothing is estimated to remove 75%–90% of the contamination (Cox 1994; Vogt and Sorrensen 2002). However, this quoted level of performance does not appear to be evidence based and is likely to be founded on the percentage of the body covered by clothing. It is also important to emphasise that clothing can also afford significant protection against chemical warfare agents. However, this protection diminishes rapidly with time (Matar 2012). In addition, it should be noted that sulphur mustard appears to penetrate various fabrics with little hindrance. The relative ease with which sulphur mustard penetrates clothing has long been known and has been observed in infantrymen, where even the heaviest types of clothing offered little protection against exposure (Porton Report 1929a; Porton Report 1929b). Moreover, water or sweat may further decrease the protection afforded by clothing (Wester et al. 2000).

It is widely accepted that contaminated clothing should not be removed over the head; in order to limit the area of exposure, clothing materials are cut off (Houston and Hendrickson 2005). If cutting implements are not available, then casualties should be instructed to hold clothing away from the face and hold their breath when removing it (Chilcott and Amlot 2015). Once clothing from contaminated individuals has been removed, it is imperative that these articles of clothing are contained, as they can be a source for re-exposure. Some fabrics are able to absorb large quantities of chemicals and, depending on its physicochemical properties, the agent may volatilise and pose a secondary inhalational hazard (Feldman 2010; Okumura et al. 1996; Gaskin et al. 2013a, 2013b).

25.11.2 Decontamination

In the UK, mass casualty decontamination is conducted by showering in bespoke units consisting of three sections: disrobing, showering and rerobing. Previously, the UK employed a 2-minute decontamination wash with a detergent solution followed by a 1-minute rinse cycle with water (Amlot et al. 2010). However, the time spent showering has recently been reduced to 45–90 seconds, with the use of a washcloth if possible (Home Office 2015). It was found that this shorter duration did not compromise decontamination efficacy, whereas it reduced the potential for any "wash-in effect".[2]

This obviously calls into question the suitability of aqueous showering for the decontamination of mass casualty incidents. Currently, the initial response is geared towards improvised decontamination methods. In improvised dry decontamination, exposed skin surfaces (typically face and hands) are blotted and rubbed with any available dry absorbent material, such as paper tissues or a clean cloth (Kassouf et al. 2017). The rationale behind this form of improvised decontamination is to quickly remove the majority of skin contamination while awaited the arrival of specialised decontamination units. In addition, there is a rudimentary improvised form of wet decontamination that can be performed following improvised dry decontamination, as there will be a finite time delay before the arrival of specialist units.

25.12 RESPONDING TO CBRN INCIDENTS: United States

The manner of response to a chemical incident in the United States is changing, and it is akin to that in the UK in that it includes disrobe and dry decontamination. However, the method of showering to decontaminate is different to the UK. The procedure in the United States relies upon the rapid *ad hoc* decontamination procedure known as the "ladder pipe" decontamination system (LPS), which involves positioning fire engines in parallel to each other to deliver a high volume, low pressure water mist into a corridor through which casualties pass (Lake et al. 2013). It can be argued that the LPS system for gross decontamination may be more effective than the specialised MDU tent deployed by the UK with regards to response and set-up time. However, the LPS doesn't account for containment (bunding) of wastewater and modesty issues as a result of disrobing. However, until recently, the way in which fire departments performed LPS decontamination varied across the United States (Power et al. 2016). This process has now changed, and US responders have adopted principles similar to those followed in the UK, in which improvised dry decontamination and disrobe is performed while waiting for deployment of LPS. The time within which casualties are decontaminated has also been reduced to 15–30 seconds. This system is very effective in removing contaminants from skin (Chilcott et al. 2018). Furthermore, this research effort also evaluated methodologies for managing contaminated scalp hair. Chemicals are able to bind to human hair, and the contaminant can still be found on the hair even after several decontamination attempts (Blank and Kidwell 1995). A greater concern is that some semivolatile chemicals can off-gas from hair, potentially becoming a source of recontamination (Chilcott et al. 2018; Spiandore et al. 2017). Therefore, at present, the two main options for managing this risk are either close-clipping and removing contaminated hair or potentially by decontaminating hair with specific solvents (Duca et al. 2014) however, the latter is currently not a viable option on scene.

NOTES

1 Lectin from castor beans.
2 Some studies have shown that there is an increase in dermal penetration of chemicals following immersion or washing with water (Idson 1978; Moody and Maibach 2006; Moody and Nadeau 1997), a phenomenon referred to as the "wash-in effect".

BIBLIOGRAPHY

Amlot, R., J. Larner, H. Matar, et al. 2010. Comparative analysis of showering protocols for mass-casualty decontamination. *Prehosp Disaster Med* 25(5): 435–9.

Arnon, S. S., R. Schechter, T. V. Inglesby, et al. 2001. Botulinum toxin as a biological weapon: Medical and public health management. *JAMA* 285(8): 1059–70.

Bismuth, C., S. W. Borron, F. J. Baud, and P. Barriot. 2004. Chemical weapons: Documented use and compounds on the horizon. *Toxicol Lett* 149(1–3): 11–8. https://doi.org/10.1016/j.toxlet.2003.12.016.

Blank, D. L., and D. A. Kidwell. 1995. Decontamination procedures for drugs of abuse in hair: Are they sufficient? *Forensic Sci Int* 70(1–3): 13–38.

Braue, E. H., J. D. Boecker, B. F. Doxzon, et al. 2005. Nanomaterials as active components in chemical warfare agent barrier creams. In *Defense Applications of Nanomaterials*, 153–69. American Chemical Society.

Brimfield, A. A., L. M. Zweig, M. J. Novak, and D. M. Maxwell. 1998. In vitro oxidation of the hydrolysis product of sulfur mustard, 2,2′-thiobis-ethanol, by mammalian alcohol dehydrogenase. *J Biochem Mol Toxicol* 12(6): 361–9.

Brown, R. F., and P. Rice. 1997. Histopathological changes in Yucatan minipig skin following challenge with sulphur mustard. A sequential study of the first 24 hours following challenge. *Int J Exp Pathol* 78(1): 9–20.

Byrne, M. P., C. A. Broomfield, and W. E. Stites. 1996. Mustard gas crosslinking of proteins through preferential alkylation of cysteines. *J Protein Chem* 15(2): 131–6.

Chan, H. P., H. Zhai, X. Hui, and H. I. Maibach. 2013. Skin decontamination: Principles and perspectives. *Toxicol Ind Health* 29(10): 955–68.

Chauhan, S., S. Chauhan, R. D'Cruz, et al. 2008. Chemical warfare agents. *Environ Toxicol Pharmacol* 26(2): 113–22.

Chilcott, R. P. 2007. Dermal aspects of chemical warfare agents. In *Chemical Warfare Agents: Toxicology and Treatment*, ed. T. C. Marrs, R. L. Maynard, and F. R. Sidell. Chichester: John Wiley and Sons.

Chilcott, R. P. 2013. Initial management of mass casualty incidents. In *Disaster Management: Medical Preparedness, Response and Homeland Security*, eds. R. Arora and P. Arora, 311–24. Oxford: CABI Press.

Chilcott, R. P., and R. Amlot. 2015. Primary Response Incident Scene Management (PRISM) guidance for chemical incidents. In *Tactical Guidance For Mass Casualty Disrobe and Decontamination*. www.medicalcountermeasures.gov/barda/cbrn/decontamination-guidance-for-chemical-incidents/ (accessed December 2, 2020).

Chilcott, R. P., R. F. Brown, and P. Rice. 2000. Non-invasive quantification of skin injury resulting from exposure to sulphur mustard and Lewisite vapours. *Burns* 26(3): 245–50.

Chilcott, R. P., C. H. Dalton, Z. Ashley, et al. 2007. Evaluation of barrier creams against sulphur mustard: (II) In vivo and in vitro studies using the domestic white pig. *Cutan Ocul Toxicol* 26(3): 235–47.

Chilcott, R. P., J. Jenner, S. A. Hotchkiss, and P. Rice. 2001. In vitro skin absorption and decontamination of sulphur mustard: Comparison of human and pig-ear skin. *J Appl Toxicol* 21(4): 279–83.

Chilcott, R. P., J. Larner, and H. Matar. 2018. *Primary Response Incident Scene Management: PRISM Guidance*, Vol. 1, 2nd ed. Office of the Assistant Secretary for Preparedness and Response, Biomedical Advance Research and Development Authority.

Cox, R. D. 1994. Decontamination and management of hazardous materials exposure victims in the emergency department. *Annu Emerg Med* 23(4): 761–70.

Czerczak, S., and L. Fishbein. 2002. *Arsine: Human Health Aspects: Concise International Chemical Assessment Document 47*, ed. World Health Organization & International Programme on Chemical Safety. Geneva: World Health Organization.

Dacre, J. C., and M. Goldman. 1996. Toxicology and pharmacology of the chemical warfare agent sulfur mustard. *Pharmacol Rev* 48(2): 289–326.

Dalton, C. H., M. P. Maidment, J. Jenner, and R. P. Chilcott. 2006. Closed cup vapor systems in percutaneous exposure studies: What is the dose? *J Anal Toxicol* 30(3): 165–70.

Diller, W. F. 1985. Pathogenesis of phosgene poisoning. *Toxicol Ind Health* 1(2): 7–15.

Duca, R. C., E. Hardy, G. Salquebre, and B. M. Appenzeller. 2014. Hair decontamination procedure prior to multi-class pesticide analysis. *Drug Test Anal* 6(Suppl 1): 55–66.

Eddleston, M., L. Szinicz, P. Eyer, and N. Buckley. 2002. Oximes in acute organophosphorus pesticide poisoning: A systematic review of clinical trials. *Qjm* 95(5): 275–83.

Evison, D., D. Hinsley, and P. Rice. 2002. Chemical weapons. *BMJ* 324(7333): 332–5.

Eyer, P., and F. Worek. 2007. Oximes. In *Chemical Warfare Agents*, eds. T. C. Marrs, R. L. Maynard, and F. Sidell, 305–29. Chichester: John Wiley & Sons.

Feldman, R. J. 2010. Chemical agent simulant release from clothing following vapor exposure. *Acad Emerg Med* 17(2): 221–4. https://doi.org/10.1111/j.1553-2712.2009.00650.x.

Fox, M., and D. Scott. 1980. The genetic toxicology of nitrogen and sulphur mustard. *Mutat Res* 75(2): 131–68.

Gal, E. M., R. A. Peters, and R. W. Wakelin. 1956. Some effects of synthetic fluoro compounds on the metabolism of acetate and citrate. *Biochem J* 64(1): 161–8.

Ganesan, K., S. Raza, and R. Vijayaraghavan. 2010. Chemical warfare agents. *J Pharm Bioallied Sci* 2(3): 166–78.

Gaskin, S., D. Pisaniello, J. W. Edwards, et al. 2013a. Application of skin contamination studies of ammonia gas for management of hazardous material incidents. *J Hazard Mater* 252–253: 338–46.

Gaskin, S., D. Pisaniello, J. W. Edwards, et al. 2013b. Chlorine and hydrogen cyanide gas interactions with human skin: In vitro studies to inform skin permeation and decontamination in HAZMAT incidents. *J Hazard Mater* 262: 759–65.

Gerecke, D. R., J. P. Gray, M. P. Shakarjiam, and R. P. Casillas. 2009. Dermal toxicity of sulphur mustard. In *Handbook of Toxicology of Chemical Warfare Agents*, ed. R. C. Gupta. San Diego: Elsevier.

Gill, D. M. 1982. Bacterial toxins: A table of lethal amounts. *Microbiol Rev* 46(1): 86–94.

Gold, M. B., R. Bongiovanni, B. A. Scharf, V. C. Gresham, and C. L. Woodward. 1994. Hypochlorite solution as a decontaminant in sulfur mustard contaminated skin defects in the euthymic hairless guinea pig. *Drug Chem Toxicol* 17(4): 499–527.

Goncharov, N. V., R. O. Jenkins, and A. S. Radilov. 2006. Toxicology of fluoroacetate: A review, with possible directions for therapy research. *J Appl Toxicol* 26(2): 148–61.

Gordon, R. K., and E. D. Clarkson. 2009. Decontamination of chemical warfare agents. In *Handbook of Toxicology of Chemical Warfare Agents*, ed. R. C. Gupta, 1069–81. London: Academic Press.

Gordon, R. K., S. R. Feaster, A. J. Russell, et al. 1999. Organophosphate skin decontamination using immobilized enzymes. *Chem Biol Interact* 119–120: 463–70.

Goswami, D. G., R. Agarwal, and N. Tewari-Singh. 2018. Phosgene oxime: Injury and associated mechanisms compared to vesicating agents sulfur mustard and lewisite. *Toxicol Lett* 293: 112–19.

Grob, D. 1956. The manifestations and treatment of poisoning due to nerve gas and other organic phosphate anticholinesterase compounds. *AMA Arch Intern Med* 98(2): 221–39.

Han, K. H., R. Walker, and M. Kuhri. 1999. An integrated response to chemical incidents—the UK perspective. *Resuscitation* 42(2): 133–40.

Hobbiger, F. 1963. Reactivation of phosphorylated acetylcholinesterase. In *Cholinesterases and Anticholinesterase Agents*, ed. G. B. Koelle, 921–88. Berlin, Heidelberg: Springer.

Home Office. 2015. *Initial Operational Response To a CBRN Incident*. www.jesip.org.uk/uploads/media/pdf/CBRN%20JOPs/IOR_Guidance_V2_July_2015.pdf (accessed December 2, 2020).

Houston, M., and R. G. Hendrickson. 2005. Decontamination. *Crit Care Clin* 21(4): 653–72, v.

Hurst, C. R. 1997. Decontamination. In *Medical Aspects of Chemical and Biological Warfare*, eds. F. R. Sidell, E. T. Takafuji, and D. R. Franz, 351–59. Washington, DC: Office of the Surgeon General United States Army.

Idson, B. 1978. Hydration and percutaneous absorption. *Curr Probl Dermatol* 7: 132–41.

Jowsey, P. A., F. M. Williams, and P. G. Blain. 2012. DNA damage responses in cells exposed to sulphur mustard. *Toxicol Lett* 209(1): 1–10.

Kassouf, N., S. Syed, J. Larner, R. Amlot, and R. P. Chilcott. 2017. Evaluation of absorbent materials for use as ad hoc dry decontaminants during mass casualty incidents as part of the UK's Initial Operational Response (IOR). *PLoS ONE* 12(2): e0170966.

Kehe, K., and L. Szinicz. 2005. Medical aspects of sulphur mustard poisoning. *Toxicology* 214(3): 198–209.

Kehe, K., F. Balszuweit, J. Emmler, H. Kreppel, M. Jochum, and H. Thiermann. 2008. Sulfur mustard research—strategies for the development of improved medical therapy. *Eplasty* 8: e32.

Koelle, G. B. 1981. Organophosphate poisoning—An overview. *Fundam Appl Toxicol* 1(2): 129–34.

Lake, W., P. Schulze, and R. Gougelet. 2013. *ECBC-SP-036 Guidelines for Mass Casualty Decontamination During a HAZMAT/Weapon of Mass Destruction Incident*, Vols. I and II. www.nfpa.org/~/media/files/news-and-research/resources/external-links/first-responders/decontamination/ecbc_guide_masscasualtydecontam_0813.pdf?la=en (accessed December 2, 2020).

Levitin, H. W., H. J. Siegelson, S. Dickinson, et al. 2003. Decontamination of mass casualties—Re-evaluating existing dogma. *Prehosp Disaster Med* 18(3): 200–7.

Madsen, J. M. 2001. Toxins as weapons of mass destruction. A comparison and contrast with biological-warfare and chemical-warfare agents. *Clin Lab Med* 21(3): 593–605.

Matar, H. 2012. *Personal Decontamination against Chemical Warfare Agents*. PhD thesis, University of Surrey.

Matar, H., A. Guerreiro, S. A. Piletsky, S. C. Price, and R. P. Chilcott. 2016. Preliminary evaluation of military, commercial and novel skin decontamination products against a chemical warfare agent simulant (methyl salicylate). *Cutan Ocul Toxicol* 35(2): 137–44.

Matar, H., J. Larner, V. Viegas, et al. 2017. Evaluation of a new topical skin protectant (RD1433) for the prevention and treatment of incontinence-associated dermatitis. *Cutan Ocul Toxicol* 36(3): 211–19.

Maynard, R. L., and F. W. Beswick. 1992. OP compounds as chemical warfare agents. In *Clinical and Experimental Toxicology of OPs and Carbamates*, eds. B. Ballantyne, T. C. Marrs, and W. N. Aldridge, 373–85. Oxford: Butterworth-Heinemann.

Maynard, R. L., and R. P. Chilcott. 2009. Toxicology of chemical warfare agents. In *Chemical Warfare Agents: Toxicology and Treatment*, eds. B. Ballantyne, T. C. Marrs, and T. Syversen, 2875–911. Chichester: Wiley-Blackwell.

Millard, C. B., G. Kryger, A. Ordentlich, et al. 1999. Crystal structures of aged phosphonylated acetylcholinesterase: Nerve agent reaction products at the atomic level. *Biochemistry* 38(22): 7032–9.

Millerioux, J., C. Cruz, A. Bazire, et al. 2009. Evaluation of in vitro tests to assess the efficacy of formulations as topical skin protectants against organophosphorus compounds. *Toxicol in Vitro* 23(1): 127–33.

Momeni, A. Z., S. Enshaeih, M. Meghdadi, and M. Amindjavaheri. 1992. Skin manifestations of mustard gas. A clinical study of 535 patients exposed to mustard gas. *Arch Dermatol* 128(6): 775–80.

Moody, R. P., and H. I. Maibach. 2006. Skin decontamination: Importance of the wash-in effect. *Food Chem Toxicol* 44(11): 1783–8.

Moody, R. P., and B. Nadeau. 1997. In vitro dermal absorption of two commercial formulations of 2,4-dichlorophenoxyacetic acid dimethylamine (2,4-D amine) in rat, guinea pig and human skin. *Toxicol in Vitro* 11(3): 251–62.

Munro, N. 1994. Toxicity of the organophosphate chemical warfare agents GA, GB, and VX: Implications for public protection. *Environ Health Perspect* 102(1): 18–38.

Musshoff, F., and B. Madea. 2009. Ricin poisoning and forensic toxicology. *Drug Test Anal* 1(4): 184–91.

Myhrer, T., and P. Aas. 2016. Pretreatment and prophylaxis against nerve agent poisoning: Are undesirable behavioral side effects unavoidable? *Neurosci Biobehav Rev* 71: 657–70.

Nagy, S. M., C. Golumbic, W. H. Stein, J. S. Fruton, and M. Bergmann. 1946. The penetration of vesicant vapors into human skin. *J Gen Physiol* 29(6): 441–69.

Nelson, P., J. R. Hancock, and T. W. Sawyer. 2006. Therapeutic effects of hypothermia on Lewisite toxicity. *Toxicology* 222(1–2): 8–16.

Nepovimova, E., and K. Kuca. 2018. Chemical warfare agent NOVICHOK—Mini-review of available data. *Food Chem Toxicol* 121: 343–50.

Okumura, T., N. Takasu, S. Ishimatsu, et al. 1996. Report on 640 victims of the Tokyo subway sarin attack. *Ann Emerg Med* 28(2): 129–35.

Olajos, E. J., and H. Salem. 2001. Riot control agents: Pharmacology, toxicology, biochemistry and chemistry. *J Appl Toxicol* 21(5): 355–91.

Orrenius, S., D. J. McConkey, G. Bellomo, and P. Nicotera. 1989. Role of Ca2+ in toxic cell killing. *Trends Pharmacol Sci* 10(7): 281–5.

Papirmeister, B., A. J. Feister, S. I. Robinson, and R. D. Ford. 1991. The sulphur mustard injury: Description of lesions and resulting incapacitation. In *Medical Defence against Mustard Gas: Toxic Mechanisms and Pharmacological Implications*, eds. B. Papirmeister, A. J. Feister, S. I. Robinson, and R. D. Ford, 13–42. Boca Raton, FL: CRC Press.

Papirmeister, B., C. L. Gross, H. L. Meier, J. P. Petrali, and J. B. Johnson. 1985. Molecular basis for mustard-induced vesication. *Fundam Appl Toxicol* 5(6 Pt 2): S134–49.

Pather, P., S. Hines, K. Kynoch, and F. Coyer. 2017. Effectiveness of topical skin products in the treatment and prevention of incontinence-associated dermatitis: A systematic review. *JBI Database System Rev Implement Rep* 15(5): 1473–96.

Pechura, C. M., and D. P. Rall. 1993. History and analysis of mustard agent and lewisite research programs in the United States. In *Veterans at Risk: The Health Effects of Mustard Gas and Lewisite*, ed. Institute of Medicine (US) Committee on the Survey of the Health Effects of Mustard Gas and Lewisite. Washington, DC: National Academies Press.

Pitschmann, V. 2014. Overall view of chemical and biochemical weapons. *Toxins (Basel)* 6(6): 1761–84.

Porton Report, 710. 1929a. *Experiments to Ascertain the Length of Time Heavy Winter Clothing Will Protect the Skin from Large Drops of Liquid Mustard Gas*. Salisbury.

Porton Report, 727. 1929b. *Further Report on Experiments to Ascertain the Length of Time Heavy Winter Clothing Will Protect the Skin from Large Drops of Liquid Mustard Gas*. Salisbury.

Power, S., C. Symons, H. Carter, et al. 2016. Mass casualty decontamination in the United States: An online survey of current practice. *Health Secur* 14(4): 226–36.

Proudfoot, A. T., S. M. Bradberry, and J. A. Vale. 2006. Sodium fluoroacetate poisoning. *Toxicol Rev* 25(4): 213–9.

Renshaw, B. 1946. *Mechanisms in Production of Cutaneous Injuries by Sulfur and Nitrogen Mustards*. Washington, DC: Office of Scientific Research and Development.

Rice, P. 2003. Sulphur mustard injuries of the skin. Pathophysiology and management. *Toxicol Rev* 22(2): 111–8.

Riches, J. R., R. W. Read, R. M. Black, N. J. Cooper, and C. M. Timperley. 2012. Analysis of clothing and urine from Moscow theatre siege casualties reveals carfentanil and remifentanil use. *J Anal Toxicol* 36(9): 647–56.

Roberts, G., and R. L. Maynard. 2007. Responding to chemical terrorism: Operational planning and decontamination. In *Chemical Warfare Agents: Toxicology and Treatment*, eds. T. C. Marrs, R. L. Maynard, and F. R. Sidell. Chichester: John Wiley and Sons.

Rotenberg, J. S. 2003. Cyanide as a weapon of terror. *Pediatr Ann* 32(4): 236–40.

Roul, A., C. A. Le, M. P. Gustin, et al. 2017. Comparison of four different fuller's earth formulations in skin decontamination. *J Appl Toxicol* 37(12): 1527–36.

Schecter, A., L. Birnbaum, J. J. Ryan, and J. D. Constable. 2006. Dioxins: An overview. *Environ Res* 101(3): 419–28.

Schep, L. J., R. J. Slaughter, and D. I. McBride. 2015. Riot control agents: The tear gases CN, CS and OC-a medical review. *J R Army Med Corps* 161(2): 94–9.

Schiermeier, Q. 2002. Hostage deaths put gas weapons in spotlight. *Nature* 420: 7.

Scott, L. 2007. Pretreatment for nerve agent poisoning. In *Chemical Warfare Agents: Toxicology and Treatment*, ed. T. C. Marrs, R. L Maynard and F. Sidell, 343–53. Chippenham, Wiltshire: John Wiley & Sons, Ltd.

Sidell, F. R. 1976. *Use of Physostigmine by the Intravenous, Intramuscular, and Oral Routes in the Therapy of Anticholinergic Drug Intoxication*. EB-TR-76012 Edgewood Arsenal Aberdeen Proving Ground, Maryland, USA.

Sidell, F. R., G. K. Aghajanian, W. A. Groff, and S. A. Cucinell. 1973. The reversal of anticholinergic intoxication in man with the cholinesterase inhibitor VX. *Proc Soc Exp Biol Med* 144(2): 725–30.

Simeonova, F. P., and L. Fishbein. 2004. *Hydrogen Cyanide and Cyanides: Human Health Aspects: Concise International Chemical Assessment Document*, ed. World Health Organization & International Programme on Chemical Safety, 61. Geneva: World Health Organization.

Simon, J. 2011. Stratagems, combat, and "chemical warfare" in the siege mines of Dura-Europos. *Am J Archaeol* 115(1): 69–101.

Singh, R. K., S. Kumar, D. N. Prasad, and T. R. Bhardwaj. 2018. Therapeutic journery of nitrogen mustard as alkylating anticancer agents: Historic to future perspectives. *Eur J Med Chem* 151: 401–33.

Sirin, G. S., Y. Zhou, L. Lior-Hoffmann, S. Wang, and Y. Zhang. 2012. Aging mechanism of soman inhibited acetylcholinesterase. *J Phys Chem B* 116(40): 12199–207.

Spiandore, M., M. Souilah-Edib, A. Piram, A. Lacoste, D. Josse, and P. Doumenq. 2017. Desorption of sulphur mustard simulants methyl salicylate and 2-chloroethyl ethyl sulphide from contaminated scalp hair after vapour exposure. *Chemosphere* 191: 721–28.

Stone, H., D. See, A. Smiley, A. Ellingson, J. Schimmoeller, and L. Oudejans. 2016. Surface decontamination for blister agents Lewisite, sulfur mustard and agent yellow, a Lewisite and sulfur mustard mixture. *J Hazard Mater* 314: 59–66.

Taysse, L., S. Daulon, S. Delamanche, B. Bellier, and P. Breton. 2007. Skin decontamination of mustards and organophosphates: Comparative efficiency of RSDL and Fuller's earth in domestic swine. *Hum Exp Toxicol* 26(2): 135–41.

Taysse, L., F. Dorandeu, S. Daulon, et al. 2011. Cutaneous challenge with chemical warfare agents in the SKH-1 hairless mouse (II): Effects of some currently used skin decontaminants (RSDL and Fuller's earth) against liquid sulphur mustard and VX exposure. *Hum Exp Toxicol* 30(6): 491–8.

Thomas, R. J., P. A. Smith, D. A. Rascona, J. D. Louthan, and B. Gumpert. 2002. Acute pulmonary effects from o-chlorobenzylidenemalonitrile "tear gas": A unique exposure outcome unmasked by strenuous exercise after a military training event. *Mil Med* 167(2): 136–9.

Thors, L., M. Koch, E. Wigenstam, B. Koch, L. Hagglund, and A. Bucht. 2017. Comparison of skin decontamination efficacy of commercial decontamination products following exposure to VX on human skin. *Chem Biol Interact* 273: 82–9.

Vale, J. A., T. C. Marrs, and R. L. Maynard. 2018. Novichok: A murderous nerve agent attack in the UK. *Clin Toxicol*: 1–5.

Vilensky, J. A., and K. Redman. 2003. British anti-Lewisite (dimercaprol): An amazing history. *Ann Emerg Med* 41(3): 378–83.

Vogt, B. M., and J. H. Sorrensen. 2002. *How Clean is Safe? Improving the Effectiveness of Decontamination of Structures and People Following Chemical and Biological Incidents*. Oak Ridge National Laboratory. ORNL/TM-2002/178.

Vogt, R. F., Jr., A. M. Dannenberg, Jr., B. H. Schofield, N. A. Hynes, and B. Papirmeister. 1984. Pathogenesis of skin lesions caused by sulfur mustard. *Fundam Appl Toxicol* 4(2 Pt 2): S71–83.

Wang, S. M., T. S. Shih, Y. S. Huang, M. R. Chueh, J. S. Chou, and H. Y. Chang. 2006. Evaluation of the effectiveness of personal protective equipment against occupational exposure to N,N-dimethylformamide. *J Hazard Mater* 138(3): 518–25.

Warthin, A. S., C. V. Weller, and A. Arbor. 1918. The pathology of the skin lesions produced by mustard gas (dichloroethylsulphide). *J Lab Clin Med* 3: 447–79.

Way, J. L. 1984. Cyanide intoxication and its mechanism of antagonism. *Annu Rev Pharmacol Toxicol* 24: 451–81.

Wester, R. M., H. Tanojo, H. I. Maibach, and R. C. Wester. 2000. Predicted chemical warfare agent VX toxicity to uniformed soldier using parathion in vitro human skin exposure and absorption. *Toxicol Appl Pharmacol* 168(2): 149–52.

Worek, F., H. Thiermann, L. Szinicz, and P. Eyer. 2004. Kinetic analysis of interactions between human acetylcholinesterase, structurally different organophosphorus compounds and oximes. *Biochem Pharmacol* 68(11): 2237–48.

Yang, Y. C., J. A. Baker, and J. Richard Ward. 1992. Decontamination of chemical warfare agents. *Chem Rev* 92(8): 1729–43.

26 Toxicology of Drugs of Abuse

Marta Vicente-Rodríguez, Gonzalo Herradón and Carmen Pérez-García

CONTENTS

26.1 INTRODUCTION

According to the World Drug Report (UNODC 2019), around 271 million people worldwide used illicit drugs at least once a year during 2017; 35 million suffer substance use disorder (SUD), and 585,000 people died as result of drug use; of those deaths, almost one-third were directly associated with SUD. Cannabis was the most widely consumed illicit drug in 2017 (188 million people), followed by opioids (53.4 million), amphetamines and prescription stimulants (29 million), ecstasy (21 million) and cocaine (18 million). The illicit use of prescription drugs is also a major public health concern, with opioids causing the most harm, especially when combined with benzodiazepines (BZD; UNODC 2019). Alcohol and nicotine (via cigarettes, vaping or other methods) are considered legal drugs in most countries and are not included in the World Drug Report. However, they are the most harmful drugs of abuse, based upon the number of deaths they cause. Alcohol use caused more than 3 million deaths and is responsible for more than 5% of the global disease burden in 2016 (WHO 2018). It is estimated that tobacco use kills over 7 million people each year, being the leading single preventable cause of death worldwide (WHO 2017). All of these drugs coexist with new psychoactive substances (NPS) that are appearing in the market at a dramatically growing rate.

SUD is described by the latest version of the *Diagnostic and Statistical Manual of Mental Disorders* (DSM-5; American Psychiatric Association 2013) as a problematic pattern of use of an intoxicating substance leading to clinically significant impairment or distress. SUD replaces the categories of substance abuse and substance dependence that were included in the previous version of DSM (DSM-IV-TR; American Psychiatric Association 2000). "Substance abuse" is a term that can be shaming; this negative connotation prevents people from asking for help, thus it is being replaced by "substance misuse", which suggests use that can cause harm to the user or their friends or

DOI: 10.1201/9780203730584-26

family (NIDA 2018a). Diagnosis of SUD is based on 11 diagnostic criteria (Table 26.1) occurring within a 1-year period. Depending on the number of criteria shown by the patient, this disorder is classified as mild (2–3 criteria), moderate (4–5 criteria), or severe (6 or more criteria). The National Institute on Drug Abuse (NIDA) defines addiction as "a chronic, relapsing disorder characterized by compulsive drug seeking and use despite adverse consequences", which is equivalent to a severe SUD in DSM-5 (NIDA 2018a).

All known addictive drugs activate the brain reward pathway, which is based in the mesolimbic dopamine system, and mainly involves the ventral tegmental area (VTA) and the nucleus accumbens. Although each type of addictive drug binds to different molecular targets (see Table 26.2),

Table 26.1 Criteria for the Diagnosis of Substance Use Disorders (SUD) according to DSM-5

The presence of at least 2 of these criteria in the past year indicates a SUD

SUD severity:
Mild: 2–3 criteria
Moderate: 4–5 criteria
Severe: ≥6 criteria

Criteria:

1. Using a substance in larger amounts for a longer time than desired
2. Inability to cut down despite desire to do so
3. Spending a great amount of time using a substance or recovering from its effects
4. Cravings or intense desires to use the substance
5. Repeated use of the substance results in an inability to fulfil important social or professional obligations
6. Persistent usage of a substance despite having frequent problems at work, school or home caused by its use
7. Giving up important social, professional or recreational activities because of use of the substance
8. Using a substance in physically hazardous situations
9. Continued use despite knowledge that the substance is causing or worsening a physical or mental problem
10. *Tolerance*: needing to use increasing amounts of a substance to obtain its desired effects
11. *Withdrawal*: a characteristic group of physical effects or symptoms that emerge as the amount of the substance in the body decreases

Table 26.2 Mechanism of Action Involved in the Addiction Produced by Drugs of Abuse

Drug	Main molecular target	Pharmacology	Effect on dopaminergic neurons of the reward pathway	Relative risk of addiction
Opioids	μ-opioid receptors	Agonist	Disinhibition	3
Alcohol (ethanol)	$GABA_AR$, 5-HT_3R, nAChR, NMDAR, Kir3 channels	–	Excitation	2
Cannabinoids	CB_1R	Agonist	Disinhibition	1
Nicotine	nAChR (α4β2)	Agonist	Excitation, disinhibition, modulates release	3
Cocaine	DAT, SERT, NET	Inhibitor	Blocks dopamine uptake	4
Amphetamines	DAT, NET, SERT, VMAT	Reverses transport	Blocks dopamine uptake, synaptic depletion, excitation	4
Benzodiazepines	$GABA_AR$	Positive modulator	Disinhibition	2

Abbreviations: 5-HT_3R, serotonin receptor type 3; CB_1R, cannabinoid receptor type 1; DAT, dopamine transporter; $GABA_AR$, γ-Aminobutyric acid type A receptor; Kir3: G protein inwardly rectifying potassium channels; nAChR, nicotinic acetylcholine receptor; NET, norepinephrine transporter; NMDAR, N-methyl-D-aspartate receptor; SERT, serotonin transporter; VMAT, vesicular monoamine transport.

Source: Modified from Lüscher and Ungless (2006).

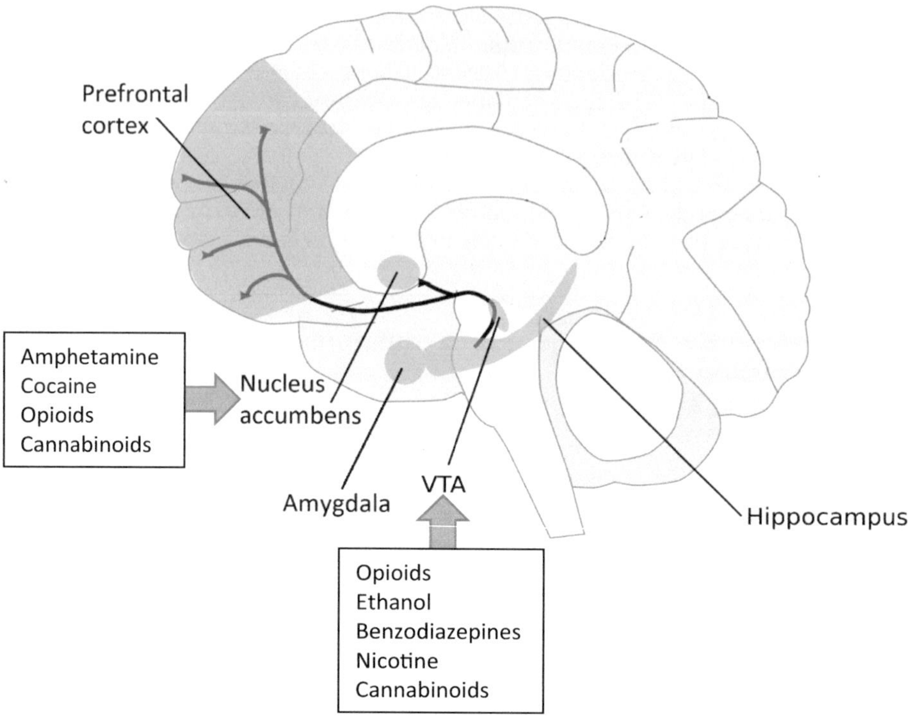

Figure 26.1 Main components of the brain reward pathway. All addictive drugs produce their initial euphoric effects (or reward) by increasing the concentration of dopamine in the projections of the ventral tegmental area (VTA) to the nucleus accumbens. To produce these effects, each drug uses specific molecular targets located either in the VTA, in the nucleus accumbens or at both locations (see Table 26.2 for further information). VTA neurons also connect with many other brain regions, including the hippocampus (memory), amygdala (emotional reactivity) and prefrontal cortex (executive function).

all of them have in common that they enhance (directly or indirectly) dopaminergic transmission in the synapsis of the VTA with the nucleus accumbens, which results in euphoria (Gardner 2011). The VTA neurons also connect with many other brain regions, including the hippocampus, amygdala, and prefrontal cortex, which are involved in learning and memory, motivation and inhibitory control over behavior (Figure 26.1). The euphoria produced by drugs may be so profound that people abandon other normal activities in favor of taking the drug. However, repeated exposure to the drug leads to changes in the brain that include the desensitization of reward circuits, which triggers tolerance to the euphoric effect; the increase of stress reactivity and the strengthening of the association of drug reward with contextual cues, which results in increased craving for the drug; and the weakening of the brain regions involved in inhibitory control, which leads to relapse (Volkow et al. 2016). These brain alterations are responsible for dose escalation (even overdose) and for the repeated use of the drug (which eventually causes addiction) and are accompanied by alterations in other organs and tissues. All of these body alterations are also responsible for the development of physical dependence that results in withdrawal symptoms when the use of the drug is reduced or abandoned.

In this chapter, we will describe the effects of the main drugs of abuse on human health with special focus on the toxic effects they produce after acute administration (intoxication or overdose) or chronic use (addiction, withdrawal or chronic toxicity). These drugs are described individually in the following sections; however, it is important to keep in mind that most drug use and SUDs involve polydrug use, some of them resulting in high toxicity (i.e. combination of central nervous system [CNS] depressants such as alcohol, opioids or BZD).

26.2 OPIOIDS

As described earlier, opioids are responsible for most of the negative health impact of illicit drug use (UNODC 2019). The mechanism of action of these drugs is well known: opioids bind and activate opioid receptors. There are different subtypes of opioid receptors, and the most widely abused opioids, such as heroin, preferentially exert their effects through mu-opioid receptors. Most opioids, like heroin, rapidly induce their effects in the brain, including the activation of the mesolimbic pathway, leading to an increase in the release of dopamine that ultimately causes the rewarding effects (Lüscher and Ungless 2006).

26.2.1 Acute Intoxication

Short-term treatment with opioids has also been reported to cause neurotoxic effects in the brain and to induce myoclonus and chorea, which refers to sudden involuntary jerking movements (Martin et al. 2018). In addition, opioids also act in different brain regions, causing important deleterious effects such as the depression of areas that control autonomic functions like breathing and heart rate, possibly leading to respiratory depression and death in case of overdose. Currently, this is critically important given the opioid overdose crisis that countries like the United States are suffering and many more are foreseeing. It is estimated that more than 40,000 persons die every year in the United States because of opioid overdose from prescription opioids, heroin, and illicitly manufactured fentanyl (National Vital Statistics System 2017). It is important to note that an opioid overdose can be treated with opioid receptor antagonists such as naloxone, which reverses the actions of opioids, including respiratory depression. Unfortunately, by the time first responders reach a person having an overdose, it is usually too late. For this reason, many efforts are being made to facilitate the access of at-risk patients to naloxone, including coprescribing naloxone and easier forms of naloxone administration, like a nasal spray and an autoinjector.

26.2.2 Chronic Toxicity

The long-term effects of opioid use are related to the neuroadaptation and changes in brain functions caused by these drugs. These include prolonged effects in different neuronal pathways affecting different neurotransmitters and neurotoxic effects in the white matter affecting the individual's decision-making, behavior and responses to stress. Importantly, these drugs are characterized by causing pronounced tolerance and physical dependence that lead to very severe withdrawal symptoms, including exacerbated muscle and bone pain, diarrhea, restlessness and insomnia. Heroin is highly addictive beyond physical dependence. Repeated heroin use often leads to opioid use disorder characterized by unchecked drug-seeking. However, not only addiction to illicit opioids but also prescription opioid misuse seems to highly contribute to the current opioid crisis. According to the Centers for Disease Control and Prevention (CDC), the economic burden of prescription opioid misuse alone in the United States is $78.5 billion a year (Florence et al. 2013). Furthermore, prescription opioid misuse seems to be a key issue in opioid addiction overall, because about 80% of people who use heroin first misused prescription opioids (Muhuri et al. 2013). A collateral effect that ultimately has become a key issue within this opioid crisis is prenatal opioid exposure, which has reached epidemic proportions; each year there are increasing numbers of children with neurodevelopmental delays and altered neural connectivity in the brain that are associated with prenatal exposure (Conradt et al. 2018).

26.2.3 Pharmacotherapy

The combination of behavioral counseling and pharmacotherapy has been successful for the treatment of opioid use disorder, as it has been proven to decrease significantly the number of deaths related to opioid overdose. Buprenorphine and methadone are key medications for the treatment of opioid use disorder. Whereas buprenorphine is a partial agonist of mu-opioid receptors and antagonist of kappa-opioid receptors, methadone is an agonist of mu-opioid receptors with a longer duration of action than other opioids such as morphine and heroin. These drugs have been shown to be efficient in controlling cravings and withdrawal symptoms, without causing a high, which is a characteristic effect of heroin. To overcome the challenge of the adherence to pharmacotherapy of patients with opioid use disorders, novel, long-acting formulations of existing therapeutics have been approved. For instance, a subdermal implant of buprenorphine that provides constant levels of this drug in plasma for up to 6 months is especially indicated for opioid-dependent individuals who have already been treated with oral buprenorphine formulations. In addition, an extended-release formulation of the opioid antagonist naltrexone has been effective in preventing relapse once detoxification has been achieved. The combination of buprenorphine and naloxone is in turn more effective to initiate the treatment of opioid dependent individuals (D'Onofrio et al. 2015). In addition to diminishing the risk of overdose, it has to be noted that these medications have been proven to be successful in reducing the risk of the patient to be involved in criminal activity and contracting HIV or hepatitis C (Marco et al. 2013). Thus, overall, these medications are useful for the recovery of patients from opioid use disorder.

26.3 ALCOHOL

According to the World Health Organization (WHO) Global Status Report on Alcohol and Health 2018 (WHO 2018), the mortality resulting from alcohol consumption is estimated to cause six deaths every minute and is higher than that caused by other diseases, such as tuberculosis, HIV/AIDS and diabetes. Despite these facts, it is difficult to sensitize society about the health consequences from alcohol consumption. Conversely, alcohol consumption is widely socially accepted; thus, a great effort is needed towards reducing the harmful use of alcohol with public health policy interventions.

26.3.1 Acute Intoxication

It is known that alcohol (or ethanol, the main component of alcoholic beverages) diffuses through fluids and tissues in the body thanks to its lipophilic properties and small molecular weight, crossing the BBB and reaching the CNS within 30 to 90 minutes. When alcohol reaches the brain, it induces behavioral and chemical alterations that lead to the well-known acute effects of alcohol such as sedation, impaired attention and emotional changes. These primary effects of alcohol are caused by the direct activation of γ-aminobutyric acid type A ($GABA_A$) receptors in the brain, enhancing the inhibitory effect of the neurotransmitter GABA and affecting neuronal excitability. Other effects of acute alcohol consumption are amnesia and sedation, which are the result of inhibition of the N-methyl-D-aspartate receptor NMDA-type glutamate receptors, leading to inhibition of the excitatory effect of glutamate. This mechanism indirectly activates the mesolimbic dopaminergic system, which explains the rewarding effect of alcohol (Lüscher and Ungless 2006).

The health and social consequences of drinking alcohol are mainly related to its potential toxicity and addiction. It has been established that alcohol is toxic to organ systems at doses above one to two drinks per day, corresponding to 14–28 grams of pure alcohol (Guo and Ren. 2010) through acetaldehyde accumulation, formation of reactive oxygen and nitrogen species, and impairment of energy homeostasis in the brain. The potential toxicity of alcohol is directly related to the dose (number of drinks) and duration of alcohol consumption; hence, heavy chronic alcohol consumption increases the risk of damage to the cardiovascular, gastrointestinal, immune and nervous systems.

Alcohol consumption in the adolescent population represents a particular serious problem. Worldwide, more than a quarter (26.5%) of all 15- to 19-year-olds are drinkers and most of them are heavy episodic drinkers (HED; WHO 2018). One of the problems related with HED is alcohol overdose, which can lead to permanent brain damage or death. Mental confusion, vomiting, difficulty remaining conscious, seizures, trouble breathing, slow heart rate and low body temperature are the main symptoms of alcohol overdose. It is known that the developing brain is especially sensitive to alcohol, hence the risk of health consequences is higher in this population. Accordingly, several publications have reported a fourfold increased risk of developing alcohol dependence with consuming alcohol before age of 13 years (Dawson et al. 2008; Hingson and Zha 2009). Furthermore, several studies have also pointed the increased risk of mental health disorders (including depression, anxiety or psychosis) and an increased risk of suicide among chronic alcohol abusers. In turn, mental health diseases often increase the risk to develop alcohol use disorders (AUD; Yang et al. 2018). Therefore, it is difficult to elucidate whether mental health issues are a risk or a consequence of alcohol abuse.

26.3.2 Chronic Toxicity

Chronic alcohol consumption decreases the release of GABA, which induces neuroadaptive responses by upregulating NMDA receptors (Clapp et al. 2008), leading to a neuronal hyperexcitability state. The rewarding effect and the neuroadaptations induced by chronic consumption of alcohol have been suggested to play an important role in the development of tolerance and withdrawal syndrome when the drug is cleared from the body or levels are considerably reduced. The withdrawal syndrome of alcohol abuse is characterized by several physical signs as tachycardia, elevated blood pressure, tremors, seizure, hallucinations, delirium tremens and psychological symptoms as distress, irritability, agitation, anxiety, dysphoria (a state of mental discomfort or suffering), sleep disturbance and anhedonia (inability to feel pleasure in normally pleasurable activities). Alcohol withdrawal syndrome is mainly treated with BZDs.

26.3.3 Pharmacotherapy

The alcohol use disorders identification test (AUDIT) is a useful screening tool for health professionals to routinely screen adults for AUD. Different pharmacological treatments, such as disulfiram, naltrexone, nalmefene and acamprosate, are available to treat AUD (Kranzler et al. 2018; Shen et al. 2018). However, only 9% of patients with AUD are prescribed such medications. Disulfiram inhibits aldehyde dehydrogenase and therefore increases the concentration of the toxic metabolite acetaldehyde, producing a reaction characterized by flushing, nauseas, vomiting, sweating, hypertension and palpitations when drinking alcohol. These adverse reactions reduce the risk of relapse. The antagonist of opioid receptors, naltrexone, reduces alcohol consumption by modulation of the rewarding effects of alcohol. Nalmefene is another opioid antagonist, with greater oral bioavailability and less liver toxicity compared to naltrexone. Acamprosate is a $GABA_A$ agonist used to prevent relapse in people that have successfully achieved abstinence from alcohol. Other drugs such as baclofen, gabapentin and topiramate also showed efficacy in the treatment of AUD. The combination of medications with intensive psychosocial intervention is essential to support the patient recovery (Kranzler et al. 2018).

26.4 CANNABINOIDS

According to the World Drug Report (UNODC 2019), cannabis continues to be the drug seized in the greatest quantities worldwide. *Cannabis sativa* and *C. indica* are the strains of plant from which different preparations are derived, such as marijuana, which refers to the dried leaves, flowers, stems and seeds from *C. sativa* (Table 26.3). The main psychoactive substance of cannabis is tetrahydrocannabinol (THC), whereas less pronounced are the psychoactive effects of other components such cannabinol (CBN) or cannabidiol (CBD).

26.4.1 Acute Intoxication

When marijuana is smoked, its active components rapidly penetrate into the brain (within 20 to 30 seconds). When it is eaten or drunk, the effects are usually observed after 30 minutes. Its active compounds are ligands of the cannabinoid receptors CB1, mainly expressed in the brain, and CB2, mainly expressed in the immune system and hematopoietic cells. Through interactions with CB1 receptors, cannabinoids overactivate the areas of the brain where they are mainly expressed such as the hippocampus, ventral tegmental area, cortex and cerebellum, which are related to memory, judgment, reward, movement, vision and coordination. Thus, the short-term effects of cannabinoids include changes in mood, altered senses, difficulty with thinking and decision-making, and impaired memory. In the case of overdose, cannabinoids may cause hallucinations, delusions or psychosis. Other effects that are commonly observed include breathing problems, cardiac dysrhythmias, and nausea and vomiting (NIDA 2018b).

26.4.2 Chronic Toxicity

The long-term effects of cannabis in teenagers are of special concern because it affects brain development. Numerous studies have shown that this drug may cause suicidal thoughts in teens and may impair thinking and learning and memory, possibly leading to permanent sequelae such as neuropsychological decline and loss of IQ points between the preteen years and adulthood (Meier et al. 2012). In adolescents, cannabis use has been related to mental illness including worsening symptoms in patients with schizophrenia and other mental health problems, such as depression or anxiety.

Table 26.3 Comparison of *Cannabis sativa* and *C. indica* strains.

C. sativa	C. indica
High THC levels	Low THC levels, high CBD levels
Psychostimulant properties	Sedative effects
Stimulates via sympathetic nervous system	Relaxes via parasympathetic nervous system
Earthy flavor	Sweet flavor
Long, thin leaves, light color	Wide, broad leaves, dark color
Hybridizing medical, creative, social use	Medical use

Cannabis is an addictive drug of abuse, and thus it may lead to a SUD. It is estimated that up to 30% of cannabis users develop this disorder (Hasin et al. 2018). Importantly, adolescents show up to a sevenfold greater risk of developing cannabis use disorder compared to adults (Winters and Lee 2008). For any drug of abuse, withdrawal plays a recognized role in relapse. Cannabis withdrawal include mental symptoms such as mood swings, sleep disruptions, cravings and anxiety, and physical symptoms such as headaches and nausea. Although not all users will experience withdrawal, cannabis withdrawal generally peaks within the first week of abstinence and predicts relapse, possibly as a result of withdrawal-related functional impairment, hampering the chances of achieving and maintaining abstinence (Balter et al. 2014; Cousijn and van Duijvenvoorde 2018). Unfortunately, little is known about the underlying mechanisms of cannabis withdrawal.

26.5 NICOTINE IN CIGARETTES AND OTHER TOBACCO PRODUCTS

As explained in the introduction, tobacco smoke is the leading cause of death related to drug use. The combustion of tobacco produces smoke that contains more than 7000 chemical compounds; among them, nicotine is the primary reinforcing component that produces rewarding and addictive effects, but some of the other chemicals, such as acetaldehyde, greatly increase the reinforcing properties of nicotine and may also contribute to tobacco addiction (Hoffman et al. 2013). Nicotine is also found in other sources such as smokeless tobacco (chewing tobacco or "spit tobacco") and e-cigarettes (electronic systems that deliver nicotine in the absence of other tobacco chemicals), however none of these nicotine sources is devoid of toxic effects, as described below.

26.5.1 Acute Effects

The cigarette is a very efficient drug-delivery system; it enables nicotine levels in the brain to peak within 10 seconds of inhalation. But the reward from of nicotine also disappears quickly, which causes the smoker to continue smoking to maintain the drug's rewarding effect. This behavior is continuously repeated and occurs in conjunction with daily routines, which remain strongly associated with nicotine's rewards. Accordingly, it has been demonstrated that smoking seems to enhance and prolong the pleasure of other activities, including the use of other drugs, which could explain the role of nicotine as a "gateway" drug to abuse of other substances (Perkins et al. 2017). When smoking, nicotine also increases memory processes and the ability to maintain attention. All the effects of nicotine in the CNS are mediated by activation of nicotinic acetylcholine receptors (NAS) located in neurons. In the periphery, nicotine activates the NAS of the adrenal glands, causing a high discharge on adrenaline that results in an increase of blood pressure and heart and respiratory rate (Benowitz 2009).

26.5.2 Chronic Toxicity

Tobacco smoking produces toxicity in most of the organs of the body. Regarding the respiratory system, tobacco is the leading cause of lung cancer and of chronic obstructive pulmonary disease; it also causes emphysema and chronic bronchitis and exacerbates asthma symptoms. Passive smoke also increases lung cancer risk by a 20% in nonsmokers and is a risk factor for asthma in children. Smoking also substantially increases the risk for cardiovascular diseases, including heart attack, aneurysm and brain stroke. In addition, long-term smoking may lead to cognitive deterioration and increases the risk of Alzheimer's disease. Cigarette use is also linked to many other diseases such as diabetes, rheumatoid arthritis, inflammation, and impaired immune function. During pregnancy, smoking is associated to poor birth outcomes including low birth weight, preterm birth and increased risk of miscarriage (NIDA 2018c). Smokeless tobacco is also a global threat to human health given the toxicological effects, in particular in the South Asian countries (Biswas et al. 2015). Smokeless tobacco comprises many toxicants and carcinogens compounds that are also found in cigarettes, such us tobacco-specific nitrosamines (TSNAs); in fact, the use of smokeless tobacco has been linked with risk of oral cancer (Janbaz et al. 2014).

26.5.3 Pharmacotherapy

Quitting smoking produces a withdrawal syndrome that begins within a few hours after the last cigarette and includes irritability, craving, anxiety, depression, attention deficit, sleep problems and increased appetite. On the other hand, quitting smoking results in immediate health benefits and the recovery of some or all of the reduced life expectancy from smoking. Most smokers would like to stop smoking, but many of them will need multiple attempts to succeed. Pharmacotherapy can help to treat nicotine dependence. Currently approved pharmacotherapies include nicotine replacement therapy (NRT), bupropion and varenicline. NRT is available in different formulations (i.e. patches, gums, inhalers, microtabs, nasal sprays and lozenges) and stimulates the brain's nicotinic receptors, which reduces the nicotine withdrawal and cravings that lead to relapse. Bupropion inhibits the reuptake and stimulates the release of noradrenaline and dopamine; it thereby increases quit rates compared to placebo and is equally effective as NRT. Varenicline is a partial agonist of α4β2 nicotinic receptors (Table 26.2) thus it attenuates the reinforcing effects of nicotine by occupying its binding sites on these receptors but also weakly stimulating nicotinic receptors, which explains its ability to reduce nicotine cravings. Varenicline increases the chances of successfully quitting compared to placebo and appears to be more effective than NRT or bupropion, according to some studies (D'Souza and Markou 2011). In addition, behavioral therapies help smokers avoid triggers of craving; therefore, combination of this type of therapies with pharmacotherapy further increases the rates of smoking cessation (NIDA 2018c).

Some smokers use e-cigarettes to quit smoking traditional cigarettes. However, it has not been proved that e-cigarettes are a valid aid for smoking cessation. In fact, a study showed that e-cigarettes produce effects in the brain similar to the ones produced by other nicotine sources (Hobkirk et al. 2018), which could indicate that e-cigarettes may also cause addiction. Users of e-cigarettes also believe that these systems are less harmful than traditional cigarettes. However, some toxicants, carcinogens and metal particles, such as TSNAs or mercury, have been detected in the liquids and aerosols of e-cigarettes; the effects of these products after repeated use are not yet completely known (Rom et al. 2015). Moreover, an increasing number of intoxications in children have been reported due to accidental ingestion of the concentrated nicotine present in e-cigarette liquid (NIDA 2018c).

26.6 COCAINE

Cocaine is a natural alkaloid extracted from the leaves of the coca plant that is cultivated mainly in Bolivia, Colombia and Peru. According to the World Drug Report (UNODC 2019), we are facing an increase in production of cocaine in Latin America, reaching the highest levels of cocaine manufacture ever recorded. In addition, although cocaine prices have remained stable, drug purity is at the highest level for over a decade in Europe; hence, not only there will be an increase in drug availability but also a higher risk of potential health consequences from cocaine abuse.

Cocaine is the most used illicit stimulant drug in Europe, according to the European Drug Report (EMCDDA 2019). In Europe, cocaine is available as cocaine powder (the salt form, cocaine hydrochloride) and as free base (generally known as crack). Cocaine is used orally, intranasally (snorted through the nostrils), intravenously or by inhalation (smoked cocaine). Cocaine addicts mostly prefer intravenous administration or inhalation, given the rapid onset (1–2 minutes) and the greatest intensity of the effects. Intranasal and oral administration are the most used routes among more socially integrated users. Intranasal use has a slower onset (5 minutes) and the effects last for 90 minutes, whereas oral administration of cocaine is the slowest route of administration (20 minutes) but the effects last for about 3 hours.

26.6.1 Acute Intoxication

Cocaine is a powerful stimulant that inhibits dopamine, noradrenaline and serotonin transporters in the presynaptic terminals, preventing the uptake of these neurotransmitters and increasing their concentration in the synaptic cleft (Huang et al. 2009). The increase in serotonin and dopamine in the CNS results in euphoria, a feeling of wellbeing and an increase in self-confidence. However, at higher doses it can produce agitation and delirium. The peripheral effects of cocaine, which are mainly due to its indirect sympathomimetic activity, include increased heart rate, blood pressure, cardiac contractility and a characteristic mydriasis. As a consequence of the sympathetic tone and the effects in the hypothalamic heat regulation center, cocaine can also produce significant hyperthermia. Overdose from cocaine can be lethal, arising from convulsions, cerebral hemorrhage, cardiac arrhythmias, ischemia and/or heart infarction (Bachi et al. 2017). An important consideration is that cocaine decreases appetite and induces sleep deprivation, alleviates fatigue and enhances mental alertness; for these reasons, adolescents have reported to use cocaine to keep going on a night out with friends, to enhance an activity, or to help stay awake. Cocaine is one of the drugs most consumed by adolescents and it is considered a club drug along with ecstasy, methamphetamine, ketamine, lysergic acid diethylamide (LSD) and gamma hydroxybutyrate (GHB), which are mainly consumed at "raves" or clandestine parties, but nowadays these are also used in college bars, house parties and even concerts (UNODC 2019). This is a major public health concern, given the evidence from different studies demonstrating that adolescents and young adults are more vulnerable to the effects of cocaine abuse (Guerri and Pascual 2018).

26.6.2 Chronic Toxicity

Chronic abuse of cocaine may lead to neurological and psychiatric complications, such as irritability, aggressiveness, stereotyped behavior (persistent mechanical repetition of speech or movement), and a paranoia-like psychosis. Interestingly, chronic cocaine–induced neurotoxicity has been linked to Parkinson's disease through different mechanisms. For instance, several studies have shown that cocaine increases the levels of alpha-synuclein, the main component of Lewy bodies in patients with Parkinson's disease, in dopaminergic neurons (Qin et al. 2005). Furthermore, free radical formation, produced by the increase in intracellular calcium in dopaminergic neurons after chronic cocaine administration, has been associated with neuronal loss (Cunha-Oliveira et al. 2014).

Cocaine withdrawal is difficult to diagnose given the nonspecific nature of the symptoms; according to the DSM-5, it includes the presence of dysphoric mood, fatigue, vivid or unpleasant dreams, insomnia or hypersomnia, increased appetite, psychomotor retardation or agitation following cessation of use. BZD and/or clonidine combined with beta-adrenoceptors and antihypertensive drugs is the treatment of choice for cocaine withdrawal. Unfortunately, there is no proven medication which shows clinically relevant effects to treat cocaine addiction (Buchholz and Saxon 2019).

26.7 AMPHETAMINE-TYPE STIMULANTS

The amphetamine-type stimulants (ATS) constitute a very broad group of synthetic stimulant substances comprising amphetamines (amphetamine and methamphetamine) and the ecstasy group of substances (e.g. 3,4-methylenedioxymethamphetamine [MDMA] and its analogues).

There has been a significant increase in the amount of ATS seized globally. Although methamphetamine accounts for the greatest global quantities of ATS seized worldwide, amphetamine is much more commonly used in Europe, according to the European Drug Report (EMCDDA 2018a). Both methamphetamine and amphetamine are available in tablet form (usually also containing caffeine and adulterants), crystal (very high purity), powder or liquid, and they are generally taken orally or intranasally. Prescription ATS are also misused. In the United States, more than 1 million people misused this type of stimulant in 2017. In the same year, 6% of high school students reported nonmedical use of Adderall, a medication used to treat attention deficit hyperactivity disorder, which contains amphetamine and dextroamphetamine (NIDA 2018d).

26.7.1 Acute Intoxication

The acute effects of amphetamines are generally called the "rush" effects and are characterized by feelings of alertness, wakefulness, energy, wellbeing, euphoria (at high doses) and suppression of appetite. The onset and duration of the effects are mainly related to the route of administration. Oral administration produces effects in 15–20 minutes and has the longest duration, whereas nasal administration only lasts 15–30 minutes but the onset of the effect is only 3–5 minutes. The use by injection increases the risk of overdose, as the rush effect starts right away and is very intense, but it only lasts a few minutes. Amphetamine overdose is generally characterized by agitation, tachycardia, hypertension and psychosis. The most recent generation of antipsychotics (also known as "atypical antipsychotics" or $5HT_{2A}/D_2$ antagonists), BZD and beta blockers are the recommended treatments for amphetamine overdose (Cao et al. 2016).

MDMA is chemically related to amphetamines but has different effects. MDMA use has increased in recent years in Europe among adults and young people. MDMA is consumed in tablet (often called ecstasy), powder and crystalline forms. The acute effects of MDMA occur within 45 minutes after administration and include an enhanced sense of wellbeing, increased extraversion, emotional warmth, empathy toward others and a willingness to discuss emotionally charged memories. MDMA overdose can be lethal with hyperthermia, hypertension, faintness, panic attacks and, in severe cases, a loss of consciousness and seizures. However, fatal overdose from MDMA is rare. Young adults or adolescents particularly use ATS in mainstream nightlife environments as well as bars and house parties. There is consistent evidence that chronic use of MDMA in adolescents induces cognitive and behavioral impairment (Montoya et al. 2002).

ATS are drugs with a high addictive potential (Table 26.2) that mainly exert their reinforcing effects through their ability to increase the levels of dopamine, serotonin and noradrenaline in the synaptic cleft. In the specific case of amphetamine, the elevation of synaptic dopamine levels occurs through several mechanisms. First, the amphetamine acts as a substrate for the dopamine transporter (DAT), competitively inhibiting the reuptake of this neurotransmitter from the synaptic cleft. Once inside the cell, amphetamine facilitates the release of dopamine from the vesicles to the cytoplasm and stimulates its release into the synaptic cleft, promoting the reverse transport by DAT (Heal et al. 2013). MDMA causes greater release of serotonin and noradrenaline than of dopamine, and thereby confers mild hallucinogen effects.

ATS induces neurotoxicity by different mechanisms, such as excitotoxicity, oxidative stress, the induction of apoptosis or hyperthermia (Cadet et al. 2009). Methamphetamine-induced excitotoxicity is mainly produced through the accumulation of the excitatory neurotransmitter glutamate; this overactivates various downstream signaling pathways triggering an increase in intracellular Ca^{2+} concentration, which leads to endoplasmic reticulum stress and activation of apoptotic pathways (Moratalla et al. 2017).

26.7.2 Chronic Toxicity

Preclinical studies have shown that chronic consumption of amphetamines produces important neurotoxic effects in nigrostriatal dopaminergic neurons; thus a greater prevalence of Parkinson's disease has been demonstrated in individuals addicted to ATS compared to both cocaine addicts and individuals not addicted to drugs (Callaghan et al. 2012). Preclinical and clinical studies showed that chronic MDMA consumption is associated with cognitive deficits including altered memory and depressed mood (Gouzoulis-Mayfrank et al. 2000), probably due to the neurotoxic effects in serotonergic neurons induced by the drug.

The "crash" effect that appears a few hours after the end of the acute effects of ATS is characterized by anhedonia, depressive symptoms right up to suicidal ideation, anxiety, irritability, fatigue and cognitive impairment. The crash effect is occasionally followed by a withdrawal period including insomnia or hypersomnia, increased appetite, psychomotor retardation (which involves a slowing-down of thought and a reduction of physical movements in an individual) and drug craving.

26.7.3 Pharmacotherapy

Despite the great efforts to develop an effective treatment for ATS dependence, none of the therapeutic options has demonstrated sufficient efficacy. Methylphenidate, naltrexone, bupropion and mirtazapine reported promising results; however, the literature suggest that overall none of them shows reproducible efficacy (Lee et al. 2018).

26.8 BENZODIAZEPINES (BZD)

Misuse of prescription drugs is a serious public health concern worldwide. The groups of prescription drugs most commonly misused are opioids, stimulants (both are described in previous sections) and, particularly, BZD. In fact, in nearly 60 countries, BZD are among the three most misused substances (UNODC 2019).

26.8.1 Acute and Chronic Toxicity

BZD are CNS depressants and include drugs as diazepam, clonazepam, alprazolam and triazolam. They facilitate the binding of the inhibitory neurotransmitter GABA on the $GABA_A$ receptor, resulting in sedative, hypnotic, anxiolytic, anticonvulsant and muscle relaxant effects. Therefore, BZD have many therapeutic uses, including the treatment of seizures, anxiety, insomnia, muscle spasm and alcohol withdrawal. BZD are safe drugs in general, and overdoses rarely produce significant morbidity or mortality. However, BZD are not prescribed for long-term use because of possible tolerance, and in some cases dependence, which can appear in just weeks. Abruptly stopping BZD use may result in withdrawal minor symptoms (irritability, anxiety, insomnia and nightmares) and even major symptoms (perceptual disturbances, psychosis, and life-threatening convulsions) that are treated with diazepam and gradual dose reduction (Soyka 2017). However, a recent study suggests that BZD use disorders are relatively rare among the adults who use prescription BZD, even if they are misusing them: in the United States, only 2% of BZD users develop SUD (Blanco et al. 2018).

The major concern regarding the misuse of BZD is their use in combination with other CNS depressants, such as opioids or alcohol. These combinations produce severe sedation, unconsciousness, and slow heart and respiratory rates, which can lead to death. In the United States, in 2015, 23% of people who died of an opioid overdose also tested positive for BZD (NIDA 2018e). This proportion is even higher in some European countries (e.g. in France, BDZ were identified in 70% of buprenorphine-related deaths; EMCDDA 2018b). In order to reduce the prevalence of these interactions, in 2016 the CDC recommended that clinicians avoid prescribing BZD concurrently with opioids, and the US Food and Drug Administration (FDA) included "black box" warnings on the labels of prescription opioids and BZD, highlighting the dangers of combining both types of drugs.

Another emerging problem related to BZD is the appearance on the illicit market of new BZD, such as phenazepam, etizolam, diclazepam and flubromazolam. As in the case of most new psychoactive substances (NPS; see Section 26.9), the pharmacology and toxicology of the new BZD are mainly unknown and they may pose higher risks to users. In Scotland, etizolam has surpassed diazepam as the BZD most frequently reported in drug-related deaths. These new BZD have been also found mixed with other new psychoactive substances, including synthetic cannabinoids. Moreover, fake diazepam tablets which contains a new potent synthetic opioid have been seized in in Europe. These tablets pose a serious risk to users, because they may be not aware they are using a potent opioid (EMCDDA 2018b).

26.9 NEW PSYCHOACTIVE SUBSTANCES (NPS)

NPS are substances of abuse that have recently become available on the market. NPS are defined by the UN Office on Drugs and Crime (UNODC) as "substances of abuse, either in a pure form or a preparation, that are not controlled under the 1961 Single Convention on Narcotic Drugs, or the 1971 Convention on Psychotropic Substances, but which may pose a public health threat" (UNODC 2018). Street names used for NPS are terms such as "legal highs", "bath salts" and "research chemicals". The appearance of NPS has turned the drug market into a dynamic entity that has grown in complexity. A total of 803 NPS were reported in the period 2009–2017 (UNODC 2018), and it is estimated that one additional new NPS is reported every week in Europe (EMCDDA 2018b).

NPS available on the market have similar effects to substances under international control. The majority are stimulants (such as aminoindanes, piperazines or synthetic cathinones), followed by synthetic opioids, synthetic cannabinoid receptor agonists and classic hallucinogens (such as phenethylamines); less abundant are sedative/hypnotics (as the new BZD mentioned in Section 26.8.1), and dissociatives (phencyclidine-type substances, which distort users' perceptions of sight and sound and create feelings of detachment or dissociation from their environment and self; UNOCC 2019). NPS present a significant challenge to public health because for most of them, little is known about potential toxicity or carcinogenic effects. Another issue is that NPS often contain several chemicals in different concentrations, making it very difficult to determine substance-specific effects. NPS users have frequently been hospitalized with severe intoxications, suffering effects ranging from seizures to agitation and aggression as well as acute psychosis. Synthetic cannabinoids have shown to produce also cardiovascular problems. Finally, another concern related to NPS is that their use occurs mostly among young adults, thus the possible long-term effects of NPS are also unknown (UNODC 2019; EMCDDA 2018b).

BIBLIOGRAPHY

American Psychiatric Association. 2000. *Diagnostic and Statistical Manual of Mental Disorders (DSM-IV-TR)*, 4th ed. Text Revision. Washington DC: American Psychiatric Association.

American Psychiatric Association. 2013. *Diagnostic and Statistical Manual of Mental Disorders (DSM-5)*, 5th ed. Washington DC: American Psychiatric Association.

Bachi, K., V. Mani, D. Jeyachandran, Z. A. Fayad, R. Z. Goldstein, and N. Alia-Klein. 2017. Vascular disease in cocaine addiction. *Atherosclerosis* 262: 154–62.

Balter, R. E., Z. D. Cooper, and M. Haney. 2014. Novel pharmacologic approaches to treating cannabis use disorder. *Curr Addict Rep* 1(2): 137–43.

Benowitz, N. L. 2009. Pharmacology of nicotine: Addiction, smoking-induced disease, and therapeutics. *Annu Rev Pharmacol Toxicol* 49: 57–71.

Beuming, T., J. Kniazeff, M. L. Bergmann, et al. 2008. The binding sites for cocaine and dopamine in the dopamine transporter overlap. *Nat Neurosci* 11(7): 780–9.

Biswas, S., K. Manna, U. Das, et al. 2015. Smokeless tobacco consumption impedes metabolic, cellular, apoptotic and systemic stress pattern: A study on Government employees in Kolkata, India. *Sci Rep* 16(5): 18284.

Blanco, C., B. Han, C. M. Jones, K. Johnson, and W. M. Compton. 2018. Prevalence and correlates of benzodiazepine use, misuse, and use disorders among adults in the United States. *J Clin Psychiatry* 79(6): pii: 18m12174.

Buchholz, J., and A. J. Saxon. 2019. Medications to treat cocaine use disorders: Current options. *Curr Opin Psychiatry* 32: 275–81.

Cadet, J. L., and I. N. Krasnova. 2009. Molecular bases of methamphetamine-induced neurodegeneration. *Int Rev Neurobiol* 88: 101–19.

Callaghan, R. C., J. K. Cunningham, J. Sykes, and S. J. Kish. 2012. Increased risk of Parkinson's disease in individuals hospitalized with conditions related to the use of methamphetamine or other amphetamine-type drugs. *Drug Alcohol Depend* 120(1–3): 35–40.

Cao, D. N., J. J. Shi, W. Hao, N. Wu, and J. Li. 2016. Advances and challenges in pharmacotherapeutics for amphetamine-type stimulants addiction. *Eur J Pharmacol* 780: 129–35.

Clapp, P., S. V. Bhave, and P. L. Hoffman. 2008. How adaptation of the brain to alcohol leads to dependence: A pharmacological perspective. *Alcohol Res Health* 31(4): 310–39.

Conradt, E., S. E. Crowell, and B. M. Lester. 2018. Early life stress and environmental influences on the neurodevelopment of children with prenatal opioid exposure. *Neurobiol Stress* (9): 48–54.

Cousijn, J., and A. C. K. van Duijvenvoorde. 2018. Cognitive and mental health predictors of withdrawal severity during an active attempt to cut down cannabis use. *Front Psychiatry* 9: 301.

Cunha-Oliveira, T., A. C. Rego, and C. R. Oliveira. 2014. Cocaine as a neurotoxin. In *Handbook of Neurotoxicity*, ed. R. Kostrzewa. New York: Springer.

Dawson, D. A., T. K. Li, and B. F. Grant. 2008. A prospective study of risk drinking: At risk for what? *Drug Alcohol Depend* 95: 62–72.

D'Onofrio, G., P. G. O'Connor, M. V. Pantalon, et al. 2015. Emergency department-initiated buprenorphine/naloxone treatment for opioid dependence: A randomized clinical trial. *JAMA* 313(16): 1636–44.

D'Souza, M. S., and A. Markou. 2011. Neuronal mechanisms underlying development of nicotine dependence: Implications for novel smoking-cessation treatments. *Addict Sci Clin Pract* 6(1): 4–16.

EMCDDA. 2018a. *European Monitoring Centre for Drugs and Drug Addiction*. European Drug Report 2018. www.emcdda.europa.eu/edr2018_en (accessed December 3, 2020).

EMCDDA. 2018b. *The Misuse of Benzodiazepines Among High-risk Opioid Users in Europe*. www.emcdda.europa.eu/topics/pods/benzodiazepines_en (accessed December 3, 2020).

EMCDDA. 2019. *European Monitoring Centre for Drugs and Drug Addiction*. European Drug Report 2018. www.emcdda.europa.eu/edr2019 (accessed December 3, 2020).

Florence, C. S., C. Zhou, F. Luo, and L. Xu. 2013. The economic burden of prescription opioid overdose, abuse, and dependence in the United States. *Med Care* 54(10): 901–6.

Gardner, E. L. 2011. Addiction and brain reward and antireward pathways. *Adv Psychosom Med* 30: 22–60.

Gouzoulis-Mayfrank, E., J. Daumann, F. Tuchtenhagen, et al. 2000. Impaired cognitive performance in drug free users of recreational ecstasy (MDMA). *J Neurol Neurosurg Psychiatry* 68(6): 719–25.

Guerri, C., and M. Pascual. 2018. Impact of neuroimmune activation induced by alcohol or drug abuse on adolescent brain development. *Int J Dev Neurosci* pii: S0736–5748(18)30251-X.

Guo, R., and J. Ren. 2010. Alcohol and acetaldehyde in public health: From marvel to menace. *Int J Environ Res Public Health* 7: 1285–301.

Hasin, D. S. 2018. US epidemiology of cannabis use and associated problems. *Neuropsychopharmacology* 43(1): 195–212.

Heal, D. J., S. L. Smith, J. Gosden, and D. J. Nutt. 2013. Amphetamine, past and present—A pharmacological and clinical perspective. *J Psychopharmacol* 27(6): 479–96.

Hingson, R. W., W. Zha, and E. R. Weitzman. 2009. Magnitude of and trends in alcohol-related mortality and morbidity among U.S. college students ages 18–24, 1998–2005. *J Stud Alcohol Drugs Suppl* (16): 12–20.

Hobkirk, A. L., T. T. Nichols, J. Foulds, et al. 2018. Changes in resting state functional brain connectivity and withdrawal symptoms are associated with acute electronic cigarette use. *Brain Res Bull* 138: 56–63.

Hoffman, A. C., and S. E. Evans. 2013. Abuse potential of non-nicotine tobacco smoke components: Acetaldehyde, nornicotine, cotinine, and anabasine. *Nicotine Tob Res Off J Soc Res Nicotine Tob* 15(3): 622–32.

Huang, X., H. H. Gu, and C. G. Zhan. 2009. Mechanism for cocaine blocking the transport of dopamine: Insights from molecular modeling and dynamics simulations. *J Phys Chem B*. 113(45): 15057–66.

Janbaz, K. H., M. I. Qadir, H. T. Basser, T. H. Bokhari, and B. Ahmad. 2014. Risk for oral cancer from smokeless tobacco. *Contemp Oncol (Poznan, Poland)* 18(3): 160–4.

Kranzler, H. R., and M. Soyka. 2018. Diagnosis and pharmacotherapy of alcohol use disorder: A review. *JAMA* 320(8): 815–24.

Lee, N. K., L. Jenner, A. Harney, and J. Cameron. 2018. Pharmacotherapy for amphetamine dependence: A systematic review. *Drug Alcohol Depend* 191: 309–37.

Linares, O. A. 2012. The Linares addictive potential model. *J Pharm Nutr Sci* 2: 132–9.

Lüscher, C., and M. A. Ungless. 2006. The mechanistic classification of addictive drugs. *PLoS Med* 3(11): e437.

Marco, A., A. López-Burgos, L. García-Marcos, C. Gallego, J. J. Antón, and A. Errasti. 2013. Is the availability of buprenorphine/naloxone therapy for opioid-dependent inmates a necessity? *Rev Esp Sanid Penit* 15(3): 105–13.

Martin, E. J., C. L. Vaughan, R. Atayee, J. M. Hirst, K. O'Donnell, and K. P. Edmonds. 2018. Hydromorphone-induced chorea as an atypical presentation of opioid neurotoxicity: A case report and review of the literature. *Palliat Med* 32(9): 1529–32.

Meier, M. H., A. Caspi, A. Ambler, et al. 2012. Persistent cannabis users show neuropsychological decline from childhood to midlife. *Proc Natl Acad Sci U S A* 109(40): E2657–E2664.

Montoya, A. G., R. Sorrentino, S. E. Lukas, and B. H. Price. 2002. Long-term neuropsychiatric consequences of "ecstasy" (MDMA): A review. *Harv Rev Psychiatry* 10(4): 212.2–20.

Moratalla, R., A. Khairnar, N. Simola, et al. 2017. Amphetamine-related drugs neurotoxicity in humans and in experimental animals: Main mechanisms. *Prog Neurobiol* 155: 149–70.

Muhuri, P. K., J. C. Gfroerer, and M. C. Davies. 2013. Associations of nonmedical pain reliever use and initiation of heroin use in the United States. *CBHSQ Data Rev.*

National Vital Statistics System. 2017. *CDC Wonder.* Atlanta, GA: US Department of Health and Human Services, CDC. https://wonder.cdc.gov (accessed December 3, 2020).

NIDA. 2018a. *The Science of Drug Use and Addiction: The Basics.* www.drugabuse.gov/publications/media-guide/science-drug-use-addiction-basics (accessed December 3, 2020).

NIDA. 2018b. *Marijuana.* Research Report Series. www.drugabuse.gov/publications/research-reports/marijuana/letter-director (accessed December 3, 2020).

NIDA. 2018c. *Tobacco, Nicotine, and E-Cigarettes.* www.drugabuse.gov/publications/research-reports/tobacco-nicotine-e-cigarettes/introduction (accessed December 3, 2020).

NIDA. 2018d. *Misuse of Prescription Drugs.* www.drugabuse.gov/publications/misuse-prescription-drugs/overview (accessed December 3, 2020).

NIDA. 2018e. *Benzodiazepines and Opioids.* www.drugabuse.gov/drugs-abuse/opioids/benzodiazepines-opioids (accessed December 3, 2020).

Perkins, K. A., J. L. Karelitz, and M. C. Boldry. 2017. Nicotine acutely enhances reinforcement from non-drug rewards in humans. *Front Psychiatry* 1: 8–65.

Qin, Y., Q. Ouyang, J. Pablo, and D. C. Mash. 2005. Cocaine abuse elevates alpha-synuclein and dopamine transporter levels in the human striatum. *Neuroreport* 16(13): 1489–93.

Rom, O., A. Pecorelli, G. Valacchi, and A. Z. Reznick. 2015. Are E-cigarettes a safe and good alternative to cigarette smoking? *Ann N Y Acad Sci* 1340: 65–74.

Shen, W. W. 2018. Anticraving therapy for alcohol use disorder: A clinical review. *Neuropsychopharmacol Rep* 38(3): 105–16.

Soyka, M. 2017. Treatment of benzodiazepine dependence. *N Engl J Med* 23;376(12): 1147–57.

UNODC (United Nations Office on Drugs and Crime). 2018. *Early Warning Advisory on New Psychoactive Substances.* www.unodc.org/LSS/Page/NPS (accessed December 3, 2020).

UNODC (United Nations Office on Drugs and Crime). 2019. *World Drug Report 2018.* (United Nations Publication, Sales No. E.18.XI.9). https://wdr.unodc.org/wdr2019/ (accessed December 3, 2020).

Volkow, N. D., G. F. Koob, and A. T. McLellan. 2016. Neurobiologic advances from the brain disease model of addiction. *N Engl J Med* 374(4): 363–71.

WHO. 2017. *WHO Report on the Global Tobacco Epidemic, 2017.* www.who.int/tobacco/global_report/en/ (accessed December 3, 2020).

WHO. 2018. *Global Status Report on Alcohol and Health 2018.* www.who.int/substance_abuse/publications/global_alcohol_report/en/ (accessed December 3, 2020).

Winters, K. C., and C. Y. Lee. 2008. Likelihood of developing an alcohol and cannabis use disorder during youth: Association with recent use and age. *Drug Alcohol Depend* 92(1–3): 239–47.

Yang, P., R. Tao, C. He, S. Liu, Y. Wang, and X. Zhang. 2018. The risk factors of the alcohol use disorders—Through review of its comorbidities. *Front Neurosci* 11(12): 303.

27 Case Studies in Human Toxicology 1

Human Health Risk Assessments for Arsenic and Beryllium in Urban Soils

Antonio Peña-Fernández, María José González-Muñoz, Gevorg Tepanosyan, Mark D. Evans and María del Carmen Lobo-Bedmar

CONTENTS

27.1 INTRODUCTION

Risk assessment (RA) is an effective tool in the prevention of risks to human health, because it allows identification of potential risks from toxic and carcinogenic substances in a fast, simple, and relatively inexpensive way (US EPA 1989). Thus, RA is broadly used to establish norms and measures for protecting human health and the environment for a range of professionals, including toxicologists and public health experts, researchers, risk assessors and regulators/policy makers. For example, RA was critical for the prohibition of the use of tetraethyl-lead and other derivatives of this hazardous metal as an antiknock agent in gasoline owing to its deleterious effects on the nervous system and child development (Carroquino et al. 2016). As described in Chapter 16, the characterisation of risks is the last step of the RA process to determine whether a substance monitored in soil, water, food and so forth will present a risk to humans (US EPA 1989), and whether implementation of more complex processes are required, such as for prevention and/or decontamination and remediation programmes (Peña-Fernández et al. 2014a; Wyke et al. 2014). In this case study chapter, you will find a description of novel guidelines, tools and resources to aid the development of appropriate and effective plans for the decontamination and recovery of environments impacted by chemical hazards.

The main aim of this case study is to briefly describe how to carry out RA as described by the US Environmental Protection Agency (US EPA). To show this, we will use previous data described by our research group in a previously well-documented RA study of metals and metalloids present in topsoils in Alcalá de Henares, Spain (Peña-Fernández et al. 2014b). This example has been selected due to the relevance of soils to human health (Poggio et al. 2009), as described in Section 27.2.

DOI: 10.1201/9780203730584-27

27.2 RELEVANCE OF SOIL IN HUMAN HEALTH

Soil is the target environmental compartment for pollutants emitted by both natural and anthropogenic sources (Lu et al. 2011), and it is strongly interrelated with the health of humans and the ecosystem. Thus, anything that affects the quality of the soil will have an impact in the short or long term on humans. Due to the economic relevance of soils, as it supports most human activities, developed countries (including the European Union [EU] in the recent EU biodiversity strategy for 2030; European Commission 2020) have started to pay special attention to soil protection, and carefully regulate how to evaluate and manage contaminated and degraded soils, as indicated in the new EU strategy about healthy soils (European Commission 2020), most of which is based on the RA previously developed by the US EPA in 1989.

As a consequence of various anthropogenic activities and rapid economic development, the presence of chemicals, including metals and metalloids, is increasing in the urban environment, representing a serious concern for human health (Liu et al. 2020). Anthropogenic or anthropic pollution is especially relevant in the urban environment because it involves most of the polluting activities and contains nearly half of the world's population, not least due to the rapid urbanisation phenomenon (Wang et al. 2020). Therefore, it is necessary to quantify substances of toxicological relevance in the environment and their exposure in humans to assess their possible risks (Wyke et al. 2014).

27.3 ARSENIC AND BERYLLIUM

Metals and metalloids are characterised by their persistence and nonbiodegradability in the environment. Unlike other environmental pollutants, metals are elements that man neither creates nor destroys but rather introduces into the environment as a consequence of anthropogenic activities and directly, chemically or biologically altered (Poggio et al. 2009). Thus, the relevance of undertaking regular monitoring programmes for the characterisation of risks to humans due to the environmental presence of these inorganic pollutants is enormous, because they can present a marked multiplicity of toxic and/or carcinogenic effects. We will carry out an RA on the levels of arsenic and beryllium monitored in urban and industrial soils in Alcalá de Henares described in Peña-Fernández et al. (2014b), as they are potent human carcinogens.

Arsenic (As) is a metalloid that has great chemical reactivity, being found in nature mainly in inorganic form. Its most common oxidation states are As(0) and As(III) (arsenite), both typical of reducing environments, and As(V) (arsenate), which is characteristic of aerobic conditions and is the dominant oxidation state in most of the soils; As(III) is more toxic and mobile than As(V) (Zhang and Selim 2008).

Arsenic is classified in group 1 of carcinogenic substances (IARC 2004). This metalloid is used in herbicides, insecticides, wood preservatives, veterinary drugs, glass manufacture, ceramics, semiconductors and so forth. Chronic exposure to inorganic As produces a variety of harmful effects on different organs (Jomova et al. 2011) and has great carcinogenic potential for the liver and lung, and it is also potentially leukemogenic. Additionally, As adversely affects the central nervous system, producing asthenia, muscle weakness, headaches and peripheral neuropathies (Aelion et al. 2008).

Beryllium (Be) and its compounds are also classified in group 1 as carcinogenic substances for animals and people (IARC 1993). Beryllium is a light metal whose main source of release to the environment is through the combustion of coal. The most important uses of this metal and its compounds are the manufacture of alloys, the phosphorous industry, the ceramic industry, defence, the space industry, jewellery, and the manufacture of electrical and electronic material (Kolanz et al. 2001). The most common toxic effect of Be is contact dermatitis. Chronic exposure can trigger what is known as berylliosis, which is characterised by the presence of cough, dyspnoea, and reduced lung function (Thomas et al. 2009).

27.4 MATERIALS AND METHODS

27.4.1 Site Characteristics

Alcalá de Henares is a UNESCO World Heritage City and municipality of the Community of Madrid. It is 31 km from Madrid (on the A-2 motorway), 15 km from Madrid-Barajas international airport and 22 km from Guadalajara. It has a population of 198,723, which makes it one of the most populated urban areas in the Community, after the capital itself and Móstoles. The outskirts of the city are characterised by numerous industrial estates with moderate to low industrial activity. The city's climate is a continental Mediterranean type characterised by hot, dry summers and cold, wet winters.

27.4.2 Soil Analysis and Characterisation

In July 2001, 120 topsoil samples were collected from urban and industrial areas across Alcalá de Henares. Only the upper part of the soils (the first 3 cm) was sampled, as it is the area that is in direct contact with humans. The spatial distribution of commercial and industrial enterprises, public parks (where topsoil samples were collected, all in the urban area), roads/streets and residential buildings in Alcalá de Henares are depicted in Figure 27.1. ArcGIS 10.1 software was used for creating this map, following previous methodologies (Tepanosyan et al. 2017).

The US EPA provides different reference levels for pollutants in both types of soils (i.e. urban and industrial), as they have different characteristics and therefore different potential implications for human health. These values can be easily compared, as indicated in Peña-Fernández et al. (2014b), for quick screening of potential risks to humans.

Samples were desiccated at 30°C and sieved to a diameter of 2 mm. Subsequently, the soils were subjected to acid mineralisation to determine the total content of As and Be, following the methodology previously described by Schuhmacher et al. (1996). Briefly, 0.70 g of each sample was digested with 5 mL of concentrated nitric acid (65%) in hermetic Teflon digesters for 8 hours at room temperature, and then 12 hours at 96°C. At this time, the solutions were filtered and made up to 25 mL with Milli-Q ultrapure demineralised water. As and Be concentrations were determined by inductively coupled plasma mass spectrometry (ICP-MS, Perkin Elmer Elan 6000) following previous methodologies.

Topsoil was also characterised by determining different physicochemical properties including organic matter content and the granulometry (percentage of clay, sand and silt), as these properties can affect the presence and bioavailability of the elements in the soil, and therefore the RA. This data, which has been previously reported by our group in Peña-Fernández et al. (2015), is not described here to simplify the aim of this case study chapter.

Carcinogenic and noncarcinogenic risks were calculated for the total concentration of As and Be monitored in Alcalá for children and adults based on US EPA Risk Assessment Guidance for Superfund (RAGS) methodology (US EPA 1989).

27.4.3 Human Health Risks

Many of the values and indices necessary for the characterisation of risk to substances present in soils are described in the Regional Screening Levels (RSLs) tables, which are available at the US EPA website[1] and are periodically modified (RSLs, Region IX, San Francisco, US EPA, May 2020), as further knowledge of different xenobiotics emitted into the environment and their health effects is acquired.

27.4.3.1 Exposure Calculation

To characterise the carcinogenic and noncarcinogenic risks from environmental pollutants, it is necessary to first determine the level to which humans are externally exposed; that is, to quantitatively determine the pollutant present in the environment that may come into contact with humans. This estimate must consider the different routes of entry for environmental pollutants: intake (oral), inhalation and dermal absorption. Although direct intake is the main route when evaluating exposure to contaminated soils, the relevance of inhalation of resuspended particles and the dermal absorption of contaminants present in the soil will depend on the physicochemical characteristics of the pollutant and the characteristics of the site. Thus, the most representative routes of exposure to metals and metalloids present in soils will be the oral route (ingestion of contaminated soil particles) and the inhalation route (particle inhalation). Dermal exposure will be very limited for metals present in soils, and it is not significant in this case.

27.4.3.1.1 Soil Intake and Oral Exposure Prediction

The soil intake value, or oral exposure, can be predicted for adults and children using the index of involuntary daily soil intake described by the US EPA (RAIS 2020; Table 27.1), which has been obtained from various epidemiological studies.

The child population has an index of involuntary daily soil intake higher than the adult population. This is attributed to their behavioural characteristics (play activities outside the home and hand-to-mouth, poor hygiene habits, etc.; Madrid et al. 2008). Adults, on the other hand, have a much lower rate of involuntary ingestion of soil because they generally spend limited time in contact with the soil.

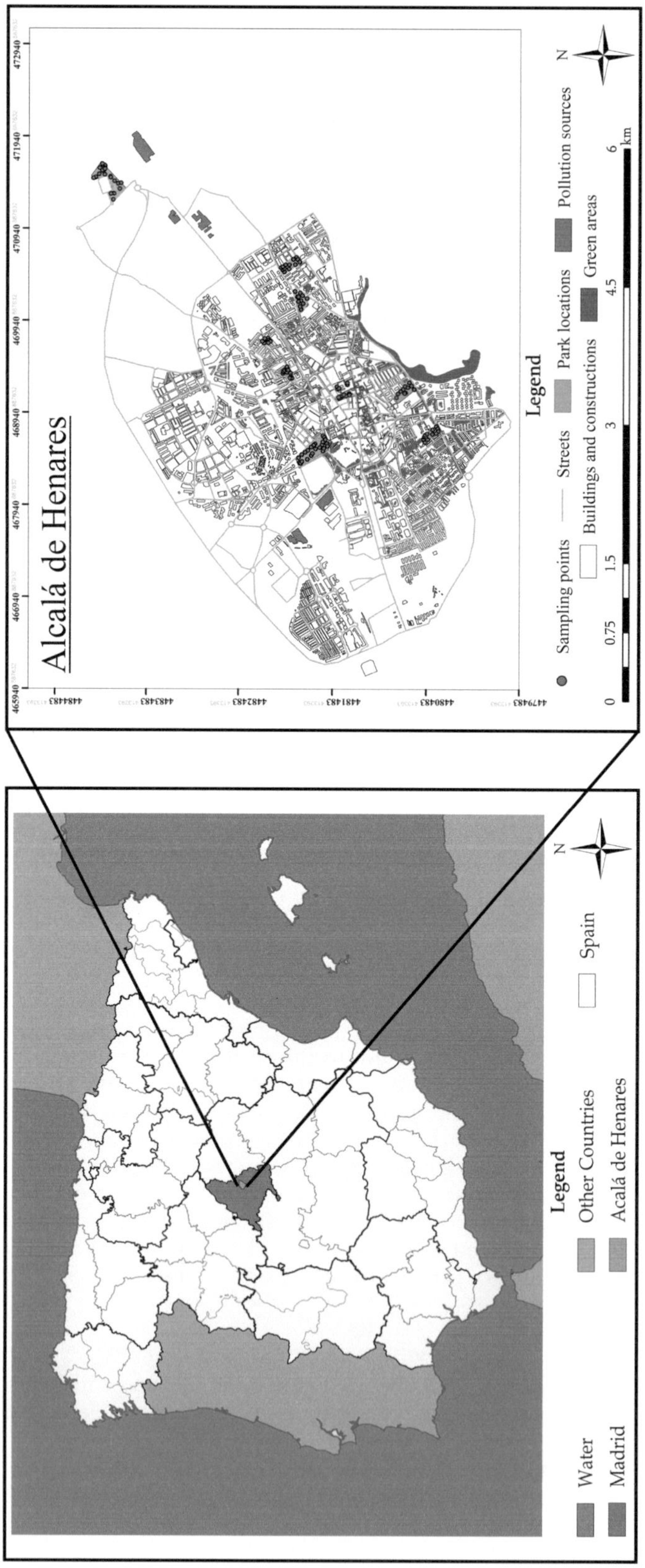
Alcalá de Henares
465940
466940
467940
468940
469940
470940
471940
472940
4484483
4483483
4482483
4481483
4480483
4479483
Legend
Sampling points
Streets
Park locations
Pollution sources
Buildings and constructions
Green areas
0 0.75 1.5 3 4.5 6 km
N
Legend
Water
Madrid
Other Countries
Acalá de Henares
Spain
N

Figure 27.1

Table 27.1 US EPA Criteria for Estimating Exposure to Pollutants in Soils

Population	Ingestion rate of soil (IRS) (mg/day)	Body weight (kg)
Adults	50	70
Children	200	15

Source: RAIS (2020).

The daily dose of chronic oral exposure or chronic daily intake (oral CDI) is expressed in milligrams of the compound/pollutant per kilogram of body weight per day; it is determined using Equation 27.1 (US EPA 2011):

$$CDI_{oral}(mg/kg \times day) = \left[(C_{soil} \times IRS \times EF \times ED \times RBA)/(BW \times AT)\right] \times CF \tag{27.1}$$

where

C_{soil} = contaminant concentration in soil (i.e. mean value of As or Be found).
IRS = ingestion rate of soil (Table 27.1). These values may change depending on information pertinent to the geographical location, the population studied, and if such empirical data are available rather than theoretical indices.
EF = exposure frequency: 365 days/year.
ED = exposure duration: 30 years for adults and 6 years for children (RAIS 2020).
RBA = relative bioavailability factor: 0.6 for As and 1.0 for Be (RAIS 2020).
BW = body weight (kg).
AT = averaging time: 70 years × 365 days for carcinogenic substances; ED × 365 days for noncarcinogenic substances.
CF = unit conversion factor (1×10^{-6} kg/mg).

However, not all the content of a pollutant in the soils will be directly bioaccessible to humans. Thus, to avoid an overestimation of risks to human health, the US EPA uses a "relative" bioavailability factor (RBA), as indicated in Equation 27.2 (Li et al. 2014; RAIS 2020). Furthermore, the risk of the working adult population (i.e. one who works in the monitored industrial area) could be also quantified. For this, an EF of 250/365 days is considered instead, which is equivalent to the time that this group remains in the workplace (5 hours a week for 50 weeks per year). This population group will not been considered in this chapter, but more information can be found in detail in Peña-Fernández et al. (2014b).

27.4.3.1.2 Inhalation of Resuspended Particles and Prediction Inhalation Exposure

The exposure through inhalation can be determined by the Inhalation Dosimetry Methodology developed by the US EPA (2009), which considers that the amount of the pollutant that reaches the target organ by inhalation would be directly related to the exposure concentration, or EC (US EPA 2009; Rovira et al. 2010). EC can be calculated using Equation 27.2:

$$EC_{inh} = (C_{air} \times ET \times EF \times ED)/AT \tag{27.2}$$

where

EC_{inh} = exposure concentration (μg/m^3)
C_{air} = contaminant concentration in air (μg/m^3)
ET = exposure time (24 hours/day)
EF = exposure frequency: 365 days/year
ED = exposure duration: 30 years for adults; 6 years for children
AT = averaging time: 70 years × 365 days × 24 hours/day for carcinogenic substances; ED × 365 days × 24 hours/day for noncarcinogenic substances.

C_{air} is site specific (it will depend on the geographical area or location), but it can be inferred by using the hypothesis of Hawley (1985), which assumes that 50% of the aerial particles that are in suspension (particles that are going to be inhaled by humans) come from the resuspension of local soils. That is, the concentration of a pollutant present in these particles would be half of that determined in the soil, and is defined as F_{res} (fraction of resuspended particles from the soil). C_{air} could then be estimated using Equation 27.3, which uses F_{res}:

$$C_{air} = C_{soil} \cdot C_{pa} \cdot F_{res} \cdot 10^{-6} \quad (27.3)$$

where
C_{air} = contaminant concentration in air (μg/m^3)
C_{soil} = contaminant concentration in soil (μg/g)
C_{pa} = concentration of particles in air (μg/m^3)
F_{res} = fraction of resuspended particles from the soil = 0.5 (dimensionless)
10^{-6} = unit converter.

The values of PM_{10} (particulate matter) estimated in the monitored area are used as the concentration of particles in air (C_{pa}). PM_{10} refers to the small particles that are in suspension that are able to penetrate the tracheobronchial section of the airways, with diameters that are 10 micrometres and smaller. Thus, owing to their impact on human health, environmental agencies and departments commonly monitor the concentration of PM_{10} in cities and regions. In this case study chapter, we use the annual average of PM_{10} of 29.04 μg/m^3 registered in the city of Alcalá de Henares in the year of soil sampling (General Directorate of the Environment of the Community of Madrid, Air Quality Network; Comunidad de Madrid 2020).

27.4.3.2 Noncarcinogenic or Toxic Risk Characterisation

Once exposure is assessed, risk characterisation can be undertaken for the substances monitored, for oral and inhalation routes, using Equation 27.4 (US EPA 1989):

$$HQ\,(\text{Hazard Quotient}) = Exp / RfD \quad (27.4)$$

where Exp = exposure (mg/kg per day)
RfD = reference dose (mg/kg per day), which is chemical specific.

When considering more than one contaminant, a hazard index (HI) can be determined using Equation 27.5:

$$HI\,(\text{Hazard Index}) = \sum HQi \quad (27.5)$$

where
HQ_i = hazard quotient for each *i* contaminant (dimensionless)

Noncarcinogenic risks derived from the oral and inhalation exposures can be characterised as follows.

27.4.3.2.1 Oral Toxicity Risk

Oral noncarcinogenic risks are quantified by comparing the chronic daily intake for each xenobiotic with its RfDo, which is chemical specific, by using Equation 27.6 (US EPA 1989):

$$HQ_{oral} = CDI_{oral} / RfDo \quad (27.6)$$

where
HQ_{oral} = hazard quotient oral (dimensionless)
CDI_{oral} = oral chronic daily intake (mg/kg per day)
RfDo = oral reference dose (mg/kg per day).

RfDo and reference concentration values (RfCi) are provided for different chemicals in the RSL tables published by the US EPA (2020); Table 27.2 collects these values for As and Be.

Table 27.2 Oral and Inhalational Reference Dose Values for Arsenic and Beryllium

Element	RfDo (mg/kg per day) (Oral exposure)	RfCi (mg/m^3) (inhalation exposure)
Arsenic (inorganic)	3.0E−04	1.5E−05
Beryllium and compounds	2.0E−03	2.0E−05

Source: US EPA (2020).

27.4.3.2.2 Inhalation Toxicity Hazard

Noncarcinogenic risks derived from the inhalation exposures are estimated by comparing the exposure by inhalation of resuspended particles with its RfCi, using Equation 27.7 (US EPA 2009):

$$HQ_{inh} = EC_{inh} / \left(RfCi \times 1000 \mu g / mg\right) \quad (27.7)$$

where
HQ_{inh} = hazard quotient through inhalation (dimensionless)
EC_{inh} = inhalation exposure concentration (μg/m^3)
RfCi = inhalation reference concentration (mg/m^3).

The RfCi values for As and Be are provided in the US EPA (2020) RSL tables (Table 27.2).

27.4.3.3 *Cancer Risk Characterisation*

Cancer risk can only be quantified for those substances for which a carcinogenic effect has been described, such as As and Be, which have been classified as carcinogenic by the International Agency for Research on Cancer (IARC 1993, 2004). Cancer risk is quantified as an excess of risk for each route of exposure (oral and inhalation), as follows.

27.4.3.3.1 Oral Carcinogenic Risk

Cancer risk through ingestion is determined for each contaminant by multiplying its chronic daily intake by the oral cancer slope factor or SFo (US EPA 1989), which is specific for each chemical (Equation 27.8):

$$Risk_{oral} = CDI_{oral} \times SFo \quad (27.8)$$

where
$Risk_{oral}$ = cancer risk through ingestion (dimensionless)
CDI_{oral} = chronic daily intake (mg/kg per day)
SFo = oral cancer slope factor (mg/kg per day)$^{-1}$.

To date, the US EPA has provided SFo values only for As (1.50E+00; US EPA 2020), so the oral carcinogenic risk for Be could not be estimated.

27.4.3.3.2 Inhalation Carcinogenic Risk

Cancer risk through inhalation is assessed by multiplying the inhalation concentration exposure by the inhalation unit risk (IUR) factor for each substance, as described by the US EPA (2009); see Equation 27.9:

$$Risk_{inh} = EC_{inh} \times IUR \quad (27.9)$$

where
$Risk_{inh}$ = cancer risk through inhalation (dimensionless)
EC_{inh} = inhalation concentration exposure (μg/m^3)
IUR = inhalation unit risk (μg/m^3)$^{-1}$.

IURs are chemical specific and are collected in the RSL tables (US EPA 2020); Table 27.3 reflects the current values for the contaminants studied in this chapter.

Table 27.3 Inhalation Unit Risk (IUR) Factor for As and Be

Element	IUR (1/(μg/m³))
Arsenic (inorganic)	4.30E−03
Beryllium and compounds	2.40E−03

Source: US EPA (2020).

Table 27.4 Chronic Daily Intake through Ingestion Values for As and Be Monitored in Urban Topsoils from Alcalá de Henares, Spain

	CDI (mg/kg per day) Noncarcinogenic		CDI (mg/kg per day) Carcinogenic	
Element	Adults	Children	Adults	Children
As	2.07E−07	3.86E−06	8.87E−08	3.31E−07
Be	5.36E−08	1.00E−06	2.30E−08	8.57E−08

Table 27.5 Inhalation Exposure Concentration Values for As and Be Monitored in Urban Topsoils from Alcalá de Henares, Spain

	EC_{inh} (μg/m³) Noncarcinogenic		EC_{inh} (μg/m³) Carcinogenic	
Element	Adults	Children	Adults	Children
As	6.31E−03	2.52E−04	1.47E−02	2.95E−03
Be	9.80E−04	3.92E−05	2.29E−03	4.57E−04

Table 27.6 Noncarcinogenic Risks through Ingestion and Inhalation for As and Be Detected in Urban Topsoils in Alcalá de Henares, Spain

	HQ_{oral}		HQ_{inh}	
Element	Adults	Children	Adults	Children
As	6.90E−04	1.29E−02	4.21E−01	1.68E−02
Be	2.68E−05	5.00E−04	4.90E−02	1.96E−03

27.4.4 Assessment of Results

There are different internationally accepted criteria to assess the results of risk characterisation. Hazard quotients/indices below the unity would suggest that there is no possibility of adverse health effect to chronic exposure to noncarcinogenic substances (US EPA 2011).

For carcinogenic substances, the tolerable limit of the increase in cancer risk must be 1.0×10^{-6}; that is, a probability of developing a cancer of 1 in 1 million. It will be unacceptable to human health if it exceeds 1.0×10^{-4} (US EPA 2011).

27.5 CASE STUDY RESULTS

We can now apply the above equations, values and hypothesis to characterise the risks to As and Be found in topsoils in Alcalá de Henares, Spain, at total concentrations (4.83 ± 2.10 μg/g and 0.75 ± 0.50 μg/g, respectively) described in Peña-Fernández et al. (2014b).

First, we can quantify the chronic daily intake for these two elements through ingestion of topsoils (CDI_{oral}) for children and adults living in Alcalá (Table 27.4).

Table 27.5 shows the inhalation exposure concentration (EC_{inh}) values obtained for each element and each population group. The C_{air} value for As is 7.01E−06 μg/m³ and that for Be is 1.09E−06 μg/m³.

27.5.1 Noncarcinogenic or Toxic Risk

The calculated hazard quotients through ingestion and inhalation are shown in Table 27.6 for each element and group of the population studied.

Table 27.7 Cancer Risk through Ingestion and Inhalation for As and Be Detected in Urban Topsoils in Alcalá de Henares, Spain

	$Risk_{oral}$		$Risk_{inh}$	
	Adults	**Children**	**Adults**	**Children**
As	1.33E−07	4.97E−07	6.33E−05	1.27E−05
Be	–	–	5.49E−06	1.10E−06

These results indicate that, for all the groups studied, there would be no potential risk of suffering any possible pathology of toxic etiology derived from the ingestion/inhalation of As and Be present in these soils and resuspended topsoils, because the risk ratios obtained are below the safety level (unity). Moreover, the hazard index HI_{oral} (7.17E−04, 1.34E−02) and HI_{inh} (4.70E−01, 1.88E−02) for adults and children, respectively, have also not exceeded the considered safety threshold (i.e. they are lower than 1).

27.5.2 Carcinogenic Risk

Cancer risks derived from the exposure to As and Be for the population living in Alcalá de Henares, Spain, are shown in Table 27.7. Risks to Be through ingestion are not shown, as an SFo has not been described for this element to date.

Although there would be no potential risk of developing a type of cancer throughout life due to oral exposure to As in the studied soils for the population groups considered, the exposure to As and Be through the inhalation of resuspended soils could *a priori* produce an increase in the cancer incidence for the population. However, the fact that the risk values obtained are within the range considered as acceptable for carcinogens (1.0E−06 to 1.0E−04; US EPA 2011), the carcinogenic risks derived from the ingestion and/or inhalation of As and Be present in soils and resuspended soils in the monitored area would be minimal.

27.6 CONCLUSIONS

In general, the risks of ingestion/inhalation of As and Be present in the selected urban soils in Alcalá de Henares, Spain, could be considered minimal, although an excess of carcinogenic risk has been observed specifically for As through the inhalation of resuspended topsoils, which would require careful monitoring to protect the public. Further potential solutions to protect the health of citizens of Alcalá de Henares could also include the implementation of public health interventions and environmental policy regulations to minimise the release of As and Be in this city and surroundings, as well as decontamination/recovery techniques.

NOTE

1 US EPA, 2020. *Region IX, Regional Screening Levels (Formerly PRGs).* San Francisco, CA 94105. www.epa.gov/region9/superfund/prg/ (accessed January 1, 2020).

BIBLIOGRAPHY

Aelion, C. M., H. T. Davis, S. McDermott, and A. B. Lawson. 2008. Metal concentrations in rural topsoil in South Carolina: Potential for human health impact. *Sci Total Environ* 402(2–3): 149–56.

Carroquino, M. J., A. Peña-Fernández, R. Duarte-Davidson, J. M. Ordóñez, and P. Martín-Olmedo. 2016. Risk characterization. In *La Evaluación de riesgos en salud. Guía metodológica. Aplicaciones prácticas de la metodología de Evaluación de riesgos en salud por exposición a químicos*, 129–52. Madrid: Sociedad Española de Sanidad Ambiental y Escuela Andaluza de Salud Pública.

Comunidad de Madrid. 2020. *Red de Calidad del Aire de la Comunidad de Madrid.* http://gestiona.madrid.org/azul_internet/run/j/AvisosAccion.icm (accessed December 9, 2020).

European Commission. 2020. New soil strategy—Healthy soil for a healthy life. In *EU Biodiversity Strategy for 2030.* Ref. Ares(2020)6391319–05/11/2020. https://ec.europa.eu/info/law/better-regulation/have-your-say/initiatives/12634-Healthy-soils-new-EU-soil-strategy (accessed December 9, 2020).

Hawley, J. K. 1985. Assessment of health risk from exposure to contaminated soil. *Risk Anal* 5(4): 289–302.

IARC (International Agency for Research on Cancer). 1993. *Monographs on the Evaluation of the Carcinogenic Risk of Chemicals to Human. Be, Cd, Hg, and Exposures in the Glass Manufacturing Industry*, Vol. 58. Lyon, France.

IARC (International Agency for Research on Cancer). 2004. *Monographs on the Evaluation of the Carcinogenic Risk of Chemicals to Human. Some Drinking-water Disinfectants and Contaminants, Including Arsenic*, Vol. 84. Lyon, France.

Jomova, K., Z. Jenisova, M. Feszterova, et al. 2011. Arsenic: Toxicity, oxidative stress and human disease. *J Appl Toxicol* 31(2): 95–107.

Kolanz, M. E. 2001. Introduction to beryllium: Uses, regulatory history, and disease. *Appl Occup Environ Hyg* 16(5): 559–67.

Li, J., Y. Wei, L. Zhao, et al. 2014. Bioaccessibility of antimony and arsenic in highly polluted soils of the mine area and health risk assessment associated with oral ingestion exposure. *Ecotoxicol Environ Saf* 110: 308–15.

Liu, L., Q. Liu, J. Ma, et al. 2020. Heavy metal(loid)s in the topsoil of urban parks in Beijing, China: Concentrations, potential sources, and risk assessment. *Environ Pollut* 260: 114083.

Lu, Y., W. Yin, L. Huang, G. Zhang, and Y. Zhao. 2011. Assessment of bioaccessibility and exposure risk of arsenic and lead in urban soils of Guangzhou city, China. *Environ Geochem Health* 33(2): 93–102.

Peña-Fernández, A., S. Wyke, N. Brooke, and R. Duarte-Davidson. 2014a. Factors influencing recovery and restoration strategies following a chemical incident. *Environ Int* 72: 98–108.

Peña-Fernández, A., M. J. González-Muñoz, and M. C. Lobo-Bedmar. 2014b. Establishing the importance of human health risk assessment for metals and metalloids in urban environments. *Environ Int* 72: 176–85.

Peña-Fernández, A., M. C. Lobo-Bedmar, and M. J. González-Muñoz. 2015. Annual and seasonal variability of metals and metalloids in urban and industrial soils in Alcalá de Henares (Spain). *Environ Res* 136: 40–6.

Poggio, L., B. Vrščaj, R. Schulin, E. Hepperle, and F. Ajmone-Marsan. 2009. Metals pollution and human bioaccessibility of topsoils in Grugliasco (Italy). *Environ Pollut* 157(2): 680–9.

RAIS. 2020. *Risk Exposure Models for Chemicals User's Guide*. https://rais.ornl.gov/tools/rais_chemical_risk_guide.html (accessed December 9, 2020).

Rovira, J., M. Mari, M. Nadal, M. Schuhmacher, and J. L. Domingo. 2010. Environmental monitoring of metals, PCDD/Fs and PCBs as a complementary tool of biological surveillance to assess human health risks. *Chemosphere* 80(10): 1183–9.

Schuhmacher, M., S. Granero, M. Bellés, J. M. Llobet, and J. L. Domingo. 1996. Levels of metals in soils and vegetation in the vicinity of a municipal solid waste incinerator. *Toxicol Environ Chem* 56: 119–32.

Tepanosyan, G., L. Sahakyan, O. Belyaeva, N. Maghakyan, and A. Saghatelyan. 2017. Human health risk assessment and riskiest heavy metal origin identification in urban soils of Yerevan, Armenia. *Chemosphere* 184: 1230–40.

Thomas, C. A., R. L. Bailey, M. S. Kent, D. C. Deubner, K. Kreiss, and C. R. Schuler. 2006. Efficacy of a program to prevent beryllium sensitization among new employees at a copper-beryllium alloy processing facility. *Public Health Rep* 2009; 124(Suppl 1): 112–24.

US EPA. 1989. *Risk Assessment Guidance for Superfund. Volume 1: Human Health Evaluation Manual (Parte A)*. EPA/540/1–89/002. https://www.epa.gov/risk/risk-assessment-guidance-superfund-rags-part (accessed September 29, 2021).

US EPA. 2009. *Risk Assessment Guidance for Superfund Volume I: Human Health Evaluation Manual (Part F, Supplemental Guidance for Inhalation Risk Assessment)*. EPA/540/R/070/002. https://www.epa.gov/risk/risk-assessment-guidance-superfund-rags-part-f (accessed September 29, 2021).

US EPA. 2011. *Exposure Factors Handbook 2011 Edition (Final)*. http://cfpub.epa.gov/ncea/risk/recordisplay.cfm?deid=236252 (accessed December 9, 2020).

US EPA. 2020. *Region IX, Regional Screening Levels (Formerly PRGs)*. San Francisco, CA 94105. www.epa.gov/region9/superfund/prg/ (accessed December 9, 2020).

Wang, X., E. Liu, Q. Lin, L. Liu, H. Yuan, and Z. Li. 2020. Occurrence, sources and health risks of toxic metal(loid)s in road dust from a mega city (Nanjing) in China. *Environ Pollut* 263(Pt A): 114518.

Wyke, S., A. Peña-Fernández, N. Brooke, and R. Duarte-Davidson. 2014. The importance of evaluating the physicochemical and toxicological properties of a contaminant for remediating environments affected by chemical incidents. *Environ Int* 72: 109–18.

Zhang, H., and H. M. Selim. 2008. Reaction and transport of arsenic in soils: Equilibrium and kinetic modeling. In *Advances in Agronomy*, Vol. 98, ed. D. L. Sparks, 45–115. San Diego: Elsevier Academic Press Inc.

28 Case Studies in Toxicology 2

Recovering Environments Affected by Chemical Incidents: The Chemical Recovery Navigation Tool

Antonio Peña-Fernández, Stacey Wyke and Raquel Duarte-Davidson

CONTENTS

28.1 INTRODUCTION

The release of a chemical substance into the environment, either deliberately or involuntarily, is termed a chemical incident. Although the use of chemicals has a significant role in our societies, chemical incidents may occur and can impact large groups of a population and affect multiple countries simultaneously, with serious implications for human health and the environment (Duarte-Davidson et al. 2014). These incidents and attacks (the use of chemicals as weapons) can require an appropriate and immediate international collaborative response from different experts, health professionals and authorities to protect the public (Olivieri et al. 2017).

Recent incidents, such as the 2017 fipronil egg contamination event, in which millions of chicken eggs were recalled from different European countries because of residual levels of the pesticide fipronil (Ruzza et al. 2020); attacks using chemicals as weapons during the recent conflict in Syria and Iraq (DeLuca et al. 2020); or the use of the nerve agent Novichok in the UK (Vale et al. 2018; Clarke and Weir 2020), have put the spotlight on the importance of having appropriate capacity, capabilities, resources and trained personnel to quickly respond to these events to protect humans and the environment. Thus, appropriate tools and technical guidance should be available as part of the preparedness and planning that all governments should establish, so first responders (central government departments, agencies and inspectorates; emergency response personnel; local authorities) can tailor robust and appropriate protection and recovery strategies should such an event occur.

Owing to their position in the community and their wide distribution, pharmacists could play a pivotal role in the response and recovery of virtually any disaster including chemical incidents, not only for the provision of medication, care products, potential antidotes and means of decontamination that can save lives if used rapidly and correctly, but also in patient care and public health (Watson et al. 2019). Pharmacists, with their multifaceted careers, could also contribute to any of the different phases of the chemical incident response phase, including the decontamination and remediation of the environment impacted, which is key to minimising the exposure to and protection of humans in the aftermath of a chemical incident (Wyke et al. 2014a).

However, despite its relevance, there is limited practical guidance or advice on how to remediate contaminated environments (e.g. for inhabited areas) affected by chemical pollutants and/or

DOI: 10.1201/9780203730584-28

how to develop a successful recovery and remediation strategy. To address this gap in knowledge, Public Health England (PHE) has led on the development of the UK Recovery Handbook for Chemical Incidents (UKRHCI; Wyke et al. 2020), which aids development of an effective recovery strategy in the aftermath of a chemical incident to protect human health and the environment. The UKRHCI, which has been updated recently (September 2020) and is freely downloadable from the PHE website (www.gov.uk/government/publications/uk-recovery-handbook-for-chemical-incidents-and-associated-publications), comprises a developed methodology and an up-to-date series of protection and recovery options or strategies (techniques or methods for environmental remediation) for different environments. In close collaboration with local authorities and relevant experts, the UKRHCI users will be able to identify the protection/recovery options that will be most applicable for their particular scenario based on the physicochemical properties of the chemical(s) involved and the characteristics of the impacted site through a multistage process based on a series of steps and decision trees.

In order to maintain and update guidance and advice for recovery and remediation to ensure that the evidence base for recovery strategies is built appropriately and effectively, PHE has also developed the Chemical Recovery Navigation Tool (CRNT; PHE 2015a) and the Chemical Recovery Record Form (CRRF; PHE 2015b) in close collaboration with the Department for Environment, Food and Rural Affairs (Defra), the Food Standards Agency, Department for Transport and the UK Government Decontamination Service (now part of Defra; Wyke et al. 2014b). The purpose of these novel PHE Chemical Recovery Tools (Figure 28.1) is to facilitate and guide recovery decision makers through the process of tailoring their recovery strategy following the innovative methodology and resources available in the UKRHCI.

Despite there being no current, standardised approach for decontaminating and restoring the environment following a chemical incident at a European level, the PHE Chemical Recovery Tools (CRT) could be used in other countries, after appropriate adaptations, following local needs, area characteristics and legislation. The purpose of this chapter is to provide an overview of these novel PHE CRT and how to use them, through a short case study.

28.2 UK RECOVERY HANDBOOKS: Where to Find Them, Content and Applicability

PHE has led the production of a complete suite[1] with methods, guidance and tools on recovery and restoration in collaboration with the Department for the Environment and Rural Affairs, Food Standards Agency, including the Government Decontamination Service, UK Home Office, Northern Ireland Environment Agency and the Scottish Government, which includes the chemical recovery guidance and tools (i.e. the UKRHCI and the Chemical RNT and associated tools) and the UK Recovery Handbooks for Biological (UKRHBI; Pottage et al. 2015) and Radiation (UKRHRI) (version 5; Nisbet et al. 2015) Incidents and the Radiation RNT. The UK Recovery Handbooks are developed methodologies to aid the recovery coordinating group, or those entitled to deal with the recovery, to design an effective and robust protection and recovery strategy after a chemical, radiation, nuclear or biological incident. The handbooks also provide a pool of protection and recovery options that are screened through a multistage process to develop a proper recovery plan for tackling events involving any of these substances, which is based on a series of steps and decision trees (Wyke et al. 2014a, 2014b).

Developing a recovery strategy requires a clear understanding of the size and complexity of the contaminated site and an assessment of the risk to the environment and public health, as depicted in Figure 28.2. Thus, the UK Recovery Handbooks also facilitate (a) identification of important issues and constraints prior to develop a remedial strategy; (b) a description of what to do and how to proceed after an incident; (c) suggestion of a series of different recovery or management options for each environment considered; and (d) selection of those recovery options potentially applicable and more effective for a particular scenario. These manuals also provide links to other government agencies and institutions that offer advice or are related with recovery in the UK. Therefore, these manuals can be used in all phases of incident management, which include planning and preparedness, incident response, and recovery and restoration. The handbooks can also be used for contingency preparedness (e.g. equipment and utilities to avoid the spread of contaminants in different environments is comprehensively described), and in training and teaching.

28.2.1 UKRHCI: Recovery Options

Recovery options are actions, techniques or strategies intended to reduce or prevent the human exposure to chemical contamination. There are three kinds of recovery options in the UKRHCI: *protection* (actions entitled to protect human health and the environment), *restoration* (actions to

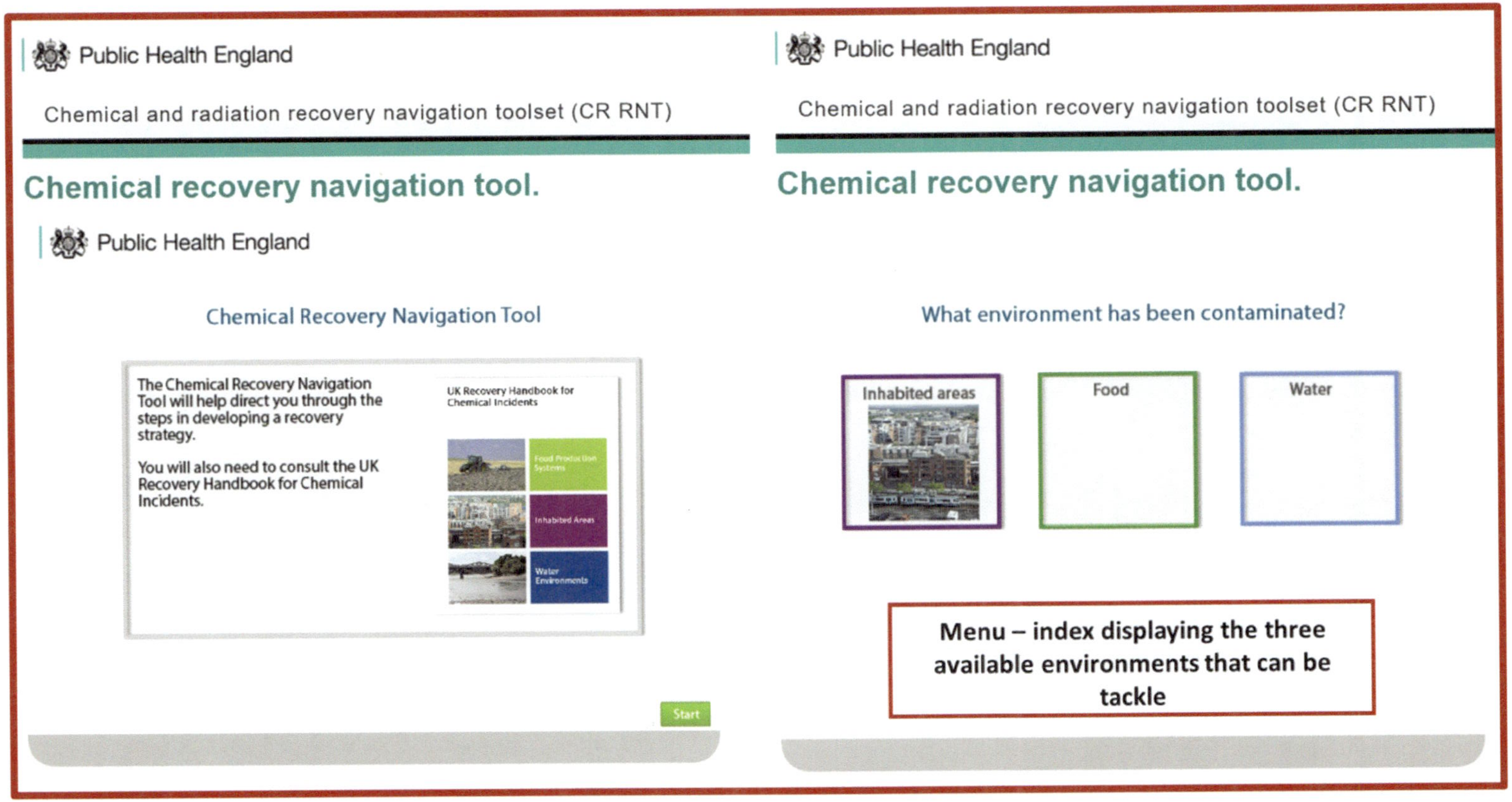

Figure 28.1 Overview of PHE's Chemical Recovery Navigation Tool

Source: http://legacyassets.phe.org.uk/tools/CRT_elearning/index.html.

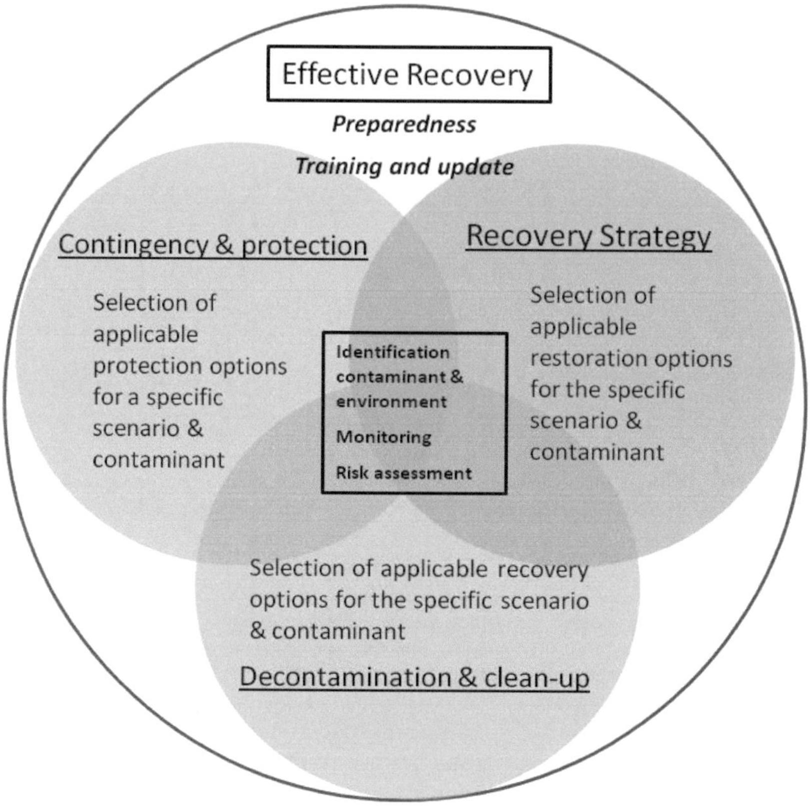

Figure 28.2 Overview of relationship between the contingency and protection, recovery strategy, and decontamination and clean-up stages in developing a robust response to a chemical incident, which includes a clear understanding of the size and complexity of the contaminated site and an assessment of the risk to the environment and the public.

get the affected environment back to normal), and *fate of affected produce or waste disposal* (options related with waste produce and appropriate disposal and treatment). Some of the restoration options recommended in this handbook, including the clean-up stage of the area impacted, will result in the generation of waste or waste by-products (e.g. water runoff), which will require consideration in any remedial plan, as these wastes can also require appropriate decontamination before processing.

Each option is provided with a summary (recovery sheet) that describes key important information, issues and constraints that should be taken into consideration when tailoring a response as part of the multistage process to evaluate and identify the potential impact of different factors that can influence the remedial response; these have been described in more detail previously by our group (Peña-Fernández et al. 2014). Thus, each recovery sheet covers the following points: (a) a description of recovery actions, objectives and potential targeted contaminants; (b) public health and environmental considerations; (c) factors influencing effectiveness; (d) specific equipment and personal requirements; (e) waste production and disposal; (f) public and worker potential exposure; (g) costs and legal constraints; and (h) references to practical/learnt experience documented in the literature with constraints and effectiveness.

28.2.2 UKRHCI: Methodology and Multistage Process

The UKRHCI provides a decision-aiding framework with decision trees for each of the four environments considered (food production systems, inhabited areas, drinking water systems and open water environments), which aids the user to select applicable recovery options during a multistage process of six steps (see Box 28.1) based on the physicochemical characteristics of the chemical(s) involved and the specific characteristics of the site. Therefore,

BOX 28.1 MULTISTAGE PROCESS FOR SELECTING RELEVANT RECOVERY OPTIONS IN THE UKRHCI

Step 1. Obtain relevant information regarding the incident
Step 2. Consult decision trees for food production systems
Step 3. Determine the effectiveness of recovery options, which includes:

a) physicochemical properties
b) effectiveness

Step 4. Review key considerations (social, economic, etc.)
Step 5. Consult recovery option sheets
Step 6. Compare remaining recovery options

⇩

Implement recovery strategy and monitor its effectiveness.

Source: Adapted from Wyke et al. (2020).

The UKRHCI also provides a specific sheet/table with the physicochemical parameters that are significant for each environment and should be carefully considered for the chemical(s) involved in order to select effective recovery options during step 3 (Box 28.1). These forms also include an interpretation of the effect of the values obtained for every physical and chemical parameter, which will allow the user to easily identify those options that will or will not work, as physicochemical properties can affect the effectiveness of the response (Wyke et al. 2014a).

Once the recovery options have been selected with the guidance of the appropriate decision tree and the physicochemical properties, they are reviewed in further steps against their effectiveness for the specific area affected (step 3b), as well as other key constraints and considerations such as public health, worker exposure, waste, time, cost or technical issues (steps 4 and 5). This review of the recovery options is intended to select only the most effective recovery option(s) from those identified in previous steps, and it is done following a screening process through the use of tables that provide a different effectiveness gradation (i.e. not applicable, none or minor, moderate, and significant or important). As part of the step 5, the user is asked to check the specific recovery sheets for the options identified, as these will inform their decisions to complete this last step successfully. The recovery options resulting from this screening process are considered to present a medium to high success to decontaminate and restore the specific scenario.

More detailed information about the UKRHCI is provided in a user-friendly e-learning module (PHE 2015c) and a user guide (Duane et al. 2016).

28.3 PHE CHEMICAL RECOVERY TOOLS

As a consequence of the UK Recovery Handbooks, and to accompany them, Chemical and Radiation Recovery Navigation Tools have been developed to facilitate and guide recovery decision makers through the process of tailoring their recovery strategy when responding to any given chemical or radiation incident.

The Chemical RNT and other tools (freely accessible from the PHE website; Figure 28.1) have been comprehensively evaluated, discussed and tested with stakeholders from a number of government departments and agencies (Department for the Environment and Rural Affairs, Food Standards Agency, UK Home Office, Government Decontamination Service, Northern Ireland Environment Agency and Scottish and Welsh Devolved Administrations) as well as with technical experts from a range of disciplines with experience of managing chemical incidents (Wyke et al. 2014a), following the same strategy used to create the UKRHCI (Wyke-Sanders et al. 2012).

28.3.1 Chemical Recovery Navigation Tool (CRNT)

The CRNT is a highly interactive resource freely available at the PHE website (PHE 2015a). This tool follows the same methodology and guidance described in the UKRHCI. Although this innovative recovery tool is intended to be used with the UKRHCI (Duane et al. 2016), the CRNT has

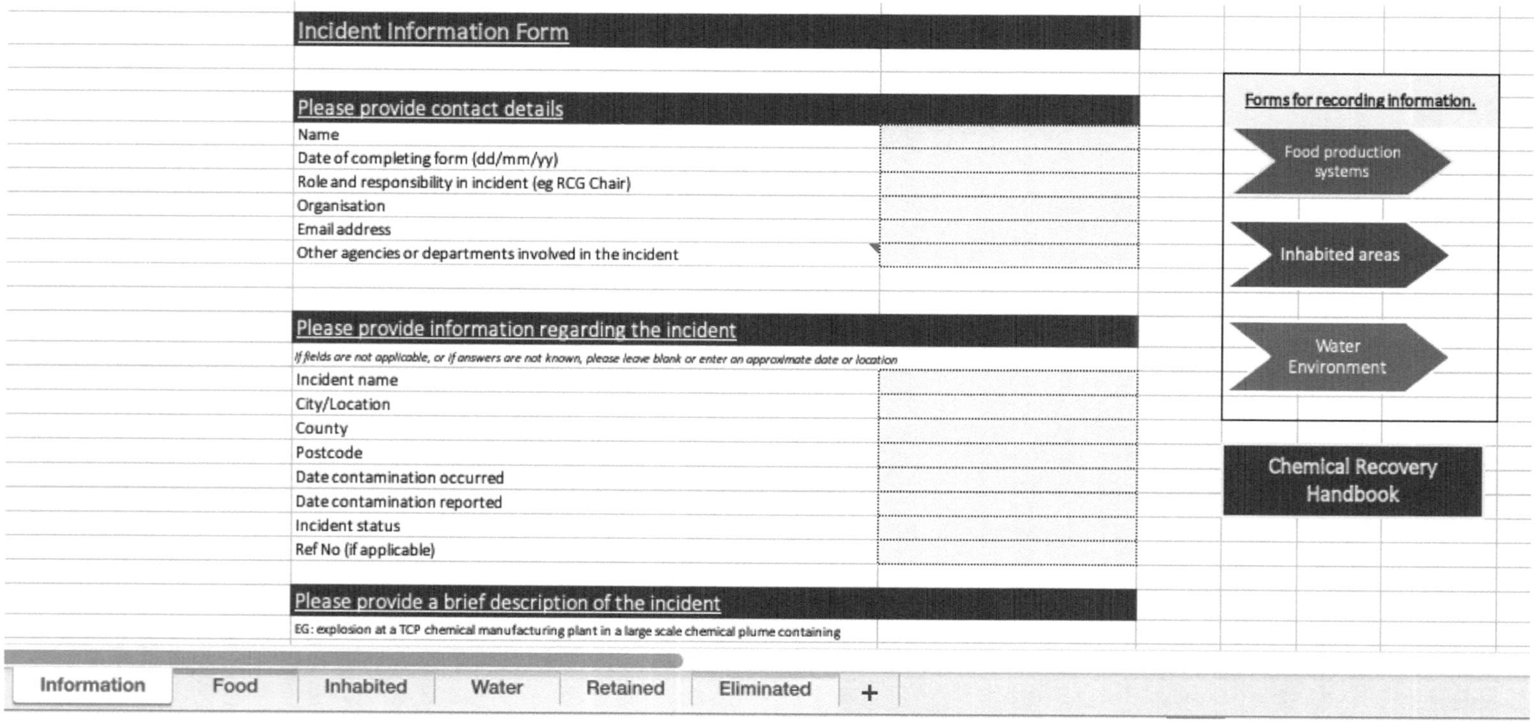

Figure 28.3 Detail of the Chemical Recovery Record Form (CRRF), displaying the three environments considered.

Source: CRRF (PHE 2015b).

been proven to be more efficient for tailoring a recovery plan in training/preparedness purposes with undergraduate and postgraduate students from different institutions (Peña-Fernández et al. 2019; Peña-Fernández 2019a, 2019b), as it has been built using standard templates within a linked framework that is easier to follow electronically than the handbook, allowing a higher level of interactivity and engagement.

28.3.2 Chemical Recovery Record Form (CRRF)

In parallel to the building of the CRNT, PHE has also developed an interactive spreadsheet, the CRRF (PHE 2015b), to aid the user with the navigating process and allow them to capture the underlying rationale for their decision on the selection of each recovery option. As described previously, the CRNT covers each of the environments considered (inhabited areas, food production systems and water environments) as displayed in Figure 28.3, which the user will need to select, as they present different decision trees and recovery options to aid development of a recovery response specifically adapted to the characteristics of the site of the scenario impacted. The CRRF can be directly downloaded from the tool itself and/or the PHE recovery website.

The CRRF has different pages/tabs to make their completion user-friendly and intuitive, which is crucial as the information recorded in the CRRF will be key when implementing the chemical incident recovery response. Thus, the user will be able to indicate why the recovery options were chosen or rejected based on the scientific evidence and constraints associated with their implementation, what containment and/or protective measures should be implemented in the first stages of the response, what equipment and capabilities are required to undertake the recovery plan, and other key issues that should be considered or monitored during the implementation of the response. In a real scenario, the CRRF will also facilitate the handover from the acute response phase to the recovery phase.

Moreover, these forms can contribute to the building of the knowledge of the lessons learnt from the chemical incident, which is important to identify what has and has not worked during the implementation of a specific recovery intervention should such an incident occur again, and could highlight gaps or failings in the policy to be addressed which in turn will enhance preparedness for future responses (Goode et al. 2018). More detail information about the CRRF, the content and how to use it is available in Duane et al. (2016).

28.3.3 Chemical Recovery: Background (CRB)

The CRNT is accompanied by an e-learning resource named Chemical Recovery: Background (CRB; PHE 2015c), which provides information on environmental decontamination and restoration and a detailed description of the CRNT. Both PHE e-learning resources, the CRB and the recovery tools e-learning module (PHE 2015d) can help enable the user to become familiar with the PHE recovery tools and acquire some basic environmental knowledge (Peña-Fernández et al. 2019).

28.3.4 How to Use the CRNT and Related Tools

Section 28.4 provides an overview of how to use the CRNT and CRRF to tailor a basic, practical recovery intervention to a chemical incident example previously developed by our group (Peña-Fernández et al. 2015). This exercise was developed as part of our training to respond to chemical incidents, which has shown to be successful in providing students with basic competencies to respond to these events to protect the public and the environment (Peña-Fernández 2019a).

28.4 SCENARIO

The scenario proposed (Box 28.2) is an industrial spill of veterinary diclofenac, benzylparaben (a preservative) and a mixture of UV filters that affect food production systems (specifically fish and shellfish) and open water systems. To develop a complete recovery response, both environments should be considered. To simplify this chapter, we will only detail here the response to target the impacts to the food production system.

BOX 28.2 CHEMICAL INCIDENT SCENARIO

A mollusc collector has reported that hundreds of mussels and several fish species have appeared dead on the shoreline. A group of experts and members of the relevant local authorities visited the area and took samples of sediment, water, fish and shellfish. The analysis showed high levels of veterinary diclofenac, benzylparaben and a mixture of UV filters in the water samples and also in different tissues of the species collected.

Later, local authorities discovered that a nearby manufacturer of pharmaceutical and personal care products had been releasing large amounts of waste products from the production of veterinary diclofenac and UV filters into a river close to the sea.

Draft a possible recovery strategy for this case.

Source: Adapted from Peña-Fernández et al. (2015).

28.4.1 Step 1: Obtain Relevant Information Regarding the Incident

To develop a response, we will open the CRNT, select the food production system environment, download the CRRF and record it in the computer/tablet. In a real situation, specific toxicological and environmental advice would be sought to assist with determining and interpreting the physicochemical properties of the chemical(s) involved in the incident after appropriate monitoring and characterisation of the impacted site. The most relevant physicochemical and toxicological information for the chemicals involved in our exercise is provided in Table 28.1, chemical information that was sought by completing a literature review. In general, the waste disposal plume of veterinary diclofenac, benzylparaben and UV filters was related to the dead of the wild marine life observed. These chemical substances can bioaccumulate in fish and shellfish tissues, presenting a risk for human health that needs to be tackled.

28.4.2 Step 2: Consult Decision Trees for Food Production Systems

An interactive window then pops up in the CRRF, in which the user can select the food production system(s) that are impacted in the chemical incident. The user will need to select all the systems affected; in our case, the only food production system impacted is "freshwater and marine fish and shellfish". After selecting the system, the form prompts another interactive window with

Table 28.1 Important Physicochemical and Toxicological Information for the Chemicals Involved in the Case Study

Physical characteristic and toxicity	Description	Interpretation
Partition coefficient between water and octanol (K_{ow})	Interpretation (units = K_{ow}) >1,000: Likely to bioaccumulate (hydrophobic): High 500–1,000: Increasing likelihood of bioaccumulating <500: Unlikely to bioaccumulate (hydrophilic): Low	Veterinary diclofenac K_{ow} = 4.9 (Lobo et al. 2014). Lipophilic, likely to be bioaccumulated (Saini et al. 2012). Poor soil sorption and relative persistence, high leaching risk that can affect groundwater systems (Lin and Gan 2011). UV filters (sunscreen agents) K_{ow} = 3000–7000 (Poiger et al. 2004). Can bioaccumulate in fish at levels similar to pesticides. Lipophilic (Brausch and Rand 2011).
Toxicity	Sum of adverse effects of the degree of danger posed by a substance to living organisms. It is expressed generally as a dose-response relationship involving the quantity of substance to which the organisms is exposed and the route of exposure skin (absorption), mouth (ingestion), or respiratory tract (inhalation).	Residues of veterinary diclofenac have caused significant decline of vulture population (Oaks et al. 2004). Benzylparabens are the most acutely toxic of the seven parabens in use (Brausch and Rand 2011) and has an oestrogenic effect (Darbre et al. 2003). The risk of parabens is currently unknown based on scarcity of data. UV filters can have oestrogenic activity (Brausch and Rand 2011).

Source: Adapted from Wyke et al. (2020).

further questions, which are based on the specific decision tree available in the handbook for each environment. The scenario provides us with monitoring data *in situ* and information about the chemicals involved. The incident involves a plume or a mix of pollutants released into open water systems. When completing the questions suggested in the specific decision tree, the CRRF is automatically populated with specific, potentially applicable recovery options for our scenario (Figure 28.4), options that should be screened against the physicochemical properties of the chemicals involved and their effectiveness and constraints in the following two steps (Section 28.4.3). Moreover, depending on the answers provided in step 2, certain recovery options may be marked as "recommended". For the food production system considered, the UKRHCI (and therefore the CRRF) only proposes two of the three types of recovery options available as relevant, specifically protection (to prevent and minimise the human exposure) and waste disposal (Figure 28.5). Thus, restoration options or options for reducing the levels of the chemical contaminants in fish and

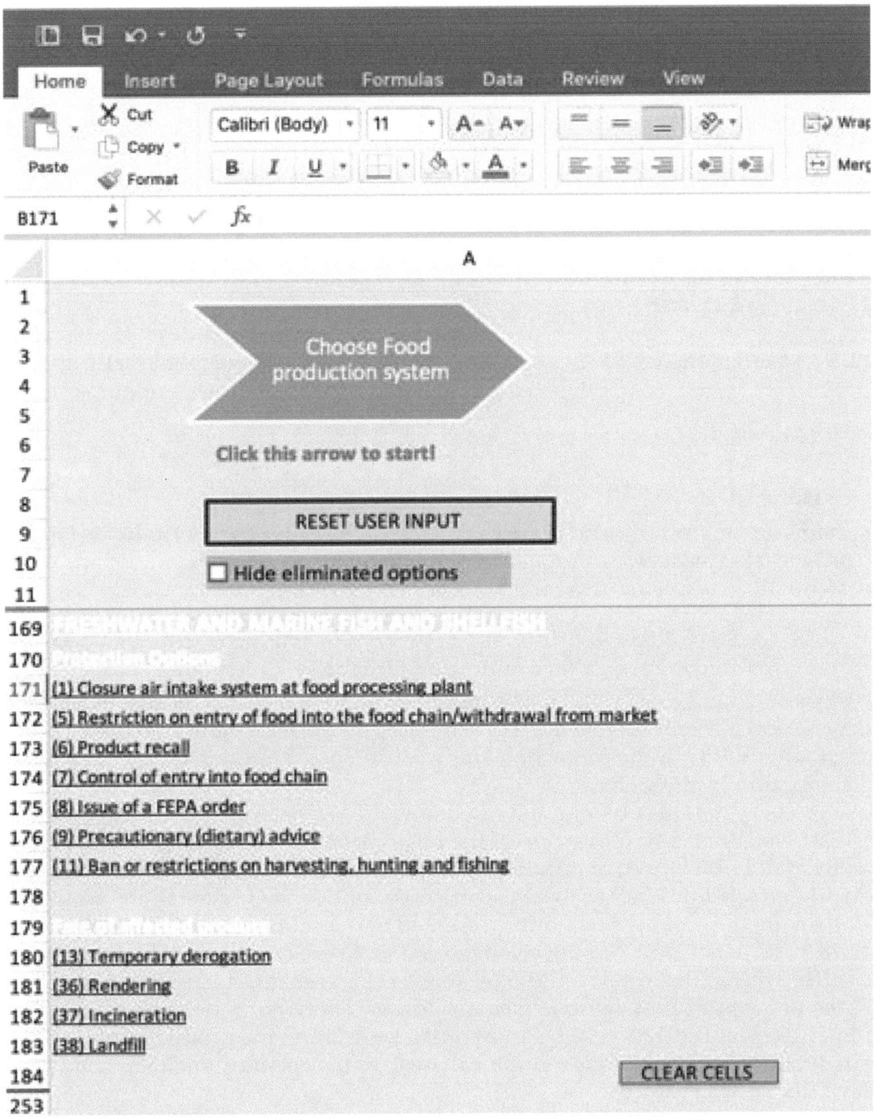

Figure 28.4 Detail of the CRRF, displaying the recovery options potentially applicable for the food production system of "freshwater and marine fish and shellfish" for our scenario.

Source: PHE (2015b).

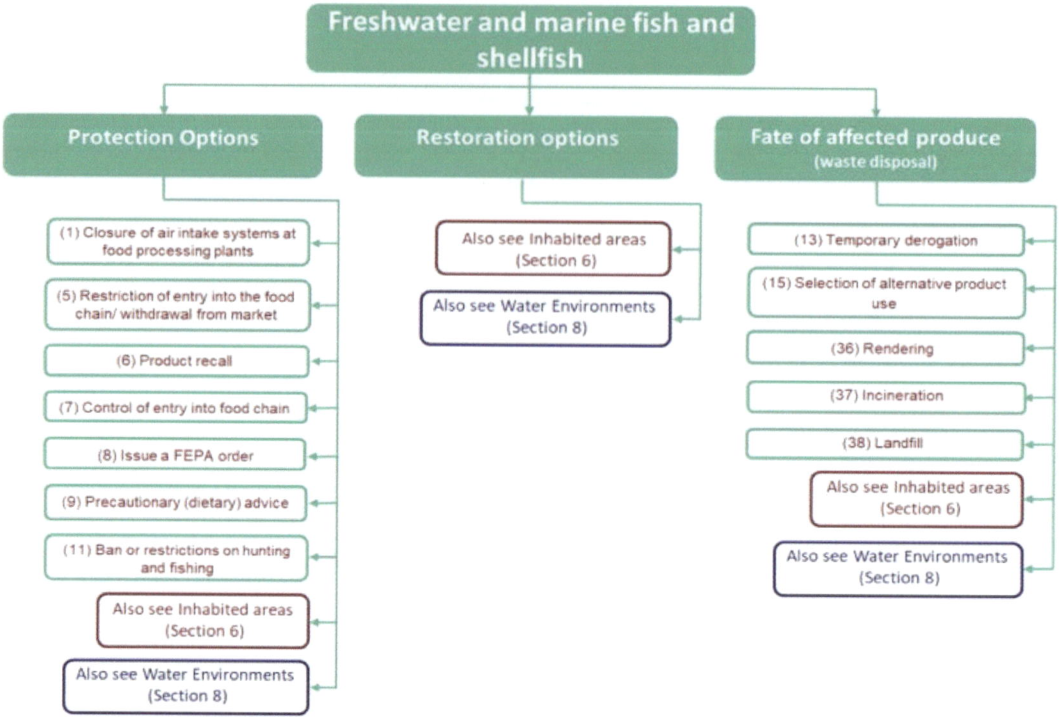

Figure 28.5 Detail of the UKRHCI, displaying the recovery options potentially applicable for the food production system of "freshwater and marine fish and shellfish" for our scenario.

Source: Wyke et al. (2020).

shellfish systems are not recommended/relevant, mainly because of practicality issues, costs and potential further risks due to, for example, incomplete removal of the chemicals and/or the by-products potentially generated during the process.

28.4.3 Steps 3 and 4: Eliminate Recovery Options according to Their Effectiveness, Constraints and Other Key Considerations

The recovery options suggested as relevant for our scenario are then comprehensively screened for appropriateness and effectiveness in the following steps. Tables 28.2 and 28.3 display the potential effectiveness/applicability of the protection and waste disposal options suggested by the CRNT and CRRF to respond to our scenario.

In step 3a, recovery options are retained or eliminated against their practicality for the chemical(s) involved; that is, as a function of the physicochemical properties of the pollutants (which are detailed in Table 28.1) and their potential for increased recovery worker exposure. Thus, the fate of affected product or the waste disposal option of "incineration" would be applicable for destroying the organic pollutants presented in the contaminated food production system (see Figure 28.5 and Table 28.3). Moreover, suggested recovery options will be checked further against the effectiveness that they would have in the environment/subenvironment impacted (step 3b). Thus, some protection options may not be very effective for our scenario (e.g. issuing a Food and Environment Protection Act [FEPA] order, preventing the production in or movement of contaminated food from a defined geographical area), as the chemical incident has affected only the local level (Table 28.2).

Suggested recovery options will be also retained/eliminated as a function of key considerations and constraints that could impact the chemical incident response and, therefore, should be considered. This screening process is contemplated in step 4 of the multistage process, and consists of checking a series of tables with a colour coding that describes the level of constraint that each recovery option could have for the following considerations: public health, waste, social

Table 28.2 Applicability of the Protection Options Suggested by the PHE Chemical Recovery Tools for Our Scenario

Protection option	Applicability
(1) Closure of air intake	Not relevant
(5) Restriction of entry into food chain	It could be useful, although it has little effectiveness for this scenario. Further monitoring is needed to decide if relevant.
(6) Product recall	This option can prevent consumers from eating contaminated fish/shellfish that they have already purchased. Therefore, local authorities can decide if relevant. The area affected is used just for the local population, so this option would have little effectiveness.
(7) Control of entry into food chain	May be relevant due to the characteristics of these chemicals: persistent and able to bioaccumulate.
(8) Issue of an FEPA order	The area affected is not a fishing production area, therefore this would be of limited effectiveness.
(9) Precautionary (dietary) advice	Applicable
(11) Ban on harvesting, hunting and fishing	Applicable

Source: Exercise adapted from Peña-Fernández et al. (2015).

Table 28.3 Applicability of the Waste Disposal Options Suggested by the PHE Chemical Recovery Tools for Our Scenario

Waste disposal option	Applicability
(13) Temporary derogation	Not applicable
(15) Selection of alternative product use	Not relevant
(36) Rendering	Not applicable
(37) Incineration	Relevant to this scenario due to the characteristics of the chemicals involved.
(38) Landfill	Applicable after appropriate treatment of the waste as some contaminants can be lixiviated and affect the groundwater.

Source: Exercise adapted from Peña-Fernández et al. (2015).

and technical issues; costs and time of applicability; and a screening process that is completed by navigating through different screens in the CRNT. Then, decisions are recorded in the CRRF by choosing, from a drop-down menu, between "no major constraints", "eliminate option due to constraints" and "major constraint but consider" for each of the recovery options remaining. Thus, for example, we can retain the waste disposal option of landfill for our scenario (Table 28.3), but we will indicate "major constraint but consider" in the CRRF as the collected waste during the recovery response will need to be appropriately treated before landfill, because the chemical substances involved can lixiviate (i.e. move through the soils layers by washing or percolation) and further affect or impact the environment.

Once step 4 is completed, the CRRF suggests reviewing other potential food production systems that could have been affected, which is not our case, as only fish and shellfish were impacted.

28.4.4 Steps 5 and 6: Polishing the Recovery Strategy

The final two steps of the multistage process consist of considering further eliminations in function of the characteristics of the impacted site and the incident, which will be done by checking the information provided in the recovery option sheets, which are available in the UKRHCI. The CRRF presents hyperlinks to the recovery sheets to facilitate this process. Step 6 will involve combining any remaining recovery options based on previous steps for tailoring a complete and relevant recovery response.

In real circumstances, recovery experts will closely monitor the effectiveness of the response and whether the levels of the chemicals involved are reduced to acceptable levels in the impacted environment. If the levels do not decrease, a new recovery strategy is necessary.

28.5 CONCLUSIONS

Training and preparedness to respond to chemical incidents and attacks is becoming critical for future pharmacists and related health professionals, as they can form part of the front line. The novel tools (Chemical RNT and associated tools) and technical guidance (UKRHCI) developed by PHE to ensure the development of a robust and effective protection and remedial response to any chemical incident or attack have also shown to be appropriate resources for delivering this increasingly relevant training.

NOTE

1 Recovery, remediation and environmental decontamination. Available at the PHE website: www.gov.uk/government/collections/recovery-remediation-and-environmental-decontamination (accessed December 11, 2020).

BIBLIOGRAPHY

Brausch, J. M., and G. M. Rand. 2011. A review of personal care products in the aquatic environment: Environmental concentrations and toxicity. *Chemosphere* 82(11): 1518–32.

Clarke, S. A., and A. G. A. Weir. 2020. UK resilience to a chemical incident. *BMJ Mil Health* 166(2): 95–8.

Darbre, P. D., J. R. Byford, L. E. Shaw, et al. 2003. Oestrogenic activity of benzylparaben. *J Appl Toxicol* 23(1): 43–51.

DeLuca, M. A., P. R. Chai, E. Goralnick, and T. B. Erickson. 2020. Five decades of global chemical terror attacks: Data analysis to inform training and preparedness. *Disaster Med Public Health Prep.* 1–12.

Duane, R., D. Cole, and S. Wyke. 2016. *Chemical Recovery Guidance and Tools: User Guide.* Version 1. ISBN 978-0-85951-781-2, PHE publications gateway number: 2015732. www.gov.uk/government/publications/chemical-recovery-guidance-and-tools-user-guide (accessed December 10, 2020).

Duarte-Davidson, R., R. Orford, S. Wyke, M. Griffiths, R. Amlôt, and R. Chilcott. 2014. Recent advances to address European Union Health Security from cross border chemical health threats. *Environ Int* 72: 3–14.

Goode, E., T. James, and S. Wyke. 2018. Chemical incident management: An overview of preparedness, response and recovery. In *Chemical Health Threats: Assessing and Alerting*, eds. R. Duarte-Davidson, T. Gaulton, S. Wyke, and S. Collins, 224–39. London: Royal Society of Chemistry.

Lin, K., and J. Gan. 2011. Sorption and degradation of wastewater-associated non-steroidal anti-inflammatory drugs and antibiotics in soils. *Chemosphere* 83(3): 240–6.

Lobo, S., H. Li, N. Farhan, and G. Yan. 2014. Evaluation of diclofenac prodrugs for enhancing transdermal delivery. *Drug Dev Indust Pharm* 40(3): 425–32.

Nisbet, A., S. Watson, and J. Brown. 2015. *The UK Recovery Handbook for Radiation Incidents.* Version 4. PHE publications gateway number: 2015067. www.gov.uk/government/publications/uk-recovery-handbooks-for-radiation-incidents-2015 (accessed December 10, 2020).

Oaks, J. L., M. Gilbert, M. Z. Virani, et al. 2004. Diclofenac residues as the cause of vulture population decline in Pakistan. *Nature* 427(6975): 630–3.

Olivieri, C., P. L. Ingrassia, F. Della Corte, et al. 2017. Hospital preparedness and response in CBRN emergencies: TIER assessment tool. *Eur J Emergency Med* 24(5): 366–70.

Peña-Fernández, A. 2019a. Teaching environmental decontamination and restoration. *Med Educ* 53(5): 523–4.

Peña-Fernández, A. 2019b. Undergraduate training to respond to chemical and biological incidents. In *Al futuro con el pasado: X Encuentro de Innovación en Docencia Universitaria*, ed. H. Del Castillo Fernández and P. Gómez Hernández, 451–60. Alcalá de Henares: Servicio de Publicaciones de la Universidad de Alcalá.

Peña-Fernández, A., M. J. González-Muñoz, and M. A. Peña. 2015. Designing training for teaching environmental toxicology to specialized pharmacists. *Curr Pharm Teach Learn* 2015; 7: 864–8.

Peña-Fernández, A., A. Nisbet, R. Duarte-Davidson, M. C. Lobo-Bedmar, and S. Wyke. 2019. Public Health England's recovery tools: Potential teaching resources? *ICERI2019 Proc*, 10520–6. https://library.iated.org/view/PENAFERNANDEZ2019PUB.

Peña-Fernández, A., S. Wyke, N. Brooke, and R. Duarte-Davidson. 2014. Factors influencing recovery and restoration strategies following a chemical incident. *Environ Int* 72: 98–108.

PHE. 2015a. *Chemical Recovery Navigation Tool (CRNT).* www.gov.uk/government/publications/chemical-and-radiation-recovery-navigation-tool-cr-rnt (accessed December 10, 2020).

PHE. 2015b. *Chemical Recovery Record Form (CRRF).* www.gov.uk/government/publications/chemical-recovery-record-form-crrf-food-inhabited-areas-and-water-environment (accessed December 10, 2020).

PHE. 2015c. *Chemical Recovery: Background (CRB).* http://legacyassets.phe.org.uk/tools/CRT_elearning/index2.html (accessed December 10, 2020).

PHE. 2015d. *Chemical and Radiation Incidents: Recovery Tools E-learning Module.* www.gov.uk/government/publications/chemical-and-radiation-incidents-recovery-handbooks-and-tools (accessed December 10, 2020).

Poiger, T., H. R. Buser, M. E. Balmer, P. A. Bergqvist, and M. D. Müller. 2004. Occurrence of UV filter compounds from sunscreens in surface waters: Regional mass balance in two Swiss lakes. *Chemosphere* 55(7): 951–63.

Pottage, T., E. Goode, C. Shieber, S. Wyke, S. Speight, and A. M. Bennett. 2015. *UK Recovery Handbook for Biological Incidents.* Version 1. PHE publications gateway number 2015406. www.gov.uk/government/publications/uk-recovery-handbook-for-biological-incidents (accessed December 10, 2020).

Ruzza, M., B. Tiozzo, V. Rizzoli, M. Giaretta, L. D'Este, and L. Ravarotto. 2020. Food Risks on the Web: Analysis of the 2017 Fipronil Alert in the Italian Online Information Sources. *Risk Anal* 40(10): 2071–92.

Saini, M., M. A. Taggart, D. Knopp, et al. 2012. Detecting diclofenac in livestock carcasses in India with an ELISA: A tool to prevent widespread vulture poisoning. *Environ Pollut* 160(1): 11–6.

Vale, J. A., T. C. Marrs, and R. L. Maynard. 2018. Novichok: A murderous nerve agent attack in the UK. *Clin Toxicol (Phila)* 14: 1–5.

Watson, K. E., J. A. Singleton, V. Tippett, and L. M. Nissen. 2019. Defining pharmacists' roles in disasters: A Delphi study. *PLoS ONE* 14(12): e0227132.

Wycisk, P., R. Stollberg, C. Neumann, W. Gossel, H. Weiss, and R. Weber. 2013. Integrated methodology for assessing the HCH groundwater pollution at the multi-source contaminated megasite Bitterfeld/Wolfen. *Env Sci Pollut Res Int* 20(4): 1907–17.

Wyke, S., N. Brooke, A. Dobney, D. Baker, and V. Murray. 2020. *The UK Recovery Handbook for Chemical Incidents*. Version 1.1, 2020. PHE publications, Gateway number: GW-1451. https://assets.publishing.service.gov.uk/government/uploads/system/uploads/attachment_data/file/912808/UKRHCI_2019_Update_v1.1__2_.pdf (accessed December 10, 2020).

Wyke, S., A. Peña-Fernández, N. Brooke, and R. Duarte-Davidson. 2014a. The importance of evaluating the physicochemical and toxicological properties of a contaminant for remediating environments affected by chemical incidents. *Environ Int* 72: 109–18.

Wyke, S., A. Peña-Fernández, A. Nisbet, V. Silvey, S. Watson, and R. Duarte-Davidson. 2014b. Developing tools to assist with evaluating a recovery strategy. *Chem Hazard Poisons Rep* 24: 52.

Wyke-Sanders, S., A. Osman, N. Brooke, A. et al. 2012. *UK Recovery Handbook for Chemical Incidents: Evaluating the Evidence Base for Recovery, Remediation and Decontamination Methods.* Chemical Hazards and Poisons Report 21: 35–7. www.gov.uk/government/publications/chemical-hazards-and-poisons-report-issue-21 (accessed December 19, 2020).

Index

Note: Information in figures and tables is indicated by page numbers in *italics* or **bold**, respectively.

F